Software Defined Radio
with
Zynq® UltraScale+™ RFSoC

Software Defined Radio
with
Zynq UltraScale+ RFSoC

Louise H. Crockett

David Northcote

Robert W. Stewart

(editors)

Department of Electronic & Electrical Engineering

University of Strathclyde

Glasgow, Scotland, UK.

January 2023

www.RFSoCbook.com

First published January 2023 by Strathclyde Academic Media (revised June 2025).

© Strathclyde Academic Media.

Cover design by Mario Stomboli, Scribble Design. Photography by Kenneth Barlee.

Warning and Disclaimer

The best efforts of the authors and publisher have been used to ensure that accurate and current information is presented in this book. This includes researching the topics covered, and developing examples. The material included is provided on an "as-is" basis in the best of faith, and neither the authors nor publishers make any warranty of any kind, expressed or implied, with regard to the documentation contained in this book. The authors and publisher shall not be held liable for any loss or damage resulting directly or indirectly from any information contained herein.

Trademarks

AMD, the AMD logo, LogiCORE, MicroBlaze, Spartan, UltraScale, UltraScale+, Versal, Vitis, Vivado, Xilinx, the Xilinx logo and Zynq, are all trademarks or registered trademarks of Advanced Micro Devices, Inc.

MATLAB and Simulink are registered trademarks of MathWorks, Inc.

Linux® is the registered trademark of Linus Torvalds in the U.S. and other countries.

Arm, Cortex, AMBA, Mali, Neon, and TrustZone are registered trademarks of Arm Limited (or its subsidiaries) in the EU and/or elsewhere. All rights reserved. This publication is independent and it is not affiliated with, or endorsed, sponsored or authorised by Arm Limited.

Dell is a trademark of Dell Inc. or its subsidiaries.

Intel and Xeon are trademarks of Intel Corporation or its subsidiaries.

Wi-Fi® is a registered trademark of the Wi-Fi Alliance®.

Android is a trademark of Google LLC.

Ettus Research and USRP are trademarks of National Instruments. Neither Strathclyde Academic Media, nor any software programs or other goods or services offered by Strathclyde Academic Media, are affiliated with, endorsed by, or sponsored by National Instruments.

All other trademarks used in this book are acknowledged as belonging to their respective companies. The use of trademarks in this book does not imply any affiliation with, or endorsement of, this book by trademark owners.

Table of Contents

Foreword

In 2019, Xilinx announced its Zynq UltraScale+ RF System on Chip (RFSoC) product portfolio, manufactured using advanced FinFET transistor technology. A confluence of compelling factors motivated the development of the RFSoC: the emergence of wideband, digitally assisted, RF data converters; the superior analog performance of FinFET transistors; the demand for large arrays of digitized RF channels in emerging 5G Massive-MIMO radios; the imperative to maximise bandwidth at low power though monolithic integration of RF and digital signal processing functions; and the success and widespread adoption of the Zynq SoC and MPSoC product families. The software adaptability, small form factor and low power operation of the RFSoC has resulted in its widespread deployment not only in 5G networks and proprietary software defined radio (SDR) implementations but also in radar systems, test and measurement equipment, and cabled networks. For some of the world's evolving RF sampling rate enabled research, design and development programmes, the RFSoC will also provide a platform for the rapidly advancing 6G communications domain, and provide capability for the high speed control and implementation requirements of quantum computing.

Building on the success of *The Zynq Book* and *Exploring Zynq MPSoC*, we were delighted to collaborate with the University of Strathclyde in this latest book 'Software Defined Radio with Zynq UltraScale+ RFSoC'. The book introduces the reader to important fundamental theory and architectures in advanced digital communications system, which can then be explored for implementation on the RFSoC platform. While the RFSoC incorporates all the features of the Zynq MPSoC including the multi-core ARM processor, complex and software programmable hardware fabric, it additionally supplies up to 16 high performance transmit and receive RF-subsystems which can synthesize and digitise wideband, RF signals all the way to-6 GHz. Traditional RF functions such as oscillators, mixers and filters are now integrated digitally as part of each RF-subsystem. Throughout the book, practical examples are provided in the form of Jupyter Notebooks on the PYNQ development platform.

Given that the RFSoC enables developers to architect a software-programmable RF radio in a single device, it is entirely appropriate that the Software Defined Radio (SDR) concept is introduced at the outset. This is followed by an overview of the RFSoC architecture, features, and available hardware development platforms. DSP and Wireless Communications fundamentals are then covered followed by an examination of SDR architectures. Digital modulation schemes in modern communications systems are explored followed by a detailed overview of typical RF subsystem architectures, all of which are supported by the RFSoC device. Later chapters deal with the more advanced aspects of communication systems including Forward Error Correction, MIMO and Beamforming, and Cognitive Radio.

Whether the reader is a professional in Communication engineering, an academic or University researcher, a first-time user or an expert seeking more insight, *Software Defined Radio with Zynq UltraScale+ RFSoC* makes the RFSoC platform easily accessible, supplying the fundamental theory, the architectures, and the practical implementation in the form of examples throughout.

The longstanding AMD collaboration in wireless technological research and innovation with the StrathSDR team at the University of Strathclyde goes from strength to strength and I would like to thank and congratulate all of the authors who have contributed to this book.

Brendan Farley,

Corporate Vice President, Wireless Engineering, AMD.

Acknowledgements

We are delighted to publish this book on Software Defined Radio Systems (SDR) with the AMD Zynq UltraScale+ RFSoC. The book represents the culmination of a few years of research, development and design activities with the latest RFSoC platforms alongside our writing efforts to integrate the theory and principles of SDR and DSP (digital signal processing) in a publication that is accessible to engineering students, as well as practising engineers and technical managers in industry. The publication of this book represents a great team effort by our StrathSDR team at the University of Strathclyde, and is built on a superb collaboration relationship with our colleagues at AMD (and as Xilinx prior to 2021), with whom we are proud to have had a fruitful relationship with since 2006.

To the researchers in our StrathSDR group who contributed as chapter authors, co-authors, reviewers, designers, even graphic artists, many thanks indeed! These individuals are (in order of 'appearance' in the book): Lewis Brown, Kenny Barlee, Josh Goldsmith, Marius Šiaučiulis, Graeme Fitzpatrick, Douglas Allan, Lewis McLaughlin, James Craig, Blair McTaggart, Tawachi Nyasulu, Andrew Maclellan, Ehinomen Atimati, and David Crawford. We greatly appreciate your enthusiasm, professionalism, and all of the knowledge and ideas that you brought to the whole process. To those who contributed in various ways to the practical materials that we have released alongside this book, please accept our thanks too — most are also chapter authors and already mentioned above, but also including Craig Ramsay and Sarunas (Shawn) Kalade. Thanks to Damien Muir for editing support and website development and to Kenny Barlee for special photography skills and setting up the book cloud and web services, and also to Jackie Malloy for highly-valued administrative support.

A huge thanks to our friends and colleagues at AMD for enabling the environment that allowed the creation of this book. To Patrick Lysaght — thanks (again) for your continuing support, hands-on engagement and guidance, and all of the opportunities we have been able to explore together, not just in this project but over many years. We are grateful to Brendan Farley and Ivo Bolsens — thanks for supporting our work, and putting your faith in us to make this book happen. To long-term colleagues Cathal McCabe, Graham Schelle, and more recently Shane Flemming, Nathan Jachimiec, and Shawn Kalade; as well as your practical help, we have also greatly valued all of the technical discussions, and the benefit of your wisdom and gentle encouragement over the course of this book project — many thanks.

A separate and significant vote of thanks is due to all those at AMD who were generous with their time in reading drafts and giving feedback. They are: Patrick Lysaght, Cathal McCabe, Nathan Jachimiec, Graham Schelle, Mario Ruiz, Shawn Kalade, Hugo Andrade, Shane Flemming, Sylvain Bertrand, Keith Lumsden, Richard Walke, and Andy Dow. We recognise the time and thought that went into all of these reviews, and your inputs have, without doubt, improved the book substantially — these efforts are greatly appreciated.

It goes without saying that the practical elements of this book project were facilitated by having easy access to software tools and development boards; therefore thanks to AMD University Program for providing us with all the resources we needed, and a special shout out to Naveen Purushotham for taking care of all the logistics. We also must acknowledge the RFSoC pioneers and engineering design teams in Ireland and in the USA (at both AMD and formerly Xilinx) for their prowess in designing the RFSoC over the last few years. On this note, a special thank you to Liam Madden for his support in the genesis and early ideas for this book project. Again, this is incredible technology and really is the ultimate SoC — system-on-chip. We are still in awe of just how flexible and powerful the computation is on the RFSoC is, and we marvel at the sheer number of RF ADC inputs and DAC output channels available, sampling at rates of several GHz — all on a single device.

Thanks also to University of Strathclyde for supporting our StrathSDR research team and providing a fantastic environment to work in. Alongside the book writing tools of PCs, desktop publishing, and graphics packages, the University provides us with a very functional lab with the latest state of the art SDR boards, instrumentation, RF facilities, test-beds as well as the academic freedom to design end to end real radio systems for standards such 5G and other applications. Our team here has more than 25 years experience in DSP, FPGAs, SDR and radio standards, so additionally thanks to the wider team at StrathSDR and our colleagues at our University SDR spin-out company, including: Malcolm Brew, Cameron Speirs, Damien Muir, Dani Anderson, Anthony Ighagbon, Samuel Yoffe, Ryan Provan, Shruthi Kumar and Marcin Mrozowski. Our wider team is not just researching and simulating, we really are building RFSoC SDR solutions with a number of industry, government and academic partners to make real SDR transceivers for 5G radios and other standards.

Therefore we hope our engineering team's design and implementation experience shines through in the book and you find it a useful.   Please do get in touch if you we can support you in your RFSoC and SDR journey.

Last but certainly not least, thanks to our families and friends for their understanding while we completed this project!

Louise Crockett, David Northcote, Bob Stewart.   *January 2023.*

Introducing the Team

There has been a whole cast of 'characters' involved in creating this book, and we'd like to take the opportunity to say hello! Here is a photo of the editing and chapter authoring team below. It was taken in late 2022, in the entrance of the Royal College building, University of Strathclyde, Glasgow, Scotland.

All names given left-to-right...
Back row: *Lewis Brown, Marius Šiaučiulis, Bob Stewart, David Northcote, Andrew Maclellan, Douglas Allan;*
Middle row: *Josh Goldsmith, David Crawford, James Craig, Graeme Fitzpatrick, and Ehinomen Atimati;*
Front row: *Louise Crockett, Tawachi Nyasulu, Lewis McLaughlin, Blair McTaggart, and Kenny Barlee.*

About StrathSDR

The University of Strathclyde Software Defined Radio team (StrathSDR) is a research, development and design team with more than 20 staff, researchers and PhD students and is supported by research funding grants and industry partnerships. Our heritage and evolution dates from the late 1990s when the then *Digital Signal Processing (DSP) Enabled Communications* group (DSPeC) was a very active part of the University's Department of Electronic and Electrical Engineering. In the late 90s it was *'all about the baseband'* with DSP providing the acceleration for digital mobile and wireless networks via baseband speech coders, MPEG audio coders, video encoding, low frequency numerically controlled oscillators (NCOs), pulse shaping and channelisation, leading to the implementation of the first digital mobile standard, GSM and emerging early Wi-Fi® / wireless implementations. However from baseband sampling rates of the orders of 10 kHz to 100 kHz in the late 80s and early 90s, and the success of oversampling converters (sigma-delta), the early 2000s saw MHz sample rate converters arrive and the first IF (intermediate frequency) digital radios in operation. In these systems signals at the RF carrier frequencies (800 MHz+ typically), could be mixed down and centred at low MHz values and then directly sampled with a few MHz sample rate ADC (analogue to digital converter). Similarly, IF-centred outputs created by MHz sampling DACs (digital to analogue converters) were available for mixing up to RF transmission bands. Tracking the increase in ADC and DAC sampling rates (not quite Moore's law growth, but perhaps doubling every 3 to 4 years since 1985), 2022 sees SDR being a fundamental core technology and methodology for the implementation of mobile and wireless networks, from 5G to 6G and other bespoke and custom implementations.

Since 2011 the StrathSDR team has been building private shared spectrum mobile and wireless networks. Early networks used TV white space (TVWS) UHF spectrum, and more recently private 5G standalone (SA) networks were implemented on shared spectrum bands from 3.4 to 4.2 GHz (n78 and n77 bands). Recent industry/academic research partnership projects using AMD FPGA and RFSoC technology for 5G SA radio networks includes UK government DCMS funded partnership projects, 5GRuralFirst, 5GNewThinking, and as a founding partner of the Scotland 5G Centre. These projects have led to the deployment of real radio networks working with a host of key partners to build both the low PHY and high PHY for 5G and other standards using AMD technology at both the remote radio head (RRH) and the baseband units (BBU). In addition to AMD, our team has collaborated recently with many tier 1 suppliers and mobile network operators. Part of our endeavour in this changing market is also around the rise of the 'tier 2s' and SMEs (small medium sized enterprises) in providing RAN (radio access networks and O-RAN), particularly for private networks.

About the University of Strathclyde

The University of Strathclyde is a public research university located in Glasgow, Scotland. Founded in 1796 as the Andersonian Institute, it is Glasgow's second-oldest university, having received its Royal Charter in 1964 as the first technological university in the United Kingdom. The name is derived from the historic Kingdom of

Strathclyde, with 'Strath' referring to valley and 'Clyde' to the river running through Glasgow. It is Scotland's third largest university by number of students, with students and staff from over 100 countries. The University operates with four main Faculties: (i) Engineering, (ii) Science, (iii) Business and (iv) Humanities and Social Sciences. The institution was named UK University of the Year 2012 by Times Higher Education and again in 2019, becoming the first university to receive this award twice.   The StrathSDR team is part of the Department of Electronic and Electrical Engineering in the Faculty of Engineering. This Department has more than 80 full time faculty, 200 PhD and research students, and 600+ undergraduates and 100+ taught masters students.

AMD and University of Strathclyde Engagement

The AMD and University of Strathclyde engagement and partnership dates back to 2002 with our first joint projects with Xilinx (who were acquired by AMD in 2021) to create educational courses using the Virtex® FPGA devices. In 2006 University of Strathclyde and Xilinx, via the Xilinx University Program (XUP) team, partnered to create university and professional teaching materials for the latest generation of FPGAs (Virtex and Spartan® generations) and jointly presented XUP DSP workshops in USA, Europe and Asia from 2008 to 2014. With the advent of the Zynq SoC in 2011, the activities extended to this device, and as well as teaching and training materials, the first text book from the partnership was published in 2014 — 'The Zynq Book: Embedded Processing with the Arm Cortex-A9 on the Xilinx Zynq-7000 All Programmable SoC' [130] in 2014. The book was (and still is!) available as a free download as well as in traditional print form. In 2019, a second book was published 'Exploring Zynq MPSoC: With PYNQ and Machine Learning Applications', [131] and again a free download version was made available alongside the print version. Across all of the joint book publications there has been more than 160,000 book downloads to date.

In the last three years, work with AMD has moved to the RFSoC, with projects focusing first on the RFSoC2x2 board, and more recently on the RFSoC4x2 board, which features in this book. With the publication of this book and the open-sourcing of the supporting hands-on notebooks and other materials, we look forward to continuing to work with AMD in this most exciting of technologies.

About the Editors

Louise H. Crockett — Louise was awarded MEng (distinction) and PhD degrees in Electronic and Electrical Engineering, both from the University of Strathclyde, in 2003 and 2008, respectively. She is currently a Senior Teaching Fellow and senior member of the StrathSDR research team where she supervises and manages researchers and key sponsored projects. Her core research interests are in the implementation of DSP systems, FPGAs and SoCs, wireless communications, and SDR. Louise has previously co-authored two books on AMD technology: *The Zynq Book* (2014), and *Exploring Zynq MPSoC* (2019). Her teaching focuses on digital systems design targeting FPGA and SoC technology, and builds practical skills to equip graduates for roles in industry.

David Northcote — Received the BEng (Hons) degree in Electronic and Electrical Engineering in 2015. He is currently a Researcher with the Department of Electronic and Electrical Engineering (EEE), University of Strathclyde, supported by AMD. His PhD research was on the efficient implementation of the Hough Transform for embedded vision systems using Zynq MPSoC. David is a co-author of the technical book *Exploring Zynq MPSoC* [43]. His research interests include efficient implementation of wireless communication and computer vision applications on Zynq, and he has published in IEEE, and at various international conferences over the last 6 years.

Robert (Bob) Stewart — Bob graduated in 1985 with a BSc (Hons) in Electronic Eng., followed by a PhD in Parallel Signal Processing in 1990. He has been an academic at Strathclyde since 1991, where he is currently a Professor, and from 2014-17 was the Department Chair. In his early career he was Design Engineer at Wolfson Microelectronics Ltd, and spent time at USC, University of Minnesota, and as a Visiting Professor at UCLA Extension (until 2017). He was co-founder and CEO of digital communications company, Steepest Ascent Ltd, which was acquired by MathWorks. Bob is head of the StrathSDR team, director of the StrathSDR spin-out company, and has published extensively and led a number of collaborative R&D projects with industry.

Chapter Authors

Douglas Allan — Douglas received his BEng (Hons) and PhD degrees from the University of Strathclyde in 2013 and 2019 respectively. His PhD and post-doctoral research involved development of novel algorithms for detection of OFDM signals in cognitive radio receivers. He has been involved in development of PHY hardware for FPGA and SoC platforms and configuration and deployment of private 5G network solutions for the broadcasting industry. Douglas is a principal SDR design engineer with Neutral Wireless Ltd in Glasgow, and also a research engineer with the StrathSDR team.

Ehinomen Atimati — Atimati is a PhD researcher with the StrathSDR team at the University of Strathclyde. Her current work, supported by Schlumberger Faculty for the Future (FFTF), is focused on exploring reinforcement learning techniques in improving coexistence management within Dynamic Spectrum Access networks. Her interests are in exploring artificial intelligence-driven shared spectrum wireless communication systems for inclusive connectivity. She has several years of international teaching experience in Electrical/ Electronic Engineering and is passionate about increased participation of females in STEM.

Kenneth W. Barlee — Kenny Received BEng (Hons) and a PhD degrees from the University of Strathclyde in 2014 and 2020 respectively. His PhD research presented FBMC-based real-time cognitive SDR transceivers that targeted vacant spectrum implemented on Zynq devices. He has worked on 5G NSA/SA network design, on the UK Government funded 5GRuralFirst and 5G NewThinking projects, where he led on RF network design and implementation of vRAN and distributed cloud + edge core networks. Kenny is a research engineer with the StrathSDR team and is also a principal 5G RAN design engineer with Neutral Wireless Ltd.

Lewis J. Brown — Lewis was awarded the MEng (Distinction) in Electronic and Electrical Engineering from the University of Strathclyde in 2020. Since 2020 he has been working on his PhD with the StrathSDR research group. In 2021, he completed an internship with AMD, where he developed embedded hardware systems and performed initial bring-up for RFSoC evaluation boards with the PYNQ team. Previously he was also an intern at Xilinx in Edinburgh. His core research interests involve reconfigurable, hardware-based implementations of 5G New Radio (5G NR) standards, focusing on RFSoC design.

James Craig — James received the MEng (Distinction) in Electronic and Electrical Engineering with International Study from the University of Strathclyde in 2021. Since then, he has been working as a PhD researcher with the StrathSDR research group. He has completed two internships at MathWorks Ltd in 2021 and 2022 with the Wireless HDL and Wireless Testbench teams. His research involves investigating 5G New Radio (NR) standard algorithms and how these can be effectively implemented on FPGAs and SoCs, with a focus on Multiple-Input Multiple-Output (MIMO) techniques.

Graeme Fitzpatrick — Graeme received the MEng (Distinction) in Electronic and Electrical Engineering from the University of Strathclyde in 2020. He is currently a PhD researcher with the StrathSDR research group. Graeme has core research interests in radio spectrum regulation and the hardware solutions that make up Dynamic Spectrum Access (DSA) techniques. The focus of his research is investigating the potential of the dynamic partial reconfiguration of FPGAs for SDR systems. Mainly, this involves using Dynamic Function eXchange (DFX) controlled via PYNQ Composable Overlays for communications solutions on the RFSoC.

Joshua Goldsmith — Josh received his BEng (Hons) degree from the University of Strathclyde in 2017, where he is currently completing his PhD degree. Integrated with his academic research, Josh completed two internships in 2019 and 2021 at Xilinx (now AMD) developing hardware systems and training material for the RFSoC. He is also a contributing author of the Exploring Zynq MPSoC book and has published a number of journal papers. His research is focused on run-time reconfigurable hardware, specifically for FPGA radio applications, and he has related interests in signal processing and embedded systems.

Andrew Maclellan — Andrew was awarded the MEng (Distinction) in Electronic and Electrical Engineering at the University of Strathclyde in 2018. He joined the StrathSDR research group in 2018 to pursue a PhD and has progressed three internships at MathWorks in 2017, 2018, and 2019 working with the Wireless HDL team. In 2020/21, he also interned at AMD in the PYNQ research team, working on the bring-up of the RFSoC 2x2 board, and experimental PYNQ features. His core research interests are in Deep Learning for Physical Layer Wireless Communications and developing FPGA Deep Learning architectures for communications SoCs.

Lewis D. McLaughlin — Lewis was awarded MEng (Distinction) in Electronic and Electrical Engineering from the University of Strathclyde in 2018. Since graduating, he has been pursuing his PhD within the StrathSDR research group. Between 2019 and 2020 Lewis completed an internship with AMD in Colorado, developing embedded hardware systems, performing development board bring-up and investigating design toolflows within the PYNQ team. His research interests include abstracted hardware design automation, communications channel emulation and short-wordlength architectures for FPGAs including RFSoC.

Blair McTaggart — Blair was awarded his MEng degree in Electronic and Electrical Engineering from the University of Strathclyde in 2018. He is currently working on his PhD with the StrathSDR research group and his research is supported by UK government funding and contracts on SDR systems. Specifically his work has focussed on the design a configurable, real-time FPGA implementation of a multi-element adaptive beamformer, using the QR algorithm. He has also developed an automatic tool that will rapidly build and implement QR based adaptive beamforming designs for an array of different antenna or FPGA characteristics.

Tawachi Nyasulu — Tawachi received the PhD degree from University of Strathclyde in 2022 and previously the MSc (Eng) degree from University of Leeds in 2015. She is currently a researcher with the StrathSDR team in the Department of Electronic and Electrical Engineering, University of Strathclyde and is supported and engaged with the Scotland 5G Centre as part of the Wave 1 Rural Testbed projects. Her research is focused on 5G PLMN/SNPN for rural/offshore connectivity and IoT solutions, modelling techniques for spectrum sharing and coexistence management, and business models for community-led network projects.

Marius Šiaučiulis — Obtained the degree of BEng (Hons) in Electronic and Electrical Engineering from the University of Strathclyde in 2019. Since then, he has pursued PhD research on 5G implementation using SDR with the support of AMD and the CENSIS (Scotland's national centre for sensing, imaging and IoT). His research interests include embedded software aspects of SDR design, high speed interfacing, and design tool development including PYNQ and GNU Radio. Marius has previously undertaken internships working with the MathWorks Glasgow office in 2019, and more recently with Xilinx in 2021.

David Crawford — David has received BSc (Hons), MSc, PhD, and MBA degrees from the University of Strathclyde. After spending several years in industry, he returned to Strathclyde in 2011 to run the then newly-formed Centre for White Space Communications, working closely with industry partners to investigate techniques and novel approaches for affordable Internet connectivity in hard-to-reach areas. He currently leads the 5G project activities of the StrathSDR team. A key aspect of this work involves spectrum sharing and affordable access to spectrum for use in under-served, difficult-to-reach areas using private 5G.

"In the new era, thought itself will be transmitted by radio."

GUGLIELMO MARCONI

(1874 - 1937)

Chapter 1

Introduction

For many people, businesses, and organisations around the world, radio communication is a vital component of everyday life. It enables everything from national radio and television broadcasts, to mobile cellular networks, private data networks; to emergency services communications, air traffic control, navigation systems, and many other technologies. While some uses have changed little over decades (such as listening to local radio stations), the bigger picture shows continuing innovations in the world of wireless communications. Cellular and private data networks are at the forefront of innovation in this field — driving towards higher data rates, more robust security and reliability, greater mobility, lower latency, and with provision for increasing densities of subscribers.

The rapid evolution of radio standards to achieve these various performance improvements, as well as evolving use-cases, and new paradigms for dynamically accessing the radio spectrum, mean that radio equipment must increasingly be upgradeable in the field, and flexible in its operation. Collectively, these requirements make Software Defined Radio (SDR) the natural solution.

In this book, we address the design and implementation of SDR systems using the AMD Zynq UltraScale+ Radio Frequency System on Chip (RFSoC) device. We review the underpinning concepts of digital signal processing and wireless communications, consider the features and capabilities of the RFSoC architecture, and demonstrate how to implement SDR systems on this exciting platform.

1.1. The Coming of SDR

Software Defined Radio (SDR) is not a new concept — the term was first introduced in the 1990s, with its origins extending back to the 1980s [264], [265]. The fundamental idea behind SDR is that one or more aspects of a radio's functionality can be controlled via software. Originally, this was a challenging aspiration, when radios were traditionally fixed-function hardware devices; however, in the decades since, developments in technology have increasingly provided platforms that enable SDR. To give a simple example, front-end SDR processing platforms are available that can be software-programmed with a local oscillator frequency for modulation and demodulation. SDR concepts are introduced further in Chapter 2; more detail on architectures follows in Chapter 8.

In the current day, it might be argued that almost all radios are SDRs, because software-based control is usually embedded somewhere within the radio architecture. Only the simplest, least expensive radios do not require any software elements — for instance, you can still buy analogue radio sets for the Amplitude Modulation (AM) and Frequency Modulation (FM) bands, tuned manually by the user rotating a dial, and watching an indicator bar shift along the frequency scale. These are not, however, the radio systems that provide the fast, reliable and secure data communications that underpin modern economy and society. Instead, think of standards such as Wi-Fi, Bluetooth, 4G, 5G — all sophisticated and complex systems that are orchestrated using software. In many cases, core processing functions are also implemented in software. As we discuss in Chapter 19, it can be argued that even more flexible and responsive radios will be needed in the future, for instance to realise Dynamic Spectrum Access (DSA), a disruptive new model of radio spectrum management.

At the time we release this book, the design and development of 6G has now commenced at pace. This will be the first generation and set of standards that is virtually all SDR at the RF (radio frequency) front end (certainly for the low and mid-band RF frequencies up to 10 GHz) and the true power and flexibility of RF sampling will be to the digital communications market in the late 2020s, as the first digital audio format of the CD was to the audio marketplace in the mid 1980s. Another digital communications revolution beckons with 6G!

Therefore, what modern communications engineers require are platforms that *empower* SDR, as they seek to deploy better solutions for today's radio standards, and to develop the new radio systems of the future.

1.2. SDR with Zynq UltraScale+ RFSoC

The focus of this book is SDR system implementation using the Zynq UltraScale+ RFSoC platform (hereafter referred to simply as 'RFSoC'). As the name suggests, the device is a System on Chip (SoC) that is targeted specifically towards Radio Frequency (RF) applications.

RFSoC is the third major Zynq SoC developed by AMD (formerly as Xilinx, which was acquired by AMD in 2022), the first being the Zynq-7000 SoC in the early 2010's. The Zynq UltraScale+ Multi-Processor SoC (MPSoC) preceded the RFSoC and, aside from radio-specific features, the two have much in common. All three SoC types are composed of a Processing System (PS) for running software system components, coupled with Programmable Logic (PL) equivalent to a Field Programmable Gate Array (FPGA), and high speed interconnections between the two parts. The hardware features of the three SoC types are compared in Chapter 3, but for now, we focus only on the RFSoC.

A high level view of the RFSoC device architecture is shown in Figure 1.1. Note the PS and PL sections, and in particular that the PL includes some highlighted features that are particularly important for SDR applications.

These features are *hardened*, meaning that they are implemented in dedicated silicon on the device, rather than programmable logic (bringing performance benefits, but still providing programmability).

The highlighted features of the RFSoC can be summarised as follows:

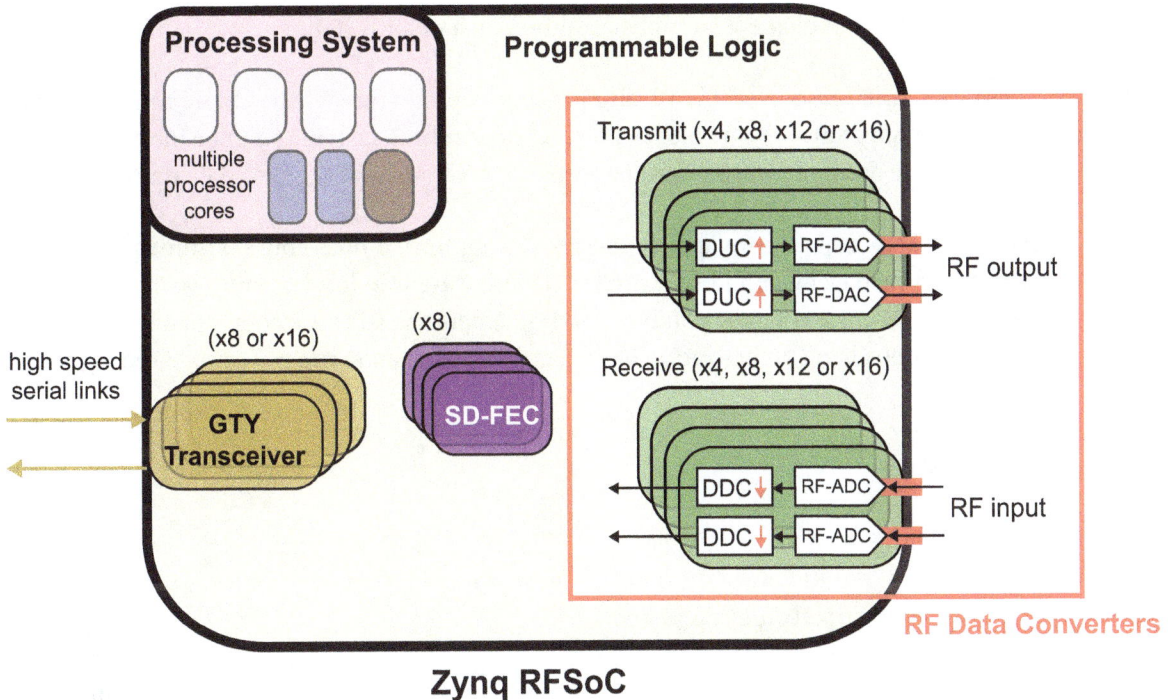

Figure 1.1: *High level illustration of the Zynq UltraScale+ RFSoC.*

- **RF Data Converter (RFDC) blocks** — Integrated Analogue to Digital Converters (ADCs) and Digital to Analogue Converters (DACs) that are capable of operating at very high sampling rates (multiple giga-samples per second, GSps, or GHz), making it possible to sample many radio signals directly. The RFDC blocks also incorporate programmable Digital Upconverters (DUCs) and Digital Downconverters (DDCs) to undertake the translation between baseband (signals close to 0 Hz) and modulated frequencies, and vice versa.

- **Soft Decision Forward Error Correction (SD-FEC) blocks** — Wireless communications schemes usually incorporate some form of Forward Error Correction (FEC) coding to mitigate against errors introduced in the radio channel (i.e. allowing the receiver to detect bit errors, and correct them where possible). The RFSoC architecture includes hardened SD-FEC blocks for implementing the coding and/ or decoding schemes used in selected mobile cellular standards. The use of these hardened SD-FEC blocks is an optimised and low power means of including FEC functionality within a radio design.

- **Gigabit Transceivers (GTY Transceivers)** — While the RFDCs represent the RF interface, in many cases the RFSoC requires a wired link to other elements of a radio system. The most prominent example is wireless cellular infrastructure: the RFSoC can be used to implement the radio front end, but high speed links back to the core network are needed, and these are usually implemented in wired or optical form. The required interface is supported on the RFSoC by hardened GTY Transceiver blocks, high rate serial interfaces that can implement a number of different standards [50].

In addition to these specialised resources, the PL also provides a programmable hardware resource for implementing custom radio architectures. From an SDR perspective, runtime flexibility can be integrated to change parameters of the hardware design from software, or even to reprogram sections of the PL design on the fly.

The PS architecture includes a quad-core application processing unit, a dual-core real-time processing unit, and features for platform management and security [131]. There are also local memories, interconnects, and peripheral interfaces. These features can be combined to implement the software components of an SDR, such as an operating system, control and orchestration of the hardware elements, and software-based algorithms.

1.3. Design Methods

SDR implementations can be realised using a variety of hardware technologies and software approaches. As noted in the previous pages, the RFSoC platform provides a powerful hardware platform for SDR, and its PS represents a capable host for the software components.

This book covers SDR implementation from a conceptual perspective, and also delves into the mathematical background where appropriate. Practical aspects of SDR system design are also presented, along with several examples and reference designs. The *Vivado*® Integrated Development Environment (IDE) is used extensively for hardware development, along with the block-based tools, *Vitis*™ *Model Composer* and *HDL Coder*, which are both used within the MATLAB® and Simulink® environment. The basis of the 'software' part of the SDRs we feature in the book is the PYNQ framework [39], an AMD open source project that aids software-hardware integration and productivity on AMD adaptive computing platforms. It achieves this by combining elements of SoC hardware design, Linux®, and Python, with a Jupyter™ environment for applications development [294].

RFSoC-PYNQ is an extension of PYNQ that incorporates support for the RFSoC platform and accelerates SDR development. It includes design support for hardened features such as the RFDCs, enabling easy control over the parameters and operation of hardware blocks. RFSoC-PYNQ is further introduced through the practical elements of the book, starting with Notebook Set A which immediately follows this chapter, and forms the basis of the reference designs provided alongside the book.

1.4. How to Use this Book

Whether you are reading in print or electronic form, this book is more than just the set of pages you see here. The intention is to provide a rich set of design resources, through the inclusion of sets of practical materials.

In the main book, you will find short sections that are enumerated as: ***Notebook Set A, B, C, D, E, F, G H, I.***

These are actually short summaries (just a few pages) of practical resources that exist separately, and can be freely obtained from the GitHub® repository accompanying the book, which is introduced in the next section. The user interface for these designs is via a set of Jupyter notebooks, hence the title 'Notebook Set'.

The Notebooks and designs can be freely obtained, downloaded to a supported RFSoC development board, and interacted with via Jupyter. Some of the notebooks sets cover related concepts on topics such as DSP and communications theory, and do not need to be run on a development board (although they can be). Notebook Set A (see page 7) provides an introduction to the practical elements of the book, and explains how to get started. These design resources can be investigated, reused, and built upon.

Of course, the rest of the book is simply a book — a set of chapters covering key topics in SDR and RFSoC technology, which are intended to be an accessible and informative read. These chapters are enumerated conventionally, from this current chapter through to Chapter 19. We hope you find them useful!

1.5. Related Work and Resources

This book aims to provide a valuable resource for getting started with RFSoC-based SDR design, and to act as a useful companion for those further along the journey. As we discovered in the process of writing the book, RFSoC-SDR design actually involves quite a number of topics! We have sought to cover each of these in an accessible style, and in a practical level of detail. There are undoubtedly limits to the scope of material that can be included here, and in many cases you can find entire textbooks on topics that are covered in a chapter, or subchapter, in this book. Sources of further information are cited where appropriate throughout, and these are listed at the end of the book in the *List of References*.

In addition, it is useful to highlight some particular sources of information and supporting resources:

- **The website for this book** https://www.rfsocbook.com.

- **RFSoC-PYNQ website** https://www.rfsoc-pynq.io/ **and GitHub repository** https://github.com/Xilinx/RFSoC-PYNQ — RFSoC-PYNQ resources, and a repository for the open source project.

- **StrathSDR GitHub repository** https://github.com/strath-sdr/RFSoC-Book — A repository hosting the practical materials accompanying this book.

- **The 'Exploring Zynq UltraScale+ MPSoC' book** [https://www.zynq-mpsoc-book.com] [131] — Our previous book on the Zynq UltraScale+ MPSoC is a useful reference, as many aspects of the architecture are in common with the RFSoC.

- **AMD support webpages** [https://www.xilinx.com/support.html] — AMD provides extensive documentation and support for RFSoC, and related development tools and design resources.

- **AMD University Program** [https://www.xilinx.com/support/university.html] — Information about university support, projects, events and other initiatives.

Materials shared via GitHub can be considered 'living' designs, and may benefit from bug-fixes, upgrades, and extensions over time.

1.6. Chapter Organisation

The book is organised into four main sections:

- **Introduction** — *Chapters 1 to 3, and Notebook Set A.* This first part of the book introduces SDR, the RFSoC platform, and provides a 'getting started' tutorial as the first practical element of the book.

- **DSP and communications concepts for SDR with RFSoC** — *Chapters 4 to 8, Notebook Sets B to E.* The fundamental concepts from Digital Signal Processing (DSP) and wireless communications theory are reviewed, along with complementary practical examples, with reference to the RFSoC as appropriate. A particular feature in this section is the introduction of an RFSoC-based spectrum analyser — if you have access to a supported RFSoC development board, you can download this application to your board, and explore the radio spectrum around you!

- **RFSoC features and practical design** — *Chapters 9 to 15, Notebook Sets F to H.* This section of the book looks more closely at the architecture and features of the RFSoC, and considers how to develop designs using the device. Particular focus is placed on the RFDCs, which represent the analogue/digital interface of the SDR architecture; we also examine the SD-FEC blocks and learn how to use them, and cover design methods for RFSoC more generally. The practical chapters also demonstrate frequency planning techniques, and present a complete transmit-receive radio design.

- **Systems and applications** — *Chapters 16 to 19, Notebook Set I.* The final chapters present a variety of systems as context for RFSoC SDR design. The multiple access method of Orthogonal Frequency Division Multiplexing (OFDM) is covered, along with a practical example. The use of RFSoC in cellular networks, and in the implementation of Multiple-Input-Multiple-Output (MIMO) and beamforming systems, is discussed. Finally, future directions in spectrum management and cognitive radio solutions are considered.

Moving forward, we continue in Notebook Set A with an introduction to Jupyter Notebooks and PYNQ, followed by a feature on SDR in Chapter 2.

Notebook Set A

Introduction to Jupyter Notebooks and PYNQ

Welcome! You have just arrived at the first of nine notebook chapters distributed throughout this book, which comprise practical exercises for your computer (or in fact, any device that can host a web browser) and RFSoC development board. The practical exercises are delivered using Jupyter [294], an interactive platform that can be accessed using a web browser. Before we begin, let's get started by installing the practical exercises on your system and introduce you to some guidelines that will be used throughout this book.

First of all, the exercises in this book can be installed on your computer or RFSoC platform, by using the set of installation and setup instructions at the GitHub repository: https://github.com/strath-sdr/RFSoC-Book. After navigating to this web page, simply scroll down to find up-to-date instructions. Additionally, if you have any questions or problems with the practical exercises and RFSoC designs featured throughout this book, you can use the GitHub issue tracker in this repository to tell us about them.

Jupyter notebooks are interactive documents that comprise executable code, documentation, and visualisation of results. Two types of Jupyter notebooks are included in the 'Notebook Set' chapters in this book. The first type can be executed on your computer or RFSoC platform; the second type must be executed on an RFSoC platform, because it uses the hardware features of the RFSoC (note that when a notebook is run on the RFSoC, it is executed natively using its Arm processors).

The following icons are used to distinguish between the types of notebook:

ALL   This icon is used when a Jupyter notebook can be executed on a computer or an RFSoC platform.

RFSoC   This icon is used when a Jupyter notebook can only be executed on an RFSoC platform.

The icons above appear at the start of every notebook chapter, and can help you distinguish between Jupyter notebooks that require an RFSoC development board, and those that can also run successfully on a computer. It is important that you have an RFSoC platform for the best experience with this book and the associated practical examples. However, if you do not have an RFSoC platform, you can still enjoy many of the supported practical exercises on your computer, via a web browser.

After following the installation and setup instructions on the RFSoC book's GitHub repository [327], you will now have everything you need to begin using the practical exercises presented in this book. To begin, the first notebook will present the Jupyter project and the key components of a Jupyter notebook, so that you can become familiar with this interactive environment. The second notebook will demonstrate key plotting and scientific computing libraries that enable visualisation and analysis in Jupyter. Notebooks three and four can only be used on your RFSoC platform. These Jupyter notebooks will introduce you to the RFSoC-PYNQ framework [44], which is a specialisation of the PYNQ framework [39] specifically for RFSoC platforms, and the use of hardware overlay designs, respectively.

There are four notebooks to explore throughout this chapter using Jupyter and your RFSoC platform. The notebooks and their relative locations are listed as follows:

ALL   01_jupyter_lab.ipynb — *rfsoc_book/notebook_A/01_jupyter_lab.ipynb*

ALL   02_visualisation_and_analysis — *rfsoc_book/notebook_A/02_visualisation_and_analysis.ipynb*

RFSoC   03_pynq_introduction.ipynb — *rfsoc_book/notebook_A/03_pynq_introduction.ipynb*

RFSoC   04_overlays.ipynb — *rfsoc_book/notebook_A/04_overlays.ipynb*

A.1. Getting Started with Jupyter

Project Jupyter is a non-profit, open source and community driven effort to create a web-based interactive computing environment with a focus on data science and scientific computing [294]. The Jupyter community is responsible for developing and governing the Jupyter Notebook standard, as well as creating a complete set of tools to experiment, develop, share, and explore all things relating to software code.

Jupyter initially started as an interactive shell for Python [295], where it was known as IPython [289]. It now supports over one-hundred different programming languages, which use their own Jupyter kernels written by the community. Throughout this series of notebooks, we will use the Python programming language to demonstrate examples and interact with the RFSoC development platform. The first notebook, which will introduce you to Jupyter, can be opened here: *rfsoc_book/notebook_A/01_jupyter_lab.ipynb*.

Jupyter has two interactive graphical environments. These are classic *Jupyter Notebooks,* and the more recent *JupyterLabs,* which is a browser-based Integrated Development Environment (IDE) that includes support for

Jupyter Notebooks. All notebooks that are used throughout this book will leverage JupyterLab, which combines several useful tools into one environment. These tools include the text editor, command terminal, workspace viewer and other useful features that improve the overall user experience when developing with this environment. After setting-up JupyterLab, you will be presented with the window shown in Figure A.1.

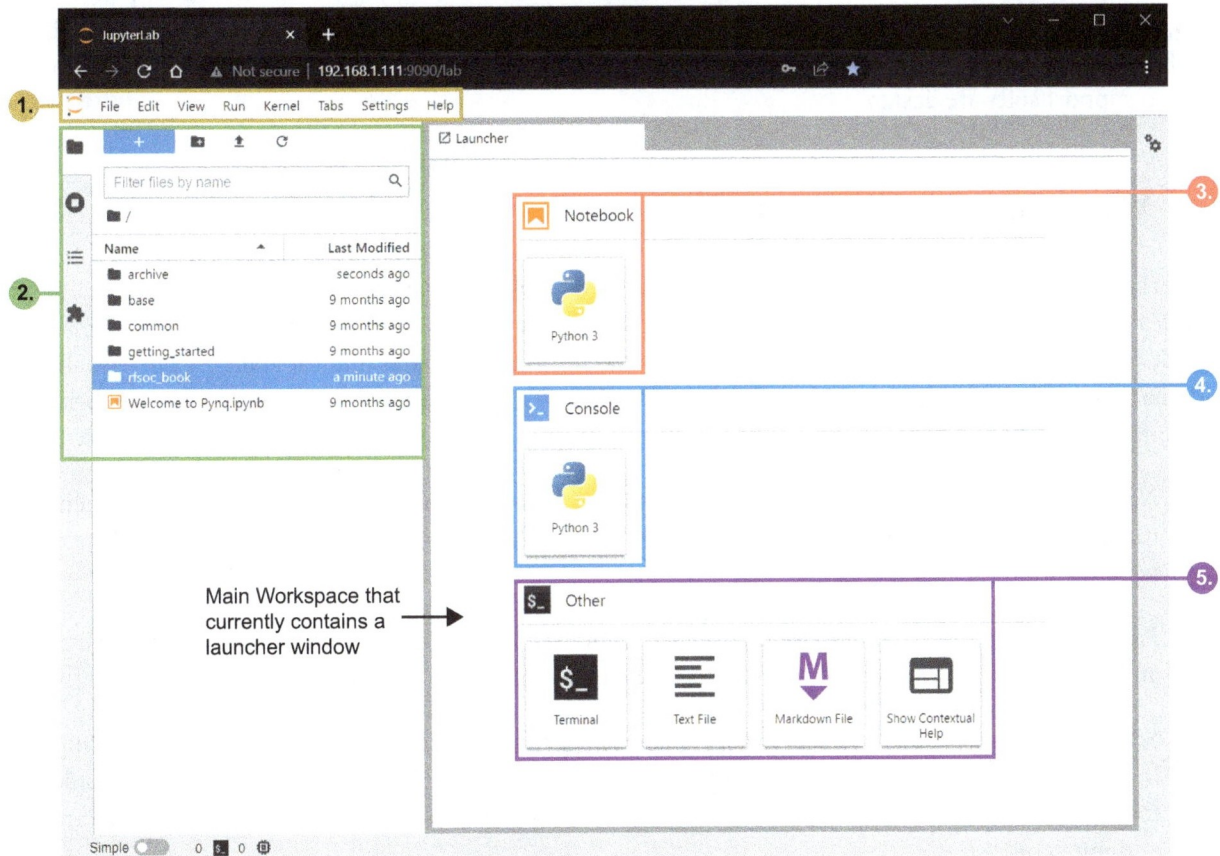

Figure A.1: The JupyterLab window with annotations.

There are three main areas on the JupyterLab interface. These are the **menu bar**, **sidebar**, and the **main workspace**, which is in the centre of the window given in Figure A.1. The menu bar at the top of the window exposes several settings and commands relating to file control and the operation of JupyterLab. The sidebar provides a set of useful tools, which includes a file browser, a workspace manager, a notebook navigation tool, and the Jupyter extension manager.

The main workspace usually contains interactive notebooks that you are able to view, modify, and run. In Figure A.1, the main workspace contains a launcher window, which will be visible when you first launch JupyterLab. The launcher window helps you to create new notebooks and other useful files.

The ***Getting Started with Jupyter*** notebook will introduce you to the JupyterLab environment and help you navigate many of its key tools and features. By the end of this notebook, you will have a better understanding of why JupyterLab is an excellent development environment, and you will also learn about useful Python libraries that can support software design.

A.2. Visualisation and Analysis

JupyterLab can leverage several Python libraries for plotting, evaluating, and manipulating data. These libraries include NumPy™, Pandas™, SciPy™, MatplotLib, and Plotly™. In the second notebook, we will introduce these libraries so that you can become familiar with their features and capabilities. The visualisation and analysis notebook is located at ***rfsoc_book/notebook_A/02_visualisation_and_analysis.ipynb***. Below is a list containing Python libraries that are used in this notebook and their descriptions.

- ***NumPy*** [281] may be one of the most commonly used Python libraries for high-level mathematical functions and multi-dimensional matrix operations. It is based on optimised C code, which results in fast execution.

- ***Pandas*** [286] is a data manipulation and analysis library, which provides developers with data structures and numerical tables (primarily sequences and data frames) for fast, flexible data processing.

- ***SciPy*** [314] is an enormous Python library consisting of various modules for scientific computing tasks that are commonly found in science and engineering disciplines. There are many different modules and tools including integration, interpolation, decimation, spatial processing, optimisation, Fourier analysis, and much more.

- ***MatplotLib*** [257] is a useful library for generating plots in Python. This library provides a vast number of options for user customisation.

- ***Plotly*** [291] provides users with a simple set of classes and methods for creating figures and plots quickly with very few lines of code. By default, the generated plots are interactive, allowing the user to hover over data points to reveal further information.

- ***ipywidgets*** [221] allows the user to create graphical interfaces in software, such as buttons and text boxes. These interfaces are commonly known as *widgets* and are very useful for facilitating user interaction with software and hardware designs.

A.3. The PYNQ Framework and RFSoC-PYNQ

Python Productivity on Zynq, known as PYNQ [39], is an open-source project from AMD. The aim of the project is to provide a framework that simplifies hardware and software development on AMD SoC platforms. The third notebook requires an RFSoC development platform to explore and use PYNQ. You can launch this notebook from the following directory: ***rfsoc_book/notebook_A/03_pynq_introduction.ipynb***. If you are not using JupyterLabs on RFSoC, this notebook will not be available in your workspace.

This notebook begins with a brief introduction to the PYNQ framework and ecosystem, which is illustrated using a simplified diagram in Figure A.2. Note the green components, which are specific to RFSoC-PYNQ.

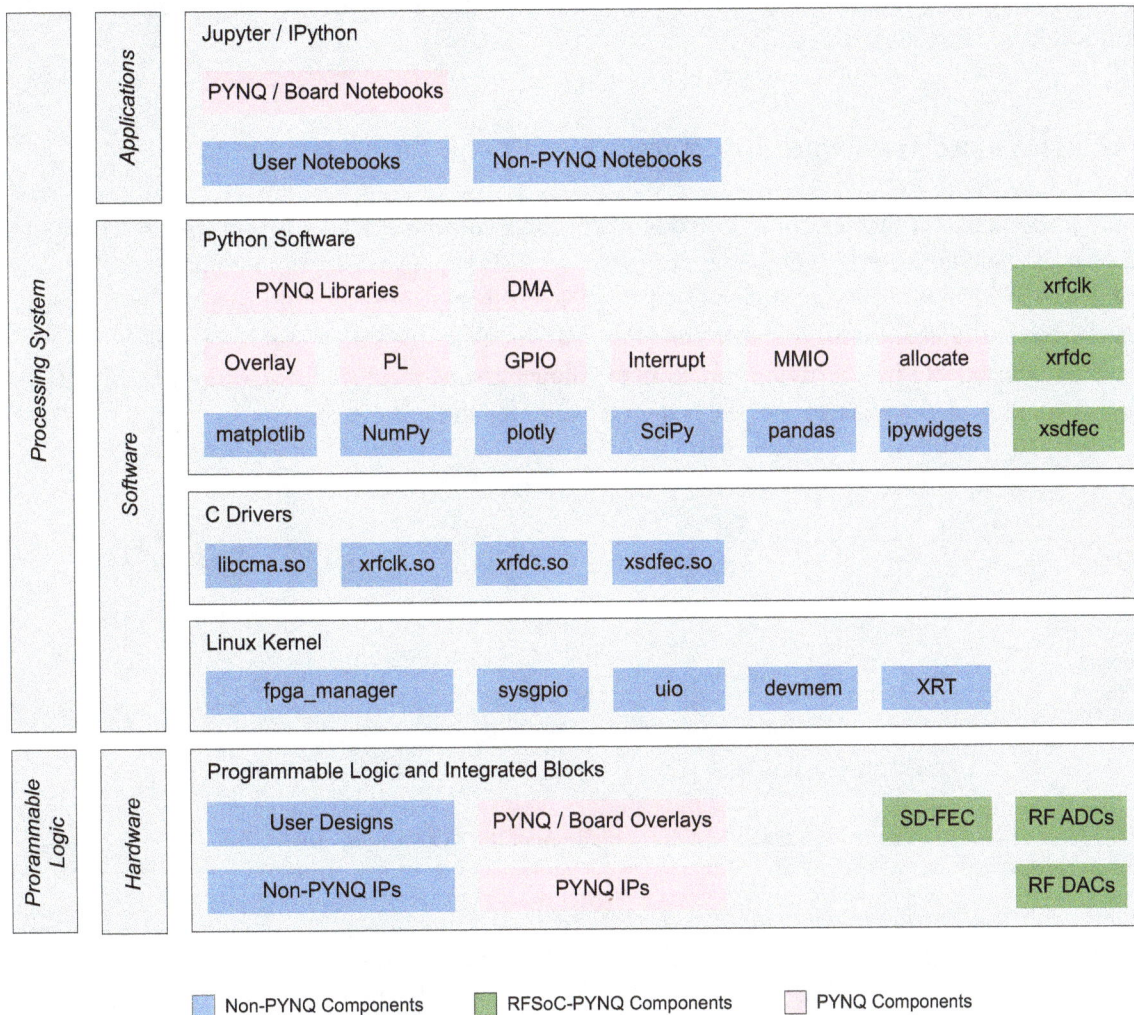

Figure A.2: Simplified diagram of the PYNQ framework and ecosystem (contains relevant RFSoC software libraries).

Online resources such as the PYNQ website [39] and the corresponding *Read The Docs* page [40], which contains useful documentation, will also be discussed. We will then explore several libraries that are provided alongside PYNQ. These libraries are generally used to control and interface to Intellectual Property (IP) cores that are implemented in the PL section of the device.

An introduction to PYNQ *overlays* will be provided. Overlays are programmable hardware configurations that are implemented on the PL [38]. They can be controlled from software, and may be used to accelerate a software application operating in the RFSoC's PS, to give one example. We will investigate the *base overlay* design using an RFSoC development board (the base overlay is the default PL configuration provided in the PYNQ distribution for your RFSoC platform).

Lastly, if you have any questions about the PYNQ framework, you can visit the support area of the PYNQ website, located at: https://discuss.pynq.io/.

A.4. Overlays and Hardware Interfacing

The final notebook in this chapter can be found in ***rfsoc_book/notebook_A/04_overlays.ipynb***. This notebook expands our understanding of PYNQ overlays by investigating a simple overlay design on RFSoC. The overlay design consists of a Numerically Controlled Oscillator (NCO) that generates a sine wave. The sine wave is sent to a data movement IP core, so that the data can be transferred from the FPGA logic fabric to the RFSoC's PS. The NCO is controllable by the user from Jupyter, allowing sine wave frequencies and amplitudes to be modified. A functional block diagram of the system is presented in Figure A.3.

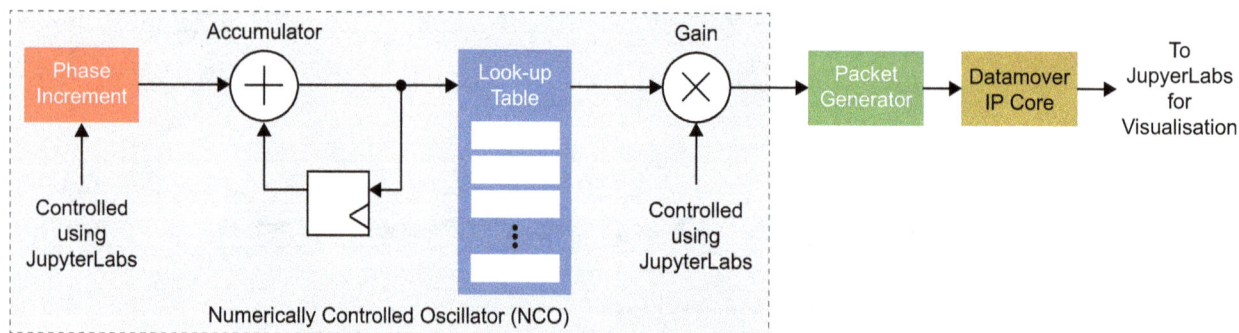

Figure A.3: Functional block diagram of the NCO overlay design.

Throughout this notebook, we will explore several PYNQ classes including the DefaultIP and DefaultHierarchy classes, which support IP core and hierarchy driver development, respectively. The Advanced eXtensible Interface (AXI) Direct Memory Access (DMA) controller [18] will also be introduced, which facilitates data movement between the RFSoC's FPGA logic fabric and PS (shown as the 'Datamover IP core' in Figure A.3). Finally, visualisation and analysis tools, such as NumPy and Plotly, will be used to process the sine wave data and plot it in a graph.

Chapter 2

Software Defined Radio

Louise Crockett

This chapter provides an introduction to SDR, prefaced by some fundamental background on radio communications in general (further review on wireless communications will be presented in Chapter 6). In particular, we consider the increasing sampling rates in modern devices, and the three main architectures for SDRs that arise from low, medium, and high sampling rates at the Analogue-Digital interface. The associated design considerations and trade-offs are also discussed.

2.1. Radio Fundamentals

Before discussing *Software Defined Radio*, it is worthwhile first defining *radio communications* in general.

2.1.1. Basic Radio Architecture

As a a starting point, a simple model of an ideal radio communications system is presented in Figure 2.1, highlighting three sections of interest: the transmitter, channel, and receiver. As we proceed through this chapter, more detail will be added to this model, and architectures for *software defined* radio will be developed.

At the transmit side, data is prepared for transmission (*baseband processing*), and the resulting signal is then converted up in frequency, such that it is centred in the allocated frequency band for transmission (i.e. the signal is *modulated*, which will be explained in more detail shortly). The modulated signal is then amplified, and transmitted by an antenna.

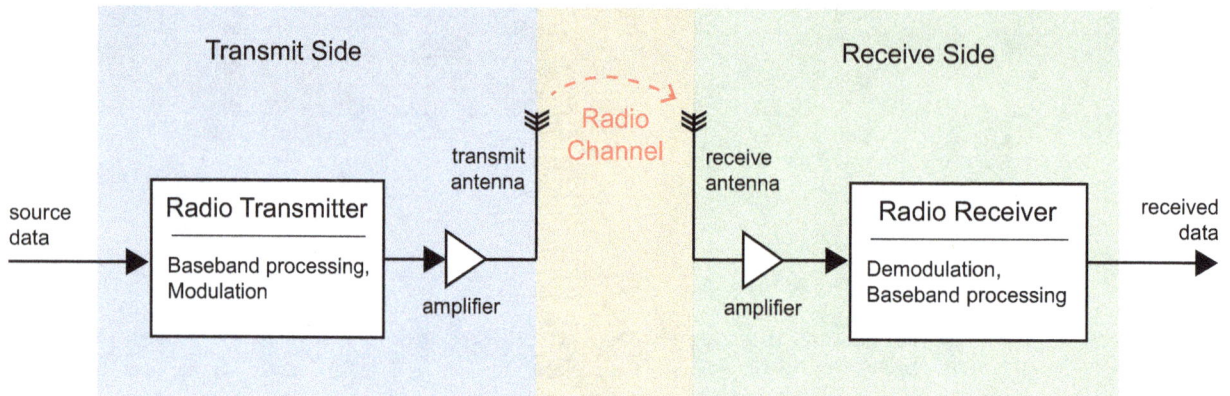

Figure 2.1: A simplified model of a radio communications system.

The physical link between the transmitter and receiver is the *channel*, and it can be wired or wireless. In radio communications, we usually consider the channel to be wireless: at the transmit side of a wireless link, an antenna is used to emit radio waves, and another antenna is used at the receive side to absorb them. It is also possible to transmit radio signals along a cable or fibre (indeed, many of the practical activities presented in this book will use a cable, to avoid emitting radio signals that could interfere with other devices using the radio spectrum). As will be discussed later, various impairments are typically introduced in the radio channel, which degrade the signal, making it more difficult for the receiver to retrieve the transmitted data.

At the receive side, the equivalent operations to the transmitter are performed in reverse. After the signal is absorbed by a receive antenna, it is amplified, *demodulated* (or mixed), and then further processed at baseband.

As we will discuss in the coming chapters, the transmitter is responsible for preparing data for transmission, and modulating it onto the allocated RF band. The receiver has the task of demodulating the received signal and recovering the information that was transmitted. As a result of impairments experienced in the channel, the receiver has the difficult task of recovering the data, and consequently it is normally significantly more complex to design the receiver than the transmitter.

2.1.2. Modulation and Demodulation

The operations of *modulation* (usually referring to shifting, or mixing, a signal up in frequency to a carrier frequency), and *demodulation* (analogously shifting a signal down in frequency) are shown in Figure 2.2. The source data for transmission is said to be *at baseband*, meaning that its constituent frequency components are located at DC (i.e. close to 0 Hz). Likewise, the demodulated received data is also *at baseband*. Modulation and demodulation are used to shift the signal between baseband, and the Radio Frequency (RF) band allocated for the transmission. As already mentioned the term mixing is often used alongside modulation and demodulation. (In Figure 2.2 we are just illustrating the general process of modulation and demodulation, or mixing to and from carrier frequencies, and the design details for required components, any filtering stages required are not shown in this introductory section - this will be detailed later.)

The mathematics of these modulation and demodulation operations will be covered later, in Chapter 7, however we can note that in each case, the signal is multiplied or 'mixed' with a cosine wave at the RF centre frequency, f_c. The value of f_c can vary across a very wide range, but some examples are ~100 MHz for FM broadcast radio stations, ~600 MHz for digital television, ~1.575 GHz for GPS signals, or 28 GHz for mm (millimetre) waves for the high bands of 5G.

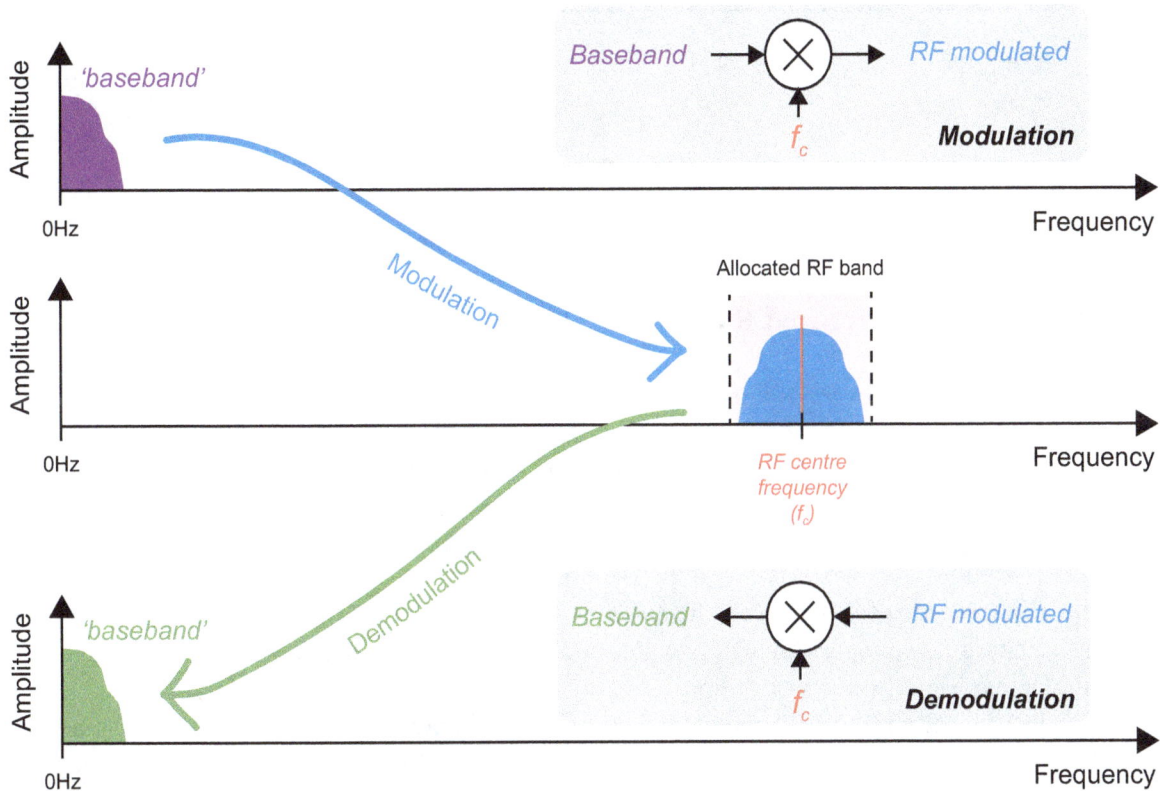

Figure 2.2: Concept of simple modulation and demodulation from baseband to a carrier frequency. (Note that the low pass filtering components for the receiver products at twice the carrier frequency are not shown.)

2.1.3. Transmission over the Radio Spectrum

Radio communication is achieved by transmitting information using radio waves or microwaves, which are within the electromagnetic spectrum as illustrated in Figure 2.3. Other parts of the electromagnetic spectrum can also be used to send information over wireless channels, for instance infra-red light, which is customarily used for television remote control handsets as well as some other short-range applications.

If a signal were not modulated, it would include frequency components around 0 Hz, which is too low in frequency for successful propagation through a radio channel. Further, by modulating signals to *different* sets of frequencies, multiple transmissions can take place simultaneously on different portions of the RF spectrum.

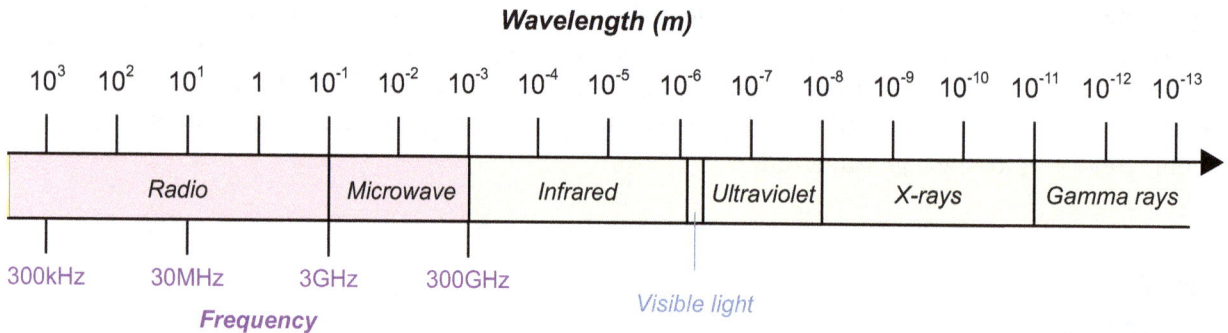

Figure 2.3: *The electromagnetic spectrum.*

2.2. What is Software Defined Radio (SDR)?

In this book, we consider a Software Defined Radio to be:

A radio in which aspects of functionality are implemented in, or controlled by, software.

Therefore, the term *Software Defined Radio* refers collectively to systems that are implemented in this manner.

An important characteristic of SDRs is their flexible functionality, which contrasts with the more traditional approach of fixed hardware designs. An SDR contains one or more programmable elements, often in the form of reconfigurable hardware. Aspects of radio processing may also be implemented in software. The major benefit of SDR is the inherent flexibility obtained: the operation of a radio can be changed without making any physical alterations to the device; making it easy to deploy systems in different territories and frequency bands, or to add new modes or features.

As will be discussed later in this chapter, there is no single, definitive architecture for an SDR. Multiple different types of devices, and combinations thereof, can be used to create SDR systems. Of course, in this book our key focus is RFSoC-based implementation, and therefore the discussion in later chapters will be centred on SDR architectures using the RFSoC platform.

2.3. Motivations for SDR: Then and Now

Software Defined Radio is not a new concept, and in fact has been around for several decades. In some ways, it was an idea that pre-dated the technology required to enable it! In this section we examine some of the early motivations for SDR, and how this has evolved to the present day.

2.3.1. The Advent of the SDR Concept

The concept of SDR originated with Joe Mitola's seminal papers on 'software radio' and 'software defined radio', published in the 1990's [242], [264], [265]. He noted that radio systems had largely migrated from analogue to digital processing in the 1970's and 1980's, and foresaw a further revolution with the introduction of more flexible architectures that would reduce the reliance on hard-wired components, and instead feature programmable elements whose functionality could be defined using software.

At that time, the rapidly increasing performance (and reducing costs) of enabling technologies such as Analogue to Digital Converters (ADCs), Digital to Analogue Converters (DACs), and embedded computing processors was seen as a key driver of the impending SDR revolution. This prediction has borne out — in the 1990's, ADCs and DACs were capable of sampling at 10's of Mega samples per second (Msps) [265], while at the time of writing (in the early 2020's), rates are entering the 10's of Giga samples per second (GSps). Meanwhile, the density of integrated circuits has expanded in line with Moore's Law, roughly doubling every 18 months since the mid 1960's [240], [268], enabling more sophisticated processing of radio signals.

The high level architecture of an *ideal* SDR was originally defined as in Figure 2.4, wherein all of the radio's functionality, other than the RF conversion stages, is implemented in software. The ADCs and DACs of that time did not enable this concept of the ideal SDR to be realised, but it was an insightful model, and one which can now be achieved using AMD RFSoC devices and other platforms. Interestingly, Mitola's ideal SDR was set in the context of voice and 'multimedia' communications, which perhaps predicted the richer set of wireless applications used in mobile networks today.

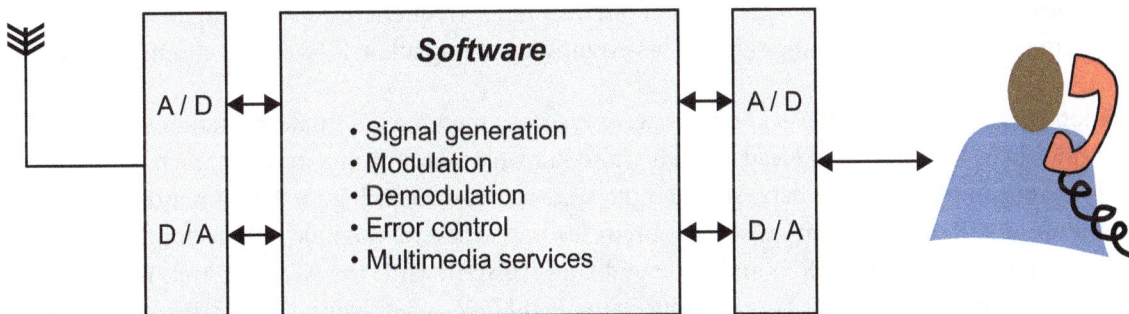

Figure 2.4: 'An Idealized Software Radio' (as defined and depicted by Joe Mitola in [265]).

2.3.2. SDR Programmes and Military Drivers

Also in the early 1990's, the US Defense Advanced Projects Research Agency (DARPA), part of the US Department of Defense (DoD), initiated research into SDR through its SPEAKeasy programme [126], which was later followed by the Joint Tactical Radio Systems (JTRS) programme. The ability to reprogram and upgrade radio equipment was seen as an attractive attribute for military systems, which often have to be maintained over long product lifecycles. Moreover, the flexibility for radios to interoperate with different systems on the battlefield was a significant motivator for military SDR.

Soon after this early work on SDR, in 1996 the Modular Multifunction Information Transfer Systems (MMITS) Forum (later renamed the *SDR Forum*... and subsequently the *Wireless Innovation Forum*) was established. A not-for-profit organisation, its purpose was to develop concepts around SDR, and to define an open, modular architecture for SDR that would enable wider adoption and innovation. Interest in SDR started to grow, with academic organisations, government, and industry seeing potential in the idea, particularly for military applications. In Europe, a number of collaborative research projects were developed to progress SDR technology. Early progress in SDR development is well reviewed in [185].

2.3.3. Modern SDR Applications

Moving forward to the present day, SDR has evolved from expensive defence programmes to a more attainable price level, even to the point that students and hobbyists can buy low cost SDR equipment (<$50) that can be interfaced to a home computer. The availability of SDR equipment at relatively low cost also provides new opportunities for amateur radio enthusiasts. Furthermore, SDRs are extremely useful for research and prototyping purposes, owing to their reprogrammable nature, and ability to operate over a wide frequency range.

SDR has found widespread application in commercial networks in recent years, largely because the enabling technology (in particular, processing platforms, DACs, and ADCs) has advanced considerably while also becoming cheaper. The early ideas of Mitola are now realisable using commodity hardware such as desktop processors, and off-the-shelf SDR front-ends. With ADC and DAC sampling rates now entering the 10's of GHz, it is possible to place the analogue / digital interface at RF frequencies, directly digitising signals without the requirement for analogue mixing stages. This evolution and impact on SDR will be discussed in Chapter 8.

Applications in the 2020's include 4G and 5G networks and emerging 6G implementations, where SDR units can be deployed as Remote Radio Heads (RRHs) (also known as Radio Units, or RUs), each handling multiple channels of transmit and receive data. An example system is shown in Figure 2.5. Functionality can be split across remote, distributed, and centralised resources, leading to several possible options for the network architecture. The *fronthaul* links the RRHs and Baseband Units (BBUs), while the *backhaul* links the BBU resources to the core network. 5G networks also introduce a *midhaul* link, which connects distributed and centralised resources (the midhaul is not depicted in Figure 2.5, but see Chapter 17 for more details).

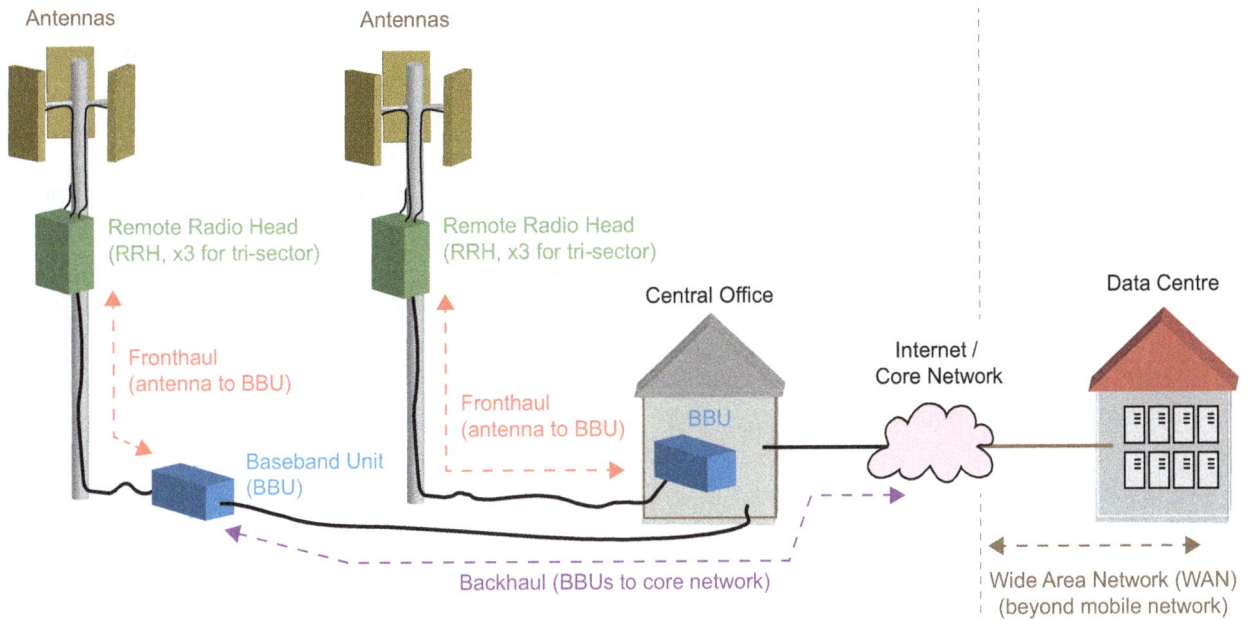

Figure 2.5: An SDR-based mobile network.

The *software defined* aspect provides operational flexibility, as well as a clear path for future upgrades; for instance, to implement new mobile standards or apply security patches.

2.4. The Radio Frequency Spectrum

The electromagnetic spectrum was introduced in Section 2.1, and it was noted that the 'radio' and 'microwave' frequency bands are used for radio communication[1]. These bands are often defined as:

- Radio: 3 kHz to 3 GHz

- Microwave: 3 GHz to 300 GHz

Most widely used radio standards use sub-6 GHz frequencies, and therefore these are the primary bands of interest for SDR. Next, we provide a brief overview of how the radio and microwave spectrum is used.

---

1. The Terahertz band, which overlaps the Microwave and Infrared regions (extending from 0.3THz to 10THz), is also emerging as a band of interest for future wireless communications [121],[275].

2.4.1. Spectrum Allocation

The radio spectrum is a valuable but finite resource, and therefore it needs to be managed for the collective good. This is traditionally achieved using a formal set of frequency allocations, i.e. specific bands of frequencies are associated with particular types of use, usually on a national basis.

Figure 2.6 shows some examples of bands that are allocated in the radio and lower microwave regions. Please note, however, that this is far from the complete picture! For a more detailed view of the radio spectrum allocations in your country or region, please refer to information published by the relevant spectrum regulator. For instance, the Office of Communications (Ofcom) in the UK, and Federal Communications Commission (FCC) in the USA both publish tables showing the allocations for all bands.

The fixed allocation of bands is useful because it establishes a very clear set of rules that can be readily followed. Bands are generally designated as either *licensed* (users apply for a licence and normally pay a fee), or *unlicensed* (meaning that no fee is required, and any user can access the band provided that they comply with the relevant rules). More recently, new paradigms for spectrum management have started to evolve, including the idea of dynamic and shared spectrum. This topic will be discussed further in Section 2.5, and Chapter 19.

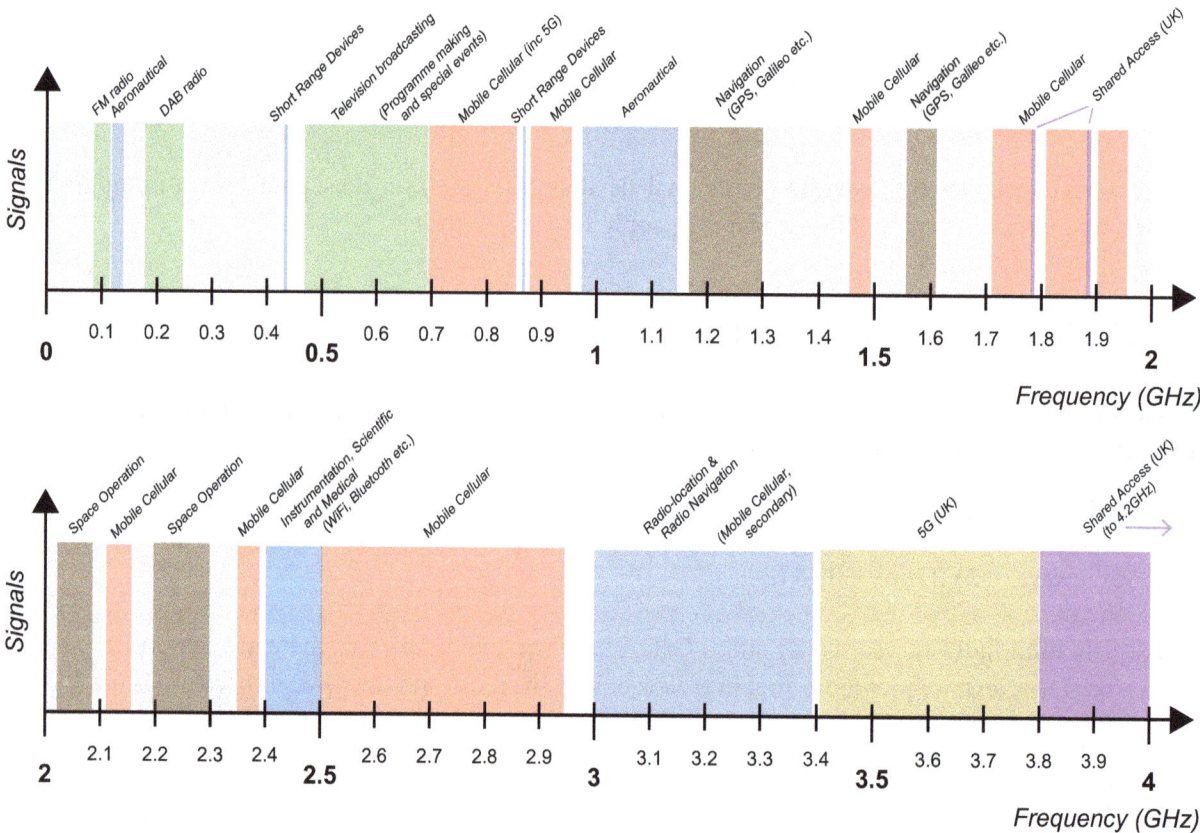

Figure 2.6: Some of the common radio bands up to 4 GHz in UK (for outline only — varies geographically).

2.4.2. Propagation and Path Loss

You might notice from Figure 2.6 that wider frequency bands tend to be available at higher frequencies. This has the advantage that more information can be carried simultaneously (faster data rates). The disadvantage of using higher frequencies is increasing path loss — even in the theoretical best case scenario, where a signal propagates through 'free space', the proportion of signal power lost increases with the square of frequency. This is defined by the Friis Free-Space Path Loss model,

$$FSPL(\text{dB}) = 20\log_{10}\left(\frac{P_r}{P_t}\right) = 20\log_{10}\left(\left(\frac{c}{4\pi df}\right)^2\right) \qquad (2.1)$$

where P_t and P_r are the transmitted and received power, respectively, d is the transmission distance, f is frequency, and c is the speed of light, i.e. $3 \times 10^8 \ ms^{-1}$.

For more realistic scenarios, where the channel is not simply 'free space', the path loss is even more significant. Higher frequencies are particularly poor at penetrating obstacles like trees, walls and buildings. A number of models have been proposed for radio propagation in different environments, including the Okumura-Hata model [190] for urban environments, the Longley-Rice model for outdoor propagation over irregular terrain [202], and the International Telecommunication Union (ITU) models covering various environments [214]. For radio links between earth and space, e.g. for satellite communications, propagation through the different layers of the atmosphere must also be taken into account [345]. We will not focus extensively on radio propagation in this book, but the interested reader may find [316] useful for further information on this topic.

Given the relationship between frequency and path loss, terrestrial radio applications that cover wide geographical areas, such as television and radio broadcasting, tend to use relatively low frequencies (up to about 700 MHz). In-building wireless systems like Wi-Fi, which are designed to deliver high data rates within small coverage areas, adopt bands at much higher frequencies (e.g. 2 GHz and above). There is even potential for *very* high frequencies (greater than 20 GHz) to be used within individual rooms to deliver ultra-high data rates, for applications like in-home media and entertainment.

2.4.3. Spectrum Harmonisation

The allocation of RF spectrum bands is managed at a country level by national regulators, such as the FCC in the USA, and Ofcom in the UK. At an international level, the ITU develops standards and regulations that promote global harmonisation in the use of the radio spectrum.

Although spectrum allocation and usage are not completely standardised across all countries and territories, many bands are common, which supports interoperability and the development of wireless technology. Also, at a practical consumer level, this means that your mobile phone continues to work when you visit other countries on holiday!

2.4.4. Spectrum Licensing and Legal Usage

In addition to allocating spectrum, national spectrum regulators are also responsible for issuing licences to users for specific frequency bands, ensuring compatibility between uses of the spectrum, and generally managing the spectrum efficiently.

When starting out in SDR, it is extremely important to note the legal implications of accessing the radio spectrum. The traditional method of managing the radio spectrum is based on fixed frequency allocations, wherein each defined frequency band is allocated for a particular class of use (e.g. public safety, military, or satellite applications), or to a specific user (such as a Mobile Network Operator, or MNO). The majority of allocated bands are *licensed*, meaning that you cannot legally transmit on these frequencies without a valid licence. Therefore, SDR developers must exercise care when building and testing SDR transmitters; it is a personal responsibility to check local spectrum regulations and abide by them.

Selected RF frequency bands are *unlicensed*, including the Instrumentation, Scientific and Medical (ISM) bands at 868MHz (in Europe), 915MHz (in the USA), and 2.4 GHz and 5.8GHz (internationally). Common uses include Bluetooth®, Wi-Fi®, and baby monitors. These bands are ready candidates for wireless transmissions using SDR equipment, provided that applicable regulations on power output are adhered to. Another set of frequency bands worth mentioning are the amateur radio bands, access to which is shared between appropriately qualified amateur operators [161].

Cabled tests are a good alternative when it is not possible to legally transmit on the desired RF frequencies.

Another important consideration relates to radio reception — while the details may vary in different jurisdictions around the world, is it generally illegal to decode radio transmissions that you are not intended to receive. For instance, it would be illegal to use an SDR to intercept others' Wi-Fi communications, or attempt to listen to their mobile phone calls, or to gather wireless sensor data from a private network. On the other hand, you may of course receive transmissions that you have generated, as well as 'public' signals including broadcasts from radio and television stations, Global Navigation Satellite System (GNSS) signals, and so on.

2.5. Spectrum Policy

While the fixed spectrum allocation model has been used very successfully for a long time, some changes are starting to take place in how spectrum is managed. The motivation for re-evaluating approaches to spectrum management derives from the pressure that radio spectrum is now under, sometimes referred to as the "spectrum crunch". In short, the demand for spectrum is accelerating, but the resource itself is finite; leading to an apparent shortage, or threat of shortage. Here, we contrast the traditional model of spectrum management, with new and emerging approaches.

2.5.1. Fixed Frequency Allocations

The fixed model of spectrum allocation, as introduced in Section 2.4.1, applies strict conditions for using each licensed band, which is robust in terms of preventing interference, but often results in spectrum being under-utilised. If a particular band is licensed but not in active use, then according to the fixed allocation model, it cannot legally be used for any other purpose. Further, if licences are issued on a country-wide basis, but the licensee only wishes to use the licensed spectrum in a particular region, the band is left fallow (yet unavailable) in other areas.

Therefore, it could be argued that there is no actual shortage of spectrum (at least, not yet!), but rather an inefficient method of controlling access to it. The fixed frequency allocation method is however well-understood, easily implemented and effective — this method is likely to persist well into the future, for much of the regulated spectrum across many regions.

2.5.2. Shared Spectrum

New approaches to spectrum management have begun to emerge. For instance, in the UK, spectrum regulator Ofcom started to make 'shared access' licences available in specific shared bands, on a geographical (local area) basis, from July 2019 [283]. These shared access bands are overlaid with existing licensees from the fixed allocation model, and permit geographical reuse of the radio spectrum, thereby increasing efficiency.

In the USA, the FCC launched Citizens' Broadband Radio Service (CBRS) as a shared spectrum service [160]. CBRS uses a different model, designating users into three tiers (Incumbent Access, Priority Access, and General Authorized Access) with decreasing levels of protection from interference. Spectrum access is coordinated by an automated system for allocating frequencies, known as a Spectrum Access System (SAS), which operates in real time. The operation of the SAS is based on a database, although it can also incorporate Environmental Sensing Capability (ESC) to detect transmissions by the Incumbent Access user. Its primary purpose is to protect incumbent user transmissions by re-allocating frequency bands to users from other tiers, to prevent interference to the incumbent user and priority users.

2.5.3. Dynamic Spectrum Access

Looking further ahead, there is significant research interest in more agile models of spectrum sharing (Dynamic Spectrum Access, or DSA). The prevailing vision of DSA will involve some combination of a database-driven approach, and local awareness at each radio terminal through active spectrum sensing. The Dynamic Spectrum Alliance (also with the acronym DSA!) is a not-for-profit organisation dedicated to advancing dynamic spectrum technology and policy around the world, and is a useful source of information on the latest developments [144].

SDR is an important enabling technology for DSA, given the inherent flexibility and software control of SDR devices. Further, SDR provides an ideal platform for *cognitive radio* systems, wherein radio terminals are additionally capable of autonomous decision-making, for instance to determine the optimum band(s) for transmission in a DSA radio. The CBRS model in particular represents a step towards DSA and cognitive radio, although its sensing and decision-making do not currently provide protection to the lowest tier (but most flexible) class of users.

2.6. Wireless Communications Standards

Most radio communications adhere to a standard, meaning that the protocol, modes and parameters of transmissions are documented and widely available. The main motivation of standards is to enable more widespread and seamless adoption, and thereby to maximise the economic benefit of the technology. Taking the example of Wi-Fi (which is based on the IEEE 802.11 group of standards for wireless local area networks), a 2021 report commissioned by the Wi-Fi Alliance® trade body estimated the economic benefit of Wi-Fi as $3.3 trillion (USD) in 2021, rising to $4.9 trillion in 2025 [356].

The alternative to published standards are proprietary schemes, where a particular company or organisation adopts its own custom protocol, without engaging more widely with other partners. There can be some distinct technical advantages of the proprietary approach, such as the ability to optimise a protocol for a particular application, thereby optimising the solution (e.g. minimising energy consumption). A practical drawback is that developing a proprietary radio is likely to be considerably more expensive than a standards-based approach, as off-the-shelf parts and solutions cannot be directly integrated.

2.6.1. Wireless Standards

By developing communications products in line with a standard, equipment from different manufacturers becomes interoperable, and this encourages competition and innovation in the market, driving quality up and prices down. For instance, you can acquire a Wi-Fi router for use in your home, without being concerned whether the manufacturer of the router matches that of your laptop — both communication devices adhere to the same standard, and as such they can work together seamlessly. Therefore, as a consumer, you can select a router based on other factors, such as price, features, brand perception, and so on.

A small selection of popular wireless communications standards are provided in Table 2.1, covering a variety of different applications. Note that in some cases, there is more than one standards body involved — for instance, the lower layers of the Zigbee® protocol stack were standardised by IEEE, and the upper layers by the ZigBee Alliance (now known as the Connectivity Standards Alliance).

Table 2.1: Examples of wireless communications standards.

Name	Standard Body	Frequency Band(s)	Applications
Bluetooth	Bluetooth Special Interest Group, IEEE (802.15.1)	2.4GHz	Short range connections (<10m). Computer peripherals, wireless headsets, in-car phone connections, etc.
Wi-Fi	IEEE (802.11)	2.4GHz 5.8GHz	Wireless connections for device-device local connectivity (computers, phones, tablets, etc. with wireless access points) up to about 100m.
5G cellular	3GPP™	numerous bands, use varies geographically	Mobile phone and mobile data networks. Coverage varies according to frequency band.
Zigbee	Connectivity Standards Alliance, IEEE (802.15.4)	868 MHz 915 MHz 2.4 GHz	Smart home, industrial Internet-of-Things (IoT), agriculture. Range: 10 - 100m (depending on the environment).

The creation and maintenance of wireless standards is a complex process, demanding specialist technical input and representations from a wide range of stakeholders such as industry trade bodies, equipment manufacturers, and end users, each of whom may have different sets of desires and requirements. As such, standards are managed by a standards body, which can maintain a position of neutrality and focus on developing the standard for the common good. The primary two standards bodies for wireless communications are the Institute of Electrical and Electronics Engineers (IEEE) [204], and the 3rd Generation Partnership Project (3GPP) [1], which works in partnership with regional and national standards bodies from around the world.

2.6.2. What's in a Standard?

A standard is a document that specifies all aspects of the communications protocol. This document can run to thousands of pages and may represent the combined efforts of hundreds of people. To provide some examples of aspects details by a standard (note, this is far from exhaustive!), it should include:

- The message formats at each layer of the protocol stack.

- The waveforms and modulation schemes used in the Physical (PHY) layer.

- The PHY layer channel configuration.

- Media Access Control (MAC) layer protocols (e.g. mechanisms for re-sending lost packets).

- Management functions.

It is perhaps equally important to outline what is *not* in a standard.

A standards document specifies <u>what</u> must be implemented, but not <u>how</u> it should be implemented. For instance, a standard might specify the baud rate and composition of a data frame, the modulation format, the pulse shaping filter applied to the transmitted signal, and the maximum permitted adjacent channel interference arising from a transmission. However, the standard would not specify how to design the transmitter to achieve these specifications; rather, many of the decisions are the responsibility of the systems developer, e.g. the company developing a standard-compliant radio, based on the specification. Importantly, there is considerable scope to innovate — for instance to develop a product that meets the specification with a more efficient design than a competitor.

2.6.3. Standards are Not Static!

One common misconception is that, once a standard is published, it is fixed. On the contrary, most standards evolve and change after their initial publication, partly to implement corrections and clarifications, but also to specify enhancements, and to add new features. For instance, 3GPP works with a series of 'Releases', which are scheduled to occur every one to two years (e.g. Rel. 15, Rel. 16 ...), as illustrated in Figure 2.7. The IEEE takes a different approach, and adds a suffix to signify the year of each new version (e.g. IEEE Std. 802.11-2020).

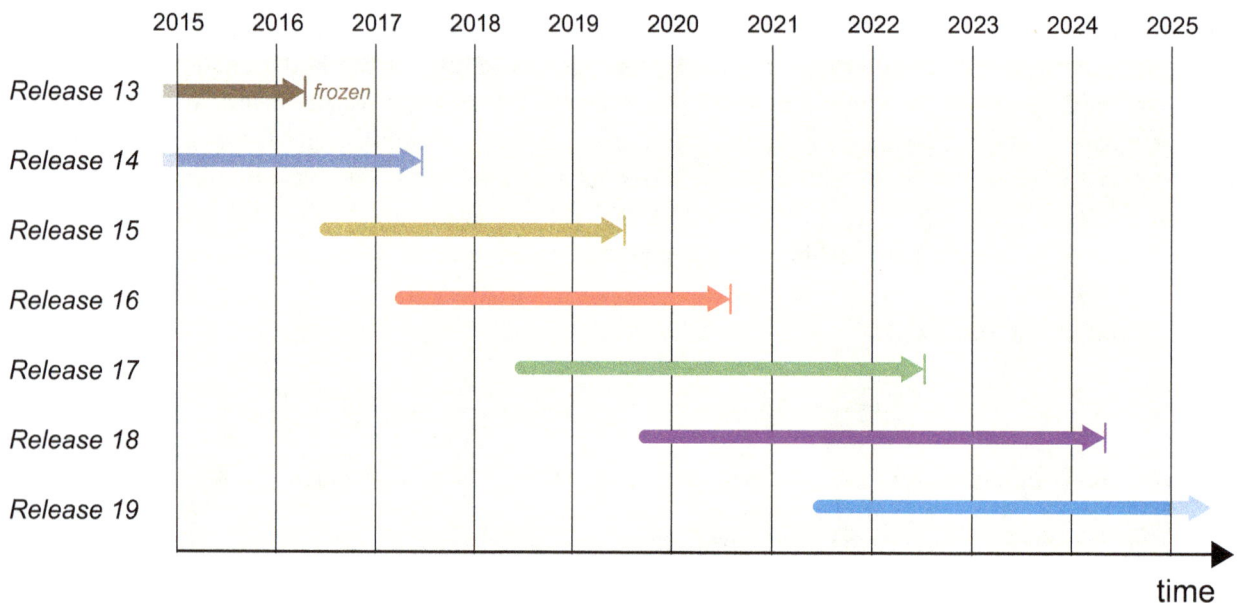

Figure 2.7: Actual and planned 3GPP Releases from 2018 - 2025.

3GPP standards are the basis of 4G and 5G cellular networks, and as shown in Figure 2.7, there is a rolling programme of Releases, with overlapping work on Release N, and the initial stages of Release $N+1$. The first full set of 5G standards was Release 15, and at the time of writing, the latest Release to be planned is Release 19.

This means that, to keep pace with evolving standards, equipment manufacturers must redesign or update their products with the latest functionality. Meanwhile network operators can upgrade infrastructure to cater for the most recent 3GPP Release, through field-upgrades. Consumer equipment such as smartphones and tablets, which typically have short lifecycles, may simply become obsolete over time.

2.6.4. Proprietary Schemes

As should be apparent from the previous discussion, communications standards are often complex, and usually become more so as they mature, as a result of new extensions or features being incorporated. These changes are a positive development on the whole, because they cater for additional use-cases, improve performance, and extend functionality. However, there can be disadvantages too, particularly if the functionality offered by available standards is insufficient to meet the requirements of an application, or conversely, is over-specified and therefore more complex to implement than necessary.

Proprietary schemes can be developed to cater for specific applications. This may include a special feature or characteristic that is not part of the standard; or it may strip out many of the complexities and features of a standard to form a more lightweight protocol. Advantages of the latter may include minimisation of power consumption or implementation cost.

Another possible use-case for proprietary schemes is for governmental, military or security use, where there is no requirement for interoperability with other systems, and perhaps even a desire to avoid interoperability (e.g. to reduce the chance of a signal being detected or received).

2.6.5. The Role of SDR

Recognising that wireless standards evolve, and also that there are clear motivations for proprietary schemes in some cases, there are obvious advantages of SDR.

As SDRs can be reprogrammed using software, they are very suitable platforms for communications network infrastructure and other radio systems. When a standard is updated, the SDR can be upgraded to implement the changes. This can be done without significant expense — even remotely (avoiding the time and expense of site visits). Moreover, SDR platforms can be reconfigured to implement multi-standard radios, wherein the same hardware components host the functionality for two or more radio standards at the same time.

While standard parts may be available for popular standards such Wi-Fi, these parts could not be used as the basis for a proprietary protocol, as their functionality is fixed and limited to the original application. SDRs, however, are well-suited — they are inherently flexible, and can be programmed with non-standard functionality. This means that SDRs can provide solutions for more specialist applications, or where adaptations or extensions to typical functionality are required.

2.7. SDR Radio Processing

Next, we briefly introduce the main approaches to implementing the modulation and demodulation aspects of a radio system. This functionality can also be mapped to analogue and digital components, which is significant in the context of SDR, as the digital elements provide scope for implementing flexible functionality (hence the *software defined* aspect of SDR).

2.7.1. Heterodyning and Superheterodyning

There are two primary architectures for radio transmitters — one involves direct modulation from baseband frequencies to RF frequencies (known as the *heterodyne*), while the second (the *superheterodyne*) achieves this transition with two modulation stages: the first from baseband to an Intermediate Frequency (IF), and the second from IF to RF. In each case, the receiver mirrors the operations of the transmitter.

Previously, in Figure 2.2 on page 15, we showed the modulation and demodulation required for a heterodyne scheme. In this case, a single stage of modulation is used to transition the signal from basedband to RF frequencies, and a single demodulation state makes the opposite transition, from RF back down to baseband.

The superheterodyne scheme is depicted in Figure 2.8, showing the frequency shifts that take place. Note in particular that the IF signal is typically modulated to a frequency in the range of 10's to 100's of MHz, whereas the eventual RF frequency band may be much higher — up to 10's of GHz.

2.7.2. The Impact of Sampling Rates

The sampling rates that can be achieved by the digital processing elements in an SDR, and in particular at the digital/analogue interfaces of a radio (the ADC and DAC), determine how a radio can be implemented — in particular, which parts can be implemented digitally, and which parts require analogue circuitry. This is because digital processing can only be used when the Nyquist criterion is met, in other words when the sampling rate is more than double the maximum frequency component present in the signal. [2]

Depending on the RF frequency band used to transmit the signal, it may be possible to implement all modulation and demodulation digitally, i.e. if

$$f_s > 2f_{rf_{max}} \tag{2.2}$$

where f_s is the DAC and ADC sampling rate, and $2f_{rf_{max}}$ is the maximum frequency present in the RF-modulated signal.

---

2. There are some notable exceptions to this statement, which we will touch on later!

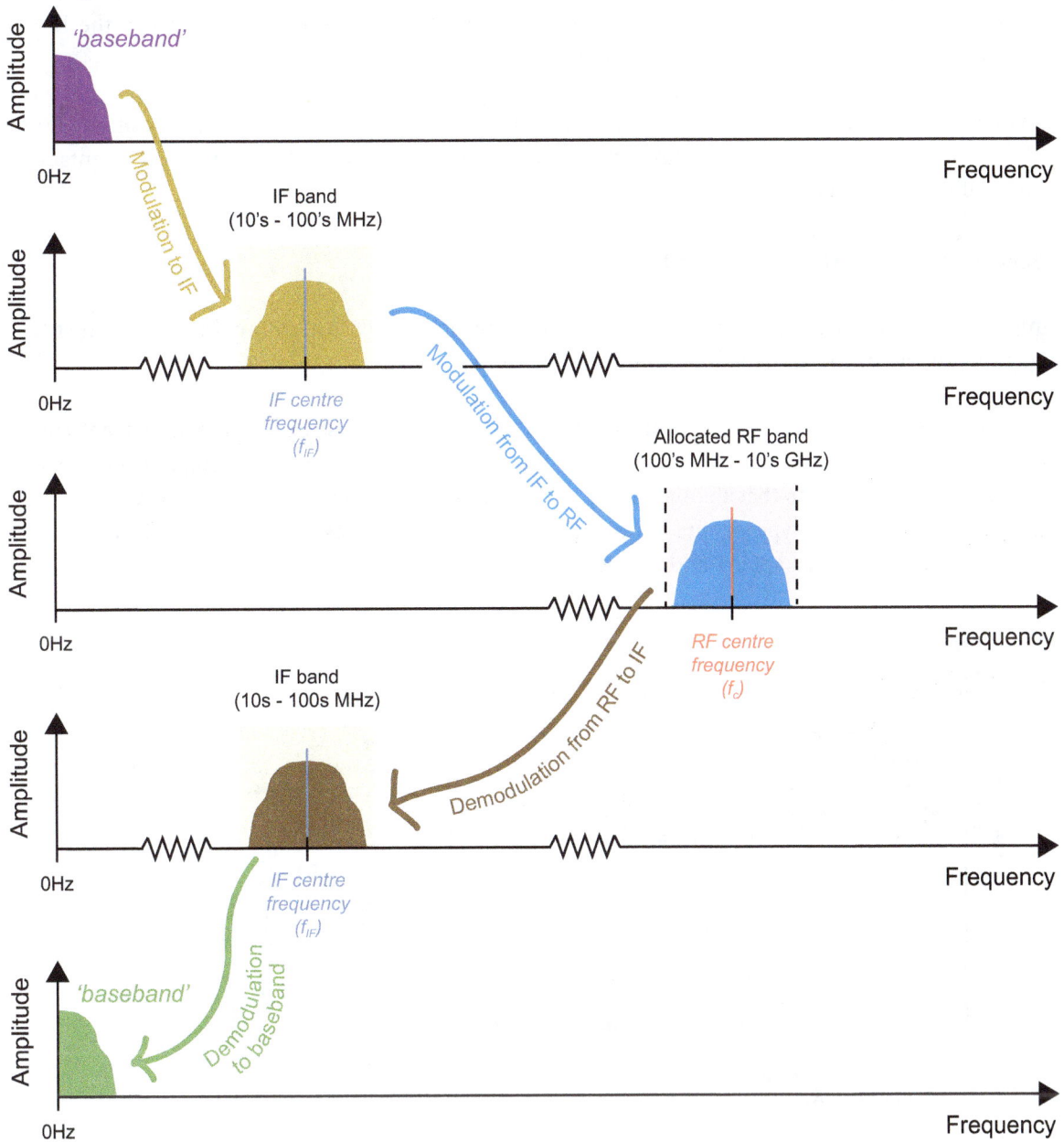

Figure 2.8: Modulation and demodulation using an Intermediate Frequency (IF) stage.

If the condition from (2.2) cannot be met, the alternative is to implement the modulation and demodulation stages that transition between baseband and IF frequencies digitally, with the analogue / digital interface at the IF stage (i.e. the second and fourth spectra shown in Figure 2.8), and modulation between the IF and RF frequency bands taking place in the analogue domain.

We can now go on to consider three different radio architectures, which arise based on the position of the DAC and ADC. Note that all of these models (as depicted in Figures 2.9 to 2.11) are simplified representations that omit some of the required filtering stages.

A Direct-RF (Almost-All-Digital) Radio

The implication of advancing DAC and ADC technology, bringing ever faster sampling rates, is that (almost) all-digital radios can be implemented for an increasing number of bands.

An almost-all-digital radio requires very little analogue processing — mostly the antenna(s), front-end filters, and amplifiers. From an SDR perspective, the fact that almost all functionality is implemented digitally is highly significant — it means that the operation of the radio can be controlled using software, as shown in Figure 2.9. (Review of this type of quadrature modulation architectures and related SDR systems will be presented in more detail in Chapter 7 and Chapter 8.)

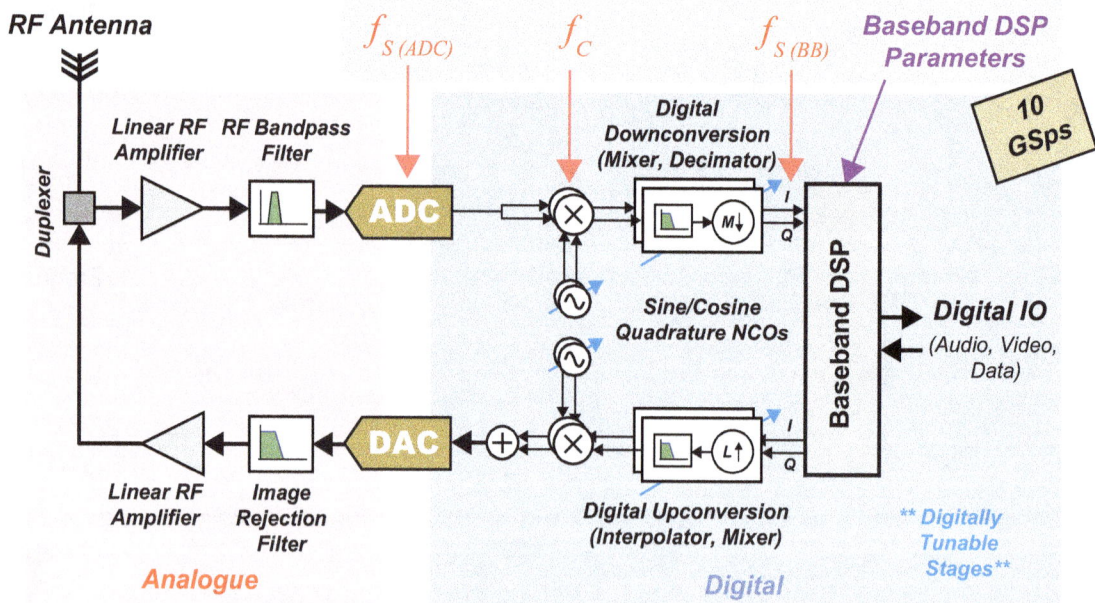

Figure 2.9: *High level architecture of a Direct-RF Software Defined Radio (the RFSoC!).*

An IF-Sampling Software Defined Radio

For situations where the RF frequency is higher than the available DAC and ADC sampling rates, a superheterodyne architecture can be used, with modulation / demodulation between baseband and IF accomplished in the digital portion of the radio, and the transition between IF and RF handled using analogue circuitry. Even so, it is often possible to exert software-based control over the analogue mixing stages. With the analogue / digital interface at IF rates, the required sampling rates may range from 10's of MHz, up to a few 100's of MHz. This type of SDR architecture is depicted in Figure 2.10.

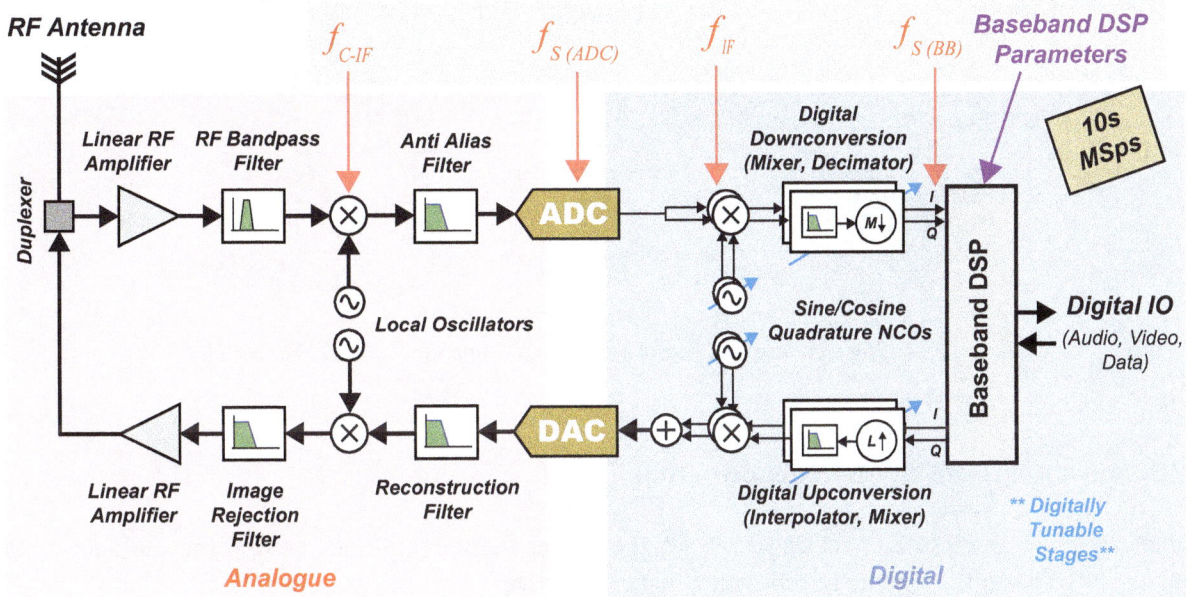

Figure 2.10: High level architecture of a Digital IF Software Defined Radio.

A Baseband-Sampling Software Defined Radio

A further class of SDR, shown in Figure 2.11, has the analogue / digital interface at baseband processing rates, with all modulation and demodulation performed in the analogue domain, either using a single stage or two stages. Historically, this approach was used where, due to the limitations of DAC and ADC technology (and in particular the achievable sampling rates), it was the only viable position for the A/D interface. This 'baseband sampling' architecture may also be adopted in low-cost, low data rate devices to minimise the requirements for the digital processor, or in applications using state-of-the-art, multi-GSps data converters, to transmit and receive extremely wide bandwidth signals.

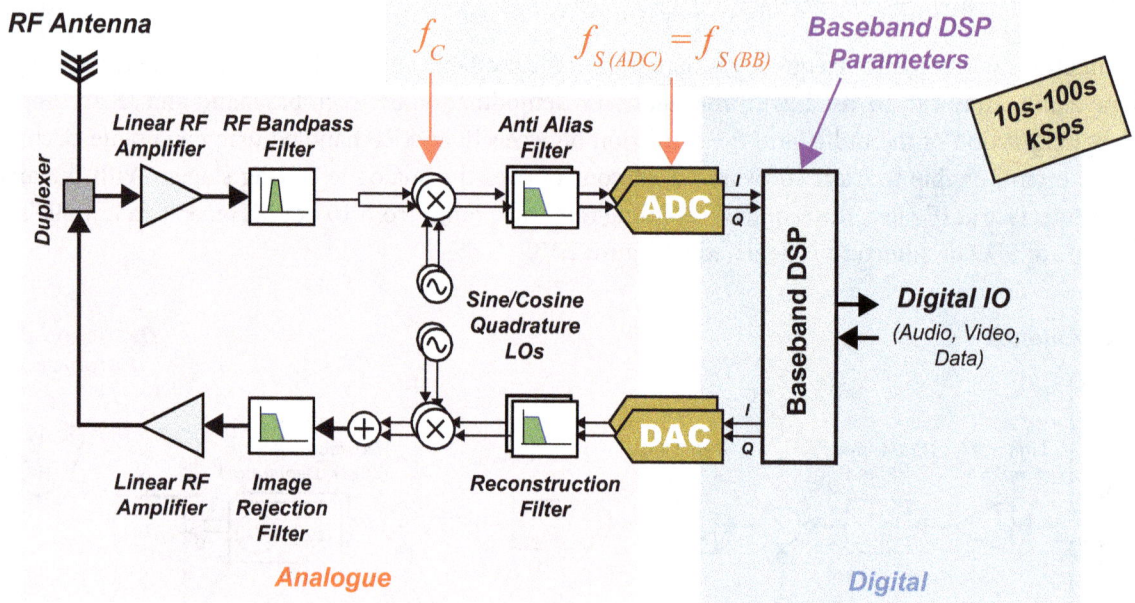

Figure 2.11: High level architecture of a Baseband-Sampling Software Defined Radio.

2.7.3. Advantages of Digital Implementation

Comparing Figures 2.9 to 2.11 makes clear that, the higher the sampling rate used at the analogue / digital interface, the greater the amount of processing that can be done digitally.

The use of digital implementations for modulation and demodulation has several advantages, including a greater accuracy of operation, reduced impact of component tolerances and greater resistance to ageing effects, a smaller physical footprint and simplified bill of materials. Power consumption may also be lower.

There are also additional benefits in the context of SDR, in particular due to the increased scope for software control over the operation of the radio, and the flexibility this brings. Where FPGAs or SoCs are used, the platform further offers reprogrammability of hardware-based processing, and therefore more fundamental upgrades and changes of functionality.

2.8. Key Radio Terminology and Parameters

Before moving on the later chapters, it is useful to define some key terms and concepts that will arise frequently in our discussions of SDR, and in the context of radio systems in general. These are illustrated in Figure 2.12 and described thereafter.

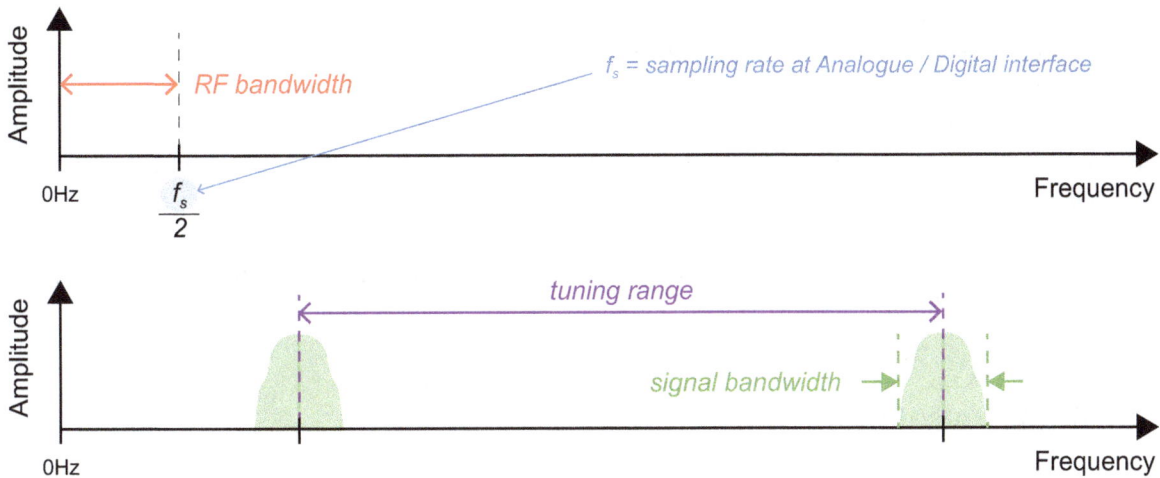

Figure 2.12: Signal and RF Bandwidths, and Tuning Range.

RF Bandwidth (of Transceiver)

The term *RF bandwidth* describes the range of frequencies that the radio transceiver can generate or capture. In the context of SDR (and digital radios in general) this is a function of the ADC and DAC sampling rates.

As shown in Figure 2.12, the RF bandwidth is half of the sampling frequency used at the ADC and DAC. This assumes that a single ADC and DAC are used, representing *real* signals (as opposed to *complex* signals). The RF bandwidth can be doubled to the full sampling frequency, if a complex input/output are used — however, this requires a pair of ADCs and pair of DACs. The topic of complex signals will be reviewed in Chapter 7.

Signal Bandwidth

When referring to a radio signal, *bandwidth* refers to the range of frequencies present in the transmitted signal. Note the signal bandwidth annotated against the green signal spectrum in Figure 2.12.

When working with SDRs, the bandwidth of the signal is a feature of the implemented design (which can be defined in software). For instance, the designed transmitter may generate a signal with a bandwidth of 10kHz, or 100MHz, depending in its configuration. In general, larger bandwidth signals can transmit data at faster rates.

The maximum signal bandwidth is limited by the RF bandwidth of the SDR, which is a property of the physical hardware device(s) used to implement the SDR. The RF bandwidth is defined next.

Tuning Range

The term *tuning range* refers to the range of frequencies across which the RF bandwidth can be shifted, when a an IF-sampling or baseband-sampling architecture is used (as depicted in Figures 2.10 and 2.11, respectively), where part or all of the modulation / demodulation is undertaken in the analogue domain. The applicable tuning range is dependent on the analogue circuitry, and consequently, the data sheet of an SDR incorporating a tuning stage will usually specify lower and an upper tuning frequencies.

For a Direct-RF architecture, there is no analogue tuning involved, as modulation and demodulation are undertaken entirely in the digital domain.

2.9. SDR Implementation

SDR activities can be pursued for a number of different purposes, including for student and hobbyist use, research and algorithm validation, for prototyping, and for the development of SDR-based radio products. Some of the simplest SDRs are single-channel, receive-only devices costing only a few 10's of dollars. At the other end of the spectrum, a professional multi-channel development platform could cost upwards of $10,000.

The major hardware elements of an SDR are:

- Antenna(s)

- Analogue front-end circuitry (signal conditioning, including amplifiers and filters)

- Analogue mixing stages (optionally, depending on the architecture as described in Section 2.7.2)

- Data converters (ADCs and DACs)

- Processing platform(s)

The choice of hardware components in an SDR typically depends on functional criteria (parameters such as the RF carrier frequency, data rates, and parameters flowing from these), performance, and cost constraints. Another important factor is the number of transmit and receive radio channels to be supported.

While it is possible to realise an SDR using discrete components for each of the above, increasingly there are integrated products available that combine some of the above functionalities. The RFSoC is a prime example, as it combines data converters with a processing platform. There are many possibilities for the hardware implementation of an SDR, especially considering that there are three candidate architectures (corresponding to the analogue-digital interface being placed at baseband, IF, or RF, as depicted in Figures 2.9 to 2.11).

Taking the example of the IF-sampling architecture, three possible hardware implementation options are depicted in Figure 2.13 (note: this is not intended to be exhaustive!). These options illustrate that, while the radio could be composed of discrete components for each of the main processing stages, there are approaches that combine two or more functional stages on the same device.

For example, integrated SDR transceiver chips include analogue front end circuitry, analogue mixing, and ADC and DAC stages. Another possibility is the integration of IF and baseband processing with software control aspects, through the use of an SoC or MPSoC device. Finally, adopting the RFSoC to implement an IF-sampling radio would additionally integrate high speed ADCs and DACs, along with IF and baseband processing, and software control.

While we have not covered software extensively here, SDRs, by their very nature, include at least software-based control of hardware elements. This software is normally hosted on a dedicated processor. Where very capable processors are used, much of the signal processing functionality may also be implemented in software.

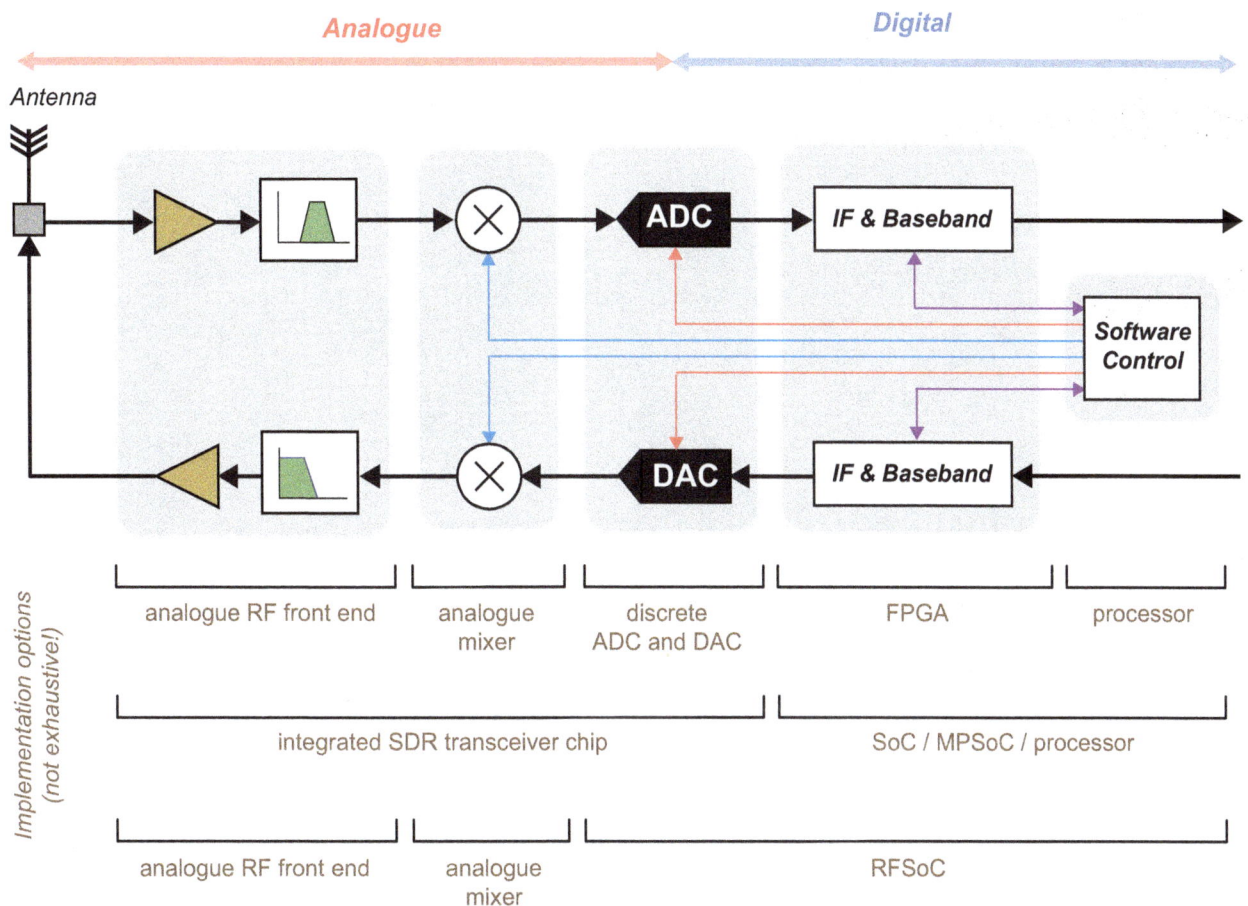

Figure 2.13: Some indicative implementation options for an IF-sampling SDR.

2.10. Chapter Summary

This chapter has introduced the concept of SDR, along with some fundamental underpinnings in radio communications. In particular, we noted an evolution in the motivating factors for SDR, with flexible and efficient use of the radio spectrum now being a key driver, due to the ever-expanding demand for data, and the emergence of new applications. New, 'dynamic spectrum' techniques promise a new approach to managing at least some of the radio spectrum, and demand a flexible, agile radio platform.

SDR is aligned both with the implementation of formal communications standards, as well as proprietary schemes. Even standards such as 5G are constantly moving and evolving, and 6G is now very much in the research, design and development phase; therefore, the option to change functionality by updating software is compelling.

Also in this chapter, we considered the three primary architectures for implementing SDRs, i.e. with the analogue-digital interface positioned at baseband, IF, or RF frequencies. The ability to sample at RF provides the option of an almost-all-digital radio, and this provides the greatest scope for implementing SDR. As was touched on here, the RFSoC device is able to perform many of required functions in an integrated manner; the next chapter will look at its capabilities for SDR in more detail.

Chapter 3

Introduction to Zynq UltraScale+ RFSoC

Lewis Brown

We now introduce Zynq UltraScale+ RFSoC as a platform for SDR systems. As will be demonstrated and discussed in this chapter, the RFSoC is a ground-breaking device with unparalleled facilities for implementing SDR, and radio systems more generally: it integrates multiple channels of transmit and receive capabilities, operating at sampling rates of up to 5.9 GSps and 10 GSps, respectively, with a Processing System (PS), FPGA Programmable Logic (PL), and optimised, hardened radio processing blocks. The integration is significant in itself, because of the benefits delivered — especially in terms of performance and power consumption, allied to the complete flexibility achieved from this synergy of programmable elements.

The chapter begins by introducing the capabilities of the RFSoC, and comparing SDRs implemented with RFSoC against other candidate architectures based on alternative devices. Later, the RFSoC device architecture is reviewed at a more detailed level, and supporting development boards are introduced.

3.1. RFSoC as an SDR Platform

It is useful to consider the architecture and capabilities of the RFSoC, as we assess its merits as a platform for implementing SDR systems. In this section, we review the features of the RFSoC at a high level, how SDR systems may be developed based on the RFSoC, and how this approach compares with other possible design approaches using alternative chips.

3.1.1. RFSoC Architecture Overview

As we start to understand the unique attributes of the RFSoC, it is worth returning to the high level diagram from the Introduction (reproduced as Figure 3.1).

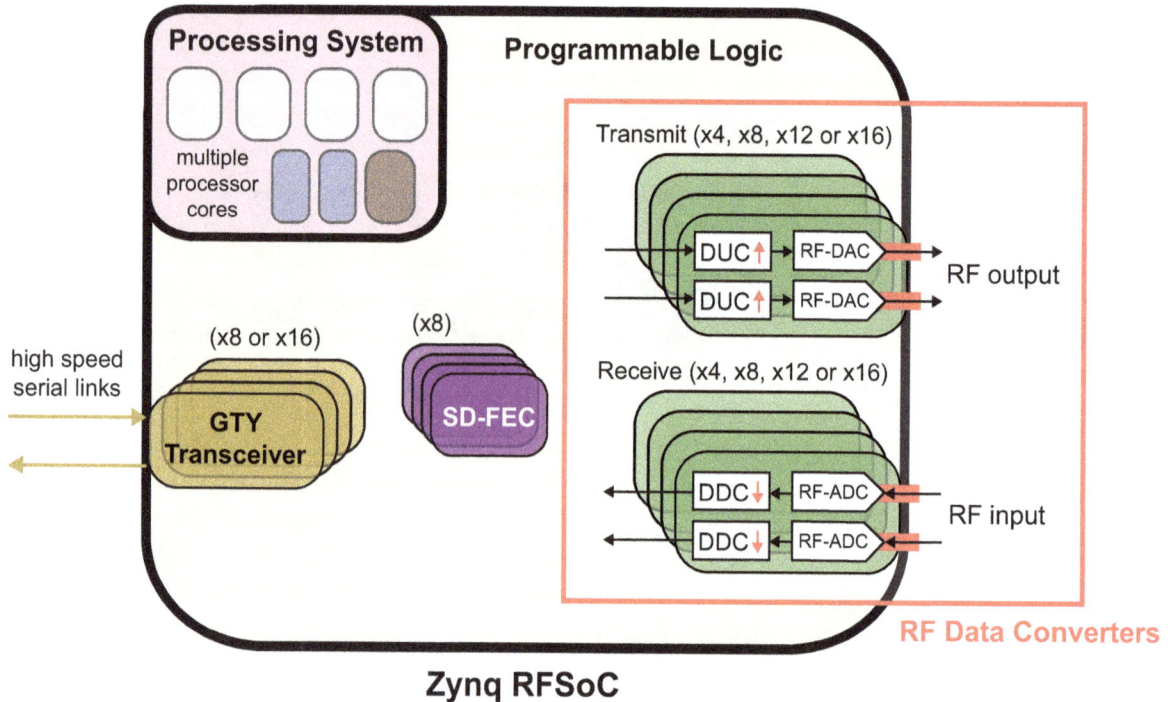

Figure 3.1: Overview of the RFSoC's integrated features (reproduced from Chapter 1).

As shown in Figure 3.1, the RFSoC platform incorporates:

- High speed ADCs and DACs (along with front-end signal processing as optimised hardened blocks);

- A processing system incorporating a set of applications and real-time processors and other features;

- FPGA programmable logic;

- Optimised, hardened blocks for Forward Error Correction (FEC) coding and decoding;

- Integrated transceivers for high speed serial communication over wired links.

Importantly, these facilities are all in *a single chip*. Therefore complete radio solutions can be realised in an extremely integrated and compact form. This has multiple benefits, as will be summarised over the next few pages (and further discussed in Section 3.5, after the RFSoC architecture has been introduced in more detail).

3.1.2. RFSoC Single-Chip Solution versus Alternatives

The RFSoC comprises all of the features needed to create a complete radio system, other than analogue components such as RF amplifiers, filters, and antennae, which must be added externally. But what are the alternatives? Figure 3.2 compares the single chip RFSoC solution against some other possible designs (other variations are possible) — in each case requiring multiple chips.

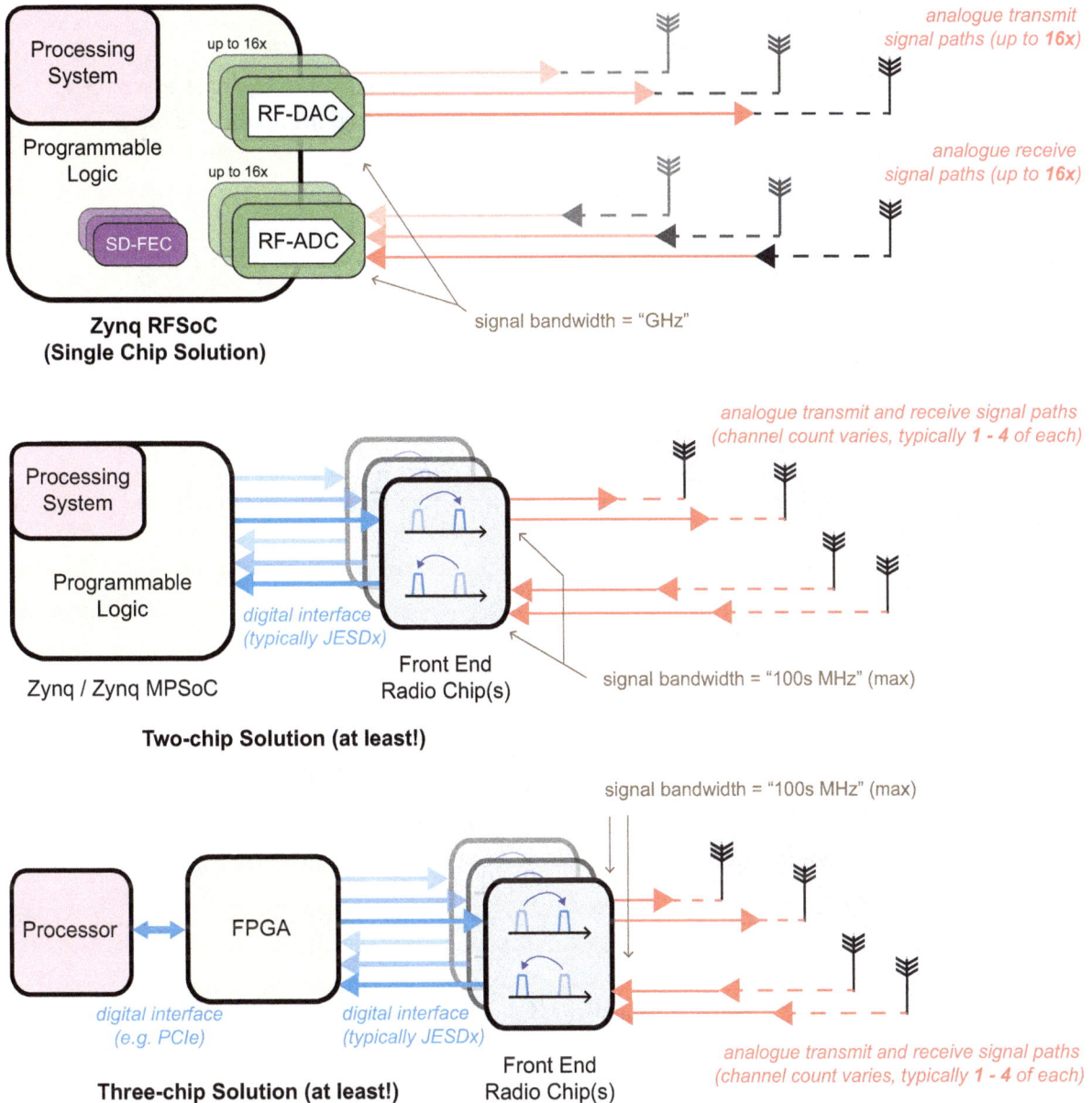

Figure 3.2: *Comparison of a single-chip RFSoC SDR solution, with multi-chip alternatives.*

Due to its highly integrated nature, the RFSoC can provide the basis for a much simpler SDR architecture than other alternatives, which would require additional components to perform Digital-to-Analogue (D-to-A) and Analogue-to-Digital conversion (A-to-D) conversion. This functionality often forms part of a front-end radio transceiver chip that also performs modulation and demodulation (instances of which are shown in Figure 3.2). Integration also means that a number of other components for signal conditioning and interfacing are not required.

3.1.3. RFSoC Highlights

The integrated nature of the RFSoC family, as well as its many features, results in multiple advantages for SDR design. We conclude our initial introduction to the RFSoC device by summarising them below.

Removed Requirement for Front-End Radio Chips

The integration of RF-DACs and RF-ADCs within the RFSoC removes the need for external front-end radio processing, including DACs and ADCs. This in turn enables:

- A smaller physical realisation of the system implementation.

- A simpler bill-of-materials and Printed Circuit Board (PCB) layout.

- Reduced power consumption, due in particular to the hardened, optimised RFDC blocks included in the RFSoC, and inter-chip communication (e.g. via JESD interfaces) not being required. The topic of power consumption will be discussed further in Section 3.5.5.

- Lower latency in both the transmit and receive paths.

Flexibility and Reprogrammability

As the RFSoC is completely integrated, the PS (which hosts software) has low-latency connections to hardware design elements operating in the PL, as well as to the RFDCs and SD-FEC blocks. This means that:

- Under the control of software operating on the PS, parameters can be changed in the PL, and in the RFDCs and SD-FEC blocks, meaning that operation can be controlled in an agile manner.

- Design effort is reduced — engineers can develop using a single device with well-documented features for parameterisation, programming, and the interfacing of internal elements.

- The integration of software and FPGA hardware enables hardware-acceleration of computationally-intensive algorithms.

- The SDR can be reprogrammed with new software and hardware functionality, including dynamically changing the hardware elements implemented on the FPGA.

Performance and Features

Of course, the performance and features of the RFSoC are vital aspects of its suitability for SDR applications. There are a number of different aspects to this, which include:

- RFSoC devices support ADC sampling rates of up to 5.9 GSps, and DAC sampling rates of up to 10 GSps, meaning that they can capture and generate extremely wide bandwidth signals[1]. RFSoC also supports operation in the 2nd Nyquist Zone, and RF input signal frequencies up to 7.125 GHz[2]. Therefore, an SDR can be developed with RFSoC that covers the majority of popularly used signal bands using a Direct-RF architecture (as previously introduced in Section 2.7.2) and thus minimal external analogue circuitry.

- The integration of up to 16 transmit and 16 receive channels in a single devices is significant, especially for Multiple Input Multiple Output (MIMO) systems, and beamforming, both of which involve multi-element antenna arrays. In both such systems, the channels must be synchronised, and RFSoC provides functionality to enable this. More generally, the availability of multiple transmit and receive channels enables SDRs to be designed that support several radio standards in the same chip.

- As well as the ability to combine and synchronise multiple channels on the same RFSoC device, two or more devices can also be combined and synchronised. This is significant because it enables integrated SDR systems with larger channel counts to be created (such as in Massive MIMO).

- The combination of a capable Arm® multi-processor system in the PS, integrated with a huge area of PL, provides huge scope to implement SDR functionality.

- Optimised, hardened blocks are integrated to provide front-end transmit and receive functionality (RFDCs, introduced further in Section 3.3.4) and SD-FEC coding and decoding (more in Section 3.3.5). The presence of these high-performance blocks reduces design effort and, due to their implementation in dedicated silicon, they also operate with low power consumption.

- The RFSoC DFE, which will be introduced further in Section 3.3.6, has an expanded set of optimised and hardened processing blocks, optimised for 5G NR and similar deployments.

- The integrated GTY transceivers provide high rate connectivity over fixed links, which is ideal for building network infrastructure.

- RFSoC devices have security and platform management functionality.

- Designers benefit from an extensive library of Intellectual Property (IP) cores to aid productivity and accelerate system design. These IP cores can be instantiated and systems developed using AMD design tools, as will be reviewed in Chapter 13.

---

1. External analogue circuitry is required to support the generated signals.
2. These figures are for the RFSoC DFE. The respective values for Gen 1, 2, and 3 differ and will be set out in Section 3.3.1.

Having now reviewed the RFSoC device at a high level, the next section explains how the device fits into the development of AMD SoCs, and provides a feature comparison.

3.2. A System on Chip for RF Applications

The Zynq UltraScale+ RFSoC is one of several System on Chip (SoC) devices that have been released by AMD (formerly Xilinx) in recent years. Announced in 2011, Zynq-7000 was the first FPGA-based SoC, combining FPGA programmable logic with a dual-core Arm applications processing system [80]. Subsequently, a more capable Zynq UltraScale+ MPSoC (Multi-Processor System on Chip) followed [84], and more recently still, the RFSoC [88], which is a relative of the MPSoC, and represents a single-chip adaptable radio platform.

3.2.1. Evolution of AMD SoCs

All AMD SoCs evolved from FPGAs, which are traditionally composed of two-dimensional arrays of reconfigurable processing elements known as Configurable Logic Blocks (CLBs) [344]. Modern FPGAs incorporate hardened blocks, optimised for specific tasks such as memory (Block RAMs), high speed arithmetic (various versions of the DSP48 block), and input/output connectivity (for instance via the 100G Ethernet CMAC subsystem [52]). Where applications require processing capabilities, a 'soft' processor such as the AMD *MicroBlaze*™ processor can be constructed using the low-level reconfigurable logic resources of the FPGA [54].

The SoC is a natural progression, because it incorporates a hard Processing System (PS) alongside FPGA Programmable Logic (PL). The main advantage of a hard processing system is that it can achieve higher performance than a soft processor, using less area, and with greater independence from the design implemented on the PL portion of the SoC. It should be highlighted that by combining PL and PS design elements, complete applications can run in an SoC, thus simplifying overall system complexity compared to a solution with separate processor and FPGA chips.

The PS is also more sophisticated than a processor in isolation; it additionally combines memory subsystems, interconnect resources for PS-PL connections, and external interfaces supporting multiple popular standards (USB, I2C, HDMI®, DisplayPort™ and others). The PS in the Zynq UltraScale+ MPSoC has more resources than the Zynq-7000, with an expanded set of hard processors (two real-time processors and up to four applications processors, compared to two applications processors in Zynq-7000). Further, the applications processors are more highly specified in the Zynq UltraScale+ MPSoC: they are 64-bit as opposed to 32-bit processors, and can operate at higher clock frequencies.

3.2.2. Comparison of SoC Families

The device of interest in this book, the Zynq UltraScale+ RFSoC, has an equivalent PS to the Zynq UltraScale+ MPSoC, but without the graphics and video coding features. For more details on Zynq UltraScale+ MPSoC devices, see [85] and [131].

The major additional features are the inclusion of 8 or 16 channels of hardened, RF-capable ADCs and DACs, which are capable of sampling at multiple Giga Samples per second (GSps). These ADCs and DACs form part of RF Data Converter (RFDC) blocks, which also contain hardened Digital Upconverters (DUCs) and Digital Downconverters (DDCs). Several RFSoC devices also include a hardened Soft Decision Forward Error Correction (SD-FEC) block. Table 3.1 provides a high level feature comparison between the three SoC families.

Table 3.1: Comparison of Key Features of AMD SoC Devices.

Feature	Zynq-7000 SoC	Zynq UltraScale+ MPSoC	Zynq UltraScale+ RFSoC
CLBs DSP Slices Memory	up to 444K up to 2020 up to 26.5 Mb	up to 1143K up to 3528 up to 70.6 Mb	up to 930K up to 4272 up to 67.8 Mb
Applications Processors	up to 2 Arm Cortex® A9 (32-bit, up to 1 GHz)	up to 4 Arm Cortex A53 (64-bit, up to 1.5 GHz)	4 Arm Cortex A53 (64-bit, up to 1.3 GHz)
Real-time Processors	-	2 Arm Cortex R5F (up to 600 MHz)	2 Arm Cortex R5F (up to 533 MHz)
Graphics Processors	-	Arm Mali™-400 MP2[1]	-
RF Data Converters	-	-	up to 16 RF-DACs up to 16 RF-ADCs
Hardened SD-FEC Blocks	-	-	up to 8 SD-FEC blocks

1. There are three Zynq UltraScale+ MPSoC sub-families (CG, EG, and EV). The graphics processor is present in EG and EV devices.

Note: There is considerably more variance in hardware resources across Zynq-7000 and MPSoC devices, compared to RFSoC devices. The major differentiation between devices in the RFSoC range is the number and specification of RFDCs and SD-FEC blocks.

3.3. Zynq UltraScale+ RFSoC Architecture Overview

As mentioned earlier, the Zynq UltraScale+ RFSoC integrates hardened RF Data Converters and SD-FEC blocks alongside PL and PS that are similar to those of Zynq UltraScale+ MPSoC devices. Therefore, the RFSoC brings all the components of a radio transceiver into a single device (an MPSoC would require an external front end radio chip with high speed DACs and ADCs, and RF signal processing). This section will review the PS and PL of the RFSoC in detail, as well as highlighting the RF Data Converters and FEC blocks which will be further explored in later chapters.

A high-level overview of an RFSoC device can be seen in Figure 3.3.

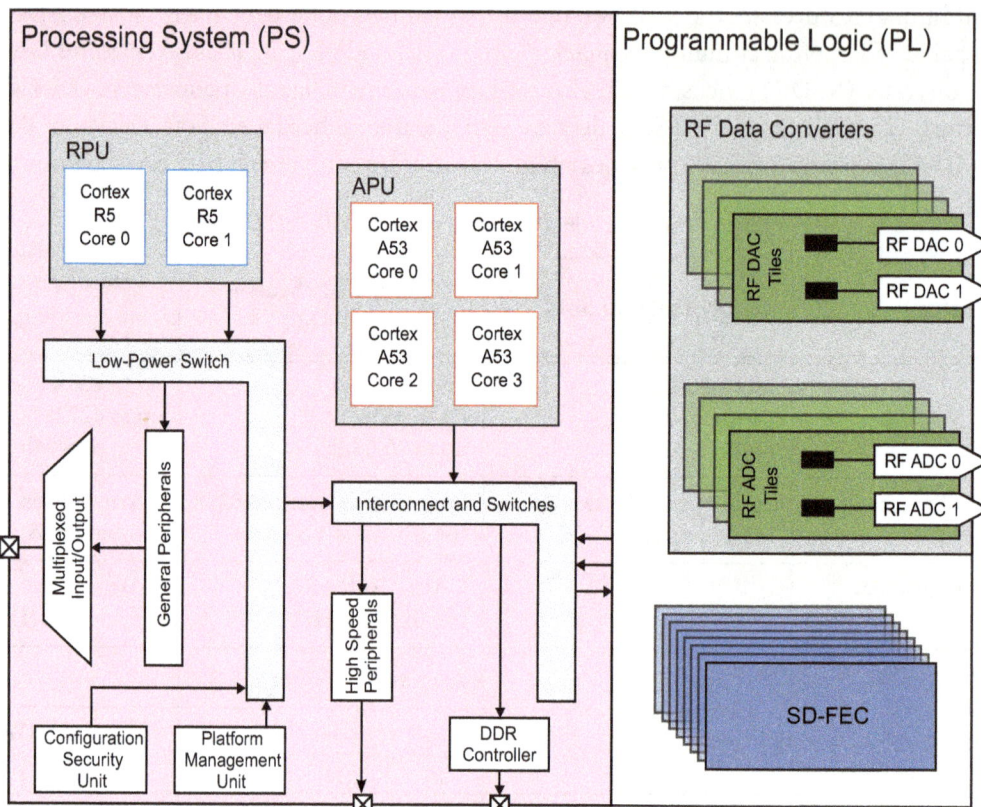

Figure 3.3: High level overview of the Zynq UltraScale+ RFSoC device architecture.

3.3.1. RFSoC Devices

At the time of writing four 'generations' of RFSoC devices are available, known as Gen 1, Gen 2, Gen 3 and RFSoC DFE devices. All devices in each generation share common features such as RFDC resolution (expressed in terms of the number of bits) and maximum supported input frequencies. The naming convention for devices within each generation is as follows:

- Gen 1   ZU2XDR

- Gen 2   ZU3XDR

- Gen 3   ZU4XDR

- RFSoC DFE   ZU6XDR

where the 'X' is replaced with a single digit to identify a particular device. Table 3.2 highlights a selection of devices across the four generations.

Table 3.2: A selection of RFSoC devices and their features.

Feature	Gen 1		Gen 2	Gen 3		RFSoC DFE
	ZU21DR	ZU28DR	ZU39DR	ZU46DR	ZU48DR	ZU67DR
Max. input frequency (GHz)	4	4	5	6	6	7.125
ADC blocks Max. rate (GSps) ADC resolution	0 0 -	8 4.096 12	16 2.220 12	8 | 4 2.5 | 5.0 14	8 5.0 14	8 | 2 2.95 | 5.90 14
DAC blocks Max. rate (GSps) DAC resolution	0 0 -	8 6.554 14	16 6.554 14	12 9.85 14	8 9.85 14	8 10.0 14
SD-FEC blocks	8	8	0	8	8	0
System Logic Cells (K)	930	930	930	930	930	489
CLB LUTs (K)	425	425	425	425	425	224
Max. distributed RAM (Mb)	13.0	13.0	13.0	13.0	13.0	6.8
Total Block RAM (Mb)	38.0	38.0	38.0	38.0	38.0	22.8
UltraRAM (Mb)	22.5	22.5	22.5	22.5	22.5	45.0
DSP Slices	4,272	4,272	4,272	4,272	4,272	1,872
Optimised hardened Digital Front End (DFE) blocks	-	-	-	-	-	Yes

Note that the number of RF-ADCs and RF-DACs varies across devices (particularly within Gen 3), up to 16 of each. The maximum available RF-ADC sample rate is related to the configuration of the RF-ADCs within the Data Converter. This RFDC structure will be discussed in Section 3.3.4.

Although most RFSoC devices incorporate RFDCs, a subset do not, such as the ZU21DR shown in Table 3.2 — these devices are targeted at elements of network infrastructure away from the front end RF interface, where the hardened SD-FEC blocks can be used to perform FEC coding and decoding. The ZU21DR can be seen in the T1 Telco Accelerator Card, which will be introduced in Section 3.4.3.

At the time of writing, the most recent of these RFSoC generations are the DFE devices — specialised RFSoCs aimed at *5G New Radio* deployment. These devices contain fewer PL resources than other devices, however they contain additional optimised hardened blocks, including a hardened Digital Pre-Distortion (DPD) block.

Speed Grade

The Speed Grade is an indication of the timing characteristics/performance of logic elements in the PL. At the time of writing, there are eight different speed grades for the PL of the RFSoC. These are 2E, 2I, 2LE, 2LI, 1E, 1I, 1M and 1LI, where the higher numerical values correspond to faster PL timing performance. An 'L' indicates low power mode, whilst 'E', 'I' and 'M' indicate operating temperatures: Extended, Industry and Military respectively. These speed grades determine the maximum clock rate of the PL, which is device-dependent, but generally up to several hundred MHz. For more information about speed grades, see [87].

3.3.2. Processing System

The PS consists of several different types of hardened processing resources, including the Application Processing Unit (APU), Real-Time Processing Unit (RPU) and Platform Management Unit (PMU). These processors are used to run the software stacks of SDR applications. Refer to Figure 3.3 for an overview of the PS and its components.

The PS can be used alongside the PL to create a complete SDR system. Some candidate examples of PS functionality in an SDR system include:

- Quad-core Arm Cortex-A53 application processors

  - *Operating System*

  - *SDR Software Applications (including PYNQ applications)*

- PS local memories and external memory control

  - *Software programs*

  - *SDR instructions and data*

- Dual-core Arm Cortex-R5 real-time processor

  - *Low-level, real-time SDR system control*

- Platform management and security functions

  - *System booting*

  - *Power management*

  - *Physical security*

  - *SDR communications security*

- External interfaces

  - *Local IoT devices*

  - *SDR backhaul*

  - *Networked system control*

We will now review some of these PL facilities in more detail.

The Application Processing Unit

An overview of the APU is provided in Figure 3.4. It contains a Quad-core Arm Cortex-A53 processor, which hosts four processing cores, each with its own dedicated computational units [96]. These include a Floating-Point Unit (FPU), Neon™ Media Processing Engine (MPE), Cryptography Extension (Crypto), Memory Management Unit (MMU) and dedicated Level 1 cache memory per core. The entire APU has access to a Snoop Control Unit (SCU) and Level 2 cache memory.

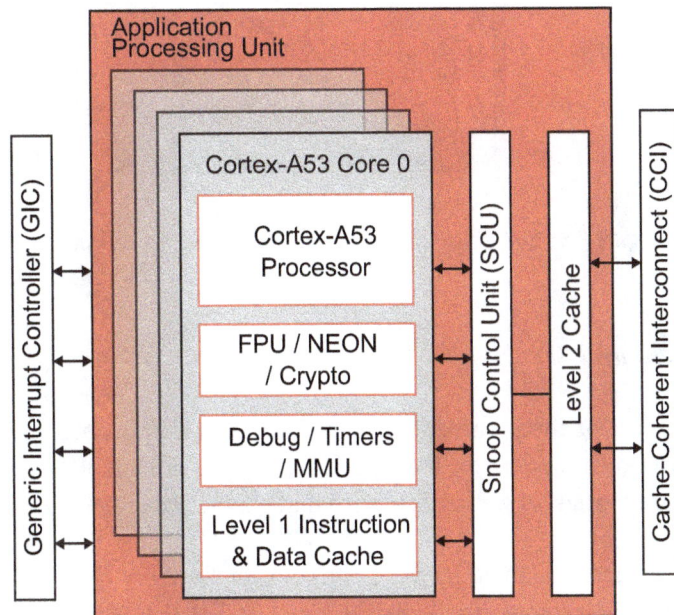

Figure 3.4: Simplified diagram of the Application Processing Unit (APU).

Most notably, the four Arm Cortex-A53 cores are capable of running a fully featured Operating System (OS) alongside software applications. As introduced in Chapter 1, this book is supported by Jupyter notebooks that can be run on an RFSoC platform using the PYNQ software framework [39], which is deployed on the APU.

The Real-Time Processing Unit

The Real-Time Processing Unit (RPU) contains two Arm Cortex-R5 cores for real-time applications and deterministic system control, and it provides low latency performance [98]. The RPU contains a number of computational units and memories, which include an FPU, Tightly Coupled Memories (TCMs), two local caches and a Memory Protection Unit (MPU). A simplified overview of the RPU architecture is presented in Figure 3.5.

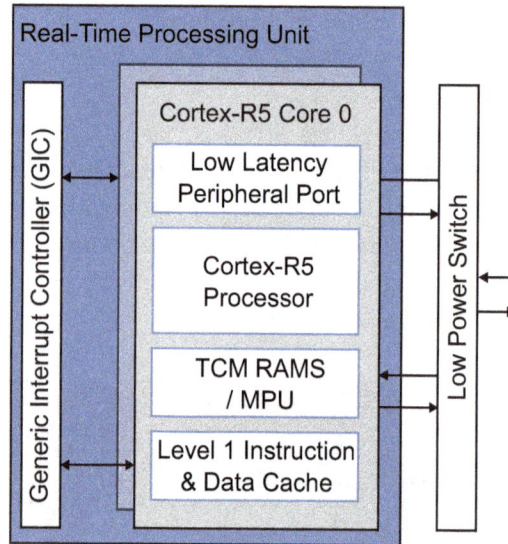

Figure 3.5: Simplified diagram of the Real-Time Processing Unit (APU).

Platform Management and Security

The PMU consists of a set of three hardened MicroBlaze processing units [37]. The MicroBlaze CPUs are configured in a majority voting system for increased reliability of critical platform management functions. The PMU contains several memories, as well as firmware that enables effective management of the RFSoC device.

Security of the RFSoC device is handled by the Configuration Security Unit (CSU) which consists of a Secure Processor Block (SPB) and Cryptography Interface Block (CIB) [24]. Similar to the PMU, the SPB contains three MicroBlaze processing units. These manage the secure boot of the Arm processors and several other security features, such as Physically Unclonable Functions (PUFs) and tamper protection. The CIB contains several cryptographic blocks for secure applications: Advanced Encryption Standard with Galois Counter Mode (AES-GCM), SHA-3 and RSA 4096.

3.3.3. Programmable Logic

The PL available on an RFSoC device is equivalent to an FPGA. The PL is an integral part of SDR design as it directly interfaces to the RF-ADCs and RF-DACs. Before examining the specialised resources available on an RFSoC device, let us begin by reviewing the fundamental building blocks of an FPGA and the key resources required for high-speed processing.

Recall Table 3.2, where we saw that most devices in the RFSoC family provide a common set of resources, including 930K CLBs, 4,272 DSP48E2 slices and 38 Mb of memory in Block RAMs (high-speed dedicated memories; UltraRAMs are similar but even larger memory blocks). These various resources are highlighted in Figure 3.6, which provides a high-level architecture overview of the RFSoC PL.

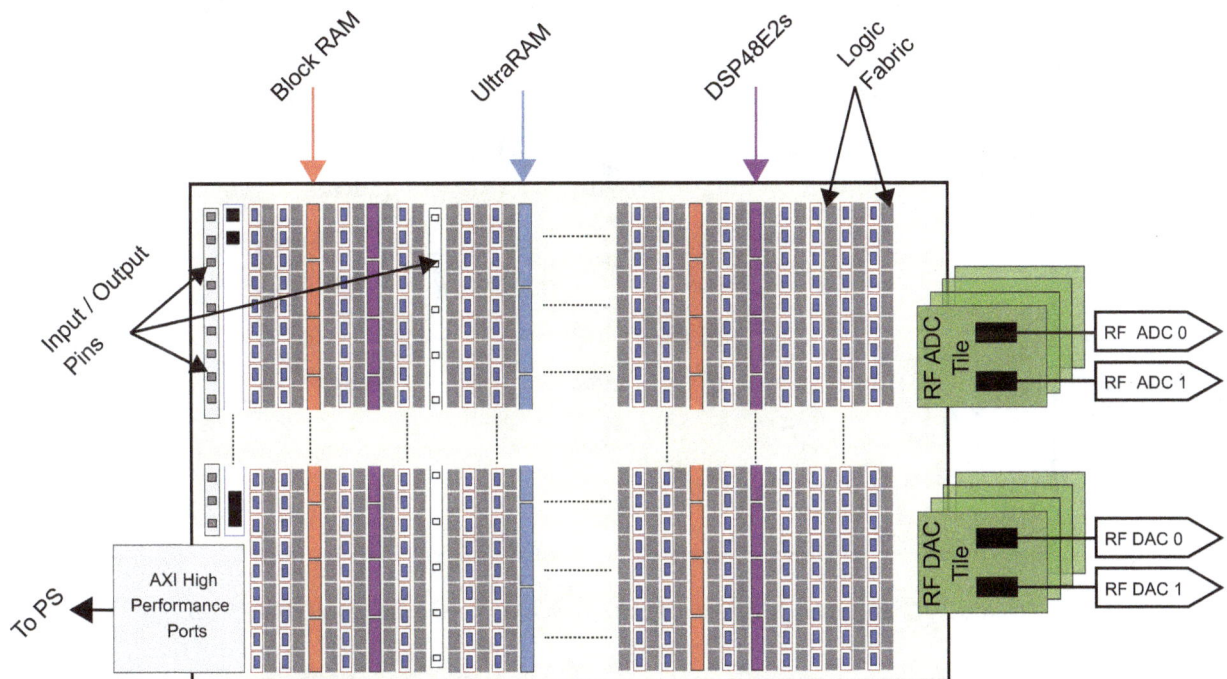

Figure 3.6: RFSoC PL architecture, highlighting PL features, RF Data Converters, and AXI ports.

The Logic Fabric

As noted when reviewing the selection of RFSoC devices in Table 3.2, each RFSoC contains a large number of CLBs. These resources are fundamental for implementing DSP algorithms, as well as logic circuits more generally. CLBs are arranged in columns in the FPGA logic fabric, and are closely aligned with switch matrices to support signal routing between neighbouring resources. The local routing logic includes multiplexers and optimised arithmetic carry logic.

The composition of a single CLB is shown in Figure 3.7; note that it contains a number of Lookup Tables (LUTs) and Flip-Flops (FFs). LUTs can implement a varied set of functionalities, including logic functions, Read Only Memory (ROM), Random Access Memory (RAM), and shift registers (SRL). Each FF represents a 1-bit register. Multiple LUTs and FFs can be combined using multiplexers to form larger logic functions, memories, registers, etc.

Of particular note, single or multiple CLBs can be used to implement small, local memories (which are known as Distributed RAM) within the PL; this contrasts with much larger, dedicated memory blocks (Block RAMs and UltraRAMs) which we describe a little later.

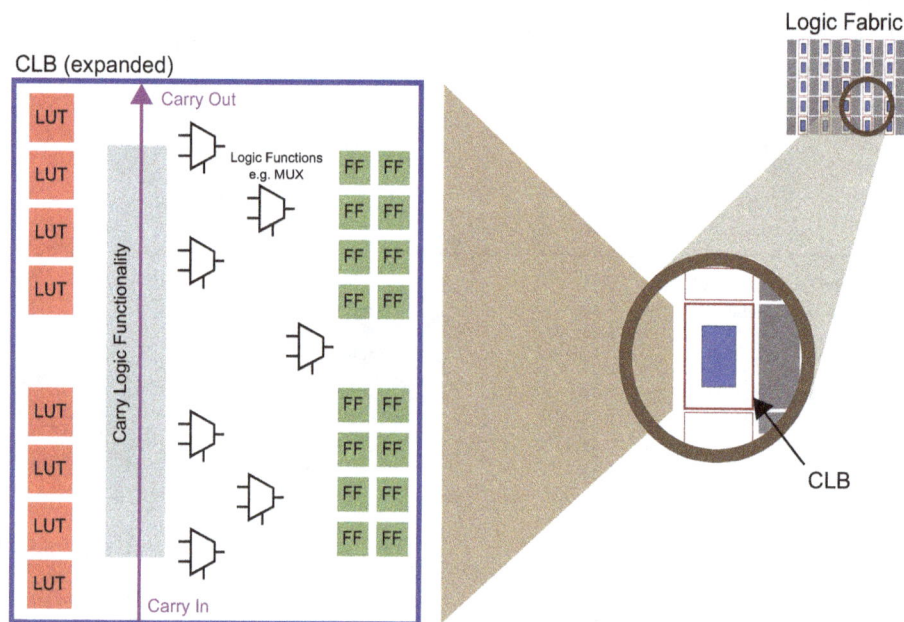

Figure 3.7: Configurable Logic Block (CLB) composition and connections (simplified).

Due to its highly programmable nature, the RFSoC logic fabric has a wide variety of applications, as almost any desired circuit can be built from these fundamental building blocks. Further detail on this highly flexible resource can be found in [49].

Advanced eXtensible Interface (AXI)

Also shown in Figure 3.6 are the Arm Advanced Microcontroller Bus Architecture (AMBA®) Advanced eXtensible Interface (AXI) ports present within the RFSoC logic fabric, which enable data transfer between the RFSoC's PL and PS. The RFSoC implements the fourth iteration of the AXI standard (AXI4, [95]) in both PL and PS.

Three types of AXI4 buses are supported in RFSoC: **AXI4** and **AXI4-Lite** (both of these are memory mapped interfaces, for burst transfers and single transactions, respectively), as well as **AXI4-Stream**, which enables the direct flow of data from source to destination. PS-to-PL connections are typically controlled through AXI4 and AXI4-Lite buses, enabling large data movement and single-burst IP core register control. For more information on AXI4 protocols within PL design, see [20].

DSP48E2 Slices

SDR transmitters and receivers commonly require processing such as Finite Impulse Response (FIR) filters (see Chapter 4), Fast Fourier Transforms (FFTs, see Chapter 5) and other computationally intensive DSP algorithms. The DSP48E2 slices within the PL are particularly valuable for implementing these design elements. DSP48E2 slices provide high speed Multiply Accumulate (MAC) hardware, supporting word lengths of up to 48 bits, optimised for high-speed and low-power operation. An example of a DSP48E2 slice is shown as Figure 3.8, highlighting the structure of arithmetic and logical operators contained within the slice.

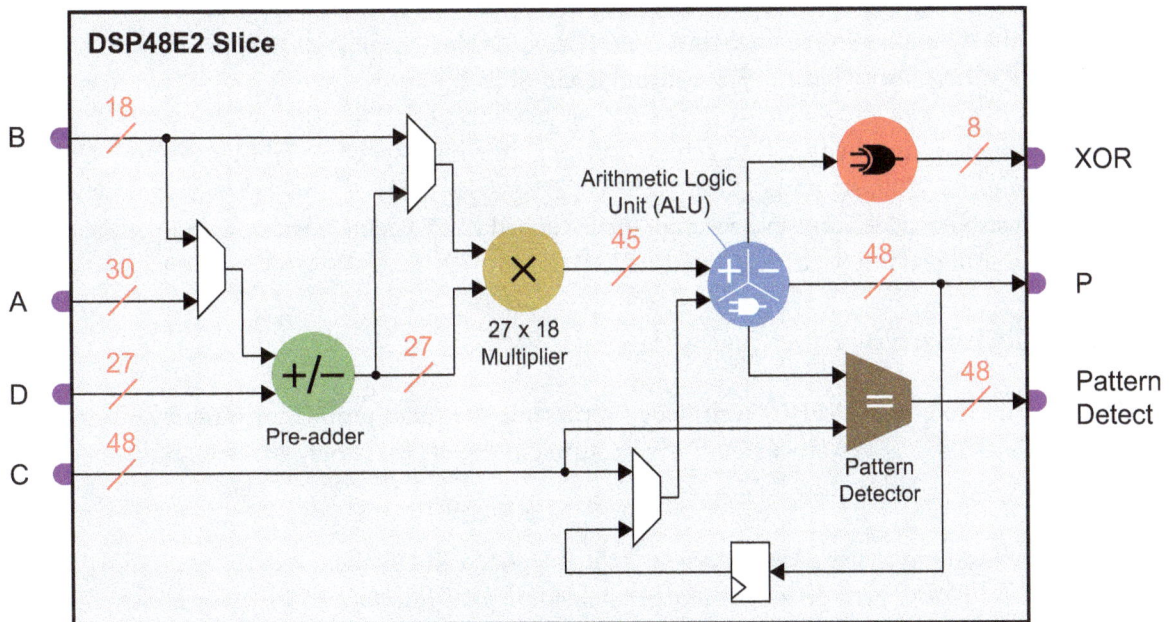

Figure 3.8: DSP48E2 slice (simplified).

Block RAM and UltraRAM

As well as being able to implement distributed memory using LUTs and FFs, the PL also contains Block RAMs and UltraRAM for data storage [51]. These are dense, high speed memories, suitable for storing larger amounts of data.

Block RAMs can be set to operate as RAMs, ROMs or First In First Out (FIFO) buffers. An individual Block RAM can be configured as either a single storage element, storing up to 36 Kb of data, or as two individual memories, each capable of storing 18 Kb of data. They are unique amongst PL memory as they have the ability to be reshaped. For example, a Block RAM could be configured to store 4,096 elements × 8 bits, or 8,192 elements × 4 bits, or several other dimensions.

UltraRAM provides significantly larger storage than Block RAM, storing up to 288 Kb of data in one tile [48]. Unlike Block RAM, an UltraRAM cannot be reshaped, however multiple UltraRAM blocks can be combined to form extremely large on-chip memories. Individual UltraRAM blocks have a fixed address configuration of 4,096 elements × 72 bits.

Connectivity

The RFSoC PL has many different interfaces for connecting to peripherals. As with MPSoC devices, this is primarily achieved using the Multiplexed Input/Output (MIO). The MIO provides a flexible interface that configures the route mapping between pins and peripheral interfaces, and is analogous to a very large multiplexer. MIO-enabled peripherals include: UART, SPI, CAN, I2C buses, General Purpose IO, Gigabit Ethernet (GigE), NAND flash memory, USB 3.0, SD card and Quad-SPI.

Additional peripheral connectivity can be achieved through the high-speed Serial Input Output Unit (SIOU) present within the PS. This supports PCIe, USB 3.0, DisplayPort, SATA and Ethernet protocols. Certain connections can also be made accessible through the Extended MIO, which creates a direct path of communication from the peripheral interfaces in PS, to the PL.

3.3.4. RF Data Converters

The RFSoC's Data Converter (RFDC) technology represents the most prominent difference between Zynq MPSoC and RFSoC devices. The other significant inclusion in RFSoC (but not in MPSoC) is the set of SD-FEC blocks, which are covered in Section 3.3.5.

A key reason for choosing to use RFSoC devices is the integrated RFDC tiles. because the RF-ADCs and RF-DACs within these blocks provide an integrated analogue/digital interface at RF frequencies. An MPSoC-based equivalent could be a viable alternative, but would require external components and interfacing; this would lead to performance limitations, increased power consumption, and larger physical system size. Extended details of the RFDCs can be found in [90], or in later chapters of this book (Chapters 9 through 11).

The PL to RF Interface

The RFSoC PL acts a gateway to the RF-ADC and RF-DAC channels. As previously shown in Figure 3.6, there are several RFDC channels that each require an interface to the PL; however, for the purpose of exploring the RF interface, we will focus on a single RF-ADC and a single RF-DAC channel. Signal data is transferred between the PL and the RFDCs using the AXI4-Stream interface. The AXI4-Stream interface features a data source (referred to as the master / manager) and a data sink (the slave / subordinate).

Taking first an RF-DAC channel, we can see that the PL fabric is responsible for transferring data onto the RF interface. In this setup, the PL is the manager and the RF-DAC is the subordinate, as shown in Figure 3.9(a). Here, the Complex-to-Real configuration is shown, as complex data in the RF-DAC channel becomes analogue real data. In contrast, an RF-ADC channel transfers data onto the RF interface for the PL to consume. The RF-ADC is the manager in this situation, and the PL is the subordinate. An illustration of this setup is provided in Figure 3.9(b), which is an example of a Complex-to-Complex configuration.

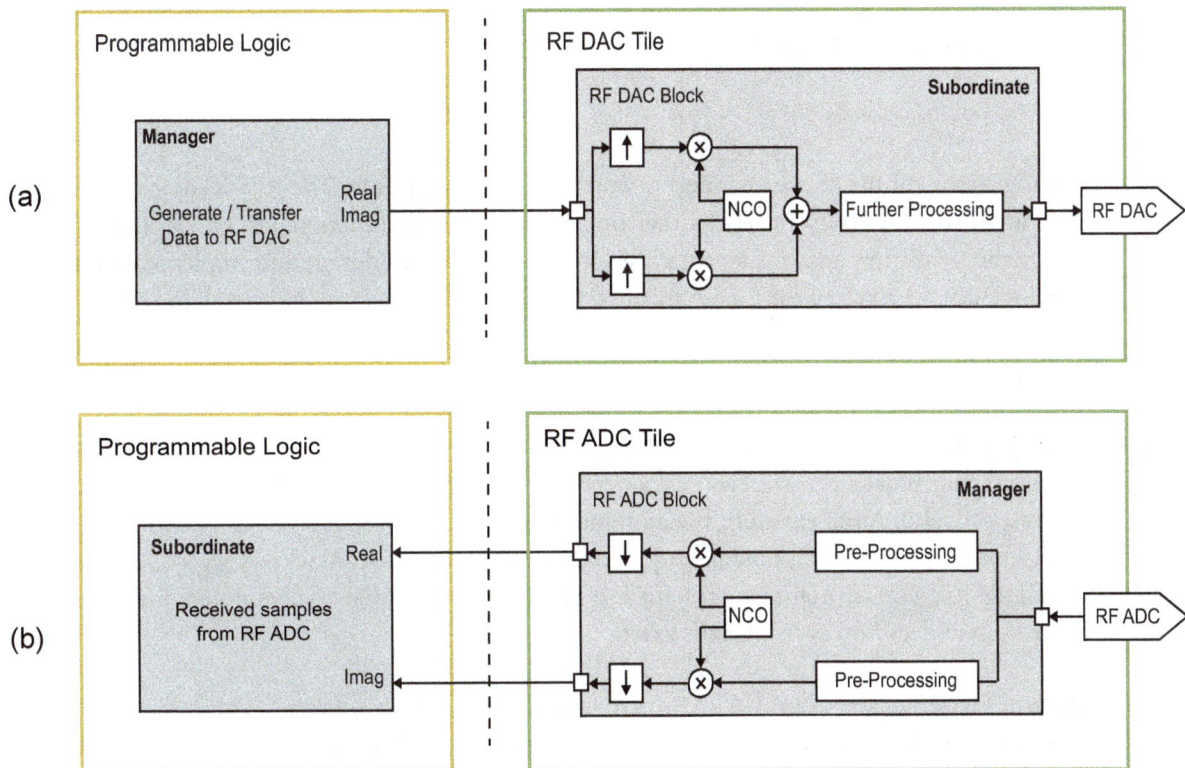

Figure 3.9: The manage / subordinate interface between: (a) the PL and the RF-DAC; (b) the RF-ADC and PL.

The RF Data Converters

As a quick review of the RFDC specifications for the Generation 3 ZU48DR device, the key parameters are:

- 8 RF-ADC channels:

    - *Max sample rate = 5 GSps*

    - *RF-ADC resolution = 14 bits*

- 8 RF-DAC channels:

    - *Max sample rate = 9.85 GSps*

    - *RF-DAC resolution = 14 bits*

The physical configuration of the above channels is worthy of particular note. The RFSoC data converters are laid out in tiles, which host a group of blocks that implement the core functionality of the associated data converter. This hierarchy of tiles and blocks simplifies the data converter design and implementation. In the remainder of this section, we explore the hierarchy of each RFDC to understand the overall layout.

RF-ADC Hierarchy

The RF-ADCs are configured in one of three different styles, which are fixed for any given device. RF-ADCs may be configured as four blocks per tile, two blocks per tile, or one block per tile (referred to as Quad, Dual and Single tiles, respectively). As an example, the aforementioned ZU48DR uses a layout of two blocks per tile (i.e. Dual tile), meaning that a total of 4 tiles are required to host all 8 RF-ADC blocks. A high-level overview of the ZU48DR Dual tile RF-ADC structure is presented in Figure 3.10; note that analogue signals enter from the right and the direction of data flow is from right to left.

Each tile contains a Phase Locked Loop (PLL), which generates the clocks required within the tile. The PLL requires an external, low-jitter, off-chip clock to operate effectively. RFSoC development boards often host these external clocks; for example, the ZU48DR device is located on the ZCU208 board, which features an add-on clocking board.

The RF-ADCs operate using differential signalling, however a received signal from an antenna will be single ended. Development boards such as the aforementioned ZCU208 contain RF baluns which convert the signal from single ended to differential, prior to the input to the RFSoC device.

The RF-ADC processing pipeline samples the received analogue signal, converting it to the digital domain. The RF-ADC block then applies DSP techniques such as threshold detection (to enable adjustments to the input signal level) and Quadrature Modulation Correction (QMC), which corrects imbalances in the analogue signal chain in a complex configuration, i.e. where the are two (I and Q) inputs subject to possible gain and phase offsets [90]. The signal is subsequently down-converted using a complex mixer and programmable decimator, which are collectively known as a Digital Down Converter (DDC).

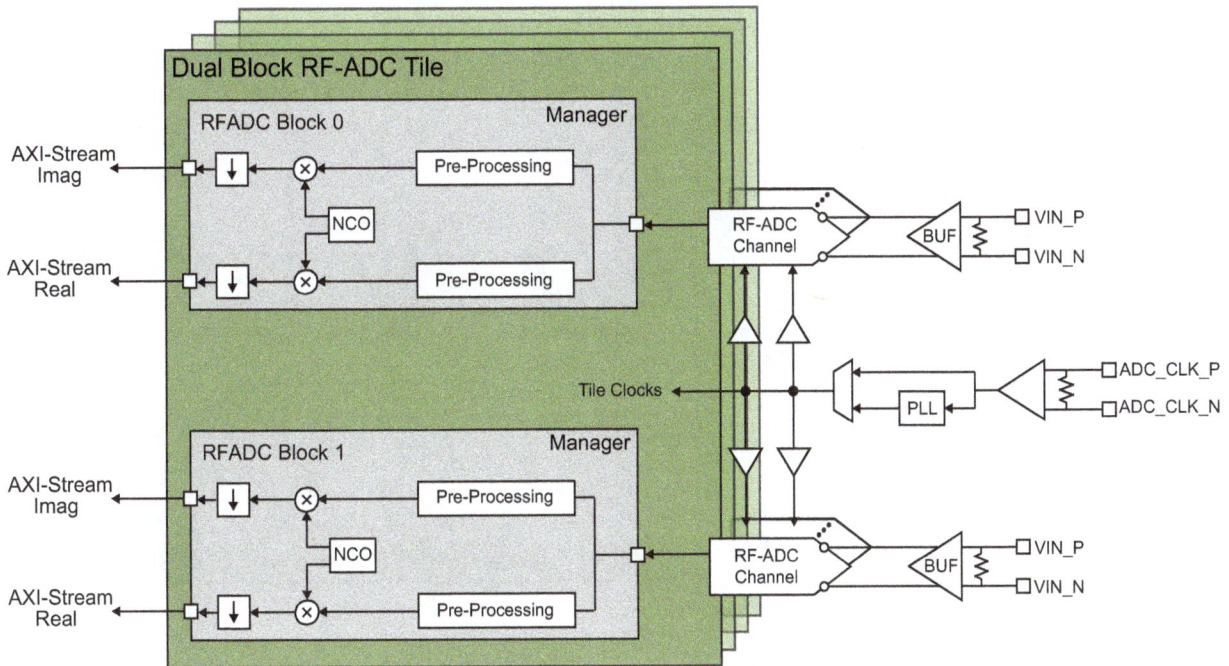

Figure 3.10: RF-ADC hierarchy for the ZU48DR RFSoC device [90].

RF-DAC Hierarchy

For Gen 1 and 2 RFSoC devices, the RF-DAC is configured as four blocks per tile, and therefore two tiles are required to support 8 RF-DAC channels. In Gen 3 devices, RF-DAC tiles can be configured as either two or four blocks per tile, matching the RF-ADC tiles. Similar to the RF-ADC tiles, an internal PLL is present on each RF-DAC tile, which can be driven from an external low-jitter clock. Figure 3.11 presents an overview of the Dual tile RF-DAC configuration found on the ZU48DR device; the direction of data flow is right to left.

As with the RF-ADC, differential signalling is used at the external interface of the RF-DAC. Baluns can be used to convert between differential signals and single-ended signals.

Each RF-DAC contains several stages including a programmable interpolator and complex mixer (collectively known as a Digital Up Converter, or DUC), a QMC block, and inverse sinc compensation filtering. The inverse-sinc compensation filter which can be optionally enabled to correct 'droop' (non-linear gain) in Nyquist Zone 1, and also Nyquist Zone 2 in Gen 3 devices, which arises from the D-to-A conversion process — this technique is further explained in Chapter 11. Finally, the digital signal is converted to the analogue domain for transmission.

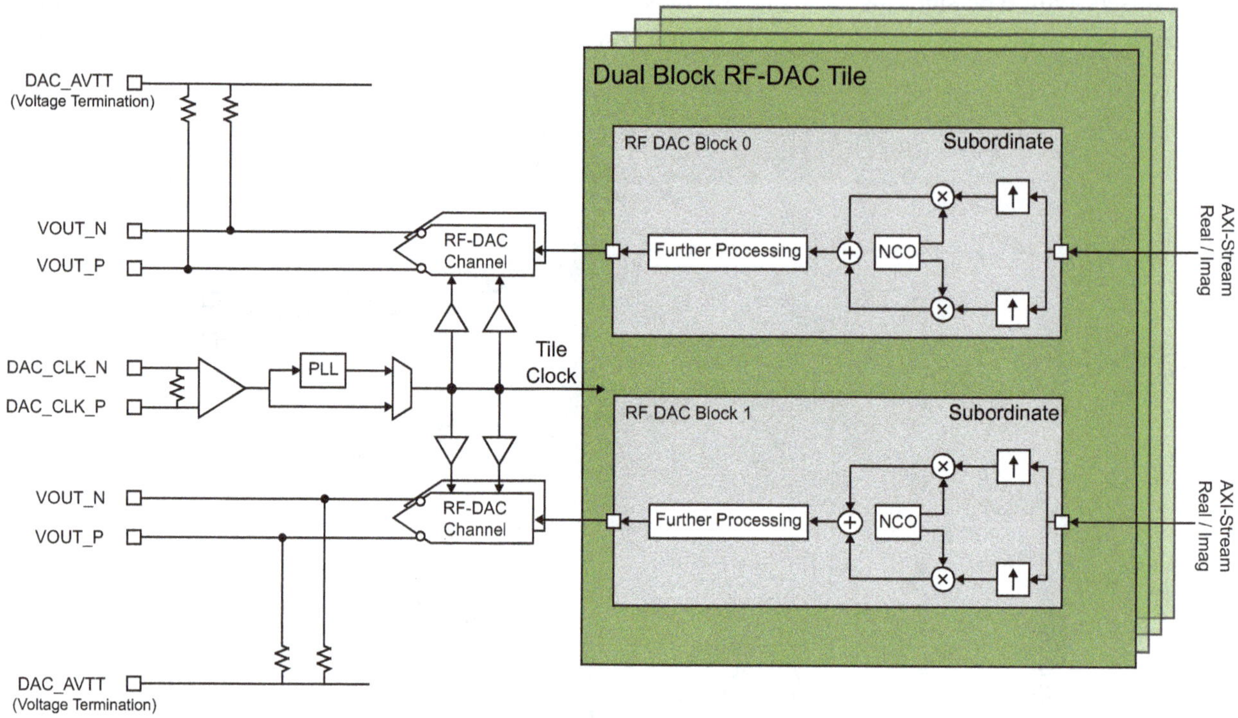

Figure 3.11: RF-DAC hierarchy for the ZU48DR RFSoC device [90].

3.3.5. SD-FEC Blocks

Forward Error Correction (FEC) coding is often applied to source data prior to modulation and transmission over the radio channel. FEC is a powerful method that increases the robustness of transmissions to channel impairments, and thus improves link quality. At the receiver, the FEC decoder is able to detect and correct any errors that occur (up to some limit that depends on the properties of the coding scheme).

When implementing a FEC scheme, redundancy is added to the transmission, as additional data is transmitted beyond the original source data. This implies that the channel has greater bandwidth and energy requirements compared to an identical system without FEC, however, the benefit of this extra encoding is that it provides more protection against noise. As a result, bandwidth and energy may actually be saved, because there would be fewer instances retransmissions required as a result of data being corrupted or 'lost' in the channel. As the need for retransmissions implies a time delay while handshaking protocols take place and the data is retransmitted, overall latency is also likely to be improved as a result of introducing FEC.

An example of a radio channel with FEC encoding can be seen in Figure 3.12, demonstrating how errors in the received signal can be corrected by a FEC decoder.

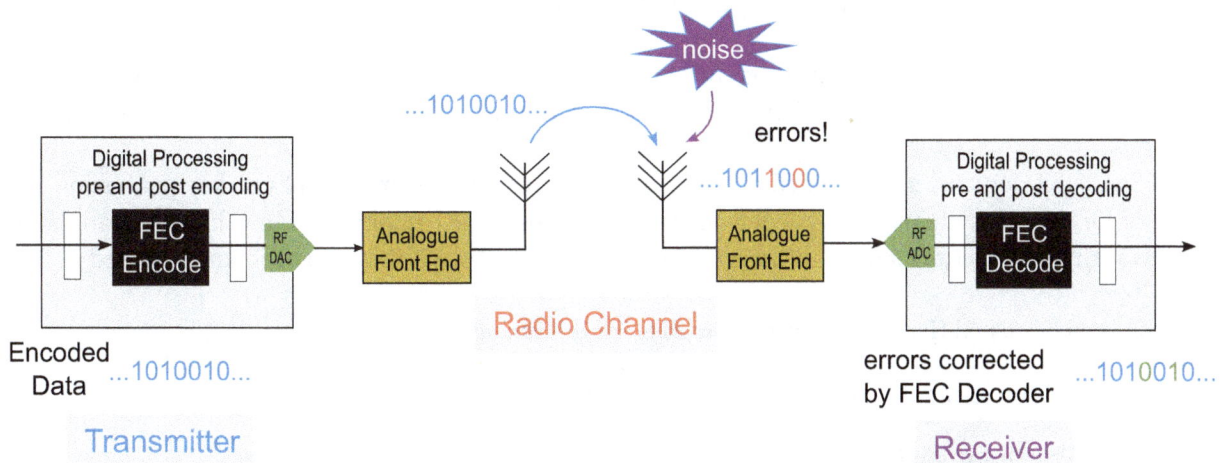

Figure 3.12: Simple model of a radio link using Forward Error Correction.

FEC coding and decoding are computationally intensive processes, and this is particularly true for the robust schemes used in 4G and 5G radio systems. Therefore, to reduce the computation required on the RFSoC PL, a hardened Soft Decision Forward Error Correction (SD-FEC) block is implemented on selected RFSoC devices[3] [21]. If needed, 'soft' SD-FEC cores can be implemented on the PL of any RFSoC device, instead of (or in addition to) hard cores.

The term 'Soft Decision' is distinct from the 'soft' / 'hard' manner of describing the implementation of the core. Rather, in this context, 'soft' reflects that the decoding of received bits is based on a computed probability of an error having occurred.

The SD-FEC core supports the following operational modes:

- Low Density Parity Check (LDPC) coding

- LDPC decoding

- Turbo decoding

Both LDPC and Turbo codes are high-performance coding schemes that have been adopted in modern communications standards. Turbo codes are used in 3G and 4G LTE™, whereas LDPC codes are used in 5G and provide improved error coding performance. Later in the book, Chapter 14 reviews each of these schemes in detail.

---

3. Hardened SD-FEC blocks are available on a range of RFSoC devices including the ZU21DR, ZU28DR and ZU48DR.

3.3.6. RFSoC DFE Architecture and Special Features

The RFSoC DFE is worth a special mention. Although our discussion throughout the book focuses mostly on Gen 1, 2 and 3 RFSoC devices, the DFE has a different architecture, tailored for mobile network applications.

The RFSoC DFE contains an expanded set of hardened IP cores that implement the key processing tasks necessary for 5G NR systems [82]. The advantage of these hardened resources is to achieve key functional elements with higher performance and lower power consumption than if they had been implemented in PL.

The additional hardened IP cores specific to the RFSoC DFE include:

- RF Signal Processing (includes resampling and equalisation functions);

- Digital Pre-Distortion (compensates for non-linearities in the external amplifiers);

- Crest Factor Reduction (reduces the Peak-to-Average-Power Ratio (PAPR) to aid amplifier efficiency);

- DUC and DDC (optimised for 5G NR, up to 8 channels);

- Channel Filtering (optimised for 5G NR, up to 8 channels).

Each of these RFSoC DFE's hardened IPs can be actively included within a system design (and configured with user parameters), or bypassed, as shown in Figure 3.13. Where a hardened IP is bypassed, the option is open to implement a custom IP to replace it, based in the PL — therefore, the designer has considerable flexibility.

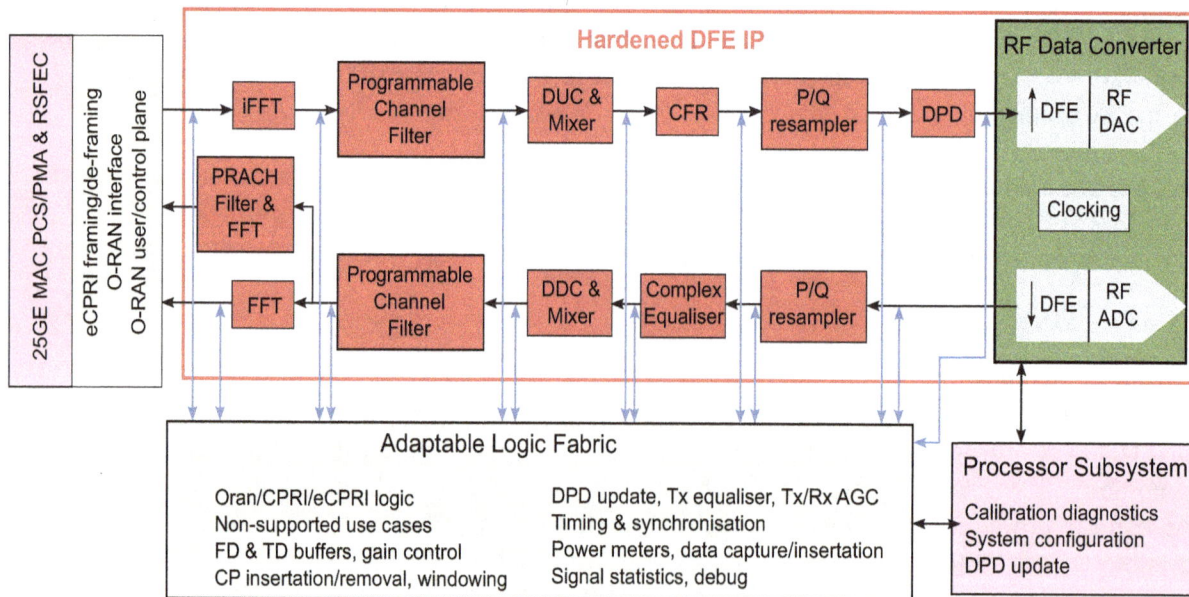

Figure 3.13: Block diagram of the RFSoC DFE architecture [79],[82].

To illustrate the value of these hardened blocks, we can consider the example of Digital Pre-Distortion (DPD). DPD is a technique that implements a corrective gain in the digital domain to compensate for the non-linear gain of an external Power Amplifier (PA) in a communications system. Amplifiers are inevitably non-linear, but have a linear region at lower power levels. The operating range of a PA can be extended (and thus efficiency improved) if the upper region can be linearised. An intuitive diagram of DPD is shown in Figure 3.14.

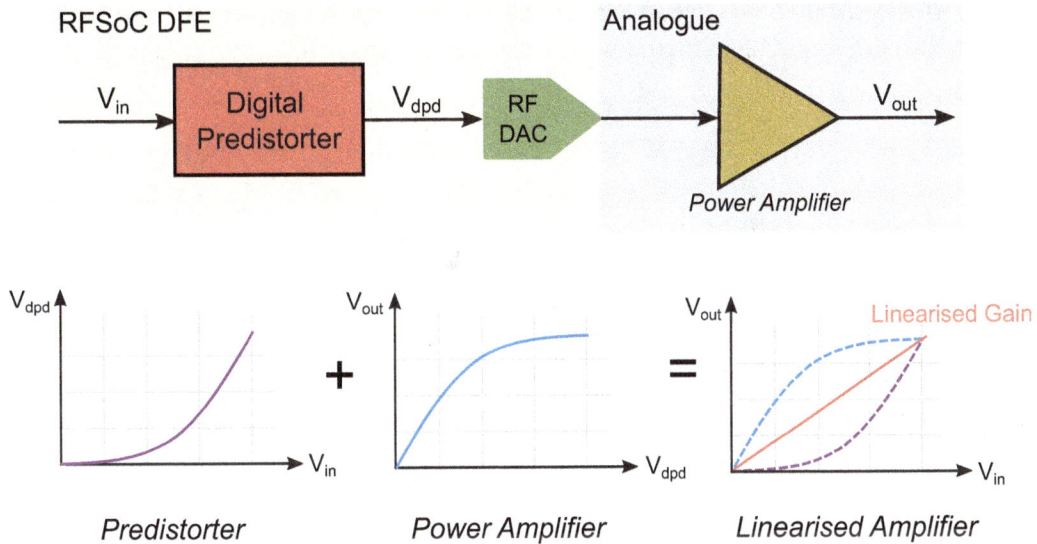

Figure 3.14: The concept of DPD.

The Digital Pre-Distorter must develop a model of the PA gain, updated over time, from which the compensating digital gain function is derived. DPD algorithms are mostly based on the Volterra series, with increasing interest in Machine Learning based techniques [105]. Architectures for DPD are non-trivial to design, and can be computationally complex. The availability of DPD of a hardened block is therefore of considerable value, because this commonly required function is pre-designed and verified (thus saving design effort), and implemented in dedicated silicon (and so is more compact, and reduces power consumption by up to 80% compared to an equivalent in PL [82]). If a different solution is needed, the DPD block can be bypassed in favour of a custom-designed alternative, and the same applies to other hardened blocks in the RFSoC DFE.

3.4. RFSoC Development Boards

Now that we have explored the key components of an RFSoC device, it is worthwhile reviewing the available development and evaluation boards. Such boards combine an RFSoC device with supporting features such as additional DDR memory, IO connectors, external clocking and baluns, which are used for external interfacing of the RFDCs. As a result, such boards provide excellent platforms for designing RF based applications.

In this book, we focus on RFSoC boards that are developed and supplied by AMD. A number of other boards are available from third party suppliers, covering a range of different devices and form factors, and the selection continues to increase over time.

3.4.1. Overview of Development Boards

At the time of writing, AMD offers five RFSoC boards for industry, as captured in Table 3.3. Extended details are available in the corresponding user guides, i.e. references [75] to [79].

Table 3.3: AMD RFSoC development boards.

Type	Feature	Board ZCU111 [75]	ZCU1285 [76]	ZCU208 [77]	ZCU216 [78]	ZCU670 [79]
RFSoC part	Generation RFSoC device	1 ZU28DR	2 ZU39DR	3 ZU48DR	3 ZU49DR	DFE ZU67DR
RFDCs	ADCs DACs	8 8	16 16	8 8	16 16	10 10
SD-FEC	SD-FEC blocks	8	-	8	-	-
Connectors	RJ-45 SFP28 USB 3.0 UART/JTAG QPSI I2C FMC+ FMC RFMC Pmod	1 4 1 1 2 2 1 - 2 x v1.0 2	- - - 2 1 2 - 2 - -	1 4 1 1 2 2 1 - 2 x v2.0 -	1 4 1 1 2 2 1 - 2 x v2.0 2	1 4 1 1 2 2 1 - 2 x v2.0 -
Memory	PL DDR4 PS DDR4 PS DDR3 SD Card	4 GB 4 GB - Yes	- 2 GB Yes	8 GB 4 GB - Yes	8 GB 4 GB - Yes	4 GB 4 GB - Yes
Add-on cards *(see Section 3.4.2)*	Balun card Loopback card Clocking card(s)	XM500 - -	RF200 - CLK101 CLK103	XM655 XM650 CLK104	XM655 XM650 CLK104	XM755 XM650 CLK104 (optional)

Table 3.3 highlights some of the key details and features of each board, including RFDCs, connectors, memory, and add-on cards. More detail about the available add-on cards is provided in Section 3.4.2. As is evident from the table, at least one development board exists for each generation of device, and a range of tile configuration options are covered.

To give an example, the ZCU208 is a popular Gen 3 board with 8 channels of ADCs and DACs, and is pictured in Figure 3.15 with selected features and interfaces highlighted.

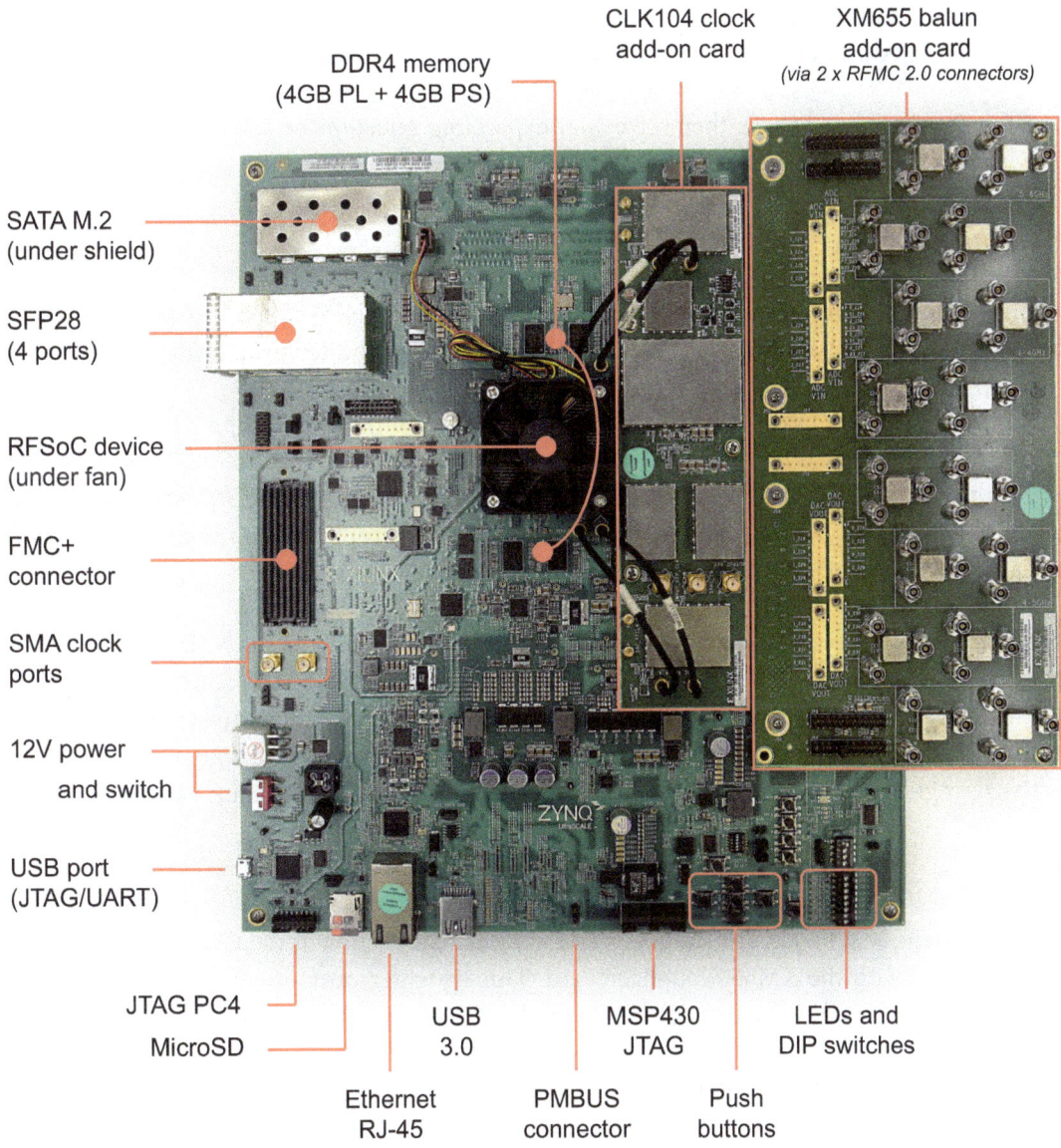

Figure 3.15: The ZCU208 development board (ZU48DR Gen 3 RFSoC).

A further class of boards are the 'Telco accelerator cards' which combine RFSoC and MPSoC devices for centralised processing in mobile wireless networks. These are briefly introduced in Section 3.4.3 of this chapter, and discussed more extensively in Chapter 17. Additional RFSoC boards are available for academic use, as reviewed in Section 3.4.4.

3.4.2. Add-On Cards

Each development board featured in Table 3.3 supports one or more add-on cards. These cards expand the functionality and/or ease-of-use of a development board. Add-on cards generally fall into three categories: balun cards; loopback cards; and clock module cards; which are reviewed in the following subsections.

Similar to the development boards themselves, an expanding selection of add-on cards are available from third-party suppliers, and so the review that follows is not exhaustive.

Balun Daughter Cards

RFDCs transmit and receive using differential signals, and conversion is required to and from a single-ended analogue signal format, in many applications. These conversions can be achieved using baluns.

To fulfil this interfacing requirement, each of the AMD boards from Table 3.3 supports a balun daughter card, which can be directly attached to the development board. Each generation of RFSoC device is supported by a different daughter card, but they all serve the same purpose of interfacing the RFDCs through a set of baluns to a respective set of SubMiniature Version A (SMA) connectors. Front-end components such as filters and antennas can thus be connected to the RFDCs via balun add-on cards.

Typically, baluns are optimised for particular frequency ranges. For example, the XM500 card for the ZCU111 supports two DACs and two ADCs routed to high frequency baluns; two DACs and two ADCs routed to low frequency baluns; and four DACs and four ADCs routed directly to SMAs for external baluns, which permits customisation. Detailed information about the balun card accompanying a development board can be found in its respective user guide, e.g. [75] for the ZCU111, which features the XM500 balun daughter card.

Loopback Cards

Along with a conventional breakout balun card (for external connections), the Gen 3 and DFE development boards are packaged with the XM650 loopback card. This daughter card also incorporates baluns, however, instead of providing external access through SMAs, each RF-DAC is internally connected to an RF-ADC.

The XM650 loopback card provides an internal 4.4 to 5 GHz band pass filter and balun between the RF-DACs and RF-ADCs, allowing applications to be tested with no external components. Additional details on the XM650 internal loopback card can be found in the user guides of the respective boards [77], [78] and [79].

Clock Module Cards

Most RFSoC development boards (excluding the ZCU111 which has on-board clocking) are packaged with a Clock Module add-on-card. These cards, such as the CLK104 for the ZCU208 and ZCU216 boards [22], provide an external ultra low-noise RF clock source for the RF-ADCs and RF-DACs.

Functionally, Clock Module cards normally consist of two stages. The first stage is an 'LMK' clock, which has two internal Phase Locked Loops (PLLs) that are combined to provide a smooth, jitter-free clock, generated from either a reference clock on the card, or from an external clock interfaced via SMA. The second stage is composed of two 'LMX' clocks, one to supply a clock signal to the RF-ADCs, and another for the RF-DACs. These second stage LMX clocks are directly driven by the LMK clock, and provide dedicated RF PLLs for the associated RFDCs on the RFSoC device. A high-level overview of the CLK104 card is provided in Figure 3.16, highlighting the LMK and LMX clock stages.

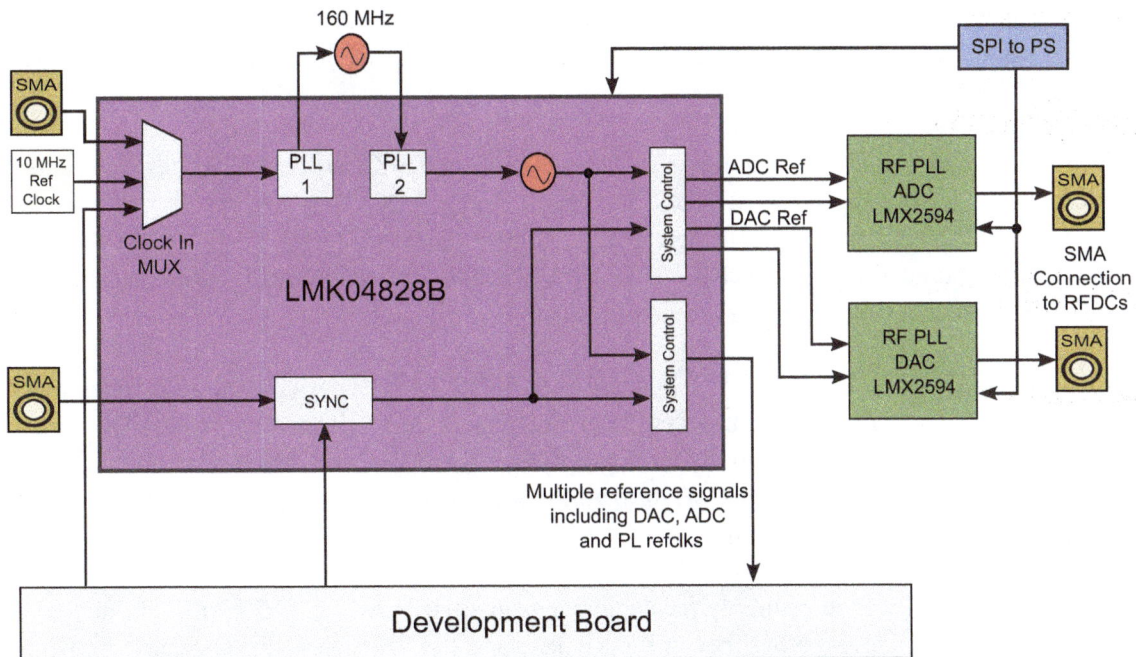

Figure 3.16: High level Overview of the CLK104 Add-On Card for ZCU208 and ZCU216 Boards.

Frequency planning, discussed in Chapter 12, is key to selecting the proper input frequency for the LMX clocks. Generally, the chosen frequency should be an integer factor of the specified RFDC sampling frequency. Within the PYNQ software framework, a clocking driver exists to ease the programming of the LMK and LMX clocks. The use of this clocking driver will be discussed in practical design examples later in the book.

Other Supported Add-On Cards: Avnet® Qorvo®

Although not included in a standard AMD development kit from Table 3.3, the Avnet Qorvo 2x2 small cell RF Front End 1,800 MHz add-on card is also worth highlighting [100]. This card is supported by the ZCU111 and acts as a radio front end, using four RF-ADCs and two RF-DAC channels of the ZCU111's ZU28DR device to provide two transmit and two receive channels. The remaining two RF-ADCs are utilised as observation channels, enabling Digital Pre-Distortion (DPD) to linearise the analogue power amplifiers in a developed system. The channels are fitted with 1,800 MHz bandpass filters, which are suited for LTE Band 3 designs.

3.4.3. Telco Accelerator Cards

AMD produces 'Telco' (Telecomms operator) accelerator cards that feature RFSoC and MPSoC devices. The aim of these cards is to enable high performance, low latency and power efficient 5G deployment, recognising that a 5G network is structured as a hierarchy of links and processing units, and that the network infrastructure must meet exacting requirements, as well as the front-end radio interface. Thus, Telco accelerator cards are targeted at applications within the network but away from the RF interface. At the time of writing, two such Telco cards are available.

Firstly, the T1 Accelerator Card [72], features a ZU21DR RFSoC and a ZU19EG MPSoC device on one card. As identified in Section 3.3.1, the ZU21DR RFSoC does not contain RFDCs — rather, the T1 card leverages the RFSoC for its SD-FEC blocks, which enable high speed encoding and decoding for 4G and 5G LDPC schemes. The MPSoC device is used alongside this RFSoC device to provide further FPGA PL resources, and software control and orchestration for fronthaul 5G protocols.

Secondly, the T2 Accelerator Card [73] is based on a ZU48DR RFSoC device, and accelerates real-time baseband processing. The T2 card assumes that fronthaul processing is performed externally to the card, allowing the RFSoC device to be used entirely for forward error correction and data offload through the PCIe Gen 4 connector. This includes LDPC FEC, Hybrid Automatic Repeat Request (HARQ), rate matching and Cyclic Redundancy Check (CRC) functions.

As these Telco cards are designed primarily for deployment in Distributed Units (DUs) in 5G networks, i.e. within the network infrastructure rather than at the RF interface, the RFDCs are less relevant, as reflected by the choice of ZU21DR RFSoC for the T1 card. Rather, the powerful PL logic fabric and hardened SD-FEC blocks are exploited to meet 5G protocol requirements. RFSoC applications in 5G networks are discussed in detail in Chapter 17, including the use of the Telco accelerator cards.

3.4.4. Academic Boards

The primary development board for academic use is the RFSoC 4x2, pictured in Figure 3.17 [42]. This Gen 3 RFSoC board replaces the (now discontinued) RFSoC 2x2, which features a Gen 1 device [41].

Figure 3.17: The RFSoC 4x2 board (Gen 3 RFSoC).

Both boards are named to reflect the number of ADC and DAC channels that are supported with external interfaces, i.e. the RFSoC 4x2 has four ADC and two DAC channels, and the RFSoC 2x2 has two of each. These academic boards are physically smaller than their mainstream counterparts, due to the reduced channel count.

The RFSoC 4x2 features a ZU48DR Gen 3 RFSoC device, and the board design routes a subset of its 8 transmit and 8 receive channels (4 RF-ADCs and 2 RF-DACs) through RF baluns to SMAs, without an add-on card.

This board also features an OLED screen, real-time clock and high speed QSFP28 interface. The RFSoC 4x2 (and the RFSoC 2x2 up to a version limit) are supported by the RFSoC-PYNQ framework, enabling applications such as the Spectrum Analyser, which will be introduced in Notebook Set C of this book.

3.5. Key Features and Capabilities

Having introduced RFSoC devices in general, as well as the portfolio of AMD development boards available to support the design process, it should be apparent that RFSoC is a unique platform. RFSoC has some significant features and capabilities that make it highly suited for SDR applications, as highlighted in this section.

3.5.1. Multi-GSps Sampling

Two defining characteristics of the RFSoC are the extremely high sampling rates that it can support — up to multiple GSps —and the wide analogue bandwidths that its RF-ADCs can capture. Similarly, the RF-DACs can generate extremely wideband signals (at least equivalent to the bandwidths of the RF-ADCs).

To give an example of RF-ADC capabilities, recall the ZU48DR RFSoC device that was discussed earlier in this chapter. This device has an RF-ADC sampling rate, of f_s = 5 GSps. Therefore, the entire band between 0 Hz and $f_s/2$ = 2.5 GHz can be directly sampled by the device[4], as illustrated by Figure 3.18(a). This region of frequencies is known as the 1st Nyquist Zone. The inclusion of an analogue low-pass filter to isolate this band, prior to the sampler, is recommended.

An equivalent bandwidth can be captured using the 2nd Nyquist Zone, i.e. the range of frequencies between $f_s/2$ and f_s, by exploiting the effect of *aliasing* (as explained further in Chapter 4). Therefore, with the RF-ADC sampling rate set to 5 GSps, the band from 2.5 to 5 GHz can be digitised. An analogue bandpass filter should be used prior to the sampler to isolate this band, as shown in Figure 3.18(b).

Therefore, the RFSoC device is effectively able to directly sample bandwidths of multiple GHz. In the context of SDR design, this is incredibly powerful; an SDR can convert between the digital and analogue domains at RF frequencies, enabling a large set of wireless standards and technologies to be supported using an 'almost all digital ratio' architecture (see Section 2.7.2, and page 30 in particular). Examples of standards that can be directly sampled include, but are not limited to, Wi-Fi bands at 2.4 GHz, 5G New Radio bands around 3.5 GHz and various cellular network bands from 800 MHz to 1.8 GHz (in the UK, other countries may vary) [35].

By comparison, many traditional 'off-the-shelf' SDRs are limited to sampling rates of 10's to 100's of MHz only, and require an analogue mixing stage to demodulate the above example signals from RF to a suitable IF frequency, prior to analogue-to-digital conversion. Increased fidelity, along with a physically compact design, can be accomplished using an RFSoC-based design approach.

---

4. The properties of external analogue circuitry (e.g. linearity of amplifiers, filters etc.) may restrict the operational range.

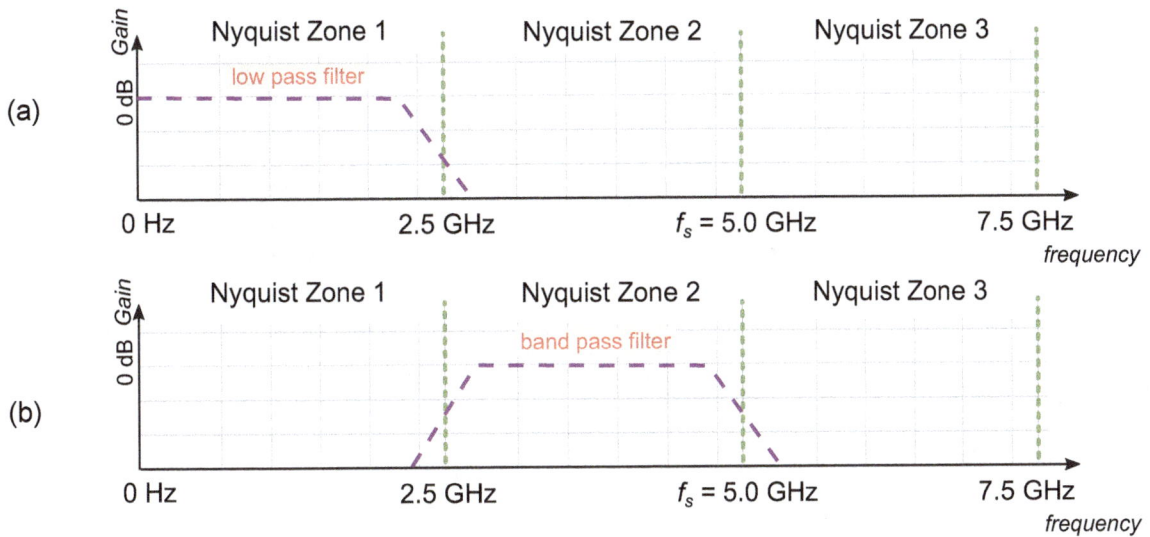

Figure 3.18: Using the ZU48DR (Gen 3 RFSoC) to sample signals in the: (a) first and (b) second Nyquist Zones.

3.5.2. Super Sample Rate

As PL operating clocks are limited to high hundreds of MHz, the Multi-GSps sampling rates used by the RF-ADCs and RF-DACs are often transitioned to a lower sampling rate at the interface with the PL. This is achieved using the hardened DDCs and DUCs integrated within the RFDCs. Such an approach is not always suitable, however, particularly for extremely wideband signals. In cases where very high sampling rates need to be maintained, samples require to be transferred from the RFDCs to the PL (at the receive side), or from the PL to the RFDCs (at the transmit side) at sampling rates higher than the maximum clock rate of the PL. In other words, digital samples cannot be clocked sequentially, as the required clock frequencies are too high. The Super Sample Rate (SSR) method addresses this issue.

SSR Interfacing

An SSR interface between the RFDC and the PL contains several time-contiguous samples per AXI-Stream clock cycle. To explain this concept, consider the conceptual diagram[5] presented in Figure 3.19. Here we consider a Gen 3 ZU48DR RF-ADC channel, sampling at the maximum rate of 5 GSps. To acquire a signal sampled at 5 GSps in the PL, the samples must first be deserialised. This process increases the signal wordlength, but also has the effect of decreasing the required AXI-Stream clock frequency — in this case, using an SSR of 8 to reduce the required clock frequency to 625 MHz. Subject to good design practices, a clock rate of 625 MHz can be supported on the PL.

---

5. The SSR interface depiction of Figure 3.19 is conceptual only; the underlying implementation differs.

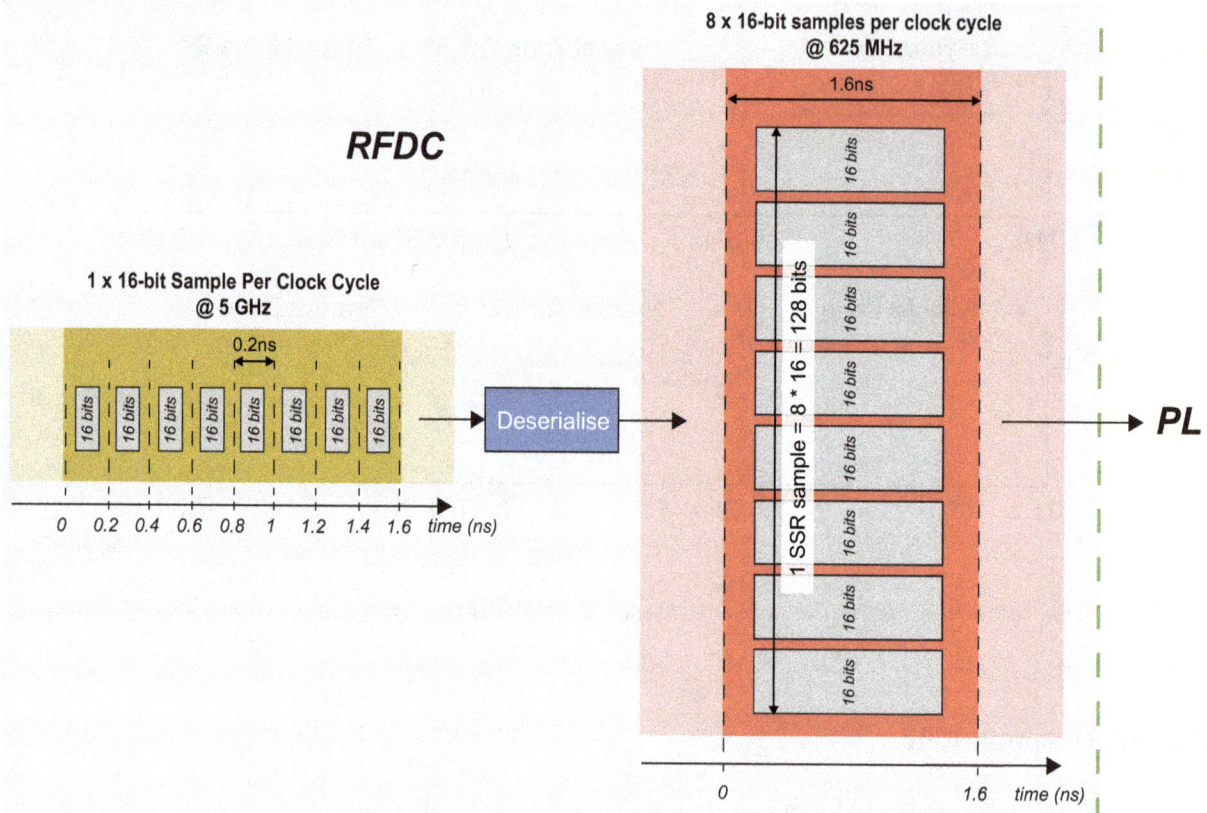

Figure 3.19: Conceptual diagram of converting from a serial interface to an SSR interface.

SSR interfaces are used regularly to transfer data between the RFDC channels and the RFSoC's PL fabric, in either direction (i.e. SSR is applicable to both transmit and receive paths). SSR is primarily required when the sample rate requirements at the RFDC / PL interface are too high for single-rate implementation. For all RFSoC devices, the interface between the RFDCs and the PL uses 16 bits to represent each sample. Therefore, when the SSR is 1 (i.e. the trivial case of one sample per clock cycle, or effectively where SSR is not used) SSR samples are 16 bits wide; with an SSR of 2 (two samples per clock cycle), SSR samples are 32 bits wide; and for an SSR of 4 (four samples per clock cycle) SSR samples are 64 bits wide.

SSR Design

The SSR interface between the RFDCs and the PL is only one part of the solution, where very high sample rates require to be processed on the PL. The other aspect is implementing highly parallel PL-based architectures that are capable of processing several samples per clock cycle. To support the creation of SSR designs, optimised SSR blocks for commonly required functions are available as IP cores in AMD design tools. This topic will be discussed further in Chapter 13.

3.5.3. Multiple Channels

The majority of RFSoC devices incorporate RFDCs, as outlined in Section 3.3.1, and such devices typically support either 8 or 16 channels of transmit and receive functionality. The integration of such a large number of channels on a single chip is a powerful enabler for various different radio applications. To highlight three examples, the RFSoC's multi-channel capabilities can be exploited to implement: Multiple Input Multiple Output (MIMO) systems; beamforming; and radios that support several different standards concurrently.

MIMO is a wireless communications technique that extends upon the fundamental Single Input Single Output (SISO) configuration, i.e. a single transmitter and receiver, to instead use multiple transmitters and receivers. Consequently, MIMO introduces spatial diversity, as each of the transmit and receive antennas occupies a different position in space, and therefore components of the signal take different physical paths through the radio channel. Sets of transmit and receive antennas are usually laid out in a linear or two-dimensional arrays, with the simplest MIMO systems involving two transmitters and two receivers. 5G helped to popularise the concept of 'Massive MIMO', which involves substantially more antennas (in this context, *massive* is not strictly defined and does not refer to a fixed minimum number of antennas, but it is indicative that commercial Massive MIMO systems may contain 64 or more elements) [244]. The greater the number of antennas, the greater the potential to exploit spatial diversity and thus improve the throughput over a wireless link.

Beamforming is another example system where multiple antennas are required. The technique can be used at the transmitter, to electronically steer the direction of emitted energy in a desired direction, and at the receiver, to electronically steer the direction of sensitivity in a desired direction. It is also possible to generate 'nulls' in particular directions and as a means of mitigating interference. One of the major benefits of multi-element antennas in this context is that the beam direction can be altered dynamically using DSP algorithms, without any need to physically move the antennas. Similar to MIMO, antennas can be arranged in different configurations in physical space, and a larger number of antennas implies a greater precision of control in a steering the beam. Beamforming and MIMO can also be combined, as they are in 5G systems, to steer beams towards specific subscribers in three-dimensional space [187]. This technique increases capacity in cellular networks by spatially reusing the available radio channel.

For both MIMO and beamforming applications, it is important that the channels comprising the arrays are synchronised with the same time and frequency reference. In RFSoC, this is supported by the Multi-Tile Synchronisation feature, which can synchronise channels within the same device, as well as across multiple devices [46],[90].

Aside from these multi-antenna systems, which are discussed further in Chapter 18, the multiple channels of the RFSoC can also be used to support several different transmit and receive signal paths, from either the same or different radio standards, on the same chip. For instance, an RFSoC device could support LTE and 5G NR radio standards simultaneously, or even enable multiple network operators to share a device [260], [288]. This approach compares to a more complex overall system design, or even an implementation requiring several separate radios.

3.5.4. Flexible SDR Design

As reviewed earlier in this chapter, the RFSoC architecture combines PL (FPGA logic); a PS integrating applications and real-time processors, and platform management functionality; hardened RF Data Converters capable of directly sampling at multiple GSps; and hardened SD-FEC blocks. Therefore, the RFSoC is an ideal platform for radio applications, and SDR in particular. It is useful to highlight key features of the RFSoC architecture that support SDR design, where the radio must be dynamic and flexible in operation, and therefore under software control.

The multi-GSps sample rates of the RFSoC enable support for wide RF bandwidth signals. By applying a suitable analogue filter to isolate a single Nyquist Zone, signals occupying multiple radio bands within that Zone can be generated or captured. In fact, using *multi-band operation* mode, several RF signals can be received through a single RF-ADC channel before being processed through separate DDCs, or similarly at the transmit side, multiple signals can be generated using different DUCs, and then combined before passing through a single RF-DAC. This flexibility is extremely useful for SDR, where dynamic operation and support for multiple signal bands are often required.

The RFSoC architecture provides the means to exert dynamic control over multiple aspects of a radio's functionality. At the front end, the RFDCs are highly programmable, both at setup and during operation. The RF sample rate can be reconfigured even during run time, catering for any number of different requirements. Similarly, the DUC and DDC parameters for each RFDC channel are flexible, allowing decimation and interpolation ratios to be programmed from software while the system is operational. Software-based control can also be exerted over other design elements implemented in PL, as well as the configuration of the hardened SD-FEC blocks. Of course, elements of SDR processing can be realised in the PS using software, and these can also be developed with flexibility in mind.

In Chapter 13, development tools and practical RFSoC design workflows will be reviewed. The use of a Python driver to reconfigure the RFDCs will also be discussed, highlighting that changes can be achieved in a straightforward manner (allowing modulation and demodulation frequencies to be dynamically altered, for instance). We also note that flexibility is further enabled through Dynamic Function eXchange (DFX) support, wherein designated areas on the PL can be reprogrammed while the rest continues to operate unaffected [63].

The integration of a capable PS with the hardware elements of the RFSoC is particularly significant for SDR implementation. The single chip architecture enables software to reside very close to the hardware that it controls, and enabling lower latency interaction than would be possible with a multi-chip solution.

3.5.5. Power Efficiency

As outlined in Section 3.1, the RFSoC is an integrated device, and integration provides a number of benefits. One of these is power efficiency. There are also a number of features of the RFSoC that support low-power operation. We will now review several aspects of the RFSoC that contribute to power-efficient SDR design.

Integration

A key advantage of the RFSoC as a single-chip solution is that it removes the need for external front-end SDR chips (or external DACs and ADCs), as previously discussed in Section 3.1.2 and illustrated in Figure 3.2. If an FPGA or MPSoC is interfaced with external chips, JESD204 links are often used, especially where high data transfer rates are required. AMD offers an IP core for implementing JESD interfaces [32].

JESD204 is a high speed serial interface standard for data converters, developed and maintained by the Joint Electron Device Engineering Council (JEDEC), with JESD204C the most recent revision at the time of writing [219]. JESD204 is implemented as set of 'lanes' which each represent a serial link. The number of lanes needed depends on the data rate (and therefore the sampling rate and resolution of the data converter), but as an example, a 14-bit converter operating at 4 GSps would generate data at a rate of 56 Gb/s. This may require as many as 16 JESD204C lanes. For each set of four lanes used, it is estimated that an additional 1W of power is used compared to the RFSoC with its integrated data converters [351], and therefore an additional power consumption of 4W would be implied.

Aside from the integration of data converters, the integration of PS and PL reduces power consumption over a comparable system developed using a separate processor and FPGA. This power efficiency arises from reducing the number of devices, avoiding the necessary external interfacing between them, and eliminating the need for various supporting components on the PCB.

Hardened Processing Blocks

In Gen 1, 2, and 3 RFSoCs, the SD-FEC blocks are hardened, and this provides a substantial power saving of approximately 80% compared to equivalent 'soft' cores implemented in PL [30].

The RFSoC DFE has a much-expanded set of hardened processing blocks, as outlined in Section 3.3.6, and similar power savings can be achieved by using these blocks rather than soft cores on the PL. Overall, a system taking advantage of all DFE hardened features is estimated to save around 50% of the consumed power, compared to an equivalent system implemented using a Gen 3 RFSoC device [82].

Platform Management Features

The PS of the RFSoC includes a Platform Management Unit (PMU), in common with the Zynq UltraScale+ MPSoC. One of the functions of the PMU is power management during operation [83], [131].

The PMU supports three different power modes: full power mode, low power mode, and battery power mode. The RFSoC can therefore be switched from full power mode (for normal operation) to low power mode when it does not need to actively transmit or receive data. This can assist in reducing the average power consumed by the device. Further, the RFSoC architecture has a defined set of power domains, which can be power gated (i.e. they can be powered down when not actively required), which also helps to reduce power consumption.

DPD Support

While also covered under the category of DFE hardened blocks earlier in this section, support for DPD is worth highlighting again, because the linearisation of the PA response improves their power efficiency, as discussed in Section 3.3.6. This advantage applies regardless of whether the DPD functionality is implemented using the hardened facilities of the RFSoC DFE, or as a soft core on RFSoC Gen 1, 2, or 3 devices, and it can be substantial. For instance, [156] reports an increase in PA efficiency from 8 - 15%, to 30 - 40% (or more) when DPD is used.

Lidless Packages

A power-related innovation is the concept of housing RFSoC devices in 'lidless' packages [33]. This means that the outer casing of the RFSoC chip does not have a top, and therefore the heatsink can be placed in direct contact with the chip. Consequently, heat is dissipated more effectively, and the device can operate in a cooler environment. Implications include increased component reliability, and an easing of cooling requirements (such as a physically smaller heatsink, and/or smaller or less powerful fans).

3.6. Chapter Summary

This chapter has introduced the Zynq RFSoC family of devices in the context of SDR applications. It was highlighted that the device architecture comprises two main components: the Processing System (PS); and Programmable Logic (PL) with hardened RF Data Converters (RFDCs) and Soft Decision Forward Error Correction (SD-FEC) blocks. Various RFSoC devices were compared, both within and between generations, noting the number of RFDCs and maximum sampling rates as key parameters.

The architectures of the PS and PL were explored, with particular emphasis on the hardened, radio-centric blocks integrated into the PL. These hardened resources, the RFDCs and SD-FEC blocks, are a key factor in selecting RFSoC devices for radio applications. Further, the main advantages of RFSoC architecture and design support were reviewed from an SDR perspective. The very high sampling rates, multi-channel support, and features for integrating software and hardware elements were highlighted as particularly powerful features. The range of development boards that are available to support the practical creation of RFSoC designs was also reviewed. As will be explored further later in the book, the RFSoC can enable an array of radio applications including radio standards for mobile networks (and many other technologies), beamforming, MIMO systems, and cognitive radio.

Chapter 4

DSP Fundamentals

Louise Crockett

Several fundamental Digital Signal Processing (DSP) concepts underpin wireless communications and the operation of the RFSoC. While DSP principles will be familiar to many, it may be difficult for readers without this background to appreciate some of the theory introduced later in the book. Feel free to dip into this chapter as appropriate to your level of prior knowledge.

We begin with sampling and quantisation, before briefly reviewing the frequency domain (noting that more extensive coverage of this topic follows in Chapter 5), and aspects of filtering. Conversion between the analogue and digital domains is fundamental to the operation of the RFSoC, and these conversions are considered in detail from a signal processing perspective. Finally, multirate signal processing is presented — increasing and decreasing the sampling rate is integral to almost any radio design we might develop for implementation on an RFSoC device.

4.1. Sampling

Sampling is one of two processes that take place when an analogue signal is converted into a digital equivalent (the other is quantisation, which will be covered in Section 4.2). We can think of sampling as converting the time axis of a signal to a set of discrete time instants, while quantisation converts its amplitude to a discrete set of representable amplitude values. This simple model of Analogue-to-Digital Conversion is depicted in Figure 4.1, where an analogue signal, $z(t)$, is passed through an ADC to create a digital equivalent, $z[k]$.

As shown at the left hand side of Figure 4.1, the analogue signal is considered to be *continuous time*, meaning that its amplitude is defined over all time, and can be measured or represented at any arbitrary point in time.

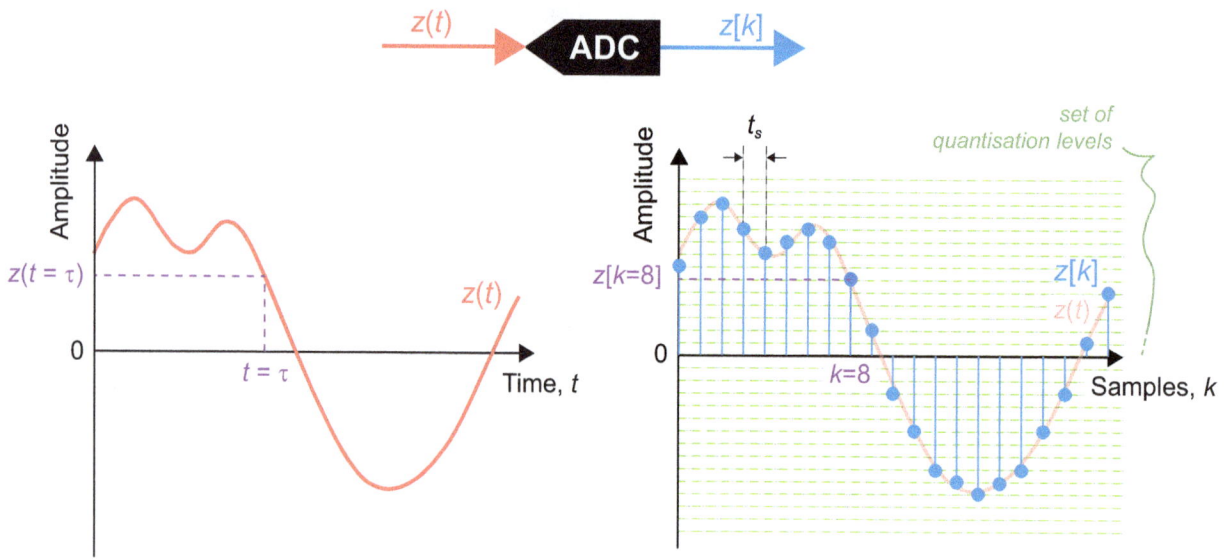

Figure 4.1: An intuitive model of Analogue-to-Digital conversion.

When converted to the digital domain, shown on the right of Figure 4.1, the signal is instead represented at discrete instants in time, which are separated by a sampling period, t_s. This conversion process is known as sampling. The sampling period and the sampling frequency, f_s, have a reciprocal relationship as shown in (4.1). Note that the term *sampling rate* is also common, and can be used interchangeably with *sampling frequency*.

$$f_s = \frac{1}{t_s} \tag{4.1}$$

Quantisation is a separate but concurrent process in the ADC that converts the amplitude of each sample to the closest representable level (the set of available quantisation levels is indicated in green in Figure 4.1).

4.1.1. Sampling Rate Selection

The selection of an appropriate sampling frequency is primarily influenced by the frequency content of the signal to be processed. Sample too slow, and information about the signal is not properly captured; sample too fast, and the operations needed to process the signal are inflated unnecessarily. To take a simple intuitive example, consider the 100Hz sine wave seen in Figure 4.2, which is sampled at three different rates.

If we sampled this sine wave at 2kHz (shown in the second plot), the signal would certainly be captured faithfully (we can clearly see that the samples correspond closely to the shape of the original sine wave), however,

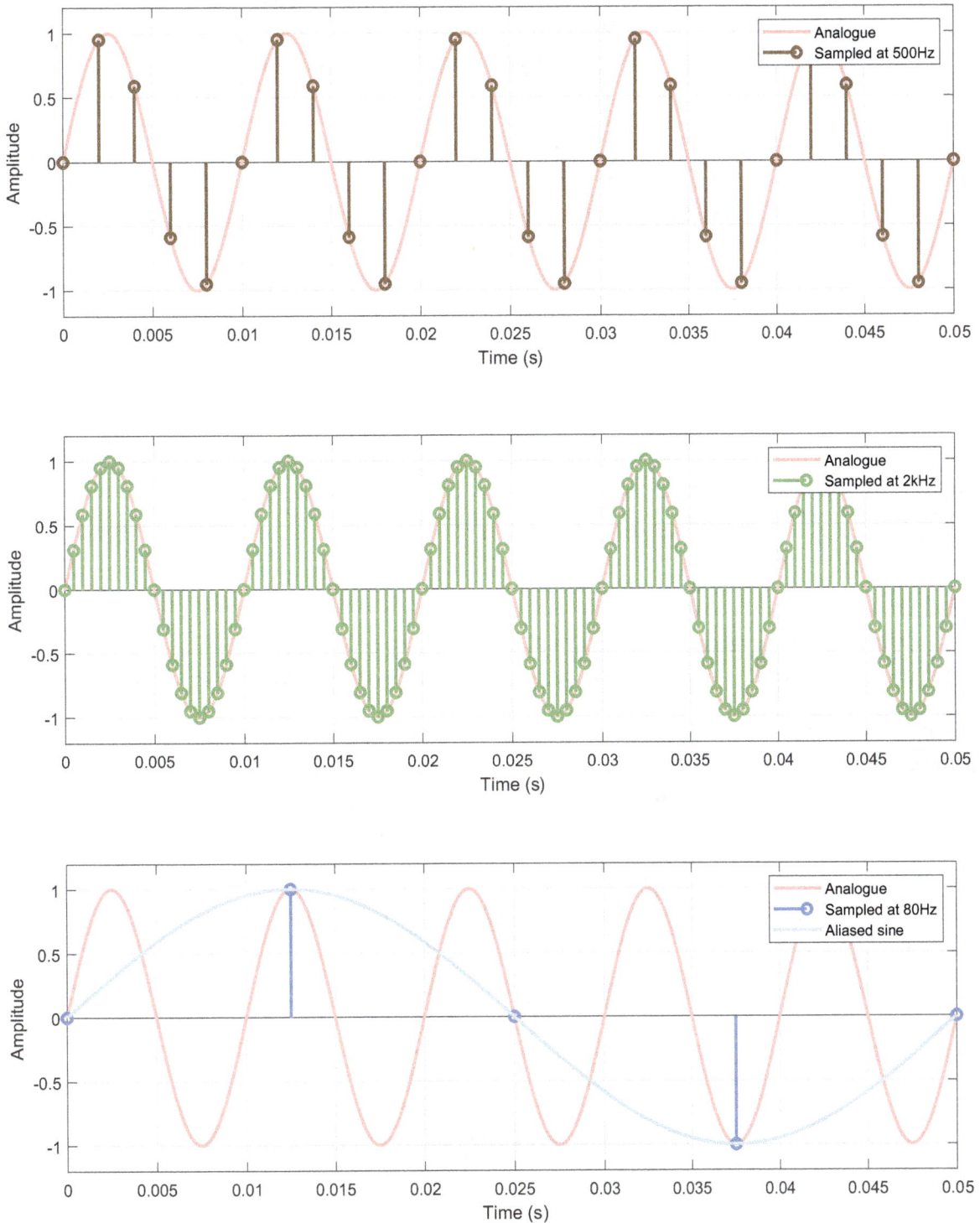

Figure 4.2: Sampling a 100Hz sine wave at three different rates: top — about right; middle — too fast; bottom — too slow.

perhaps this is a higher rate than needed? Twenty samples per sine wave period is actually far in excess of the minimum required, and implies a computational overhead for any subsequent processing operations applied to the signal. This might commonly include operations like digital filtering, which is introduced in Section 4.5.

At the other extreme, a sampling rate of 80Hz (as shown in the third plot) seems insufficient. This would provide fewer than one sample per period of the sine wave, and in fact the sine wave cannot be accurately reconstructed from samples taken at this rate. If sampled at a rate of 80Hz, then actually a lower frequency sine wave would be perceived. This effect is commonly known as *aliasing*. In our current example, the aliased signal has a frequency of 20Hz, which is related to the sine wave frequency and sampling rate as

$$f_{alias} = f_{sine} - f_s = 100\,\text{Hz} - 80\,\text{Hz} = 20\,\text{Hz}. \tag{4.2}$$

Although aliasing is generally considered an undesirable effect, there are occasions where it can be deliberately exploited, as will be discussed later.

Finally, a sampling rate of 500Hz (the first plot) might seem "about right". In this case, there are five samples per sine wave period, resulting in a good balance of accurately capturing the sine wave, and keeping the computation rate reasonably low.

4.1.2. Signal Frequency Range Terminology

In this context of sampling rates, it is useful to define some key terms that describe the range of frequencies present in a signal. Figure 4.3 shows the frequency domain content of two signals in spectrum sketches.

Firstly, when referring to a signal whose lowest frequency component is at, or close to, 0 Hz, the signal is said to be "at *baseband*", or "a *baseband* signal". Examples of baseband signals include sensor data, such as measured voltage and current from a power system, audio signals, and unmodulated communications signals.

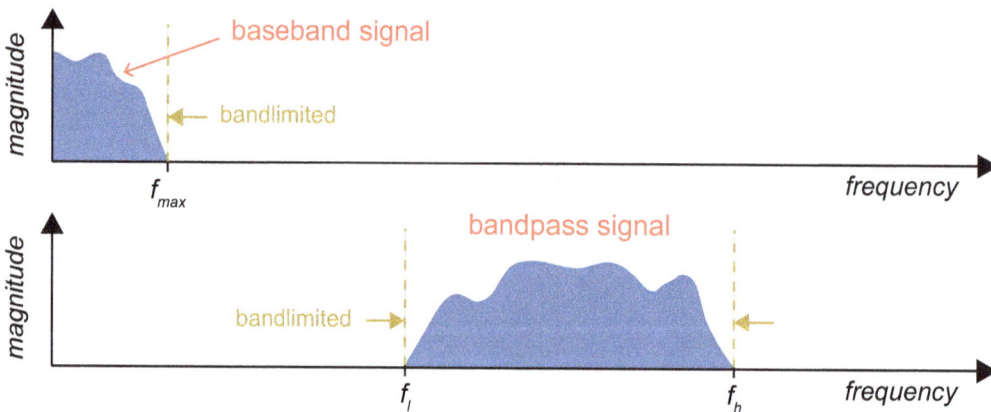

Figure 4.3: Clarification of 'baseband', 'bandpass', and 'bandlimited' signals.

Often in communications, we consider signals that are in the process of being modulated or demodulated by a transmitter or receiver, respectively, and which do not have any components close to 0Hz. A term that may be applied to these signals is *bandpass*, reflecting that the signal occupies a range of frequencies, not adjacent to 0Hz (akin to a signal that has been filtered by a bandpass filter). Both baseband and bandpass signals may also be referred to as *bandlimited*, meaning that the signal energy is contained within a particular range of frequencies.

4.1.3. Nyquist Sampling Theorem

A more formal definition of the minimum possible sampling frequency, in order to avoid aliasing, is given by Nyquist Sampling Theorem. The Nyquist Sampling Theorem states that a baseband, bandlimited signal must be sampled at greater than twice the maximum frequency component present in the signal, i.e.

$$f_s > 2f_{max} \tag{4.3}$$

in order to accurately preserve all of the frequency content. If the above condition is not met, then all frequency components above $0.5f_s$ are subject to aliasing. To aid in describing this lower limit on the sampling rate, the Nyquist frequency is commonly defined as $2f_{max}$.

Returning for a moment to our 100Hz sine wave example, in that case a minimum sampling frequency of just over 200Hz would be the required according to Nyquist Sampling Theorem. Interestingly, if we observe the samples taken at a sampling rate of 250Hz (as an example), then the correspondence with a sine wave may be unclear; from a human inspection point of view, these samples do not "look like" a sine wave in a time domain waveform. Importantly, however, the original sine wave is fully mathematically captured by these samples, and an analogue version of the signal could be perfectly reconstructed.

Nyquist sampling theory can also be applied to bandpass, bandlimited signals. The bandwidth of the signal is defined as the difference between the highest and lowest frequency components present, denoted by f_h and f_l, respectively. Provided that sampling takes place at greater than twice the bandwidth, i.e.

$$f_s > 2(f_h - f_l) \tag{4.4}$$

with reference to Figure 4.3, then it is possible to retain all information in the signal. This is accomplished using the technique of *undersampling*, implicitly relying on aliasing effects, and alignment of the signal within the applicable Nyquist Zones.

4.1.4. Nyquist Zones and Aliasing

Aliasing occurs in a regular pattern based on *Nyquist Zones*, which are partitions of bandwidth $0.5f_s$ in the frequency domain. Any arbitrary number of Nyquist Zones may be defined, but only sampled signals in the 1st Nyquist Zone can be directly represented, according to Nyquist Sampling Theorem. Any signal components present in higher Nyquist Zones are 'folded' down into the 1st Nyquist Zone as a result of aliasing.

To consider an example with some simple numbers, let us assume a sampling rate of 200Hz. Nyquist Zones are 100Hz wide, with the first Nyquist Zone spanning the range from 0 to 100Hz, the second from 100Hz to 200Hz, and so on, as shown in Figure 4.4.

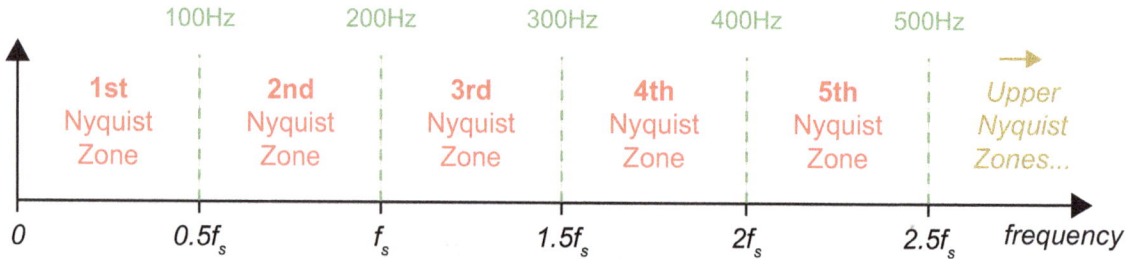

Figure 4.4: Definition of Nyquist Zones.

With an ideal ADC sampling at 200Hz, any input signals of frequency less than 100Hz would therefore be perfectly captured, residing in the 1st Nyquist Zone. If signal frequencies between 100Hz and 200Hz were applied to the ADC, these components would experience aliasing, resulting in terms in the range 0 to 100Hz (in other words, components in the 2nd Nyquist Zone would be folded into the 1st Nyquist Zone). The frequencies of the aliased terms are predictable due to a point of symmetry at $0.5f_s$, or 100Hz, as seen in Figure 4.5, which illustrates the effect of aliasing.

In Figure 4.5 shows some examples of tones input at different frequencies within the first, second, third, and fourth Nyquist Zones. Notice the pattern of folding that applies, depicted here by the gradiated bar below the frequency axis.

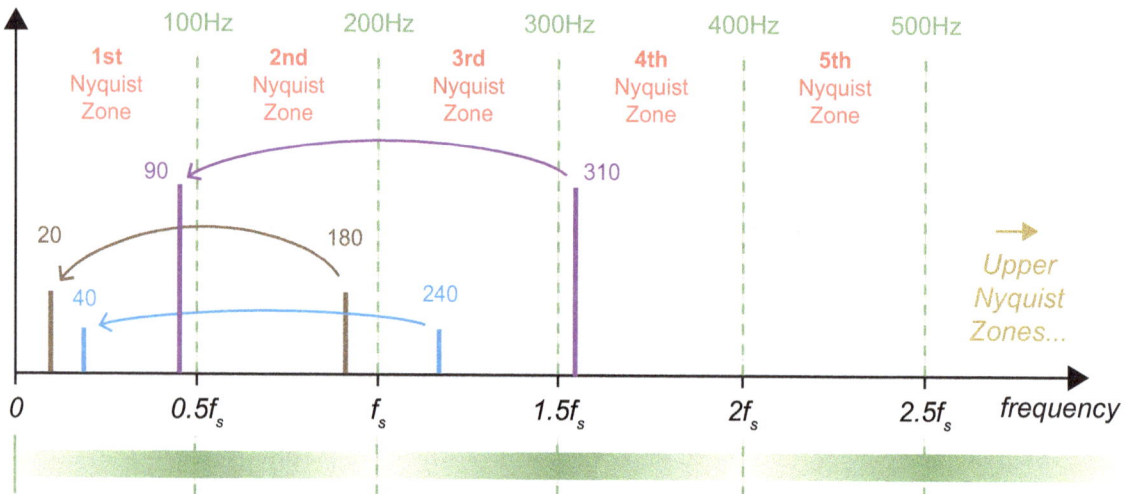

Figure 4.5: Examples of aliasing with reference to Nyquist Zones.

The technique of undersampling can be used to exploit aliasing, and deliberately fold a bandpass signal down to Nyquist Zone 1. In the context of communications, this technique can be used in the receiver as a means of converting an IF- or RF-modulated signal to baseband (often this is referred to as *direct downconversion*). An example of this technique is shown on the upper axis of Figure 4.6, in purple.

One of the major advantages of downconverting the signal is that the sampling rate required for subsequent processing stages may be substantially lower, which can reduce power consumption and computational load. The technique relies on the original signal being entirely contained in a single Nyquist Zone, otherwise the resulting spectrum is corrupted by the superposition of aliased components, as depicted in the lower axis of Figure 4.6. Notice that, as the received signal straddles two Nyquist zones, two separate portions of the signal are both aliased into the upper part of the first Nyquist zone, and are superimposed upon each other.

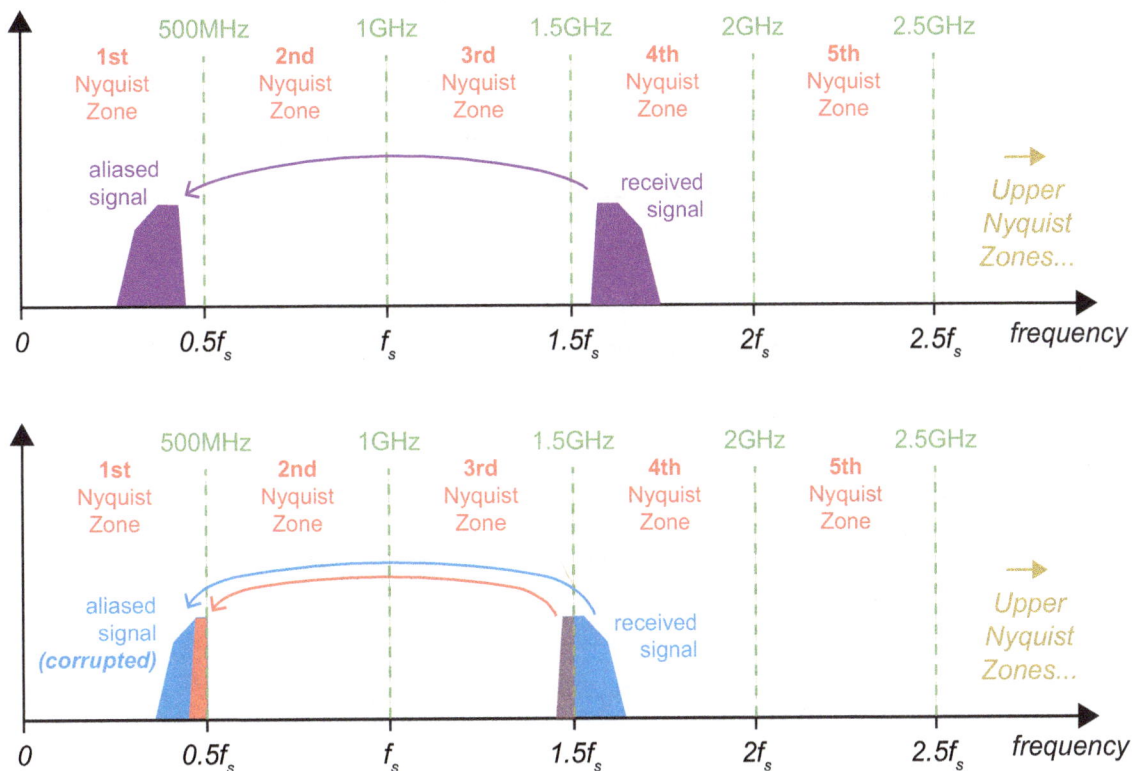

Figure 4.6: Successful and unsuccessful direct downconversion using Nyquist Zones.

These examples have not considered any other frequency components present in the spectrum. However, prior to downconversion using this method, it would also be prudent to *bandpass filter* the signal (more detail on filters coming up in Section 4.5!). This would remove any noise or spurious frequency components present in any other Nyquist Zones, all of which would otherwise alias to the 1st Nyquist Zone and degrade the quality of the desired signal.

The RFSoC device exploits Nyquist Zones for signal reception in its RF-ADC processing, as will be discussed further in Chapter 9. At the transmit side, signals can be generated in upper Nyquist Zones using the RFSoC's RF-DACs, and this will be covered in Chapter 11.

4.1.5. Sampling Jitter

All of our coverage of sampling so far has assumed that samples are perfectly timed, with a consistent separation in time of t_s. In practice, ADCs do not operate in this perfect way, and some variation in the periods between samples should be expected, as illustrated in Figure 4.7 (note, the degree of variation in sampling periods has been exaggerated for visualisation purposes).

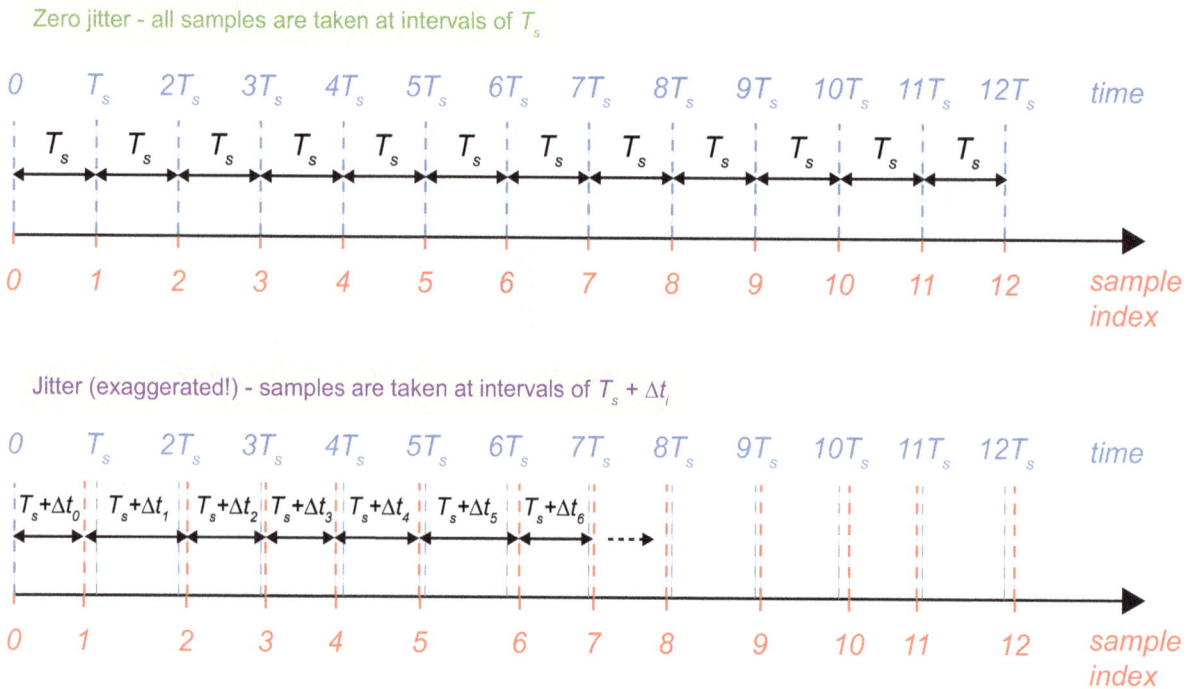

Figure 4.7: Sampling jitter.

This effect of variable sampling periods is referred to as *jitter*, and it is undesirable because samples of the input signal are not taken at the ideal time instants. Jitter is also a characteristic of DACs, wherein the desired output samples are subject to errors in the sampling period, and therefore are not generated at exactly the correct time instants. As a result of samples being misplaced in time compared to the ideal sampling instants, the amplitude values experience a corresponding error. Therefore, jitter can be modelled as a source of noise.

High fidelity DACs and ADCs minimise jitter to a great degree, and often other effects such as quantisation noise dominate, i.e. they contribute significantly more to the overall level of noise experienced. At very high sampling frequencies, however, jitter can be a notable issue (a small deviation in sampling time is significant in proportion to the sampling period). We will not consider jitter extensively in the remainder of this book, however it is worthwhile being aware of the phenomenon and its effects, particularly as the Zynq RFSoC can operate at extremely high (GHz) sampling rates which are naturally more likely to experience jitter.

4.2. Quantisation

Quantisation is one of the two processes that takes place when an analogue signal is converted to a digital one (the other being sampling, as previously covered in Section 4.1). When a signal is quantised, the amplitude value taken at each sampling instant of an ADC is mapped to one of a discrete set of possible amplitude levels. At the output of the sampling and quantisation processes, therefore, the analogue signal has been discretised in both time and amplitude — in other words, it has been converted to a digital signal.

4.2.1. The Quantisation Process

To define the process of quantisation in a little more detail, we consider that the ADC has N bits of resolution, and therefore can represent 2^N different values. These correspond to the set of discrete quantisation levels to which samples of the input signal will be mapped. Usually this is done in a linear fashion as shown in Figure 4.8[1], although a realistic ADC would have many more bits (and hence quantisation levels) than this example.

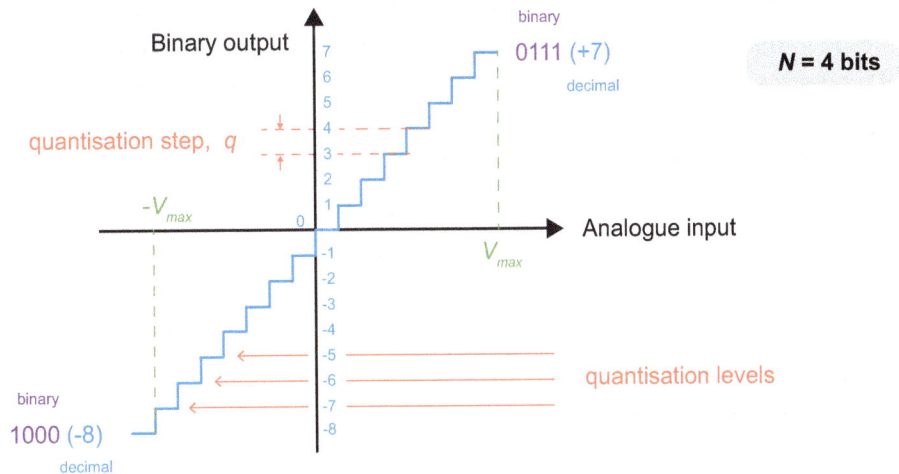

Figure 4.8: An example linear quantiser.

The larger the value of N, the greater the number of quantisation levels that are spread across the operating range of the ADC. We note a slight imbalance (there is one more negative quantisation level than positive), but this effect becomes negligible for realistic values of N.

---

1. Non-linear quantisers are adopted in some specific applications, for instance the A-law and μ-law companding methods used for audio quantisation in telephony systems.

When analogue samples are quantised, they are 'moved' to the nearest available quantisation level; the conversion is more accurate when N is large and the *resolution* is said to be higher. More formally, the changes in represented amplitude between the analogue and quantised samples are referred to as quantisation errors.

Where V_{max} is the maximum input voltage of the ADC, the quantisation step (or interval), q, is expressed as

$$q = \frac{V_{max}}{2^{N-1}}.$$ (4.5)

The dynamic range of an N-bit converter, a commonly used figure of merit that expresses the ratio of largest to smallest representable numbers, is often given in dBs. It is defined as

$$\text{Dynamic Range} = 20log_{10}(2^N) = 20Nlog_{10}(2) = 6.02N.$$ (4.6)

Therefore, a 12-bit converter has a dynamic range of approximately $6 \times 12 = 72\,\text{dB}$.

An example of quantising a sine wave is shown in Figure 4.9, using two different quantisers: first using 4-bit precision, and then with 6-bit precision. The reference (sampled by not quantised) and quantised sine waves have been reproduced in large format here, for ease of inspection — the errors introduced by quantisation are relatively obvious in the 4-bit case, but they are more difficult to see when 6-bit quantisation is used. Thus we can say that quantising the signal with 6-bit resolution produces smaller amplitude errors than the 4-bit equivalent. Although the errors arising from quantisation become difficult to observe in the time domain, at 6-bits this still represents a considerable degradation in signal quality when viewed in the frequency domain.

Practical ADCs and DACs for most applications use at least 8-bits of precision. All RFSoC devices have at least 12-bit ADCs (Gen 1 and 2 devices have 12-bit ADCs, and Gen 3 have 14-bit ADCs); all have 14-bit DACs [88].

4.2.2. Quantisation Errors

When the quantisation of an individual sample is considered, and its amplitude is moved to the nearest quantisation level, it follows that the worst-case error is half of one quantisation interval. Using the symbol q to represent the amplitude of a quantisation interval, the maximum error is therefore $\pm q/2$.

If we assume that the amplitudes of samples input to the quantiser are random, then a Probability Density Function (PDF) can be formed as shown in Figure 4.10. The area of this PDF is 1, meaning that when a particular sample is quantised, a quantisation error in the range $-q/2$ to $q/2$ is certain to occur.

The quantisation noise associated with an ADC can therefore be analysed based on this statistical expectation of quantisation errors. As we will see later, the assumption that errors across the range are equiprobable does not always hold, as the distribution of errors depends on the properties of the signal being quantised. However the PDF from Figure 4.10 provides a useful mechanism for quantifying the noise introduced by the quantisation process. We will now go on to consider how quantisation noise arises and is quantified.

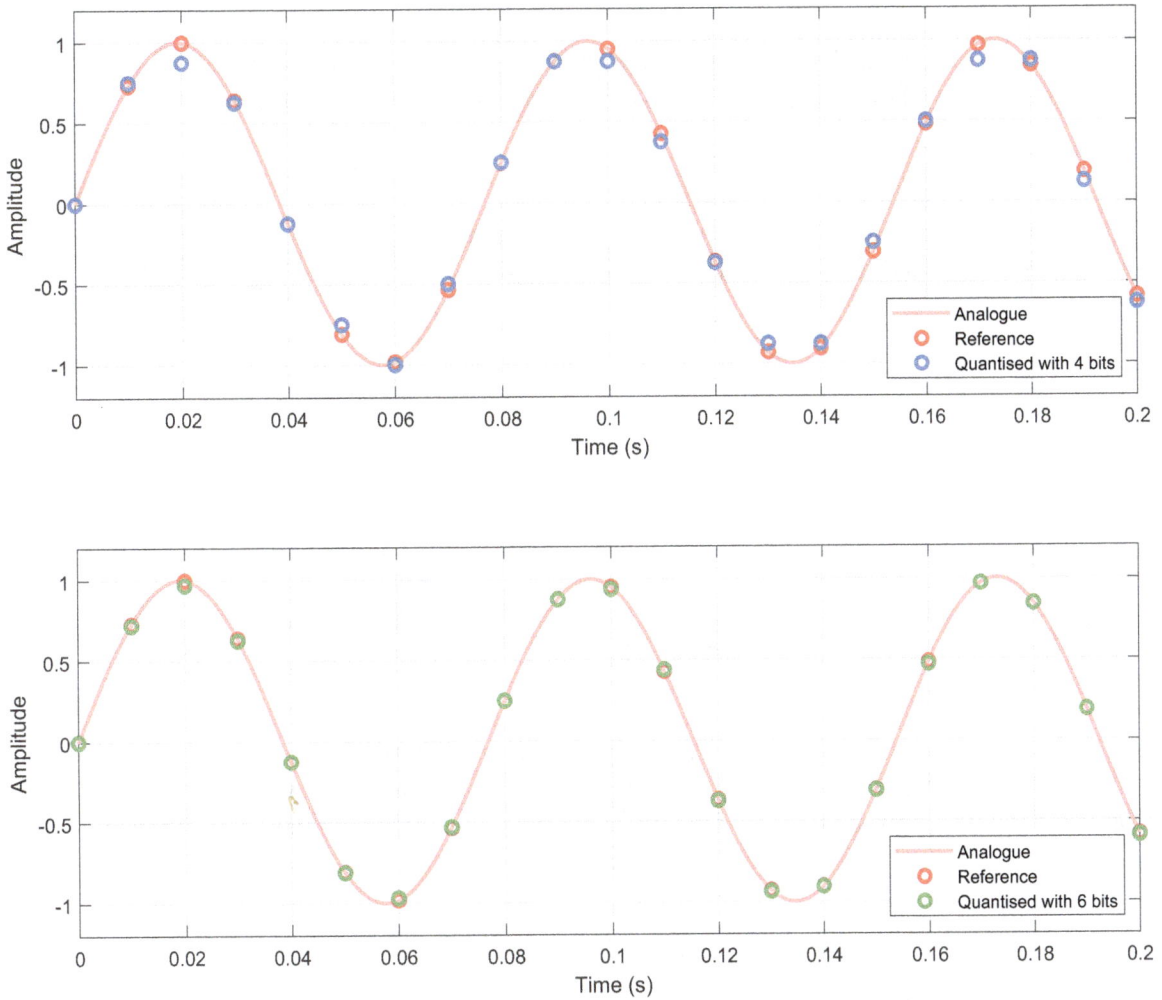

Figure 4.9: An analogue sine wave, sampled and then quantised with 4 bits (upper plot), and 6 bits (lower plot).

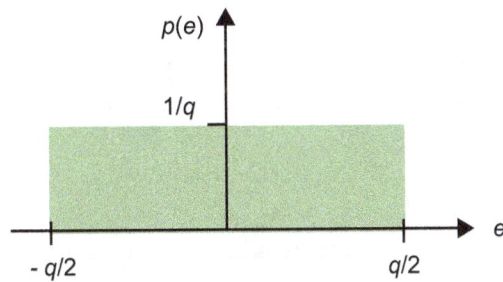

Figure 4.10: Probability density function (PDF) of quantisation noise (for quantisation interval q).

4.2.3. Quantisation Noise

The quantisation process discussed in the previous section can also be modelled as the addition of noise. Each quantised sample is subject to an error, and the quantised sample at index k can be expressed as

$$v[k] = s[k] + e[k] \tag{4.7}$$

where $v[k]$ is the sample after quantisation, $s[k]$ is the true (unquantised) sample value, and $e[k]$ is the difference between the two (i.e. the quantisation error). This process is illustrated Figure 4.11.

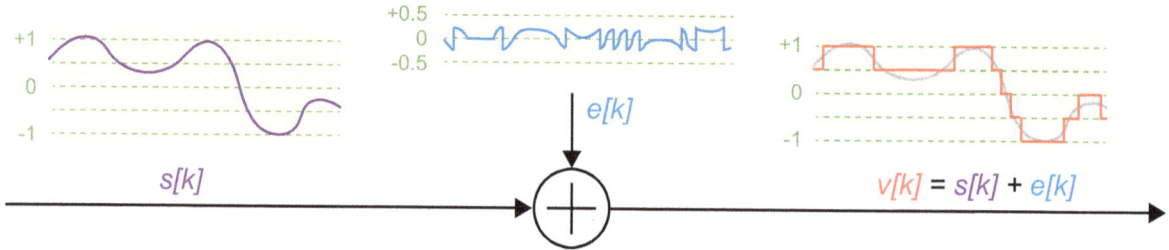

Figure 4.11: Modelling of the quantisation process as the addition of quantisation noise.

The power of the quantisation error signal (equivalently, the quantisation noise power of the ADC) can be derived based on the error PDF from Figure 4.10. The square of the error signal, weighted by the probability of the error, is integrated across all possible error values. As the range of possible error values is limited to the range $-q/2$ to $q/2$, the integral can be evaluated over those limits,

$$n_{ADC} = \int_{-\infty}^{\infty} e^2 p(e)\, de = \int_{-q/2}^{q/2} e^2 p(e)\, de. \tag{4.8}$$

Noting that $p(e) = 1/q$ for all values of e, the error power is therefore expressed as

$$n_{ADC} = \frac{1}{3q} e^3 \Big|_{-q/2}^{q/2} = \frac{q^2}{12}, \tag{4.9}$$

where q is the quantisation interval. We can therefore confirm that the noise power introduced by an ADC reduces as the number of quantiser bits is increased (as this results in a smaller quantisation interval, q).

Theoretically, the quantisation error signal extends across the full baseband region from 0Hz to $f_s/2$, Therefore, there is a chance that low level components present in the signal of interest can be 'masked' by the presence of quantisation noise, as depicted in Figure 4.12. Based on our analysis above, this problem is more acute for low resolution ADCs (i.e. those with few bits), and therefore the motivation for higher resolution ADCs is clear. With current technology, ADCs operating at very high (Gsps) sampling rates do not provide as many bits as ADCs for lower frequency applications, such as audio processing (operating at tens of ksps).

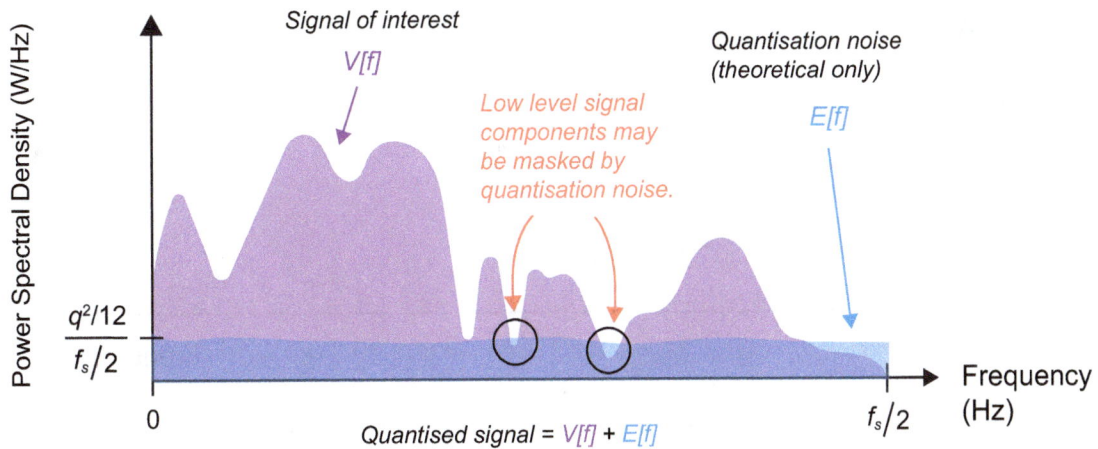

Figure 4.12: A sketch of quantisation noise overlaid on a signal of interest.

4.2.4. Periodic Signals and Frequency Spurs

Not all input signals to the ADC quantiser produce the random quantisation errors discussed in the previous section. Where an input signal is periodic, the sequence of quantisation errors follows a repeating pattern, and therefore the quantisation error signal is also periodic. We may also find that only a subset of possible quantisation error values are ever generated (in contrast to the PDF from Figure 4.10, which assumes randomness).

The digitisation of a sine wave is a good example. We can observe that the sequence of quantisation errors repeats after some time; the repetition period is defined by the relationship between the sampling period and the sine wave period. The resulting quantisation error signal is therefore also periodic, and its periodic components correspond to unwanted tones ('spurs') in the frequency domain. In practice, however, the signal being quantised is usually more complex than a single sine wave (e.g. in communications, a baseband signal is composed of a range of frequency components), and consequently the issue of spurs is less pronounced.

Frequency spurs are a form of harmonic signal distortion, and are undesirable. The extent of this distortion is quantified by the metric *Spurious Free Dynamic Range* (SFDR), which is the ratio between the fundamental component (e.g. the sine wave) and the most significant spur, expressed in dBs. The concept of SFDR is illustrated in Figure 4.13.

Aside from the most direct solution of using a higher resolution ADC (i.e. increasing the number of bits), another common approach is to add a 'dither' signal to the ADC input, prior to quantisation. The dither signal is low level noise, which introduces sufficient randomness to prevent the strict periodicity of the quantisation errors, and thus avoids the creation of frequency spurs. While adding noise to improve signal quality may seem counter-intuitive, the use of dithering can be beneficial in suppressing spurious frequency components, and thus improving SFDR. This is often a primary concern in communications systems.

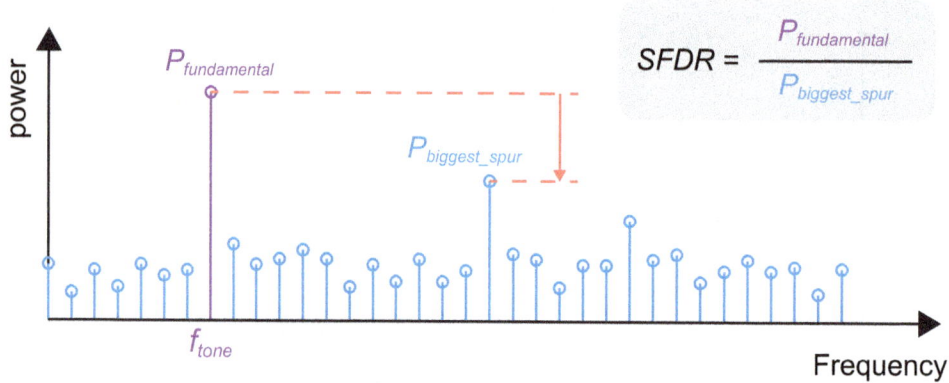

Figure 4.13: Measuring Spurious Free Dynamic Range (SFDR) with reference to a sine wave input at frequency f_{tone}.

4.3. Time and Frequency Domains

So far in this introduction, we have touched on both time and frequency domain signal analysis, and shown sketches of both. Time domain plots are probably the more intuitive of the two — the amplitudes of a measured signal are simply plotted against time. Frequency domain analysis involves converting sets of consecutive time domain samples (which may be referred to as a *window* of samples) into a frequency domain representation, such that the magnitude or power of the signal can be plotted against frequency. Being able to analyse signals in this way is a very useful tool — for instance, a frequency spectrum plot enables an engineer to verify that the signal emitted by a communications transmitter occupies only the allocated bandwidth.

A simple depiction of conversion from the time domain to the frequency domain is provided in Figure 4.14. The Discrete Fourier Transform (DFT), which is often implemented in a more efficient form as the Fast Fourier Transform (FFT), is used to perform the conversion [127]. The Inverse DFT (IDFT) or Inverse FFT (IFFT) can be used to make the opposite transition, from the frequency domain to the time domain.

There is considerably more to describe on this topic, including parameter choices, and interpretation of the resulting spectra, and therefore frequency domain analysis has its own dedicated chapter (see Chapter 5). We will leave further discussion of the frequency domain until then.

The various frequency domain plots and sketches presented in this chapter allow us to view the energy content of a signal against frequency, over the interval from 0 Hz to $f_s/2$ Hz (assuming that the signal is *real*). Later, we will also consider *complex* signals, which are generally plotted over the frequency range from $-f_s/2$ Hz to $f_s/2$ Hz, and the relevant background to this is covered in Chapter 7.

Next in this chapter, we consider digital filtering. Digital filters are designed to change the frequency content of a signal in some desired way.

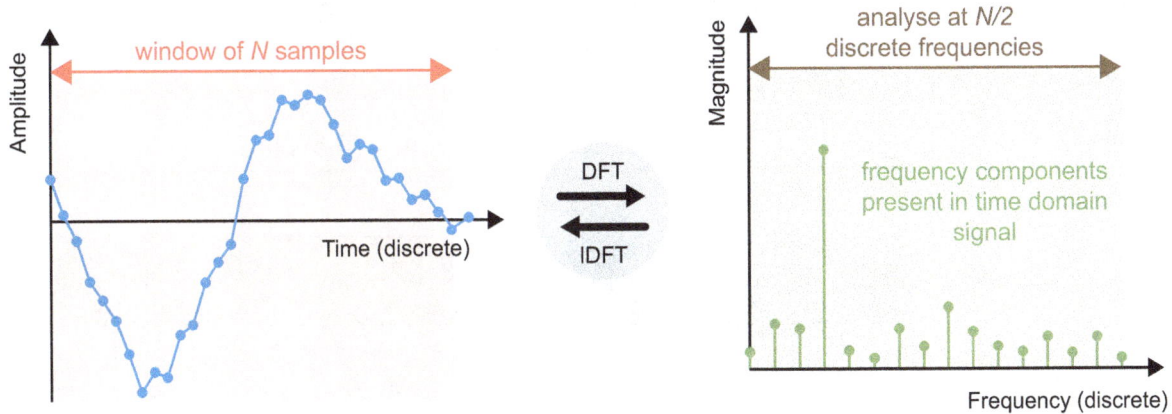

Figure 4.14: Conversion from the time domain to the frequency domain to analyse the frequency composition of a signal.

4.4. Fixed Point Arithmetic

In Section 4.2, we discussed quantisation and the process of mapping analogue input values to a discrete set of representable amplitude values, with the number of bits defining the number of available values, or in other words, the *resolution*. More generally, signals within the hardened processing blocks of the RFSoC device, along with custom algorithms developed in the PL are, in the vast majority of cases, represented using a fixed point number format. In this section, we review the fundamentals of fixed point numbers and arithmetic.

Two fixed point number formats are predominantly used in hardware designs: unsigned, and 2's complement. In both cases, numbers are composed of n integer bits and b fractional bits, separated by a binary point. Design tools often specify formats in the form $(s, n + b, b)$, where s represents the format (0 = unsigned, 1 = signed 2's complement). For instance, (0,8,5) indicates an unsigned number with 3 integer and 5 fractional bits.

4.4.1. Unsigned Fixed Point Format

Unsigned fixed point format is appropriate in cases where a signal can only take on positive values (or zero); in other words, where negative values are not required. In signal processing for radio and other applications, negative numbers are usually needed (think sinusoidal signals), however there are occasions where this is not the case: unsigned fixed point can be used in counters for control circuits, for example.

The unsigned fixed point number is composed of n integer bits and b fractional bits, separated by a binary point. The bits forming the number are arranged from Most Significant Bit (MSB) on the left, to the Least Significant Bit (LSB) on the right, and each has an individual bit weighting, which is derived from its index, i.e. its position in the array. In unsigned fixed point, all bits have a positive weighting, including the MSB. Figure 4.15 depicts the structure of a generic fixed point number; the MSB is positive for unsigned fixed point.

To provide an example, the range of an unsigned fixed point number with 5 integer bits and 3 fractional bits is 0 to 31. 875. If we wanted to represent the arbitrary value of 14.75 using this unsigned fixed point format, it would be expressed as: **01110 · 110**.

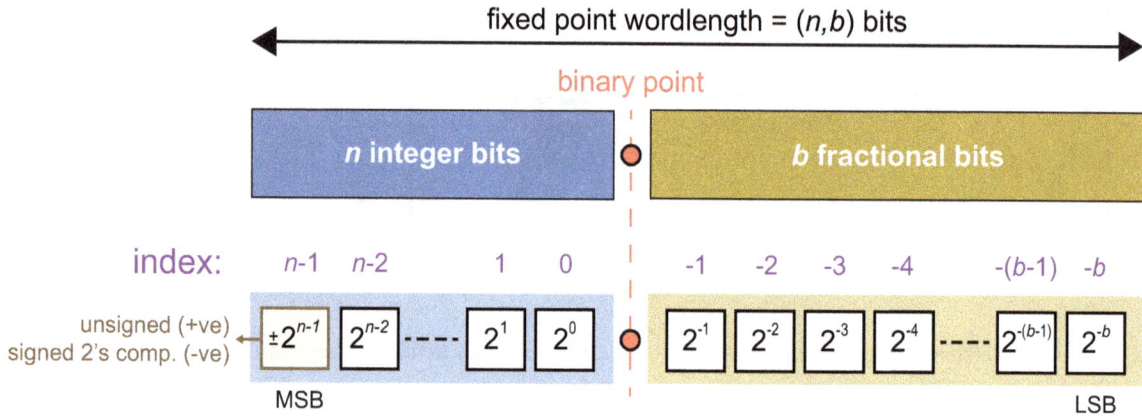

Figure 4.15: Generic fixed point number format (covers both unsigned, and signed 2's complement).

4.4.2. 2's Complement (Signed Fixed Point) Format

2's complement is composed similarly to unsigned format, the difference being that the MSB has a negative weighting, as indicated in Figure 4.15, and therefore 2's complement can express negative numbers. Taking the example parameters of $n = 5$ and $b = 3$, the range is -16 to +15.875. Notice that the range is not quite symmetric around 0; this occurs because there is one more negative level than positive level.

The smallest representable number occurs when the MSB (which has a negative weighting) is 1, and all other bits (which all have positive weightings) are 0; the largest representable number is the opposite. Therefore, it is convenient to look at the MSB to confirm the sign of a 2's complement number (0 = positive, 1 = negative).

The range and precision of both fixed point formats are summarised in Table 4.1.

Table 4.1: Summary of unsigned and signed 2's complement fixed point number formats

Format	Lower End of Range		Upper End of Range		Smallest Representable Number
Unsigned	0	$00\ldots00.00\ldots00$	$11\ldots11.11\ldots11$	$2^n - 2^{-b}$	2^{-b}
Signed 2's Complement	$-2^{(n-1)}$	$10\ldots00.00\ldots00$	$01\ldots11.11\ldots11$	$2^{(n-1)} - 2^{(-b)}$	2^{-b}

4.4.3. Wordlength Growth

It is worth noting that, when fixed point numbers are the subject of arithmetic operations, they should normally be allowed to grow, in order to avoid overflow[2] (i.e. where a number exceeds the available range, and 'wraps around' to the opposite of the range), or underflow (where there is insufficient resolution to represent the number, and it is rounded to zero). Longer wordlengths result in more costly implementation of arithmetic circuits, however, and therefore we often require to manage wordlength growth carefully.

As a simple rule of thumb, when two numbers are added or subtracted, the integer part of the number should be increased by 1 bit. This doubles the available range and ensures that all possible answers can be represented.

Multiplication can result in a larger change in magnitude. To avoid overflow and underflow, the integer sections of the two multiplicands should be added, as should the fractional sections. For instance, multiplying two numbers using our earlier format of $n = 5$ and $b = 3$ would give largest positive and negative results of $-16 \times -16 = 256$ and $15.875 \times -16 = -254$, respectively. The smallest magnitude result can be calculated as $0.125 \times 0.125 = 0.015625$. Therefore, the multiplier output should have 10 integer and 6 fractional bits.

4.5. Filtering

Filtering is a fundamental signal processing operation that is normally employed to change the frequency composition of a signal. In this section, we briefly review some important aspects of filtering, including the basic filter response types, the roles of analogue and digital filters in a wireless communications system, and the design and implementation of digital filters.

4.5.1. Filter Response Types

Signals are commonly filtered when there is a desire to change their frequency content in some way, for instance to remove a certain band of frequencies. Although arbitrary responses are possible, most filters can be categorised into four response types, which are: low pass, high pass, band pass, and band stop. These four responses are depicted in Figure 4.16.

A fifth class of filter, which we will not focus on here, is the all-pass. The magnitude of the signal is not altered by filtering (or it is altered very little), and instead the purpose of filtering is to modify the phase of the signal.

Most filters have fixed responses that are set at design time. Some applications in wireless communications, such as channel equalisation, require adaptive filters. This a distinct class of filter whose response is adjusted continuously while the system operates, based on observed signals. Again, we will not consider adaptive filters extensively in this book; the interested reader might wish to refer to textbooks such as [136], [191], and [331].

---

2. Overflow is generally undesirable, although it can be mitigated using saturation, a technique whereby a positive or negative overflow is detected, and the closest representable number is substituted.

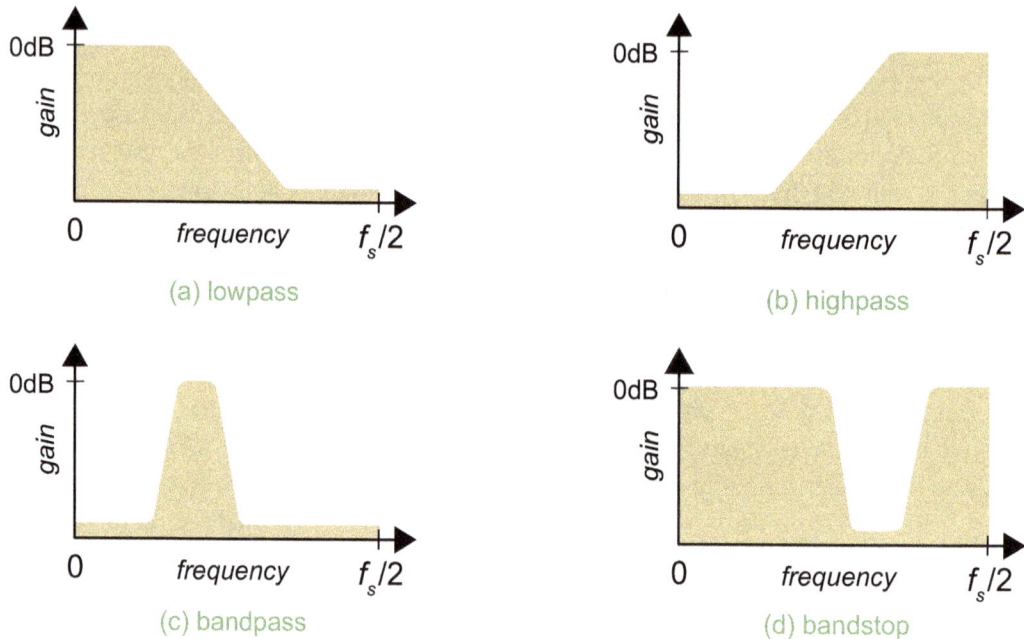

Figure 4.16: *The four main filter response types.*

4.5.2. Analogue and Digital Filters

Filters can be implemented in both the analogue and digital domains, and wireless communications systems require both, as was evident from the three main radio architectures presented in Chapter 2 (see Figures 2.9 - 2.11, starting from page 30). For example, the front end analogue section of a receiver should normally include a band pass filter to isolate the signal(s) of interest, and remove components at other frequencies, prior to conversion to digital at the ADC. This maximises the Signal-to-Noise Ratio (SNR), i.e. the ratio between the signal and noise powers, and thus improves the quality of the signal passing through the digital section of the receiver. For a more flexible SDR implementation, the front-end analogue filter in the receiver should at minimum reject any frequency components outside the Nyquist Zone of interest. The default scenario is to capture signals from Nyquist Zone 1, which implies the use of a low pass filter; as we will discuss later in the book, however, the RFSoC can also capture signals from higher Nyquist Zones, which would require an appropriate band pass filter.

Analogue filters are necessary as part of the overall architecture of an SDR, even where the conversion between analogue and digital takes place at RF frequencies and there are no analogue modulation and demodulation stages. Analogue filters play an important role in selecting RF bands, as mentioned above, and are also required as anti-image and anti-alias filters for DACs and ADCs, respectively. For instance, spectral images are generated by the digital-to-analogue process, and these unwanted images should be removed by applying an analogue filter to the signal, prior to transmission.

For SDR applications, it is generally considered desirable to implement as much of the radio processing in the digital domain as possible. With particular reference to filtering, this has several advantages, including savings on cost and power consumption (compared to implementing analogue filters), and avoiding the issues where temperature or component ageing effects alter the response of an analogue filter. Depending on the implementation platform, a digital filter can also be reprogrammed with a new response while the system is operational. However, there are some tasks that cannot be performed digitally, and analogue filters are vital components in the overall system architecture.

4.5.3. Filter Magnitude Response Features

Turning our attention towards digital filters (although similar principles apply for their analogue counterparts), a filter response can be described by a set of key features. These are in addition to the response type as outlined in Figure 4.16.

Figure 4.17 illustrates these filter features, for the example of a low pass filter response. Note that three regions of the response are defined: the passband, transition band, and stopband. Signal frequency components in the passband experience approximately 0dB of gain (a linear gain of 1), although there is some 'ripple' in the passband, meaning that there is some variability in gain across the passband region. In the stopband, the signal is subject to attenuation, which has the effect of diminishing or even effectively eliminating these signal frequency components. A transition band with sliding gain is present between the passband and stopband.

When specifying the requirements for a filter, the following parameters are commonly used:

- Passband edge frequency (in Hz, or on a normalised frequency scale)

- Stopband edge frequency (as above)

- Passband ripple (dB)

- Stopband attenuation (dB)

The transition bandwidth corresponds to the difference between the passband and stopband edge frequencies.

An ideal filter response, sometimes referred to as a 'brick wall' filter, has exactly 0dB gain across the passband, infinite attenuation in the stopband, and a 0 Hz transition bandwidth, as shown in Figure 4.18. However, such a filter cannot be realised in practice, and there is always a trade-off between the response characteristics of a filter, and its computational cost. The closer a filter design is to the brick wall response, the more "expensive" its implementation.

For bandstop and band pass filters, there are two passband regions and two stopband regions, respectively (refer back to Figure 4.16 as needed), and the parameters of these additional features of the response can be set individually. For instance, the attenuation in the two stopband regions of a bandpass filter need not be equal.

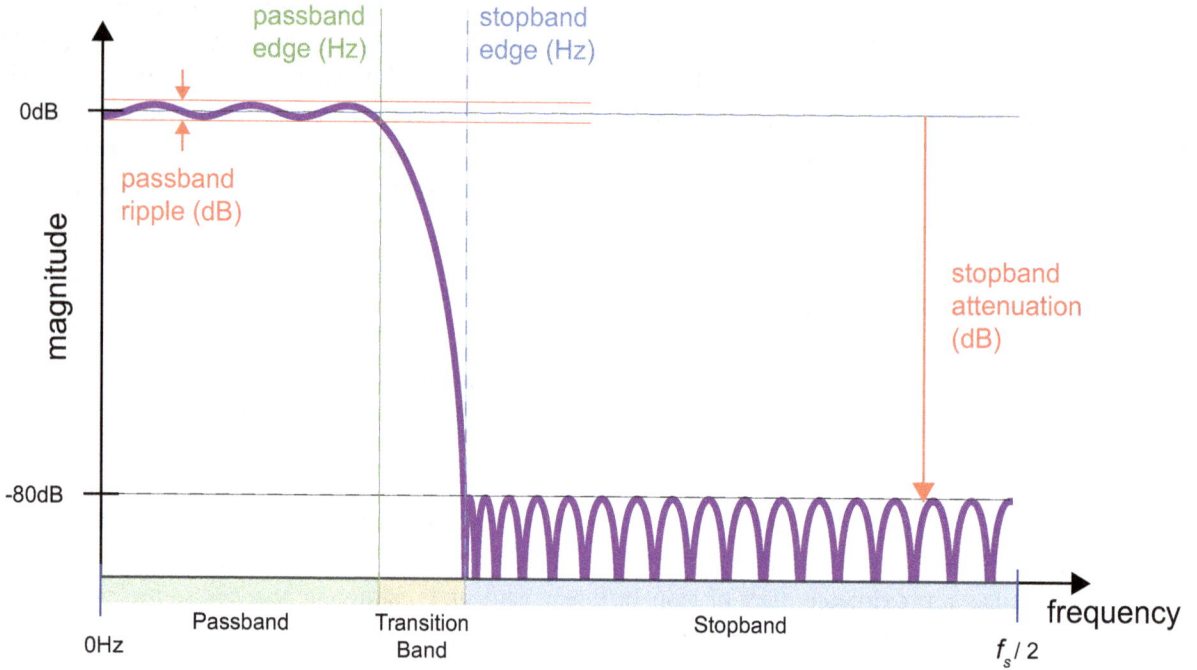

Figure 4.17: Features of an example low pass filter response.

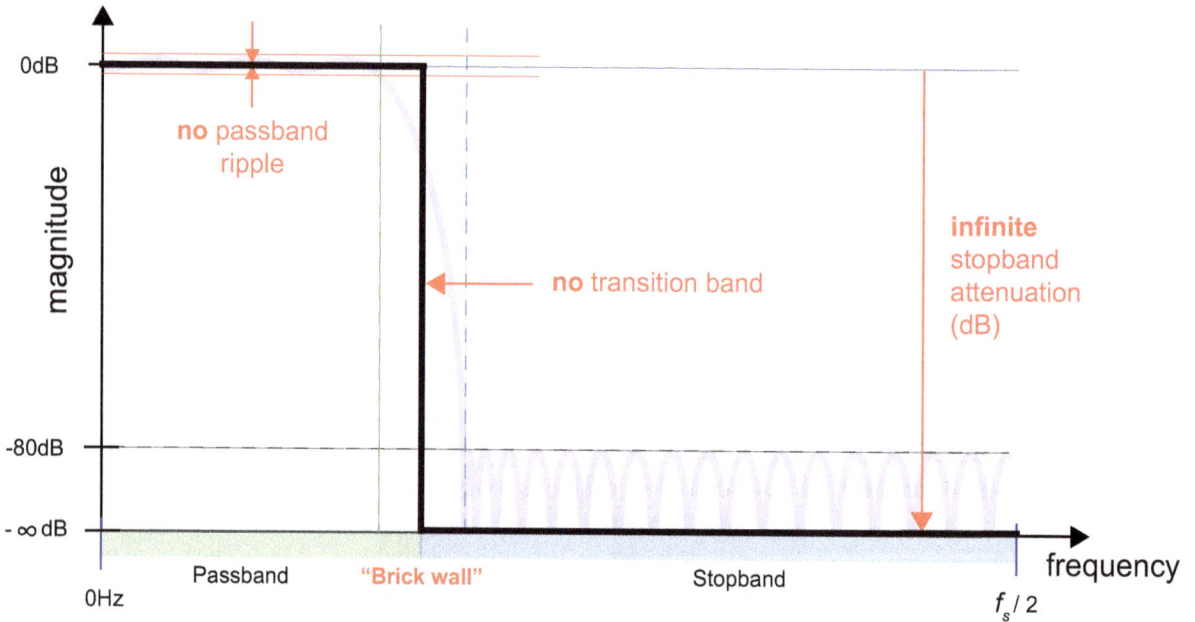

Figure 4.18: Features of an ideal 'brick wall' low pass filter response (unrealisable in practice).

4.5.4. Infinite and Finite Impulse Response Filters

One of the fundamental design choices to be made is the selection of a Finite Impulse Response (FIR) or Infinite Impulse Response (IIR) filter. As the names suggest, the response of an FIR to an impulse (i.e. a single input sample of amplitude 1, followed by zero-valued samples), lasts only for a finite number of samples. By contrast, the response of an IIR filter to an impulse could in theory continue forever. An illustration comparing the two filter response types is provided in Figure 4.19.

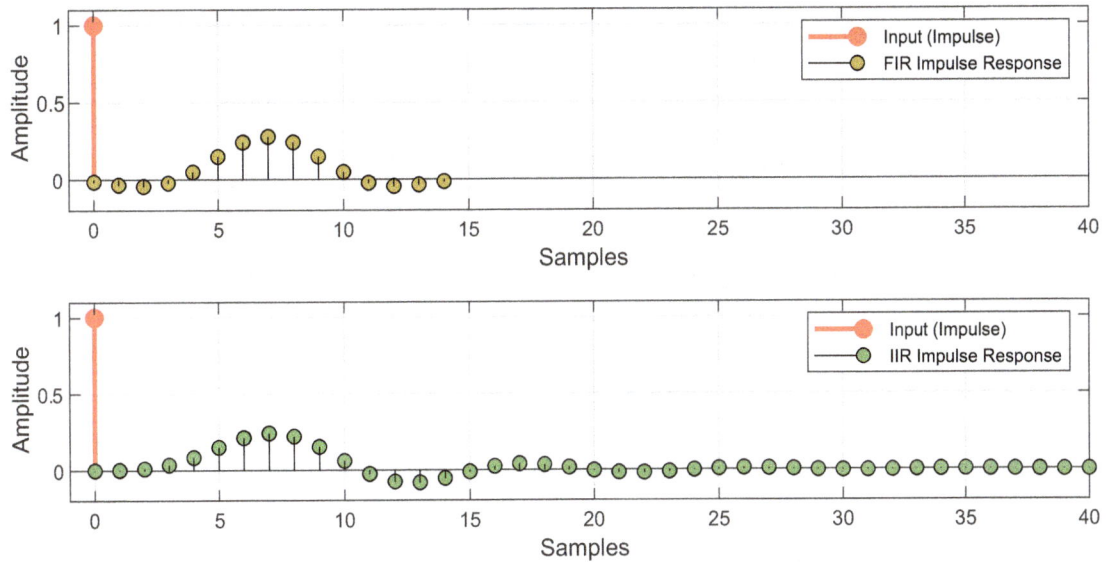

Figure 4.19: Example FIR and IIR filter impulse responses.

The FIR filter performs a weighted average (convolution) on a window of N data samples, to produce

$$y[k] = \sum_{n=0}^{N-1} w_n x[k-n],\tag{4.10}$$

where w_n represents the nth weight, and $x[k]$ is the input at sample index k.

For the example of a 5-weight FIR filter, we could alternatively write the difference equation

$$y[k] = w_o x[k] + w_1 x[k-1] + w_2 x[k-2] + w_3 x[k-3] + w_4 x[k-4],\tag{4.11}$$

which highlights the computation that is required to calculate each output. For this 5-weight example, five multiplications (each between a weight value and a delayed input sample) are required, and four additions are needed to sum these products to form the filter output, $y[k]$.

It is relatively intuitive to map the FIR difference equation from (4.11) to a graphical *signal flow graph* representation, as depicted in Figure 4.20. In this signal flow graph depiction of the 5-weight filter, we can confirm that five multipliers and four adders are required. The blocks labelled with 'delta' symbols represent one-sample delays (we will shortly replace them using the more formal Z-notation).

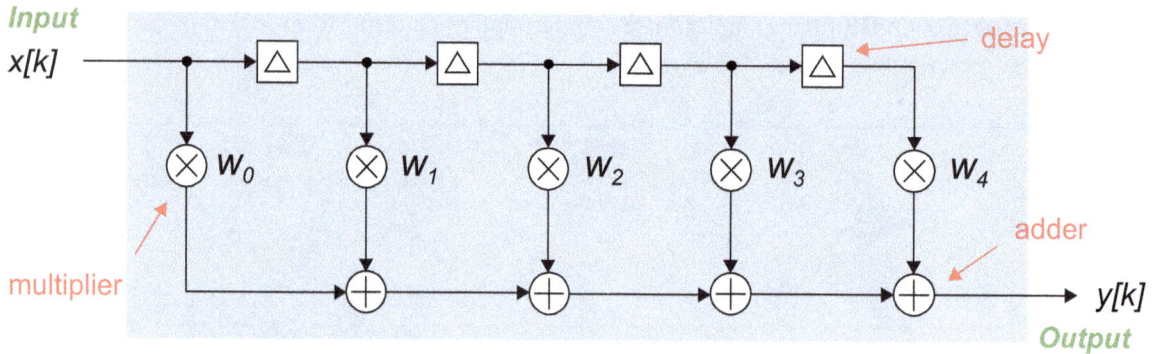

Figure 4.20: Signal flow graph for an example 5-weight FIR filter.

While an FIR filter computes the weighted sum of past inputs (a feedforward path in the signal flow graph shown in Figure 4.20), the IIR filter also includes a feedback path, and so its output is composed of weighted sums of past inputs *and* past outputs. The operation of an IIR filter can therefore be described by the equation

$$y[k] = \sum_{n=0}^{N-1} b_n x[k-n] + \sum_{m=1}^{M} a_m y[k-m] \qquad (4.12)$$

where N and M are the numbers of feedforward and feedback weights, respectively; and where b_n is the nth feedforward weight, and a_m is the mth feedback weight. Notably, the feedforward section of an IIR filter is equivalent to an FIR filter. An example IIR filter with four feedforward and three feedback weights is shown in Figure 4.21.

The primary motivation to select IIR filters is that they can achieve the same magnitude response as an FIR filter, using fewer weights, and therefore they require less computation and are less costly to implement in hardware. However, FIR filters have two advantageous characteristics which make them the preferred choice for communications applications:

- FIR filters are intrinsically stable, as they have no feedback path (IIR filters, on the other hand, have feedback paths and carry the risk of becoming unstable).

- Any FIR filter designed by one of the standard algorithms will have symmetric weights, and therefore possess the attribute of *linear phase response*. This means that all frequencies passing through the filter are delayed by the same amount of time, which corresponds to a linearly increasing phase difference.

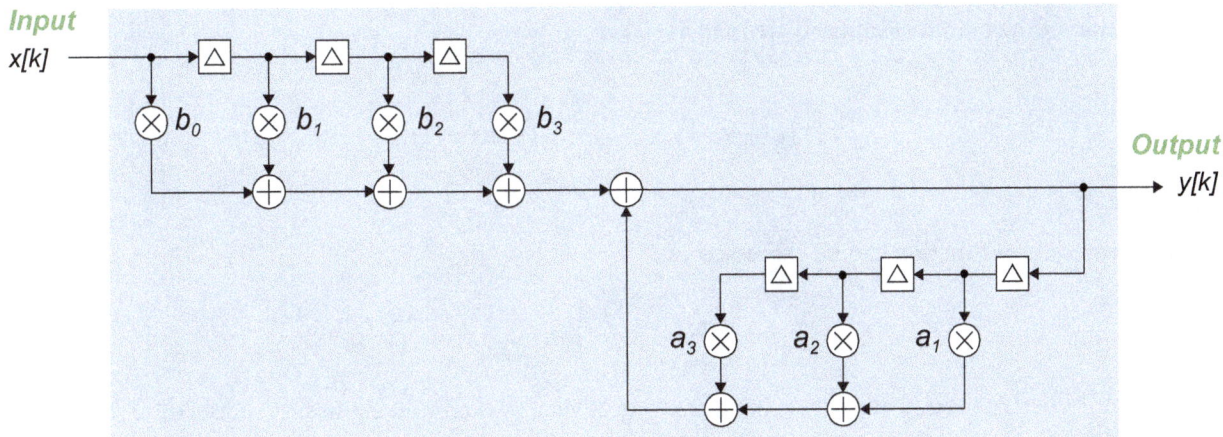

Figure 4.21: Signal flow graph for an example IIR filter with 4 feedforward and 3 feedback weights.

The significance of the *linear phase response* attribute is to preserve the phase relationship between different frequency components — and this is particularly important in communications, where information is carried by 'pulses', which are composed of several different frequencies. The pulse shapes can be destroyed by non-linear phase filters, but are preserved by linear phase filters.

In this book, therefore, we focus on FIR filters exclusively[3]. For more background on phase response, and the linear phase response property of FIR filters, the reader may wish to refer to textbooks such as [203] or [234].

4.5.5. Z-Domain Filter Representation

Digital filters and other DSP algorithms are commonly described using Z-domain representation, which provides a convenient and mathematically tractable form that is amenable to analysis. In the context of digital filtering, Z-domain representation (or simply 'Z-notation') can be used to express input and output signals, and also the transfer function of a filter. For the sake of brevity, we do not derive the Z-transform here, but rather demonstrate how it can be applied.

The one-sided Z-transform of a causal signal, $x[k]$ can be defined as:

$$z\{x[k]\} = X[z] = \sum_{k=0}^{\infty} x[k]z^{-k}. \tag{4.13}$$

---

3. IIR filters can be used in communications where they are constrained to have approximately linear phase, but that is an advanced topic not considered here.

If we consider $x[k]$ to represent the input to a filter, and $y[k]$ to represent its output, then the Z-domain representation of the output is similarly defined as

$$z\{y[k]\} = Y[z] = \sum_{k=0}^{\infty} y(k)z^{-k},$$

(4.14)

and the filter transfer function can be expressed as

$$H[z] = \frac{Y[z]}{X[z]}.$$

(4.15)

It is possible to derive the Z-transform representations of various common signals, such as a discrete impulse, delayed discrete impulse, unit step, and ramp functions.

One of the most important is the discrete impulse, denoted as $\delta[k]$, given that we are often interested in the impulse response of a filter or other system (an impulse contains all frequencies, and therefore applying an impulse as the filter input excites it across its entire range of operation).

It is also useful to note the Z-transform representation for the delayed impulse, $\delta[k-m]$, where m is the number of samples delay. These two signals are illustrated in Figure 4.22, for the example delay of $m = 3$.

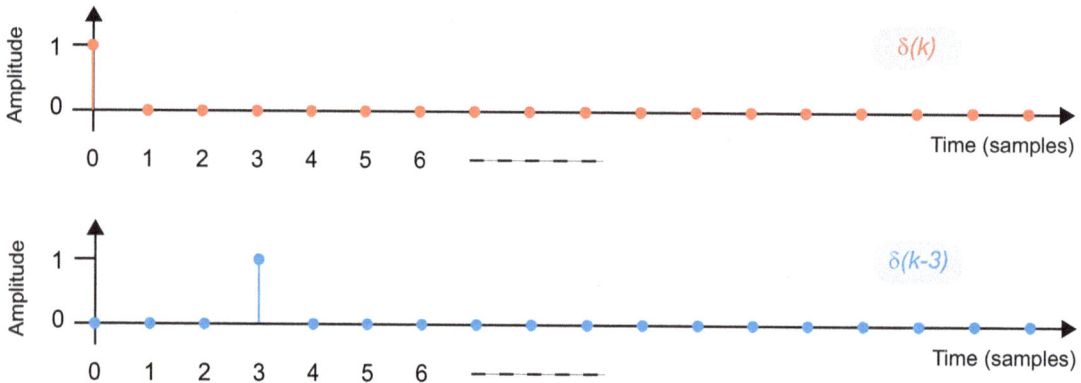

Figure 4.22: discrete impulse (top), and discrete impulse delayed by 3 samples (bottom).

Starting with the discrete impulse, we can define the Z-domain representation as

$$z\{\delta[k]\} = \sum_{k=0}^{\infty} \delta[k]z^{-k}.$$

(4.16)

Noting the amplitude of $\delta[k]$ at sample indices 0, 1, 2, 3... , i.e. $\delta[0] = 1$, and $\delta[k] = 0$ for all other sample indices, we can confirm that the value of $z\{\delta[k]\} = 1$,

$$z\{\delta[k]\} = (1 \times 1) + (0 \times z^{-1}) + (0 \times z^{-2}) + (0 \times z^{-3})... = 1. \tag{4.17}$$

Similarly, for the delayed impulse, we obtain

$$z\{\delta[k-3]\} = \sum_{k=0}^{\infty} \delta[k-3]z^{-k} \tag{4.18}$$

$$z\{\delta[k]\} = (0 \times 1) + (0 \times z^{-1}) + (0 \times z^{-2}) + (1 \times z^{-3}) + (0 \times z^{-4}) + (0 \times z^{-5})... = z^{-3}. \tag{4.19}$$

By applying the Z-transform to convert a time domain signal to the Z-domain, we therefore observe that one-sample delays can be represented by z^{-1} (or more generally, z^{-d} for a d-sample delay). This notation is commonly used in signal flow graphs to depict delays (for instance, the delta symbols in Figures 4.20 and 4.21 could be replaced by z^{-1} annotations).

One further observation relating to the Z-transform of the discrete impulse, is that if $X[z] = 1$, then we can easily determine the transfer function of a filter, $H[z]$, from the output $Y[z]$, using (4.15),

$$H[z] = \frac{Y[z]}{X[z]} = \frac{Y[z]}{1} = Y[z]. \tag{4.20}$$

In other words, the impulse response of a filter corresponds directly to its transfer function.

4.5.6. Digital Filter Design

The desired characteristics of a filter response represent inputs to the design process. For instance, to design a low pass filter, the parameters of passband edge frequency, stopband edge frequency, passpand ripple, and stopband attenuation are typically supplied. The filter design algorithm then attempts to meet these requirements, and assuming it can do so, outputs a set of filter weights that realise the desired response. In general, unless constrained to generate a filter of a specific length, the design algorithm outputs a filter that achieves the required response with the minimum number of weights, as this represents the most computationally efficient solution.

There are several different methods for filter design, including the Windowing, Frequency Sampling, Least Squares, and Parks-Mclellan (Equiripple) methods for FIR filters. Each method is likely to produce a slightly different response for any given set of input parameters. It is outside the scope of this book to describe such filter design algorithms in detail, however there is plentiful support for them in signal processing design tools like MATLAB, Numpy etc. There is also useful theoretical background available in books such as [92] and [196]. Additionally, 'special' types of filters, such as halfband filters, raised cosine filters, and inverse sinc compensation filters, all of which are commonly used in communications systems, are generally supported by dedicated functions within these environments.

From a practical perspective, we note simply that the filter design process produces a set of weights that are used in the implementation of the filter. The weight values are converted to the chosen fixed point format, which is specified as part of the design process; note that the filter response becomes non-ideal if the weights are represented with insufficient resolution. As previously shown in Figures 4.20 and 4.21, all weights of an FIR filter are in the feedforward path (there is no feedback path), while the IIR has both a set of feedforward weights, and a set of feedback weights. As we are now focusing on FIR filters, we additionally note that all of the above mentioned design methods produce filters with symmetric weights. Not only does this achieve the property of linear phase response outlined in Section 4.5.4, but it is also convenient for efficiently implementing filters on FPGAs and other platforms, as we touch on next.

4.5.7. FIR Filter Implementation

Having designed a filter response and generated the required set of weights to realise the filter, the next step is to implement it. In the context of communications design using the RFSoC platform, the natural target resource is the PL (it is also possible to implement filters in software code running on a processor). The PL provides a flexible, highly parallel, and low power option, and is capable of supporting the high sampling rates often needed in communications systems. Therefore, we focus our discussion on the PL-based implementation of FIR filters.

Taking the example of an 11-weight symmetric FIR filter, we exploit the symmetry by pre-adding those samples from the input delay line that should be multiplied with the same weight value. For instance, the first and last weights in the filter have exactly the same numerical value, and so the first and last samples from the delay line are pre-added together, and then multiplied by the weight value w_0, which is also equal to w_{10}. This reduces the total number of multipliers needed to implement the filter by approximately half — a significant saving, as the multipliers are the most costly element within the filter architecture. The efficient, symmetric structure is shown in Figure 4.23.

The basic unit of computation in a filter is a Multiply Accumulate (MAC), and the hardware required to implement a MAC is referred to as a MAC unit. An example MAC unit is annotated in Figure 4.23. The PL is optimised for implementing FIR filter structures, and each of its DSP48E2 slices can support one MAC unit, which includes the weight multiplier and post-adder, along with the pre-adder and all associated delay elements. Therefore, this 11-weight example can be implemented using six DSP48E2 slices, without requiring any additional PL logic.

The registers within the DSP48E2 slice can be programmed with several configurations, and thus support different retimed versions of the FIR, such as the transpose and systolic forms (which perform equivalent computation as the standard form presented here, but are optimised to maximise clock rate). We will not focus on retiming here, but the reader may be interested to refer to [226] for more information on this topic.

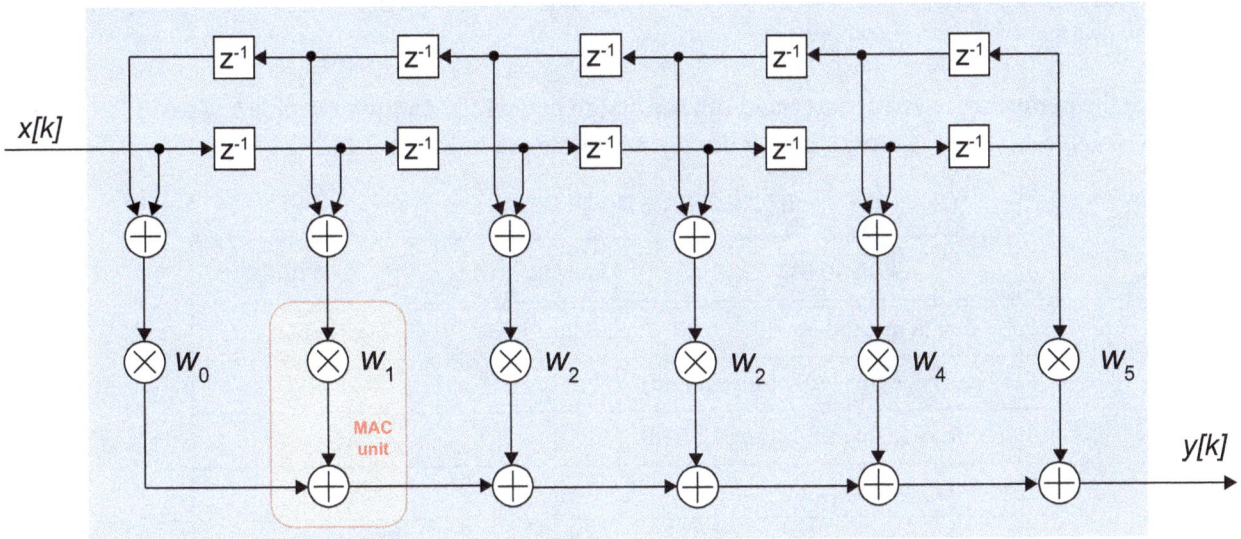

Figure 4.23: Example of a symmetric 11-weight FIR filter.

Further, it is worth noting that the implementation discussed above is a fully parallel one, wherein each multiplier in the signal flow graph (for instance) corresponds to a physical hardware multiplier in the PL. The amount of hardware required to implement the filter therefore scales within the number of filter weights, or in other words, the filter length. Another possibility is to serialise, or partially serialise, the filter implementation such that its resource footprint is reduced — this is achieved by time-sharing MAC units, so that each MAC unit calculates two or more MAC operations for each execution of the filter. This is only possible where the filter hardware can run at least twice as fast as the sampling rate to be supported, and therefore it is not suitable for processing at very high sampling rates, but it can be a very useful tool for filter implementations operating at lower sampling rates.

4.5.8. Filter Design Trade-Offs

The 11-weight FIR filter presented in the previous section represents a modest filter length, and the response achievable by such a filter is also likely to be modest. For instance, it may not be able to achieve a particularly sharp transition between the passband and stopband.

In general, the following design choices tend to increase the number of weights needed to implement a filter:

- Reducing the amount of permissible passband ripple.

- Increasing the required stopband attenuation.

- Specifying a narrower transition band between the passband and stopband.

For instance, if you were to reduce the transition bandwidth by a factor of two, this would approximately double the number of filter weights required.

To explore the performance versus cost trade-off, the design parameters for two example low pass filter designs are given in Table 4.2, and the resulting filter designs are presented in Figure 4.24.

Table 4.2: Filter Design Parameters

Parameter	Design A	Design B
Filter design method	Equiripple	Equiripple
Passband edge frequency (normalised)	0.1	0.1
Stopband edge frequency (normalised)	0.2	0.125
Transition bandwidth (normalised)	*0.1*	*0.025*
Passband ripple	3dB	0.1dB
Stopband attenuation	30dB	60dB

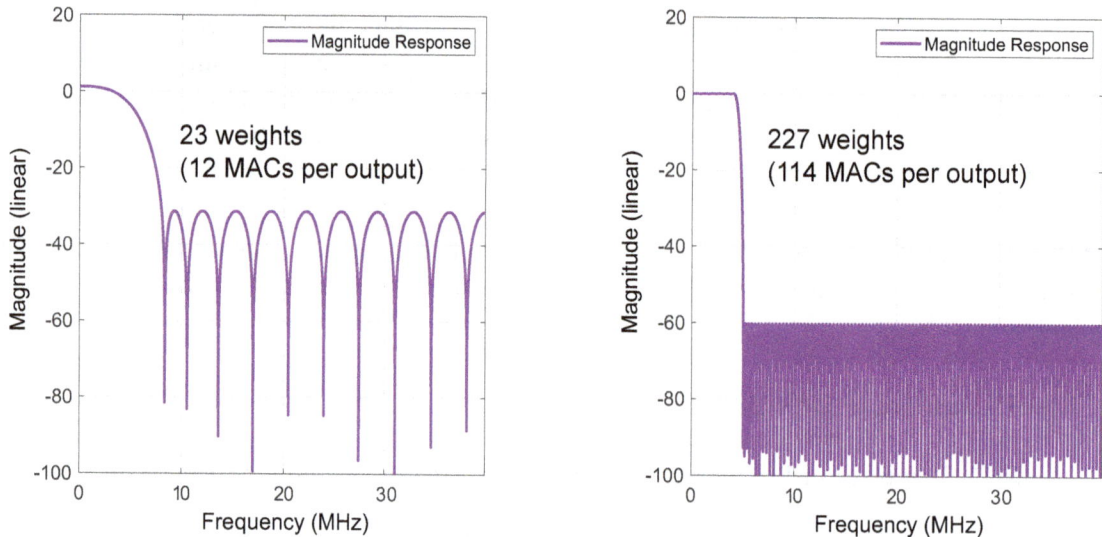

Figure 4.24: Two example filter designs (left — Design A; right — Design B).

By inspection of the filter responses, it might be interpreted that the filter on the right (Design B) is "better", but it will require almost 10 times more computation than Design A, on the left. For a parallel implementation, this also implies 10 times more resources on the PL, higher power consumption, and so on. Therefore, if both filters meet the application requirements, it would be preferable to choose Design A, the 23-weight filter.

The fixed point parameters used to represent the filter weights also play a part — as mentioned previously, the filter response starts to deviate from the ideal, the shorter the wordlengths used. On the other hand, shorter wordlengths imply less expensive arithmetic components (the adders and multipliers that comprise a filter).

We can therefore conclude that there is a trade-off between the performance of a filter, and the computation and processing hardware required to implement it. A prudent approach is to design each filter to minimise the number of weights and the fixed point wordlengths, while still achieving the required functional specification.

4.6. The Analogue-Digital Interface

In this section, we briefly review the operations involved in analogue-to-digital (A-to-D), and digital-to-analogue (D-to-A) conversion. With respect to the former, we have already discussed the sampling and quantisation processes that are inherent to A-to-D conversion, but it is useful to confirm the overall structure and the role of the anti-alias filter.

4.6.1. Analogue to Digital Conversion

As reviewed in Section 4.1 and Section 4.2, a continuous-time analogue signal can be converted to a digital equivalent by sampling and quantising the signal. The key parameters linked to these operations are the sampling rate, in Hz, and the number of bits used in the quantiser.

A conventional ADC operating in Nyquist Zone 1 is preceded by an analogue low pass anti-alias filter, to retain only the frequency components in Nyquist Zone 1, and attenuate all higher frequency signal components that are present at the ADC input. Without the anti-alias filter, these higher frequency components would alias into Nyquist Zone 1, at minimum degrading the SNR, and potentially even destroying the signal of interest.

As noted earlier in this section, the ideal filter response is a 'brick wall', but this is not realisable in practice, either in the analogue or digital domains. Just like digital filters, analogue filters also become more expensive and difficult to design, the more exacting the specification. It is therefore realistic to expect that:

- The anti-alias filter cuts off below the upper limit of Nyquist Zone 1, with the effect that some frequency components of interest are lost; and/or

- The transition band extends a little into Nyquist Zone 2, meaning that some level of aliasing may occur.

The above highlighted issues may be addressed by oversampling at the ADC, i.e. sampling at a higher rate than is necessary, according to Nyquist Sampling theorem. As a result, a more relaxed analogue anti-alias filter can be used, which is less expensive to implement. Any aliased components will not impact the frequency band of interest, and can be removed after the ADC using digital filters.

A comparison between these two different approaches is presented in Figure 4.25.

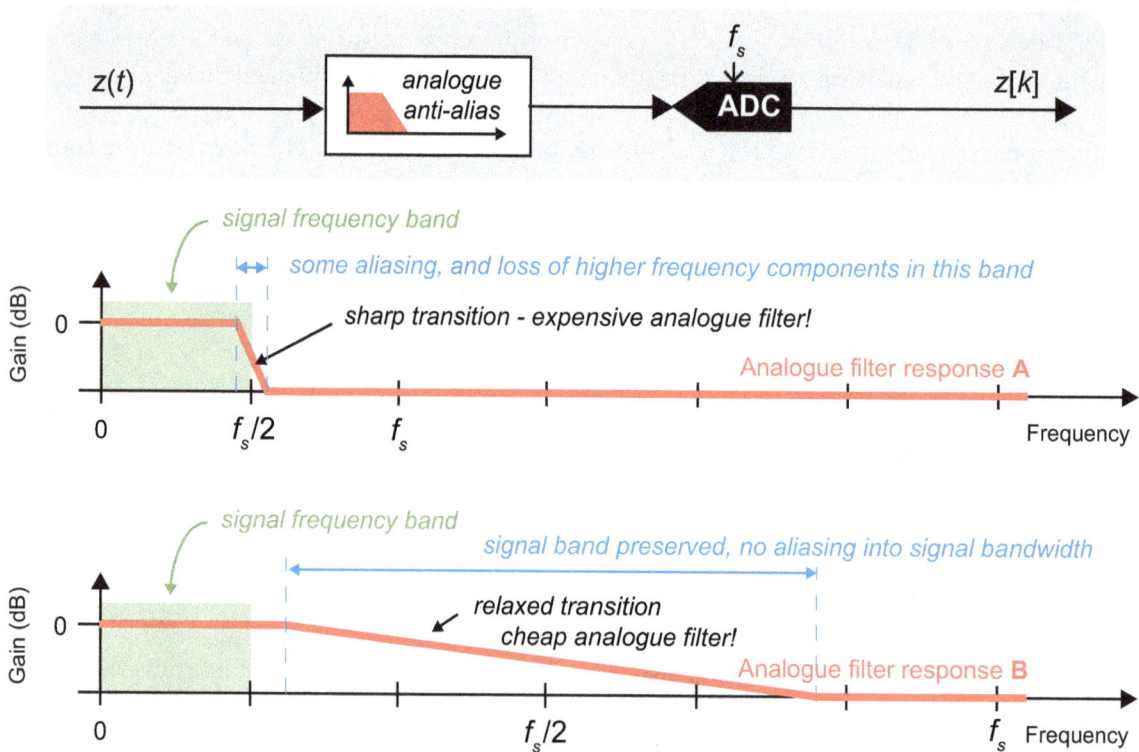

Figure 4.25: Comparison between Nyquist rate (top) and oversampled (bottom) ADCs and anti-alias filters.

One possible disadvantage of the oversampled method is that the sampling rate is higher than necessary at the input to the digital domain, which implies a higher computation load for subsequent processing stages. This issue can be easily solved by using a multirate filter to reduce the sampling rate soon after the ADC, such that the signal is sampled at a rate closer to the Nyquist rate. Multirate filters will be covered in Section 4.7.

4.6.2. Digital to Analogue Conversion

The conversion from Digital to Analogue (D-to-A) has received relatively little attention so far, but is also a significant theme within this book, as DACs are required to generate the communications signals to be transmitted across the wireless link.

The conventional D-to-A process involves regenerating a continuous time signal from discrete samples and amplitudes. This is conventionally done using a zero-order-hold techniques, which creates a "steppy" signal as illustrated in Figure 4.26. Analytically, we could consider that the discrete time samples have been passed through a rectangular filter of duration T_s, i.e. the sampling period.

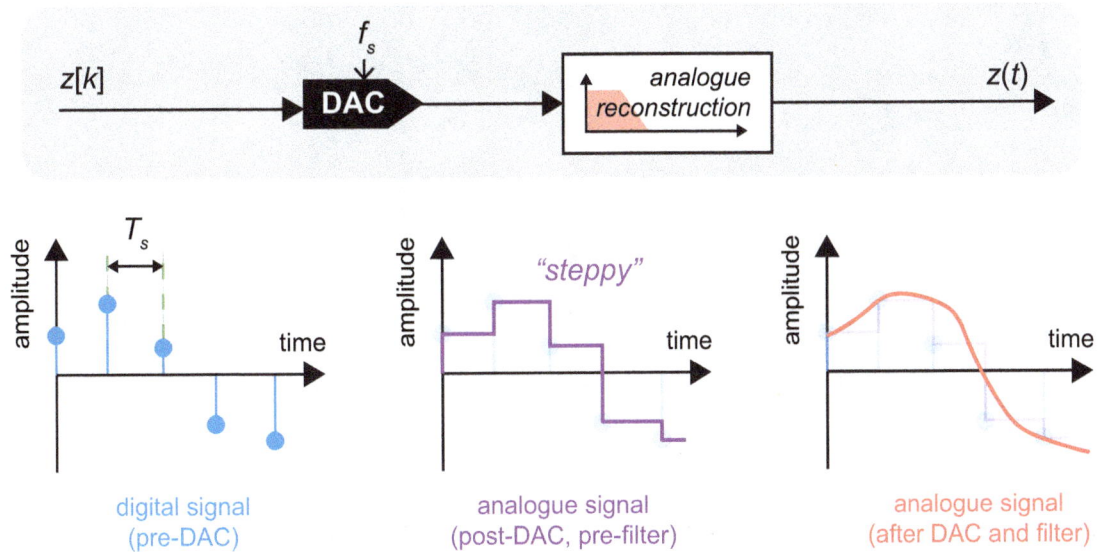

Figure 4.26: DAC time domain signals and reconstruction filtering.

The D-to-A conversion process produces image spectra at integer multiples of the sampling frequency, which need to be removed by low pass filtering — if the signal of interest occupies the whole of Nyquist Zone 1, i.e. frequencies up to $f_s/2$, this demands an expensive analogue filter with a sharp cut off.

The signal output by the DAC, which includes the signal of interest as well as the image spectra, follows the sinc shape of the rectangular filter response, shown in Figure 4.27. The profile of the DAC response means that the spectral images are only partially attenuated, leaving remnants with significant energy that require to be further attenuated by an analogue reconstruction filter. Their effective removal is especially important in wireless communications, where the bandwidth of transmitted signals must be tightly controlled to meet spectral emission masks (this topic will be discussed further in Chapter 6).

To address this issue, a low pass analogue reconstruction filter is included immediately following the DAC. This acts to attenuate the significant frequency components present in upper Nyquist Zones, and in doing so, smooths out the time domain signal (as shown at the right hand side of Figure 4.26) to produce a more intuitively 'analogue' waveform.

Even so, the application of such a reconstruction filter does not address the non-linear gain across Nyquist Zone 1 — this can distort the signal band of interest, especially if it extends all the way up to $f_s/2$. We also encounter the same issue as with the anti-alias filter in the ADC, in the sense that a sharp transition in the analogue filter response is required, which is expensive and difficult to implement. Once again, oversampling techniques can be used to manage these issues: firstly, by increasing the sampling rate at the DAC, the require-

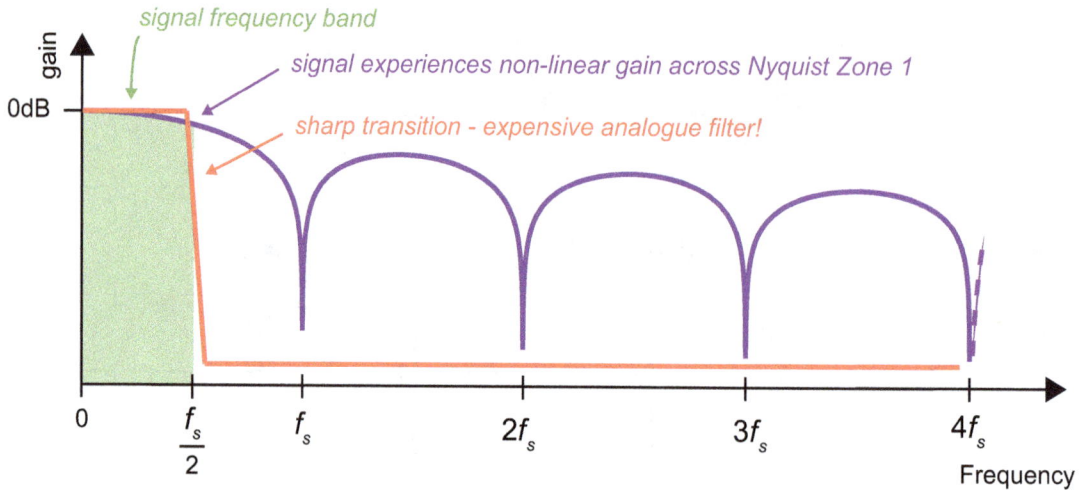

Figure 4.27: DAC frequency response, and near-ideal reconstruction filter response.

ments for the reconstruction filter are lessened; and secondly, as the signal of interest occupies a smaller portion of Nyquist Zone 1, it experiences less severe sinc-shaped distortion resulting from the frequency response of the DAC.

Another mitigation is possible, which is especially relevant if the oversampling ratio is relatively low (or the DAC is not oversampled). In this case, there is a significant non-linear gain across the signal bandwidth, and it would be desirable to compensate for this effect. To do so, an inverse sinc filter is introduced in the digital domain, prior to the DAC, and corrects for its sinc-shaped response.

4.6.3. Upper Nyquist Zones

Our discussion in this section has focused on 'conventional' ADCs and DACs that operate in Nyquist Zone 1. One of the significant innovations included in the RFSoC device, is the ability of the RF-ADCs and RF-DACs to additionally operate in upper Nyquist Zones. In other words, in some cases the signal of interest does not actually reside in Nyquist Zone 1, but rather might be present in Nyquist Zone 2, or even higher Nyquist Zones in some cases.

In situations where the ADC or DAC operates in an upper Nyquist Zone, the analogue filters discussed in this section (anti-alias for the ADC, and reconstruction for the DAC) should no longer be low pass filters, but must instead be designed as band pass filters targeting the desired range of frequencies.

The operation of the RF-ADC and RF-DAC, including in upper Nyquist Zones, will be discussed in further detail in Chapters 9 and 11, respectively.

4.7. Multirate Operations

Multirate operations are required to change the sampling rate in a DSP system. For instance, if a signal was sampled at 10 MHz, a multirate operation could be used to increase the sampling rate to 40 MHz (*interpolation by a factor of 4*), or alternatively, *decimation by a factor of 2* could be used to reduce the sampling rate to 5 MHz. Further types of multirate operations are used if the sampling rate needs to be changed by a more difficult factor, for instance a ratio expressed as a rational fraction (like 3/2), where the ratio cannot be readily expressed as a rational fraction (e.g. an arbitrary factor such as 1.178134...), or even where the resampling ratio needs to change dynamically over time.

For the remainder of this chapter, we focus our discussion on the most fundamental multirate operations, where the sampling rate is increased or decreased by an integer factor. These operations underpin the major tasks of *Digital Upconversion* and *Digital Downconversion* that are integral to almost any Direct-RF or IF Sampling SDR design and make the transition between baseband and IF or RF sampling rates. Hardened DUCs and DDCs are available in RFSoC devices, as discussed in Chapters 3, 9 and 11. For extended coverage of applications, theory and implementations, the interested reader is referred to one of the available textbooks on multirate signal processing [129], [261], [348], and in particular [188] which focuses on multirate DSP as it applies to digital communications systems.

4.7.1. Motivations for Multirate Processing

There are several reasons why it may be desirable to change the sampling rate in a system; however the overarching motivations are to optimise computational efficiency, and often, to ease the analogue filtering requirements at the digital/analogue interface.

Multirate processing is often used to maintain a sampling rate that is not much greater than the minimum rate according to Nyquist, which is defined as

$$f_n > 2f_{max},$$ (4.21)

where f_{max} is the maximum frequency component present in the signal. With the sampling rate set in this way, the computation involved in processing the signal takes place at a minimal (or close to minimal) rate, and the overall computational effort of processing the signal is optimised. This may have positive implications for the implementation cost and power consumption of a system design. Where the signal bandwidth changes, then it may be desirable to increase or decrease the sampling rate using multirate techniques, such that a similar relationship between the sampling rate and signal bandwidth is maintained.

Some example scenarios that require sampling rate changes are:

- **To match the sampling rates of two signal paths that will be combined** — Two signals can only be added if they have the same sampling rate. An example might be an audio mixing desk, where voice captured from a phone call is combined with high quality music from a media file. The sampling rate from the voice call is 8 kHz and the music is sampled at 48 kHz. Therefore, the voice can be interpolated by a factor of 6 to achieve the 48 kHz sampling rate, before being added to the music.

- **To adjust the sampling rate closer to Nyquist when the signal bandwidth changes** — In wireless communications, the bandwidth of a received signal can be reduced after it has been demodulated to baseband, as the higher frequency components (which contain mostly noise and signals on other bands) are not required. The benefit of doing so is to reduce the computation rate as outlined earlier.

- **To match the sampling rate of an external interface, such as a DAC** — Rate conversion may be required to match the sampling rate specified at a DAC or other external interface. Sometimes this can involve a rate transition by an awkward factor, if the local system sampling rate and interface sampling rate are not conveniently related. In wireless communications, a common scenario is to interpolate by an integer factor as part of the upconversion process, where a signal is modulated onto a carrier.

- **To ease analogue anti-alias or image-rejection filter requirements** — It is often desirable to operate a DAC or ADC at a higher sampling rate than is required, according to Nyquist sampling theorem (often referred to as *oversampling*), in order to achieve less exacting filtering requirements for the anti-alias or anti-image filter. This implies an additional rate change in the digital domain, so that the sampling rate can then be reduced closer to the Nyquist rate, enabling efficient computation for subsequent DSP tasks.

The next section will explore the last of these examples in a little more detail.

4.7.2. Oversampled ADCs and DACs

Taking the example of A-to-D conversion, if a signal occupied frequencies between 0 MHz and 480 MHz, and the ADC was operated at a sampling frequency of 1 GHz, then the band of frequencies between 520 MHz and 1 GHz would alias into the band of interest, as shown in the upper part of Figure 4.28. An analogue anti-alias filter with a very tight transition band (cutting off between 480 MHz and 520 MHz) would therefore be required — such a filter design can be difficult and costly to achieve.

As an alternative, adopting an ADC sampling rate of 2 GHz would allow a much more relaxed analogue anti-alias filter response, as shown in the lower section of Figure 4.28. A multirate operation (decimation by 2) could be introduced after the ADC to reduce the sampling rate to 1 GHz, and perform the remainder of the anti-aliasing task digitally (noting that an components present between 1 GHz and 1.5 GHz will alias into the band from 0.5 - 1 GHz. We will review the mechanics of decimation in the next section.

At the DAC, similar rationale applies regarding oversampling, and analogue filtering requirements. The introduction of oversampling at the DAC can be used to lessen the requirements for the analogue image-rejection filter. The higher sampling rate is achieved via interpolation prior to the DAC; interpolation will be covered in Section 4.7.4.

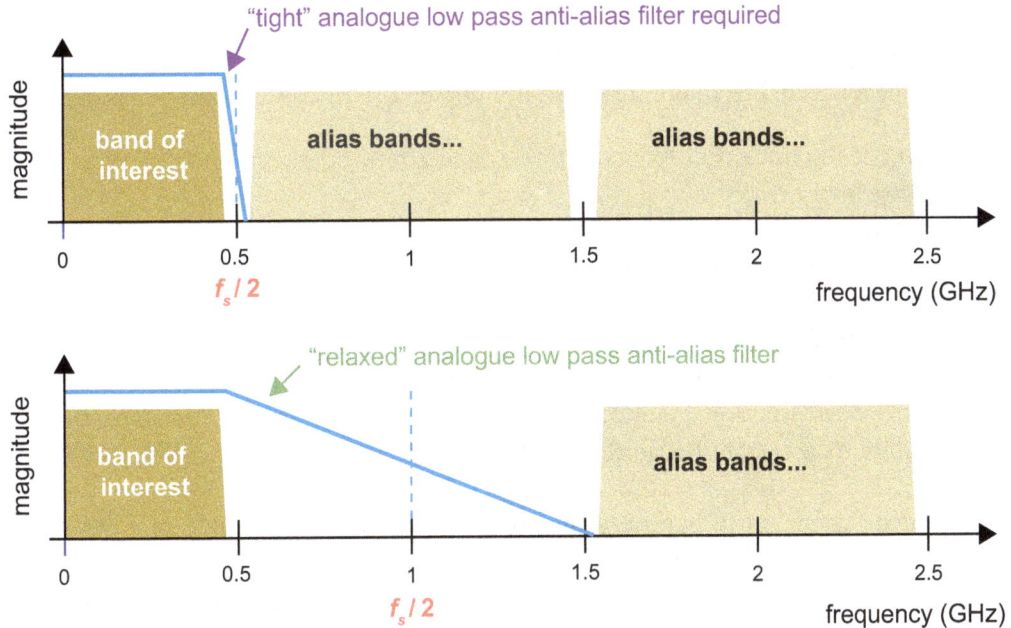

Figure 4.28: Alternative ADC sampling rates and analogue anti-alias filter requirements.

4.7.3. Decimation

Decimation is the process of reducing the sampling rate. In the simplest and most common cases, this is done by an integer factor, which is usually denoted by M, for instance a decimation factor of $M = 3$ would be used to reduce the sampling rate from 300 MHz to 100 MHz.

Decimation involves two processes: anti-alias low pass filtering, followed by downsampling, as illustrated in Figure 4.30. Here, we assume an input sampling rate of f_s, and denote the output sampling rate (after decimation by M), as f_d.

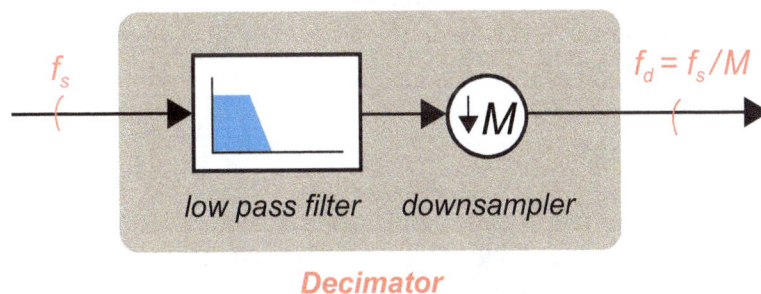

Figure 4.29: Decimator for a factor of M.

These two processes that comprise decimation can be outlined as follows:

- **Low pass filtering** is used to remove frequency components that would otherwise alias, when the sampling rate is reduced. FIR filters are generally adopted for this task, as will be assumed in this book.

- **Downsampling** is the process of reducing the sampling rate, by retaining only every M^{th} sample and discarding those in between.

The process of decimation is shown in both the time and frequency domains, in Figure 4.30.

In the upper part of the diagram, we note that the original signal, $x[k]$, has most of its energy at low frequencies, but there are some components extending up to $f_s/2$, i.e. half of the original sampling rate. These higher frequency components, above $f_d/2$, or half of the new sampling rate, are attenuated by a low pass anti-alias filter. The required frequency response of this filter is shown in blue in the middle right diagram.

After the signal has been filtered to remove terms that would alias, the sampling rate can be reduced by a downsampler. In the time domain, notice that the period between samples is increased by a factor of 3, due to the downsampler discarding 2 out of every 3 input samples. In the diagram, we denote sample indices after downsampling using a different index term, n as opposed to k, reflecting that there are fewer samples; in this case only every third sample is retained, i.e.

$$y[n] = v[3n].$$ (4.22)

In the frequency domain, the spectral images move closer together as a result of downsampling. However, as the low pass filter has successfully removed frequency components above $f_d/2$, no aliasing occurs.

In implementation terms, the 'direct' method of decimation shown in Figure 4.30, while intuitive, is computationally wasteful. The downsampler immediately discards $M-1$ out of every M samples output by the filter, and therefore the computation required to generate these discarded samples is redundant. Moreover, the proportion of redundant computation increases with the decimation ratio.

To address this issue, decimators are normally implemented in the *polyphase* form, which exploits the Noble Identities to reorder the computation and remove the redundancy, while generating identical outputs to the direct method. It is outside the scope of this book to derive the polyphase form, however this topic is well covered elsewhere, e.g. [188] and [261].

4.7.4. Interpolation

Interpolation increases the sampling rate, usually by an integer factor denoted by L. As an example an interpolator could increase the sampling rate by a factor of $L = 3$ from 200 MHz to 600 MHz. Similar to decimation, interpolation also requires a low pass filter along with the rate change operation, although the two operators

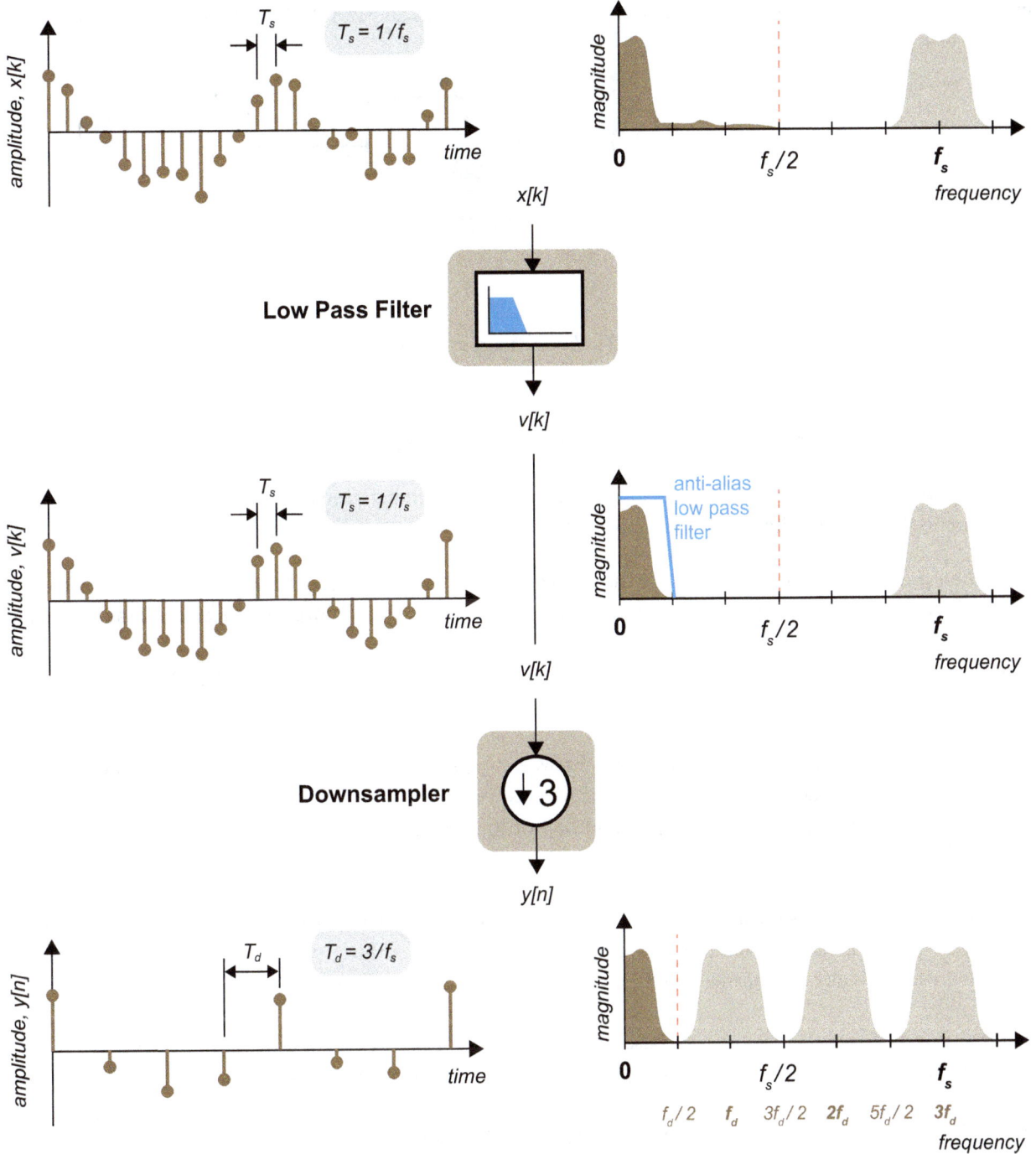

Figure 4.30: The process of decimating by a factor of M = 3.

occur in the opposite order. An interpolator is composed of an upsampling operation, followed by a low pass image rejection filter (assumed in this book to be implemented as an FIR filter), as shown in Figure 4.31.

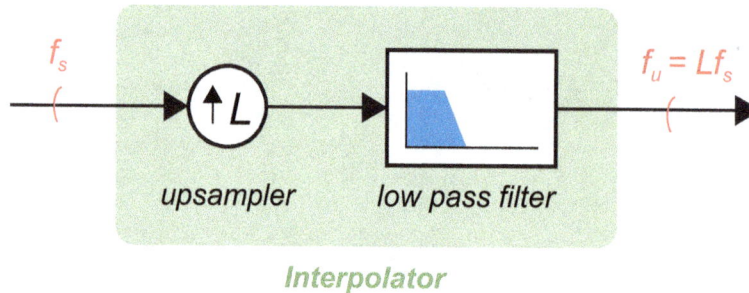

Figure 4.31: Interpolator for a factor of L.

The two operations comprising the interpolator can be summarised as:

- **Upsampling** involves inserting $L - 1$ zero-valued samples in between each pair of original samples, thus increasing the sampling rate by a factor of L.

- **Low pass filtering** is used to remove the spectral images that are present between 0 Hz and $f_u/2$, where f_u is the upsampled sampling rate. These images occur at integer multiples of the original sampling rate.

Like the decimator, we can visualise the process of interpolation in both the time and frequency domains, as shown in Figure 4.32 for the example of $L = 3$.

The upsampler inserts two zero-valued samples between the original ones, raising the sampling rate by a factor of 3. In the frequency domain, the two spectral images that are symmetric around f_s, the original sampling rate, now exist in the region between 0 Hz and half of the new, upsampled sampling rate, i.e. $f_u/2$. This is shown in the centre right portion of Figure 4.32.

These two spectral images must then be removed using a low pass image rejection filter, as a final stage in the decimation process. This manifests in the time domain as a 'smoothing' of the signal, i.e. the zero-valued samples are interpolated to intermediate amplitudes between the original samples.

Interpolation is an inefficient process, based on the model shown in Figure 4.31: upsampling inserts $L - 1$ zero-valued samples for every original sample, and these zero-valued samples are then processed by the filter, even though they contribute nothing to the computed output samples (zero multiplied by any weight value is still zero!). This implies that there is a degree of redundant computation in the interpolation filtering process, which is more significant for higher interpolation ratios.

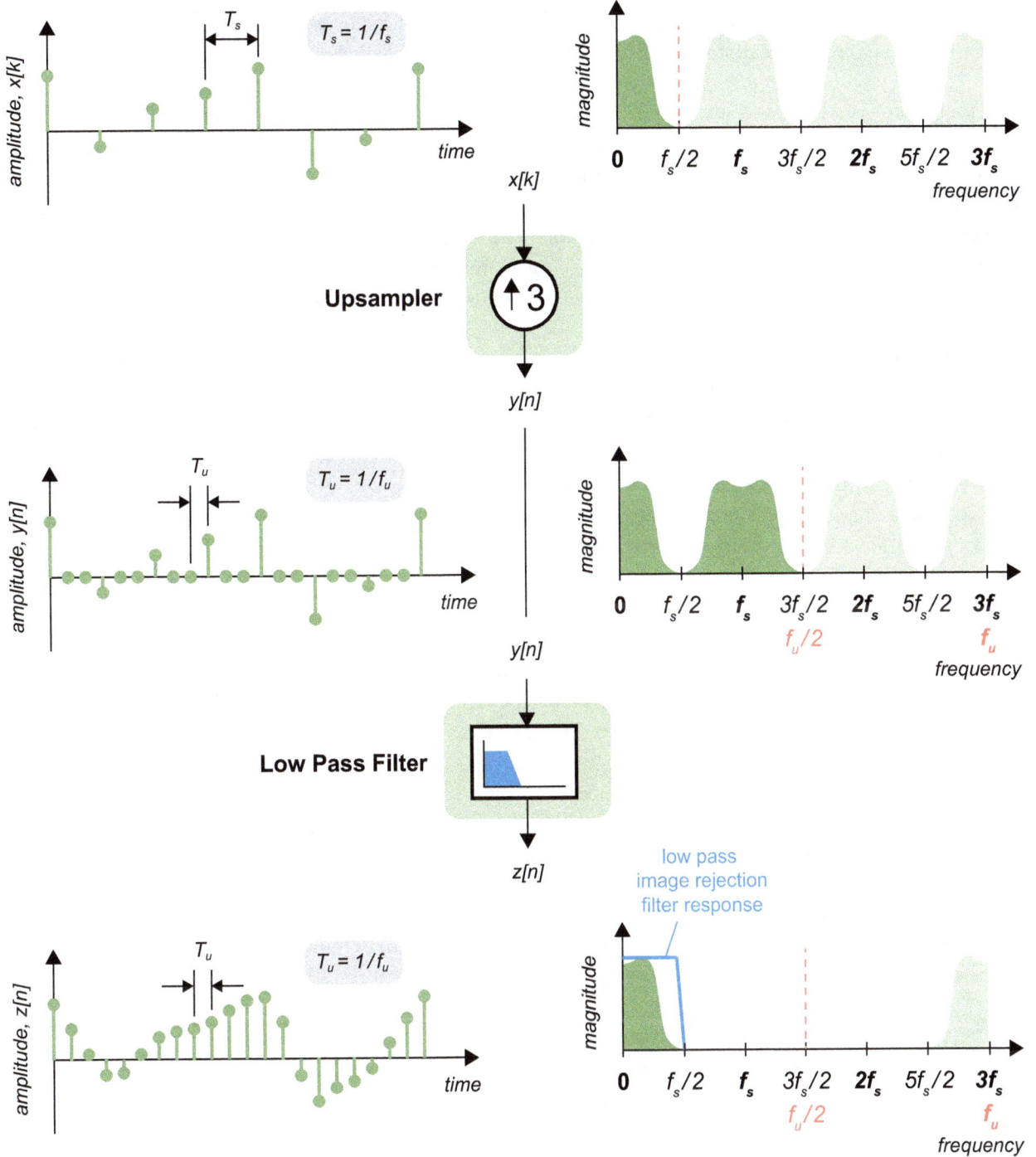

Figure 4.32: Interpolation by a factor of L = 3.

As for the decimator, a set of optimisations can be made to develop a polyphase form of the interpolator, which produces identical results to the direct form described here, but removes the redundant computation. The reader is again referred to textbooks such as [188] and [261] for extended treatment of the topic.

The premise of the polyphase method, for both interpolation and decimation, is "only calculate what you need!". There is no real reason <u>not</u> to use the polyphase method, as it requires significantly less computation compared to the direct method of implementing the decimator and interpolator (as shown in Figures 4.29 and 4.31, respectively), and generates identical results. Polyphase interpolators and decimators are amenable to implementation in PL, and indeed the reduced computational requirements can be exploited to produce very resource-efficient designs. In this book, we therefore assume that any multirate designs for DUCs, DDCs and other interpolation or decimation tasks are implemented using the polyphase method.

4.7.5. Halfband and L-Band Filters

On the theme of efficient implementation, it is worth highlighting a special class of filter that is especially suited for multirate applications. We note that a rate change by an integer factor of R requires a low pass filter passing $1/R^{th}$ of the band, regardless of whether the rate is being increased or decreased; for instance, factor-of-2 decimators and interpolators both require a low pass filter that passes 1/2 of the band. *Nyquist* filters, also known as *Lth-band* filters, satisfy this brief: they pass $1/R^{th}$ of the band, as shown in Figure 4.33 for example filters from 2-band to 6-band.

Nyquist filters designs have a characteristic impulse response, where every R^{th} weight is exactly zero, other than the central one (which has a value of $1/R$). Therefore, from an implementation perspective, these zero-valued weights do not need to be computed as MAC operations, and thus the filter implementation can be optimised to save computation and hardware. The saving is most prominent for lower values of R, with the *halfband* filter (i.e. $R = 2$) offering the largest benefit and being more commonly used (for higher order bands, other filter design methods may produce less expensive designs, i.e. with fewer weights).

As shown in Figure 4.34, the impulse response of the halfband filter is symmetric (which can be exploited as previously demonstrated in Figure 4.23) and can be further optimised by not computing the zero-valued weight multiplications. The central weight multiplication by 0.5 can be achieved by an arithmetic shift to the right by one place, which costs almost nothing — only routing resources to rewire the bits. Halfband filters can be therefore implemented particularly cheaply.

The requirement to interpolate or decimate by factor of 2 arises frequently, particularly as large interpolation and decimation tasks can be efficiently implemented as a cascade of smaller ones (to be further discussed in Section 4.7.6), and 2 is a factor of any even-valued rate change.

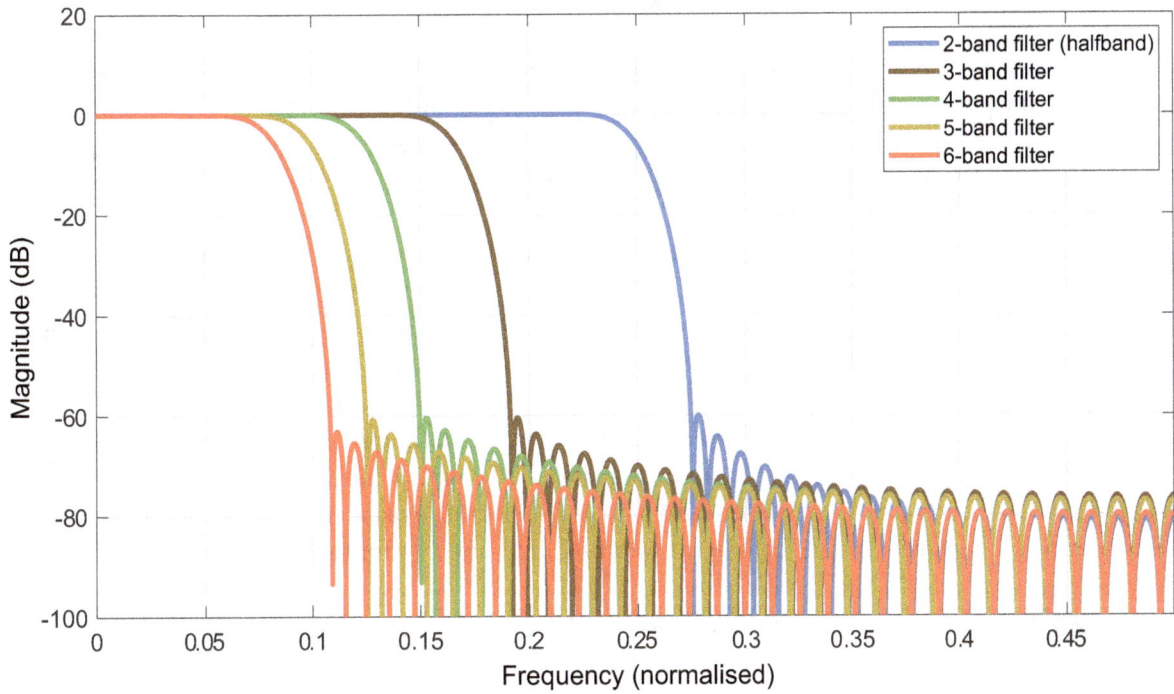

Figure 4.33: *Example Nyquist filter magnitude responses (normalised).*

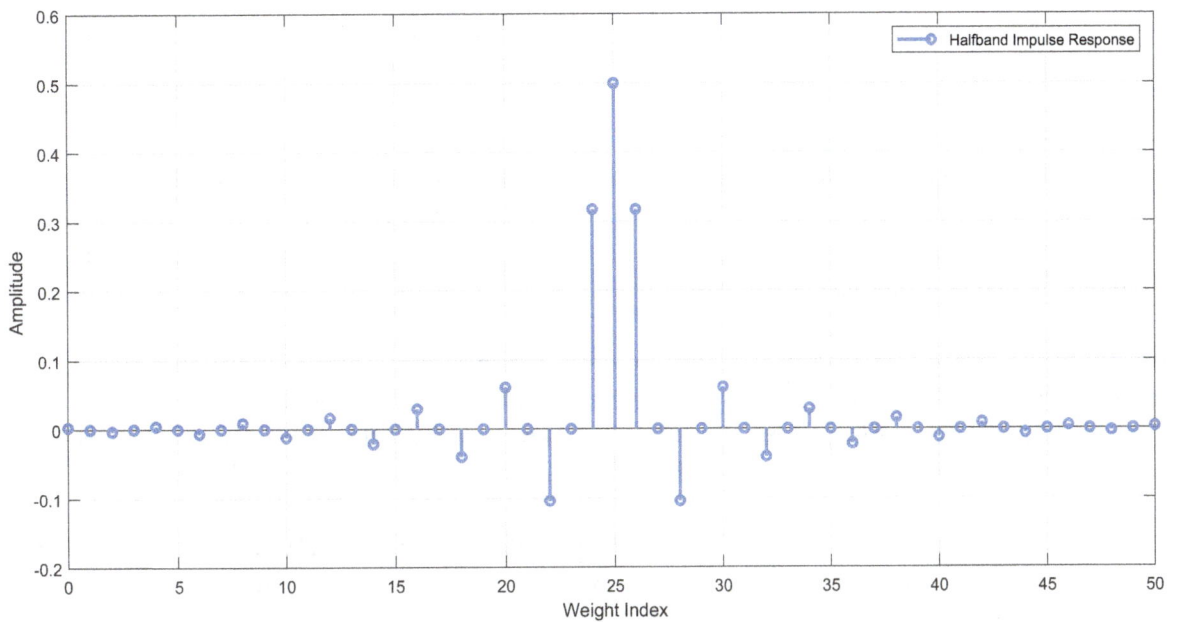

Figure 4.34: *Impulse response of an example halfband filter.*

4.7.6. Decimation and Interpolation Cascades

When the decimation or interpolation task extends to larger integer factors, the low pass response required for the anti-alias or image rejection filter, respectively, can become expensive to implement. The expense arises because the required transition band becomes progressively narrower (with respect to the sampling rate), the higher the rate change factor, and this leads to longer filters requiring more computation.

For a more computationally efficient solution, it is common to partition a larger rate change factor into several smaller rate change tasks. For instance, rather than interpolating by a factor of 18 using a single stage, an alternative approach is to cascade three interpolators, performing smaller rate changes of 2, 3, and 3 (noting that $2 \times 3 \times 3 = 18$). This results in simpler filter designs, and less computation overall.

The method of cascading smaller interpolators / decimators is used in the hardened DUCs and DDCs present in RFSoC devices. Cascades of interpolators and decimators are used to achieve the set of supported interpolation and decimation ratios, with the aid of multiplexers to select or bypass stages as required. This topic is covered in detail in Chapter 9 for the DDC decimators, and in Chapter 11 for the DUC interpolators.

Example: Rate Change of 6, as (i) a Single Stage, and (ii) a Cascade of Two Stages

For a simple example, let us consider a signal occupying the band from 0 to 45 MHz, which is to be interpolated by a factor of 6, from a sampling rate of 100 MHz to an interpolated rate of 600 MHz. We assume that a passband ripple of 0.2 dB is permissible, and that the stopband attenuation (image rejection) must be at least 60 dB. We assume the direct method of interpolator implementation (polyphase methods have not been covered in sufficient detail to permit analysis, but could be used to optimise the computation rate of both of the solutions presented here).

A single-stage approach, interpolating by a factor of 6, would require a low pass filter response that passes 1/6 of the band, cutting off between 45 MHz and 55 MHz. With the filtering operating at 600 MHz, the transition band is relatively narrow in a normalised sense, leading to an expensive filter. This requirement could be met by a symmetric filter design with 151 weights. The filter operates at 600 MHz and therefore the computation rate is 45.6 GMACs/s (exploiting coefficient symmetry).

Alternatively, the task could be partitioned into two simpler interpolators with individual rate changes of 2 and 3 (in either order), making an overall interpolation ratio of 6. For the purposes of this illustrative example, we will assume that Stage 1 interpolates by 3, and Stage 2 interpolates by 2. These two stages are depicted in the upper and lower halves of Figure 4.36, respectively

Notice the filter designs shown in Figure 4.36 — the Stage 2 filter can be more relaxed than you might imagine! This is because Stage 1 has created a significant portion of 'empty' spectrum, by removing the spectral images arising from the Stage 1 upsampler. The transition band of the Stage 2 filter can therefore be much wider, extending from 45 to 255 MHz, leading to a cheaper filter design.

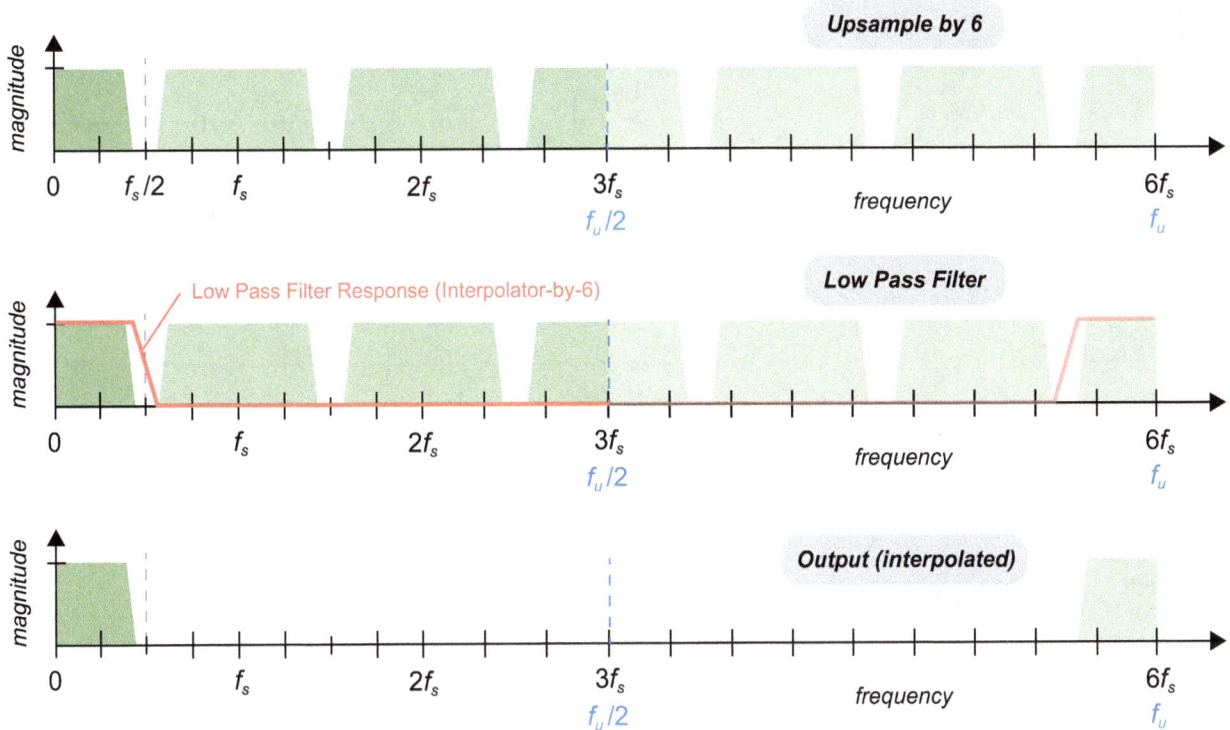

Figure 4.35: Interpolation by 6 using a single interpolator.

We can therefore design filters to suit the requirements for both approaches, i.e.

- A single stage interpolator, and

- A cascade of two interpolator stages;

and compare the computation involved. All filters are specified with the parameters mentioned earlier (0.2dB passband ripple and 60dB stopband attenuation), with the frequency specifications derived from the positions of the spectral images. The results are given in Table 4.3.

Notably, the single stage interpolator design requires a 151 weight filter, which translates to 76 MAC operations per calculated output sample, taking into account coefficient symmetry. As the filter operates at 600 MHz, the computation rate of this interpolator option is 45.6 GMACs/s. The cascaded design has two stages, with one being able to operate at an intermediate sampling rate of 300 MHz. The design of both of the filters — but in particular the second one — can be relaxed significantly. The cascaded design therefore yields a considerable saving in terms of computation, requiring 15 GMACs/s, less than one third of the original value.

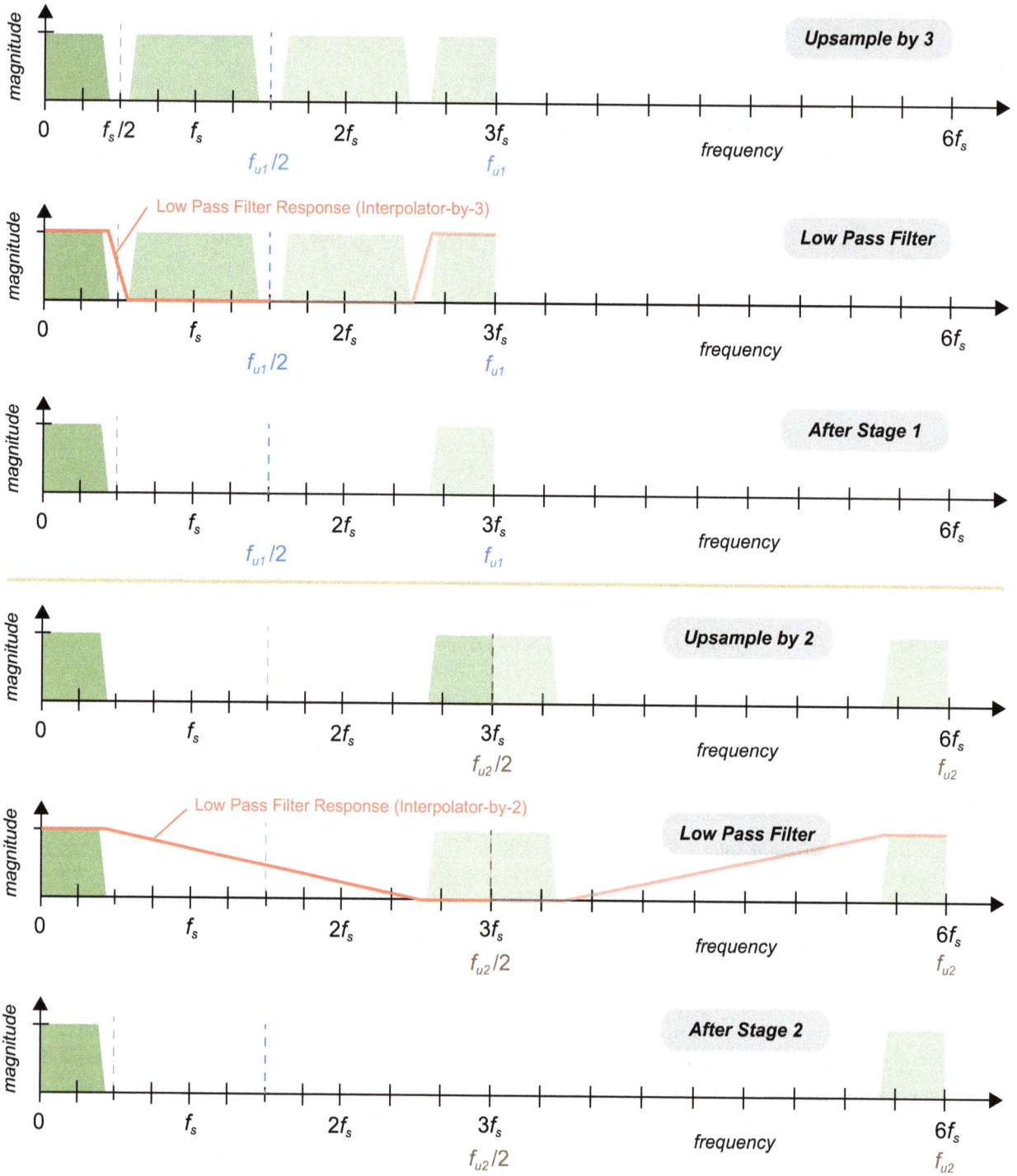

Figure 4.36: Interpolation by 6, using two interpolators in cascade (stage 1 interpolates by 3, stage 2 interpolates by 2).

Table 4.3: Comparison of two alternative interpolator-by-6 designs (single stage, and a cascade of two stages).

	Filter sampling rate	**Filter length (no. weights)**	**MAC operations per output sample**[a]	**Computation rate**
Single stage	600 MHz	151	76	76 * 600 MHz **= 45.6 GMACs/s**
Cascade: Stage 1	300 MHz	79	40	40 * 300 MHz = 12 GMACs/s
Cascade: Stage 2	600 MHz	15	5[b]	5 * 600 MHz = 3 GMACs/s
Cascade: Total				**= 15 GMACs/s**

a. Taking into account coefficient symmetry to reduce the number of MAC operations.
b. Of the 15 weights, there are only 5 individual weight values (due to symmetry and zero-valued weights).

This example demonstrates that a cascaded interpolator can reduce the computational requirements, compared to a single stage interpolator. A good rule of thumb is to choose smaller rate change factors where possible, as these result in simpler filter designs, and to maximise the use of factor-of-2 decimators and interpolators, where the efficiency of the halfband filter type can be exploited. The concepts explored here can be applied to interpolation cascades with additional stages, and also to cascaded decimators, and can be further optimised through the use of polyphase methods.

The RFSoC's integrated DUCs and DDCs make use of the same concepts to achieve efficient interpolator and decimator designs in dedicated silicon; through analysis of the individual filter responses comprising these cascades, we will later observe a similar relaxation of frequency specifications, depending on the position of each filter in the cascade.

4.7.7. Cascade Integrate Comb (CIC) Filters

The Cascade Integrate Comb (CIC) is a particularly efficient class of filter, often used for multirate operations [197]. As its name suggests, the CIC is composed of several stages of integrators and combs (which are very simple, multiplierless filters) in cascade. An individual stage is derived from the moving average, i.e. an FIR filter of length W weights, where each weight value is $1/W$. The trick of the CIC is to reformulate the computation of the moving average, such that instead of the conventional FIR structure, a reduced-complexity architecture can be used to achieve the same output.

The CIC filter, while inexpensive in hardware terms, has the disadvantage that its magnitude response is very sub-optimal: the passband is flat only at very low frequencies (and quickly starts to 'droop'), while stopband attenuation is acceptable only when several stages are cascaded — which makes the droop worse. There are also some implementation challenges in terms of specifying the various internal arithmetic wordlengths. Nevertheless, the CIC filter often finds utility in multirate systems such as DUCs and DDCs. The role it tends to play is the final stage of interpolation, or the first stage of decimation, where the sampling rates are highest and its computational efficiency can be enjoyed the most, but where the characteristic CIC 'droop' tends not to affect the signal bandwidth of interest[4].

We elect not to focus on CIC filters in this book, because the first stages of the post-ADC decimation chain, and the final stages of the pre-DAC interpolation chain, are both already provided in hardened and optimised form as part of the RFSoC's RFDCs [90]. These decimators and interpolators are implemented as cascades of polyphase FIR filters; their functionality will be reviewed in Chapters 9 and 11, respectively. While there may be some specific applications where it is useful to implement CIC filters in addition to the RFDC's decimators and interpolators, or even in place of them (the RFDCs permit their internal decimators and interpolators to be bypassed if desired), it is anticipated that most RFSoC SDR designs will not involve CIC filters. Design tool support for CICs is however available if needed [21].

4.7.8. Resampling and Other Multirate Operations

In addition to the integer rate changes discussed over the last few pages, there are other types of multirate operations, as briefly mentioned in the introduction to Section 4.7. These can be referred to collectively as *resampling* operations, i.e. where the amplitudes of a signal are generated at a new set of sampling instants. This is analogous to passing a digital signal through a DAC to convert it to analogue, and then through an ADC to convert it back to digital, using a different sampling frequency or phase. However, the entire process can be undertaken digitally using multirate signal processing techniques, provided that signals are sampled above Nyquist rate.

Briefly, there are three main types of operation to be aware of, beyond simple decimation and interpolation by integer factors. These are:

- **Resampling a signal by a rational fraction** — If the sampling rate is to be changed by the ratio of two integers, e.g. a rate change from 100 MHz to 150 MHz could be expressed as $R = 3/2$. Rational fractional rate changes can be achieved using a cascade of an interpolator and decimator, e.g. $L = 3$ and $M = 2$ in this example. The resulting structure can be optimised using polyphase methods.

- **Resampling a signal by an irrational fraction, or by a factor that changes over time** — Where there is no convenient integer-based expression for the resampling ratio, or where it is dynamic, a different type of approach is required. Popular methods include highly oversampled polyphase filters, and Farrow structures [158].

- **Changing the sampling phase, or introducing a fractional delay** — Occasionally it may be desired to delay a signal by a duration less than one sample period. This can be achieved using similar methods to the second category.

---

4. The droop does not significantly impact the signal bandwidth of interest, provided that the signal bandwidth is small with respect to the sampling rate at the output of the CIC. If desired, droop can be compensated by inserting a correction filter, although this adds to the overall cost of implementation.

These more complex multirate techniques are often required in communications receivers, which must synchronise to the frequency and timing parameters of incoming signals. It is beyond the scope of the current book to cover these topics in detail; books such as [188], [230], and [301] may be useful for further study.

4.8. Chapter Summary

In this chapter, we have reviewed some of the key DSP concepts underpinning the SDR theme of this book. In particular, the processes of sampling, quantisation, digital-to-analogue and analogue-to-digital conversion were covered, and the frequency domain was briefly touched upon (much more on this to follow in Chapter 5).

The topic of digital filtering was introduced. Filtering is used to change the frequency content of a signal, and forms a key building block in many DSP systems, including in wireless communications. The PL portion of the RFSoC is particularly well-equipped for implementing high speed filters, via its hardened DSP48E2 arithmetic blocks.

The last part of the chapter focused on multirate operations, in particular interpolation and decimation to raise or lower the sampling rate by an integer factor. The RFSoC architecture includes hardened support for interpolation and decimation at the DAC and ADC interfaces, respectively, which is discussed in Chapters 9 and 11. The concepts presented here should prove useful background for understanding the functionality of these dedicated, hardened RFSoC features, or building your own custom designs using the PL.

Notebook Set B

DSP Fundamentals

For those without a background in DSP, or who would like a refresher, a set of Jupyter notebooks is provided to review key concepts from a simulation perspective, consolidating on the theory presented in Chapter 4. These notebooks review the DSP fundamentals of sampling and quantisation, ADCs and DACs, and digital filtering. In all cases, you can investigate the impacts of different parameters by making adjustments and re-running the simulations.

There are four notebooks to explore on the fundamentals of DSP. Their relative locations are listed as follows:

ALL   01_sampling.ipynb — *rfsoc_book/notebook_B/01_sampling.ipynb*

ALL   02_quantisation.ipynb — *rfsoc_book/notebook_B/02_quantisation.ipynb*

ALL   03_adcs_and_dacs.ipynb — *rfsoc_book/notebook_B/03_adcs_and_dacs.ipynb*

ALL   04_digital_filter_design.ipynb — *rfsoc_book/notebook_B/04_digital_filter_design.ipynb*

B.1. Sampling

Sampling is the process of converting a continuous-time signal to a discrete-time signal representation, and it is an important aspect of analogue-to-digital conversion. The first notebook in this series named *01_sampling.ipynb* provides an interactive example of sampling, similar to the one presented in Chapter 4. You can vary the sampling rate and the frequencies of the input sine waves, and view the resulting changes in the time and frequency domains.

This notebook also contains an example that demonstrates the effect of aliasing, which shows how signals above $f_s/2$ are aliased (or 'folded' down) into the frequency band between 0 Hz and $f_s/2$ Hz. As noted in Chapter 4, this occurs based on Nyquist Zones in a predictable way. A simple time domain plot that illustrates aliasing is presented in Figure B.1. Notice that aliasing has occurred due to an insufficient sampling rate, resulting in a lower-frequency sinusoid being interpreted from the set of available samples.

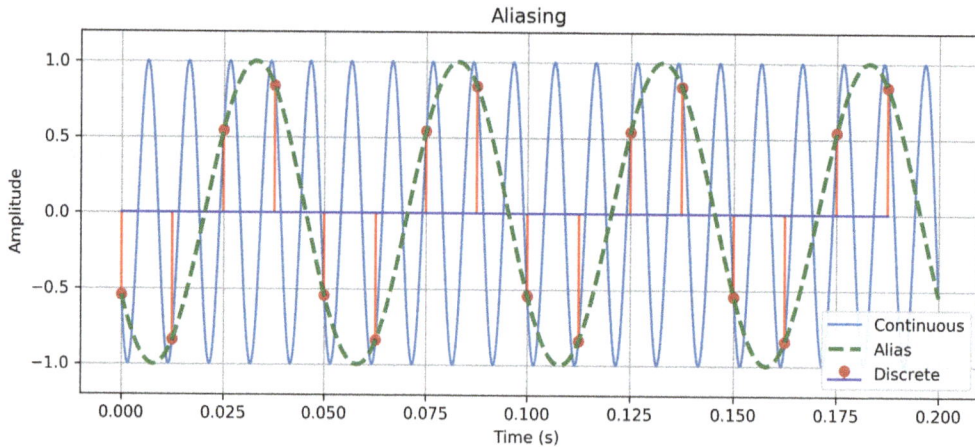

Figure B.1: Time domain plot illustrating aliasing of a sampled waveform.

B.2. Quantisation

Quantisation is the process of converting the amplitude of a signal to a discrete set of representable levels. The second notebook to explore in this series is named ***02_quantisation.ipynb***. This notebook will investigate the effect of quantisation in the time and frequency domains for a set of test input signals, and will confirm that quantising with a greater number of bits produces a more accurate representation of the signal.

The modelling of quantisation as noise is also explored. The notebook confirms that this noise can be *tonal*, i.e. where energy is concentrated at a number of discrete frequencies. Additionally, the use of dithering to "whiten" the quantisation noise (in other words, to reduce or remove the tonal nature of the noise), is demonstrated.

Dithering is useful in wireless communications applications, because (in the transmitter) tonal noise can generate significant out-of-band emissions that interfere with other users, and (in the receiver) tonal noise can create harmonics that compromise signal integrity and make signal reception more difficult — white noise (even with higher overall energy) is preferable.

Through this notebook, you can experiment with different parameters to gain an appreciation of the effects on the quantised signal.

B.3. ADCs and DACs

Notebook three reviews the process of converting analogue signals to digital signals, and vice versa. The sampling and quantisation operations reviewed in the first two notebooks are integral to these conversions, but there are also aspects of filtering and oversampling to consider too, as investigated here.

To begin exploring this example, open the notebook named ***03_adcs_and_dacs.ipynb***.

In this notebook, we review the use of an anti-aliasing filter and oversampling techniques when acquiring a signal using an ADC. We also explore signal reconstruction with a DAC, by investigating the Zero Order Hold (ZOH) technique and the reconstruction filter (as previously reviewed in Section 4.6.2). A time domain plot of the digital to analogue conversion process is provided in Figure B.2. As shown, a ZOH is applied to a digital signal, resulting in a 'stepped' output signal. This signal is smoothed using a reconstruction (lowpass) filter.

As for the other examples in this set, you can experiment by applying your own custom parameters for the ADC and DAC processes, and observing how they effect the results.

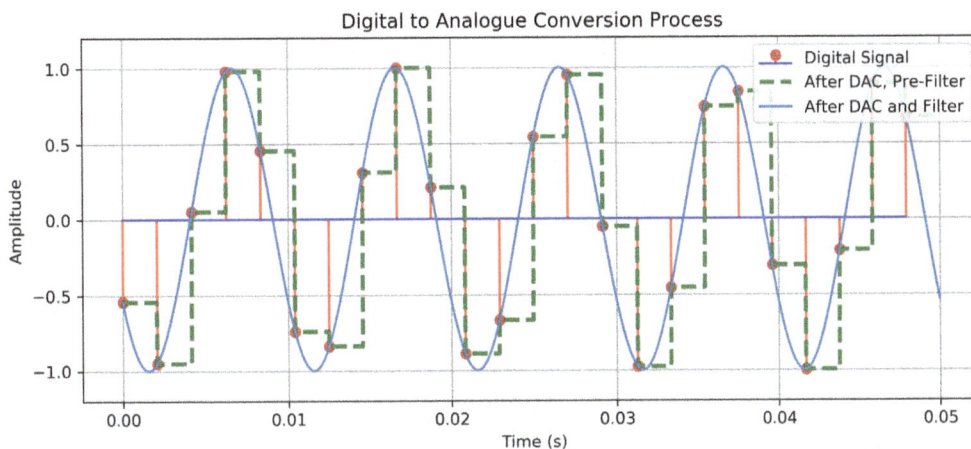

Figure B.2: Time domain plot illustrating the digital to analogue reconstruction process.

B.4. Filtering and Filter Design

The final notebook in this set, ***04_digital_filter_design.ipynb***, introduces the fundamentals of digital filters, which are ubiquitous components in digital communications systems, and DSP applications in general. By working through this notebook, you will gain an understanding of how to design your own digital filters.

The design of filters to change the frequency content of a signal is reviewed, covering lowpass, highpass, bandstop, and bandpass filters. An example plot of the magnitude response for each type is shown in Figure

B.3. The filter phase response is also explored. We also consider the key parameters involved in specifying a filter design, an example of which is shown in Figure B.4. Filter design methods are also explored, in particular the Window Method, Parks McClellan Algorithm, and the Least Squares Method.

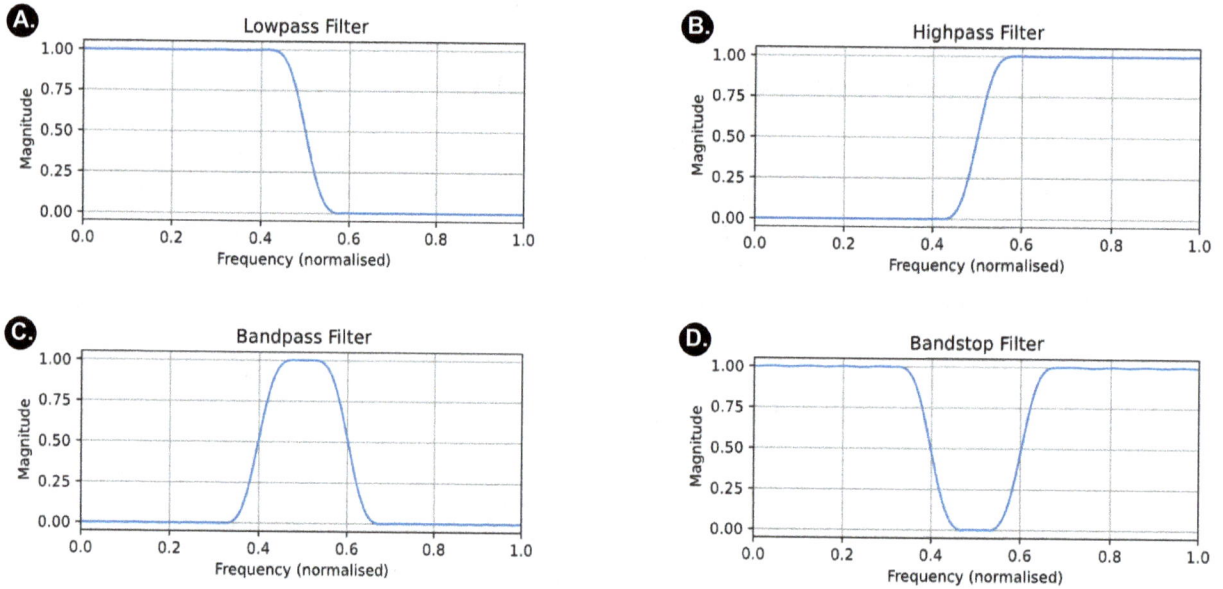

Figure B.3: *Magnitude response plots of a lowpass filter (a), highpass filter (b), bandpass filter (c), and bandstop filter (d).*

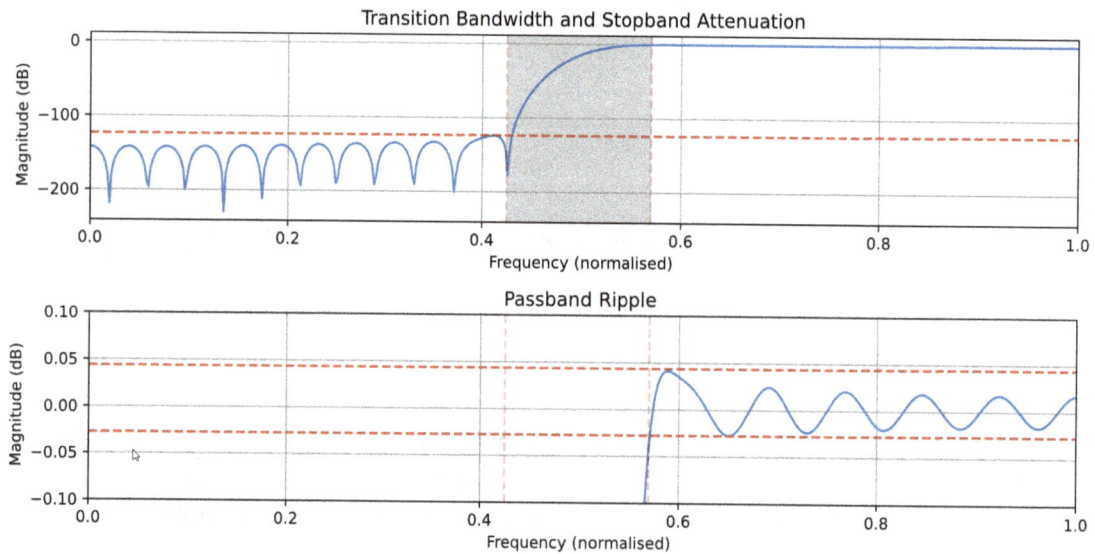

Figure B.4: *An example of a highpass filter design.*

Chapter 5

Spectral Analysis

David Northcote and Robert W. Stewart

In this chapter, we will review standard spectral analysis techniques, including the Fourier Series, the Fourier Transform, the Discrete Fourier Transform (DFT), and its more efficient version, the Fast Fourier Transform (FFT). We will also explore common techniques for analysing and displaying the frequency spectrum as both the one sided and two-sided forms. This discussion will also include spectrograms, which are created using the Short-Time Fourier Transform (STFT). Later in Chapter 7 we will review the complex signal frequency representation of baseband I and Q signals, and also use the FFT and inverse FFT for orthogonal frequency domain modulation and demodulation in Chapter 16.

Electromagnetic waves are everywhere around us, from light waves to radio waves, to X-rays, and so on. Most modern communications use frequency bands from VHF (a few 10s of MHz) to mmWave (a few 10s of GHz). Measuring the frequency content and power levels of these radio waves is a process known as spectral analysis. Instrumentation and specialised equipment, such as a spectrum analyser tool, can measure essential properties of electromagnetic waveforms and display these measurements to a user. With the likes of the RFSoC sampling at a few GHz, we can digitise wide radio spectrum bands. Using Fourier or frequency domain analysis, we can measure and then present acquired waveforms' magnitude and phase spectra.

5.1. Fourier Theorem

Many modern communication systems and related measurement instrumentation use Fourier analysis to investigate a waveform's frequency content and orthogonal Fourier transform methods for signalling. Jean-Baptiste Joseph Fourier was a French mathematician who studied in Grenoble, France, around the end of the French Revolution (1789-1799). Fourier experimented with heat propagation, and in 1822 he published a book on heat flow named La Théorie Analitique de la Chaleur (The Analytic Theory of Heat) [165]. In his book, Fourier presented the process for decomposing a signal into a set of harmonically related periodic waveforms, which is essentially the Fourier Theorem or the Fourier Series.

In modern-day Fourier Series for time domain signals, we can take any periodic waveform of fundamental period T and decompose it into a sum of sine and cosine waves of varying amplitudes at the fundamental frequency ($f_0 = 1/T$) of the signal, and harmonically related components $2f_0$, $3f_0$, $4f_0$. This summation is known as a Fourier Series and can be used to decompose a waveform into a harmonically related sum of *any* arbitrary base waveform. For example, triangular waves, square waves (known as the Walsh-Hadamard decomposition) or some defined wavelet.

In the world of time domain signal processing, decomposing a signal into a sum of sine waves is always the way forward. The reason is simple. If a sine wave is input to a linear system, then the output is a sine wave at *exactly* the same frequency — the amplitude might be modified, and the phase might be changed, but the frequency remains the same. Given that we invariably work with linear systems in DSP, and often assume radio environments to be linear (or nearly linear), then by presenting the characteristics of a signal or a system in terms of sine waves means that we can characterise the frequency content or the transfer function across the frequency spectrum. This is not true of any other base waveform. If we input a square wave into a linear system, it will NOT always output a square wave of the same frequency with modified amplitude and phase.

In the following sections, we will review the fundamental Fourier series using some simple periodic waveforms, and illustrate how we can calculate, present and analyse the outputs of the Fourier Series calculation. We will also show how the more general Fourier Transform is derived.

5.1.1. Fourier Series of a Square Wave

The classic Fourier series example is to decompose a periodic square wave into a sum of harmonically related sine and cosine waves. We can begin by defining our square wave to have a period of 2π and a peak-to-peak amplitude of 6. You can inspect the square wave on the left of Figure 5.1. Notice that the square wave does not oscillate around 0 as it contains a DC bias of +3 (or in frequency terms we could define this as the 0 Hz cosine term of amplitude 3). The DC bias is equivalent to the average value of the square wave and is very important for our upcoming analysis.

We can approximate the square wave using a sinusoid consisting of similar characteristics. This sine wave is defined mathematically as $3 + (12/\pi)\sin(x)$, where x is the phase in radians. You can inspect a plot of the sine wave on the right of Figure 5.1. Notice that we gave the sinusoid a DC bias of +3 and an amplitude coefficient of $(12/\pi)$. Don't worry about how the amplitude coefficient of the sinusoid has been calculated as we will discuss this later in Section 5.1.2. It is not a great approximation of the square wave, but it is a start.

The Fourier Theorem states we should add several harmonically related sine and cosine waves together to improve the Fourier series approximation of our periodic waveform. The next sine wave can be defined mathe-

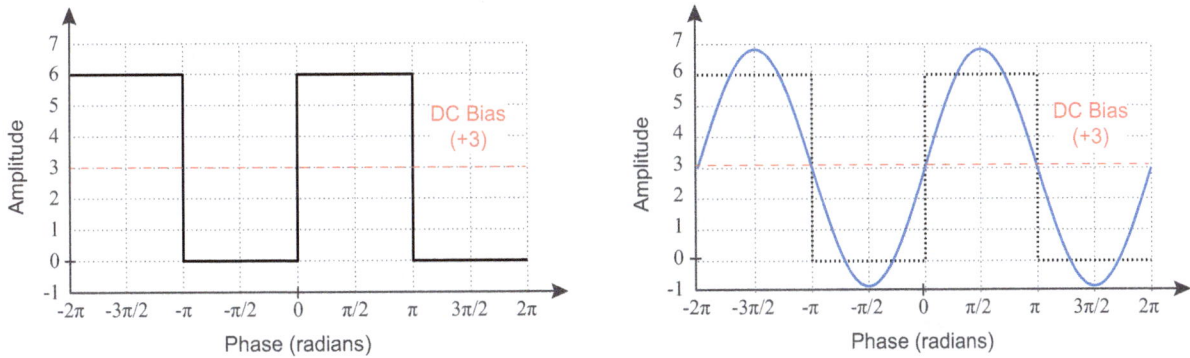

Figure 5.1: A square wave (left). A sinusoid used to approximate the square wave (right).

matically as $(12/(3\pi))sin(3x)$ and can be inspected on the left of Figure 5.2. This sinusoid oscillates three times as fast in comparison (its the 3rd harmonic) to the fundamental sine wave and has a lower amplitude coefficient. This sine wave can be added to our existing approximation resulting in the waveform shown on the right of Figure 5.2.

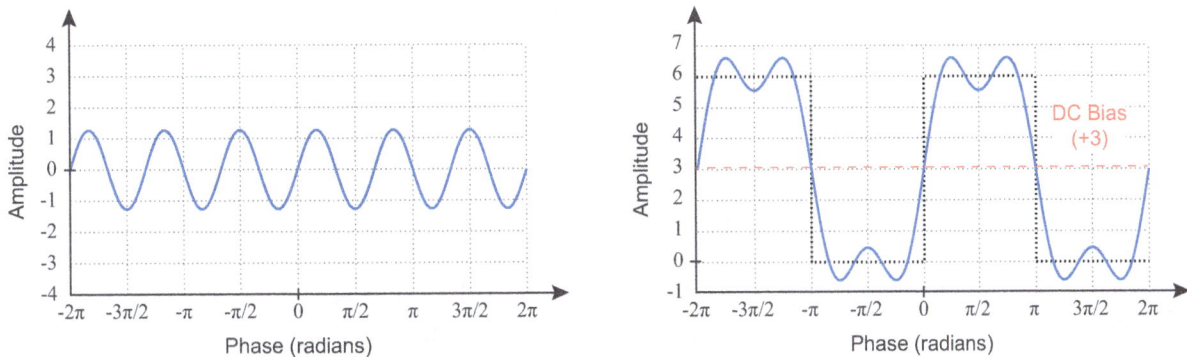

Figure 5.2: The second sine wave (left). Accumulation of two sine waves that are approximating a square wave (right).

Notice that our approximation is beginning to look more like a square wave. Let's add another sinusoid defined as $(12/(5\pi))sin(5x)$. This (5th order harmonic) sine wave oscillates faster and has a lower amplitude coefficient than the previous two. We can inspect our new square wave approximation on the right of Figure 5.3.

It should be clear now that the approximation of the square wave will improve as we sum more harmonically related sine waves together. For example, consider the plots on the left and right of Figure 5.4. These square wave approximations are created using a summation of 13 and 26 sine waves, respectively.

The main point of the example above is to explore the underlying principles of Fourier's Theorem, where we can represent a periodic waveform by summing sine and cosine waves together. In this example, if we summed

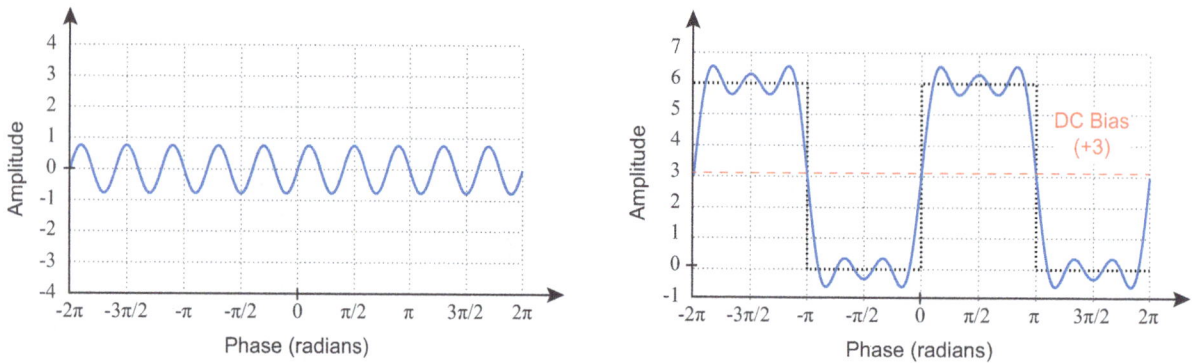

Figure 5.3: *The third sine wave (left). Accumulation of three sine waves that are approximating a square wave (right).*

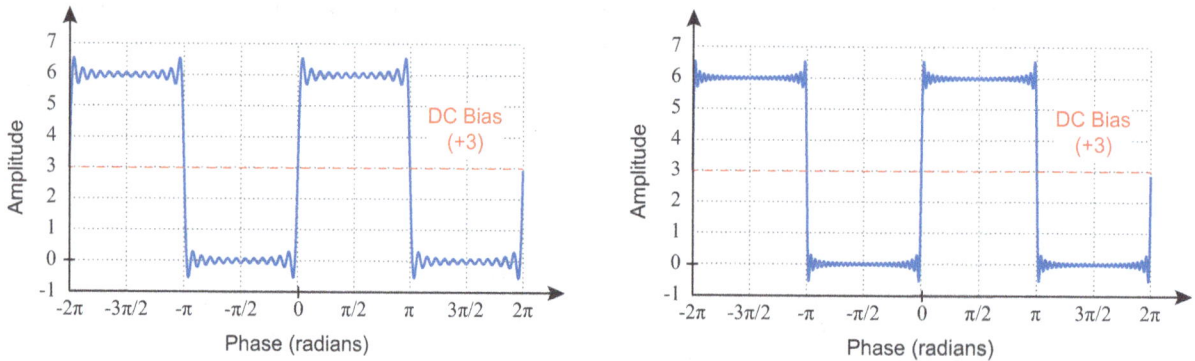

Figure 5.4: *Square wave approximation using a sum of 13 sine waves (left) and 26 sine waves (right).*

together an infinite number of harmonically related sine waves, with the correct amplitudes, we would be able to reproduce the square wave. We can write this summation mathematically,

$$g(x) = 3 + \frac{12}{\pi}\sin(x) + \frac{12}{3\pi}\sin(3x) + \frac{12}{6\pi}\sin(5x) + \dots \text{ (infinte sum of weighted sine waves)}, \qquad (5.1)$$

where $g(x)$ is the periodic square wave. The sum of sine waves given in (5.1) is known as a Fourier Series. We can take a step further and describe a generalised Fourier Series mathematically as,

$$\begin{aligned} g(x) = \ & a_0 + a_1\cos(1x) + b_1\sin(1x) \\ & + a_2\cos(2x) + b_2\sin(2x) \\ & + a_3\cos(3x) + b_3\sin(3x) \\ & + \dots \\ & + a_n\cos(nx) + b_n\sin(nx). \end{aligned} \qquad (5.2)$$

As you can see, the periodic waveform $g(x)$ is equal to a weighted sum of sine and cosine waves (albeit the first example only has sine waves and the amplitudes of the cosines were zero) where:

- a_0 is the DC bias,

- a_1 to a_n are the cosine amplitude coefficients,

- b_1 to b_n are the sinusoid amplitude coefficients,

- and $n = 1, 2, 3, \ldots$.

We can also represent $g(x)$ using a series summation:

$$g(x) = a_0 + \sum_{n=1}^{\infty} a_n \cos(nx) + \sum_{n=1}^{\infty} b_n \sin(nx), \qquad n = 1, 2, 3, \ldots . \tag{5.3}$$

In this form, we should be able to represent any (real valued) periodic waveform as a sum of sine and cosine waves with defined amplitudes. In the following section, we will present the classical Fourier Series to calculate the DC bias (a_0) and amplitude coefficients a_n and b_n of the Fourier Series expansion.

5.1.2. Fourier Series Coefficients

The Fourier Series coefficients provide the weighting for the DC bias and each sine and cosine wave required to expand a periodic function. It will be necessary to use appropriate mathematics in the form of definite integrals to compute these coefficients. We can begin by defining three fundamental equations that allow us to calculate the DC bias a_0 and the amplitude coefficients a_n and b_n.

$$a_0 = \frac{1}{2\pi} \int_{-\pi}^{\pi} g(x)\,dx, \tag{5.4}$$

$$a_n = \frac{1}{\pi} \int_{-\pi}^{\pi} g(x)\cos(nx)\,dx, \qquad n = 1, 2, 3, \ldots, \tag{5.5}$$

$$b_n = \frac{1}{\pi} \int_{-\pi}^{\pi} g(x)\sin(nx)\,dx, \qquad n = 1, 2, 3, \ldots . \tag{5.6}$$

We will now demonstrate how to compute the DC bias a_0 for the periodic square wave we defined previously in Section 5.1.1. This square wave is periodic over the interval $[-\pi, \pi]$. Two statements can be made about the square wave across this interval.

1. The square wave amplitude is 0 between the interval $[-\pi, 0]$.

2. The square wave amplitude is 6 between the interval $[0, \pi]$.

To compute a_0, we expand the definite integral in (5.4). The amplitude of the square wave across each interval is used to compute the DC bias as follows.

$$a_0 = \frac{1}{2\pi} \int_{-\pi}^{\pi} g(x)\,dx = \frac{1}{2\pi}\left(\int_{-\pi}^{0} 0\,dx + \int_{0}^{\pi} 6\,dx \right)$$

$$= \frac{1}{2\pi}(6x)\Big|_{0}^{\pi}$$

$$= \frac{1}{2\pi}(6\pi - 0) = 3.$$

The DC bias result makes sense as we can clearly see that the square wave oscillates around the value of +3 in Figure 5.1. We can use this value later when defining our Fourier Series of the square wave. We will now evaluate the cosine coefficients. There may be an infinite number of cosine coefficients as $n \to \infty$. Therefore, we should create an expression that can be used to calculate a cosine coefficient a_n for any value of n.

$$a_n = \frac{1}{\pi} \int_{-\pi}^{\pi} g(x)\cos(nx)\,dx = \frac{1}{\pi}\left(\int_{-\pi}^{0} 0\cos(nx)\,dx + \int_{0}^{\pi} 6\cos(nx)\,dx \right)$$

$$= \frac{1}{n\pi}(6\sin(nx))\Big|_{0}^{\pi}$$

$$= \frac{6}{n\pi}(\sin(n\pi) - 0) = 0$$

You may be surprised that $a_n = 0$. In essence, our particular square wave does not contain a cosine component for any value of n. We will explain this result in a moment. Next we move on to derive a similar expression for the sine wave coefficients b_n.

$$b_n = \frac{1}{\pi} \int_{-\pi}^{\pi} g(x)\sin(nx)dx = \frac{1}{\pi}\left(\int_{-\pi}^{0} 0\sin(nx)dx + \int_{0}^{\pi} 6\sin(nx)dx \right)$$

$$= \frac{1}{-n\pi}\left(6\cos(nx)\right)\Big|_{0}^{\pi}$$

$$= \frac{6}{-n\pi}(\cos(n\pi) - 1)$$

In the above expression, the term $\cos(n\pi)$ evaluates to 1 when n is even, or -1 when n is odd. We can write these two cases mathematically to simplify our expression for b_n as below,

$$b_n = \begin{cases} 0, & \text{if } n \text{ is even,} \\ \dfrac{12}{n\pi}, & \text{if } n \text{ is odd.} \end{cases} \tag{5.7}$$

If we plug-in the expressions for a_0, a_n and b_n into (5.3), we arrive at the Fourier Series expansion for the square wave,

$$g(x) = 3 + \sum_{n=1}^{\infty} \frac{12}{n\pi}\sin(nx), \qquad n = 1, 3, 5, \dots . \tag{5.8}$$

We can calculate the first four terms of (5.8) and plot the results for comparison, as shown in Figure 5.5. Notice that the square wave approximation improves as the number of accumulated sine waves increase. Also, it is important at this stage to mention the "ringing" that occurs at discontinuities (sudden changes in amplitude). This "ringing" is known as Gibbs Phenomenon. We won't cover Gibbs Phenomenon in this chapter, but you can read more about it in [195].

The square wave above is a specific case that expands into a Fourier Series consisting of only sine waves. Figure 5.6 presents two similar square waves, but with different phase, and the first sinusoidal term of their corresponding Fourier Series expansions. The square wave on the left is another unique case whose Fourier Series consists of only cosine waves (by virtue of being an even symmetric square wave). Notice that the square wave has a similar amplitude and duty cycle as before. However, the phase of the waveform has shifted by $\pi/2$.

The Fourier Series expansion for this square wave is given as,

$$g(x) = 3 + \sum_{n=1}^{\infty} \left(-1^{\frac{n-1}{2}}\right)\frac{12}{n\pi}\cos(nx), \qquad n = 1, 3, 5, \dots . \tag{5.9}$$

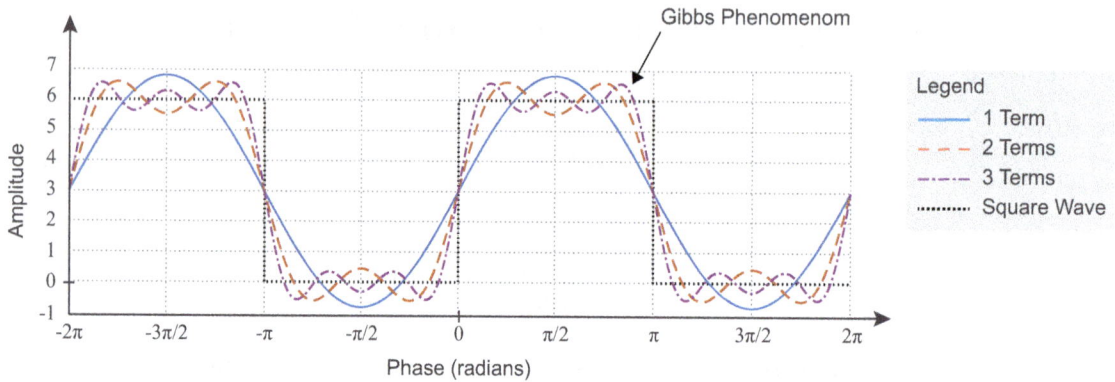

Figure 5.5: First three significant sine wave terms with amplitude (b_1, b_3, b_5) with DC bias (a_0) of a square wave Fourier Series expansion.

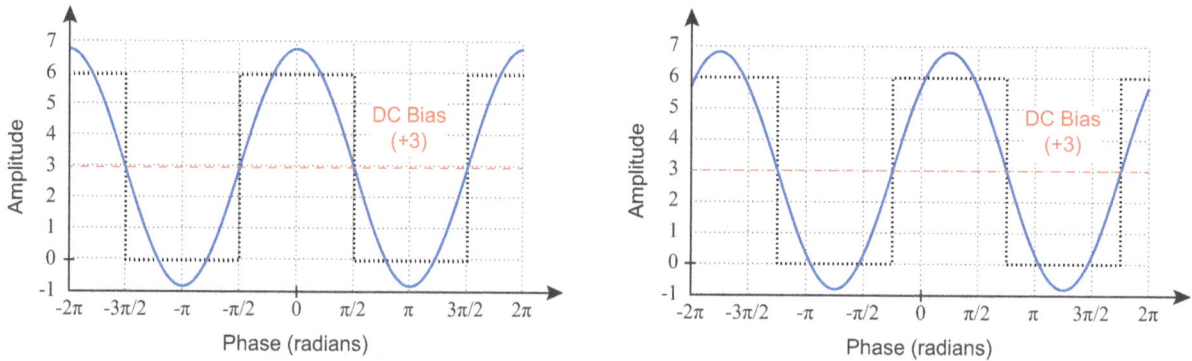

Figure 5.6: Two square waves and the first term of their Fourier Series expansions.

The square wave on the left of Figure 5.6 shares the same phase relationship with a cosine wave, which means its Fourier Series expansion will only use cosines. Many practical Fourier Series expansions will use a sum of sine and cosine waves. For instance, the square wave on the right of Figure 5.6 does not share the same phase relationship with a sine or cosine wave as it exhibits a phase shift of $\pi/4$ (in comparison to the square wave in Figure 5.5). Therefore, its corresponding Fourier Series expansion will require a weighted sum of both sine and cosine waves to represent the square wave correctly.

Up until now, we have only explored the Fourier Series expansion of a square wave. The same process can be used to obtain the Fourier Series expansions of other periodic waveforms, such as a sawtooth wave. Figure 5.7 contains a plot of a sawtooth wave and the first three terms of its Fourier Series expansion for you to inspect.

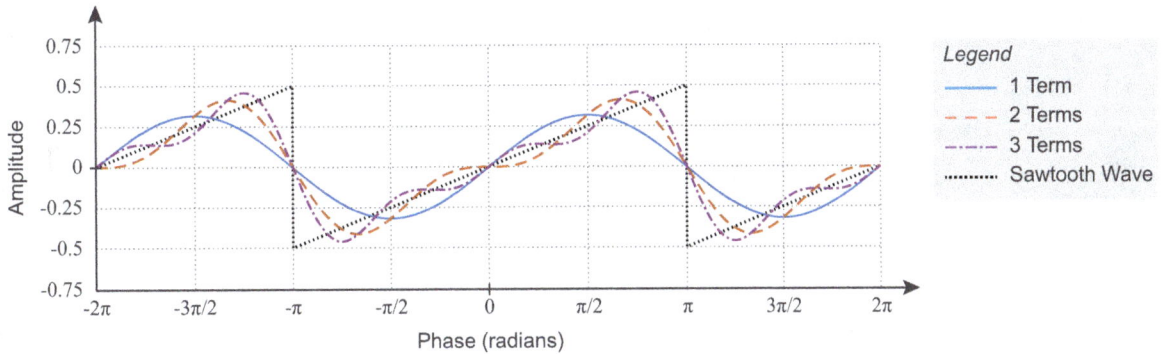

Figure 5.7: *First three terms of a sawtooth Fourier Series expansion.*

5.1.3. Time, Frequency, and Harmonics

The Fourier Series expansions we have performed so far are periodic across the interval $[-\pi, \pi]$. We can also expand waveforms that are periodic over an interval of time, such as the waveform given in Figure 5.8. This waveform is periodic over the interval $[-T/2, T/2]$, where T is the period of the waveform in seconds (s).

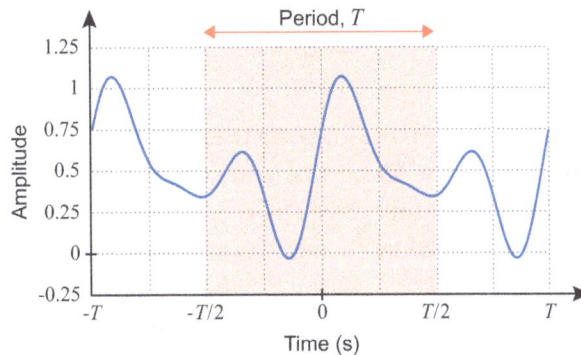

Figure 5.8: *A waveform that is periodic over time.*

We can compute the Fourier Series over an interval of time by ensuring the sine and cosine waves, which expand a periodic function, are still periodic over the interval $[-T/2, T/2]$. Begin by modifying (5.3) to express the Fourier Series expansion over any time interval by substituting in $x = 2\pi t/T$ as below,

$$g(t) = a_0 + \sum_{n=1}^{\infty} a_n \cos\left(\frac{2\pi nt}{T}\right) + \sum_{n=1}^{\infty} b_n \sin\left(\frac{2\pi nt}{T}\right), \qquad (5.10)$$

where t is the time in the range $[-T/2, T/2]$. We can rewrite the Fourier Series coefficients as,

$$a_0 = \frac{1}{T} \int\limits_{-T/2}^{T/2} g(t)dt, \tag{5.11}$$

$$a_n = \frac{2}{T} \int\limits_{-T/2}^{T/2} g(t)\cos\left(\frac{2\pi nt}{T}\right)dt, \qquad n = 1, 2, 3, \ldots, \tag{5.12}$$

$$b_n = \frac{2}{T} \int\limits_{-T/2}^{T/2} g(t)\sin\left(\frac{2\pi nt}{T}\right)dt, \qquad n = 1, 2, 3, \ldots. \tag{5.13}$$

Now that we are expressing the Fourier Series across an interval of time, we can make several observations about the frequency content of a Fourier Series expansion. We can use the waveform given in Figure 5.8 as an example. The DC bias and the first three sine and cosine terms for this waveform are plotted in Figure 5.9.

Notice that the first sine and cosine waves have the same period T as the original waveform. The period of the subsequent terms are shorter, i.e. $T/2$ and $T/3$. We can express the waveforms in the Fourier Series expansion in terms of the fundamental frequency f_0 of the original periodic waveform as,

$$f_0 = \frac{1}{T}. \tag{5.14}$$

We can deduce that the sine and cosine waves that make up the expansion are a positive integer multiple of the original waveform's fundamental frequency. These sine and cosine waves are known as harmonics and are present at frequencies nf_0. For instance, the first three terms of a Fourier Series expansion may consist of sine and cosine waves at multiples of the fundamental frequency, i.e. f_0, $2f_0$ and $3f_0$.

The Fourier Series can expand a periodic function into harmonically related sine and cosine waves of the fundamental frequency. As we will explore later in Section 5.2, the Fourier Series cannot be used to derive the frequency content of an aperiodic waveform. The Fourier Transform is normally applied to continuous aperiodic waveforms to reveal their frequency content.

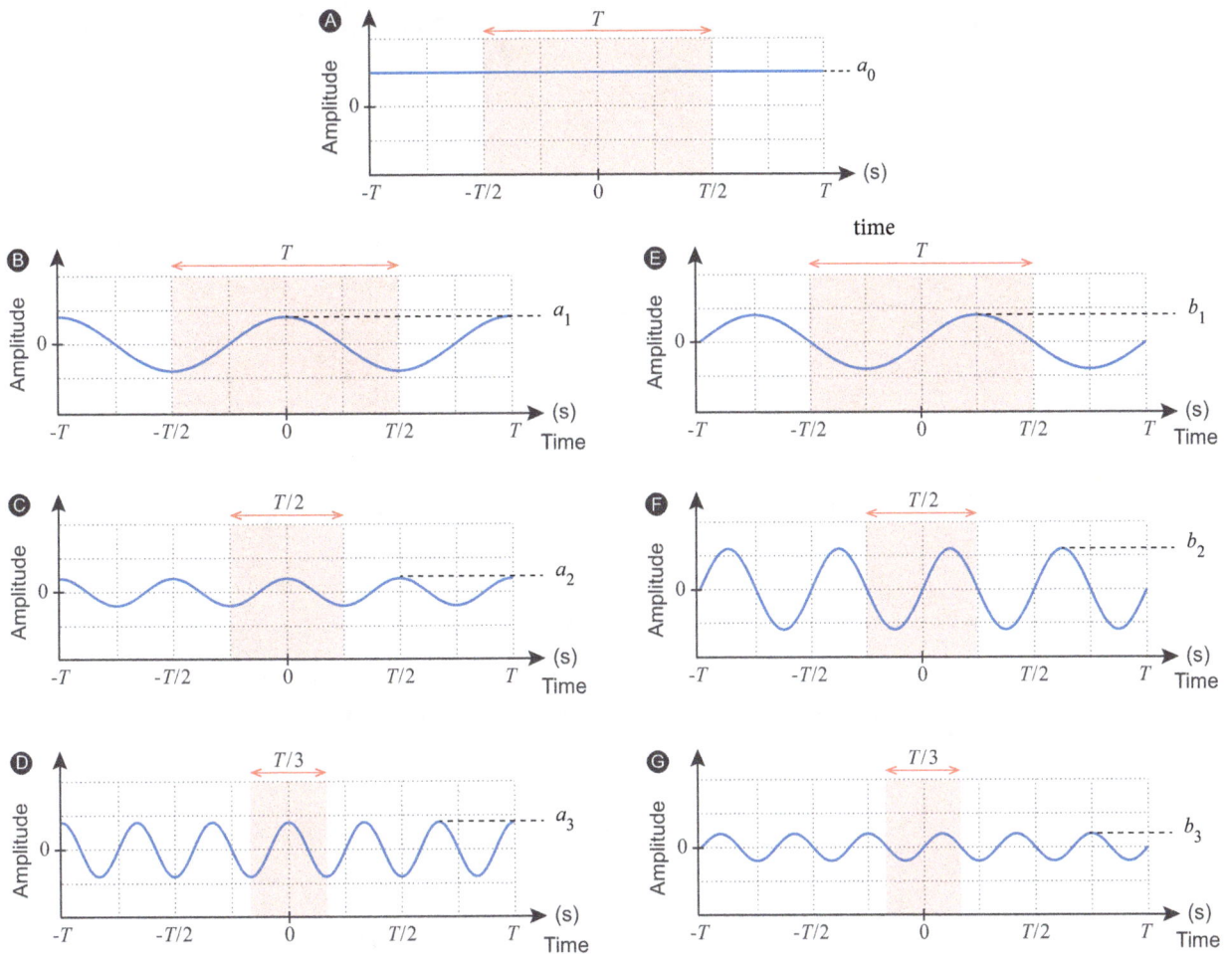

Figure 5.9: The plots above illustrate the DC bias and the first three terms in the Fourier Series expansion for the waveform given in Figure 5.8. Plot A contains the DC bias, plots B to D show the first three cosine waves, and plots E to G contain the first three sine waves.

5.1.4. Euler's Formula

We will soon investigate a complex definition of the Fourier Series in Section 5.1.5. Before progressing further, we should review Euler's formula, which is widely known across many science, engineering, and mathematical disciplines. Euler's formula describes the relationship between a complex exponential function and associated trigonometric functions. It can be expressed as

$$e^{j\theta} = cos(\theta) + j\,sin(\theta), \tag{5.15}$$

where e is the base of the natural logarithms (also known as Euler's constant and is equal to approximately 2.718...), θ is an angle, and $j = \sqrt{-1}$. We can plot Euler's formula on the complex plane by cycling through values of θ in the range $[0, 2\pi]$, as shown in Figure 5.10.

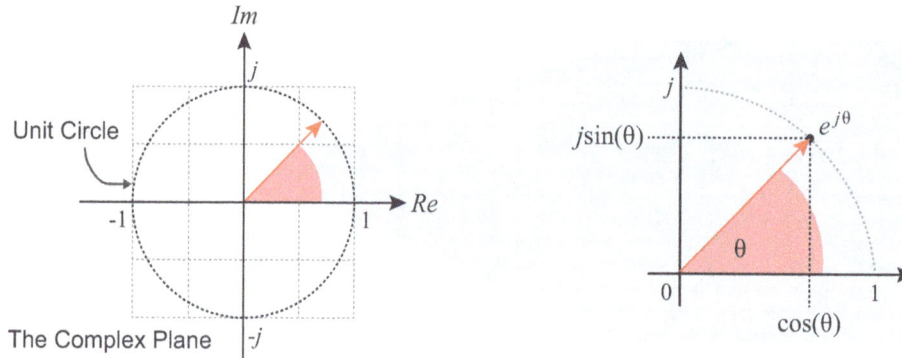

Figure 5.10: Sketch of Euler's formula showing the unit circle and the relationship between trigonometric functions and the complex exponential.

The plot of Euler's formula has produced a circle of radius 1 on the complex plane, which is commonly known as the unit circle. The real and imaginary co-ordinates are equal to $cos(\theta)$ and $sin(\theta)$, respectively. We can easily compute a point on the complex plane for any angle of θ. For example, when $\theta = \pi/2$, we obtain,

$$e^{j(\pi/2)} = cos(\pi/2) + j sin(\pi/2) = 0 + j \cdot 1. \tag{5.16}$$

If the complex exponential is negative, $e^{-j\theta}$, the equation can be expressed as

$$e^{-j\theta} = cos(\theta) - j sin(\theta), \tag{5.17}$$

since $cos(-\theta) = cos(\theta)$ and $sin(-\theta) = -sin(\theta)$. Equation (5.17) is the complex conjugate of (5.15), which is useful to remember.

Although Euler's formula may appear intimidating, it is a very valuable identity. Complex exponentials can reduce the difficulty of performing mathematical operations on complex numbers. For instance, multiplying exponentials can be as simple as summing their powers,

$$e^{(A+B)} = e^A \cdot e^B. \tag{5.18}$$

We can represent sine and cosine functions in terms of positive and negative complex exponentials. Let us begin by defining the equations for the real (Re) and imaginary (Im) parts of a complex number denoted as c, where $\bar{c}$ is the complex conjugate.

$$\text{Re}(c) = \frac{c + \bar{c}}{2} \tag{5.19}$$

$$\text{Im}(c) = \frac{c - \bar{c}}{2j} \tag{5.20}$$

Then, substitute $c = e^{j\theta}$ and $\bar{c} = e^{-j\theta}$ into (5.19) to produce an expression for $cos(\theta)$.

$$cos(\theta) = \frac{e^{j\theta} + e^{-j\theta}}{2} \tag{5.21}$$

Repeat the substitution with (5.20) to obtain an expression for $sin(\theta)$.

$$sin(\theta) = \frac{e^{j\theta} - e^{-j\theta}}{2j} \tag{5.22}$$

It is worth mentioning that both trigonometric and complex exponential notations are valid representations of complex numbers. We will use both notations throughout this chapter as required.

5.1.5. The Complex Fourier Series

Up to this point, we have obtained Fourier Series expansions of real valued waveforms using trigonometric terms. However, it is useful to represent the Fourier Series in terms of complex exponentials. This technique is known as the Complex Fourier Series expansion, which allows the Fourier series of a complex waveform to be calculated i.e. a waveform with both real and imaginary terms. In this section, we will derive the Complex Fourier Series, which will prove useful later when investigating the Fourier Transform.

Begin by substituting the identities in (5.21) and (5.22) into (5.3). Then rearrange the expression as,

$$
\begin{aligned}
g(x) &= a_0 + \sum_{n=1}^{\infty} \left[a_n \left(\frac{e^{jnx} + e^{-jnx}}{2} \right) + b_n \left(\frac{e^{jnx} - e^{-jnx}}{2j} \right) \right] \\
&= a_0 + \sum_{n=1}^{\infty} \left[\frac{a_n e^{jnx}}{2} + \frac{a_n e^{-jnx}}{2} + \frac{b_n e^{jnx}}{2j} - \frac{b_n e^{-jnx}}{2j} \right] \\
&= a_0 + \sum_{n=1}^{\infty} \left[\left(\frac{a_n}{2} + \frac{b_n}{2j} \right) e^{jnx} + \left(\frac{a_n}{2} - \frac{b_n}{2j} \right) e^{-jnx} \right].
\end{aligned}
\tag{5.23}
$$

We can use $1/j = -j$ to simplify the expression further.

$$g(x) = a_0 + \sum_{n=1}^{\infty} \left(\frac{a_n - jb_n}{2}\right) e^{jnx} + \sum_{n=1}^{\infty} \left(\frac{a_n + jb_n}{2}\right) e^{-jnx} \tag{5.24}$$

Now, reduce the expression by simply reversing the limits of the second summation so that n operates over the range $[-\infty, -1]$. Each summation now contains the same complex exponential, e^{jnx}.

$$g(x) = a_0 + \sum_{n=1}^{\infty} \left(\frac{a_n - jb_n}{2}\right) e^{jnx} + \sum_{n=-\infty}^{-1} \left(\frac{a_{-n} + jb_{-n}}{2}\right) e^{jnx} \tag{5.25}$$

Let us set a new variable c_n in terms of the Fourier Series coefficients as,

$$c_n = \begin{cases} \dfrac{a_{|n|} + jb_{|n|}}{2}, & \text{for } n < 0, \\ a_0, & \text{for } n = 0, \\ \dfrac{a_n - jb_n}{2}, & \text{for } n > 0. \end{cases} \tag{5.26}$$

Substituting c_n into (5.25) we obtain the following.

$$g(x) = c_0 + \sum_{n=1}^{\infty} c_n e^{jnx} + \sum_{n=-\infty}^{-1} c_n e^{jnx} \tag{5.27}$$

Finally, we can rewrite (5.27) to reveal a definition for the Complex Fourier Series over $n = [-\infty, \infty]$.

$$g(x) = \sum_{n=-\infty}^{\infty} c_n e^{jnx}, \qquad n = 0, \pm1, \pm2, \ldots \tag{5.28}$$

The variable, c_n, is a complex number that contains the original Fourier Coefficients a_0, a_n, and b_n. We can rearrange (5.28) to solve for c_n, which is provided in Appendix A. The expression for c_n over $[-\pi, \pi]$ is,

$$c_n = \frac{1}{2\pi} \int_{-\pi}^{\pi} g(x) e^{-jnx} dx, \qquad n = 0, \pm1, \pm2, \ldots . \tag{5.29}$$

If $n = 0$, (5.29) reduces to the expression for a_0 given in (5.4). As above for the trigonometric Fourier series, we can generalise the Complex Fourier Series so it operates over any interval in time. Substitute $x = 2\pi t/T$ into (5.28) to produce,

$$g(t) = \sum_{n=-\infty}^{\infty} c_n e^{j2\pi nt/T}, \qquad n = 0, \pm1, \pm2, \dots . \tag{5.30}$$

We can now create a new expression for c_n over the interval $[-T/2, T/2]$ as follows.

$$c_n = \frac{1}{T} \int_{-T/2}^{T/2} g(t) e^{-j2\pi nt/T} dt, \qquad n = 0, \pm1, \pm2, \dots \tag{5.31}$$

Conventionally, we normally write the complex exponential using the fundamental frequency of the periodic waveform, where $f_0 = 1/T$. The final expression for c_n becomes,

$$c_n = \frac{1}{T} \int_{-T/2}^{T/2} g(t) e^{-j2\pi f_0 nt} dt, \qquad n = 0, \pm1, \pm2, \dots . \tag{5.32}$$

5.1.6. Positive and Negative Frequencies

Notice that the Complex Fourier Series in (5.30) operates on negative values of n, which means our Fourier Series expansion will contain negative harmonics of the fundamental frequency. These negative frequencies must be addressed in a similar way that we approach positive frequencies. The most effective way to discuss both positive and negative frequencies is through an example that uses the Complex Fourier Series.

Let us begin by considering a waveform $g(t)$ that is composed of three cosine waves. We will represent each cosine wave using the notation $A\cos(2\pi ft)$, where A is the amplitude, f is the frequency, and t is an interval of time. The waveform is defined as,

$$g(t) = 8\cos(2\pi 100t) + 3\cos(2\pi 200t) + 2\cos(2\pi 300t). \tag{5.33}$$

For simplicity, the frequency of the cosine waves are 100Hz, 200Hz, and 300Hz and they each have different amplitudes to make our analysis interesting. The time domain plot of this waveform is given in Figure 5.11.

Using the relationship given in (5.21), we can rewrite $g(t)$ as a sum of complex exponentials.

$$g(t) = 8\left(\frac{e^{j2\pi 100t} + e^{-j2\pi 100t}}{2}\right) + 3\left(\frac{e^{j2\pi 200t} + e^{-j2\pi 200t}}{2}\right) + 2\left(\frac{e^{j2\pi 300t} + e^{-j2\pi 300t}}{2}\right) \tag{5.34}$$

We can expand and rewrite the expression above to separate the positive and negative complex exponentials into their own groups.

$$g(t) = 4e^{j2\pi 100t} + \frac{3}{2}e^{j2\pi 200t} + e^{j2\pi 300t} + 4e^{-j2\pi 100t} + \frac{3}{2}e^{-j2\pi 200t} + e^{-j2\pi 300t} \tag{5.35}$$

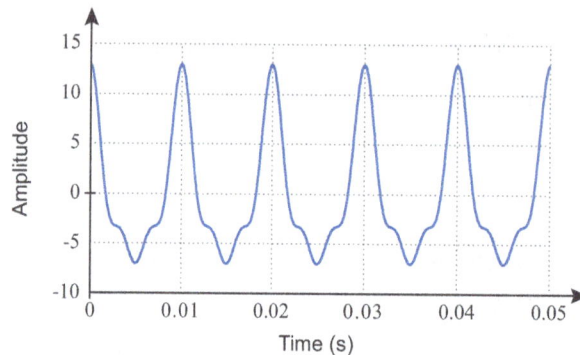

Figure 5.11: *A waveform created by summing three cosine waves together as defined in (5.33).*

As you can see, the first three terms of (5.35) are positive complex exponentials and the last three terms are negative complex exponentials. We can plot the amplitude of each complex exponential against its frequency, as shown in Figure 5.12.

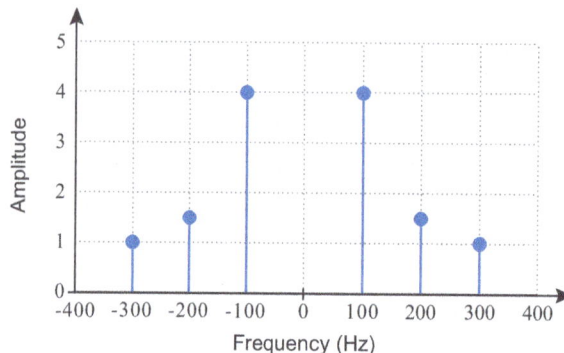

Figure 5.12: *Complex frequency spectrum showing negative and positive frequency components.*

The plot given above is known as a complex frequency spectrum. You can clearly see that each cosine wave has its own equal contribution of positive and negative frequency components (this statement is true for all real waveforms).

In essence, we can express a cosine wave as being a sum of a positive-frequency complex exponential, and a negative-frequency complex exponential. We can sketch an illustration of this sum for inspection as given in Figure 5.13. Notice that the positive-frequency complex exponential is a helix rotating on the complex plane over time in the anti-clockwise direction, while the negative-frequency complex exponential rotates in the clockwise direction over time.

Each helix is characterised as having a peak amplitude of 0.5. When we sum each helix together, as above, we produce the original cosine wave. The main point of this discussion is to recognise that all real waveforms have a positive and negative frequency component. In contrast, a complex waveform defined as $Ae^{j2\pi ft}$ only has

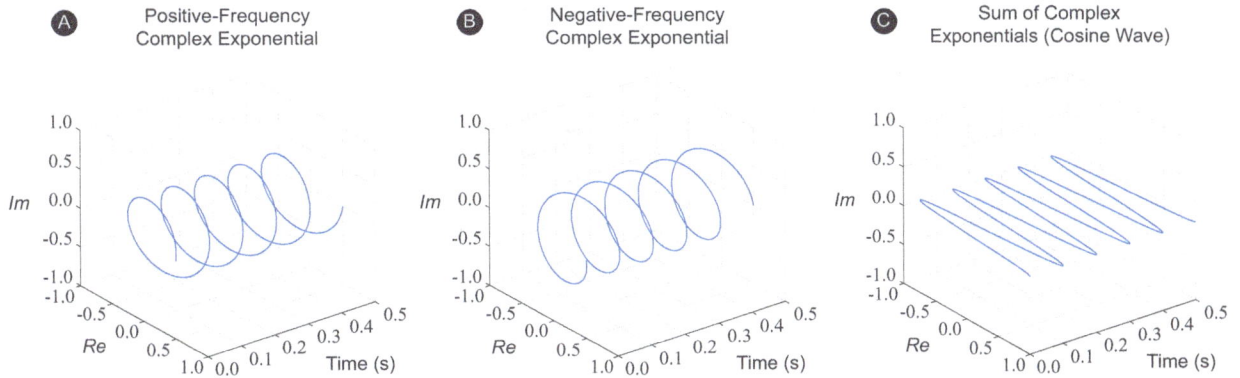

Figure 5.13: Plot A contains a positive-frequency complex exponential rotating anti-clockwise over time. Plot B contains a negative-frequency complex exponential rotating clockwise over time. Plot C is the sum of each complex exponential, resulting in a cosine wave. Notice the cosine wave only oscillates over the real axis. The imaginary axis is always zero.

one frequency component, which can be either positive or negative. We will explore positive and negative frequencies again when investigating the Discrete Fourier Transform in Section 5.3.

5.2. Understanding The Fourier Transform

Until this point, we have investigated the Fourier Series expansion, which is advantageous when obtaining the frequency content of a periodic wave. However, most practical waveforms are aperiodic and are unable to expand into a Fourier Series (without significant caveats). How do we interpret the frequency content of a waveform that is aperiodic? Of course, the answer is the Fourier Transform, which is an extension of the Fourier Series for aperiodic waveforms. The Fourier Transform converts aperiodic waveforms between the continuous time and frequency domains and is one of the most well known tools in spectral analysis.

This section will explore the Fourier Transform by diving into its operation. Before proceeding, ensure that you have read and understood the Complex Fourier Series in the previous section, as we will use this to derive an expression for the Fourier Transform.

5.2.1. Continuous Frequencies

Previously in Section 5.1.3, the Fourier Series of a periodic waveform contained harmonics of the fundamental frequency, f_0. We can plot the harmonics of a time domain waveform using a frequency magnitude plot as shown on the right of Figure 5.14. Notice that the harmonics are evenly spaced across multiples of f_0.

Real world waveforms are never truly periodic as they contain transient and random components pertaining to information content. Aperiodic waveforms cannot be expanded into a Fourier Series as it is not possible to

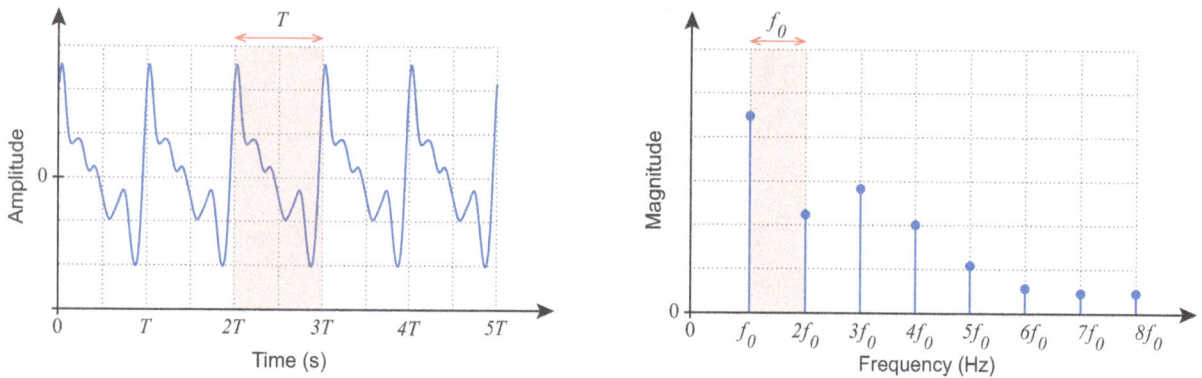

Figure 5.14: A continuous time waveform (left), a frequency magnitude plot of Fourier harmonics (right).

obtain the period of the waveform (since it is not actually periodic). However, in order use the Fourier series, we could assume the waveform repeats after an 'infinite' time.

If we assume that the period of the waveform increases such that $T \to \infty$, then the fundamental frequency and the spacing between the associated harmonics tends to zero. To observe the effect of the period tending to infinity, $T \to \infty$, we can analyse the effects of increasing the size of the period T. Figure 5.15 presents the frequency magnitude plots for a square pulse using increasing pseudo-periods. Notice that as the pseudo-period increases, the number of harmonics also increase (causing the space between harmonics to become smaller) and the amplitude of the harmonics decreases, since the total energy in the time domain and Fourier Series representations must remain the same.

The final example in Figure 5.15 (labelled D) depicts a square wave with a period tending to infinity, $T \to \infty$. The resulting frequency magnitude response is one of continuous frequency i.e. as $T \to \infty$, then $f_0 \to 0$. The outcome of the Fourier series with $T \to \infty$ is the Fourier Transform and the frequency spacing between the harmonics is a continuous frequency spectrum. You may have noticed that as $T \to \infty$, the magnitude response of the harmonics decrease given the total energy in the waveform remains the same in both time and frequency representations. The magnitude of the harmonics are of the order $1/T$, which requires scaling of the y-axis by T if we are to practically plot and inspect the frequency magnitude response. Hence, in evolving the Fourier Series equation to the Fourier Transform, we will require to scale the Fourier Series coefficients by a factor of T.

5.2.2. The Fourier Transform

In this section, we will derive the equation for the Fourier Transform. We begin with the Complex Fourier Series expression for c_n given in (5.31), and scaling both sides by T and rearranged as

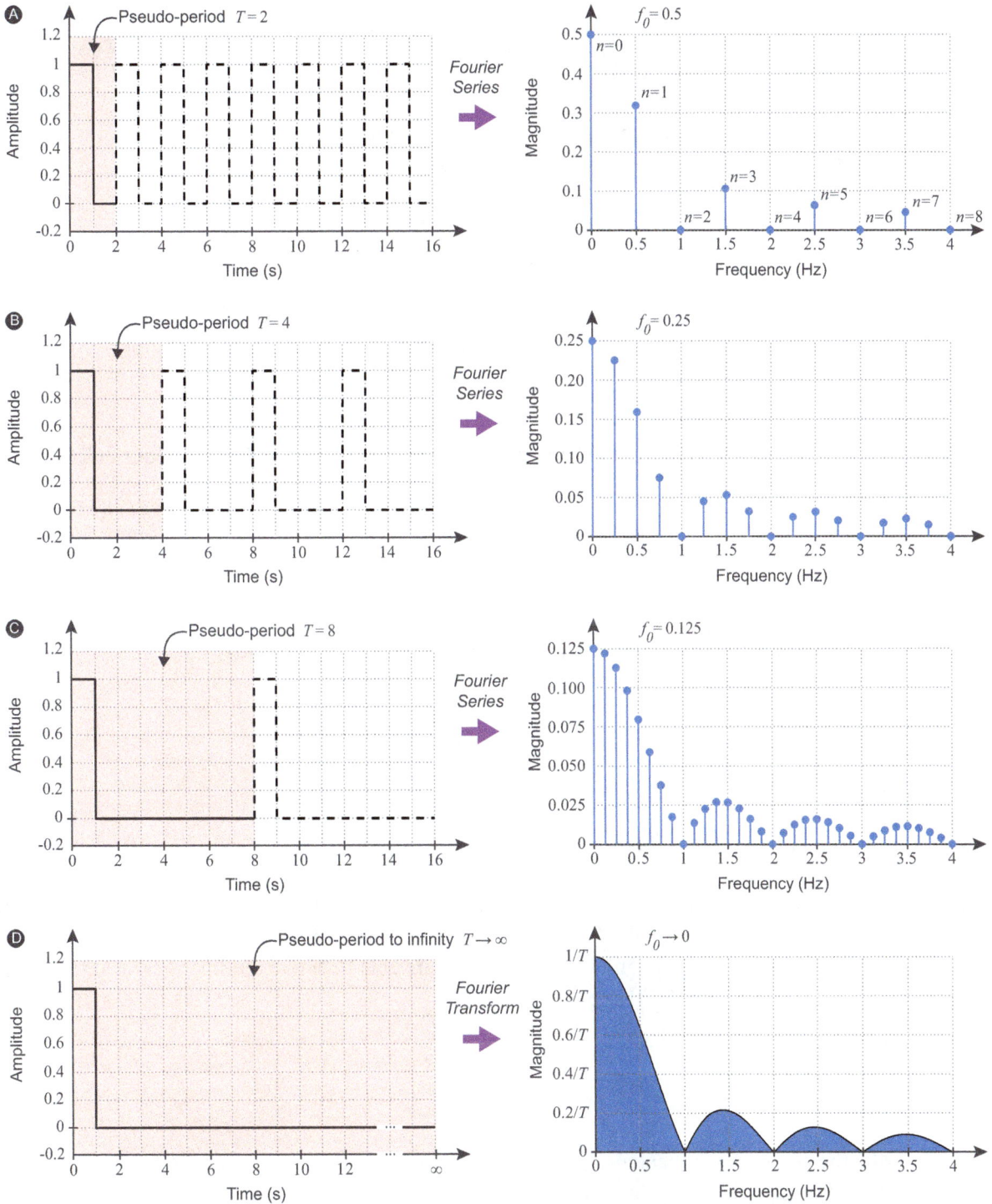

Figure 5.15: *Four square waves labelled A, B, C, and D that use pseudo-periods of 2, 4, 8, and infinity. Each square wave has a corresponding magnitude plot showing the frequency response.*

$$Tc_n = \int_{-T/2}^{T/2} g(t)e^{-j2\pi f_0 nt}\,dt, \qquad n = 0, \pm 1, \pm 2, \ldots . \tag{5.36}$$

As $T \to \infty$, the discrete Fourier spectrum $f_0 n$ tends to a continuous variable, f, as $f_0 \to 0$ and $n \to \infty$. The interval of integration is now between $[-\infty, \infty]$. Our expression becomes,

$$Tc_n = \lim_{T \to \infty} \int_{-T/2}^{T/2} g(t)e^{-j2\pi f_0 nt}\,dt \tag{5.37}$$

$$= \int_{-\infty}^{\infty} g(t)e^{-j2\pi ft}\,dt.$$

We now replace Tc_n with $G(f)$ to realise the Fourier Transform (also called the Forward Fourier Transform),

$$G(f) = \int_{-\infty}^{\infty} g(t)e^{-j2\pi ft}\,dt . \tag{5.38}$$

Note that many engineers and physicists commonly write the Fourier Transform in terms of angular frequency, which is defined as $\omega = 2\pi f$ and given below.

$$G(\omega) = \int_{-\infty}^{\infty} g(t)e^{-j\omega t}\,dt \tag{5.39}$$

5.2.3. The Inverse Fourier Transform

It is possible to convert a continuous frequency spectrum (obtained using the Forward Fourier Transform) back to a continuous time waveform. The Inverse Fourier Transform is a tool that transforms a function of continuous frequency, $G(f)$, to a function of time, $g(t)$. We are able to derive the Inverse Fourier Transform by substituting $c_n = G(f)/T$ into (5.30) as below.

$$g(t) = \sum_{n=-\infty}^{\infty} \frac{1}{T}G(f)e^{j2\pi nt/T} \tag{5.40}$$

We can describe the complex exponential in terms of the fundamental frequency of the periodic waveform, where $f_0 = 1/T$. The expression now becomes,

$$g(t) = \sum_{n=-\infty}^{\infty} \frac{1}{T} G(f) e^{j2\pi n f_0}. \qquad (5.41)$$

As before with the Forward Fourier Transform, the period $T \to \infty$. Therefore, $f_0 n$ tends to a continuous variable, f, as $f_0 \to 0$ and $n \to \infty$. We can also replace the term $1/T$ with Δf as follows.

$$g(t) = \sum_{n=-\infty}^{\infty} G(f) e^{j2\pi f t} \Delta f \qquad (5.42)$$

Finally, we use a special mathematical relationship to swap the above summation to an integral. This relationship is known as a Reimann sum, which allows us to approximate an integral given a summation as $\Delta f \to 0$. The Inverse Fourier Transform can then be expressed as,

$$g(t) = \int_{-\infty}^{\infty} G(f) e^{j2\pi f t} df. \qquad (5.43)$$

We can also compute the Inverse Fourier Transform in terms of angular frequency ω, as shown in (5.44). Note that a scaling factor of 2π is required when integrating with respect to the angular frequency.

$$g(t) = \frac{1}{2\pi} \int_{-\infty}^{\infty} G(\omega) e^{j\omega t} d\omega \qquad (5.44)$$

This concludes our investigation into the Fourier Series and Fourier Transform. These techniques are for manipulating continuous waveforms. The next section will introduce the Discrete Fourier Transform, which is a version of the Fourier Transform that yields the discrete frequency spectrum of a sampled waveform, which is of key relevance to the RFSoC and operates on discrete time signals.

5.3. The Discrete Fourier Transform

We seen previously in Chapter 4 that a continuous waveform can be periodically sampled when its time axis is divided into discrete points, which are separated by a sampling period t_s. When working with digital systems or devices, such as an FPGA, you will likely be performing arithmetic operations on sampled (digital) waveforms. The Discrete Fourier Transform (DFT) is a special version of the Fourier Transform that will yield a discrete frequency spectrum of a sampled waveform.

In this section, we will begin by deriving the DFT and then work through an example to demonstrate its uses. We will also explore various properties of the DFT including its frequency domain symmetry and periodicity. Finally, we will investigate spectral leakage, windowing, scalloping loss, and zero padding.

5.3.1. The DFT Equation

The DFT will operate on a digital waveform that has been discretely sampled by a regular sampling period t_s. Take a moment to inspect the digital waveform that is illustrated in Figure 5.16. We denote digital waveforms with the variable $x(n)$, where n is used to index a discrete sample. The variable N, shown in the sketch, indicates the number of samples in $x(n)$, which is $N = 10$.

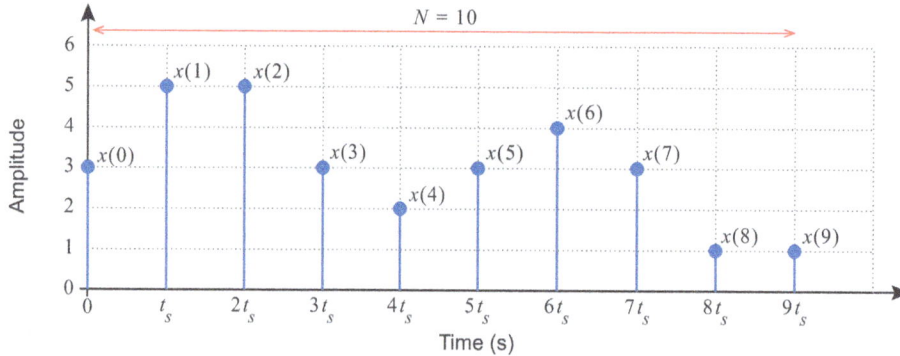

Figure 5.16: A digital waveform that has been discretely sampled by a regular sampling period.

We can obtain the time index of any discrete sample in $x(n)$ by simply computing nt_s. Alternatively, we can express this term using the sample frequency f_s of $x(n)$, such that the time index is obtained using n/f_s.

We can derive the DFT by starting with the equation for the Fourier Transform in (5.38) and substitute the continuous variable t for the discrete term n/f_s. Additionally, we also exchange the continuous waveform $g(t)$ for our discretely sampled waveform $x(n)$. Lastly, we rename $G(f)$ as $X(f)$ so that our mathematical notation remains consistent.

$$X(f) = \sum_{n=-\infty}^{\infty} x(n)e^{-j2\pi fn/f_s} \tag{5.45}$$

The discrete waveform $x(n)$ must obey the laws of causality, where the amplitude of its data points depend on past and current inputs only (and not future inputs). Therefore, assuming our discrete waveform is causal then the lower limit of the summation can be changed so $n = 0$, which is the first sample point of $x(n)$. Additionally, it is highly unlikely that there are an infinite number of data samples in our discrete waveform. Therefore, we can set the upper limit of the summation to $N-1$. The expression in (5.45) now becomes,

$$X(f) = \sum_{n=0}^{N-1} x(n)e^{-j2\pi fn/f_s}. \tag{5.46}$$

We are getting very close to the final expression for the DFT, with only one step remaining. By using a finite number of data samples, N, we are no longer evaluating a waveform over an infinite period of time, which is required by the continuous Fourier Transform. Instead, we evaluate $x(n)$ with the implicit assumption that the waveform contains pseudo-periods every N samples, or Nt_s seconds.

Assuming $x(n)$ is pseudo-periodic, we can reduce the continuous frequency f to a specific set of discrete frequencies. We can evaluate (5.46) across integer multiples of the fundamental frequency, f_0, which is commonly referred to as the frequency resolution Δf. The frequency resolution is calculated using

$$\Delta f = \frac{1}{Nt_s} = \frac{f_s}{N}. \tag{5.47}$$

To reduce the expression in (5.46), we can introduce a discrete variable k that is used to index the discrete frequencies. The variable k operates in the range $[0, N-1]$. We can exchange the continuous frequency f in (5.46) for the discrete frequencies kf_s/N. The final equation for the DFT becomes,

$$X(k) = \sum_{n=0}^{N-1} x(n)e^{-2\pi kn/N}. \tag{5.48}$$

It is common to think about the operation of the DFT defined above as a Discrete Complex Fourier Series, given that the periodicity of the signal was over the window length of N samples. In the next section, we will work through an example of applying the DFT to a discrete waveform.

5.3.2. A Closer Look at the DFT

It can be easier to understand the DFT equation when we use its rectangular form, which can be created by separating the complex exponential of (5.48) into its real and imaginary components using Euler's formula. We can rewrite the DFT equation as,

$$X(k) = \sum_{n=0}^{N-1} x(n)[cos(2\pi kn/N) + jsin(2\pi kn/N)]. \tag{5.49}$$

We can apply the DFT to an input waveform $x(n)$ to reveal its frequency spectrum. Our input waveform will be composed of two frequency components at 1200Hz and 2400Hz. These frequency components can be created using sine waves, which are added together to create the input waveform as follows.

$$x(n) = sin(2\pi 1200 nt_s + \pi/4) + 0.5 sin(2\pi 2400 nt_s) \tag{5.50}$$

The waveform is regularly sampled at discrete time intervals nt_s, where the sampling frequency is 9600Hz (sometimes defined as 9600 samples/second). Figure 5.17 presents a plot of the input waveform and its sinusoidal components for you to inspect.

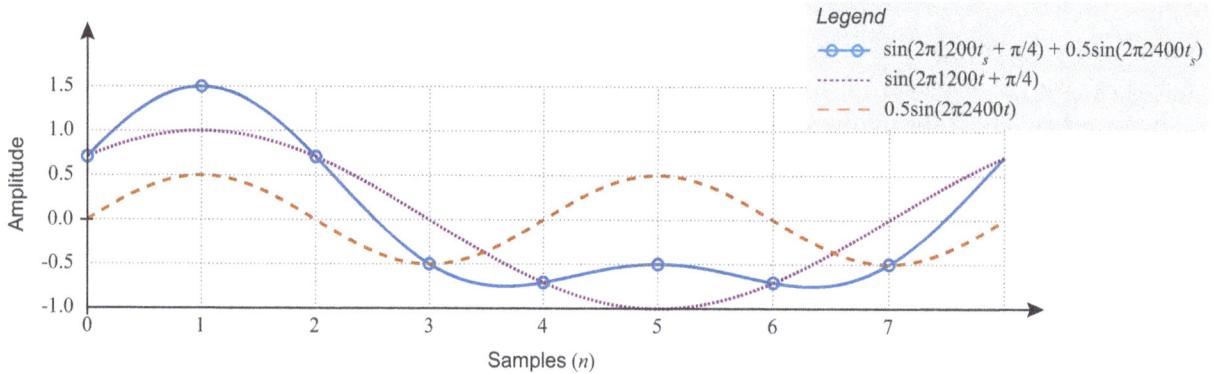

Figure 5.17: Plot of a discretely sampled waveform that was created by summing two sine waves (also shown).

Our input waveform contains exactly 8 samples, which means we will be able to compute an 8-point DFT i.e. the discrete variable $k = 0, 1, 2, ..., 7$. The amplitude of each sample in our waveform is,

$$
\begin{aligned}
x(0) &= 0.7071, \\
x(1) &= 1.5000, \\
x(2) &= 0.7071, \\
x(3) &= -0.5000, \\
x(4) &= -0.7071, \\
x(5) &= -0.5000, \\
x(6) &= -0.7071, \\
x(7) &= -0.5000.
\end{aligned}
\tag{5.51}
$$

The first frequency term of the DFT, when $k = 0$, is a unique case. It is defined as,

$$
X(0) = \sum_{n=0}^{7} x(n)[\cos(2\pi \cdot 0 \cdot n/8) - j\sin(2\pi \cdot 0 \cdot n/8)].
\tag{5.52}
$$

The term $\cos(2\pi \cdot 0 \cdot n/N) - j\sin(2\pi \cdot 0 \cdot n/N)$ reduces to $1 - j0$, which means (5.52) becomes,

$$
X(0) = \sum_{n=0}^{N-1} x(n).
\tag{5.53}
$$

The special relationship in (5.53) evaluates to the sum of the input samples, $x(n)$. When evaluating the DFT using $k = 0$, we are actually computing the frequency component of $x(n)$ that does not vary in time, which is commonly referred to as the DC bias. Our input waveform would exhibit a DC bias if $X(0)$ returned a non-zero value. Using the samples in (5.51) and (5.53), we can demonstrate that our input waveform does not contain a DC bias.

$$
\begin{aligned}
X(0) \; = \; & 0.7071 \cdot 1.0 && -j(0.7071 \cdot 0.0) \\
+ \; & 1.5000 \cdot 1.0 && -j(1.5000 \cdot 0.0) \\
+ \; & 0.7071 \cdot 1.0 && -j(0.7071 \cdot 0.0) \\
- \; & 0.5000 \cdot 1.0 && -j(-0.5000 \cdot 0.0) \\
- \; & 0.7071 \cdot 1.0 && -j(-0.7071 \cdot 0.0) \\
- \; & 0.5000 \cdot 1.0 && -j(-0.5000 \cdot 0.0) \\
- \; & 0.7071 \cdot 1.0 && -j(-0.7071 \cdot 0.0) \\
- \; & 0.5000 \cdot 1.0 && -j(-0.5000 \cdot 0.0)
\end{aligned}
$$

$$
\begin{aligned}
= \; & 0.7071 - j0.0 \\
+ \; & 1.5000 - j0.0 \\
+ \; & 0.7071 - j0.0 \\
- \; & 0.5000 - j0.0 \\
- \; & 0.7071 - j0.0 \\
- \; & 0.5000 - j0.0 \\
- \; & 0.7071 - j0.0 \\
- \; & 0.5000 - j0.0
\end{aligned}
$$

$$
= \; 0.0 - j0.0
$$

This result can be verified by summing the amplitude of each sample in $x(n)$. We can write the result given above using magnitude and phase representation, which gives us $0.0\angle 0°$. The magnitude and phase equations for the DFT are covered later in Section 5.3.3.

Since we know the waveform's sampling frequency, we can also compute the frequency resolution of the DFT using (5.47), which gives us,

$$
\Delta f = \frac{f_s}{N} = \frac{9600 \text{ samples/second}}{8 \text{ samples}} = \frac{1200}{\text{second}} = 1200 \text{ Hz}. \tag{5.54}
$$

The value of $X(0)$ corresponds to the DC component of $x(n)$. The next component $X(1)$ evaluates to the complex amplitude of the first frequency component at 1200Hz. Similarly, $X(2)$ and $X(3)$ will equate to the complex amplitude of the harmonics, $2\Delta f$ and $3\Delta f$, respectively. We proceed with computing the 8-point DFT by evaluating the second frequency term, $X(1)$ as,

$$
X(1) = \sum_{n=0}^{7} x(n)[\cos(2\pi \cdot 1 \cdot n/8) - j\sin(2\pi \cdot 1 \cdot n/8)]. \tag{5.55}
$$

Our input waveform contains a sine wave that has a frequency of 1200Hz. We should expect to obtain a non-zero value for $X(1)$.

$$
\begin{aligned}
X(1) = \;& 0.7071 \cdot 1.0 && -j(0.7071 \cdot 0.0) \\
& +\, 1.5000 \cdot 0.7071 && -j(1.5000 \cdot 0.7071) \\
& +\, 0.7071 \cdot 0.0 && -j(0.7071 \cdot 1.0) \\
& -\, 0.5000 \cdot -0.7071 && -j(-0.5000 \cdot 0.7071) \\
& -\, 0.7071 \cdot -1.0 && -j(-0.7071 \cdot 0.0) \\
& -\, 0.5000 \cdot -0.7071 && -j(-0.5000 \cdot -0.7071) \\
& -\, 0.7071 \cdot 0.0 && -j(-0.7071 \cdot -1.0) \\
& -\, 0.5000 \cdot 0.7071 && -j(-0.5000 \cdot -0.7071)
\end{aligned}
$$

$$
\begin{aligned}
= \;& 0.7071 - j0.0 \\
& +\, 1.0607 - j1.0607 \\
& +\, 0.0 - j0.7071 \\
& +\, 0.3536 + j0.3536 \\
& +\, 0.7071 + j0.0 \\
& +\, 0.3536 - j0.3536 \\
& -\, 0.0 - j0.7071 \\
& -\, 0.3536 - j0.3536
\end{aligned}
$$

$$
= 2.8285 - j2.8285 = 4.0\angle -45°
$$

The result given above for $X(1)$ indicates that there is a frequency component in $x(n)$ at a frequency of 1200Hz. Notice that the phase of the frequency component is $-45°$, which is equivalent to $-\pi/4$ in radians. As you can see in (5.50), the 1200Hz sine wave contains a phase shift of $\pi/4$. The phase appears to have changed sign from positive to negative, which we will explain in a moment. For now, we will evaluate $X(2)$ as,

$$
X(2) = \sum_{n=0}^{7} x(n)[\cos(2\pi \cdot 2 \cdot n/8) - j\sin(2\pi \cdot 2 \cdot n/8)]. \tag{5.56}
$$

A 2400Hz sine wave is present in $x(n)$, so the complex amplitude of $X(2)$ should return a non-zero value.

$$
\begin{aligned}
X(2) = \;& 0.7071 \cdot 1.0 && -j(0.7071 \cdot 0.0) \\
& +\, 1.5000 \cdot 0.0 && -j(1.5000 \cdot 1.0) \\
& +\, 0.7071 \cdot -1.0 && -j(0.7071 \cdot 0.0) \\
& -\, 0.5000 \cdot 0.0 && -j(-0.5000 \cdot -1.0) \\
& -\, 0.7071 \cdot 1.0 && -j(-0.7071 \cdot 0.0) \\
& -\, 0.5000 \cdot 0.0 && -j(-0.5000 \cdot 1.0) \\
& -\, 0.7071 \cdot -1.0 && -j(-0.7071 \cdot 0.0) \\
& -\, 0.5000 \cdot 0.0 && -j(-0.5000 \cdot -1.0)
\end{aligned}
$$

$$
\begin{aligned}
= \;& 0.7071 - j0.0 \\
& +\, 0.0 - j1.5000 \\
& -\, 0.7071 - j0.0 \\
& -\, 0.0 - j0.5000 \\
& -\, 0.7071 + j0.0 \\
& -\, 0.0 + j0.5000 \\
& +\, 0.7071 + j0.0 \\
& -\, 0.0 - j0.5000
\end{aligned}
$$

$$
= 0.0 - j2.0 = 2.0\angle -90°
$$

The magnitude and phase of $X(2)$ has evaluated to a non-zero value, meaning there is a frequency component in $x(n)$ at $2\Delta f$, or 2400Hz. The phase of the frequency component is $-90°$, which is equivalent to $-\pi/2$ using radians. This result may appear a little strange as the 2400Hz sine wave in (5.50) does not have a phase shift. We can explain these results very easily by recognising that the phase returned by the DFT is relative to the phase of a cosine wave. The phase of $X(2)$ is correct since $cos(\theta - \pi/2) = sin(\theta)$, where θ is an angle in radians. This relationship is proven below.

$$\begin{aligned} cos(\theta - \pi/2) &= cos(\theta)cos(\pi/2) + sin(\theta)sin(\pi/2) \\ &= cos(\theta) \cdot 0 + sin(\theta) \cdot 1 \\ &= sin(\theta) \end{aligned} \qquad (5.57)$$

This relationship can also be used to explain the phase result for $X(1)$ i.e. $cos(-\pi/4) = sin(\pi/4)$.

The third frequency term of the DFT, $X(3)$, should evaluate to zero as $x(n)$ does not have a frequency component at $3\Delta f$. We begin with the expression for $X(3)$ as given below.

$$X(3) = \sum_{n=0}^{7} x(n)[cos(2\pi \cdot 3 \cdot n/8) - j sin(2\pi \cdot 3 \cdot n/8)] \qquad (5.58)$$

Now, we compute $X(3)$ as follows.

$$\begin{aligned} X(3) = \quad & 0.7071 \cdot 1.0 && -j(0.7071 \cdot 0.0) \\ &+ 1.5000 \cdot -0.7071 && -j(1.5000 \cdot 0.7071) \\ &+ 0.7071 \cdot 0.0 && -j(0.7071 \cdot -1.0) \\ &- 0.5000 \cdot 0.7071 && -j(-0.5000 \cdot 0.7071) \\ &- 0.7071 \cdot -1.0 && -j(-0.7071 \cdot 0.0) \\ &- 0.5000 \cdot 0.7071 && -j(-0.5000 \cdot -0.7071) \\ &- 0.7071 \cdot 0.0 && -j(-0.7071 \cdot 1.0) \\ &- 0.5000 \cdot -0.7071 && -j(-0.5000 \cdot -0.7071) \end{aligned}$$

$$\begin{aligned} = \quad & 0.7071 - j0.0 \\ &- 1.0607 - j1.0607 \\ &+ 0.0 + j0.7071 \\ &- 0.3536 + j0.3536 \\ &+ 0.7071 + j0.0 \\ &- 0.3536 - j0.3536 \\ &- 0.0 + j0.7071 \\ &+ 0.3536 - j0.3536 \end{aligned}$$

$$= 0.0 - j0.0 = 0.0 \angle 0°$$

This result shows that $x(n)$ does not contain a frequency component at $3\Delta f$, or 3600Hz.

Rather than calculate each frequency component of $X(k)$, we will simply list the remaining results in (5.59). As an exercise, you should try to compute these frequency components on your own. You will be able to see that the results we obtained for $X(1)$ and $X(2)$ will re-emerge (with the phase values negated).

$$\begin{aligned}
X(0) &= 0.0 - j0.0 &&= 0.0\angle 0° \\
X(1) &= 2.8285 - j2.8285 &&= 4.0\angle -45° \\
X(2) &= 0.0 - j2.0 &&= 2.0\angle -90° \\
X(3) &= 0.0 - j0.0 &&= 0.0\angle 0° \\
X(4) &= 0.0 - j0.0 &&= 0.0\angle 0° \\
X(5) &= 0.0 - j0.0 &&= 0.0\angle 0° \\
X(6) &= 0.0 + j2.0 &&= 2.0\angle 90° \\
X(7) &= 2.8285 + j2.8285 &&= 4.0\angle 45°
\end{aligned} \tag{5.59}$$

We have successfully computed the DFT of our input waveform $x(n)$. In the following sections, we will explore the operation of the DFT in more detail and describe techniques to plot and manipulate the discrete frequency waveform $X(k)$.

5.3.3. Magnitude and Phase

We often plot the magnitude and phase spectra of $X(k)$ to illustrate its frequency content. You can visualise the relationship between a complex number in $X(k)$ and its associated magnitude and phase in Figure 5.18.

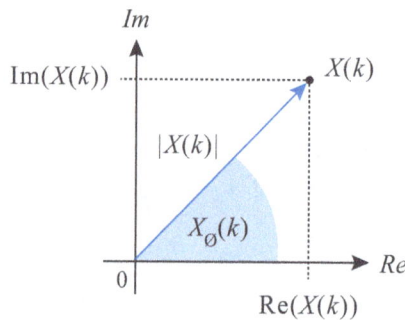

Figure 5.18: Relationship between a complex number and its magnitude and phase properties.

We can compute the magnitude of a complex number by initially extracting the real and imaginary coefficients using (5.19) and (5.20), respectively. The magnitude $|X(k)|$ is then obtained using Pythagoras theorem.

$$|X(k)| = \sqrt{\text{Re}(X(k))^2 + \text{Im}(X(k))^2} \tag{5.60}$$

The phase spectra of $X(k)$, denoted as $X_\varnothing(k)$, is calculated using,

$$X_\varnothing(k) = \tan^{-1}\left(\frac{\text{Im}(X(k))}{\text{Re}(X(k))}\right). \tag{5.61}$$

We have already used the above definitions to compute the magnitude and phase spectra in our example of the DFT in Section 5.3.2. We will proceed by plotting the magnitude and phase spectra against frequency to reveal interesting symmetric and scaling properties of the DFT. We will also plot the real and imaginary parts of $X(k)$ in a similar way for analysis. Each plot is available for inspection in Figure 5.19.

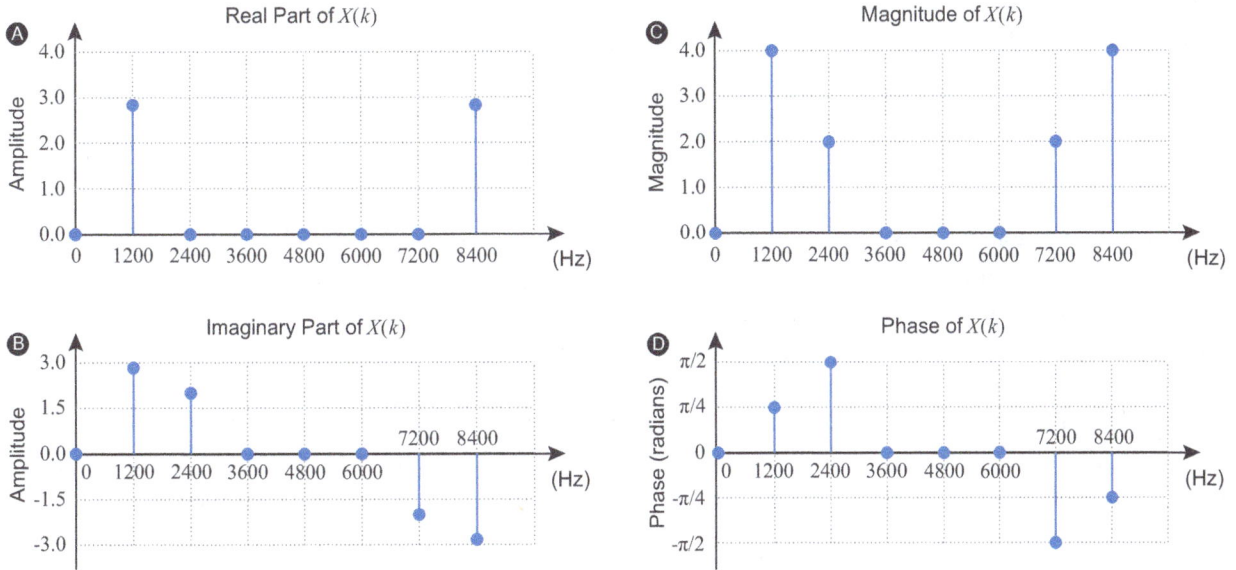

Figure 5.19: Stem plots showing the DFT results from Section 5.3.2: (A) the real part of X(k), (B) the imaginary part of X(k), (C) the magnitude of X(k), and (D) the phase of X(k).

The magnitude plot shows that there is a frequency component at 1200Hz and 2400Hz. Each component's magnitude appears to be four times larger than the peak amplitude of the original sine waves given previously in (5.50). These results are caused by the DFT and need to be scaled, as we will discuss later.

Further study of the magnitude and phase plots reveal symmetry. We can clearly see that the magnitude plot exhibits symmetry across the y-axis. The phase spectra also contains symmetry around the origin. DFT symmetry is explored further in the next section.

5.3.4. Symmetry and Periodicity

The input waveform $x(n)$ is assumed to be periodic over an interval N. In a similar way, the output of an N-point DFT is also periodic over an interval N. Let us prove output periodicity by substituting $k = k + N$ into the DFT equation in (5.48), as follows.

$$X(k + N) = \sum_{n=0}^{N-1} x(n)e^{-j2\pi n(k+N)/N} \tag{5.62}$$

We can expand the complex exponential above using the rule in (5.18). The expression becomes,

$$X(k + N) = \sum_{n=0}^{N-1} x(n)e^{-j2\pi kn/N}e^{-j2\pi nN/N}. \tag{5.63}$$

Notice that the second complex exponential can be reduced, since $e^{-j2\pi nN/N} = 1$ for any value of n. The periodic property of the DFT is proved since $X(k) = X(k+N)$, as given below.

$$X(k) = X(k+N) = \sum_{n=0}^{N-1} x(n)e^{-j2\pi kn/N} \tag{5.64}$$

Previously in Section 5.1.6, we investigated positive and negative frequencies with respect to the Complex Fourier Series. An example showed that a real waveform consists of positive-frequency and negative-frequency complex exponentials. A similar idea can be extended to the DFT. The periodicity property of the DFT in (5.64) is useful, as it means we can reorganise our DFT results to accommodate negative frequencies.

The sketch on the left of Figure 5.20 contains the magnitude spectra of the DFT example in Section 5.3.2. We can rearrange the magnitude spectra to include negative frequencies. Consider the plot on the right of Figure 5.20. We can see that $X(0)$ is still the DC component and $X(1)$, $X(2)$ and $X(3)$ correspond to positive frequency components.

By using the periodicity property in (5.64), the frequencies $X(5)$, $X(6)$, and $X(7)$ can been reinterpreted as negative frequencies $X(-3)$, $X(-2)$, and $X(-1)$, respectively. Note that the frequency component at $X(4)$ straddles between the positive and negative frequencies. For our purposes, $X(4)$ will remain on the positive side of the spectra. Take a moment to inspect the new magnitude spectrum on the right of Figure 5.20.

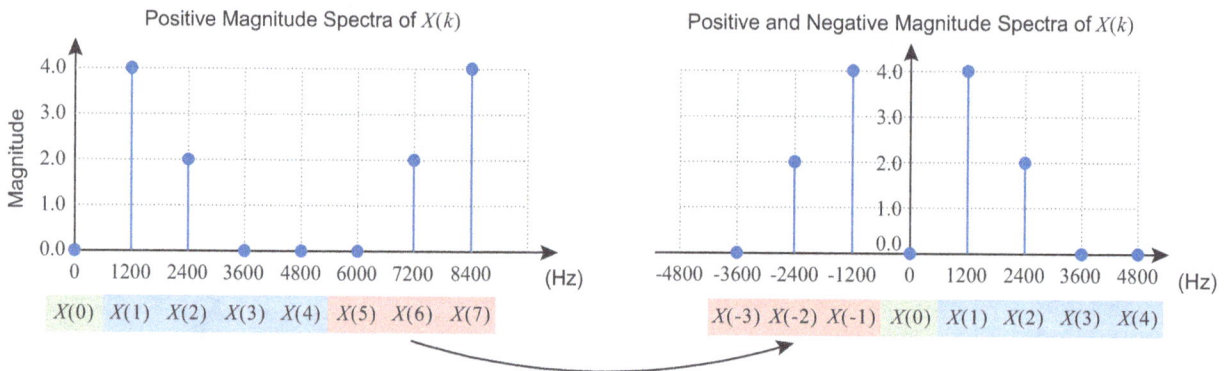

Figure 5.20: Rearranging the positive magnitude spectra into positive and negative magnitude spectra.

We can see that there is symmetry between the positive and negative frequencies in the rearranged plot of the magnitude spectrum. This symmetry corresponds to the real waveform's positive and negative frequency components (described previously in Section 5.1.6).

For most practical applications where the input waveform is real, it is not necessary to compute both halves of the DFT as it contains redundant values. For instance, if we are applying the DFT to a real waveform and N is even, then it is only necessary to compute the first $N/2+1$ values of $X(k)$. Similarly, if N is odd, we only need to compute $(N+1)/2$ values of $X(k)$ instead.

Remember that this redundancy is only true for input waveforms that are real. When the input waveform is complex (contains non-zero real and imaginary parts), then the DFT does not feature symmetry as a complex waveform consists of frequency components that can only be positive or negative (not both).

5.3.5. Normalisation

The magnitude response $|X(k)|$ of the DFT example in Section 5.3.2, contained values that were larger than the peak amplitudes of sine waves in $x(n)$. For instance, $|X(1)| = 4$ and $|X(2)| = 2$, while the peak amplitudes of their sine wave counterparts were 1 and 0.5, respectively. These magnitude values appear to have grown over four times in size.

We can normalise the complex output of the DFT by dividing $X(k)$ by the length of the input waveform, N. This produces the normalised complex output $X'(k)$, as below.

$$X'(k) = \frac{1}{N} \sum_{n=0}^{N-1} x(n)e^{-j2\pi kn/N} \tag{5.65}$$

The normalised frequency magnitude values become $|X'(1)| = 0.5$ and $|X'(2)| = 0.25$. You may still be confused by this result, but it can be explained by considering the positive and negative counterparts of a real waveform. We seen in Section 5.1.6, that a real waveform is actually a sum of a positive-frequency complex exponential and a negative-frequency complex exponential. When the DFT is applied to a real waveform, the corresponding frequency spectrum contains positive and negative frequency components. For instance, $|X'(1)| = |X'(-1)| = 0.5$ and $|X'(2)| = |X'(-2)| = 0.25$.

The frequency magnitude of a real waveform is evenly distributed between its positive and negative frequency components. If we sum $|X'(1)|$ and $|X'(-1)|$ together, we obtain 1. Similarly, summing $|X'(2)|$ and $|X'(-2)|$ together gives us 0.5. These normalised frequency magnitudes are now equivalent to the peak amplitude of their sine wave counterparts in $x(n)$. It is not necessary to sum each half of the spectrum to obtain this result. In the future, we can compute one half of $|X(k)|$ (as discussed in the previous section) and multiply it by 2. For the remainder of this chapter, we will plot the magnitude response of a real waveform in this way.

Note that the DFT of a complex waveform can also be normalised using (5.65). However, (5.65) is only used for rectangular windows, as we will discuss later in Section 5.3.10.

5.3.6. Magnitude, Power, and the Logarithmic Scale

In this section, we will demonstrate a variety of different ways that you can plot the output of a DFT. We have already seen plots of the real and imaginary parts of $X(k)$ and corresponding magnitude and phase plots in Figure 5.19. There are other ways engineers, scientists, and mathematicians like to present and measure the output frequency spectrum of a DFT. Before we begin this investigation into frequency representation, take a moment to inspect the sine wave given in Figure 5.22.

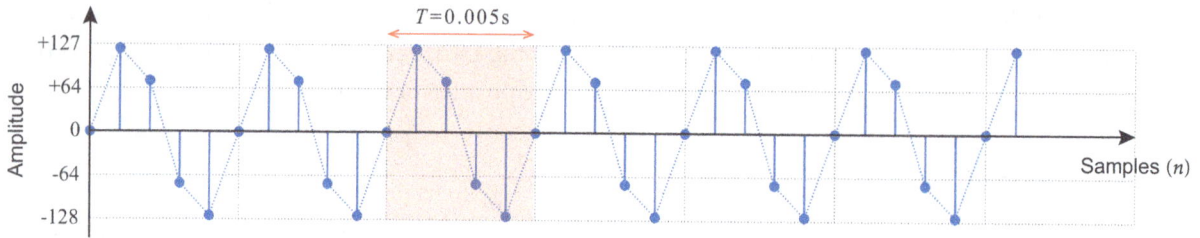

Figure 5.21: A 200Hz sine wave that was sampled using a frequency of 1kHz, and quantised using 256 levels.

We can see that the sine wave has a fundamental frequency of 200Hz (and period of 0.005s) and its amplitude has been quantised using 256 levels (8 bit in the 2's complement range of -128 to +127). The sine wave was acquired using a sampling frequency of 1000Hz and there are a total of 32 samples. As previously discussed in Section 5.3.5, we are able to compute the normalised magnitude of the DFT output. We can then perform a calculation to derive the power spectrum, $|X_{ps}'(k)|$, by using,

$$X'_{ps}(k) = |X'(k)|^2. \tag{5.66}$$

An example plot showing the power spectrum of the sine wave is given on the left of Figure 5.22. The power spectrum was computed using a 32-point DFT (the same number of samples as in the sine wave, $N = 32$).

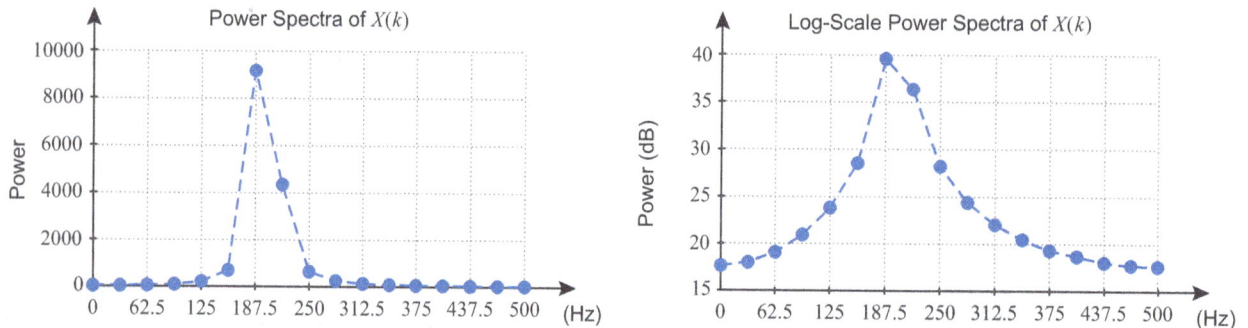

Figure 5.22: Power spectrum plot (left) and log-scale power spectrum plot (right).

Power spectrum plots often allow small (but significant) frequency components to be seen and that are not always apparent using a linear y-axis scale. We can more readily see a wider scale by using a logarithmic scale for the y-axis. The right of Figure 5.22 contains a log-scale magnitude plot. This plot should make it easier to inspect small and large magnitude spectra at the same time. The log-scale power spectrum is calculated as:

$$X'_{dB}(k) = 10\log_{10}(|X'(k)|^2) \text{ dB } . \tag{5.67}$$

It is possible to exchange the square operation above for a multiplication as shown in (5.68). This exchange may reduce the complexity of an FPGA architecture design that uses the log-scale power spectrum.

$$X'_{dB}(k) = 20\log_{10}(|X'(k)|) \text{ dB} \tag{5.68}$$

There is another log-scale representation approach known as *decibels relative to full-scale*, which is abbreviated as dBFS. Recall in Chapter 4 that an ADC will quantise the amplitude of a continuous time waveform into discrete levels known as quantisation levels. When a digital waveform reaches its maximum quantisation level, it is said to be at full-scale. The *decibels relative to full-scale* method of representing the spectrum is defined as the ratio of the input waveform's magnitude versus the waveform's full-scale representation, as given below.

$$X'_{dBFS}(k) = 20\log_{10}\left(\frac{|X'(k)|}{\text{full-scale}}\right) \text{ dBFS} \tag{5.69}$$

Lets use a few example to show how the dBFS technique works (rounded to 2 significant figures).

- If $|X'(k)| = 128$ and full-scale $= 128$, then $X'_{dBFS}(k) = 0.00$ dBFS.

- If $|X'(k)| = 64$ and full-scale $= 128$, then $X'_{dBFS}(k) = -6.02$ dBFS.

- If $|X'(k)| = 1$ and full-scale $= 128$, then $X'_{dBFS}(k) = -42.14$ dBFS.

Notice that it isn't possible for *decibels relative to full-scale* to produce a value higher than 0.00 dBFS. A plot of the log-scale magnitude spectra for the sine wave, which uses dBFS, is given on the left of Figure 5.23.

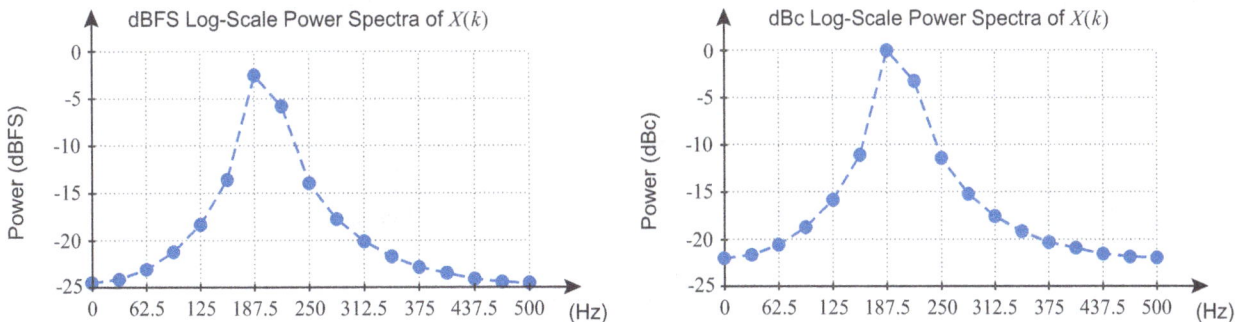

Figure 5.23: Decibels relative to full-scale plot (left) and decibels relative to carrier plot (right).

The final representation we will discuss is another log-scale approach known as *decibels relative to the carrier*, which is denoted as dBc. This approach presents the spectrum as a ratio of the input waveform power versus the carrier waveform power. If the output ratio is positive, then the power of the input waveform is greater than the carrier. Alternatively, if the output ratio is negative, then the carrier has more power than the input waveform. The *decibels relative to the carrier* can be computed as,

$$X'_{dBc}(k) = 10\log_{10}\left(\frac{|X'(k)|^2}{\text{carrier power}}\right) \text{ dBc}. \tag{5.70}$$

An example magnitude spectra plot for the sine wave, which uses dBc, can be see on the right of Figure 5.23. Notice that the 200Hz sine wave, which we have set to be the carrier, has a value of 0.00dBc. The surrounding spectra is scaled relative to the carrier. This representation is useful when we need to compute measurements such as the Spurious Free Dynamic Range (SFDR), which was discussed previously in Section 4.2.4.

5.3.7. Frequency Bins

In the previous section, we investigated several ways to plot the frequency magnitude spectra of a discrete time waveform. In these plots, we can clearly see that the frequency axis is divided into several discrete frequencies. Engineers commonly name a point on the frequency axis a "bin", or "bins", if referring to all of them collectively. Each bin is regularly spaced by a frequency resolution Δf (also known as the fundamental frequency). Take a moment to inspect the sine wave and its corresponding frequency magnitude plot given below in Figure 5.24. The sine wave was acquired using a sampling frequency f_s of 1000Hz.

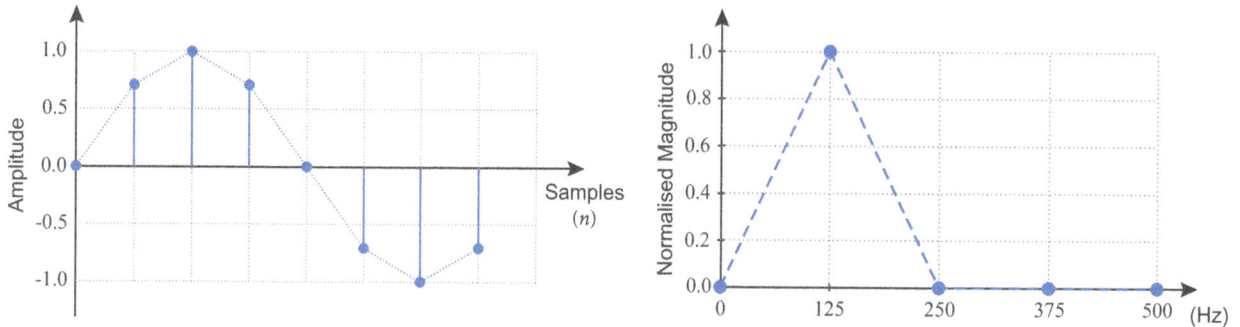

Figure 5.24: Discrete time waveform of a 125Hz sine wave (left) and its normalised frequency magnitude plot (right).

We can see that the sine wave has 8 samples ($N = 8$) and the corresponding magnitude plot has 5 discrete frequency bins. The frequency axis in this example uses 5 bins because the input waveform is real, and the negative frequencies contain redundant information (see Section 5.3.4). We can compute the frequency resolution (or fundamental frequency) of the DFT output using (5.47) as below.

$$\Delta f = \frac{f_s}{N} = \frac{1000 \text{ samples/second}}{8 \text{ samples}} = 125 \text{ Hz} \tag{5.71}$$

Notice that the frequency axis of the magnitude plot increases in steps of 125Hz. The maximum frequency we can represent is 500Hz, which is equivalent to $f_s/2$. Similarly, the minimum frequency that we can represent is $-f_s/2$. The bins on the frequency axis are spaced in multiples of Δf between $[-f_s/2, f_s/2]$. For this example, the bins correspond to the frequencies 0Hz, 125Hz, 250Hz, 375Hz, and 500Hz (excluding negative frequencies). It is worth noting that our input waveform has an exact frequency of 125Hz. We can see that all of the bins in the frequency magnitude plot are zero except for the 125Hz bin, which has all of the energy.

We can make several observations about the frequency resolution of the DFT. For example, if we increase the number of input samples, N, then we can improve the frequency resolution of the DFT (making it finer). We have provided an example below in Figure 5.25 that applies a 16-point DFT to a 125Hz sine wave that was acquired using a sampling frequency of 1000Hz. We can see that the frequency resolution has now become 62.5Hz and the bins are closer together. It is worth mentioning that increasing the size of N will also increase the number of arithmetic operations required to compute the DFT.

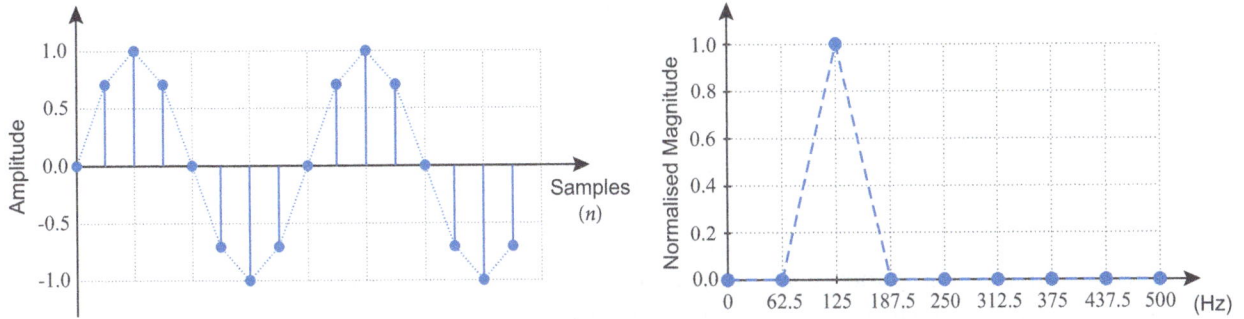

Figure 5.25: The discrete 125Hz sine wave with 16 samples (left) and its normalised frequency magnitude plot (right).

Another way of obtaining a finer frequency resolution is by reducing the sampling frequency of the input waveform. This change will not have a considerable impact on the computational requirements of the DFT. To demonstrate the effects of reducing the sampling frequency, we have provided an example in Figure 5.26 that applies an 8-point DFT to a sine wave that has been acquired using a sampling frequency of 500Hz.

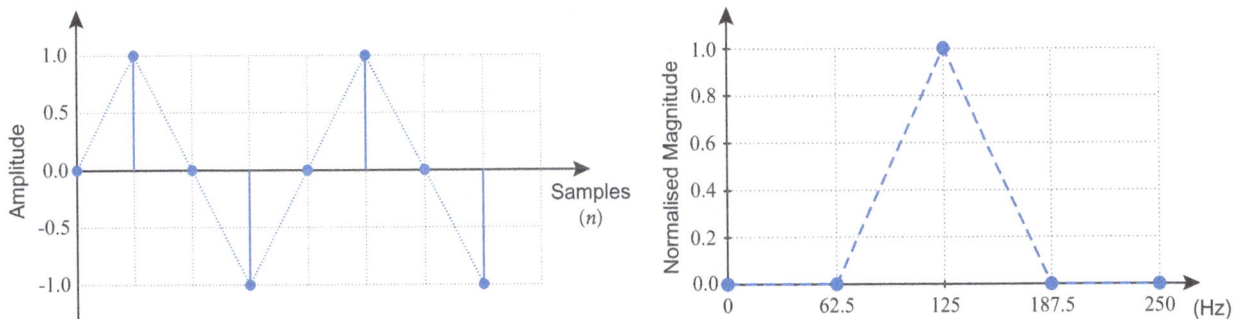

Figure 5.26: The discrete 125Hz sine wave sampled at 500Hz (left) and its normalised frequency magnitude plot (right).

Notice that the resolution of the frequency axis for this example is equal to 62.5Hz. We are still able to see energy in the 125Hz bin on the frequency magnitude plot. However, the range of the frequency axis has now decreased, which may be a problem for applications that require a specific range of frequencies. Later in Section 5.3.12, we will investigate the zero padding technique, which can also improve frequency resolution.

5.3.8. The Inverse DFT

The DFT is a tool to convert a discrete time waveform into a discrete frequency representation. We can reverse the operation by using the Inverse DFT, which is commonly abbreviated as the IDFT. You can see the equation for the IDFT below.

$$x(n) = \frac{1}{N} \sum_{k=0}^{N-1} X(k) e^{j2\pi kn/N} \tag{5.72}$$

159

To demonstrate the functionality of the IDFT, we will use it to simply reverse the DFT performed previously in Section 5.3.2. The values for $X(k)$ are given in (5.59). As before, we will use the trigonometric notation to perform the IDFT, which is given as,

$$x(n) = \frac{1}{N} \sum_{k=0}^{N-1} X(k)[\cos(2\pi kn/N) + j\sin(2\pi kn/N)]. \tag{5.73}$$

We will demonstrate how to compute the first sample of $x(n)$. We begin by substituting $n = 0$ and $N = 8$ into the above equation, which gives the expression for $x(0)$.

$$x(0) = \frac{1}{8} \sum_{k=0}^{7} X(k). \tag{5.74}$$

This expression computes the average value of $X(k)$, as follows.

$$
\begin{aligned}
x(0) = (&0.0 - j0.0 \\
+ &2.8285 - j2.8285 \\
+ &0.0 - j2.0 \\
+ &0.0 - j0.0 \\
+ &0.0 - j0.0 \\
+ &0.0 - j0.0 \\
+ &0.0 + j2.0 \\
+ &2.8285 + j2.8285)/8 \\
= \ &0.7071 + j0.0
\end{aligned}
$$

We have successfully obtained the same value for $x(0)$ as given previously in (5.51).

We could progress by computing the remainder of the IDFT. However, this would involve a lot of complex multiplications, which consumes a lot of page space. Instead, you should try converting from the frequency domain to the time domain on your own. We recommend using a software tool or programming language (such as Python) to assist you with your calculations.

It is worth mentioning that the IDFT is only applied to the discrete complex waveform, $X(k)$. The IDFT equation in (5.72) is not applied to the normalised discrete complex waveform $X'(k)$.

5.3.9. Spectral Leakage

The DFT is a marvel of applied signal processing and an indispensable tool for analysing the radio frequency electromagnetic spectrum. However, it has various characteristics that users need to be aware of and interpret accordingly, or appropriately mitigate against if possible. One such issue is spectral leakage. We will begin by discussing how spectral leakage occurs and its impact on DFT results. Then, we will explore ways to reduce spectral leakage via windowing in Section 5.3.10.

We previously established in Section 5.3.7 that the DFT operates across a discrete set of frequencies. These frequencies are integer multiples of the frequency resolution, f_s/N. When we apply the DFT to a discrete-time waveform that contains frequency components equal to one or more of these discrete frequencies, we obtain a perfect DFT result. We have already seen this outcome several times now in this chapter. For instance, the original DFT example in Section 5.3.2 had carefully selected sine waves that produced a perfect DFT output.

Take a moment to inspect the sine wave and its frequency magnitude plot given in Figure 5.27. The sine wave has 16 samples, a frequency of 80Hz, and was acquired using a sampling frequency of 1000Hz. Notice that the sine wave completes one whole period but does not complete its second period. The magnitude plot was obtained by applying a 16-point DFT to the sine wave. There does not appear to be a discrete bin for an 80Hz frequency so the energy has spread to other neighbouring frequency bins, which is a common problem known as spectral leakage. The bins with the most energy are 62.5Hz, 125Hz, and 0Hz.

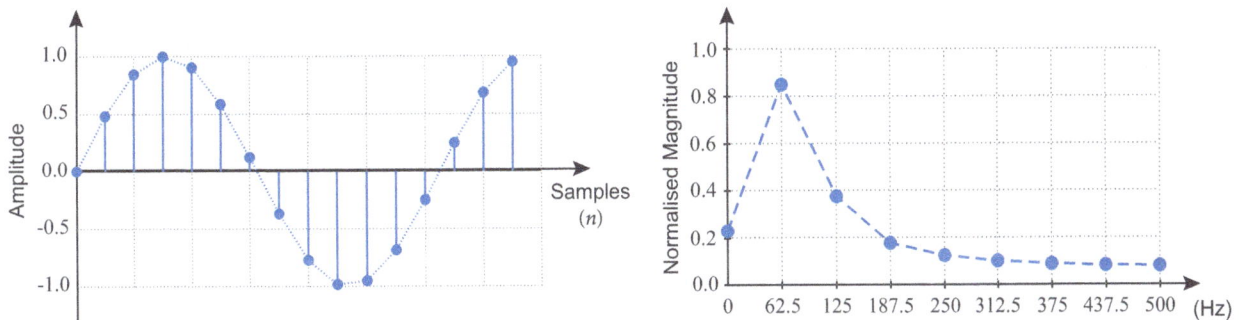

Figure 5.27: A discrete sine wave with a frequency of 80Hz (left) and its normalised frequency magnitude plot (right).

We can explain this behaviour by recalling a very important property of the DFT. Previously, in Section 5.3.1 we derived the DFT equation using the continuous Fourier Transform. An important observation we made for the input waveform $x(n)$ is given below.

"By using a finite number of data samples, N, we are no longer evaluating a waveform over a infinite period of time, which is required by the continuous Fourier Transform. Instead, we evaluate $x(n)$ with the implicit assumption that the waveform contains pseudo-periods every N samples, or Nt_s seconds."

When the DFT is applied to the sine wave in Figure 5.27, we are actually assuming the waveform is periodic as illustrated in Figure 5.28. As you can see, there is an abrupt transition between one period of the waveform and another. These points are known as discontinuities and are only found at the endpoints of the waveform's sampling interval. Discontinuities occur because the endpoints of the input waveform do not align with the period of the fundamental frequency, or frequency resolution, of the DFT.

We can plot the discrete magnitude response that was given previously in Figure 5.27, alongside an equivalent continuous frequency response of the 80Hz sine wave. This plot can be seen in Figure 5.29, with negative frequencies excluded. The main lobe of the continuous magnitude response is directly over 80Hz.

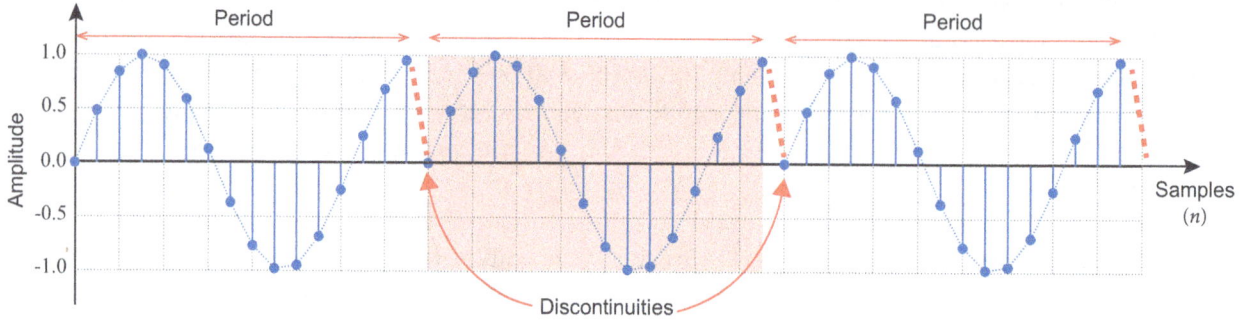

Figure 5.28: A discrete time plot showing how discontinuities can occur when using the DFT.

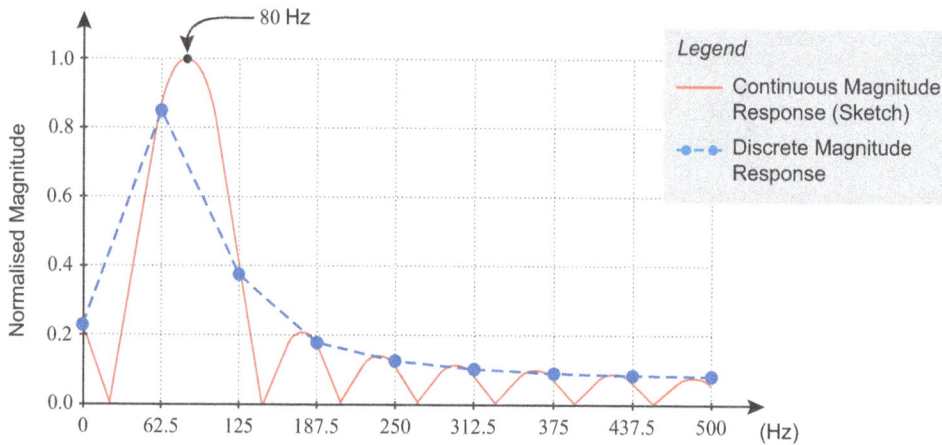

Figure 5.29: A plot of the continuous magnitude spectra, overlapped on the discrete magnitude spectra.

The plot in Figure 5.29 shows that the DFT is a sampled version of the equivalent continuous frequency magnitude response. In other words, each bin of the discrete magnitude response has been awarded a value that corresponds to an equivalent point on the continuous magnitude spectrum. We do not have a frequency bin for 80Hz, so the energy leaks into neighbouring bins and causes the DFT output to inaccurately represent the frequency content of the discrete time waveform.

5.3.10. Windowing

We can reduce the effect of spectral leakage by applying particular windows to a discrete waveform before using the DFT. Many windows have the effect of "tapering" the endpoints of a discrete waveform to reduce abrupt transitions, such as those seen previously in Figure 5.28. There are many different types of windows including Hamming, Hann, Blackman-Harris and Bartlett. In this section, we will demonstrate how to apply a window to a discrete waveform. Then, we will present different types of windows and briefly explore and discuss their magnitude responses.

Windowing is the process of extracting a subset of data from a larger set so that it can be processed by another function. We also have the option to modify the data before it is passed on. If the data remains unmodified, then we have simply truncated the dataset across a specified window (or interval). This type of window is known as a rectangular window. Figure 5.30 contains a plot that illustrates the process of using a rectangular window on a discrete 80Hz sine wave.

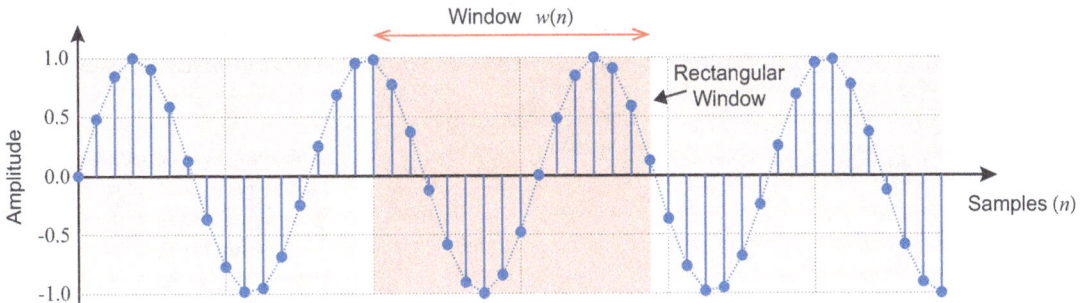

Figure 5.30: A rectangular window applied to a discrete sine wave of 80Hz.

The rectangular window is simply used to extract an interval of data from the discrete waveform without affecting the data. As we have already suggested, windows can be used to modify data too. For instance, we can taper the endpoints of the window before applying the DFT to reduce spectral leakage. A common window used for tapering the endpoints of a discrete waveform is the Hann window. Take a moment to see the effects of applying this window to the discrete 80Hz sine wave in Figure 5.31.

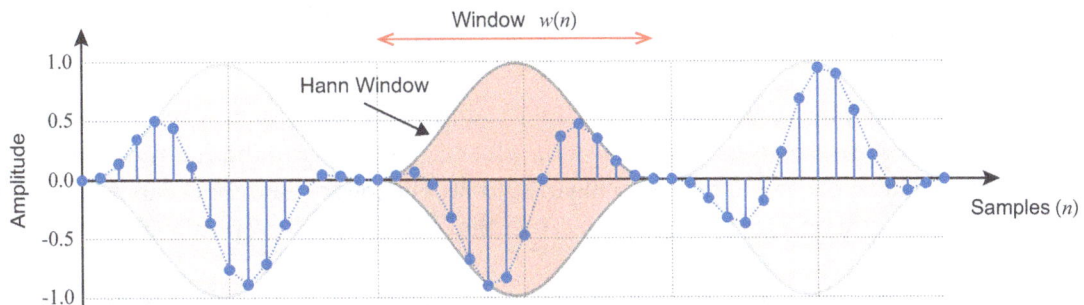

Figure 5.31: A Hann window applied to a discrete sine wave of 80Hz.

We can mathematically express the process of using a window in the DFT equation. For a discrete window denoted as $w(n)$, and a discrete waveform $x(n)$, we simply perform element-wise multiplication before applying the DFT. This operation is given as,

$$X_w(k) = \sum_{n=0}^{N-1} w(n) \cdot x(n) e^{-j2\pi kn/N}. \tag{5.75}$$

As we will see in a moment, tapered windows can reduce spectral leakage in the DFT. However, there are some caveats. Windowing has the effect of widening the main lobe of the peak frequency. However, the side lobes that cause spectral leakage are reduced.

Let us take a look at the frequency magnitude plots for the windowed discrete waveforms given in Figure 5.30 and Figure 5.31. Each wave was sampled using 1000Hz and a 16-point DFT was applied to obtain their frequency representations. We will not normalise the frequency magnitude responses, as we also want to inspect the impact that the window has on scaling. The frequency magnitude responses are each plotted below in Figure 5.32.

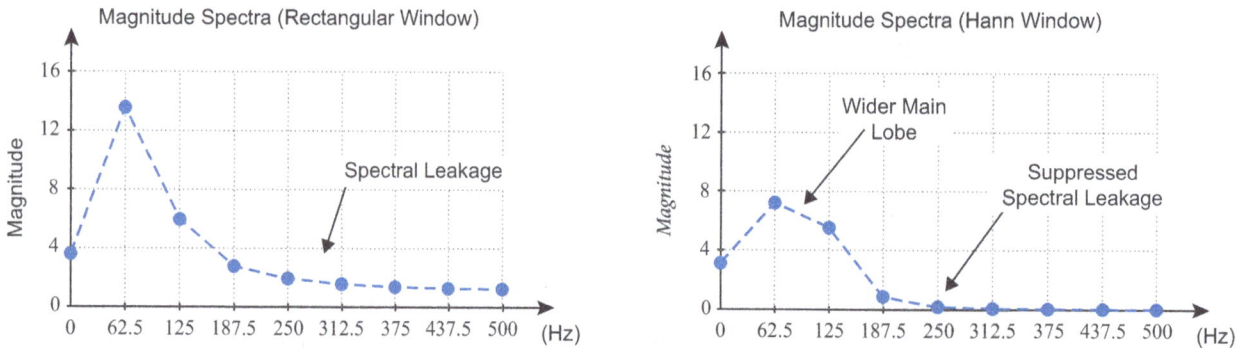

Figure 5.32: Frequency magnitude plot for the rectangular waveform (left) and the Hann waveform (right).

We can see that the main lobe has widened in the magnitude plot corresponding to the Hann window (in comparison to the rectangular window). The side lobes have also reduced in magnitude, which effectively suppresses the spectral leakage.

We deliberately did not normalise the plots (using (5.65)) so that you can visualise the frequency magnitude response on the y-axis of each plot. As you can see, the plot corresponding to the Hann window has a lower magnitude response in comparison to the rectangular window. The lower response occurs because we scaled the amplitude of the discrete time waveform, which causes the DFT output to proportionately decrease.

As described in Section 5.3.5, we usually normalise the DFT output by dividing it by the number of samples in the discrete waveform, N. We only normalise in this way for rectangular windows. When we use other windows that modify the amplitude of the discrete waveform, we need to scale (or normalise) the DFT output by the sum of the window samples, as given below.

$$X_w{}'(k) = \frac{1}{\sum_n w(n)} X_w(k) \tag{5.76}$$

For example, a rectangular window, $w(n) = 1$, means (5.76) reduces to $1/N$. We can plot each frequency magnitude response again. This time using the scaled (or normalised) form, as shown in Figure 5.33.

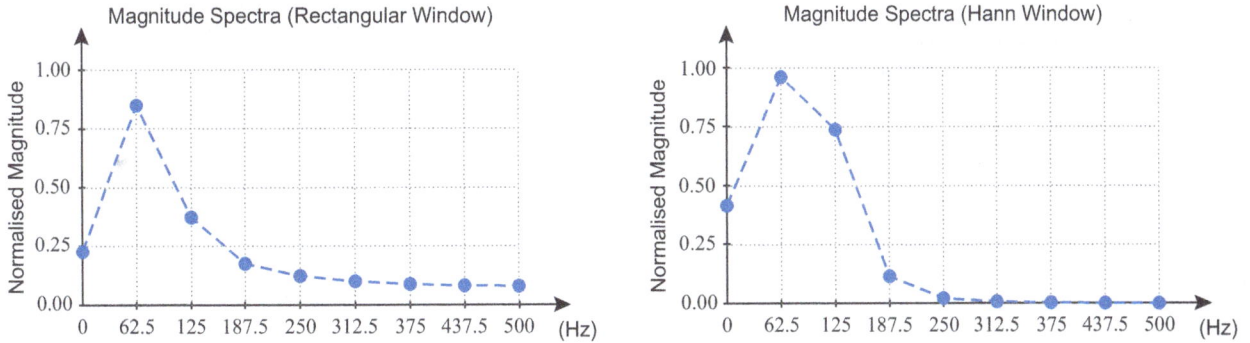

Figure 5.33: Normalised frequency magnitude plot for the rectangular waveform (left) and the Hann waveform (right).

There are many different types of windows. Every engineer, scientist, and mathematician has their favourite (and you will too). There are various reasons for selecting one window over another. For the remainder of this section, we will introduce several windows that you will commonly see in literature including, Bartlett, Blackman, Hamming, and Hann windows. We will express each window mathematically and describe their features. You can inspect the time representation of each window in Figure 5.34.

At times, we will be discussing the attenuation of side lobes in terms of dB (log-scale magnitude). These values correspond to the window's frequency magnitude response. We have provided a plot of each window response in Figure 5.35 for you to inspect. This magnitude response was created by applying a 16384-point DFT to the windows. We will refer to this plot when describing each window below.

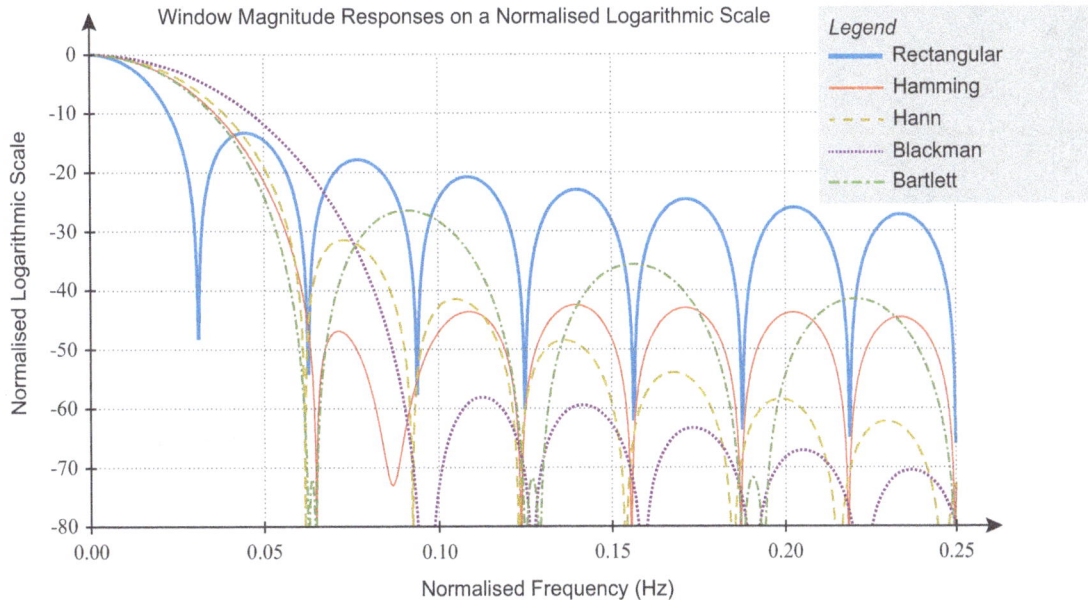

Figure 5.35: Window magnitude responses on a normalised logarithmic scale (16384-point DFT).

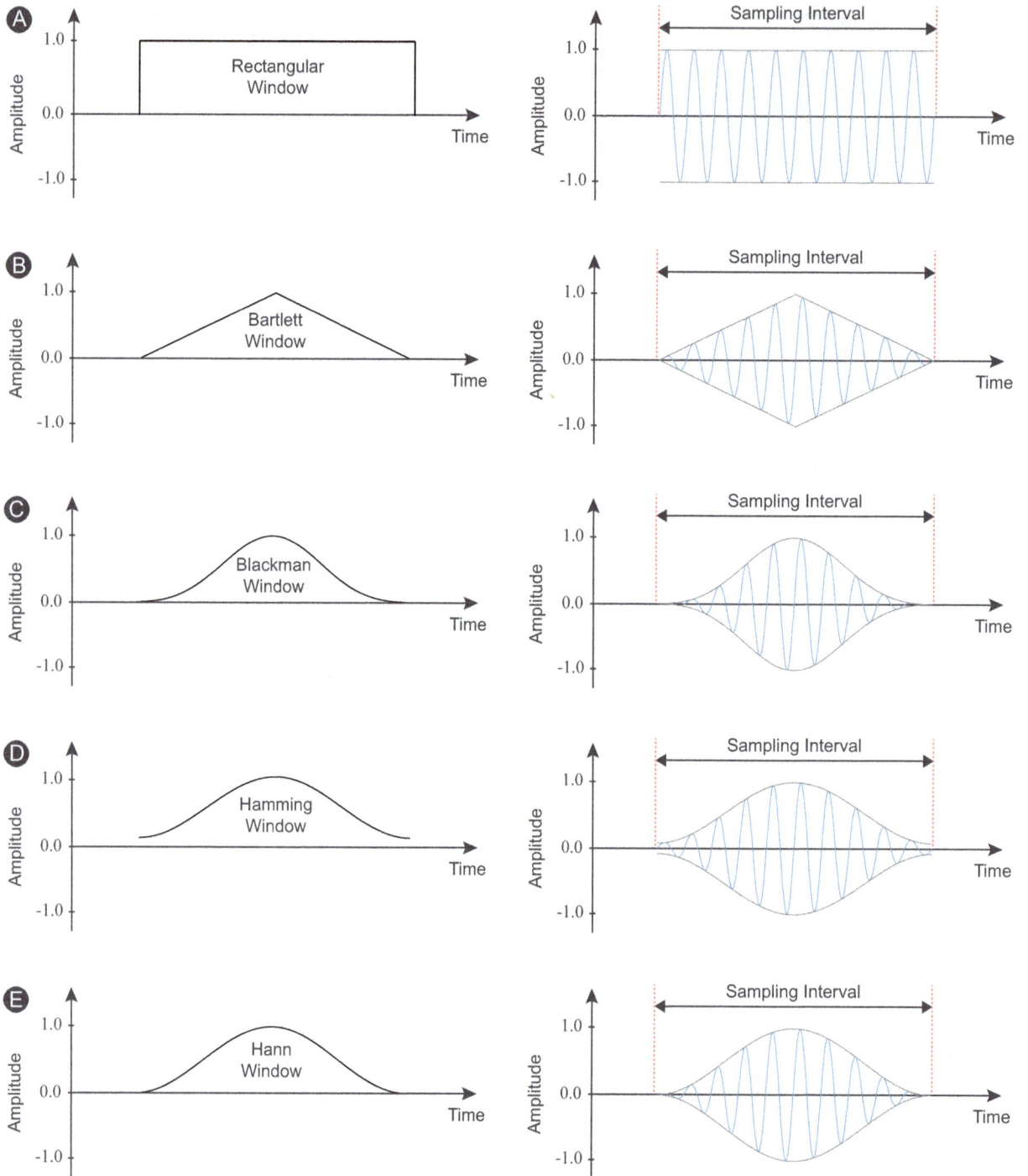

Figure 5.34: Rectangular window (A), Bartlett window (B), Blackman window (C), Hamming window (D), and Hann window (E). Each window is accompanied by a plot demonstrating their endpoint suppression capabilities.

Rectangular Window

The rectangular window is commonly known as the uniform, or boxcar window. It is the most basic window that can be used before applying the DFT to a discrete waveform. It is expressed as,

$$w(n) = 1, \quad (0 \leq n \leq N-1). \tag{5.77}$$

Remember that using a rectangular window allows spectral leakage to occur unsuppressed (depending on the input waveform to the DFT). We typically use the rectangular window as a reference to compare other different types of windows.

Bartlett Window

In comparison to the rectangular window, the Bartlett window doubles the width of the main lobe. The first side lobe is attenuated by 26dB, which is double that of the rectangular window. For $N + 1$ data samples, the Bartlett window is express as,

$$w(n) = \begin{cases} \dfrac{2n}{N}, & \left(0 \leq n \leq \dfrac{N}{2}\right) \\ 2 - \dfrac{2n}{N}, & \left(\dfrac{N}{2} \leq n \leq N\right). \end{cases} \tag{5.78}$$

Note that a Bartlett window is very similar to a triangle window. The exception is a Bartlett window has zeros at the endpoints.

Blackman Window

The Blackman window has superior spectral leakage rejection of all the windows discussed in this chapter. It boasts approximately 58dB of attenuation for the first side lobe. The Blackman window is defined as,

$$w(n) = 0.42 - 0.5\cos\left(\frac{2\pi n}{N}\right) + 0.08\cos\left(\frac{4\pi n}{N}\right), \quad (0 \leq n \leq N-1). \tag{5.79}$$

The Blackman window also has a fairly wide main lobe, which causes the frequency peak to spread across several bins. If you are interested, there is a window in the same family as the Blackman window, which is known as the Blackman-Harris window. It has approximately 71.48dB of side lobe rejection.

Hamming Window

In comparison to the rectangular window, the Hamming window doubles the width of the main lobe. However, it attenuates the primary side lobe by 46dB. The Hamming window is expressed as,

$$w(n) = 0.54 + 0.46\cos\left(\frac{2\pi n}{N}\right), \quad (0 \leq n \leq N-1). \tag{5.80}$$

Notably, the Hamming window suppresses the first adjacent side lobe more than the subsequent side lobes, as shown in Figure 5.35. The other side lobes can only be attenuated by around 42dB.

Hann (or Hanning) Window

Finally, the Hann window features a main lobe that is around double the width of the rectangular window. The primary side lobe attenuation for this window is around 32dB. The Hann window is defined by,

$$w(n) = 0.5 + 0.5\cos\left(\frac{2\pi n}{N}\right), \quad (0 \leq n \leq N-1). \tag{5.81}$$

We have now finished our discussion on windowing and will be moving on to briefly discuss how to measure the error introduced by spectral leakage, which is called scalloping loss. We will also introduce a clever technique known as zero padding that can improve the frequency resolution.

5.3.11. Scalloping Loss

A discrete waveform may contain a frequency that resides precisely between two DFT bins. When this happens, the energy from that frequency leaks into the neighbouring bins in the DFT output response. We can measure the worst-case reduction in the waveform's level. Formally, this is known as the scalloping loss and is usually abbreviated as SL. Take a moment to inspect Figure 5.36, which contains a plot of the continuous frequency magnitude response of a discrete 93.75Hz sine wave, acquired at a sampling frequency of 1000Hz.

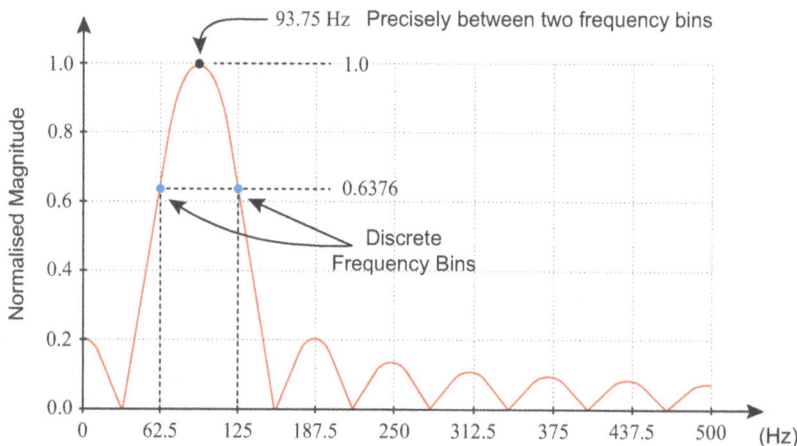

Figure 5.36: *The continuous frequency magnitude response of a 93.75Hz sine wave that was acquired using a sampling frequency of 1000Hz. The maximum point of the response falls precisely between two discrete frequency bins of a 16-point DFT. The energy leaks into the neighbouring bins.*

The maximum point of the response is centred on 93.75Hz. If we applied a 16-point DFT to the sine wave then the energy would be precisely between two discrete frequency bins (62.5Hz and 125Hz). We do not have a frequency bin for 93.75Hz, so the energy falls into the neighbouring bins instead.

The scalloping loss in the plot above is 0.6376, as this is the worst case reduction in the waveform's frequency magnitude response. The frequency output was computed without applying a window to the discrete sine wave, meaning a rectangular window is inferred. The type of window used and the DFT size impact the severity of the scalloping loss. We can define the scalloping loss as the coherent gain located half way between two DFT bins, divided by the coherent gain at a DFT bin [189]. The coherent gain is simply the sum of all samples in a discrete waveform, divided by the number of samples in the waveform (similar to the average value of a waveform). We can express the scalloping loss mathematically as,

$$\text{SL} = \frac{\left| \sum_n w(n) e^{-j\pi n/N} \right|}{\sum_n w(n)} . \tag{5.82}$$

If we substitute in the example above, where $w(n) = 1$ and $N = 16$, we obtain,

$$\text{SL}_\text{Rec} = \frac{|1.0000 - j10.1532|}{16} = \frac{10.2023}{16} = 0.6376 . \tag{5.83}$$

The above expression in (5.82) can be used to compute the scalloping loss for any window. When we use a window that is not rectangular, we will reduce the severity of scalloping loss as the main lobe will be wider. For comparison, we will compute the scalloping loss of a Hann window using the same constraints as the above example i.e. a DFT size of 16.

$$\text{SL}_\text{Han} = \frac{|0.6367 - j6.4646|}{7.5} = \frac{6.4959}{7.5} = 0.8661 . \tag{5.84}$$

We can see that the Hann window has less severe scalloping loss than a rectangular window. This is expected as the Hann window has a wider main lobe than a rectangular window.

5.3.12. Zero Padding

We will wrap-up our investigation of the DFT by exploring a common method of improving the frequency resolution of the DFT plot, known as zero padding. When we studied the continuous Fourier Transform in Section 5.2.1, we seen that increasing the period of the time domain waveform to infinity created an infinitesimally small spacing between harmonics in the frequency response. It turns out that this technique can also be extended to the DFT (without needing to sample a waveform to infinity!).

Zero padding is a technique that involves inserting zero-valued samples at the end of a discrete waveform to improve the frequency resolution of the DFT plot. The effect of zero padding is essentially an interpolation of the frequency sample points in the DFT and as such no extra 'information' is created on the signal.

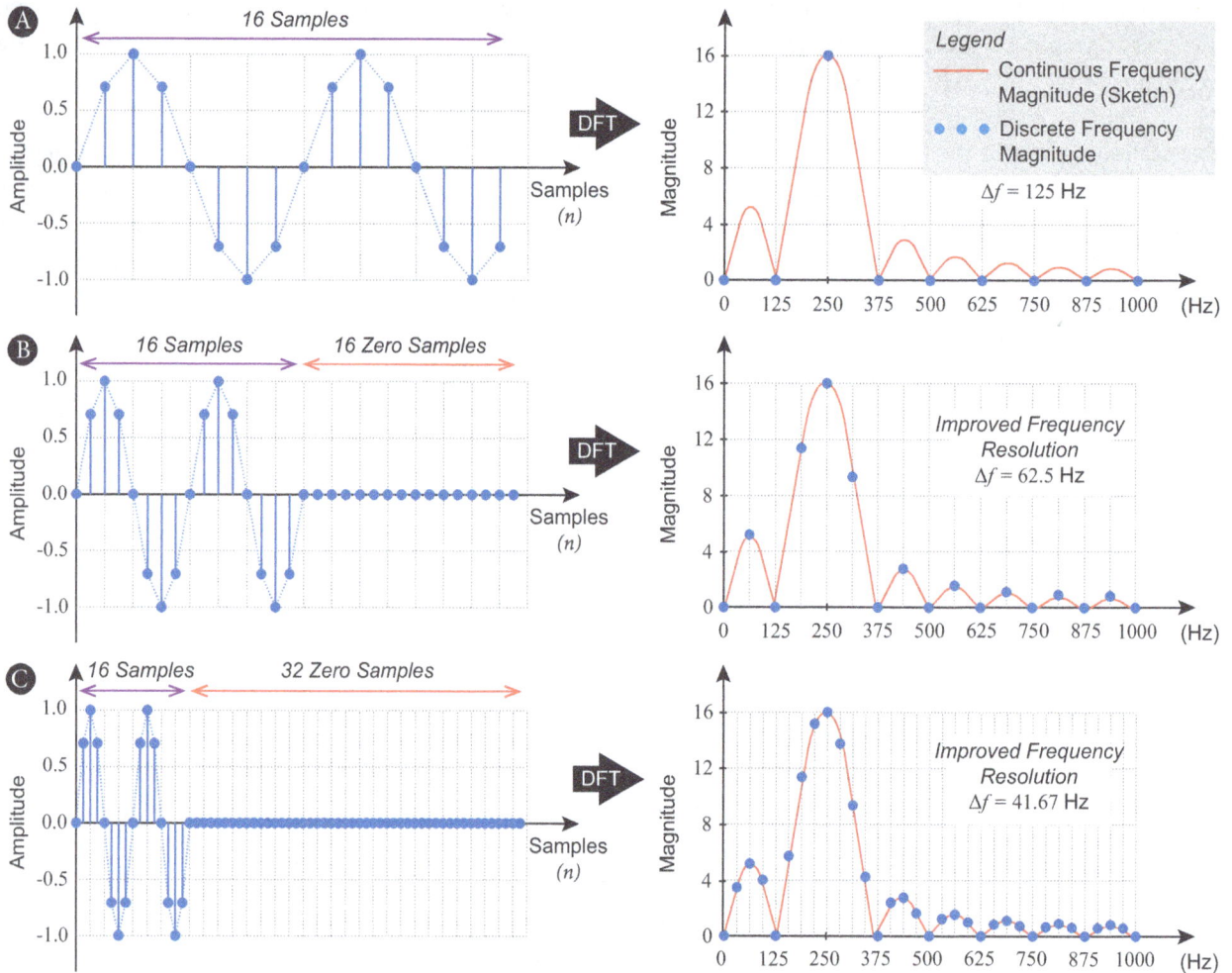

Figure 5.37: A discrete sine wave consisting of 16 samples (A), 32 samples (B), and 48 samples (C). Each discrete sine wave has a corresponding frequency magnitude plot. As zero padding increases, the frequency resolution of the DFT is finer.

The effect of time domain zero padding is shown in Figure 5.37 where each of examples above contain a discrete 250Hz sine wave at a sampling frequency of 2000Hz. The sine wave has 16 samples. The first example (at label A) demonstrates a normal 16-point DFT of the sine wave (using a rectangular window). We have provided the discrete and continuous plots of the frequency magnitude. The sine wave is perfectly represented in the frequency domain as a 250Hz peak. Also, we can see that other frequency bins do not contain any energy. The frequency resolution of this DFT is 125Hz.

We pad the discrete sine wave with 16 zeros at label B in Figure 5.37, which creates a new discrete time waveform consisting of 32 samples. Applying a 32-point DFT to this waveform improves the frequency resolution of the DFT. The resolution is now 62.5Hz. Notice that the peaks of the side lobes are now visible in the discrete spectra on the frequency magnitude plot.

At label C in Figure 5.37, we append 32 zero samples onto the end of the original sine wave. We now apply a 48-point DFT, which improves the frequency resolution of the output spectrum so that it is now 41.67Hz.

We can see that zero padding the input waveform has the effect of achieving a finer frequency resolution. This technique is particularly useful for peak finding. For instance, consider a frequency magnitude plot that has spectral leakage since the input waveform is not periodic over the sample period. We could use zero padding to improve the frequency resolution and obtain the frequency that corresponds to the peak value.

5.4. The Fast Fourier Transform

Large DFTs computations are of the order of N^2 computations. Therefore, a $\sim 10^3$ or 1024-point DFT will require $\sim 10^6$ arithmetic operations. In 1965, a paper published by Cooley and Tukey [127] described an efficient DFT implementation, commonly referred to as the Fast Fourier Transform (FFT). The FFT is an indispensable tool, as it can efficiently compute very large DFTs by reducing the total number of arithmetic operations. This is achieved by exploiting the periodicity of the DFT calculation. The FFT is somewhat more difficult to code/program compared to the DFT (which is in essence two nested FOR loops and a lookup table of cosine and sine values) and the FFT is a challenging programming effort, as will be outlined below! But once coded, its $\log_2 N$ times faster than an N-point DFT. For a 1024 point FFT, the number of arithmetic operations reduces by a factor of $\log_2(1024) \approx 10$. So its 10 times faster to do the FFT rather than the DFT, and *exactly* the same answer is produced by both computations.

Hence, it is important to highlight that the FFT is an efficient implementation of the DFT and is not an approximation. The FFT will produce the same results as the DFT and also abides all of its characteristics and features that we have discussed so far. For many pre-coded FFT computations, the FFT size is usually a power of 2 number i.e. $N = 2^m$, where m is a positive integer. This defines the window length and the spectral resolution. Non-power of 2 FFTs are indeed possible (such as prime factor FFT).

In this section, we will explore the FFT algorithm. Firstly, we will investigate the Danielson-Lanczos lemma, which presented a method of computing the DFT, recursively. Then, we will analyse the *decimation-in-time* solution that underpins the basic operating principles of the FFT. We will explain *decimation-in-time* using several diagrams and mathematical expressions for an 8-point FFT. Finally, we will work through an 8-point FFT example and then describe other notable FFT algorithms and architectures that are worth exploring in your own time.

5.4.1. Danielson-Lanczos Lemma

Danielson and Lanczos described a method of exploiting the periodicity of the DFT [132] to reduce the computational requirements i.e. reduce the number of complex multipliers required to compute the DFT. Their work is very interesting, as it describes a method of separating an N-point DFT into two smaller size DFTs. For this method to operate correctly, N must be an even number of samples.

Consider the expression below for $X(k)$ that separates the input waveform $x(n)$ into two sequences containing even and odd samples. An $N/2$-point DFT is applied to each sequence individually and they are summed together. Notice that k still operates over the discrete values $0, 1, 2, ..., N-1$.

$$X(k) = \sum_{n=0}^{N/2-1} x(2n)e^{-j2\pi(2n)k/N} + \sum_{n=0}^{N/2-1} x(2n+1)e^{-j2\pi(2n+1)k/N} \qquad (5.85)$$

The complex exponential in the second DFT above, can be moved outside of the summation, which produces

$$X(k) = \sum_{n=0}^{N/2-1} x(2n)e^{-j2\pi(2n)k/N} + e^{-j2\pi k/N} \sum_{n=0}^{N/2-1} x(2n+1)e^{-j2\pi(2n)k/N}. \qquad (5.86)$$

A common method of simplifying notation is to rewrite the same complex exponential as

$$W_N^k = e^{-j2\pi k/N}, \qquad (5.87)$$

which is referred to as the twiddle factor. We can also rewrite the even and odd DFTs in (5.86) as $X^{even}(k)$ and $X^{odd}(k)$, respectively. The Danielson-Lanczos lemma to compute an N-point DFT as two smaller DFTs, is now expressed as

$$X(k) = X^{even}(k) + W_N^k X^{odd}(k). \qquad (5.88)$$

Earlier in Section 5.3.4, we looked at the periodic property of the DFT. The same rules apply for the even and odd DFTs given in (5.88). For example, we can generate an expression for $X^{even}(k=0)$ as

$$X^{even}(k=0) = \sum_{n=0}^{N/2-1} x(2n)e^{-j2\pi(2n)0/N} = \sum_{n=0}^{N/2-1} x(2n) \qquad (5.89)$$

We can also create an expression for $X^{even}(k=N/2)$ as

$$X^{even}(k=N/2) = \sum_{n=0}^{N/2-1} x(2n)e^{-j2\pi(2n)(N/2)/N} = \sum_{n=0}^{N/2-1} x(2n). \qquad (5.90)$$

Both expressions in (5.89) and (5.90) are the same and a similar result is obtained for the odd DFT, such that $X^{odd}(k=0) = X^{odd}(k=N/2)$. Therefore, we can conclude that there is no increase in the computational complexity when calculating $X^{even}(k)$ and $X^{odd}(k)$, due to the fact that $k = 0, 1, 2, ..., N-1$ and $n = 0, 1, 2, ..., N/2-1$.

We can also observe how the twiddle factor is affected by the periodicity of k. For example, when $k = 0$, the twiddle factor becomes,

$$W_N^{(k\,=\,0)} = e^{-j2\pi(0)/N} = 1. \tag{5.91}$$

When $k = N/2$, the twiddle factor's sign is changed as

$$W_N^{(k\,=\,N/2)} = e^{-j2\pi(N/2)/N} = -1. \tag{5.92}$$

This relationship is true for all values of k that are more than or equal to $N/2$.

Let us now graphically represent the Danielson-Lanczos lemma to aid our understanding. For example, we could create a diagram that would allow us to compute $X(k = 0)$ and $X(k = N/2)$ using (5.88). This diagram is illustrated in Figure 5.38.

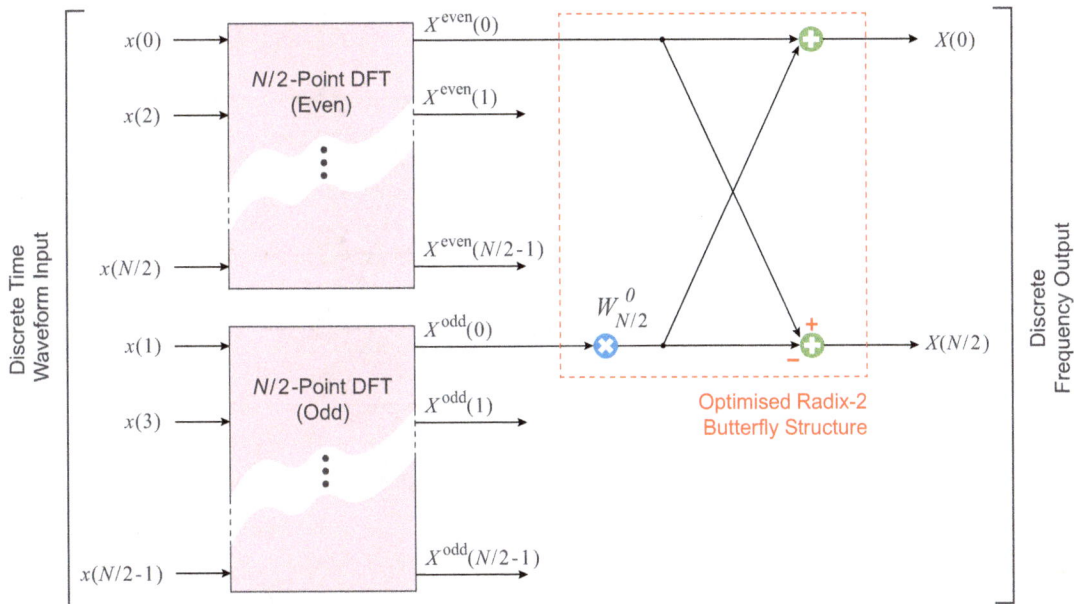

Figure 5.38: Graphically the Danielson-Lanczos lemma can be sketched as a signal flow graph (as above). Here, we can see that two smaller DFTs have been applied to the even and odd samples of a discrete time waveform. The DFT outputs are used in (5.88) to compute the final output values when k=0 and k=N/2.

We can see from this diagram that two $N/2$-point DFTs have been implemented to compute the frequency representation of the even and odd discrete time sequences. We have only shown the $X^{\text{even}}(0)$ and $X^{\text{odd}}(0)$ connections, but it is worth noting that the other DFT outputs would be connected in a similar way. These DFT outputs are connected to a structure known as a radix-2 butterfly, as shown in Figure 5.38.

We can compute $X(0)$ by simply multiplying the twiddle factor $W_{N/2}^0$ with $X^{\text{odd}}(0)$ and then adding the result to $X^{\text{even}}(0)$, as given in (5.88). Similarly, $X(N/2)$ is calculated by subtracting $W_{N/2}^0 X^{\text{odd}}(0)$ from $X^{\text{even}}(0)$. The subtraction is necessary as the twiddle factor is negative when $k \geq N/2$.

The Danielson-Lanczos lemma can be extended to an 8-point DFT. Take a moment to inspect the diagram given in Figure 5.39. It is not necessary to use trigonometry (as required by the DFT) to compute $X(k)$. The twiddle factors are used instead. We can see that we only need to perform four multiplications to compute the twiddle factors in this example. The even and odd DFTs both require $N^2/4$ complex multipliers each. This means a total of 40 multipliers are required for this design. An equivalent 8-point DFT uses 64 multipliers. As we can see, the computational complexity has reduced. Additionally, we obtain the exact same results as an 8-point DFT (there are no approximations).

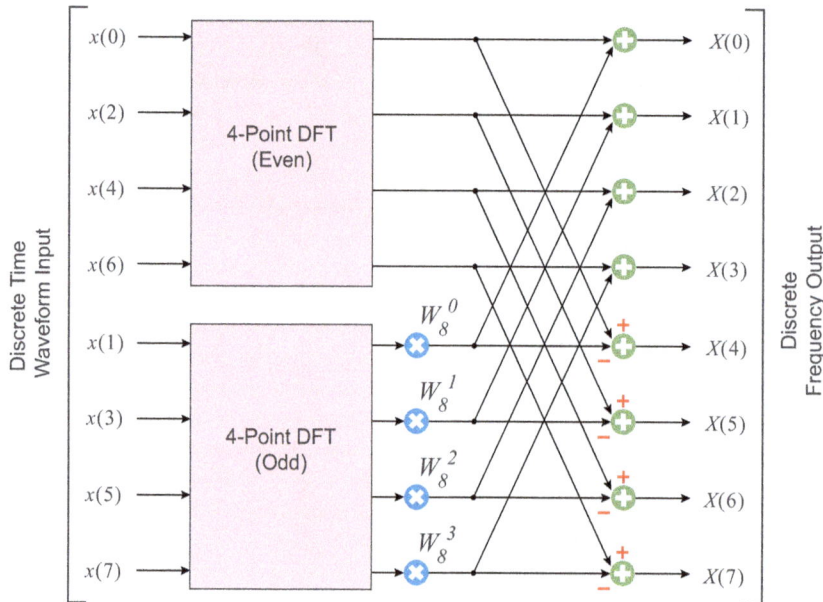

Figure 5.39: *The diagram above is an 8-point DFT that has been implemented using two 4-point DFTs and four multipliers.*

5.4.2. Decimation in Time

When we separate a discrete waveform into two sequences, such as odd and even samples, we are actually decimating the waveform in time. In the previous section, we used the Danielson-Lanczos lemma to *decimate-in-time*, which is a well known and commonly used FFT design technique.

The Danielson-Lanczos lemma can be used to recursively *decimate-in-time* so that we use four 2-point DFTs rather than two 4-point DFTs from the previous section. See Figure 5.40 for an idea of how we can further reduce the computational complexity of the DFT.

We can start by separating the even DFT, $X^{\text{even}}(k)$, into two DFTs of its even and odd samples, as

$$X^{\text{even}}(k) = \sum_{n=0}^{N/4-1} x(4n)e^{-j2\pi(4n)k/N} + \sum_{n=0}^{N/4-1} x(4n+2)e^{-j2\pi(4n+2)k/N}. \tag{5.93}$$

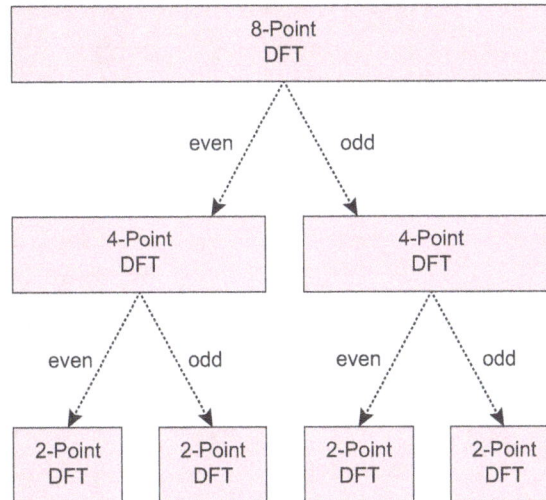

Figure 5.40: An abstract diagram that shows the Danielson-Lanczos lemma being applied recursively.

Similar to before in (5.86), we extract the complex exponential from the second summation. We then replace the complex exponential with the notation for the twiddle factor, which gives

$$X^{\text{even}}(k) = \sum_{n=0}^{N/4-1} x(4n)e^{-j2\pi(4n)k/N} + W_{N/2}^{k} \sum_{n=0}^{N/4-1} x(4n+2)e^{-j2\pi(4n)k/N}. \tag{5.94}$$

We can repeat the process for the odd DFT, $X^{\text{odd}}(k)$, so that it can be expressed as

$$X^{\text{odd}}(k) = \sum_{n=0}^{N/4-1} x(4n+1)e^{-j2\pi(4n)k/N} + W_{N/2}^{k} \sum_{n=0}^{N/4-1} x(4n+3)e^{-j2\pi(4n)k/N}. \tag{5.95}$$

Now the even and odd DFTs have both been rewritten so they use two 2-point DFTs each. We can continue our 8-point FFT design and replace the 4-point DFTs shown previously in Figure 5.39 with two 2-point DFTs. We can see this rearrangement in Figure 5.41. Notice that the twiddle factor after the odd 2-point DFT, operates over four points (which corresponds to a 4-point DFT).

We are almost finished our 8-point *decimation-in-time* FFT implementation of the DFT. We just have one step remaining, which is to replace the 2-point DFT, with a simple butterfly structure. The 2-point DFT can be efficiently computed by recognising that $W_N^0 = 1$ and $W_N^{N/2} = -1$. For example, we can replace the first 2-point DFT block using (5.88), which evaluates to $x(0) + x(4)$ for the even side, and $x(0) - x(4)$ for the odd side. See the final architecture of an 8-point FFT in Figure 5.42.

The final design is the 8-point FFT implementation of the Cooley & Tukey algorithm we mentioned at the start. In the next section, we will validate the design of this FFT in an example.

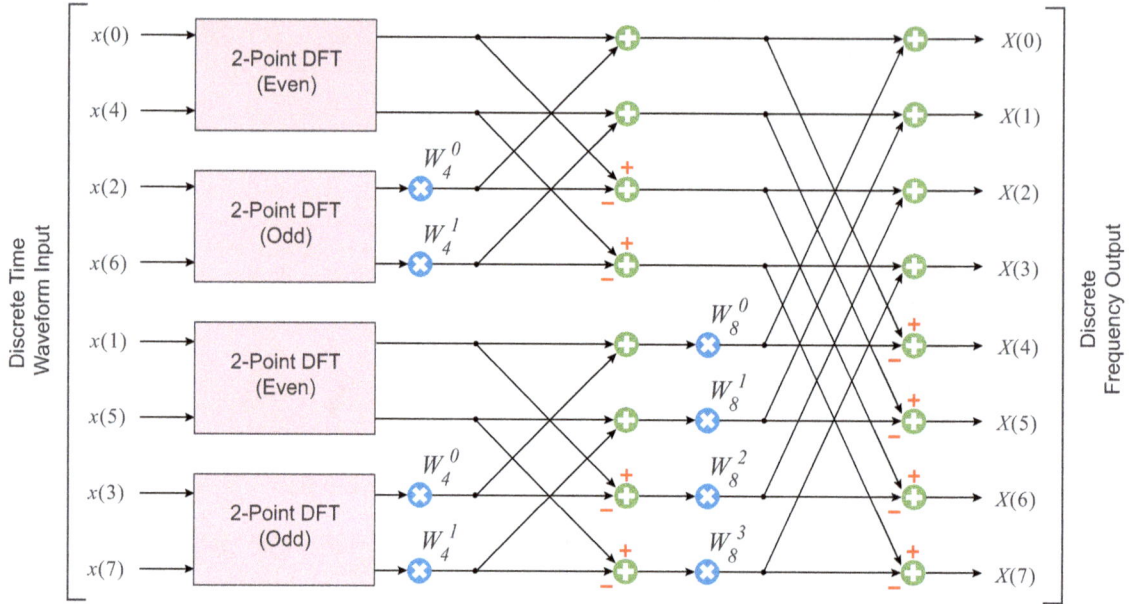

Figure 5.41: *The diagram above uses the Danielson-Lanczos lemma to recursively to implement of an 8-point FFT using two 4-point DFTs and four 2-point DFTs.*

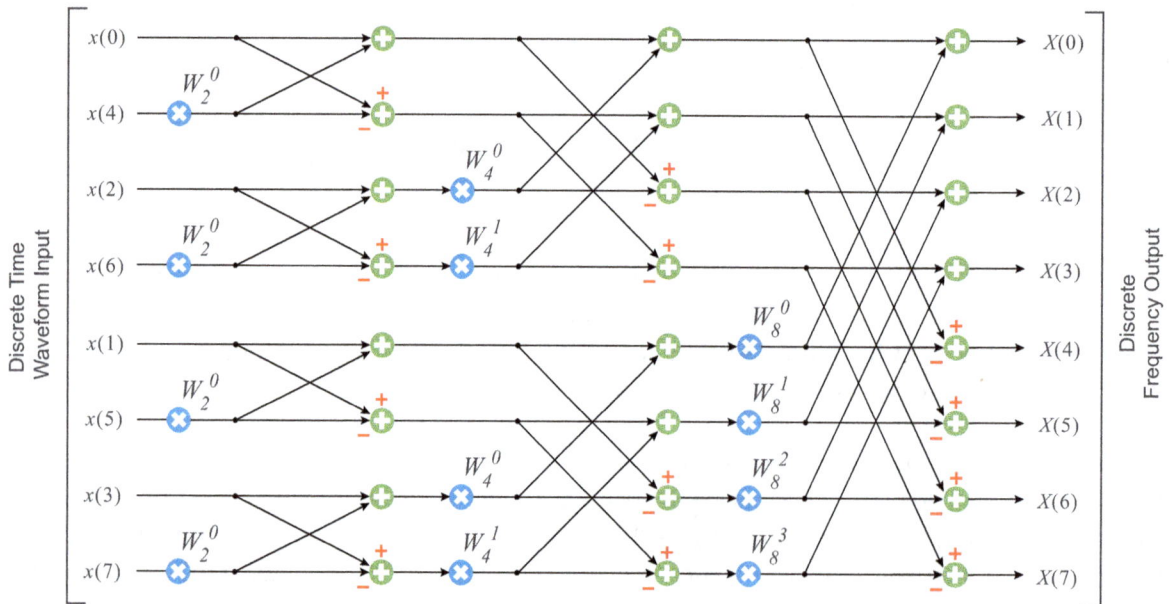

Figure 5.42: *The final architecture of an 8-point DFT using two 4-point DFTs and four 2-point DFTs. This implementation is generally known as the Radix-2 decimation-in-time FFT architecture.*

5.4.3. An FFT Example

In Section 5.3.2, we demonstrated an 8-point DFT in a simple example using a discrete input waveform. Let us use the same waveform to validate the 8-point FFT architecture we created in the last section. To remind ourselves, the input samples for the previous example was,

$$
\begin{aligned}
x(0) &= 0.7071, \\
x(1) &= 1.5000, \\
x(2) &= 0.7071, \\
x(3) &= -0.5000, \\
x(4) &= -0.7071, \\
x(5) &= -0.5000, \\
x(6) &= -0.7071, \\
x(7) &= -0.5000.
\end{aligned}
\tag{5.96}
$$

Figure 5.43 contains the results of the 8-point FFT using the input samples above. We have annotated the diagram with intermediate calculations as required. You can see the discrete frequency representation on the right of the diagram. The results match the DFT output given previously in (5.59).

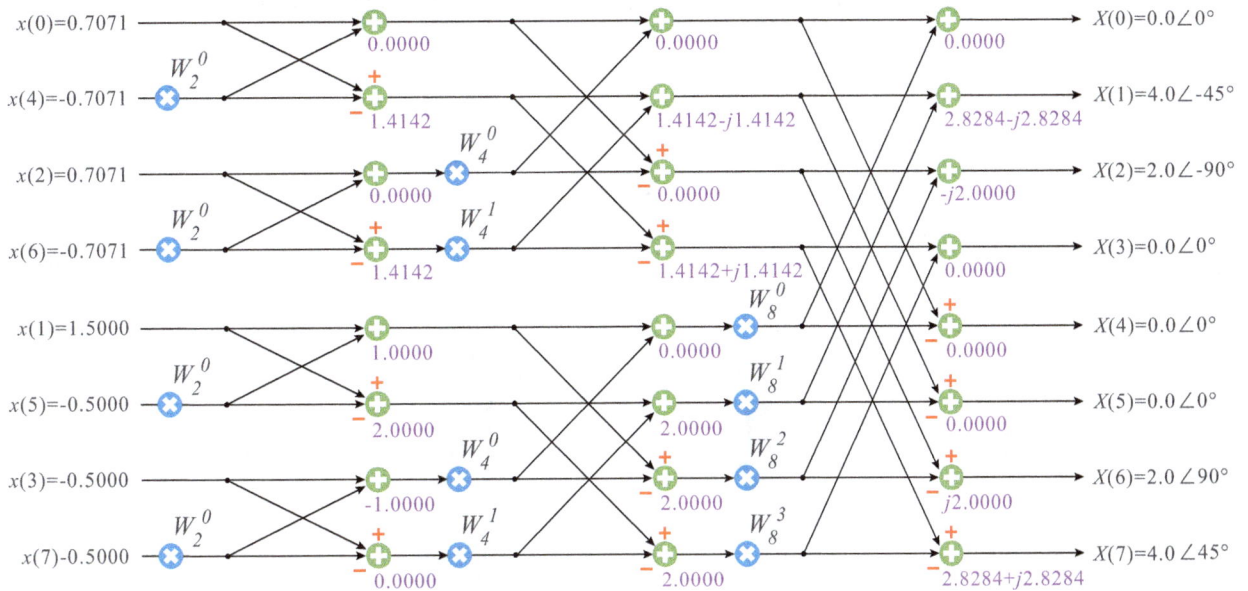

Figure 5.43: An 8-point FFT example using the data stimulus from Section 5.3.2. The results are the same as the 8-point DFT example.

We can evaluate and compare the number of multipliers that would be required to implement the design in Figure 5.43. Originally, an 8-point DFT would require 64 complex multipliers to operate successfully. The design above only requires 8 complex multipliers (noting that $W_2^0 = 1$). The FFT *decimation-in-time* method is very efficient and can be used in real-time on modern computer systems.

5.4.4. FFT Algorithms and Architectures

The FFT is a vast and interesting subject area that has a lot to discuss and explore. There are numerous resources that you should investigate on your own, which describe various FFT algorithms and designs. We will list two of these here to get you started.

- The *decimation-in-frequency* FFT technique [243].

- Fixed-point and floating-point FFT design considerations [28].

5.5. Short-Time Fourier Transform

FFTs are very efficient and have low computational complexity, allowing us to achieve real-time performance on modern computing systems. It is possible to consecutively perform FFT operations across an interval of time, which is useful for a variety of reasons. For instance, we can separate a discrete waveform into small equal size segments and use the FFT to obtain the frequency representation of each segment. This process is known as the Short-Time Fourier Transform (STFT). A simple example is given in Figure 5.44.

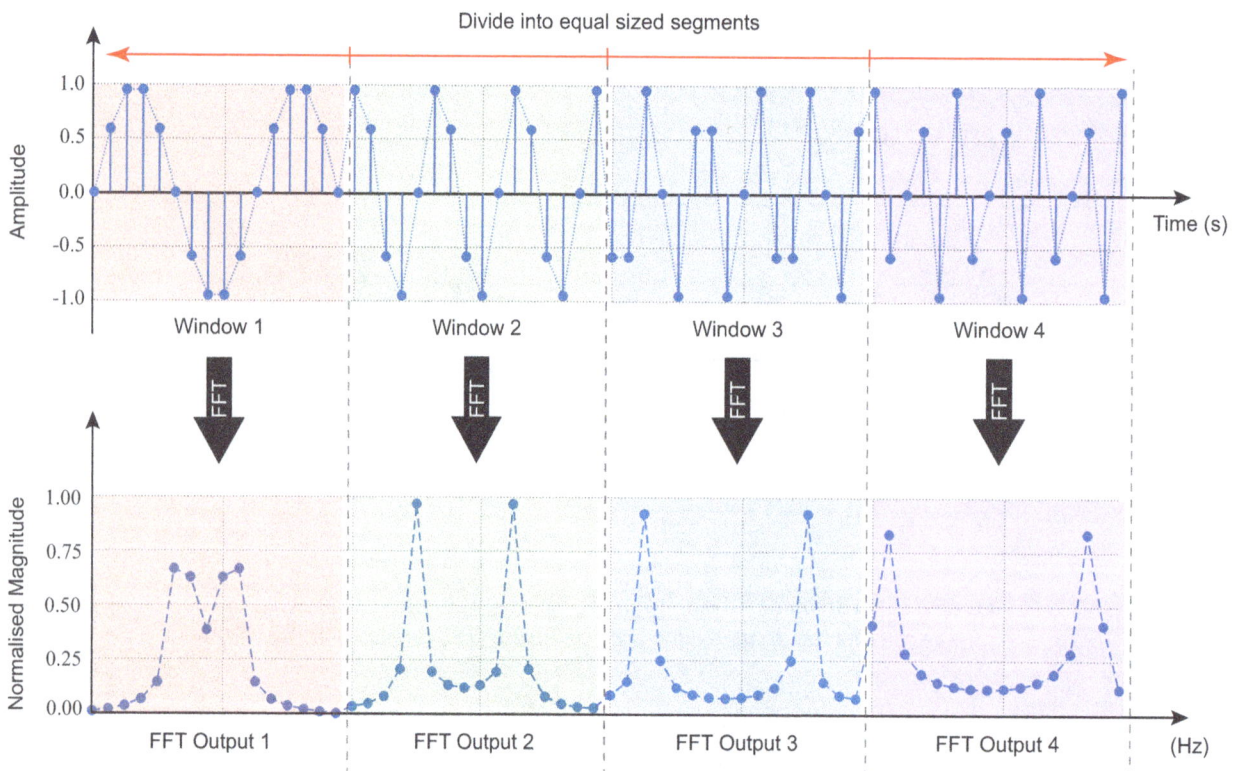

Figure 5.44: This diagram shows a discrete time waveform being divided into four equal sized segments. A 16-point FFT is applied to each segment, resulting in four FFT outputs.

The STFT can be used to explore how the frequency spectra of a discrete waveform changes over time. For example, we can plot each of the FFT outputs as a function of time, which is known as a spectrogram or waterfall plot.

5.5.1. STFT of a Chirp

A chirp is a waveform whose frequency increases or decreases over time. It is quite a valuable waveform for demonstrating the STFT in action. Take a moment to inspect the up-chirp wave given in Figure 5.45. Notice that it increases in frequency as a function of time.

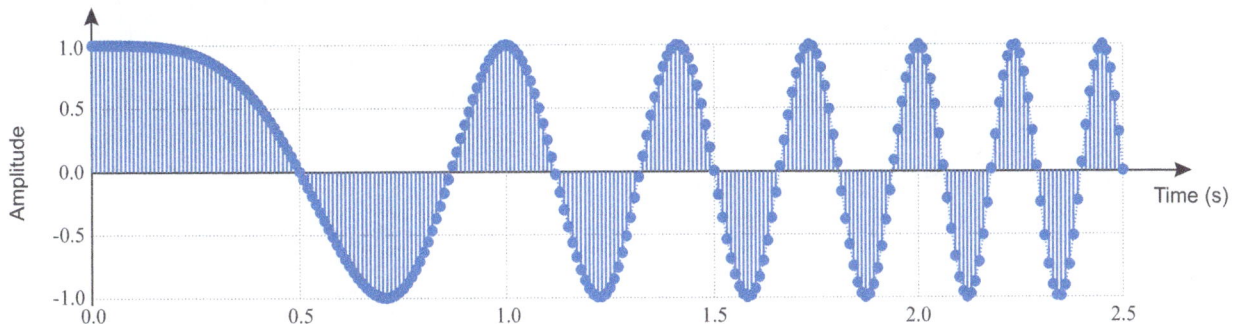

Figure 5.45: A chirp waveform of increasing frequency.

To perform the STFT, we divide the chirp into 8-equal size segments and applied a hamming window to each segment. We then use a 256-point FFT to obtain each segment's frequency representation. Plotting the FFT output frequency against time reveals an increasing frequency response, as shown in the 2-dimensional and 3-dimensional plots in Figure 5.46. Note that we have used log-scale to display the plot.

The 3-dimensional plot uses the x-axis for frequency, the y-axis for time, and the z-axis for the log-scale magnitude. Each colour on the plots represent a frequency magnitude value as shown on the scale to the right of the plot. It is worth mentioning that there are other ways to implement the STFT. For instance, if the time axis is particularly coarse in the spectrogram output, the adjoining segments can be overlapped to smooth out sharp fluctuations between FFT frames.

5.5.2. Time Versus Frequency Resolution

An issue with the STFT is deciding on the time and frequency resolution. Recall that the frequency resolution of a DFT is given by f_s/N, where f_s is the sampling frequency and N is the number of samples. To obtain a finer time resolution, we divide the input waveform into small segments. Partitioning the waveform in this way causes the frequency resolution to be coarse, since the frequency resolution becomes larger i.e. N is small. In contrast, we can divide the input waveform into larger segments, which means each segment has a large amount of samples. This results in a fine frequency resolution, but the time resolution is coarse as the value of N is large.

3-Dimensional View　　　　　　　　　2-Dimensional View

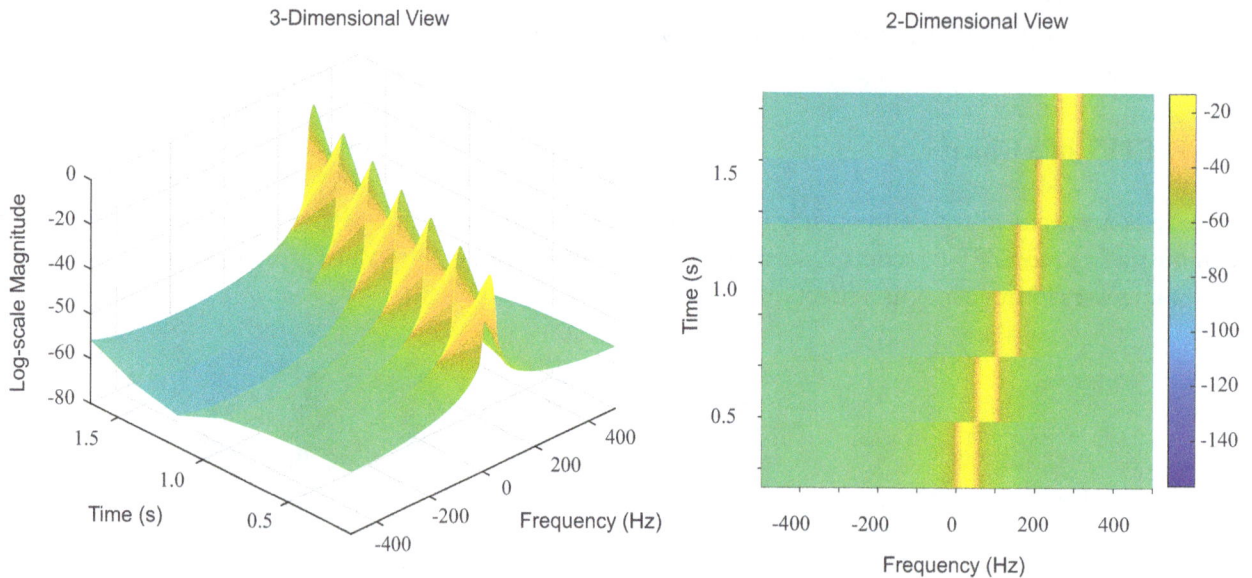

Figure 5.46: A 3-dimensional spectrogram of the chirp (left), and a 2-dimensional spectrogram of the chirp (right).

In summary, there is a trade-off between the time and frequency resolution. There are other techniques to address this issue that you can explore on your own, such as the wavelet transform [183].

5.6. Chapter Summary

We began our deep dive into spectral analysis with Fourier's Theorem, which is the foundation for all Fourier analysis techniques. We explored the Fourier Series expansion by demonstrating that any continuous periodic waveform could be expanded as a weighted sum of harmonically related sine and cosine waves. We then rearranged the Fourier Series to use complex exponentials. A simple example was given, which demonstrated a Complex Fourier Series expansion of a real waveform, revealing positive and negative frequency components. Then, we derived the continuous Fourier Transform, which obtains the frequency representation of aperiodic continuous time waveforms.

We then derived the equation for the DFT. The DFT operates in a similar way to a Discrete Complex Fourier Series and can be used to obtain the frequency representation of discrete-time waveforms. We explored a simple example for an 8-point DFT and established that the DFT output has symmetry for real input waveforms. Our investigation into symmetry revealed periodic properties of the DFT that could be used to rearrange the discrete frequency representation as a positive and negative complex frequency spectrum. Lastly, we investigated various features of the DFT, including normalisation, log-scale plots, frequency bins, spectral leakage, scalloping loss, and zero padding.

In the final part of this chapter, we derived the FFT and explored its architecture and resource consumption for an 8-point DFT. We found that the FFT is not an approximation of the DFT; rather, it is the DFT and is effective when reducing computational complexity. We established that the FFT technique could only be used with DFT sizes that are a power of two. We applied the FFT to the same stimulus as the 8-point DFT example and obtained the same results. Finally, we investigated the STFT by looking at a simple chirp example. We discussed issues regarding the time and frequency resolution of the STFT, and introduced 3-dimensional and 2-dimensional spectrogram plots.

Notebook Set C

Exploring the Spectrum with the RFSoC

Following Chapter 5 on Spectral Analysis, we can begin to explore the spectrum with the RFSoC. There are two notebooks on this topic. The first notebook will allow you to explore the ambient radio spectrum. You will be able to inspect several radio bands including those that belong to the broadcast, license-exempt, and mobile and wireless broadband sectors. The second notebook will launch a spectrum analyser tool on your RFSoC platform [328]. The spectrum analyser tool is capable of inspecting multiple channels, and also integrates features for generating test signals. The spectrum analyser tool is an excellent way of learning about RFSoC and its capabilities.

The notebooks and their relative locations are listed below. Each notebook must be used on an RFSoC platform. Section C.3 contains additional spectral analysis notebooks that can be used on a computer or RFSoC device. These notebooks cover topics on Fourier's theorem, the DFT, and the FFT.

RFSoC 01_exploring_the_spectrum.ipynb — *rfsoc_book/notebook_C/01_exploring_the_spectrum.ipynb*

RFSoC 02_rfsoc_spectrum_analyser.ipynb — *rfsoc_book/notebook_C/02_rfsoc_spectrum_analyser.ipynb*

C.1. Exploring the Spectrum

Begin by opening the notebook named *01_exploring_the_spectrum.ipynb*. This notebook contains several code cells that will automatically tune your RFSoC device to various frequency bands so that you can begin exploring the spectrum. Follow the instructions in the notebook to correctly setup your RFSoC platform.

As you progress through the notebook, you should be able to see power in various frequency bands. These radio signals are broadcast from nearby sources such as TV, radio, or mobile transmitters. Note that the presence of signals in different bands will depend on your location. If you are unable to detect a signal at some of the radio frequencies provided, this may be because no transmissions are currently occurring at that particular frequency, or because the frequency band is not used in your geographical location. Listed below are the frequency bands you will explore in this notebook.

- (88 to 108) MHz — Frequency Modulated (FM) radio broadcasts.

- (210 to 230) MHz — Digital Audio Broadcast (DAB) radio communications.

- (470 to 700) MHz — Ultra-High Frequency (UHF) digital television broadcasts.

- (700 to 1000) MHz — Spectrum access for mobile communications.

- (1700 to 2200) MHz — Spectrum access for mobile communications.

It may be useful to consult the frequency allocation tables from your national or regional spectrum regulator to obtain a better understanding of frequency allocations in your area. For example, the spectrum regulator for the United Kingdom (UK) is the Office of Communications (Ofcom) [282] and the regulator for the United States (US) is the Federal Communications Commission (FCC) [159].

Many of the frequency ranges provided above are approximate bands in the spectrum where you should be able to find radio activity in your geographical area. For the best reception of radio signals, try moving your antenna close to an external window or door, or even move your antenna outside! Be careful if the weather is not in your favour, i.e. don't place the antenna outside if it is raining or there are high winds.

Some readers may be following this book without access to an RFSoC platform. If this is the case, then you can view some images of the spectrum acquired by the RFSoC in Figure C.1. These screenshots are each labelled with the content of the radio spectrum.

C.2. The RFSoC Spectrum Analyser

The spectrum analyser tool is available to use on your RFSoC platform by navigating to the following directory in Jupyter Labs: ***rfsoc_book/notebook_C/02_rfsoc_spectrum_analyser.ipynb***. When you execute this notebook, the RFSoC spectrum analyser tool will launch and you will be presented with the analyser's graphical interface. A diagram highlighting parts of the spectrum analyser interface is presented in Figure C.2.

- You can use the control panel highlighted in Figure C.2 to switch on the spectrum analyser and spectrogram. Centre frequency selection and bandwidth control can also be configured using the control panel's drop down widgets.

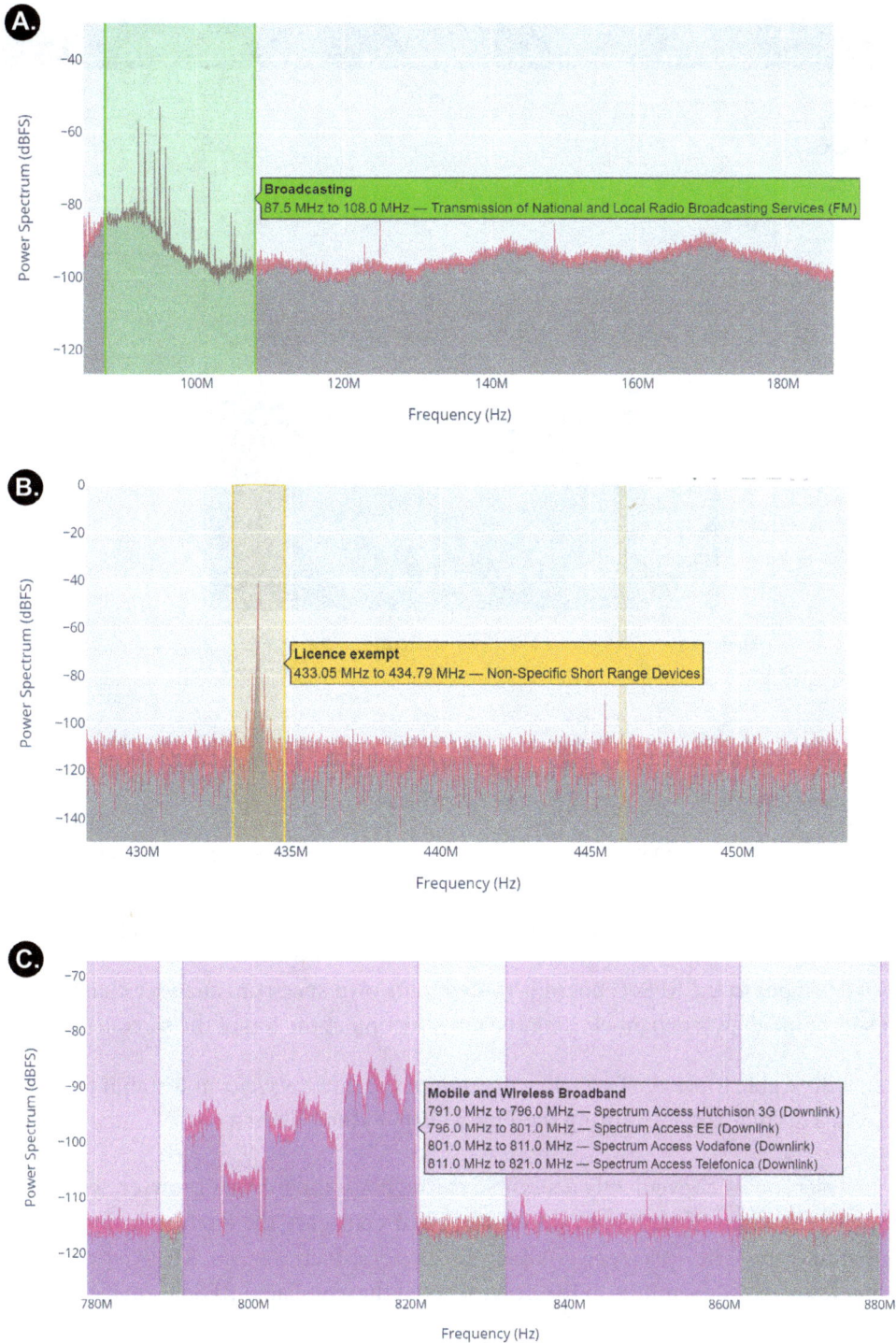

A.

Broadcasting
87.5 MHz to 108.0 MHz — Transmission of National and Local Radio Broadcasting Services (FM)

B.

Licence exempt
433.05 MHz to 434.79 MHz — Non-Specific Short Range Devices

C.

Mobile and Wireless Broadband
791.0 MHz to 796.0 MHz — Spectrum Access Hutchison 3G (Downlink)
796.0 MHz to 801.0 MHz — Spectrum Access EE (Downlink)
801.0 MHz to 811.0 MHz — Spectrum Access Vodafone (Downlink)
811.0 MHz to 821.0 MHz — Spectrum Access Telefonica (Downlink)

Figure C.1: Screenshot A: viewing the FM radio spectrum; Screenshot B: spectrum of a licence-exempt band (UK only); Screenshot C: spectrum from the mobile and wireless broadband sector.

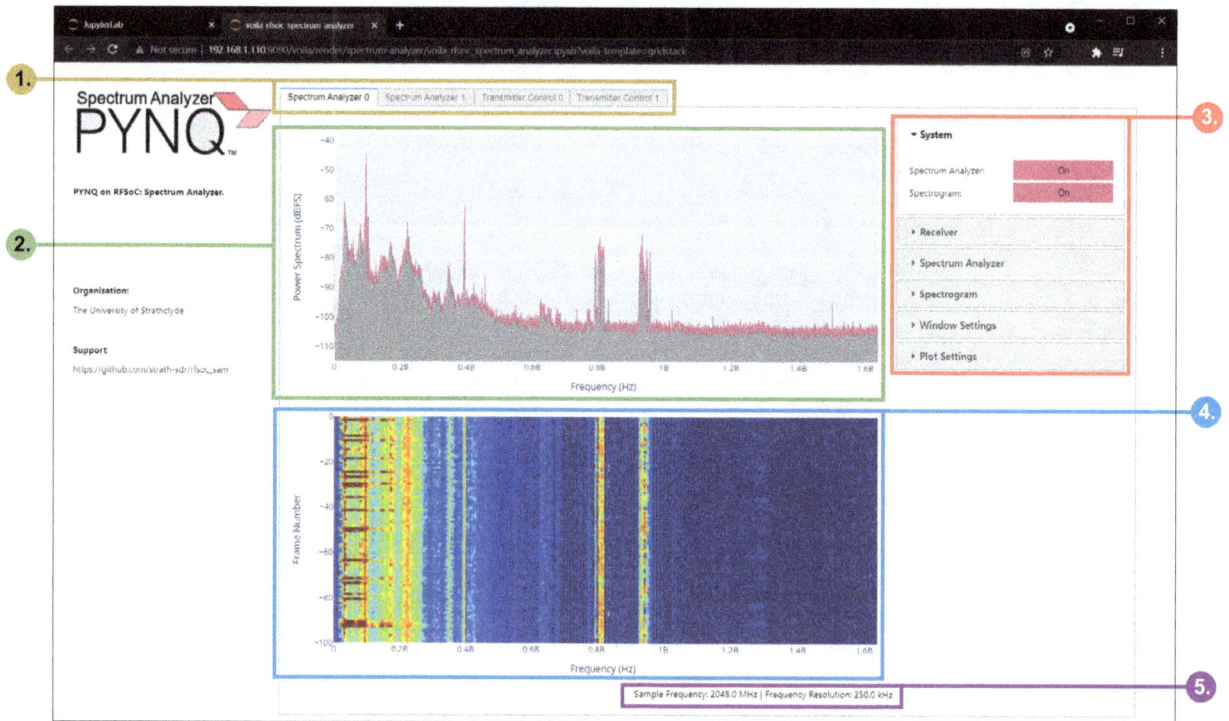

Legend / Key

1. Spectrum Analyser Tab Selection   **2.** Frequency Domain Plot   **3.** Control Panel

4. Spectrogram / Waterfall Plot   **5.** Information Panel

Figure C.2: The RFSoC spectrum analyser interface displayed in a web browser.

- The RF sampling frequency and resolution are reported in the information panel.

- Each RF-ADC input to the RFSoC board is routed to its own spectrum analyser channel. You can view a different spectrum analyser channel using the tab selection at the top of the screen.

- Lastly, you will be able to see the frequency domain of the input waveform through the plot in the centre of the screen. This plot is created using the Plotly [291] Python library.

The spectrum analyser tool is conveniently accessible through a standard web browser, where it is possible to inspect the frequency domain of acquired radio signals and configure the analyser using various properties and operating parameters. The spectrum analyser is supported on several RFSoC platforms and uses a different sampling frequency depending on the generation of the associated RFSoC device.

- Zynq UltraScale+ RFSoC Gen 1 devices are able to achieve a sampling frequency up to 4,096 MSps.

- Zynq UltraScale+ RFSoC Gen 3 devices are able to achieve a sampling frequency up to 4,915.2 MSps.

If you do not have an antenna or amplifier connected to the input of the RF-ADC channel, then you may notice that the spectrum analyser does not contain significant power in any of the frequency bands. See Section 13.7 for some useful tips that will help improve signal acquisition and visualisation of the spectrum.

As an example, it is possible for you to generate Wi-Fi signals and inspect them on the spectrum analyser. Wi-Fi signals can be generated using a Wi-Fi router or a mobile phone that uses 2.4GHz Wi-Fi channels. You should be able to inspect Wi-Fi signals by configuring the spectrum analyser to a centre frequency in the range 2,400 MHz to 2,500 MHz. It is important that the spectrum analyser is not set to apply any averaging techniques, as Wi-Fi signals only transmit across a short period of time. The spectrogram / waterfall plot can also be used to display the frequency spectrum over time.

In order to acquire Wi-Fi signals effectively, the RF-ADC front-end must be connected to an appropriate anti-aliasing filter to suppress unwanted frequency components. The RF-ADC input requires a bandpass filter, such as the Mini-Circuits® VBF-2435+ [263], to suppress the spectrum around the frequency band of interest.

You may also wish to try viewing the spectrum generated when you use your mobile phone (this has two components: the uplink (phone transmitting to basestation) and downlink (phone receiving from basestation). This is best done by switching off your phone's Wi-Fi connection (so as to direct all traffic over the mobile network — however, note that this may incur data charges, so be careful if necessary!). For instance, if you stream a video from YouTube or another source, then you should be able to view traffic in the downlink.

C.3. Additional Spectral Analysis Notebooks

We have also provided three notebooks that cover spectral analysis, which was previously introduced in Chapter 5. These notebooks can be used on your computer. The topics include Fourier's theorem, the DFT, and the FFT. See below for an enumeration of these extra notebooks and where you can find them.

ALL	03_fouriers_theorem.ipynb — *rfsoc_book/notebook_C/03_fouriers_theorem.ipynb*
ALL	04_investigating_the_dft.ipynb — *rfsoc_book/notebook_C/04_investigating_the_dft.ipynb*
ALL	05_fast_fourier_transform.ipynb — *rfsoc_book/notebook_C/05_fast_fourier_transform.ipynb*

Notebook 3 explores practical implementation of the Fourier Series and the Complex Fourier Series. The user will be able to interact with functions that perform harmonic analysis on square and sawtooth waves. Additionally, the functions can operate reveal the spectral content of any arbitrary periodic waveform. Euler's formula is also introduced.

In Notebook 4, we apply the DFT to time domain waveforms and explore the fundamental concepts of frequency domain analysis using Python. These concepts include investigating spectral resolution and leakage, computing the power spectrum, and implementing zero-padding techniques to improve spectral resolution.

Notebook 5 introduces practical implementations of the FFT. The user will investigate an implementation of the FFT using the *decimation-in-time* technique. The DFT and FFT will be compared for speed of operation and arithmetic resources. We will also describe and use the FFT support provided in the NumPy library. Finally, spectrogram plots and their implementations will also be explored.

Chapter 6

Wireless Communications Fundamentals

Louise Crockett

The aim of this chapter is to provide suitable wireless communications background for the SDR discussion that follows later in the book. We begin by introducing the multi-layered models used to describe communications protocols, focusing on the OSI model and TCP/IP models (protocols are sets of rules that are used for sending data and exchanging messages between communications nodes). The idea of a communications 'stack', i.e. the implementation of a layered set of communications protocols, is then introduced; we highlight some examples of wireless communications stacks, and consider how such a stack might be realised using SDR (in particular, the partitioning of different aspects to hardware or software).

Most of the design and implementation work covered in the book is at the Physical (or 'PHY') layer, which is the lowest layer in the OSI layered model and refers to the transmission of physical signals through the communications medium. The remainder of the chapter therefore focuses on various aspects of PHY layer radio principles, performance metrics, and design considerations.

6.1. Layered Models for Communications Systems

Several decades ago, communications systems began to shift from analogue to digital format and evolved towards data networks. New protocols were required to manage the transfer of data. Rather than dedicated point-to-point links, for instance to convey voice traffic between telephone users, communications systems were created to handle information sent in 'packets', which would share infrastructure with traffic from other users. Packets could potentially pass through a large number of intermediate nodes between source and destination. To aid the development of these more complex systems, conceptual models were developed to describe the different aspects of functionality required in communications terminals and network infrastructure, arranged into a vertical set of 'layers'. The purpose of these layered models is to provide commonly understood frameworks for the development of communications standards. Layers are defined such that they are independent from one another, and have clearly defined interfaces between layers.

In this section, we review the most widely used communications systems models, and discuss how the layers in these models map to the implementation of protocol 'stacks' for wireless communications.

6.1.1. The Open Systems Interconnection (OSI) Model

The Open Systems Interconnection (OSI) Model was first published in 1984 by the International Standards Organisation (ISO) and International Electrotechnical Commission (IEC), with the stated purpose *"to provide a common basis for the coordination of standards development for the purpose of systems interconnection, while allowing existing standards to be placed into perspective within the overall Reference Model"* [212]. It defines a set of seven layers, each describing the functionality of a communications system at a different level of abstraction. The OSI model does not provide any detail or guidance on how any of the seven layers should be implemented — rather, that is considered the remit of standards developed with reference to the model.

The OSI model is illustrated in Figure 6.1, which includes a brief outline of each of the seven layers (for a much fuller review, see [332]). In this diagram, Nodes A and B are communicating. The top layer is the Application layer, which represents the end use/user interface and is the highest level of abstraction in the model — an example might be the web browser on a tablet. The lowest level is the Physical (PHY) layer, which is concerned with the physical transmission and reception of data across a communications channel, or *medium* (whether as voltages on a wire, RF signals through a radio channel, optical pulses via an optical fibre, or another method). The PHY layer also includes the required hardware connections and signal processing operations.

The Data Link layer is distinct, in the sense that it contains two defined sub-layers. The lower is the Media Access Control (MAC) sub-layer, which manages the access of nodes to the physical transmission medium, and the upper is the Logical Link Control (LLC) sub-layer, which is responsible for the point-to-point links established at the MAC layer, including the coexistence of different types of traffic, and error mitigation.

In terms of interactions between layers, OSI protocols are conceived such that each layer in Node A can communicate with the equivalent layer in Node B; for instance, the Transport layer in Node A could interact directly with the Transport layer in Node B. In practice, however, the system implementation is likely to mean that only the PHY layers are directly connected (via the Physical Medium, as indicated in Figure 6.1). Under that assumption, for the Application layer in Node A to communicate with the Application layer in Node B, data passes all the way down the layers (7 through 1) of Node A, along the Physical Medium to Node B, and then all the way back up layers 1 to 7 of Node B, eventually reaching the Application layer. Network nodes that act as intermediaries (rather than producers or consumers of data) may only implement layers 1 - 3.

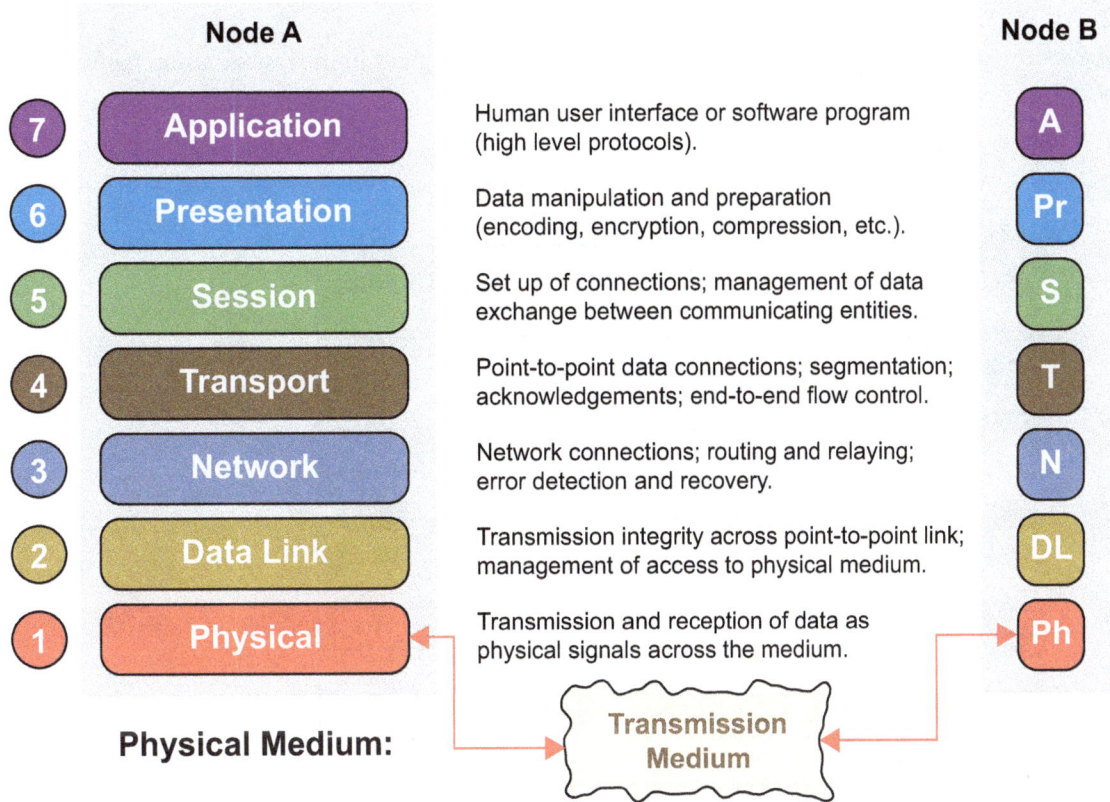

Figure 6.1: The OSI 7-layer model.

In traversing the layers at the transmit side of the link, data is *encapsulated* (and *de-encapsulated*) as depicted in Figure 6.2. In general terms, data from layer L is segmented into blocks, and prepended with protocol information to create a larger block, known as a *Protocol Data Unit* (PDU). Each PDU is therefore composed of a header (for the protocol information) followed by a payload (carrying the data), and in some cases also a trailer (with further protocol information). The PDUs from layer L are subsequently passed to layer $L-1$, which adds its own header information to each, forming new (larger) PDUs, and so on. At the receive side, in layer L, the header information in each PDU is extracted and used to implement the layer L protocol, and the payload is passed to layer $L+1$, above. Notice that the lowest layers must handle the greatest amounts of data.

PDUs are known by different terms, depending on the OSI layer. They are defined as:

- Layers 7 - 5:   Data
- Layer 4:   Segments
- Layer 3:   Packets

- Layer 2:   Frames

- Layer 1:   serialised into bits, and then physical signals in the transmission medium

OSI Layer	Protocol Data Unit (PDU) Composition	PDU Name

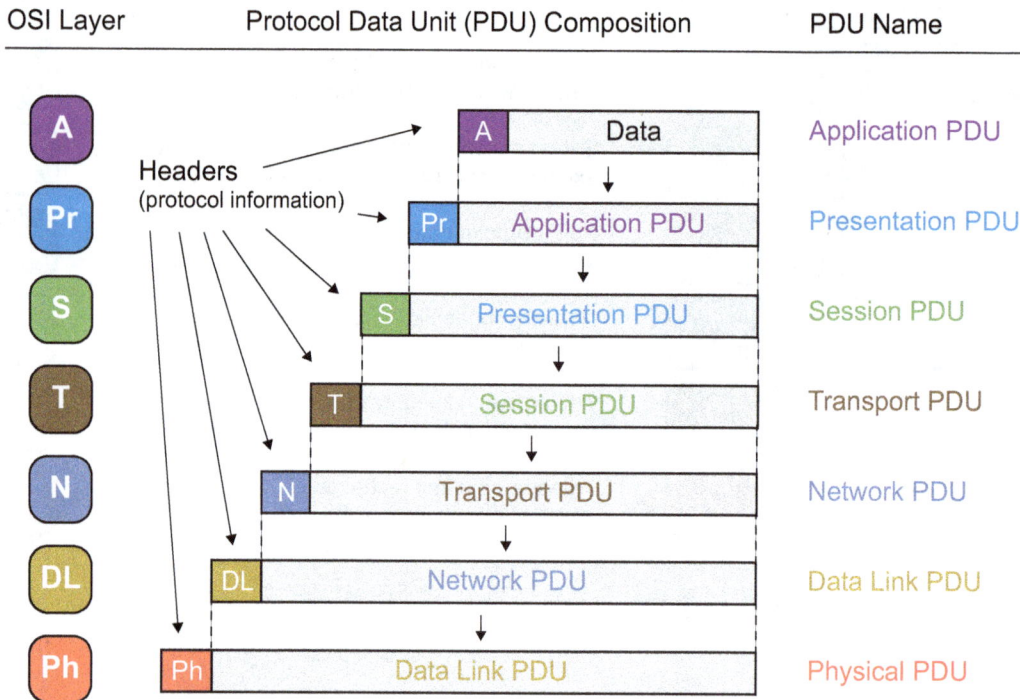

Figure 6.2: Data encapsulation in the OSI 7-layer model.

The protocol information included in the PDU headers (and trailers, where applicable), varies according to the layer, but can include fields such as addresses, sequence numbers, and error check information.

The OSI model can be successfully applied to a variety of communications systems. As such, the model is well known and studied; it provides a frame of reference for standards development, and a fairly universal context for understanding the structure of communications protocols. The OSI model has shown great endurance and remains a well used and referenced framework.

6.1.2. The TCP/IP Model

The Transmission Control Protocol / Internet Protocol (TCP/IP) model is defined with four or five layers[1], compared to the seven layers of the OSI model [322]. Its origins date back to the 1960's, when the US Defense Advanced Projects Research Agency (DARPA) began the computer networks research that created *the internet*.

The TCP/IP model therefore pre-dates the OSI model. It has been widely adopted, including in many wireless communications systems, and underpins the internet as we know it today.

The TCP/IP model is shown in Figure 6.3, which also clarifies how it maps to the OSI model. The TCP/IP model uses a similar method of encapsulating and de-encapsulating PDUs (not illustrated here), adding protocol information as the data moves down the layers at the transmit side, then extracting and utilising the protocol information at the receive side.

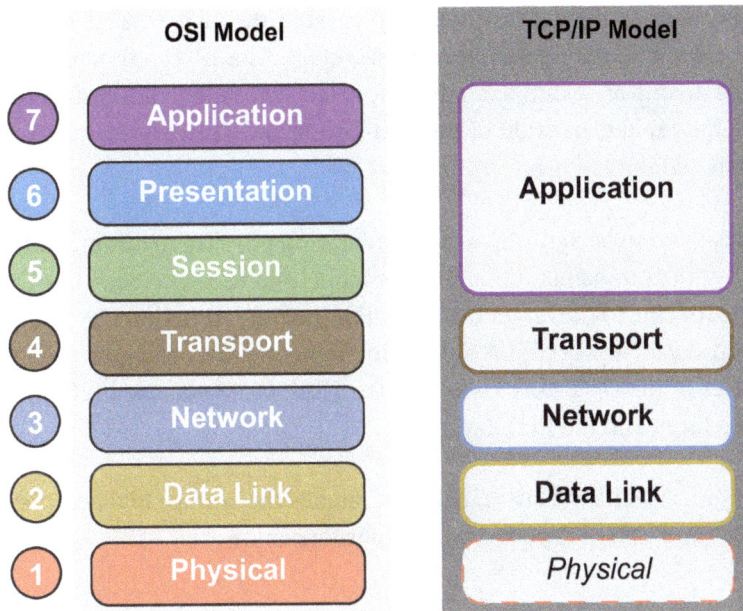

Figure 6.3: The TCP/IP model, and comparison to the OSI 7-layer model.

The term 'TCP/IP' also refers to the suite of internet-related protocols that are used to implement the model. There are several such protocols, which serve various functions throughout its four principal layers, the most important ones being TCP and IP (as the name suggests), along with User Datagram Protocol (UDP). These can be briefly summarised as follows:

- **TCP** [216],[217] — The TCP *Transport* layer protocol provides guaranteed delivery of data. Packets generated at the source are numbered, sent across the network, and reordered at the destination. A scheme of packet acknowledgements, timeouts and selective retransmissions is used to recover from any losses that occur, and this provides the guarantee of delivery, albeit at the cost of significant complexity.

---

1. The TCP/IP model can be presented in different ways. The model has five layers when the PHY layer is included. Alternative interpretations are that the TCP/IP model refers only to the four layers above the Physical layer, which is assumed to exist beneath it, or that the Data Link and PHY layers are combined into a single Network Access layer. We assume the five layer version here.

- **UDP** [292] — UDP is also a *Transport* layer protocol, but unlike TCP, does not provide any guarantees of successful delivery. This results in a far simpler protocol, where messages are only sent in one direction, and there are no acknowledgements and retransmissions. UDP is appropriate for applications where some level of data loss can be accommodated and low latency is an important attribute, such as audio and video calls.

- **IP** [215] — Both TCP and UDP assume that the *Internet* layer implements the IP protocol. The IP protocol enables data transmission from a source node on one network, to a desired destination node which may be on another network, potentially passing through several other networks on the way (hence the term *internet*, derived from *inter-networking*). The IPv6 protocol superseded IPv4, the most significant change being an extended address space (providing addresses for a larger number of endpoints, due to the rapid expansion of internet-connected devices). At the time of writing, both IPv4 and IPv6 remain in widespread use.

The TCP/IP model is designed to be agnostic to the underlying physical medium, and is therefore compatible with wireless, wired, or optical transmission. The protocols used to implement the Data Link layer, and the PHY layer beneath it, are specific to the target transmission medium, functionality and performance attributes. For instance, wired Local Area Networks (LANs) are implemented according to the ISO/IEC 8802[2] standard [213] for the LLC sublayer of the Data Link layer, and the IEEE 802.3 standard [206], which defines the MAC sublayer of the Data Link layer, and the PHY layer.

This very brief summary of TCP/IP, of course, barely scratches the surface of a complex and interesting topic. For further reading, [157] and [225] are good sources of information.

6.1.3. Communications Protocol Stacks

Considering the conceptual OSI model alongside the TCP/IP model for internet traffic, the major common feature is the use of layers to represent a 'stack' of protocols, each dealing with a different level of abstraction. This general idea of layers forming a stack is reflected by the standards defining most data communications systems, and their practical implementations. In this book, we are primarily concerned with wireless communications systems, and therefore it is pertinent to outline a few examples of wireless communications stacks. Three types are featured: Wi-Fi (for wireless LANs), Digital Enhanced Cordless Telecommunications (DECT™, primarily used for cordless telephony), and 5th Generation New Radio (5G NR, for cellular networks). Even as new 6G standards and paradigms emerge, the adherence to a layer and stack model will continue.

Wi-Fi

Wireless Local Area Networks (WLANs) are ubiquitous: they are widely used in the workplace, at home, and in public spaces, to provide wireless data connectivity inside buildings, and outdoors up to a range of ~100m.

---

2. Previously published as IEEE standard 802.2.

The name 'Wi-Fi' is the well known and widely used, however the standards underpinning the technology are maintained by the IEEE, and known as IEEE 802.11 [207]. There are several variants of this standard, e.g. targeting different frequency bands, and supporting different data transfer rates; these are denoted with lettered suffixes, as IEEE 802.11a, IEEE 802.11b, and many others.

IEEE 802.11 is used in the lower part of the Data Link layer (the MAC sublayer), and the PHY layer, as shown in Table 6.1. They are specified such that the sublayer above (the LLC Data Link sublayer) can co-operate with them in the same way as other network implementations, such as wired Ethernet (IEEE 802.3).

The upper layers are not considered part of Wi-Fi, although Wi-Fi is typically used in the context of TCP/IP networks, where protocols from the TCP/IP model (such as IP at the Network layer, TCP at the Transport layer, and HTTP at the Application layer) would be adopted.

Table 6.1: Wi-Fi communications stack.

Layer	OSI Model	TCP/IP Model	Wi-Fi Stack
-	Application	Application	
-	Presentation		
-	Session		
-	Transport	Transport	
-	Network	Network	
2	Data Link	Data Link	LLC sublayer: IEEE/ISO/IEC 8802 MAC sublayer: **IEEE 802.11**
1	Physical	Physical	**IEEE 802.11**

Digital Enhanced Cordless Telecommunications (DECT)

The primary purpose of DECT is to convey speech signals for digital cordless telephony. The DECT standard defines four protocol layers (Network, Data Link Control, MAC, and Physical), which map to the lowest three levels of the OSI model, and two planes — the Control Plane and the User Plane [153]. The mapping of these DECT planes and layers to the OSI model are summarised in Table 6.2.

The DECT PHY layer defines a Time Division Multiple Access (TDMA) protocol which is across multiple RF carriers. A frame structure is defined, which repeats every 10ms and contains 24 timeslots; each timeslot provides the opportunity to transmit one packet generated and passed down from the MAC layer. The MAC layer also selects physical channels, and sets up and closes connections across those channels.

Above the MAC layer, the protocol separates into two planes at the Data Link Control layer: a user plane for data traffic, and a control plane for management traffic and tasks. The Data Link Control Layer is responsible for link reliability and data integrity, while the Network layer is for signalling, e.g. to establish and release calls. Note that DECT does not define any layers above OSI layer 3 [153].

Table 6.2: DECT communications stack [153].

Layer	OSI Model	DECT Stack (*Control Plane*)	DECT Stack (*User Plane*)
-	Application		
-	Presentation		
-	Session		
-	Transport		
3	Network	Network layer	
2	Data Link	Data Link Control layer	Data Link Control layer
		(most of) MAC layer	
1	Physical	(part of) MAC layer	
		Physical layer	

5th Generation New Radio (5G NR)

5G NR is the most advanced wireless cellular standard at the time of writing, and it offers enhanced capabilities over its predecessor, 4th Generation Long Term Evolution (4G LTE), in terms of performance and functionality. In particular, 5G NR targets a triumvirate of use-cases: Massive Machine-to-Machine type Communications (MMMC) and Ultra Reliable Low Latency Communications (URLLC), as well as Enhanced Mobile Broadband (eMBB). Although a new set of standards, 5G builds and evolves the designs and architectures of 4G. A similar evolution from 5G to 6G is now also in play, with 6G standards, applications and new paradigms moving forward and building on the SDR designs from 5G.

5G NR defines three layers (denoted as Layers 1 - 3), of which Layers 2 and 3 are composed of sublayers. The mapping of these 5G NR layers and sublayers to the OSI model is captured in Table 6.3.

Variations on the protocol stacks apply for the control and user planes, and also for different components within the network, the detail of which are outside the scope of the current discussion. We will also omit to review the functionality of each of the protocols given in Table 6.3 (for more information, see [11]). The main point of interest is that, similar to the Wi-Fi and DECT examples, the 5G NR standard only specifies protocols equivalent to the lowest layers of the OSI model, primarily the Data Link and PHY Layers.

The Network layer, directly above the Data Link layer, contains the Layer 3 5G NR Radio Resource Control (RRC) protocol for the control plane, while the user plane of 5G NR Layer 2 interfaces with IP in the Network layer, which is defined separately from the 5G NR standard. Applications, in this context, may ultimately be 'apps' on a smartphone, supported by the appropriate Transport layer protocols (such as TCP, UDP, etc.), and these sit above the 5G NR specification.

Table 6.3: 5G NR communications stack [11].

Layer	OSI Model	5G NR Stack (layers)	5G NR Stack (sub layers)
-	Application		
-	Presentation		
-	Session		
-	Transport		
3	Network	-	*Internet Protocol (IP) — user plane*
		Layer 3	Radio Resource Control (RRC) — control plane
2	Data Link	Layer 2	Service Data Adaptation Protocol (SDAP)
			Packet Data Convergence Protocol (PDCP)
			Radio Link Control (RLC)
			Medium Access Control (MAC)
1	Physical	Layer 1	Physical layer (PHY)

Other Standards and Proprietary Schemes

There are many other standards-based wireless stacks that could be mentioned here, from broadcast networks such as Digital Audio Broadcast (DAB) [359] and Digital Video Broadcast (DVB) [143]; to Internet of Things (IoT) standards like LoRaWAN®, SigFox®, Zigbee and NBIoT [101]; to Personal Area Networks (PANs) including Bluetooth and Near Field Communication (NFC) [106]; and many others. And as 5G becomes an integral part of wireless and mobile deployments, we are already now seeing the design and development of the next evolved wireless generation — the new 6G standards, applications and paradigms — that can be built on RFSoC technologies.

There are also industry-specific standards for transportation, public safety, and military uses, and innumerable proprietary protocols for different applications. Even simple wireless communications systems, such as wireless doorbells, car keyfobs and so on, have defined protocols which could be mapped to the OSI model as a protocol stack.

Observations

From considering these examples of protocol stacks, the layer that is intrinsically 'wireless' is the PHY layer, as it interfaces with the physical medium of the radio channel. The MAC sublayer of the OSI Data Link layer is designed for the wireless medium, too, as its role is to implement the necessary protocols for accessing the radio spectrum, and the requirements for wireless and wired media differs — for instance, in radio systems there may be other concurrent transmissions in the target band, arising from the same or a third party system, which affect channel access.

The degree to which wireless standards specify the upper Data Link Layer (the Logical Link sublayer) and above may differ. In the three examples considered here, i.e. Wi-Fi, DECT, and 5G NR, only the bottom two or three layers of the OSI model are specified.

To form a complete system, the upper layers are tailored to the types of services and applications offered. Taking 5G NR as an example, there are many possible applications (think of all of the apps on your smartphone!) and these would be facilitated by the appropriate higher layer protocols.

6.1.4. SDR Implementation of Communications Stacks

In designing an SDR, consideration must be given to how the communications stack should be implemented, and in particular, how the different layers of the stack should be mapped to the facilities of the target platform. This section provides some comments on possible approaches, with reference to the layers of the OSI model.

In the context of RFSoC, we conceive a general wireless transceiver model as shown in Figure 6.4. This may not be suitable for all communications standards or scenarios; rather it is presented as a starting point for thinking about SDR implementation. As shown here, the PHY layer and parts of the MAC sublayer can be targeted to the PL portion of the RFSoC, as these algorithms are suited to hardware implementation, and benefit from low latency, deterministic operation. Upper layers are often be implemented in software on the PS, with optional hardware acceleration of suitable algorithms. Given the RFSoC architecture, the partitioning of the system implementation across hardware and software is flexible.

Lower Layers (MAC and PHY)

Based on the OSI model considered previously, and the associated encapsulation and de-encapsulation of data, we observed that the PHY layer must handle the greatest amount of data, as it includes all of the protocol information from the upper layers. The computational demands on the PHY layer implementation are the most challenging of all the layers, not just because of the volume of data, but also the algorithms and operations needed to prepare data for transmission across the physical radio channel, and recover it at the receiver. This includes various stages of filtering, synchronisation, coding/decoding, and so on.

The PHY layer requires high throughput operation, but takes the form of a constant flow of data requiring deterministic signal processing operations. In this context, we use the term deterministic to mean that processing stages have consistent latencies, and that the relative timing of data paths is maintained, both of which are important factors in successfully implementing the designed DSP algorithms. Therefore, the PHY layer maps well to hardware implementation on an FPGA, or equivalently the PL portion of an RFSoC device[3].

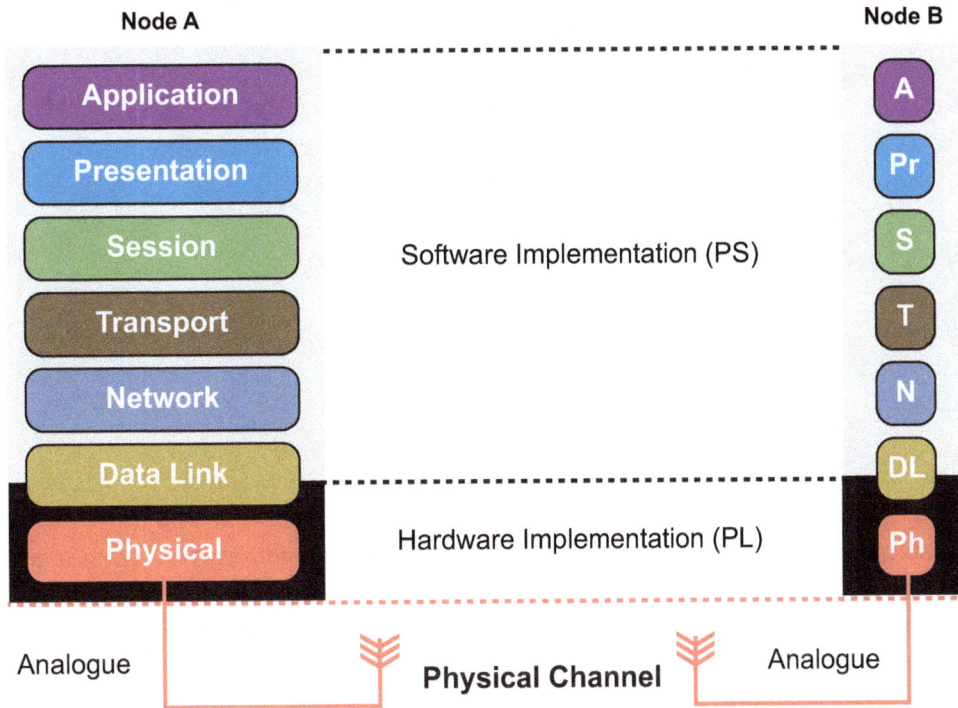

Figure 6.4: An indicative model for software / hardware partitioning of a radio onto RFSoC resources (alternative partitioning across hardware and software is possible).

The MAC layer (i.e. the lower sublayer of the Data Link layer) is also required to handle high data throughputs. At this layer, data is partitioned into frames, which in some standards may be associated with different logical channels (e.g. user data and control channels) in the layers above; one of the tasks of the MAC layer is therefore to multiplex and demultiplex these frames.

The MAC layer must also determine when to access the radio channel for transmitting data, for instance using the Carrier Sense Multiple Access / Collision Avoidance (CSMA/CA) algorithm [125] in Wi-Fi networks. This method involves "sensing" the radio channel to check if it is busy with transmissions from other users, before attempting to send a frame when the channel appears to be free (recall that Wi-Fi operates in a shared band). Whether operating in shared bands or licensed bands, the MAC layer often implements a scheme of selectively retransmitting frames, based on the knowledge (or assumption) of failed frame transmissions. This is commonly achieved using an Automatic Repeat reQuest (ARQ) handshaking scheme involving acknowledgements of successfully received frames, non-acknowledgements of corrupted frames, and/or timeouts [227].

---

3. Depending on the data rates involved, some aspects of the PHY layer can be implemented on a processor if desired.

Hybrid-ARQ (HARQ) schemes are a variation on ARQ: they add an element of FEC to reduce the requirement for retransmissions at the cost of additional redundancy [11].

Aspects of the MAC layer therefore need to operate at high rates, for frame processing, and in HARQ schemes to implement FEC coding and decoding. There is also a degree of determinism required. Therefore, some MAC layer protocols can also be mapped successfully to PL; this is generally considered to be the 'lower MAC'.

Upper Layers

As is evident from the review of wireless standards in the previous section, the upper layers of wireless communications stacks can vary considerably. They also require to process data at lower rates (more so towards the top of the stack), given the formation of PDUs as shown in Figure 6.2.

The functions implemented in the uppermost layers are naturally implemented in software. Examples of Application and Presentation layer protocols are, for instance, web browsers and other GUIs, and protocols like HyperText Transfer Protocol (HTTP), and XML. The Transport and Network layers involve tasks that are also best implemented in software; for instance, the implementation of TCP and routing protocols. Therefore, collectively, the upper layers map better to a processor running software, than to a hardware architecture implemented in PL. There are, however, some upper layer tasks that may be suitable for acceleration using a hardware co-processor, such as coding and decoding.

A Focus on the PHY Layer

In this book, attention is generally focused on the PHY layer. These are the elements of the radio that deal with physical signals for radio transmission and reception, which are most closely associated with the wireless medium, and where we make the most specific use of the PL and the RFSoC's hardened blocks (particularly the RFDCs). We additionally touch on the MAC layer via the RFSoC Radio Demonstrator (see Notebook Set G), and discuss the various layers of the 5G NR communications stack in Chapter 17.

6.2. The Wireless Physical Layer

Focusing on the PHY layer of the radio, we can present a basic overview model of Quadrature Amplitude Modulation (QAM) transmitter and receiver architectures in Figure 6.5, which provides a basis for the remainder of the chapter.

The major parts of the **transmitter** architecture are:

- Baseband modulation (bit-to-symbol mapping)

- Pulse shaping

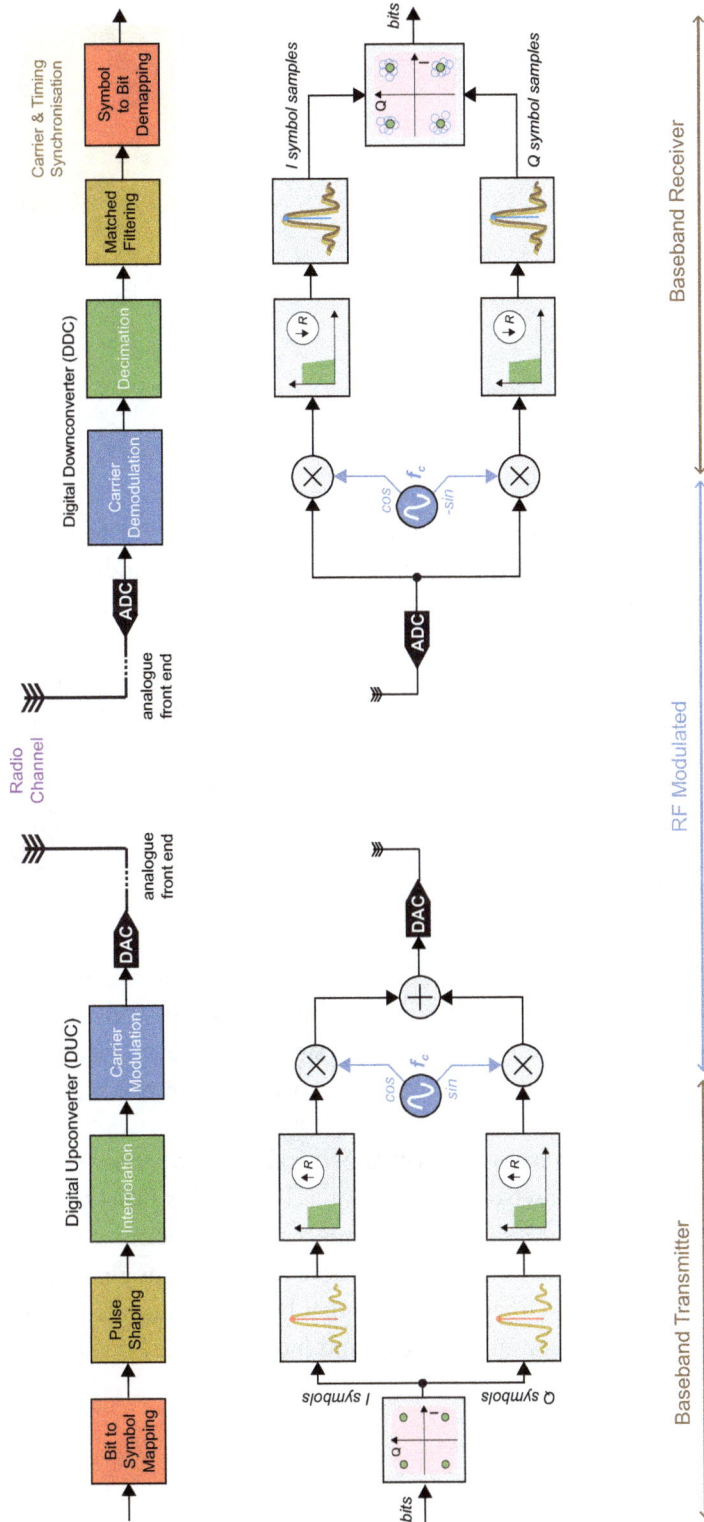

Figure 6.5: Overview of PHY layer transmitter and receiver architectures.

- Interpolation

- Carrier modulation

The major parts of the **receiver** architecture are:

- Carrier demodulation

- Decimation

- Matched filtering

- Baseband demodulation (symbol to bit demapping)

- Synchronisation (carrier and timing)

This chapter will concentrate on baseband modulation and demodulation (bit-to-symbol mapping, and vice versa), and the processes of pulse shaping and matched filtering. Carrier modulation and demodulation is covered in Chapter 7, while interpolation and decimation are discussed in Chapter 8.

Also within this chapter, we discuss the wireless channel, and the various degradations that radio signals are subject to as they pass through the channel. One of the issues encountered is that the transmitter and receiver are not synchronised in frequency and time, and we discuss the effects of this (although not the solutions — please see a textbook such as [301] or [326] for more information on implementing synchronisers).

6.3. Baseband Modulation (Bits to Symbols)

As previously noted in Section 2.1.2, a communications signal is said to be at *baseband* if it is composed of frequency components close to 0 Hz. At the transmit side of the wireless link, this corresponds to a signal that has not yet been modulated with a carrier to form a bandpass signal centred at IF or RF. At the receive side, the signal is at baseband after it has been demodulated from an IF or RF carrier, and is again close to 0 Hz.

In order to send digital data over the physical wireless channel, and prior to carrier modulation, it must first be converted from bits to symbols using a baseband modulation scheme. This process is referred to as *baseband modulation*. The modulation scheme defines the symbol mapping, i.e. the number of symbols, their amplitude levels or phases, and how groups of bits are converted to these symbols. The resulting pattern of symbols is often referred to as a *constellation*. It is also worth clarifying that baseband modulation, and modulation onto a carrier signal, are two distinct processes (sometimes, baseband modulation is known as *bit-to-symbol mapping*, which is arguably clearer!).

The remainder of this section reviews the baseband modulation schemes that are used to map bits to symbols. Digital modulation schemes can convey data by changing the amplitude, phase, or frequency of the signal. We focus on the first two of these in the discussion that follows.

6.3.1. Quadrature Modulation and Symbol Space Dimensions

Before proceeding further, a quick note on the symbol space that symbol constellations occupy, which may be depicted as one-dimensional or two-dimensional.

The majority of SDR architectures considered in this book modulate data onto both sine and cosine carriers simultaneously (*quadrature modulation*), as opposed to a single cosine carrier. The motivation is bandwidth efficiency — twice as much information can be carried within the same bandwidth, if sine/cosine orthogonal carriers are used.

This topic will be covered in more detail in Chapter 7, but for now, the main point of interest is how it defines the mapping of bits to symbols: with quadrature schemes, the symbol space is defined in two dimensions, as depicted in Figure 6.6. The x-axis represents the amplitude of the *In Phase* component (also known as the *Real* component), while the y-axis represents the amplitude of the *Quadrature Phase* component (or *Imaginary* component). The resulting two-dimensional space is therefore often referred to as the *I-Q plane* (i.e. the In Phase / Quadrature plane). If there is no Quadrature Phase, the signal space is one-dimensional and all symbols are conveyed by a position on the x-axis.

Example symbols are depicted on the axes in Figure 6.6 as an indication of how the spaces are used — more on this shortly.

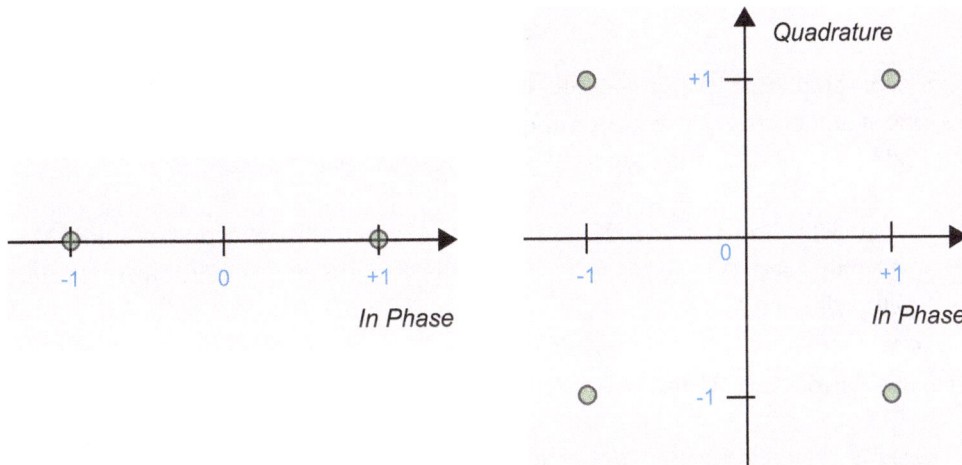

Figure 6.6: Symbol space for: (left) single carrier and (right) quadrature carrier baseband modulation schemes.

6.3.2. Amplitude Shift Keying (ASK)

Amplitude Shift Keying (ASK) conveys data on a single phase (i.e. where there is a single cosine carrier) by mapping the amplitude of a baseband signal to a discrete level from a defined set. In the simplest case, there are two such levels (+1V and -1V), which may be denoted as 2-ASK. If four levels are used, the scheme is referred to as 4-ASK, and the applicable levels are +1V, +1/3V, -1/3V, and -1V, and so on.

Taking the example of 2-ASK, each bit is mapped to one of two symbols, corresponding to amplitudes of +1 and -1, as shown in Figure 6.7. The symbol mapping diagram shows that both of these symbols belong to the In Phase component (in fact there is no Quadrature Phase component in this case). Using the 2-ASK modulation scheme, one bit corresponds to one symbol, and therefore the bit and symbol rates are the same.

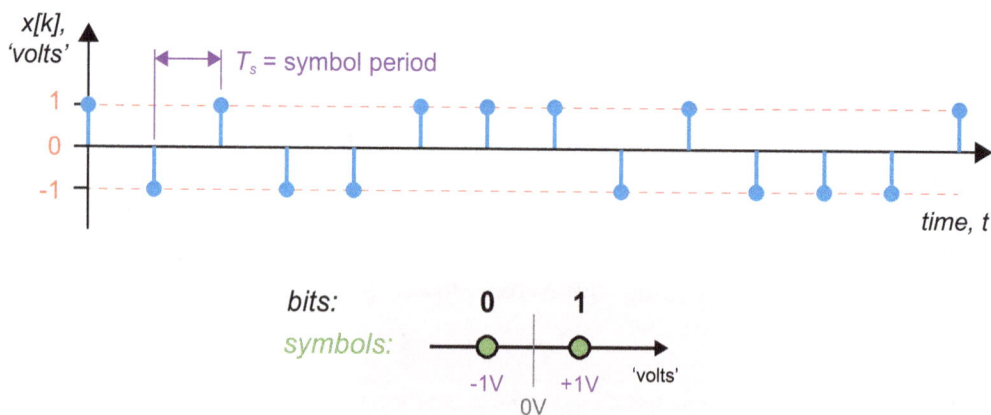

Figure 6.7: Example time domain waveform for 2-ASK (top), and 2-ASK symbol mapping diagram (bottom)

There are four possible amplitude levels in 4-ASK, and therefore two bits are needed to represent each symbol. In other words, one symbol conveys two bits of information, as can be confirmed by inspection of Figure 6.8. For a given baud rate (symbol rate), the supported bit rate is twice that of 2-ASK.

If we consider two parallel channels of 2-ASK, and conceive that one channel corresponds to the x-axis of the symbol mapping diagram, and the other to the y-axis, the concept of Quadrature Amplitude Modulation (QAM) can be developed.

6.3.3. Quadrature Amplitude Modulation (QAM)

In QAM, two baseband channels are used: one is denoted the *In Phase* or *Real* channel and is modulated onto the cosine carrier, while the other is denoted the *Quadrature Phase* or *Imaginary* channel as is modulated onto the sine carrier. As the carrier signals are sine and cosine at the same frequency, they are separated by 90°, and

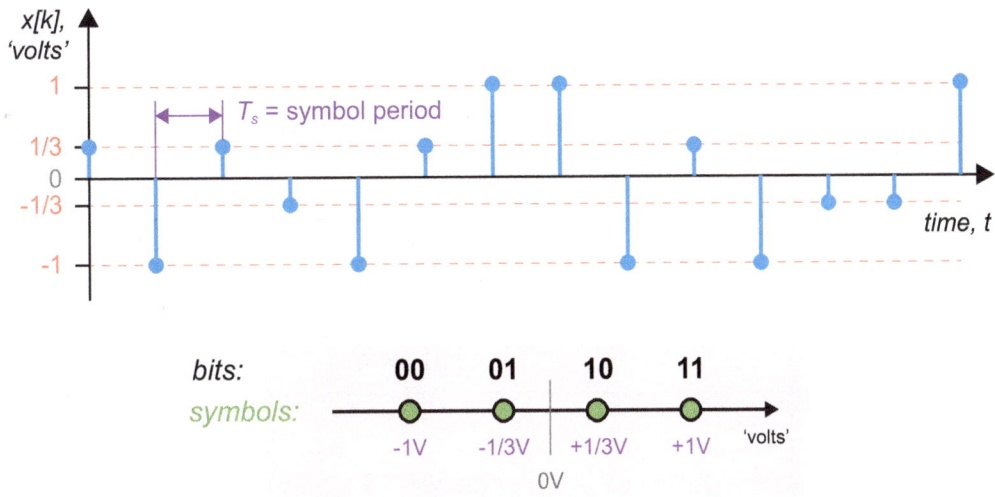

Figure 6.8: Example time domain waveform for 4-ASK (top), and 4-ASK symbol mapping diagram (bottom).

are orthogonal to each other. This property of orthogonality ensures that the data sent on the two channels remains separate, and can be perfectly recovered, i.e. one channel does not interfere with the other.

An example of two baseband signals for QAM are shown in Figure 6.9. In this case, each of the two channels (or phases) transmits data equivalent to the 2-ASK modulation scheme. In total, there are four possible symbols in the resulting symbol mapping, meaning that two bits are transmitted per symbol. This modulation scheme is therefore known as 4-QAM (and is equivalent to 4-PSK, as will be discussed in the next section).

To clarify the terminology, the following terms are equivalent: *In Phase, Real Component / Phase, Channel 1*; and the following terms are also equivalent: *Quadrature Phase, Imaginary Component / Phase, Channel 2*.

Similar to ASK, the number of levels on each of the two QAM phases can be increased, leading to a larger set of symbols. Conventionally, a power-of-two number of evenly spaced amplitude levels are used per phase. The next largest QAM symbol mapping is 16-QAM, where there are four amplitude levels on each of the two phases, and this scheme conveys four bits per symbol. A diagram of 16-QAM is provided in Figure 6.10.

Table 6.4 provides a summary of QAM modulation schemes, extended to include larger symbol mappings and restricted to square patterns (32-QAM, 128-QAM etc., are also possible and these produce non-square shapes).

Note that larger constellations convey a greater number of bits per symbol; therefore, for any given symbol rate, selecting a larger QAM scheme will result in a higher bit rate.

Table 6.4: Summary of QAM modulation scheme sizes.

Scheme	Levels per phase	Bits per phase	Total bits per symbol
4-QAM	2	$\log_2(1) = 1$	$1 + 1 = 2$
16-QAM	4	$\log_2(4) = 2$	$2 + 2 = 4$
64-QAM	8	$\log_2(8) = 3$	$3 + 3 = 6$
256-QAM	16	$\log_2(16) = 4$	$4 + 4 = 8$
1024-QAM	32	$\log_2(32) = 5$	$5 + 5 = 10$
4096-QAM	64	$\log_2(64) = 6$	$6 + 6 = 12$

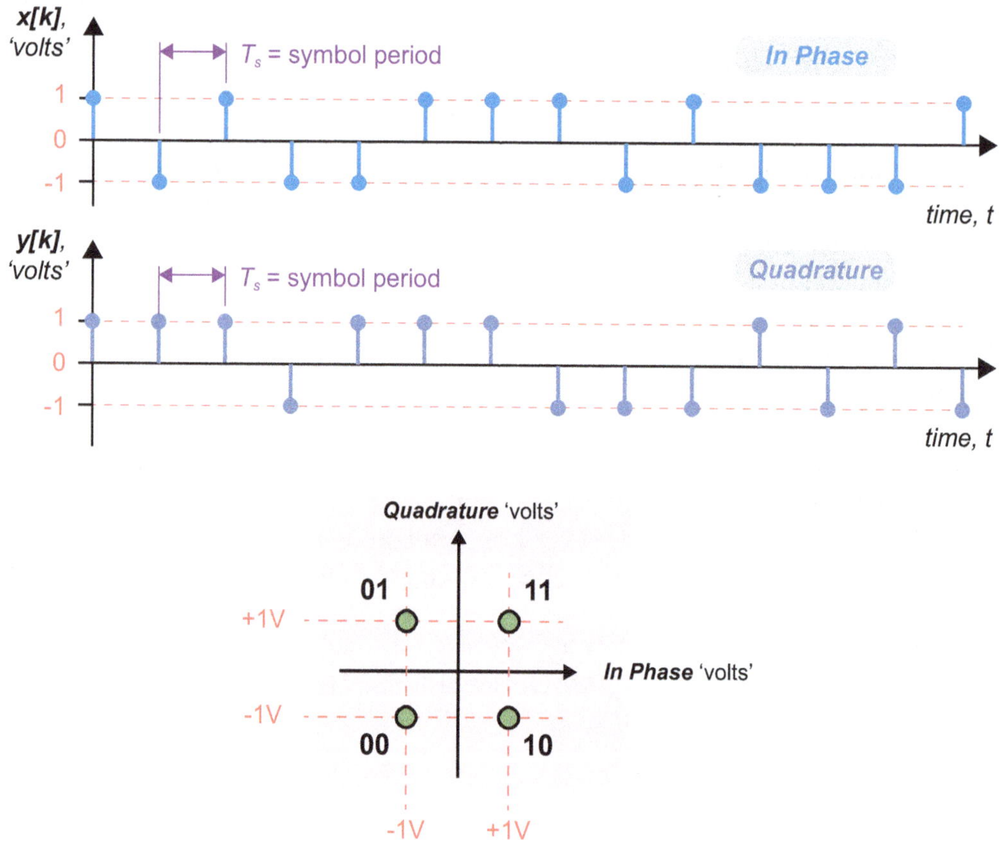

Figure 6.9: Example time domain waveform for 4-QAM (top), and 4-QAM symbol mapping diagram (bottom).

As will be discussed in Section 6.4, the disadvantage of larger QAM schemes is that their performance is more susceptible to degradation due to noise. Therefore, smaller QAM schemes such as 4-QAM and 16-QAM are preferred in noisy environments.

6.3.4. Phase Shift Keying (PSK)

Another method of digital modulation involves modulating the phase of the signal. For *m*-PSK, data bits are encoded by symbols that are placed at a set of *m* evenly spaced phases around the 360° degree range. Figure 6.11 illustrates the two lowest order PSK schemes, 4-PSK (also known as Quaternary Phase Shift Keying, or QPSK), and 8-PSK. Both larger and smaller PSK symbol mapping sizes are also possible, as summarised in

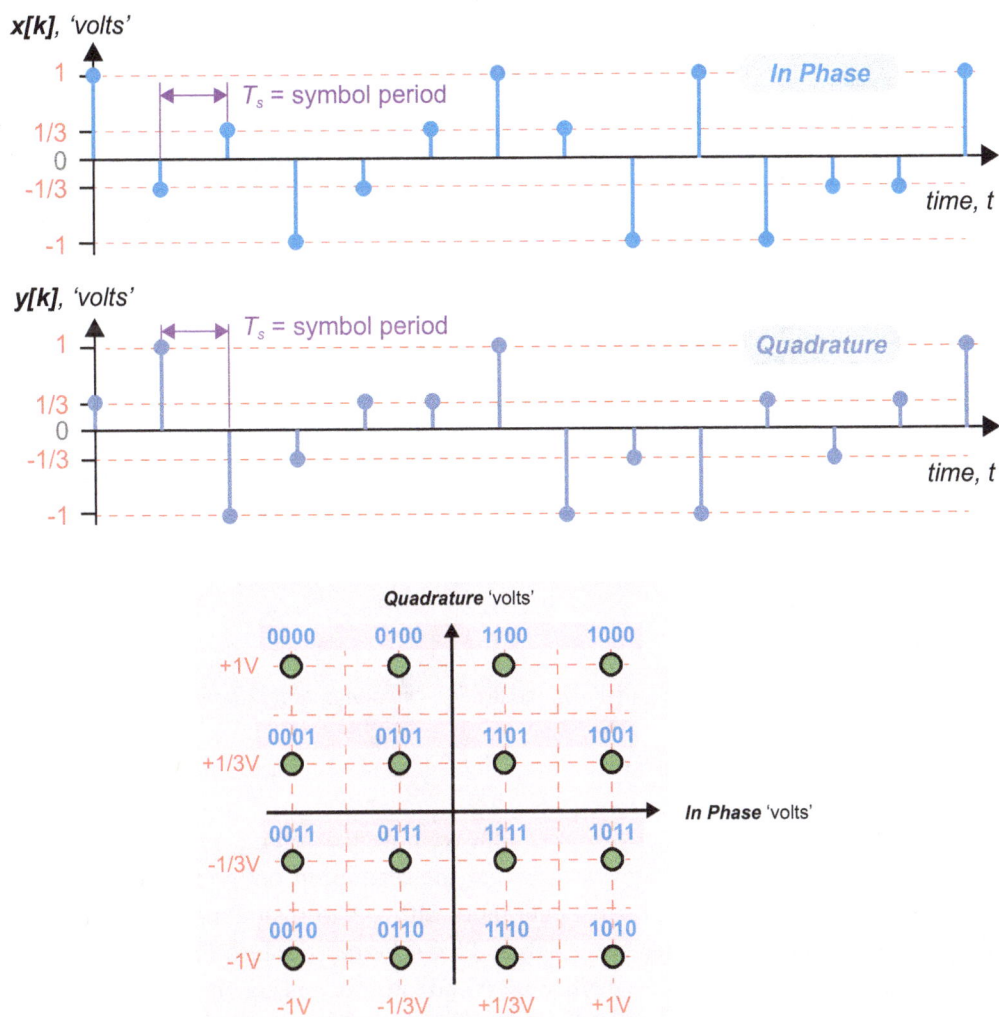

Figure 6.10: Example time domain waveform for 16-QAM (top), and 16-QAM symbol mapping diagram (bottom).

Table 6.5; Binary PSK (BPSK) is a special variation which does not have a quadrature component (its two symbols are placed at -1 and +1 on the *In Phase* axis).

Table 6.5: Summary of PSK modulation scheme sizes.

Scheme	Number of Phases	Phase Separation	Bits per symbol
BPSK (Binary PSK)	2	$360^{\circ} / 2 = 180^{\circ}$	$\log_2(2) = 1$
4-PSK	4	$360^{\circ} / 4 = 90^{\circ}$	$\log_2(4) = 2$
8-PSK	8	$360^{\circ} / 8 = 45^{\circ}$	$\log_2(8) = 3$
16-PSK	16	$360^{\circ} / 16 = 22.5^{\circ}$	$\log_2(16) = 4$
32-PSK	32	$360^{\circ} / 32 = 11.25^{\circ}$	$\log_2(32) = 5$

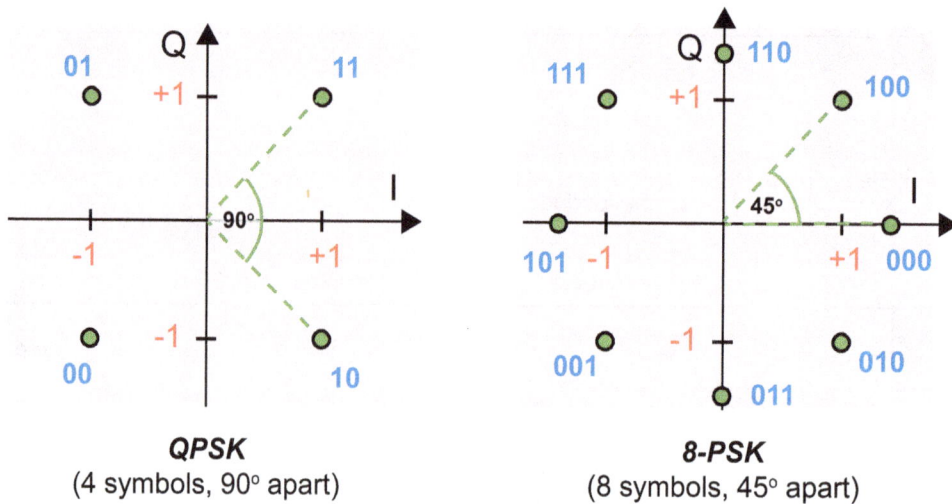

QPSK
(4 symbols, 90° apart)

8-PSK
(8 symbols, 45° apart)

Figure 6.11: QPSK (4-PSK) and 8-PSK symbol mapping schemes.

Similar to the QAM schemes from the previous section, we note that higher order PSK schemes convey a greater number of bits per symbol. The drawback is that the separation between symbols becomes smaller (here in terms of phase), and therefore they become more difficult to distinguish in the presence of noise. Larger QAM schemes are generally preferable to larger PSK schemes because they perform better in noisy conditions — this is because the symbols are further apart and fewer errors are encountered for any given level of AWGN.

The process of converting bits to symbols using a PSK scheme involves generating the required *In Phase* and *Quadrature* amplitudes. The amplitudes may be scaled compared to those shown in Figure 6.11, to normalise the output power.

6.3.5. Other Modulation Schemes

In addition to the QAM and PSK modulation schemes presented in the preceding sections, there are several other types of digital modulation schemes. These include:

- **On Off Keying (OOK)** — A very simple scheme wherein a carrier signal is multiplied with a '0' or '1', depending on the bit value, to transmit one bit per symbol. This has the effect of switching the carrier signal on or off, as the name suggests. A disadvantage of OOK is the sharply changing power envelope, which can create challenges in the analogue parts of the transmitter and receiver, such as power amplification and active gain control.

- **Frequency Shift Keying (FSK)** — Symbols are mapped to discrete frequencies from a defined set. For instance, 4-FSK comprises four symbols, corresponding to a set of four different frequencies, and conveys two bits per symbol. As the power envelope is constant, this avoids the issues of OOK, however there may be sharp phase transitions between symbols, which can expand the occupied bandwidth considerably compared to the set of frequencies defined by the modulation scheme.

- **Minimum Shift Keying (MSK)** — MSK is a form of FSK, which addresses the issue of phase discontinuities experienced in FSK by ensuring that transitions between symbols occur at the zero-crossing points. Variations of MSK include Gaussian MSK (GMSK).

- **Offset Schemes, e.g. Offset-QAM** — Offset QAM schemes involve changing the *In Phase* and *Quadrature* components of the transmitted symbols in a staggered manner, i.e. the *Quadrature* symbol transitions occur half a symbol period after the *In Phase* transitions. Offset QPSK is another offset scheme, whereby alternate symbols in the sequence are transmitted using mappings offset by 45°.

These schemes are mentioned here for background interest, but QAM (in particular) and PSK are the dominant digital baseband modulation techniques, especially for high value systems, and will receive the majority of attention in the remainder of the book.

6.4. Baseband Demodulation (Symbols to Bits)

Next, we turn our attention to the receiver, which has the opposite task to perform: translating the received symbols back to bits. For the purposes of our current discussion, we abstract away the modulation and demodulation processes that take place between baseband modulation and demodulation in a real radio system, and concentrate on the retrieval of bits from received symbols. As well as considering the process of converting received symbol samples back to bits, performance metrics are also reviewed.

The plotted received symbol samples, in the two-dimensional I-Q plane, is often called the received symbol *constellation*. The term *constellation* can also be used to describe the original symbol mapping, e.g. *reference constellation*.

6.4.1. Symbol Decisions

First, we consider the symbol-to-bits demapping process under ideal conditions where the radio channel does not degrade the signal. The receiver determines which is the closest symbol to the received sample — under perfect conditions, the transmitted and received symbols are the same, and so the process is trivial.

More generally, symbol decision boundaries can be conceived between the symbols defined in the original mapping, such that the closest possible symbol to each received symbol sample is assumed. Symbol decision boundaries are depicted in Figure 6.12 for the examples of 4-QAM / QPSK (recalling that these are equivalent), and 8-PSK. An example of 16-QAM will be included a little later. Note that in QAM schemes, the lines are drawn horizontally and vertically between symbols, and in PSK, at equally spaced angles between symbols.

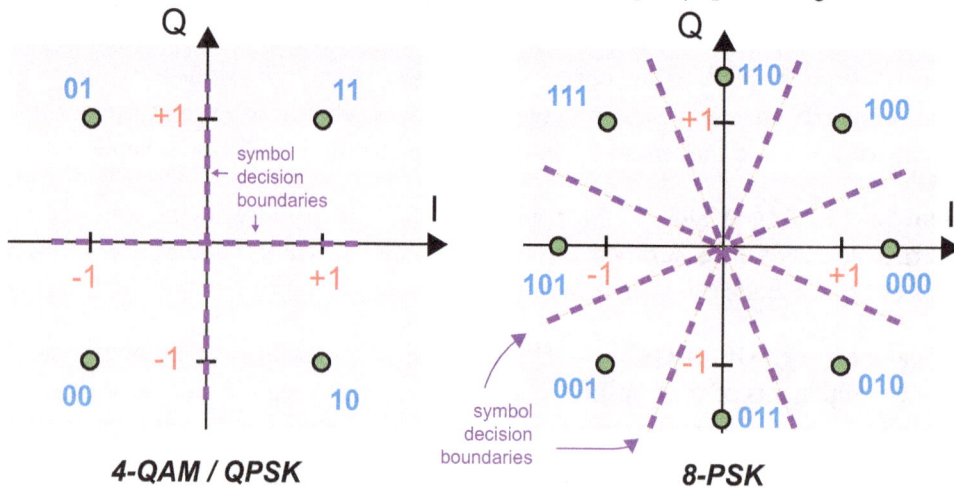

Figure 6.12: Decision boundaries for symbol-to-bit demapping.

6.4.2. Additive White Gaussian Noise (AWGN) Channel

In an ideal channel, there would be no degradation of the signal between the transmitter and receiver, but of course this is not realistic. Even in an *excellent* channel, some level of thermal noise must be expected at minimum, and this is normally modelled as Additive White Gaussian Noise (AWGN). Note that noise is described as 'white' if it contains approximately equal energy across all frequencies.

To consider the effect of an AWGN channel, we assume a simplified model of a communications link, shown in Figure 6.13. In this model, the modulation and demodulation processes (between baseband and the carrier frequency) are abstracted. This is reasonable as, in theory, a modulated signal can be perfectly reconstructed at the output of the demodulator. The channel is modelled at baseband, where it is assumed that AWGN is introduced at the input to the receiver, and that samples are taken at the ideal time instants (more on this later, in Section 6.6.4).

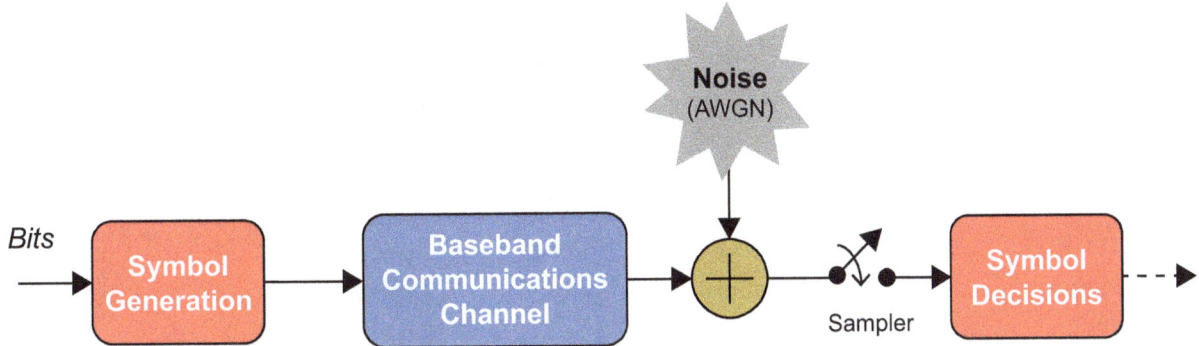

Figure 6.13: Communications link modelled at baseband with an AWGN channel.

The received signal conditions can be quantified using the metric of Signal to Noise Ratio (SNR), given by

$$SNR = 10log_{10}\left(\frac{P_{signal}}{P_{noise}}\right) \qquad (6.1)$$

where P_{signal} and P_{noise} are the powers of the signal and noise, respectively.

For a received signal of some arbitrary power, then, the greater the amount of AWGN introduced, the lower the SNR. Also, intuitively, the more difficult it is to correctly retrieve the transmitted signal.

In the time domain, AWGN has a Gaussian Probability Density Function (PDF), meaning that small magnitude errors (both positive and negative) are most likely. The variance (σ) of the AWGN describes the degree of spread; AWGN with a higher variance has a greater likelihood of large magnitude errors. Examples of AWGN with difference variances are shown in Figure 6.14.

The effect of AWGN is to spread the received symbol samples, so that they form a 'cloud' around the ideal positions. Up to a point, this can be tolerated, because the clouds are contained within the decision boundaries and the correct symbol decisions are still made. However if the level of noise is too high, some of the symbol samples forming these clouds extend beyond the decision boundaries, resulting in incorrect symbol decisions. Examples of 16-QAM in the presence of AWGN are provided in Figure 6.15: in the left hand example, the received symbols are close enough to the reference positions to prevent any errors occurring; however in the right hand example, where more noise is added, some of the received symbols stray into adjacent regions and therefore a proportion of the symbol decisions are likely to be incorrect.

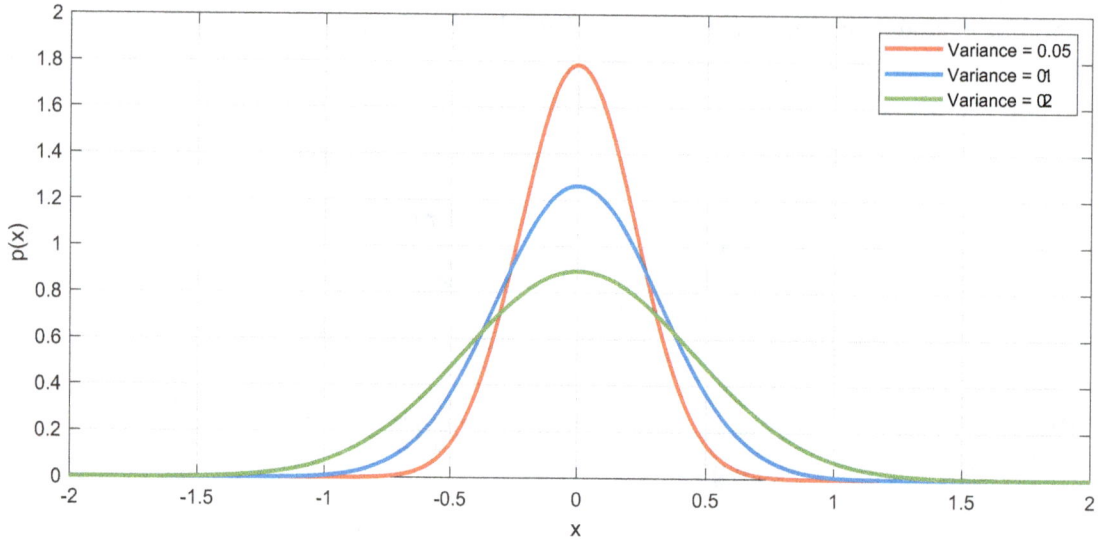

Figure 6.14: AWGN profiles for different variances.

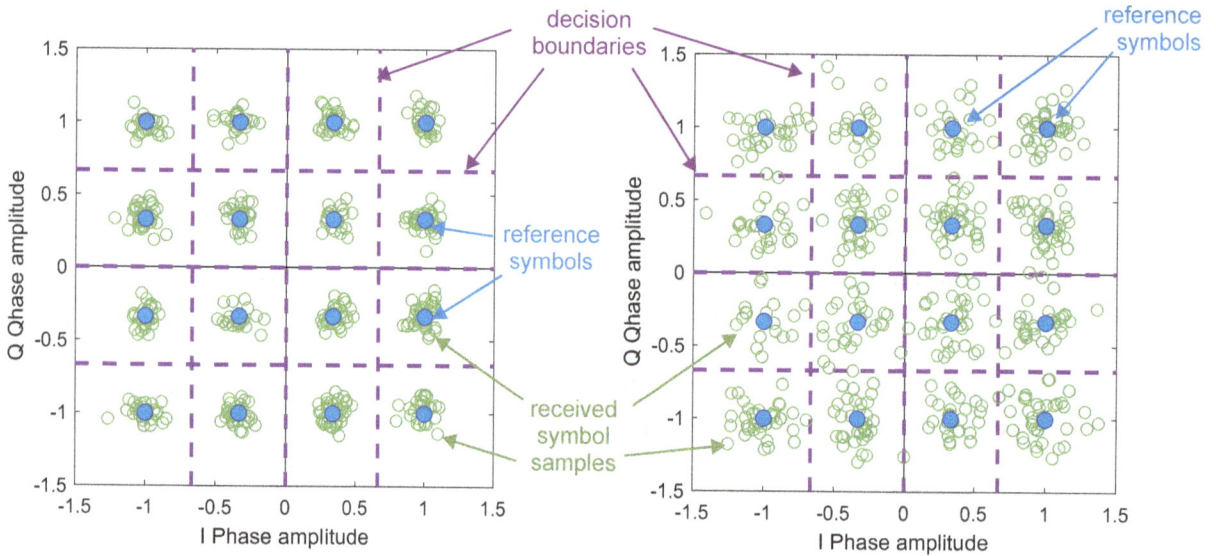

Figure 6.15: Different amounts of AWGN added to 16-QAM (left: $\sigma = 0.25$; right: $\sigma = 0.35$).

For any given level of noise, there is a practical limit to the size of the QAM scheme that can be used, because the reference symbols are more closely spaced for larger QAM schemes, and therefore the scheme is more susceptible to incorrect symbol decisions. To give an example, 16-QAM might experience very few symbol decision errors at a certain noise level, whereas 64-QAM would suffer from an intolerable level in the same noise conditions. Therefore, some communications standards support multiple baseband modulation schemes, and switch between them according to the experienced environment. Low noise implies that a higher order modulation scheme can be used, which translates to a higher data rate because each symbol conveys more bits; conversely, when the noise level is high, a low order modulation scheme is necessary to provide more robust performance, but at a reduced data rate.

6.4.3. Error Vector Magnitude (EVM)

It is useful to quantify the degree of received symbol spread around the reference symbol points, and conventionally this is done using the Error Vector Magnitude (EVM) metric. For one individual symbol, the EVM is defined as the distance between the reference and received symbol points, in the I-Q symbol space, as a ratio with respect to the magnitude of the reference sample point. Figure 6.16 illustrates how the EVM between one received sample and the reference point is calculated. This can be expressed as,

$$EVM = \frac{\sqrt{(I_{rx} - I_0)^2 + (Q_{rx} - Q_0)^2}}{\sqrt{I_0^2 + Q_0^2}} \qquad (6.2)$$

where the symbols are as defined in Figure 6.16.

When used as a quality metric, EVM is commonly expressed as a percentage of the peak signal level, and computed as an RMS value over some time window (N samples), i.e.

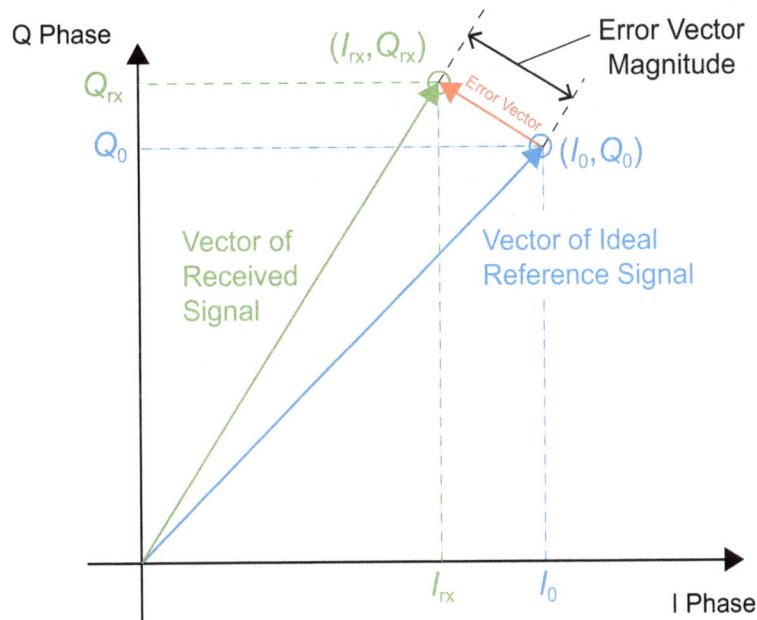

Figure 6.16: Definition of EVM.

$$EVM = \frac{\sqrt{\frac{1}{N} \sum\limits_{n=1}^{N} (I_{rx}(n) - I_0(n))^2 + (Q_{rx}(n) - Q_0(n))^2}}{\sqrt{I_0^2 + Q_0^2}} \quad (\%) \tag{6.3}$$

Note that, in the case of 16-QAM or higher, the I_0 and Q_0 values used to calculate the reference signal power are those of a corner point in the constellation, as this corresponds to the peak signal level.

Clearly, a larger EVM value indicates a greater degree of noise in the received constellation.

6.4.4. Bit Error Rate (BER)

While EVM is a useful indication of how the received symbol constellation has been affected by noise, it does not directly capture the success of accurately transmitting data across the radio channel (although the two are undoubtedly linked).

More useful measures of successful data transmission are Bit Error Rate (BER) and Symbol Error Rate (SER). SER can be related intuitively to the discussion from Section 6.4.2 — when a received sample is closer to a different reference symbol than the transmitted one (e.g. due to the effects of AWGN), an incorrect symbol decision is made, leading to a symbol error. The SER is simply the rate of occurrence of such errors, with respect to the total number of transmitted symbols. BER considers the rate of bit errors, after the received symbols have been converted back to bits.

For any given SER, the BER can be minimised by strategically allocating symbols, such that the group of bits represented by adjacent symbols in a constellation differ minimally (i.e. the *Hamming distance* is minimised). This is commonly done using Gray coding (as shown in Figure 6.10 — compare the sequences of bits represented by adjacent symbols, and you should notice that they differ by only one bit). Therefore, each symbol error results in a minimum number of bit errors.

BER is perhaps the key metrics for assessing link quality, and will be discussed further in Section 6.7.

Next, we will consider the part of the communications system not yet discussed — the channel! The channel is the 'real-world' part of the system that inevitably poses various difficulties for successful communication across a wireless link, and which is the cause of the bit errors that we have discussed here.

6.5. The Radio Channel

When a wireless signal is transmitted across a radio channel, it can experience a variety of different effects and degradations, depending on the physical environment, and factors such as the carrier frequency of the signal, and the degree of mobility involved (i.e. the relative motion between the transmitter and receiver). This section reviews commonly encountered channel effects and the reasons that they arise; these effects are also depicted graphically in Figure 6.17. We also mention some of the techniques that can be used to manage or mitigate the adverse effects introduced in the channel, although it is outside this scope of this book to review them in detail.

6.5.1. Channel Effects

First, we review the various wireless channel effects that may occur, as illustrated in Figure 6.17. The impacts of selected channel effects are shown in Figure 6.18, for the example of a QPSK / 4-QAM symbol constellation.

The Ideal Channel and Path Loss

The ideal wireless channel between a transmitter and receiver may be modelled by a simple wire, meaning that the signal is transferred perfectly. Often this method is used in simulations as a first step in validating a receiver design. To introduce a degree of realism, however, signals are subject to attenuation as they cross the radio channel: the signal power reduces in proportion to the distance travelled.

The attenuation of signal power in the channel can modelled by a path loss model. The simplest such model is the Friis free space path loss model [167], which defines that

$$\frac{P_r}{P_t} = \frac{A_r A_t}{d^2 \lambda^2},$$ (6.4)

where P_t and P_r are the transmitted and received signals power, respectively; A_t and A_r are the effective areas of the transmit and receive antennas, respectively; d is the distance between antennas; and λ is the wavelength of the transmitted signal.

A common adaptation of (6.4) replaces the antenna areas A_t and A_r with gain values G_t and G_r, which are defined with respect to an isotropic radiation pattern, i.e. where power is emitted equally in all directions. This modification results in

$$\frac{P_r}{P_t} = G_t G_r \left(\frac{\lambda}{4\pi d}\right)^2.$$ (6.5)

Bearing in mind that the RF carrier wavelength is the reciprocal of frequency, it is observed that power loss varies with the square of frequency, and the square of distance. In other words, the higher the RF carrier frequency, and the further the signal must travel, the lower the signal power will be on reaching the receiver.

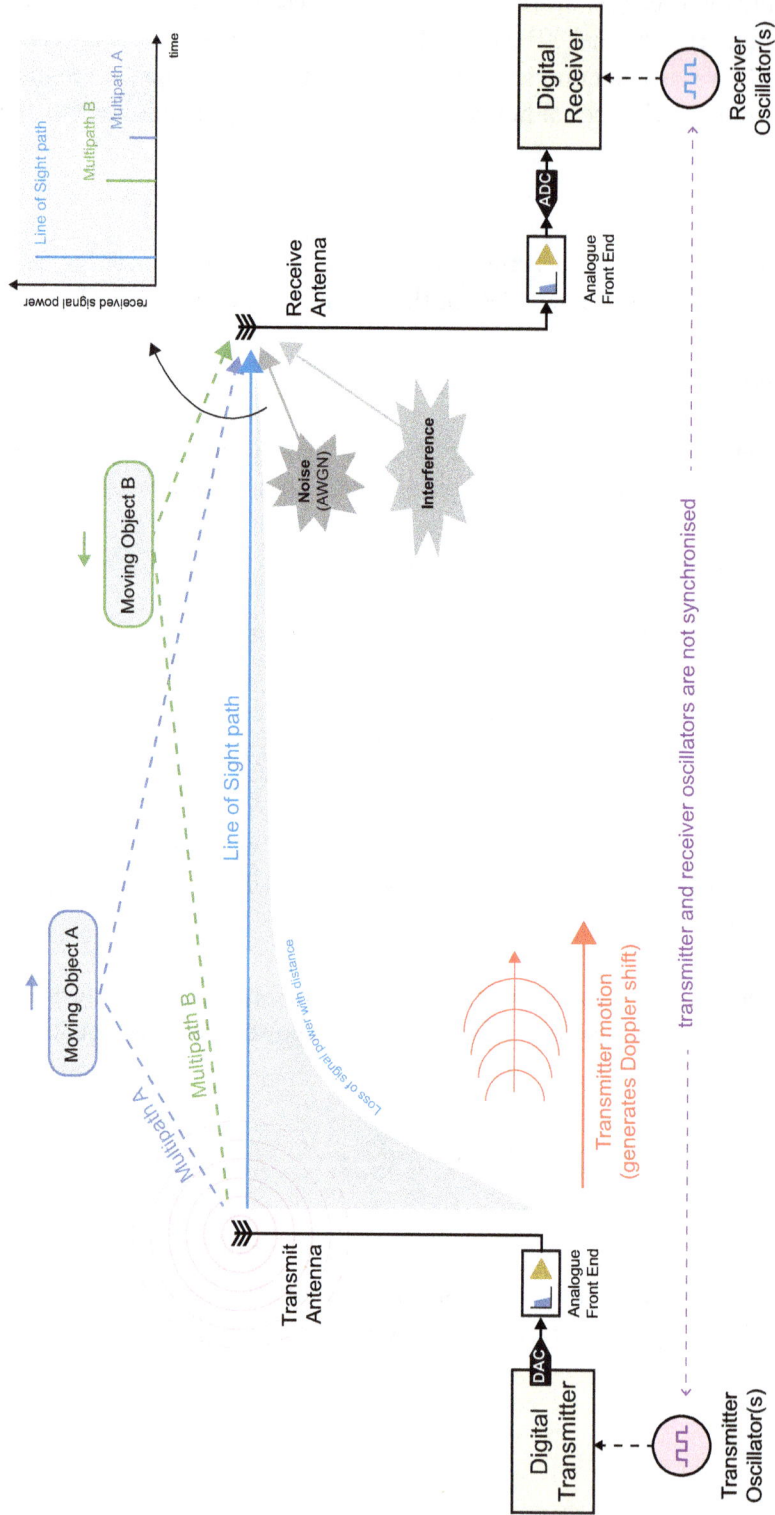

Figure 6.17: Overview of wireless channel effects.

Additive White Gaussian Noise (AWGN)

It is normal for a signal to experience additive noise in a radio channel. This is an inevitable effect, which always arises to some extent, as a result of the transmission environment and radio hardware in use (at minimum, thermal noise is generated in the analogue components of the radio front end). There can also be low level interference in the environment that arises from other uses of radio spectrum, such as spectral leakage from other bands, and distant transmissions in systems that reuse frequencies geographically (for example, mobile networks [370]). Such interference is often modelled as noise.

AWGN is usually modelled as being added to the signal at the point of reception. The impact of AWGN on the constellation of received QAM signals, i.e. to spread the received symbol samples around the reference points, as already been noted in Section 6.4.2.

Interference and Jamming

Aside from low level background interference (modelled as noise as mentioned above), there may also be more significant interference presented by other users of the radio spectrum. This is most pertinent when operating in one of the unlicensed bands, as there are no restrictions on how many other radios can seek to use these bands simultaneously, although all must comply with transmitted power limits. As unlicensed bands are shared, users must contend for access to the radio spectrum, typically using a MAC protocol such as Carrier Sense Multiple Access / Collision Avoidance (CSMA/CA) [125].

Interference can also arise from other sources, such another radio malfunctioning or otherwise not operating within its out-of-band emission limits, or perhaps even transmitting illegally in a licensed band. It is also possible for non-radio devices to generate RF interference, such as microwave ovens. Interference may be limited within one or more frequency band(s), and it may vary over time (for instance, bursts of interference may occur if another user transmits intermittently). As such, its characteristics may be dissimilar to AWGN, although it is usually handled in the same way in simulation models, i.e. as an additive signal at the receiver.

Jamming is a type of deliberate interference: it refers to malicious emission of radio signals, with the aim of disrupting other transmissions and users of the radio spectrum. Its signal characteristics can very considerably, but may include very wide bandwidths or high powers. Jamming is most relevant to military radio applications and will not be a focus in this book.

Doppler Effect

The Doppler Effect occurs when the transmitter and receiver of a radio link are in relative motion. For instance, when a mobile subscriber is travelling in a train, towards a trackside basestation, the distance travelled by the radio waves is constantly decreasing. The receiver experiences this as the wavefronts of the carrier signal arriving more frequently, or in other words, it perceives the received signal at a higher frequency than its transmitted frequency. There is therefore an error at the receiver between the expected signal frequency, and the actual signal frequency, which is known as a Doppler shift. The Doppler shift can be expressed as

$$\Delta f = \frac{f \cdot \Delta v}{c} \tag{6.6}$$

where f is the transmitted signal frequency, c is the propagation velocity of a radio signal in air (which is approximately equal to the speed of light), and Δv represents the relative motion of the transmitter and receiver (a positive value when they are moving towards each other, and negative when they are moving apart).

In wireless communications, Doppler effect causes the received symbol constellation to spin, as depicted in Figure 6.18(c). It also affects the timing parameters of the received signal, i.e. the symbol period is perceived as slightly shorter or longer than it should be.

The most challenging Doppler conditions are when the rate of relative motion is extreme, where the environment is dynamic and there are frequent changes, or where there are a variety of signal paths each experiencing different Doppler shifts (known as *Doppler spread*) [346]. Some of the largest Doppler shifts occur in satellite communications systems, where the satellites are not geo-stationary, and the relative motion between transmitter and receiver can result in several kHz of Doppler shift [305].

Fading

Fading refers in general to the loss of signal power experienced in a channel. This occurs due to distance-dependent path loss (even in free space) as discussed earlier. Another possible cause is *shadowing*, where the presence of a large object such as a hill or a building causes less signal power to reach the receiver. In dynamic environments, for instance where the transmitter and receiver are moving relative to each other, the loss of signal power can also vary over time. The general concept of fading is illustrated in Figure 6.18(a), i.e. the constellation points move closer together, although retain the desired shape.

In very broad terms, the effect can be categorised as either 'slow-fading', where the channel characteristics vary only gradually over time, or 'fast-fading', where rapid temporal variations in the receive signal power can occur. Fluctuations in the received power level can often be compensated (or at least, partially compensated) by a dynamically changing amplifier gain in the receiver (Active Gain Control, or AGC). However, a deep fast-fade can cause a loss of ability to receive the signal, e.g. temporarily when traveling through a railway tunnel.

Multipath Propagation and Fading

If the propagation environment is cluttered, it may include various objects that reflect or diffract radio signals. The result is multipath propagation, where components of the transmitted signal take several different paths between the transmitter and receiver, with each one arriving with a different time delay and loss of signal power, depending on the distance travelled and the reflectivity of the object.

Where there is a Line-of-Sight (LoS) path between the transmitter and receiver, this implies that one dominant signal component is received, along with other lower-power multipath components. The situation is more difficult when the channel is said to be Non-Line-of-Sight (NLoS), and only a collection of multipath components arrive at the receiver.

The impact of multipath propagation may vary, but in general it causes multipath fading, which may be either 'flat', or 'frequency-selective'. In the former case, the entire signal bandwidth experiences the same general effect. Therefore, at any arbitrary time instant, the received power is relatively constant across the signal bandwidth, although the degree of fading is still likely to change over time. This type of fade can be compensated with a simple gain (equivalent to a one-tap filter). The effects of fading are more challenging in the 'frequency-selective' case, when the signal bandwidth exceeds the *coherence bandwidth*. This is defined as

$$W_c = \frac{1}{2T_d} \tag{6.7}$$

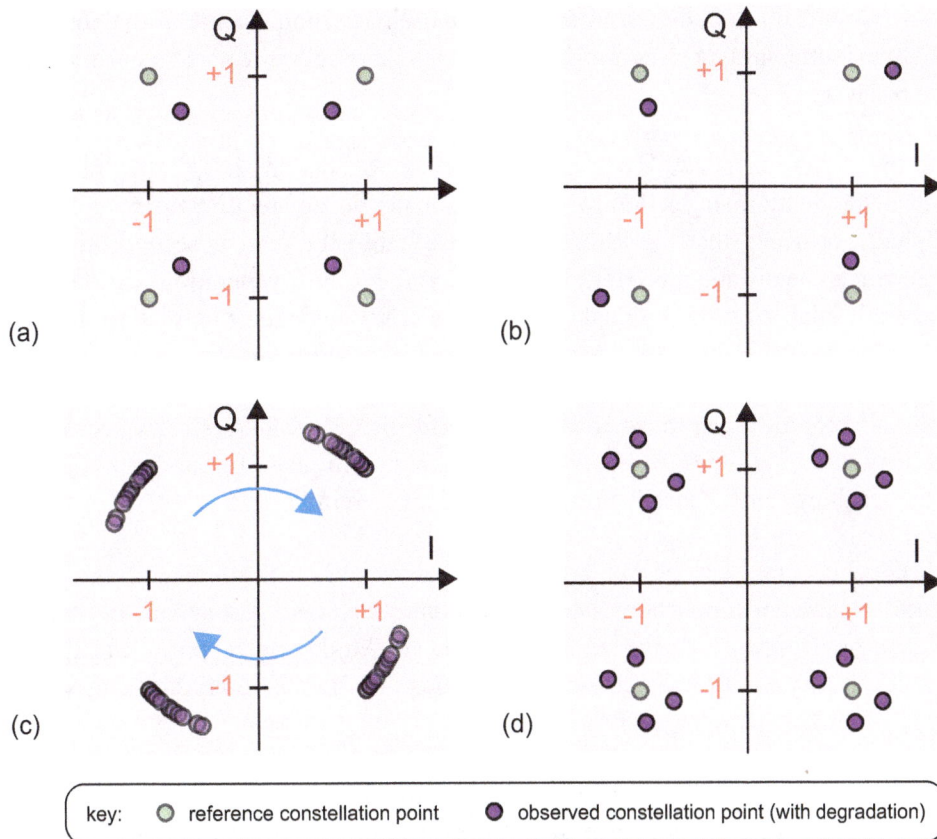

Figure 6.18: Illustration of the impacts of selected channel effects on a received QPSK / 4-QAM symbol constellation: (a) flat fading, (b) I/Q imbalance, (c) Doppler / oscillator mismatch, and (d) multipath frequency selective fading.

where T_d is the delay spread of the channel (i.e. the maximum time difference between signal components of significant energy that arise from the multipath channel) [346].

In this case, the channel exhibits frequency-selective fading, i.e. the received signal power varies with frequency, and therefore the response cannot be represented by (or compensated by) a single-tap filter. This is much more difficult from a signal reception perspective — in the time domain, frequency-selective fading manifests as a distortion of the received symbol constellation which, if severe enough and left unaddressed, may make it impossible to correctly receive the transmitted data. A (tolerable) example is shown in Figure 6.18(d); the patterns produced due to multipath propagation can however vary considerably.

An added difficulty is that multipath effects are often time-varying, for instance due to the transmitter and/or receiver moving through a cluttered environment. Fortunately, adaptive DSP techniques can be used to compensate for the effects of multipath channels (discussed further in Section 6.5.2).

Transmitter and Receiver Artefacts: I/Q Imbalance, Oscillator Mismatch

Although not strictly part of the radio channel, it can be useful to model some of the possible imperfections arising from the transmitter and receiver implementations, in particular the analogue front end sections of the transmitter and receiver.

Depending on the architecture employed (referring back to Section 2.7), some radios have analogue modulation and demodulation stages. RFSoC designs actually do not require them, and can modulate and demodulate digitally, provided that the RF signal resides within the first or second Nyquist Zones (which translates to frequencies below about 6 GHz, depending on the RFSoC generation). At the other extreme, for very high bandwidth, high frequency signals, quadrature (I/Q) analogue modulation and demodulation is needed, which is achieved using sine and cosine carriers at the carrier frequency, f_c.

For modulation and demodulation to be undertaken perfectly, the oscillators must provide sine and cosine waves with exactly equal amplitudes, and they must be separated in phase by exactly 90 degrees. If this is not the case, then the resulting signal constellation is distorted, as seen in Figure 6.18(b).

Direct-RF radio architectures are particularly beneficial here, as all modulation and demodulation is performed in the digital domain, where sine/cosine amplitudes and phases can be controlled precisely. However, analogue oscillators may have some degree of amplitude and/or phase imbalance. A benefit of RFSoC-based radios is that they only require external modulation and demodulation stages for extremely high frequencies, so I/Q imbalance can largely be avoided.

Another issue (an almost inevitable one) is that the oscillators in the transmitter and receiver are subject to component tolerances, and do not generate exactly the same frequencies. In wireless systems, the transmitter and receiver are not connected in any way, and therefore do not share a common frequency reference from which to generate sample and symbol timing parameters, or carrier signals for modulation and demodulation.

This means that, for instance, the transmitter believes that it is producing symbols at a rate of 1 Msymbol/s, but the actual rate is 1.00083 Msymbols/s; meanwhile the receiver expects symbols to arrive at 1 Msymbol/s, but actually the rate is 0.99936 Msymbols/s. Similarly, the nominal RF carrier frequency is 2.45 GHz, but the transmitter actually modulates the signal to 2.45203 GHz, and the receiver believes the received signal should be located at 2.44843 GHz. This issue occurs even in Direct-RF systems, because all radios rely on a frequency reference provided by an oscillator.

Fortunately, the resulting deviations in timing and frequency parameters tend to be relatively small, provided that high quality oscillators are specified for the radio design. They also generate the same effect as Doppler-induced timing and frequency shifts (such as that illustrated in Figure 6.18(c)). The combined effects of both Doppler and oscillator tolerances can be compensated in the radio receiver using synchronisation techniques [230], [301].

6.5.2. Mitigation Techniques

Having reviewed the problems that can be introduced as the radio signal propagates through the channel, we will now briefly review some of the techniques used to address these issues at the PHY layer. It should be highlighted that, in addition to PHY layer processing, higher layer protocols usually apply further mitigations against data loss, e.g. error coding, buffering, and selective retransmissions.

Analogue Receiver Front End Filtering

Perhaps the most obvious mitigation against channel effects is bandpass filtering, around the signal of interest, to remove as much of the incident AWGN and interference as possible. This should be done initially in the analogue domain, to ensure that the SNR is as high as possible when reaching the ADC. Further filtering can subsequently be undertaken in the digital section of the receiver (acknowledging that there are practical limits to the performance of the analogue filters).

Analogue front-end filtering is especially useful when there are high-power signals present on adjacent or nearby frequency bands, as these could otherwise saturate the ADC.

Quadrature Modulation Correction

Noting that the symbol constellation is distorted when a signal is subject to modulation and/or demodulation with imperfect quadrature modulators (an example of which was shown in Figure 6.18(b)), it may be desirable to apply corrections for these effects in the receiver. This can be achieved in two ways:

- **I/Q gain imbalance** — The effect of unequal I and Q amplitudes can be corrected by applying a compensating gain to the I and/or Q branches in the receiver.

- **I/Q phase error** — Where the separation between the two phases is not the required 90°, and therefore the signals are not orthogonal, an element of mixing between the two phases takes place. This can be corrected by adding a scaled version of the received Q signal to the I signal.

Support for the above corrections, and also offset compensation (to correct for an incorrect DC level) is available in RFSoC devices, via the QMC block, a hardened resource on the RFSoC [90]. This block provides the facility to apply corrections to the I and Q signal paths, based on user-developed algorithms.

Forward Error Correction (FEC)

FEC is one of the key techniques used in wireless communications to protect against bit errors. By applying a coding scheme which adds redundancy to the transmitted data, the receiver is able to detect when bit errors have occurred (up to some limit), and also in most cases to correct them (again, up to some limit). The degree of protection depends on the coding scheme used, and its parameters. FEC is reviewed in considerable detail in Chapter 14, and so further discussion is not included here.

Synchronisation

As noted earlier, and depicted in Figure 6.17, the transmitter and receiver do not have a common frequency or timing reference. Therefore, the frequency and timing parameters of signals arriving at the receiver are almost certain not to correspond to the nominal values expected by the receiver. The Doppler effect will add to these offsets and in total there may be considerable deviations in the actual and expected signal characteristics.

Synchronisation systems are used in the receiver to estimate frequency and timing offsets, and apply adjustments to correct for them. There are two main synchronisation tasks involved:

- **Carrier synchronisation** — The receiver must adjust the frequency and phase of its local oscillator, to match the frequency and phase of the carrier within the received signal. The outcome of successful carrier synchronisation is that the symbol constellation ceases to spin.

- **Symbol timing synchronisation** — The receiver must take samples of the incoming symbols at the correct rate, which is determined from observation of the incoming signal. Ideally it should position them at the *maximum effect points*, i.e. the optimum timing instants to achieve the best possible SNR.

Depending on the structure of transmitted data, *frame synchronisation* is additionally required in the receiver MAC layer to determine when the start of each frame occurs, and correctly extract the payload.

Equalisation

As noted in the previous section, and depicted in Figure 6.18(d), fading caused by multipath propagation can distort the received symbol constellation.

Where the fading is frequency-selective, the implication is that the channel acts like a filter, and causes varying gain across the signal bandwidth. In compensating for this effect, we consider a baseband equivalent channel, which includes the entire signal chain between applying input symbols at the transmitter, and retrieving output symbols in the receiver. In other words, the baseband equivalent channel includes aspects of the transmitter and receiver architectures, as well as the wireless channel itself. Adaptive DSP techniques can be used to generate an inverse of the baseband channel. If the output of the baseband channel is passed through this inverse, it is equalised, which achieves an approximately constant gain across the signal bandwidth, and in doing so, corrects the distortion witnessed in the constellation.

When the channel introduces flat fading, the equalisation only involves applying a gain, and is therefore trivial.

The popular technique of Orthogonal Frequency Division Multiplexing (OFDM), employs an interesting approach that converts a wideband signal into a set of smaller sub-channels to simplify the equalisation task; more information on this, along with further review on the subject of equalisation, is presented in Chapter 16.

6.6. Pulse Shaping and Matched Filtering

Our discussion from Section 6.3 presented baseband modulation, i.e. conversion of bits to symbols, and then Section 6.4 focused on converting symbol values back to bits; however, neither touched on how the symbols are represented as physical signals for transmission across the channel, or how the symbols are retrieved from the received signal. The purpose of this section is to address these issues.

6.6.1. Symbols as Impulses

The representation of symbols as impulses, for instance as shown in the time domain waveforms from Figures 6.7 to 6.10, is problematic from the perspective of physical signal transmission across the radio channel. The reason is that an impulse contains all frequencies, and therefore, transmitting impulse across the radio channel would create extremely wide bandwidth radio signals, which would intrude into adjacent bands (and beyond), causing interference to other users of the radio spectrum.

The solution to this issue is to apply pulse shaping — equivalent to passing each of the pulses through a filter. In the following sections, we set out the requirements for pulse shaping, in both the time and frequency domains, and outline how the task can be decomposed into a pair of matched filters.

6.6.2. Pulse Shaping Requirements and Implementation

Pulse shaping can be easily achieved in the digital domain by upsampling the symbol waveform, and passing it through an FIR filter with the desired filter response. The design of the pulse shaping filter must be specified carefully — there are two key requirements:

- The filter should appropriately contain the signal energy within the desired bandwidth.

- The filter should enable symbols to be accurately recovered when sampled at the ideal timing instants.

Taking these requirements in turn, the containment of signal energy within a specific bandwidth is usually needed to comply with a spectrum licence, and/or the spectral mask specified by the wireless standard. An example spectral mask is illustrated in Figure 6.19. The performance of a radio transmitter implementation with respect to this mask can be measured experimentally: Adjacent Channel Leakage Ratio (ACLR), i.e. the degree to which emissions are suppressed in the adjacent band, is often quoted as a figure of merit.

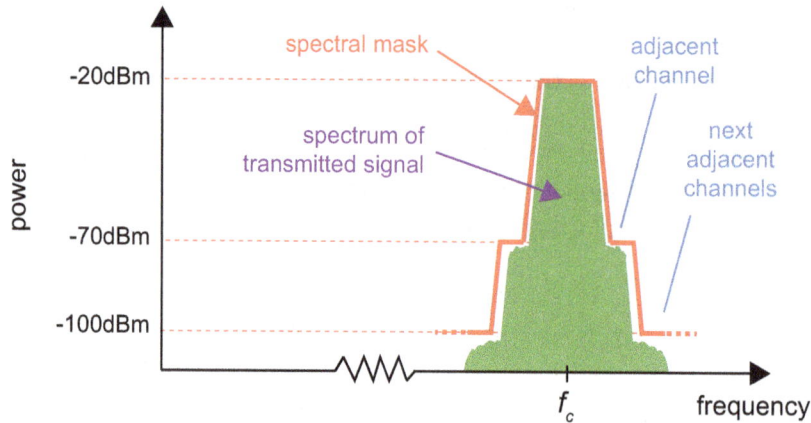

Figure 6.19: Example of a spectral mask specification.

In both cases, limits are placed on the power that can be emitted both within the allocated band, and also in the adjacent and next adjacent bands. It is usually necessary to achieve a high degree of suppression in adjacent bands, which protects other users of the radio spectrum from interference effects.

The second requirement refers to the time domain process of sampling the received symbols at their *maximum effect points*, i.e. the instants when the best SNR is achieved, which corresponds with the centre of the shaped pulses. The chosen pulse shape should not produce any interference between successive symbols (known as *inter-symbol-interference*) at these ideal sampling points. The most reliable way to achieve this is using a pulse shape whose impulse response is one symbol period (or less); however, such a constraint limits the frequency domain performance of the pulse shaping process[4]. Possible pulse shapes include the rectangular, half-sine, Gaussian, and probably the most popular — the Raised Cosine (RC).

---

4. The most basic pulse shape is a rectangle. In this case, all of the filter coefficients are 1, and the filter length is equal to the upsampling ratio — it is therefore a particularly simple filter to implement. Unfortunately, the frequency response of the square rectangular pulse shape is sinc-shaped, which equates to considerable spectral leakage into adjacent bands. It is therefore not preferred.

6.6.3. Square Root Raised Cosine Matched Filtering

The RC is a desirable pulse shape, because it satisfies both of the requirements set out in Section 6.6.2. It is not necessary, however, to implement this response entirely via transmit filtering; rather, the important point is to apply the pulse shape across the link. Therefore it is possible to split the filtering task into two sections, and implement one each in the transmitter and receiver. This is actually preferable because it enables the receive-side filter to also filter out some of the noise introduced in the channel.

The RC filter response can be split into two Square Root Raised Cosine (RRC) filters, which in cascaded form correspond to the RC response. When implemented in this separated form, the filters are referred to as the Pulse Shaping Filter (in the transmitter) and the Matched Filter (in the receiver).

When designing an RC filter (or equivalently, an RRC filter), three parameters need to be specified:

- The **oversampling ratio** (or upsampling ratio, i.e. the number of samples per symbol period)

- The filter **span**, in symbol periods

- The filter **roll-off** parameter, usually denoted by α. This controls the excess bandwidth, i.e. how much additional bandwidth is occupied by the signal when pulse-shaped. In the time domain, it determines how quickly the 'tails' of the RC impulse response diminish.

We explore RC design further in Notebook Set D (which immediately follows this chapter).

6.6.4. Maximum Effect Points

Although the RC impulse lasts for several symbol periods, it avoids Inter-Symbol Interference (ISI) provided that the signal is sampled at the ideal timing instants, i.e. *the maximum effect points*, in order to retrieve the symbols. Referring to Figure 6.20, which shows the RC responses arising from successive pulses, it is clear that at the maximum effect points (i.e. the peak of each impulse response, where the amplitude is greatest), the amplitude contribution of all other pulses is exactly zero. Therefore, no ISI is experienced if the samples are ideally timed.

The difficulty is timing the samples correctly at these exact instants, especially given the timing frequency offset issues discussed in Section 6.5.1, which may arise from oscillator mismatches in the transmitter and receiver, and/or the Doppler effect in the channel. In the receiver, symbol timing synchronisation is required to adjust the timing parameters of the symbol sampler, such that samples are taken as closely as possible to the maximum effect points. We do not cover timing synchronisation in this book, but the interested reader can find more information in textbooks such as [230], [301], and [326].

If symbol samples are not correctly timed, this results in a contribution to the overall error, or in other words, to the spreading of received symbol samples around the reference constellation points. As discussed in Section

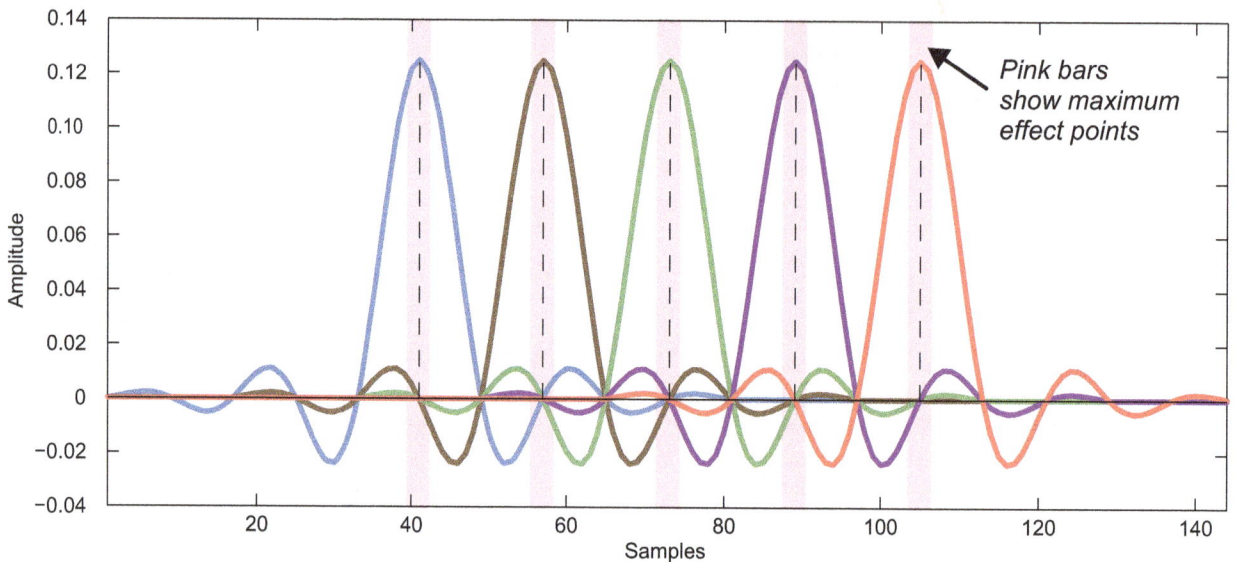

Figure 6.20: Successive RC pulses, demonstrating zero-ISI at the maximum effect points.

6.4, spreading is measured in terms of EVM, and a larger EVM leads towards incorrect symbol decisions (and hence, bit errors), with higher order modulation schemes being more susceptible. The symbol timing synchroniser is therefore an important part of the receiver.

6.7. Bit Error Rate (BER) Analysis

The Bit Error Rate (BER) metric is used to quantify the ability of a communications link to convey data accurately from the transmitter to the receiver. It expresses the average incidence of measured errors, or the rate of expected errors; for instance, a BER of 10^{-3} indicates that 1 bit in every 1000 transmitted bits will, on average, be in error. A transmission that incurs a significant number of bit errors is unable to deliver high quality video, audio, or other data reliably across the link.

In our earlier review from Section 6.4.2, we observed that increasing levels of AWGN made it more difficult to accurately receive transmitted symbols: the received symbols were seen to spread out around the reference points, and began to cross decision boundaries, and therefore some of the received symbols were misinterpreted. Incorrect symbol decisions lead directly to incorrect reception of the bits, as each symbol conveys one or more bits. Intuitively, therefore, as the level of AWGN increases, so does the BER.

Other 'real world' effects from the radio channel also impact the received symbol constellation; such as the Doppler effect, multipath propagation, and so on. The receiver, however, normally includes circuitry to compensate for these effects (in the form of carrier and timing synchronisation circuits, equalisers and so on).

The experienced BER therefore becomes a function of the radio channel environment *and* the ability of the receiver to mitigate channel effects.

FEC schemes are incorporated into many communications systems, and these methods can correct bit errors up to some defined threshold, which depends on the parameters of the coding scheme. The uncoded BER (i.e. without error correction), and coded BER (after FEC decoding) may both be of interest in such cases. For instance, an uncoded BER of 1e-2 may be acceptable, if the coded BER achieves a BER of 1e-4. This is because the latter is the effective one for later stages of receiver processing.

BER performance is often characterised and visualised using a BER curve; a two-dimensional plot that graphs BER on the y-axis against E_b/N_0 on the x-axis. E_b/N_0 is a normalised digital measure of SNR, which is explained very clearly in [319]; our explanation borrows from this. As the E_b/N_0 value increases, the transmission environment becomes less challenging, and the BER is seen to decrease.

In the E_b/N_0 term, the symbol E_b represents energy per bit, which is equivalent to the signal power, S, multiplied by the time period per bit, T_b. The N_0 term is the noise spectral density, i.e the noise power per Hz, which is equivalent to the total noise power, N, divided by the bandwidth, W.

E_b/N_0 can therefore be expressed as

$$\frac{E_b}{N_0} = \frac{ST_b}{N/W} = \frac{S/R_b}{N/W},$$

(6.8)

noting that T_b is the reciprocal of the bit rate, R_b.

It can therefore be shown that E_b/N_0 is a measure of Signal-to-Noise ratio, normalised by bit rate and bandwidth.

$$\frac{E_b}{N_0} = \frac{S}{N} \cdot \frac{W}{R_b}$$

(6.9)

Theoretical BER curves exist for commonly used modulation schemes (BPSK, QPSK, 16-QAM, 64-QAM etc.), described by equations [293],[319], and a selection are plotted in Figure 6.21. Note that modulation schemes with fewer symbols in the constellation can achieve any target BER at a lower level of E_b/N_0 (in other words, noisier conditions) than those with a greater number of symbols in the constellation. This is in line with our discussion from Section 6.4.2.

During development work, the relevant theoretical curve can be compared with the measured or simulated BER curve of a radio design to assess its relative performance.

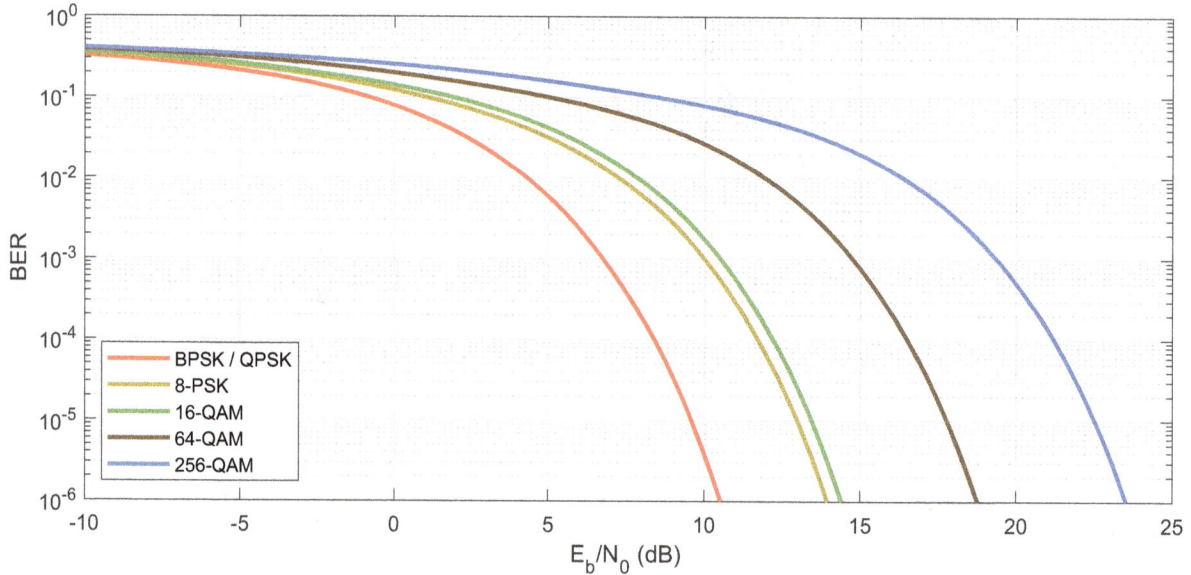

Figure 6.21: Example BER curves.

6.8. Chapter Summary

This chapter has reviewed various aspects of communications, and in particular radio communications. We began by discussing the 7-layer OSI model, which all communications systems can be compared to, and the various levels of abstraction, tasks and protocols that must be undertaken by a communications system. This review was expanded to include the popular TCP/IP model, and examples of the 'stacks' defined in three well-used standards for wireless communication.

Our subsequent review focused primarily on the lower layer in each of these stacks, the PHY layer, which has the role of transferring data across the physical radio medium. We noted that the RFSoC is a particularly powerful implementation platform for the PHY layer, due to its PL and features such as the RFDCs. The RFSoC also has the facility to integrate well with higher layers of the communication stack, due to its PS and the tight coupling between the PS and PL (which enables predictable, low latency interaction between the two sections).

Theoretical concepts useful to later chapters were reviewed, including baseband modulation and demodulation, the effects of the radio channel, pulse shaping and matched filter, and BER analysis. Many of these topics are further investigated in the example notebooks immediately following this chapter.

Notebook Set D

Wireless Communications Fundamentals

The notebooks in this section are focused towards baseband wireless communications, i.e. the signal processing that takes place when the signal frequency is close to 0 Hz. In other words, before a signal has been modulated (shifted up in frequency) to IF or RF, and after it has been demodulated (shifted back down).

A little confusingly, the term *modulation* is commonly used in communications to describe two different processes. One of them is the mapping of bits to symbols (commonly known as *baseband modulation*), and the other is to shift a signal up in frequency by mixing it with a carrier (*carrier modulation*). In this chapter, we focus only on the former — modulation and demodulation with a carrier will be covered in Notebook E.

The following notebooks are briefly introduced in this chapter:

| ALL | 01_baseband_modulation — *rfsoc_book/notebook_D/01_baseband_modulation.ipynb* |

| ALL | 02_evm_and_ber.ipynb — *rfsoc_book/notebook_D/02_evm_and_ber.ipynb* |

| ALL | 03_pulse_shaping.ipynb — *rfsoc_book/notebook_D/03_pulse_shaping.ipynb* |

D.1. Baseband Modulation Schemes

In digital communications systems, there are various different ways of mapping bits to symbols, and these are commonly referred to as *modulation schemes*. Modulation schemes can convey different symbols by varying the frequency, amplitude or phase of a signal.

The first notebook, named ***01_baseband_modulation.ipynb***, explores some of the modulation schemes that are widely used in wireless communications, in particular Binary Phase Shift Keying (BPSK), Quadrature Phase Shift Keying (QPSK) and various sizes of Quadrature Amplitude Modulation (QAM) scheme.

D.2. Noise and Errors

The task of recovering symbols in the receiver is made more difficult because of impairments encountered in the radio channel. In the best-case scenario, the channel introduces some degree of Additive White Gaussian Noise (AWGN), and it may also involve other degradations such as multipath propagation, and the Doppler effect. The second notebook concentrates on AWGN, and explores the impact of additive noise on the receiver's ability to correctly recover transmitted bits. To explore this example, open the notebook named ***02_evm_and_ber.ipynb***.

Initially, we will explore a useful measurement known as the Signal to Noise Ratio (SNR), which is the ratio of the signal power and the noise power. Several simulations will be performed that add noise to QAM signals. Thes noisy signals plotted using constellation diagrams different values of SNR. These plots allow us to inspect the impact noise has on a QAM signal and can help determine the amount of error that has been introduced. For instance, see the constellation diagrams in Figure D.1, where a QAM-16 signal has been plotted with an SNR of 20dB and 10dB. It is clear that a lower SNR introduces more errors.

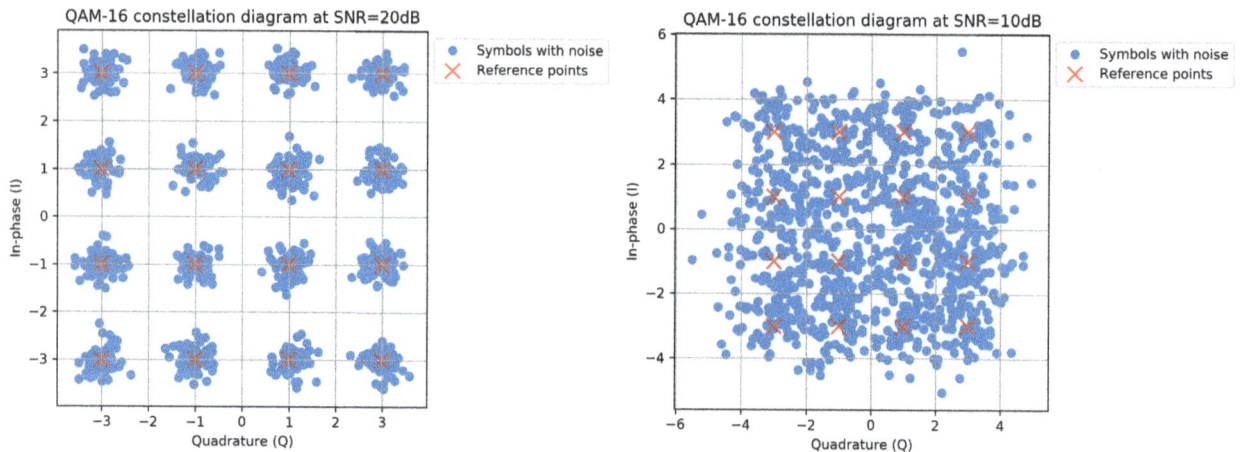

Figure D.1: QAM-16 symbol constellations: (left) tolerable noise, and (right) too much noise!

After understanding how noise impacts QAM signals, we will then review two important metrics. Firstly, we will explore *Error Vector Magnitude (EVM)*, which is commonly used to characterise the impact of channel degradations on received symbol constellations. An example of EVM is sketched on the left hand side of Figure D.2, where the magnitude of an error for a generic sample, k, has been labelled. The EVM measurement is the mean of the measured error magnitudes.

The second metric we will investigate is the *Bit Error Rate (BER)*, which quantifies the amount of errors encountered in the received data. We also inspect BER curves, where BER is plotted against E_b/N_0, which is a normalised SNR used for digital communications analysis that provides an overall characterisation of communications link quality (refer back to Section 6.7 for further discussion of BER and E_b/N_0). An example BER curve is sketched on the right of Figure D.2.

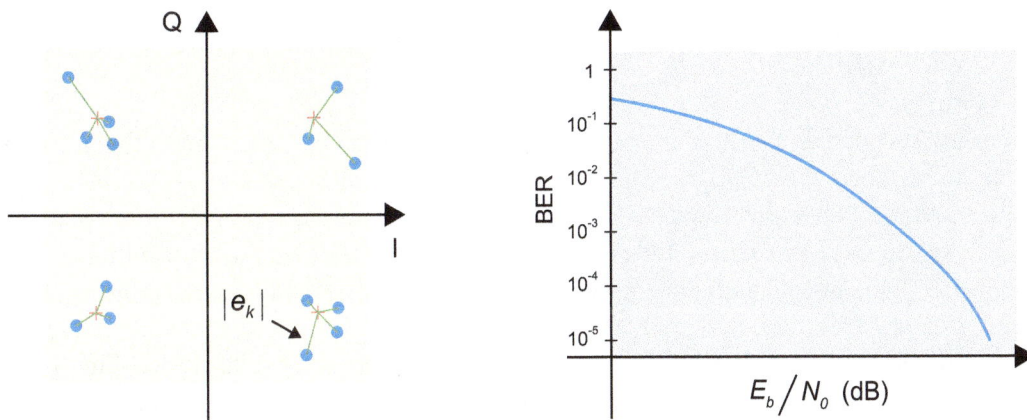

Figure D.2: Error Vector Magnitude (left) and a BER curve (right).

D.3. Pulse Shaping and Matched Filtering

This final notebook in this set, *03_pulse_shaping.ipynb,* demonstrates the technique of pulse shaping. The purpose of pulse shaping is to constrain the bandwidth occupied by a transmitted signal, helping to ensure that it complies with a spectral mask (i.e. the limits placed on transmitted power across the allocated and adjacent spectrum). Pulse shaping is performed at baseband, before the signal is modulated onto a carrier.

Several types of pulse shapes can be used and they have different properties, such as the sinc and raised cosine filters presented in Figure D.3. Here we confirm the undesirable spectrum generated by transmitting impulses, and compare the properties of square, sinc, and raised cosine pulse shapes. Note that the α parameter shown in the right hand plot refers to the roll-off of the raised cosine pulse shaping filter, a design parameter that determines the excess bandwidth occupied by the filtered signal (in the frequency domain), and how quickly the 'tails' of the impulse response diminish (in the time domain).

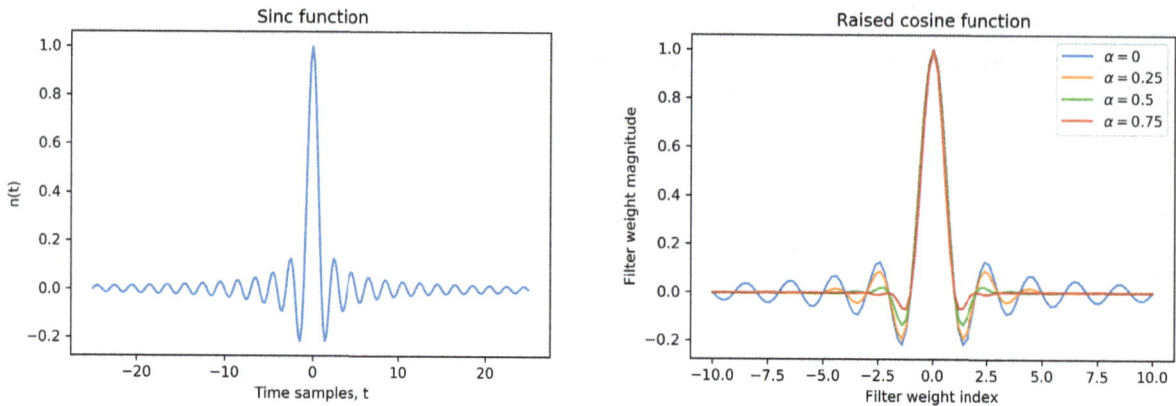

Figure D.3: Example plot of a sinc function (left) and a raised cosine (right).

The raised cosine is a popular choice as the pulse shape — the notebook demonstrates that it can be successfully implemented as a pair of matched, Square Root Raised Cosine (SRRC, or simply RRC) filters, one each in the transmitter and receiver, which results in the raised cosine response across the communications link as a whole. This filter topology introduces Inter-Symbol Interference (ISI) at the transmitter, which is when successive symbols interfere with each other, however the matched filter compensates at the receiver.

Plots of the notebook simulation can be inspected in Figure D.4, where the results of RRC-filtering transmitted and received symbols are presented on the left and right, respectively. Notice that the zero-ISI property is restored after matched filtering, where the symbol maximum effect point of each symbol (i.e. where its amplitude is greatest) occurs simultaneously with the zero crossings of all symbols that are transmitted before and after. .

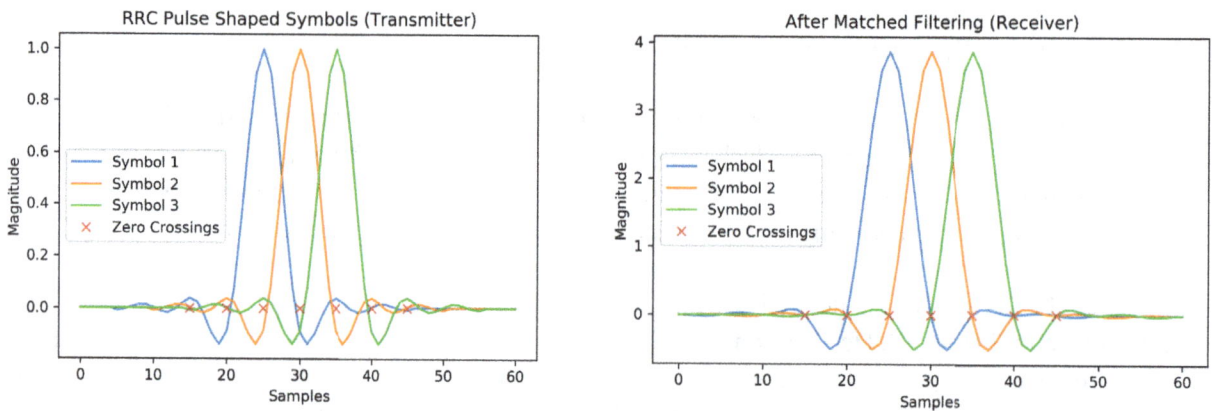

Figure D.4: Root raised cosine pulse shaped symbols at the transmitter (left), and symbols after matched filtering at the receiver (right).

Chapter 7

Quadrature Modulation & Complex Exponentials

Robert W. Stewart and Louise Crockett

When working with communications systems, it is common to encounter signals that are described as *complex signals* (i.e. with real and imaginary components) and communication systems being presented with complex exponentials rather than with modulating real sine and cosine waveforms. Hence in this chapter, we explain what *complex* signals really are in terms of representing the I and Q channels of a quadrature modulator, and specifically how they relate to the mixing (or modulation and demodulation) of quadrature signals in communications. As a way of explanation we will present the quadrature mixing or modulation/demodulation using both standard trigonometric notation (sines and cosines), and also with complex exponential mathematics. Both of these representations are precisely equivalent — they are simply alternative ways of describing the same signal processing — and, as will be demonstrated in this chapter, the complex signal representation and notation will actually make the maths and system design that little bit 'easier'!

Before presenting an analysis of classic quadrature modulation and demodulation, the next two sections provide context and motivation for choosing to work with a complex signal representation. We will review the mathematical background for presenting real signals using a complex signal spectrum representation, the so-called two-sided spectrum.

7.1. Signal Representations

As a preliminary step in advance of analysing modulation and demodulation, we first review the methods used to represent signals in this chapter.

7.1.1. Analogue and Digital Signals

In this chapter, our review will be based on continuous time real signals rather than sampled discrete digital signals. This will minimise the mathematical notation required and allow the principles of quadrature modulation, demodulation, mixing etc. to be emphasised and more clearly presented. For example, a 100 Hz sine wave with amplitude 5, over time, t, is given by:

$$x(t) = 5\cos(2\pi 100 t). \tag{7.1}$$

If we were to consider a sampled version of a 100 Hz sine wave with sampling frequency $f_s = 10{,}000$ Hz, then we can write this as:

$$x(nT) = 5\cos\left(\frac{2\pi 100 n}{fs}\right) \tag{7.2}$$

with n the discrete sample index, and noting that $t = nT$ and $T = 1/f_s$. While not an arduous addition of more variables (n, f_s & T), in order to focus on the trigonometry and complex notation in this chapter, we will just work with the continuous time mathematics and signals.

The RFSoC is of course a DSP device, and therefore signals are always sampled and quantised, as was reviewed earlier in Chapter 4. In Chapter 10 we will review the RF-ADC quadrature and complex implementations for various receivers. and these designs will be presented in terms of discrete-time signals.

7.1.2. Real and Complex Signals

A complex signal is composed of 'real' and 'imaginary' signal components. When working with quadrature modulators or mixers the in-phase signal path and quadrature phase signal path in a transmitter and receiver can be conveniently represented as a complex signal. More mathematically, as the quadrature oscillator path is 90° offset (or *orthogonal*) to the in-phase oscillator path, then it becomes possible to represent one channel as a real signal path and one channel as an imaginary signal path (multiplied by $j = \sqrt{-1}$)[1]. Therefore, when working with quadrature modulated versions of two real signals, we can depict one of these signals as *real* and the other as *imaginary*. The motivation for doing so is to simplify the mathematics of spectrum 'shifting' that comes later. If the trigonometric equations for modulation and demodulation can instead be expressed using complex exponentials (sometimes called complex sinusoids) then, as we will see later, the mathematics is easier to work with compared to directly using the quadrature sine /cosine representations.

---

1. We assume that readers are familiar with the imaginary number, $j = \sqrt{-1}$. (Also represented by i by mathematicians.)

As a very simple illustration of the 'ease' of working with complex exponentials (or complex sinusoids), suppose that you wished to express the multiplication of two cosines,

$$cos(A)cos(B) \tag{7.3}$$

as a sum of sine and cosine terms. You may remember (or be able to consult a list of trigonometric identities!), but otherwise, the result of

$$cos(A)cos(B) = 0.5\,cos(A+B) + 0.5\,cos(A-B) \tag{7.4}$$

might be difficult to derive from first principles (although we show this done with complex exponentials in Eq (7.11) on page 236). [2]

On the other hand, if you were asked to express the product of two complex exponential terms (e^{jA} and e^{jB}) as a single exponential, that task is much easier, just add the indices, i.e.

$$e^{jA}e^{jB} = e^{j(A+B)}. \tag{7.5}$$

Therefore, we might anticipate that this simplicity can be exploited. Indeed, the use of complex mathematics to describe signals, and the processes of modulation and demodulation, could be most convenient if we the trigonometric notation for quadrature modulation and demodulation could be replaced with complex exponentials and complex notation.

7.1.3. Euler's Formula

When implementing wireless communications systems, the signals transmitted and received via antennas are *real* voltages, which change over time, and may take on positive or negative values. We often choose to represent these signals in our receivers in a way that yields a complex signal, i.e. a signal which includes both *real* and *imaginary* parts. These are known as *analytic* signals, i.e. signal representations that are used for analysis purposes only. However by using complex numbers as a notation, we can describe the operation of quadrature modulators and demodulators in a more convenient way, and make the associated mathematics much more tractable.

The basis of this translation into the complex world is Euler's formula,

$$e^{j\omega t} = cos(\omega t) + j\,sin(\omega t), \tag{7.6}$$

where e is the base of the natural logarithm (a constant approximately equal to 2.71828...), ω is an angular frequency given by $\omega = 2\pi f$, $j = \sqrt{-1}$, and t represents time.

When we have a negative exponential, $e^{-j\omega t}$, this can be written as

---

2. A list of trigonometric identities is provided at the back of this book. See page 703.

$$e^{-j\omega t} = \cos(\omega t) - j\sin(\omega t), \tag{7.7}$$

because $\cos(-\omega t) = \cos(\omega t)$ and $\sin(-\omega t) = -\sin(\omega t)$.

By finding the sum and difference of (7.6) and (7.7), we note that

$$2\cos(\omega t) = e^{j\omega t} + e^{-j\omega t} \tag{7.8}$$

and

$$j2\sin(\omega t) = e^{j\omega t} - e^{-j\omega t}, \tag{7.9}$$

and with some further reorganisation, we can represent both sine and cosine terms using positive and negative powered complex exponential notation, i.e.

$$\cos(\omega t) = \frac{e^{j\omega t} + e^{-j\omega t}}{2} \quad \text{and} \quad \sin(\omega t) = \frac{e^{j\omega t} - e^{-j\omega t}}{2j} \tag{7.10}$$

Now, returning to the problem of multiplying $\cos(A)\cos(B)$ from a little earlier, it is possible to derive (7.4) using the complex exponential terms given in (7.10), without too much difficulty:

$$
\begin{aligned}
\cos(A)\cos(B) &= \left[\frac{e^{jA} + e^{-jA}}{2}\right]\left[\frac{e^{jB} + e^{-jB}}{2}\right] \\[2mm]
&= \frac{e^{jA}}{2}\left[\frac{e^{jB} + e^{-jB}}{2}\right] + \frac{e^{-jA}}{2}\left[\frac{e^{jB} + e^{-jB}}{2}\right] \\[2mm]
&= \frac{1}{2}\left\{\left[\frac{e^{jA+jB} + e^{jA-jB}}{2}\right] + \left[\frac{e^{-jA+jB} + e^{-jA-jB}}{2}\right]\right\} \\[2mm]
&= \frac{1}{2}\left\{\frac{e^{j(A+B)}}{2} + \frac{e^{j(A-B)}}{2} + \frac{e^{-j(A-B)}}{2} + \frac{e^{-j(A+B)}}{2}\right\} \\[2mm]
&= \frac{1}{2}\left\{\frac{e^{j(A+B)}}{2} + \frac{e^{-j(A+B)}}{2}\right\} + \frac{1}{2}\left\{\frac{e^{j(A-B)}}{2} + \frac{e^{-j(A-B)}}{2}\right\} \\[2mm]
&= \frac{1}{2}\cos(A+B) \quad + \quad \frac{1}{2}\cos(A-B)
\end{aligned}
\tag{7.11}
$$

In fact, using Euler's formula, all of the sine and cosine terms arising from the standard trigonometric analysis of quadrature systems can be expressed using exponentials.

As stated earlier, both trigonometric and complex representations are equivalent, and equally valid — the use of complex notation is entirely optional, and it is perfectly valid to stay in the real domain (in other words, to use only real numbers, and trigonometric representations). Given the omnipotence of quadrature modulation in communications systems, however, it is useful to have at least an awareness of complex exponentials and complex notation.

In the coming sections we will show visually the relationship between presenting the frequency spectrum of a signal in terms of its 'sine' and 'cosine' representations (the so-called *real* spectrum, or sometimes the *one-sided* spectrum) and as complex exponentials (the *complex* spectrum, or *two-sided* spectrum).

7.1.4. Viewing Real Signals in the Frequency Domain using Complex Spectra

As presented earlier in the book, it is useful, in fact essential, to view and analyse signals in the frequency domain. When the signal to be analysed is complex, i.e. in communications terms, it has both *In Phase* (I) and *Quadrature Phase* (Q) parts, then it becomes necessary to understand the complex frequency spectrum representation. A set of example signals are analysed over the next few pages, as a means of introducing this method, starting with a simple sum of sine waves (in cosine form), i.e. a real signal. Note than Chapter 7 will look at computation methods for frequency domain calculation from a discrete sampled time domain signal (via discrete Fourier series and transforms), whereas in this chapter we will create frequency spectra from simple observation of the constituent sinusoidal components in the signals.

Simple Real Signal in the Frequency Domain: A Sum of Three Tones (One-Sided Spectrum)

In this first example, we consider a real signal composed of three cosine terms, which have frequencies of 100 Hz, 200 Hz, and 300 Hz, respectively, and amplitudes of 10, 1, and 4,

$$s_1(t) = 10\cos(2\pi100t) + \cos(2\pi200t) + 4\cos(2\pi300t) \tag{7.12}$$

A time domain plot of this signal, covering the first few tens of milliseconds, is provided in Figure 7.1. By inspection, we can ascertain that the signal is periodic, and perhaps pick out that the lowest frequency term has a period of 0.01s (i.e. 100 Hz), however it is difficult to analyse the frequency content of the signal in any detail.

An alternative method of displaying the signal is using magnitude and phase spectra, as shown in Figure 7.2, and this permits easy interpretation of the sinusoidal frequency components and, as we show the cosine frequency values only (which are positive), we often refer to this as a the *one-sided spectrum*. Of course, the current example is an artificially simple one where we know exactly what we are looking for, but the same principle applies for any arbitrary signal (more generally, analysis would involve computing an FFT and generating the required plots — in the case of this straightforward example, we can simply sketch the spectra).

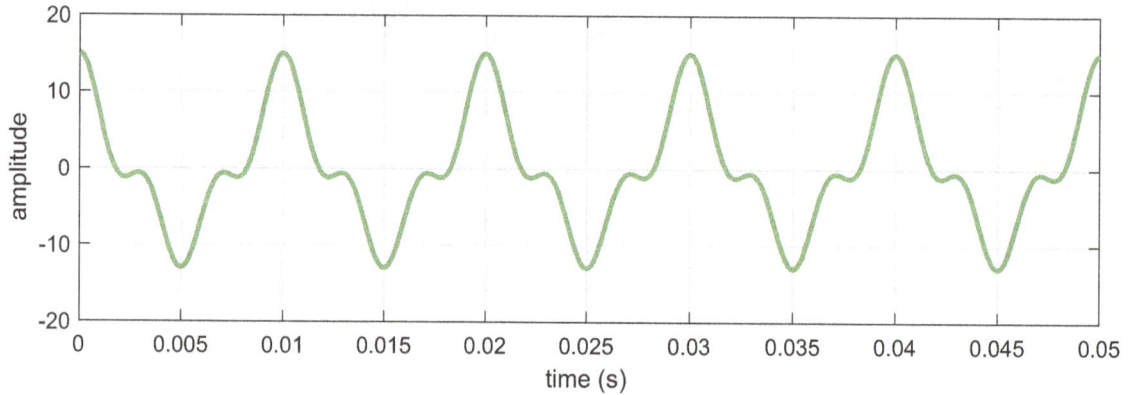

Figure 7.1: *Time domain plot of the signal $s_1(t)$ as defined in (7.12).*

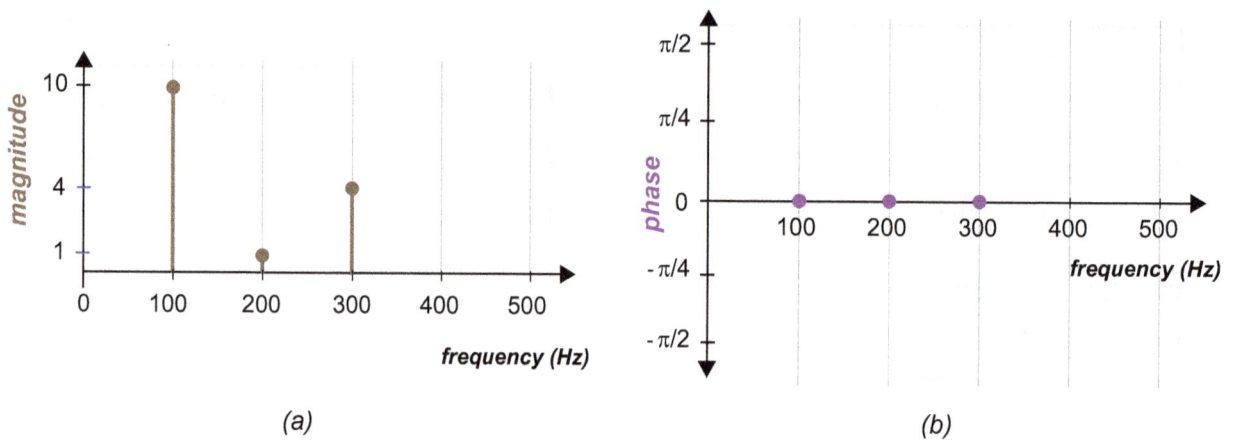

(a)

(b)

Figure 7.2: *Spectra of the signal $s_1(t)$ from (7.12): (a) magnitude spectrum and (b) phase spectrum.*

Sum-of-Three-Tones Signals in the Complex Frequency Domain (Two-Sided Spectrum)

Next, we will analyse the signal $s_1(t)$ from (7.12) in the complex frequency domain. The signal can be converted to a complex exponential representation by substituting the complex version of $cos(\omega t)$ from (7.10), i.e. $cos(\omega t) = (e^{j\omega t} + e^{-j\omega t})/2$. This produces a sum of exponential terms:

$$s_1(t) = 10\,cos(2\pi 100t) \quad + \quad cos(2\pi 200t) \quad + \quad 4\,cos(2\pi 300t)$$

$$= 10\left(\frac{e^{j2\pi 100t} + e^{-j2\pi 100t}}{2}\right) + 1\left(\frac{e^{j2\pi 200t} + e^{-j2\pi 200t}}{2}\right) + 4\left(\frac{e^{j2\pi 300t} + e^{-j2\pi 300t}}{2}\right) \qquad (7.13)$$

$$= 5e^{j2\pi 100t} + 5e^{-j2\pi 100t} + \frac{1}{2}e^{j2\pi 200t} + \frac{1}{2}e^{-j2\pi 200t} + 2e^{j2\pi 300t} + 2e^{-j2\pi 300t}$$

We can then group the positive (blue ☐) and negative (pink ☐) exponentials to give

$$s_1(t) = 5e^{j2\pi100t} + \frac{1}{2}e^{j2\pi200t} + 2e^{j2\pi300t} \quad + \quad 5e^{-j2\pi100t} + \frac{1}{2}e^{-j2\pi200t} + 2e^{-j2\pi300t} \quad , \quad (7.14)$$

and these terms can be plotted in a complex frequency spectrum, as the amplitudes of the complex exponential terms. The resulting plot is sketched in Figure 7.3 and we now have both significant components in the positive and negative axes, and will refer to this as a *two-sided spectrum*.

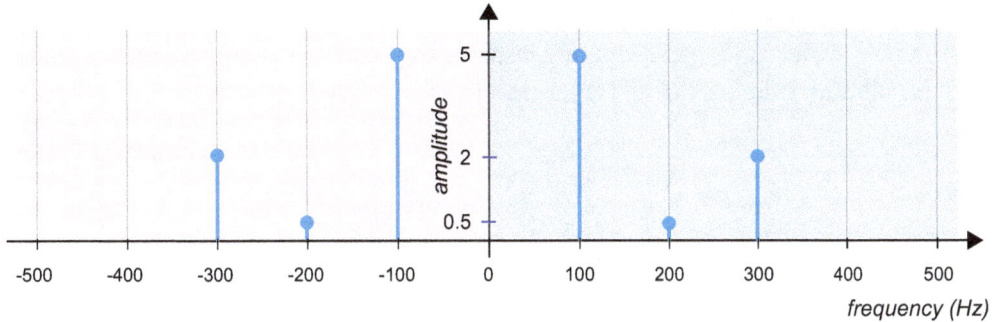

Figure 7.3: Complex frequency spectrum of the signal $s_1(t)$ as defined in (7.12)

In this plot, we can see the positive exponential terms plotted against the blue ☐ background, and the negative exponential terms plotted against the pink ☐ background. Sometimes these are referred to as 'positive' and 'negative' frequencies, although the terms 'positive frequency' and 'negative frequency' are arguably confusing and misleading, since a frequency refers to the *cycles per second* of a real-valued, varying quantity. A more accurate term would be 'negative complex exponential', in reference to components such as $e^{-j\omega t}$. In practice, however, the somewhat imprecise language of 'positive and negative frequencies' is commonly used rather than the correct 'positive and negative (power) complex exponentials'.

Notice that the right hand side of the complex spectrum in Figure 7.3 corresponds to the real magnitude spectrum from Figure 7.2(a), with a change in amplitude being the only difference. Noting once more that $cos(\omega t) = (e^{j\omega t} + e^{-j\omega t})/2$, the symmetry of the complex spectrum means that the components at, for instance, -200 Hz and 200 Hz, add together to form a real cosine at a frequency of 200Hz. Hence this two-sided spectrum is a complex notation representation of a real signal, because it is symmetric around 0 Hz.

The phase of the signal, which previously was shown explicitly as a phase spectrum plot (Figure 7.2(b)), is now captured in complex spectra plots for the real and imaginary components, although in this particular example there is no imaginary spectrum. When we introduce a phase shift away from a standard cosine, things become a little more complicated and we will have both real amplitude complex exponentials and imaginary amplitude complex exponentials, and hence will require two spectra (to be presented in Figure 7.7).

Sum-of-Three-Tones Signals in the Complex Frequency Domain (with Phase Shifts)

Our examples so far have considered a signal comprised of cosine terms, all with phase terms of zero (i.e. no phase shift from standard cosine waves). The next step is to introduce phase offsets into these cosine components. We now define a new signal,

$$s_2(t) = 10\cos\left(2\pi 100t + \frac{\pi}{4}\right) + \cos\left(2\pi 200t + \frac{\pi}{6}\right) + 4\cos(2\pi 300t) \tag{7.15}$$

which includes phase shifts in the 100 Hz and 200 Hz components. Figure 7.4 shows the time domain waveform for the new signal, $s_2(t)$, and Figure 7.5 shows the corresponding frequency domain (magnitude and phase spectra) plots, in subplots (a), and (b), respectively.

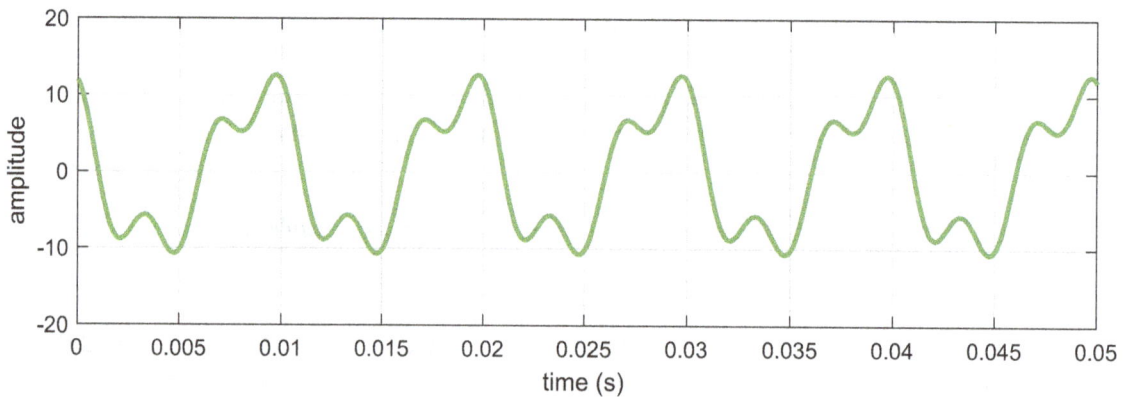

Figure 7.4: Time domain plot of the signal $s_2(t)$ as defined in (7.15).

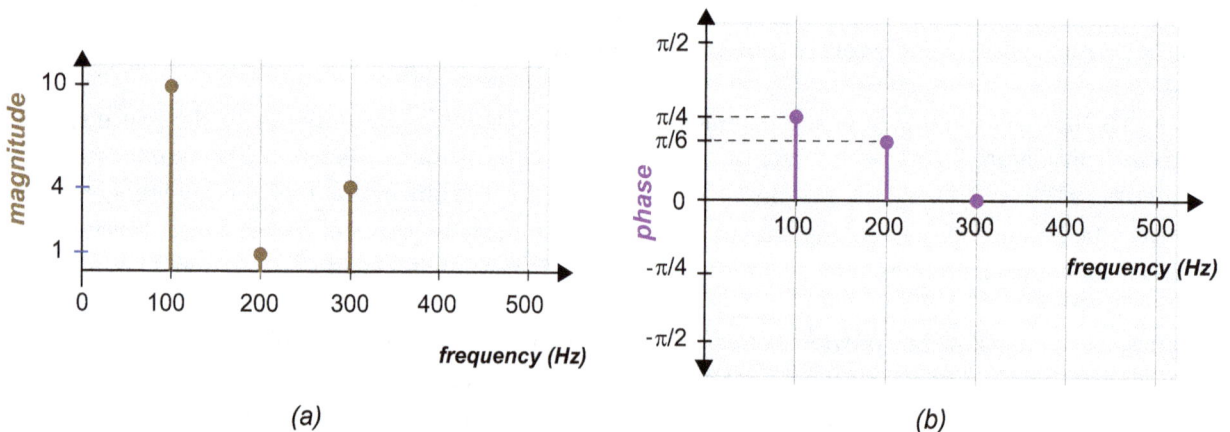

(a)   (b)

Figure 7.5: Spectra of the signal $s_2(t)$ as defined in (7.15): (a) magnitude spectrum; (b) phase spectrum.

First, notice by comparing the time domain waveforms from Figure 7.1 and Figure 7.4 that they look distinctly different from each other — this is due to the 100 Hz and 200 Hz cosine terms having different (non-zero) phases in the second plot.

By comparing the frequency plots in Figure 7.5 with those from Figure 7.2, we note that the magnitude-frequency plot for $s_2(t)$ looks identical to that for $s_1(t)$ — this is to be expected, as the two signals contain the same cosine terms with the same amplitudes. The phase-frequency plots differ, and non-zero phases for the 100 Hz and 200 Hz terms of $s_2(t)$ are now shown.

Additionally, the same $s_2(t)$ signal can be analysed and plotted using a complex frequency spectrum. The non-zero phase components cause the derivation to be rather more complicated than was presented in (7.13) for $s_1(t)$, which resulted in the spectrum shown in Figure 7.3. This derivation from first principles is presented in Figure 7.6 using the complex exponential form, with (7.15) as the starting point (note that this figure showing the maths spans across two pages!).

Note that the final complex expression for $s_2(t)$ (shown at the final stage in Figure 7.6) includes imaginary amplitude components, whereas for $s_1(t)$, there were only real components (as seen in Figure 7.3).

The real and imaginary spectra for $s_2(t)$ can be plotted using colours to represent the different components, as shown in Figure 7.7:

- ☐ real amplitude positive complex exponentials ('positive' frequencies),

- ☐ real amplitude negative complex exponentials ('negative' frequencies),

- ☐ imaginary amplitude positive complex exponentials ('positive' frequencies) and

- ☐ imaginary amplitude negative complex exponentials ('negative' frequencies).

It is not necessary to generate a phase spectrum because phase is inherently captured in the real / imaginary spectra. The phase information can be extracted using the $tan^{-1}(\ )$ function, if required, as indicated in Figure 7.8.

A key point to note is that, if a signal is real-valued only (i.e. it is not complex, and has no imaginary or j component), its two-sided spectrum is also even-symmetric (as this is formed from any cosine terms present in the signal), and its imaginary-valued spectrum is always odd-symmetric (this is formed from any sine terms present in the signal). We can therefore confirm by inspection of the complex two-sided spectrum in Figure 7.7 that the analysed signal $s_2(t)$ is real-valued (which of course we knew, as it is a real signal composed of a sum of real-world, real-valued sine waves).

It is generally more convenient to work with a magnitude spectra, which plots the magnitude of the components at each complex exponential value, as shown in Figure 7.8. This actually returns us back to the

$$s_2(t) = 10\cos(2\pi 100t + \pi/4) \quad + \quad \cos(2\pi 200t + \pi/6) \quad + \quad 4\cos(2\pi 300t)$$

First, simplifying the notation, denoting $A = 2\pi 100t$, $B = 2\pi 200t$ and $C = 2\pi 300t$:

$$= \quad 10\cos(A + \pi/4) \quad + \quad \cos(B + \pi/6) \quad + \quad 4\cos(C)$$

Expressing the cosines as complex exponentials yields:

$$= 10\left[\frac{e^{j(A+\pi/4)} + e^{-j(A+\pi/4)}}{2}\right] + 1\left[\frac{e^{j(B+\pi/6)} + e^{-j(B+\pi/6)}}{2}\right] \quad + \quad 4\left[\frac{e^{jC} + e^{-jC}}{2}\right]$$

$$= 5\left[e^{j(A+\pi/4)} + e^{-j(A+\pi/4)}\right] \quad + \quad 0.5\left[e^{j(B+\pi/6)} + e^{-j(B+\pi/6)}\right] \quad + \quad 2\left[e^{jC} + e^{-jC}\right]$$

Noting that $e^{j(u+v)} = e^{ju}e^{jv}$, we can re-write as:

$$= 5e^{j(\pi/4)}e^{jA} + 5e^{-j(\pi/4)}e^{-jA} \quad + \quad 0.5e^{j(\pi/6)}e^{jB} + 0.5e^{-j(\pi/6)}e^{-jB} \quad + \quad 2e^{jC} + 2e^{-jC}$$

And using Euler's formula; $e^{j(\pi/4)} = \cos(\pi/4) + j\sin(\pi/4)$ and $e^{j(\pi/6)} = \cos(\pi/6) + j\sin(\pi/6)$

$$= 5(\cos(\pi/4) + j\sin(\pi/4))e^{jA} \quad + \quad 0.5(\cos(\pi/6) + j\sin(\pi/6))e^{jB} \quad + \quad 2e^{jC} + 2e^{-jC}$$
$$+ 5(\cos(\pi/4) - j\sin(\pi/4))e^{-jA} \quad + \quad 0.5(\cos(\pi/6) - j\sin(\pi/6))e^{-jB}$$

Noting $\cos(\pi/4) = 1/\sqrt{2}$, $\sin(\pi/4) = 1/\sqrt{2}$, $\cos(\pi/6) = \sqrt{3}/2$, & $\sin(\pi/6) = 1/2$

$$= 5\left(\frac{1}{\sqrt{2}} + j\frac{1}{\sqrt{2}}\right)e^{jA} + 5\left(\frac{1}{\sqrt{2}} - j\frac{1}{\sqrt{2}}\right)e^{-jA} + 0.5\left(\frac{\sqrt{3}}{2} + j\frac{1}{2}\right)e^{jB} + 0.5\left(\frac{\sqrt{3}}{2} - j\frac{1}{2}\right)e^{-jB} + 2e^{jC} + 2e^{-jC}$$

$$= \frac{5}{\sqrt{2}}e^{jA} + j\frac{5}{\sqrt{2}}e^{jA} + \frac{5}{\sqrt{2}}e^{-jA} - j\frac{5}{\sqrt{2}}e^{-jA} + \frac{\sqrt{3}}{4}e^{jB} + j\frac{1}{4}e^{jB} + \frac{\sqrt{3}}{4}e^{-jB} - j\frac{1}{4}e^{-jB} + 2e^{jC} + 2e^{-jC}$$

Grouping all of the positive exponentials, e^{jX}, and negative exponentials, e^{-jX} together:

$$= \left[\frac{5}{\sqrt{2}}e^{jA} + j\frac{5}{\sqrt{2}}e^{jA} + \frac{\sqrt{3}}{4}e^{jB} + j\frac{1}{4}e^{jB} + 2e^{jC}\right] + \left[\frac{5}{\sqrt{2}}e^{-jA} - j\frac{5}{\sqrt{2}}e^{-jA} + \frac{\sqrt{3}}{4}e^{-jB} - j\frac{1}{4}e^{-jB} + 2e^{-jC}\right]$$

Now grouping the real and imaginary scaled terms in each of the positive and negative exponentials:

$$= \left[\frac{5}{\sqrt{2}}e^{jA} + \frac{\sqrt{3}}{4}e^{jB} + 2e^{jC}\right] + j\left[\frac{5}{\sqrt{2}}e^{jA} + \frac{1}{4}e^{jB}\right] + \left[\frac{5}{\sqrt{2}}e^{-jA} + \frac{\sqrt{3}}{4}e^{-jB} + 2e^{-jC}\right] + j\left[-\frac{5}{\sqrt{2}}e^{-jA} - \frac{1}{4}e^{-jB}\right]$$

Replacing the scaling terms with numbers to two decimal places:

$$= \left[3.53e^{jA} + 0.43e^{jB} + 2e^{jC}\right] + j\left[3.53e^{jA} + 0.25e^{jB}\right]$$
$$+ \left[3.53e^{-jA} + 0.43e^{-jB} + 2e^{-jC}\right] + j\left[-3.53e^{-jA} - 0.25e^{-jB}\right]$$

$$= \left[3.53e^{jA} + 0.43e^{jB} + 2e^{jC}\right] + \left[3.53e^{-jA} + 0.43e^{-jB} + 2e^{-jC}\right]$$
$$+ j\left[3.53e^{jA} + 0.25e^{jB}\right] + j\left[-3.53e^{-jA} - 0.25e^{-jB}\right]$$

Figure 7.6: continues on next page...

Substituting back for $A = 2\pi100t$, $B = 2\pi200t$ and $C = 2\pi300t$,

$$s_2(t) = \left[3.53e^{j2\pi100t} + 0.43e^{j2\pi200t} + 2e^{j2\pi300t} \right] + \left[3.53e^{-j2\pi100t} + 0.43e^{-j2\pi200t} + 2e^{-j2\pi300t} \right]$$
$$+ j\left[3.53e^{j2\pi100t} + 0.25e^{j2\pi200t} \right] + j\left[-3.53e^{-j2\pi100t} - 0.25e^{-j2\pi200t} \right]$$

Figure 7.6: First principles calculation of the complex spectrum of Eq. (7.15).

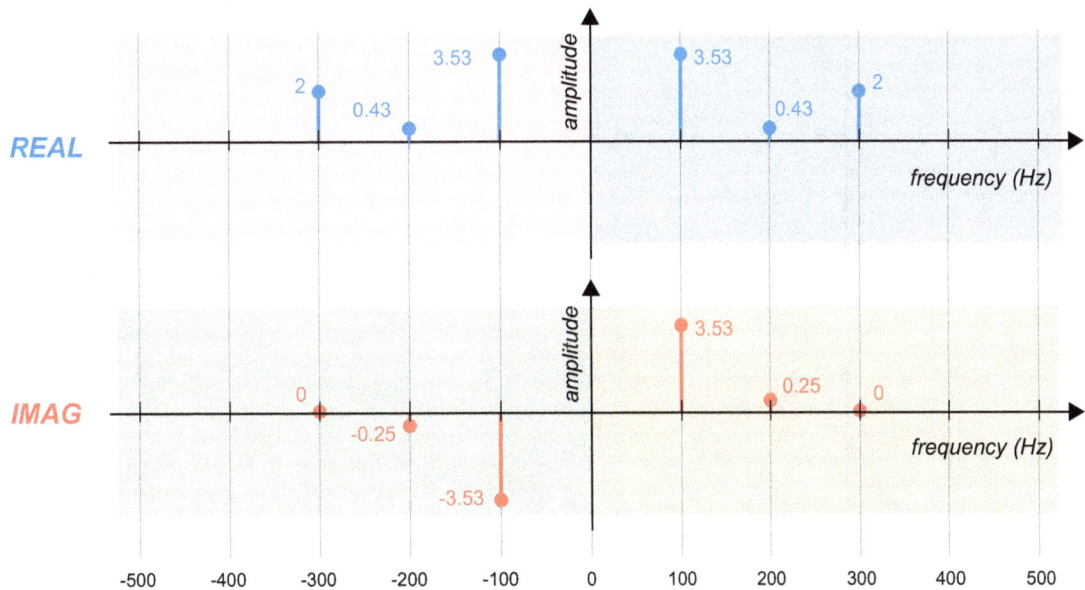

Figure 7.7: Complex-valued spectrum of the real signal $s_2(t)$ as defined in (7.15).

magnitude / phase spectrum from Figure 7.5, when we viewed just the right hand side of the frequency axis. As noted previously, for real-valued signals, the magnitude spectrum is always even-symmetric and therefore the convention is to plot only the positive frequency values.

As will be shown shortly, when working with quadrature communications signals, it becomes very useful to use complex number notation and to view complex spectra. The mathematical analysis using complex exponentials and the viewing of complex (two-sided) spectra will make the receiver design more straight-forward in terms of mixing the right signal to baseband with a quadrature mixer/demodulator. A number of examples with complex number notation and two sided spectrum design will also be derived for RFSoC RF-ADC receivers in Chapter 10.

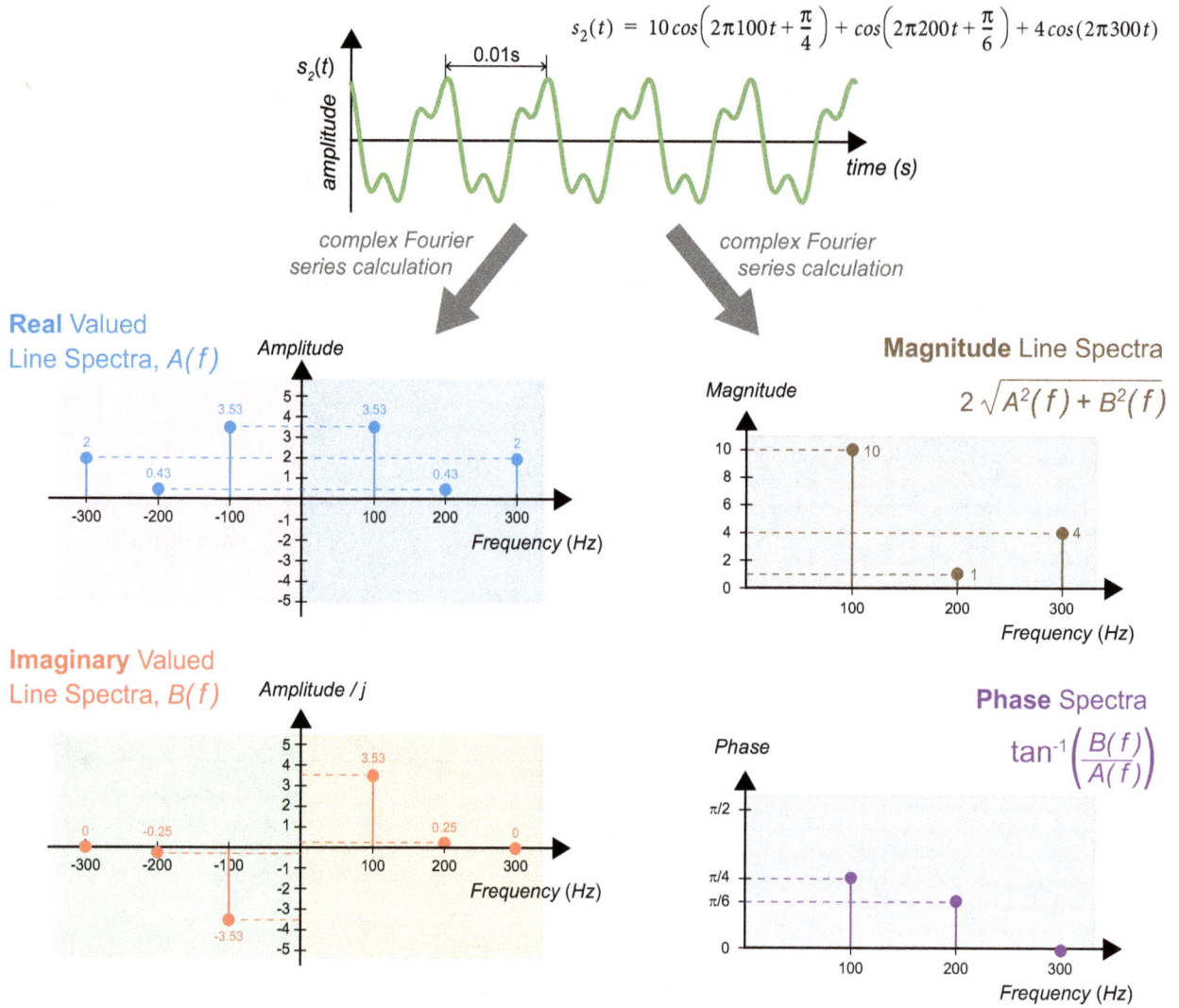

$$s_2(t) = 10\cos\left(2\pi 100t + \frac{\pi}{4}\right) + \cos\left(2\pi 200t + \frac{\pi}{6}\right) + 4\cos(2\pi 300t)$$

Figure 7.8: *A real signal as complex spectra (left hand side) and as a magnitude and phase spectra (right hand side).*

7.2. Amplitude Modulation and Demodulation

The use of complex signal notation is associated with quadrature modulation, where two independent streams or channels of baseband data are modulated onto the same carrier. A quadrature carrier refers to when these two baseband information signals are modulated onto cosine and sine carriers at the same frequency (i.e. 90° out of phase) — this creates two orthogonal components that coexist within the same frequency band, without causing interference to one another. As we will note in Section 7.4, this can then be elegantly represented with a complex notation, and considering that one of the channels is real-valued and the other is imaginary-valued.

As a prelude to quadrature modulation or mixing (or sometimes called QAM — Quadrature Amplitude Modulation) in this section we will review the standard AM (Amplitude Modulation) modulator and demodulator using real valued trigonometric analysis. We will first review the process of AM modulation using a single carrier signal, and then extend the discussion to quadrature modulation, before introducing the complex exponential representation in Section 7.3.

7.2.1. Double Sideband Suppressed Carrier Amplitude Modulation (DSB-SC AM)

Consider the amplitude modulation of a low frequency baseband signal, $g(t)$, with a higher frequency carrier, $c(t)$. To form the simplest possible example as a starting point, we assume that the baseband signal is a single cosine wave, i.e.

$$g(t) = A\cos(2\pi f_b t),\tag{7.16}$$

where f_b is the frequency of the baseband signal, and A represents its amplitude.

Note that the amplitude A is somewhat arbitrary, given that in an RFSoC implementation, modulation with a carrier is achieved by binary multiplication of the signal with the output of a Numerically Controlled Oscillator (NCO), as will be reviewed in Chapter 10. Denoting the amplitude as A allows us to more easily note the outcome of amplitude scaling to $A/2$ that happens later in the receiver stage.

The carrier signal is of frequency f_c and is expressed as

$$c(t) = \cos(2\pi f_c t).\tag{7.17}$$

Using simple trigonometry, we can determine that the modulated signal is given by

$$s(t) = A\cos(2\pi f_b t)\cos(2\pi f_c t) = \frac{A}{2}\Big[\cos(2\pi(f_c - f_b)t) + \cos(2\pi(f_c + f_b)t)\Big]\tag{7.18}$$

The modulation of the information signal $g(t)$ onto carrier $c(t)$ is depicted in the time domain in Figure 7.9, with a corresponding frequency domain representation shown in Figure 7.10. Notice that two spectral components are generated by the modulation process, at frequencies $f_c - f_b$ and $f_c + f_b$. The 'modulation' might also be referred to as 'mixing' and indeed is accomplished by simple multiplication.

More generally, baseband signals are not simple cosines, but spectrally rich signals occupying the frequency band from 0 Hz to f_b Hz. We denote such a signal $G(f)$. The result of modulating this more realistic baseband signal onto a carrier at f_c Hz is shown in Figure 7.11, where it is seen to produce upper and lower sidebands. At this stage we can see some spectral inefficiency here — the RF transmission bandwidth required, $2f_b$, is twice the baseband signal bandwidth, f_b.[3]

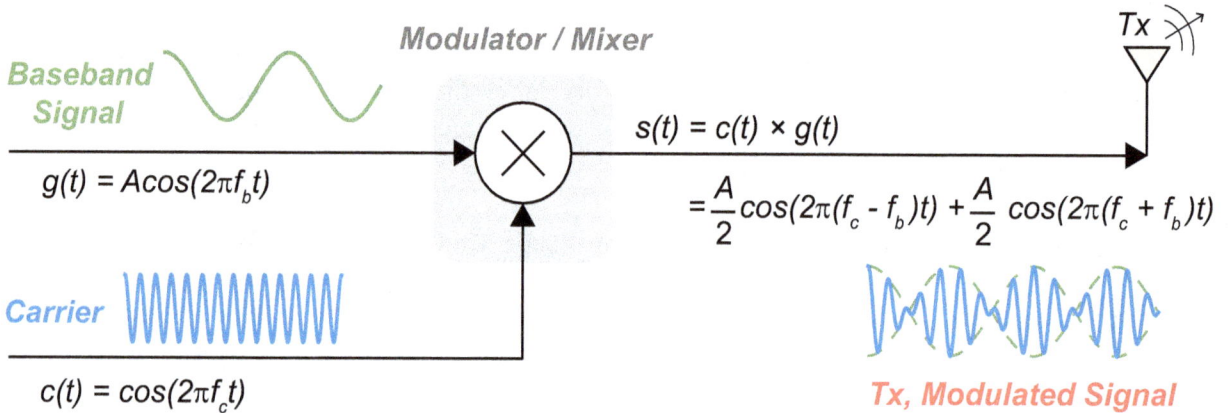

Figure 7.9: *Modulation of a low frequency cosine onto a high frequency carrier (time domain view).*

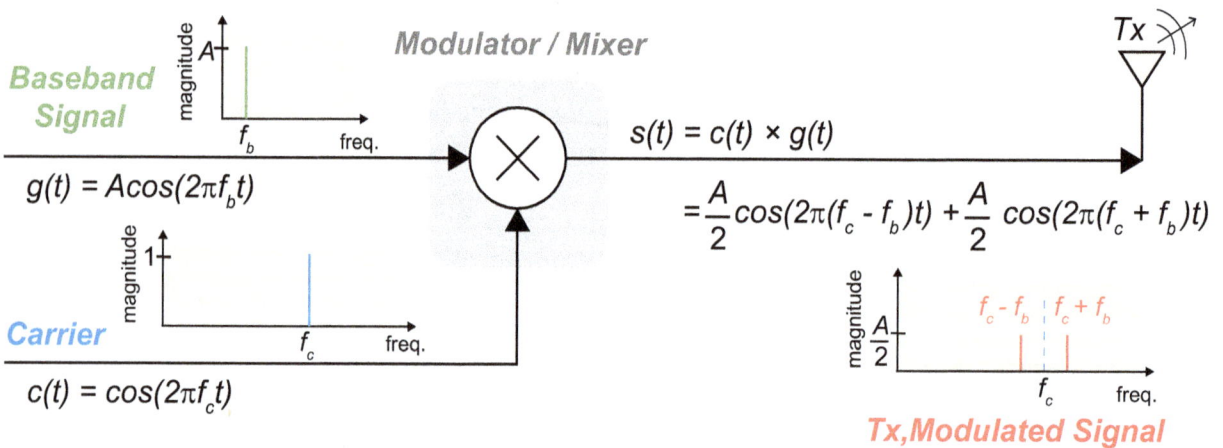

Figure 7.10: *Modulation of a low frequency cosine onto a high frequency carrier (frequency domain view).*

---

3. Of course, there is a type of communications called *single sideband* which is spectrally efficient, but QAM helps solve this problem as we will see in later sections.

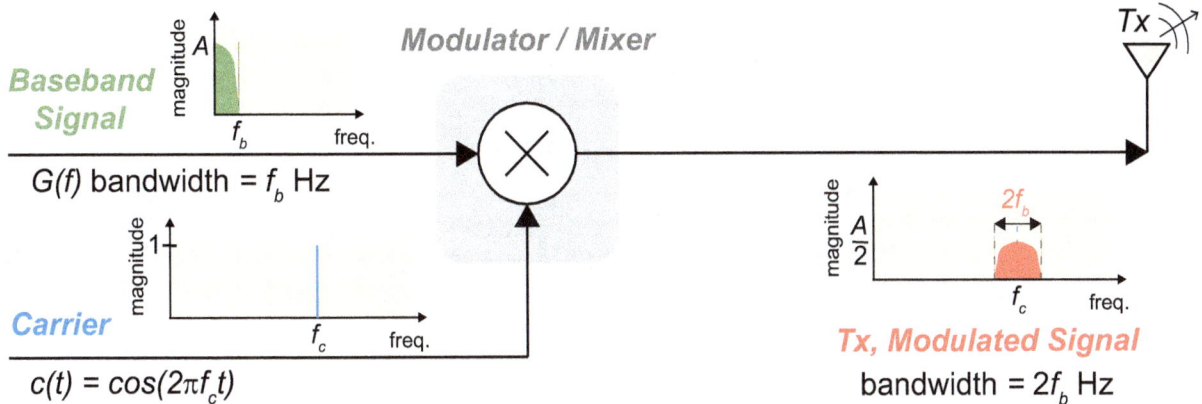

Figure 7.11: Modulation of a baseband signal onto a carrier, producing upper and lower sidebands.

7.2.2. Amplitude Demodulation

The process of demodulating a simple AM signal is straightforward, under the assumption that the carrier signal generated in the receiver, with which to demodulate the signal, has exactly the same frequency and phase as the incoming signal. (In practice this is almost never the case, which is why synchronisation systems are needed in receivers, but that is beyond the scope of the current discussion, more detail on this can be found in [176], [230]!).

The demodulation process involves multiplying (or mixing) the received signal with the locally generated carrier. This produces two sets of frequency components: one at baseband, and a second at twice the carrier frequency, which can be conveniently removed with a low pass filter. An example of demodulation is shown in Figure 7.12, for the example of receiving the modulated cosine wave used in our initial modulation example in Figure 7.9.

The signal received at the antenna in Figure 7.12 is the same transmitted signal from Figure 7.9, and is:

$$s(t) = \frac{A}{2}\cos\left(2\pi(f_c - f_b)t\right) + \frac{A}{2}\cos\left(2\pi(f_c + f_b)t\right) \tag{7.19}$$

Within the receiver, it can then be perfectly demodulated by multiplying with a locally generated carrier, given by $c(t) = \cos(2\pi f_c t)$.

$$s(t) = \frac{A}{2}\cos(2\pi(f_c - f_b)t) + \frac{A}{2}\cos(2\pi(f_c + f_b)t)$$

Rx

Modulated Signal

Demodulator / Mixer

$x(t)$

Low pass filter

$$u(t) = \frac{A}{2}\cos(2\pi f_b t)$$

Signal

Local Oscillator

$$c(t) = \cos(2\pi f_c t)$$

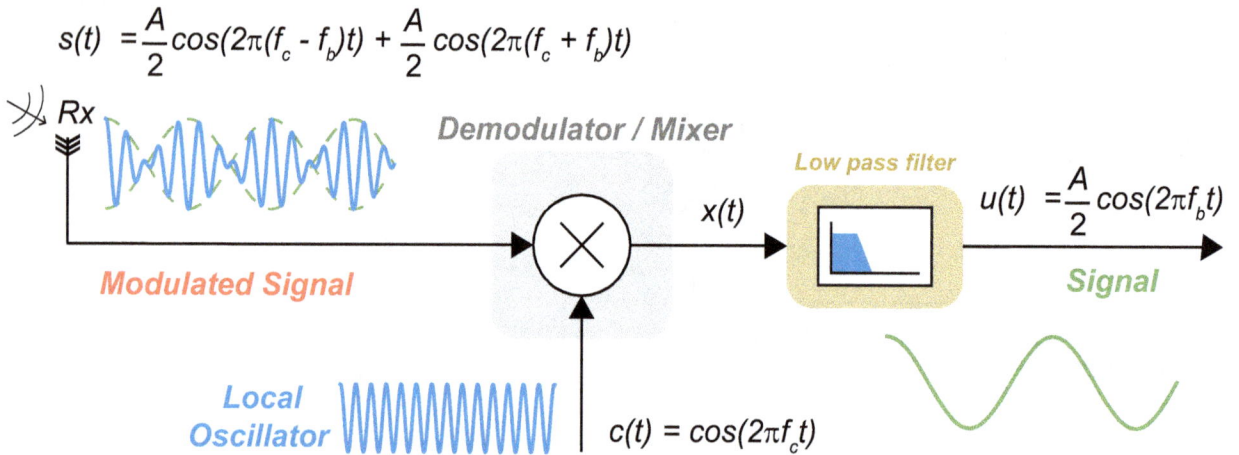

Figure 7.12: Amplitude demodulation of the AM-modulated, transmitted cosine signal from Figure 7.9.

The output of the demodulation process is given by

$$
\begin{aligned}
x(t) &= c(t) \times s(t) \\[6pt]
&= \frac{A}{2}\cos(2\pi f_c t)\left[\cos(2\pi(f_c - f_b)t) + \cos(2\pi(f_c + f_b)t)\right] \\[6pt]
&= \frac{A}{2}\cos(2\pi f_c t)\cos(2\pi(f_c - f_b)t) + \frac{A}{2}\cos(2\pi f_c t)\cos(2\pi(f_c + f_b)t) \\[6pt]
&= \frac{A}{4}\cos(2\pi(2f_c - f_b)t) + \frac{A}{4}\cos(2\pi f_b t) + \frac{A}{4}\cos(2\pi(2f_c + f_b)t) + \frac{A}{4}\cos(2\pi f_b t) \\[6pt]
&= \frac{A}{2}\cos(2\pi f_b t) \quad + \quad \left[\frac{A}{4}\cos(2\pi(2f_c - f_b)t) + \frac{A}{4}\cos(2\pi(2f_c + f_b)t)\right] .
\end{aligned}
\tag{7.20}
$$

Note that the trigonometric identity

$$\cos(A)\cos(B) = 0.5\cos(A + B) + 0.5\cos(A - B) \tag{7.21}$$

is used in the third and fourth lines of (7.20).

The two high frequency terms that are present in the demodulated output (sitting around $2f_c$) are attenuated by the low pass filter, which leaves only a scaled version of the signal that was originally transmitted ($g(t)$, as shown in Figure 7.9).

The resulting signal is given by

$$\Rightarrow u(t) = \frac{A}{2}\cos(2\pi f_b t) \quad + \quad \left[\frac{A}{4}\cos 2\pi(2f_c - f_b) + \frac{A}{4}\cos 2\pi(2f_c + f_b)\right]$$

low pass filtered terms

$$= \frac{A}{2}\cos(2\pi f_b t) \tag{7.22}$$

$$= \frac{g(t)}{2}$$

This represents perfect modulation and demodulation of a simple cosine wave — but what about the more realistic baseband signal introduced via Figure 7.11?

Actually the same process of demodulation can be shown to produce two copies of the originally transmitted signal, one of them at baseband, and another centred at $2f_c$, the latter of which is removed using a low pass filter. Demodulation of this signal is depicted in Figure 7.13.

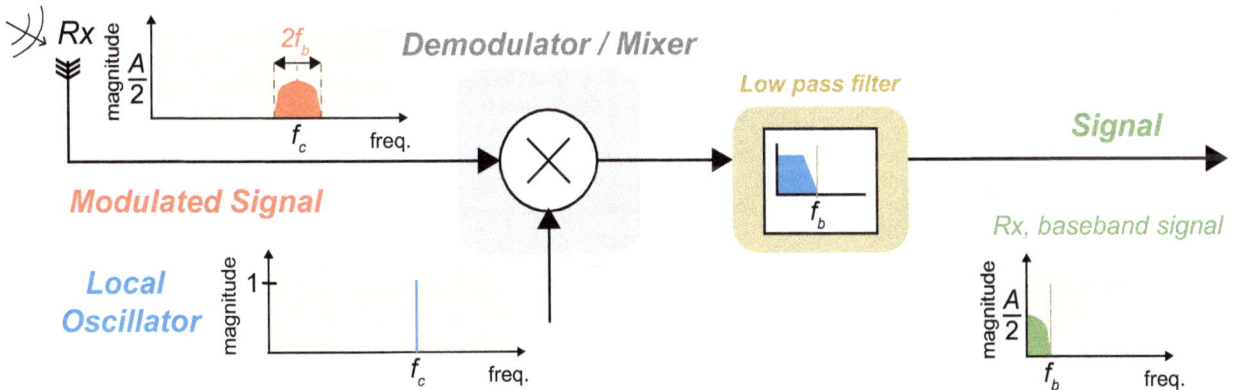

Figure 7.13: Amplitude demodulation of the AM-modulated, transmitted baseband signal from Figure 7.11

7.2.3. Amplitude Demodulation with a Phase Error

As an extension to the original analysis, we now consider the case where the local oscillator used to demodulate the signal is phase offset from the received signal. As mentioned earlier, it is more realistic to expect a phase error, or even a frequency error, to exist between the two carrier signals, and if present this produces a deviation from the ideal received signal.

To illustrate the effect of a carrier phase error, we add a phase offset term, θ, to the local oscillator output shown in Figure 7.12. We consider that the signal being modulated is $g(t) = A\cos(2\pi f_b t)$, as previously analysed for perfect modulation and demodulation.

The demodulation that was previously presented in (7.20) now becomes:

$$
\begin{aligned}
x(t) &= c(t) \times s(t) \\
&= \cos(2\pi f_c t + \theta)\left[\frac{A}{2}\cos(2\pi(f_c - f_b)t) \quad + \quad \frac{A}{2}\cos(2\pi(f_c + f_b)t)\right] \\
&= \frac{A}{2}\left[\cos(2\pi f_c t + \theta)\cos(2\pi(f_c - f_b)t) \; + \; \cos(2\pi f_c t + \theta)\cos(2\pi(f_c + f_b)t)\right] \\
&= \frac{A}{4}\cos(2\pi(2f_c - f_b)t + \theta) + \frac{A}{4}\cos(2\pi f_b t + \theta) + \frac{A}{4}\cos(2\pi(2f_c + f_b)t + \theta) + \frac{A}{4}\cos(2\pi f_b t - \theta) \\
&= \frac{A}{4}\cos(2\pi f_b t + \theta) + \frac{A}{4}\cos(2\pi f_b t - \theta) + \left[\frac{A}{4}\cos(2\pi(2f_c - f_b)t + \theta) + \frac{A}{4}\cos(2\pi(2f_c + f_b)t + \theta)\right]
\end{aligned}
\tag{7.23}
$$

Following this, the low pass filter removes the high frequency components centred around $2f_c$, and the remaining terms are simplified using a trigonometric identity, to obtain

$$
\begin{aligned}
u(t) &= \frac{A}{4}\cos(2\pi f_b t + \theta) + \frac{A}{4}\cos(2\pi f_b t - \theta) + \left[\frac{A}{4}\cos(2\pi(2f_c - f_b + \theta)t) + \frac{A}{4}\cos(2\pi(2f_c + f_b + \theta)t)\right] \\
& \text{\textit{low pass filtered terms}} \\
&= \frac{A}{4}\cos(2\pi f_b t + \theta) + \frac{A}{4}\cos(2\pi f_b t - \theta) \\
&= \frac{A}{2}\cos(2\pi f_b t) \, \cos(\theta) \\
&= \frac{A}{2} g(t)\cos(\theta)
\end{aligned}
\tag{7.24}
$$

Thus, we can confirm that the output is scaled by $\cos(\theta)$, which must have a value in the range $-1 \le \theta \le 1$. However in the absolute worst case, $\theta = \pm\pi/2$, or equivalently $\pm 90^o$, and this produces an output of zero!

Any time-varying phase error can be represented as $\theta(t)$, and where the phase error increases or decreases at a constant rate, this represents a frequency offset error (noting that the frequency is the derivative of phase). In the time domain, this will cause the amplitude of the received signal to fluctuate. Therefore, again we recognise that synchronisation is required to compensate for such errors. We do not cover synchronisation in any detail this book, but many references and support texts are available (see textbooks such as [230], [301], and [326] for further reading on this topic).

Having reviewed amplitude modulation and demodulation for basic AM signals, the next logical step is to progress to QAM. First, it is important to establish the rationale for using QAM, and then we move on to analyse QAM modulation and demodulation (initially using trigonometric methods, and then with complex notation). We will demonstrate that one of the key rationales for QAM is spectral efficiency. The earlier example for AM transmission (see Figure 7.11) requires twice the baseband frequency for transmission — QAM will make this more efficient.

7.3. Quadrature Amplitude Modulation and Demodulation

We can now begin to review quadrature amplitude modulators (also generically referred as QAM, or just quadrature modulation) and present its complex exponential (i.e. real and imaginary) representation.

The motivation to use QAM, as opposed to modulating with a single carrier, is bandwidth efficiency. Referring back to Figure 7.11, a bandwidth of $2f_b$ is required to transmit the standard AM signal, i.e. double the bandwidth of the baseband signal. Therefore, it could be said that AM is only 50% efficient. QAM allows the efficiency to be improved (all the way back up to 100%!), because two signals are transmitted using orthogonal carriers at the same frequency, thus occupying the same bandwidth. As the carrier phases are 90° apart (in quadrature), i.e. a sine wave and a cosine wave, these two signals are orthogonal and do not interfere with one another, and can be perfectly separated and recovered at the receiver.

7.3.1. Trigonometric Representation of Quadrature Modulation

The quadrature modulator is shown in Figure 7.14. Notice that the sine carrier has a negative amplitude here, i.e. $-sin(2\pi f_c t)$, which is often stated for mathematical notational convenience (the quadrature modulation and demodulation works with orthogonal carriers of $\{cos(2\pi f_c t), -sin(2\pi f_c t)\}$ and equally with carriers of $\{cos(2\pi f_c t), sin(2\pi f_c t)\}$.

Note that if the carriers were $\{cos(2\pi f_c t), sin(2\pi f_c t)\}$, then we could negate the Q channel input, giving $-g_2(t)$, which would achieve the same result as the carriers of $\{cos(2\pi f_c t), -sin(2\pi f_c t)\}$ and $g_2(t)$.

Therefore, there is no need to worry about the negative / positive setting of the oscillator amplitude; in the examples that follow, the amplitude is set to achieve the same polarity of modulated outputs when presenting both the quadrature and the complex exponential versions of the modulators/mixers.

In this model of quadrature modulation, we now denote two independent baseband signals, $g_1(t)$ and $g_2(t)$. By convention, the baseband channel modulated by the cosine carrier is referred to as *In-Phase*, or the *I-Channel*, or *Real Channel*; while the channel modulated by the sine carrier is known as the *Quadrature-Phase*, *Q-Channel* or *Imaginary Channel*. As mentioned above, the use of the term 'quadrature' arises because the negative sine carrier is 90° separated from the cosine carrier, i.e. a phase difference of one quadrant.

When the information signals, $g_1(t)$ and $g_2(t)$, are modulated onto the cosine and sine carriers, the resulting quadrature modulated signal is given by

$$y(t) = g_1(t)cos(2\pi f_c t) - g_2(t)sin(2\pi f_c t) . \tag{7.25}$$

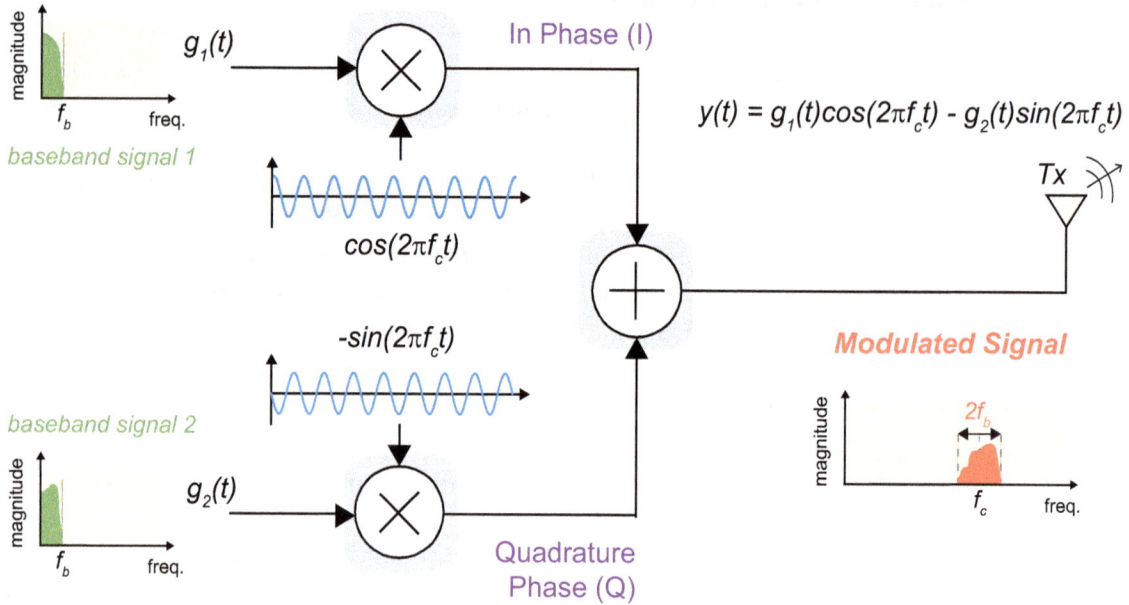

Figure 7.14: Quadrature modulation of two independent baseband signals.

7.3.2. Trigonometric Representation of Quadrature Demodulation

At the receive side, the QAM demodulator likewise uses two local carriers to demodulate the signal, separated by 90°. The demodulator is shown in Figure 7.15, and assumes the ideal case where no signal degradation occurs in the channel (i.e. the transmitted signal from Figure 7.14 and Eq. (7.25) is the received signal here).

We can now demonstrate with some trigonometric analysis that the transmitted baseband information signals, $g_1(t)$ and $g_2(t)$, can be successfully recovered from the $y(t)$ signal at the receiver.

First of all taking the I-Phase, the output after demodulation with the local cosine oscillator is given by

$$
\begin{aligned}
x_1(t) &= y(t)\cos(2\pi f_c t) \\
&= \Big[g_1(t)\cos(2\pi f_c t) - g_2(t)\sin(2\pi f_c t)\Big]\cos(2\pi f_c t) \\
&= g_1(t)\cos^2(2\pi f_c t) - g_2(t)\sin(2\pi f_c t)\cos(2\pi f_c t) \\
&= \frac{1}{2}g_1(t)\Big[1 + \cos(4\pi f_c t)\Big] - \frac{1}{2}g_2(t)\sin(4\pi f_c t)
\end{aligned}
\tag{7.26}
$$

and after low pass filtering the resulting signal, we obtain

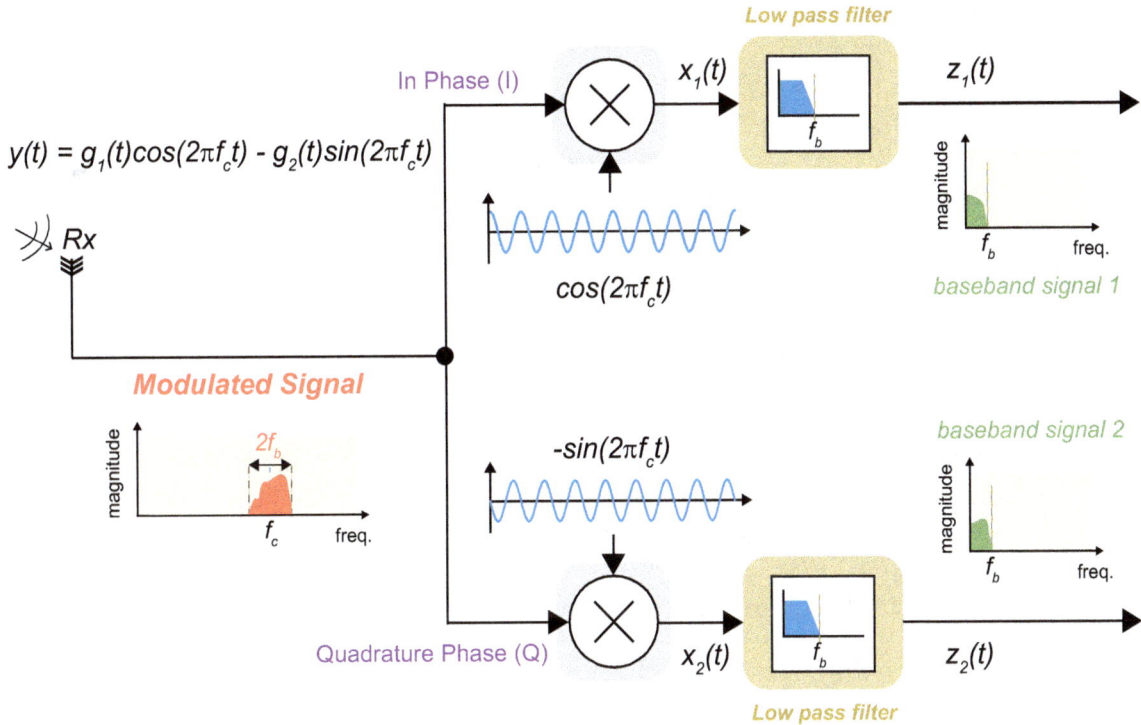

Figure 7.15: Quadrature demodulation of a quadrature-modulated signal

$$z_1(y) \;=\; \frac{1}{2}g_1(t) + \left[\frac{1}{2}g_1(t)\cos(4\pi f_c t) - \frac{1}{2}g_2(t)\sin(4\pi f_c t)\right] \;\text{low pass filtered terms} \qquad (7.27)$$

Note that the above manipulation relies on trigonometric identities as listed on page 703.

The equivalent process applies for demodulation of the Q-Phase, where the input to the sine wave demodulator is identical to that supplied to the cosine demodulator. The output of the sine demodulator is given by

$$
\begin{aligned}
x_2(t) &= y(t)(-\sin(2\pi f_c t)) \\[6pt]
&= \Big[g_1(t)\cos(2\pi f_c t) - g_2(t)\sin(2\pi f_c t)\Big](-\sin(2\pi f_c t)) \\[6pt]
&= \Big[-g_1(t)\cos(2\pi f_c t)\sin(2\pi f_c t)\Big] + g_2(t)\sin^2(2\pi f_c t) \\[6pt]
&= -\frac{1}{2}g_1(t)\sin(4\pi f_c t) + \frac{1}{2}g_2(t)\Big[1 - \cos(4\pi f_c t)\Big]
\end{aligned}
\qquad (7.28)
$$

and then after passing through the low pass filter, the output is

$$z_2(t) \;=\; \frac{1}{2}g_2(t) - \left[\frac{1}{2}g_1(t)\sin(4\pi f_c t) + \frac{1}{2}g_2(t)\cos(4\pi f_c t)\right] \qquad \text{low pass filtered terms} \qquad (7.29)$$

Therefore we can confirm that after modulation and demodulation with the quadrature carriers, both of the information signals, $g_1(t)$ and $g_2(t)$, are recovered perfectly. The only difference between the transmitted and received signals is a scaling factor of 0.5, which can be easily compensated.

7.3.3. Quadrature Demodulation with a Phase Shift

The analysis from the previous section assumed the perfect scenario, where there is no phase shift between the transmit and receive oscillators. We now repeat the analysis of demodulation, in the presence of a θ phase shift in the receiver's local oscillator. As will be demonstrated, this causes the I and Q channels to become mixed together (the severity varies according to the value of θ).

Rather than the signal obtained in (7.26), the demodulated I-Channel instead contains:

$$
\begin{aligned}
x_1(t) &= y(t)\cos(2\pi f_c t + \theta) \\[6pt]
&= \left[g_1(t)\cos(2\pi f_c t) - g_2(t)\sin(2\pi f_c t)\right]\cos(2\pi f_c t + \theta) \\[6pt]
&= g_1(t)\cos(2\pi f_c t)\cos(2\pi f_c t + \theta) \quad - \quad g_2(t)\sin(2\pi f_c t)\cos(2\pi f_c t + \theta) \\[6pt]
&= \frac{1}{2}g_1(t)\Big[\cos(-\theta) + \cos(4\pi f_c t + \theta)\Big] - \frac{1}{2}g_2(t)\Big[\sin(-\theta) + \sin(4\pi f_c t + \theta)\Big]
\end{aligned}
\qquad (7.30)
$$

Noting that $\cos(-x) = \cos(x)$ and $\sin(-x) = -\sin(x)$, we can simplify to

$$= \frac{1}{2}g_1(t)\Big[\cos(\theta) + \cos(4\pi f_c t + \theta)\Big] - \frac{1}{2}g_2(t)\Big[-\sin(\theta) + \sin(4\pi f_c t + \theta)\Big] \qquad (7.31)$$

and then filter out the high frequency terms, to obtain

$$z_1(t) \;=\; \frac{1}{2}\Big[g_1(t)\cos(\theta) + g_2(t)\sin(\theta)\Big] + \left[\frac{1}{2}g_1(t)\cos(4\pi f_c t + \theta) - \frac{1}{2}g_2(t)\sin(4\pi f_c t + \theta)\right] \qquad (7.32)$$

low pass filtered terms

where we note that the demodulated signal contains a component of $g_2(t)$, as well as $g_1(t)$.

Using the same analysis, the demodulated Q-Channel is expressed by

$$x_2(t) = y(t)(-sin(2\pi f_c t + \theta))$$

$$= \left[g_1(t)cos(2\pi f_c t) - g_2(t)sin(2\pi f_c t) \right](-sin(2\pi f_c t + \theta))$$

$$= -g_1(t)cos(2\pi f_c t)sin(2\pi f_c t + \theta) + g_2(t)sin(2\pi f_c t)sin(2\pi f_c t + \theta)$$

$$= -\frac{1}{2}g_1(t)\left[-sin(-\theta) + sin(4\pi f_c t + \theta)\right] + \frac{1}{2}g_2(t)[cos(-\theta) - cos(4\pi f_c t + \theta)]$$

(7.33)

and can be simplified to

$$= -\frac{1}{2} g_1(t)\left[sin(\theta) + sin(4\pi f_c t + \theta)\right] + \frac{1}{2} g_2(t)[cos(\theta) - cos(4\pi f_c t + \theta)]$$

(7.34)

using the property that $cos(-x) = cos(x)$ and $sin(-x) = -sin(x)$.

After low pass filtering, this yields

$$z_2(t) = \frac{1}{2}\left[-g_1(t)sin(\theta) + g_2(t)cos(\theta)\right] - \left[\frac{1}{2}g_1(t)sin(4\pi f_c t + \theta) + \frac{1}{2}g_2(t)cos(4\pi f_c t + \theta)\right]$$

(7.35)

low pass filtered terms

and similarly, we see that the demodulated signal contains not only the desired $g_2(t)$, but also a component of $g_1(t)$.

Therefore in summary, when the receive local oscillator has a phase shift of θ degrees, the demodulated I and Q Channels are mixed versions of the two transmitted signals, $g_1(t)$ and $g_2(t)$, scaled by 0.5,

$$z_1(t) = 0.5\left[g_1(t)cos(\theta) + g_2(t)sin(\theta)\right]$$

$$z_2(t) = 0.5\left[-g_1(t)sin(\theta) + g_2(t)cos(\theta)\right]$$

(7.36)

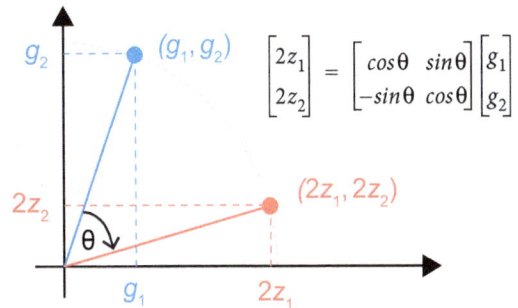

Figure 7.16: A Cartesian plane rotation applied to a transmitted symbol, $(g_1(t), g_2(t))$

and of course the scaling can be very easily compensated by applying a gain of 2 to both $z_1(t)$ and $z_2(t)$.

Considering (7.36) further, if $(g_1(t), g_2(t))$ represents a point in the Cartesian (x-y) plane at a given sample time, then the point recovered after demodulation, $(2z_1(t), 2z_2(t))$, is simply $(g_1(t), g_2(t))$ rotated about the origin by θ degrees, as shown in Figure 7.16. This represents one of the classic problems in wireless communications, where a symbol constellation is rotated and has to be 'de-rotated' by a synchroniser in the receiver.

Consider the example of a digital communications link where there is a phase shift at the receiver. If the phase shift was constant, then the received constellation would be rotated by θ degrees. For example, referring to the 4-QAM of QPSK transmitted constellation shown in Chapter 6 (in Figure 6.9 on page 206), the received constellation points would be rotated by θ degrees, as shown in Figure 7.17(a).

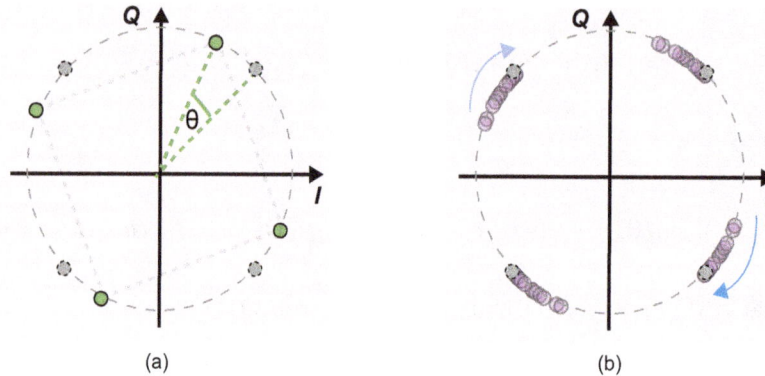

Figure 7.17: (a) Rotation of received symbol positions caused by phase error of θ. (b) Receiver constellation with frequency error f_Δ, which is a continuously changing phase and is seen in the receiver as a rotating constellation.

7.4. Quadrature Modulation and Demodulation with Complex Notation

The QAM system model considered in the previous section uses real signals only, and therefore it is absolutely not complex and does not involve any complex numbers or notation. A major benefit can be achieved from introducing complex arithmetic to represent this model, however, as it makes the mathematics more tractable. We will now redefine the equations describing the QAM system reviewed earlier, using complex exponential notation.

7.4.1. Quadrature Modulation with Complex Exponential Notation

First, consider the block diagram in Figure 7.18, which depicts the modulation of a complex baseband signal, using a complex exponential at the carrier frequency. Note that complex number signal paths are indicated by the double arrow style: ⟹ . For the purposes of analysis, the two baseband signals $g_1(t)$ and $g_2(t)$ are now represented by a single complex signal which is composed of a real part, $g_1(t)$, and an imaginary part, $jg_2(t)$.

$$g(t) = g_1(t) + jg_2(t) \tag{7.37}$$

Similarly the quadrature carrier pair is represented by another complex signal, but this time it is more convenient to represent it in its complex exponential form:

$$e^{j2\pi f_c t} = cos(2\pi f_c t) + j sin(2\pi f_c t) \tag{7.38}$$

The modulator creates a signal that consists of real and imaginary components, and is therefore complex.

$$
\begin{aligned}
v(t) = g(t)e^{j2\pi f_c t} &= \left[g_1(t) + jg_2(t) \right] e^{j2\pi f_c t} \\
&= \left[g_1(t) + jg_2(t) \right] [cos(2\pi f_c t) + j sin(2\pi f_c t)] \\
&= g_1(t)cos(2\pi f_c t) + jg_2(t)cos(2\pi f_c t) + jg_1(t)sin(2\pi f_c t) - g_2(t)sin(2\pi f_c t) \\
&= [g_1(t)cos(2\pi f_c t) - g_2(t)sin(2\pi f_c t)] \quad + \quad j[g_1(t)sin(2\pi f_c t) + g_2(t)cos(2\pi f_c t)]
\end{aligned}
\tag{7.39}
$$

Real **Imaginary**

After the modulation stage, we simply retain the real part and discard the imaginary part. Therefore, only the real part of the signal $v(t)$ is transmitted.

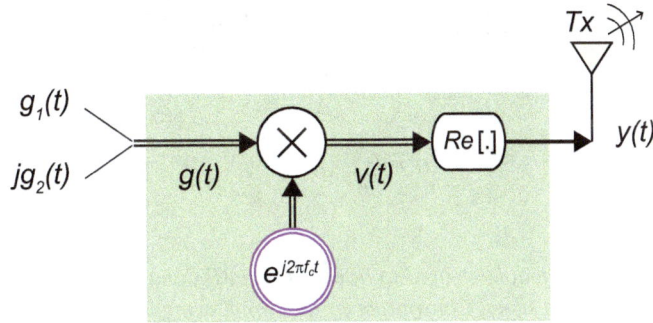

Figure 7.18: Baseband signals set as real and imaginary and then mixed with a complex exponential carrier. After the mixer we retain only the real component for transmission as a real world signal.

This last stage before transmission can be expressed mathematically as

$$
\begin{aligned}
\Re\{v(t)\} &= \Re\left\{ \left[g_1(t)cos(2\pi f_c t) - g_2(t)sin(2\pi f_c t) \right] + j\left[g_1(t)sin(2\pi f_c t) + g_2(t)cos(2\pi f_c t) \right] \right\} \\
&= g_1(t)cos(2\pi f_c t) - g_2(t)sin(2\pi f_c t) \\
&= y(t)
\end{aligned}
\tag{7.40}
$$

(Just for note we could have achieved the same output as Eq. (7.40) by mapping the complex baseband as $g(t) = g_1(t) - jg_2(t)$ and mixing with a negative complex exponential, i.e. $e^{-j2\pi f_c t}$. The design choices for selecting the complex exponential as a positive or negative value will be presented via a few examples in Chapter 10.)

Referring back to Figure 7.14, recall that $y(t)$ is also the output of the quadrature modulation when analysed using trigonometry. Therefore, we have now successfully used complex exponential notation to create the final real signal for real-world transmission.

As mentioned a number of times earlier in this chapter, the benefit of complex notation and the use of complex exponentials is that it requires simpler mathematics and lessens the need to use (and remember) trigonometric identities. Going forward, we can more elegantly design systems and work with two-sided spectra to design quadrature receivers using the RFSoC RF-ADCs! A number of examples of such receiver designs will be presented in Chapter 10.

7.4.2. Quadrature Demodulation with Complex Exponential Notation

Complex exponential notation can also be used to described the process of demodulation. A model of a complex demodulator is shown in Figure 7.19.

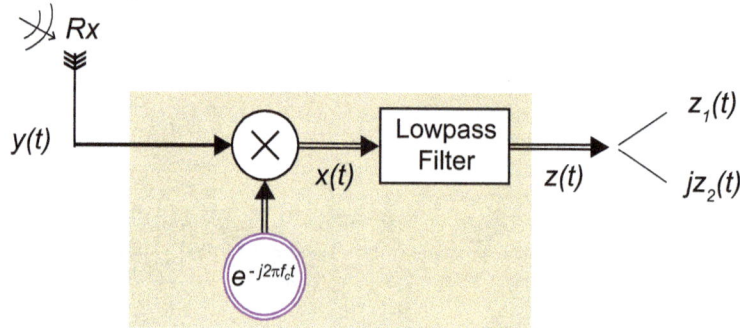

Figure 7.19: A complex demodulator used to receive a real RF signal and demodulate it to baseband (thus receiving the signal transmitted by the complex modulator in Figure 7.18).

The input to this complex demodulator, i.e. the signal received at the antenna, is a real signal, as it originates from the real world, and it is the same $y(t)$ as was generated in Figure 7.14. As in our previous discussion from Section 7.3, we assume a perfect radio channel, where the signal received is exactly the signal transmitted.

After multiplication of the received signal, $y(t)$, with the complex exponential, $e^{-j2\pi f_c t}$ (note this requires to be a negative to allow the equivalence to the earlier quadrature modulator. The signal $x(t)$ is

$$x(t) = y(t)e^{-j2\pi f_c t} \tag{7.41}$$

$$= \left[g_1(t)\cos(2\pi f_c t) - g_2(t)\sin(2\pi f_c t) \right] e^{-j2\pi f_c t}$$

$$= \left[g_1(t)\cos(2\pi f_c t) - g_2(t)\sin(2\pi f_c t) \right] \left[\cos(2\pi f_c t) - j\sin(2\pi f_c t) \right]$$

To make the equations a little more concise and easier to read, we elect to make the substitutions $A = g_1(t)$, $B = g_2(t)$, and to set $\phi = 2\pi f_c t$. The equation for $x(t)$ can therefore be expressed as

$$x(t) = \Big[g_1(t)\cos(2\pi f_c t) - g_2(t)\sin(2\pi f_c t) \Big] \Big[\cos(2\pi f_c t) - j\sin(2\pi f_c t) \Big] \tag{7.42}$$

$$= \Big[A\cos(\phi) - B\sin(\phi) \Big] \Big[\cos(\phi) - j\sin(\phi) \Big]$$

$$= A\cos(\phi)\Big[\cos(\phi) - j\sin(\phi) \Big] - B\sin(\phi)\Big[\cos(\phi) - j\sin(\phi) \Big]$$

Next, we multiply out the brackets and re-order to give

$$x(t) = A\cos^2(\phi) - jA\cos(\phi)\sin(\phi) - B\sin(\phi)\cos(\phi) + jB\sin^2(\phi) \tag{7.43}$$

$$= A\cos^2(\phi) + jB\sin^2(\phi) - jA\cos(\phi)\sin(\phi) - B\sin(\phi)\cos(\phi)$$

Trigonometric identities can then be applied to achieve

$$= \frac{A}{2}\Big[1 + \cos(2\phi) \Big] + j\frac{B}{2}\Big[1 - \cos(2\phi) \Big] - j\frac{A}{2}\sin(2\phi) - \frac{B}{2}\sin(2\phi)$$

$$= \frac{A}{2} + \frac{A}{2}\cos(2\phi) + j\frac{B}{2} - j\frac{B}{2}\cos(2\phi) - j\frac{A}{2}\sin(2\phi) - \frac{B}{2}\sin(2\phi) \tag{7.44}$$

$$= \frac{A}{2} + j\frac{B}{2} + \frac{A}{2}\cos(2\phi) - j\frac{B}{2}\cos(2\phi) - j\frac{A}{2}\sin(2\phi) - \frac{B}{2}\sin(2\phi)$$

and, substituting back for $g_1(t)$, $g_2(t)$, and $2\pi f_c t$, the signal $x(t)$ in Figure 7.19 is

$$x(t) = \frac{1}{2}\Big[g_1(t) + jg_2(t) \Big] + \frac{1}{2}g_1(t)\cos(4\pi f_c t) - j\frac{1}{2}g_2(t)\cos(4\pi f_c t) \tag{7.45}$$

$$- j\frac{1}{2}g_1(t)\sin(4\pi f_c t) - \frac{1}{2}g_2(t)\sin(4\pi f_c t)$$

low pass filtered terms

Therefore, after the low pass filter in Figure 7.19, the output $z(t)$ is the same as the complex modulator input Eq. (7.37) in Figure 7.18, i.e.

$$z(t) = \frac{1}{2}[g_1(t) + jg_s(t)]. \tag{7.46}$$

This is the same signal as that obtained from the standard quadrature (cosine sine oscillator) demodulator, shown in Figure 7.15.

With this review, we have established the mathematical equivalence between the standard quadrature demodulator of Figure 7.15, and the complex exponential based demodulator of Figure 7.19. The equivalence of these models is further illustrated in Figure 7.20.

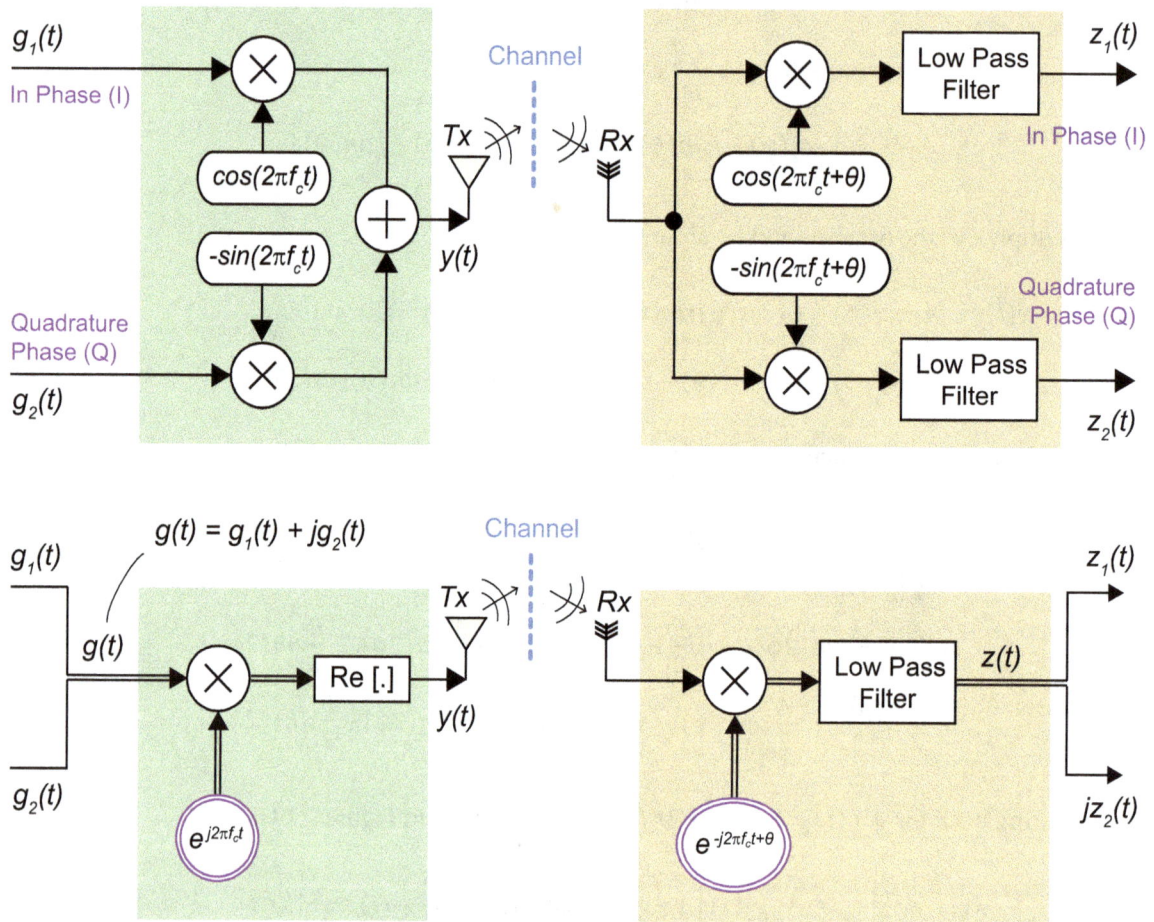

Figure 7.20: Equivalence of (top) standard QAM architecture expressed using trigonometry, and (bottom) complex model of the QAM architecture.

7.5. Spectral Representation of Complex Exponential Demodulation

It is informative to sketch the complex spectra for the 'complex' modulation and demodulation processes, as presented in Figure 7.21. The two independent signals to be transmitted are represented by $g_1(t)$ and $g_2(t)$. Therefore, if we represent them as

$$g(t) = g_1(t) + jg_2(t) \tag{7.47}$$

then $g(t)$ is a complex signal that can be represented in the complex frequency domain, by taking the Fourier transform. As this is not a real signal, then the spectrum is not symmetric. This is illustrated in the lower part of Figure 7.21, which shows the spectrum from point Ⓐ in the complex exponential system (note this style of diagram is commonly referred to as a *signal flow graph*).

The spectrum at Ⓑ shows the complex baseband signal after it has been modulated by the complex carrier,

$$v(t) = g(t)e^{j2\pi f_c t} \tag{7.48}$$

which has the effect of shifting the baseband signal spectrum to be centred at frequency f_c. As the two-sided spectrum at Ⓑ is asymmetric, then the signal is of complex form (real and imaginary components).

Taking the real part, using the $\Re[.]$ operator, we obtain the transmitted signal $y(t)$ shown in Figure 7.18, i.e.

$$\Re\left\{g(t)e^{j2\pi f_c t}\right\} = g_1(t)\cos(2\pi f_c t) - g_2(t)\sin(2\pi f_c t). \tag{7.49}$$

Therefore the signal at Ⓒ is $y(t)$, the signal that is transmitted over the channel. As this signal is real, its two-sided spectrum is symmetric as shown in Ⓒ.

At the receive side, the demodulator multiplies the received signal $y(t)$ with the complex exponential term, $e^{-j2\pi f_c t}$. In the frequency domain, this has the effect of shifting both the positive and negative spectra by f_c, to create the analytic complex signal spectra shown at Ⓓ. So once again, the two-sided spectrum is asymmetric and therefore the signal at Ⓓ is complex (with real and imaginary components).

Finally, both the complex signal at Ⓓ passes through a low pass filter (which is a real value filter only). This can be implemented as separate low pass filters on each of the two channels, or the complex signal can be passed through a single, real-valued low pass filter — producing the same output from both.

At the output of the filtering stage, the complex baseband received signal, $z(t)$, is obtained at Ⓔ. In an ideal model, with perfect correspondence between the carrier frequency and phase at the transmitter and receiver, and high quality filters, we can therefore determine that the output $z(t)$ is given by

$$z(t) = 0.5g(t) = 0.5g_1(t) + j0.5g_2(t), \tag{7.50}$$

and therefore the transmitted signal is received successfully via the complex exponential architecture.

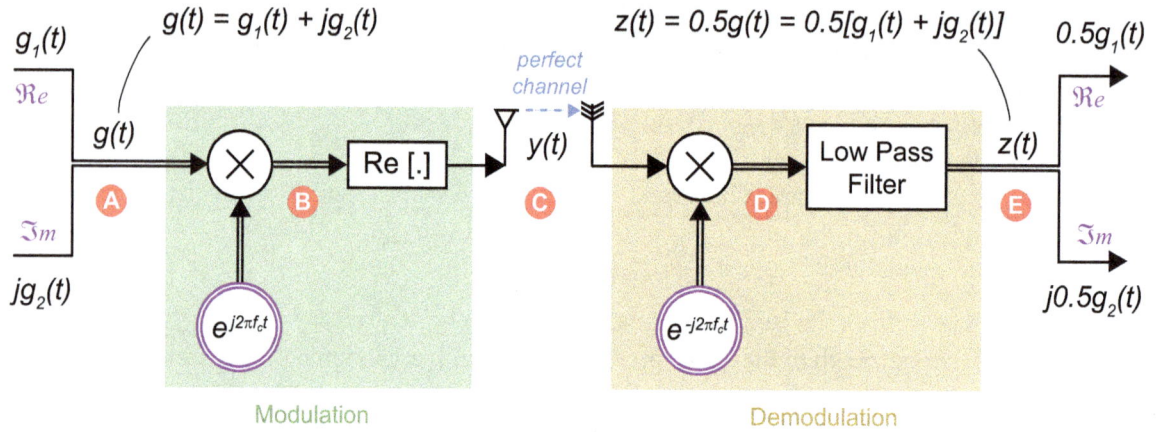

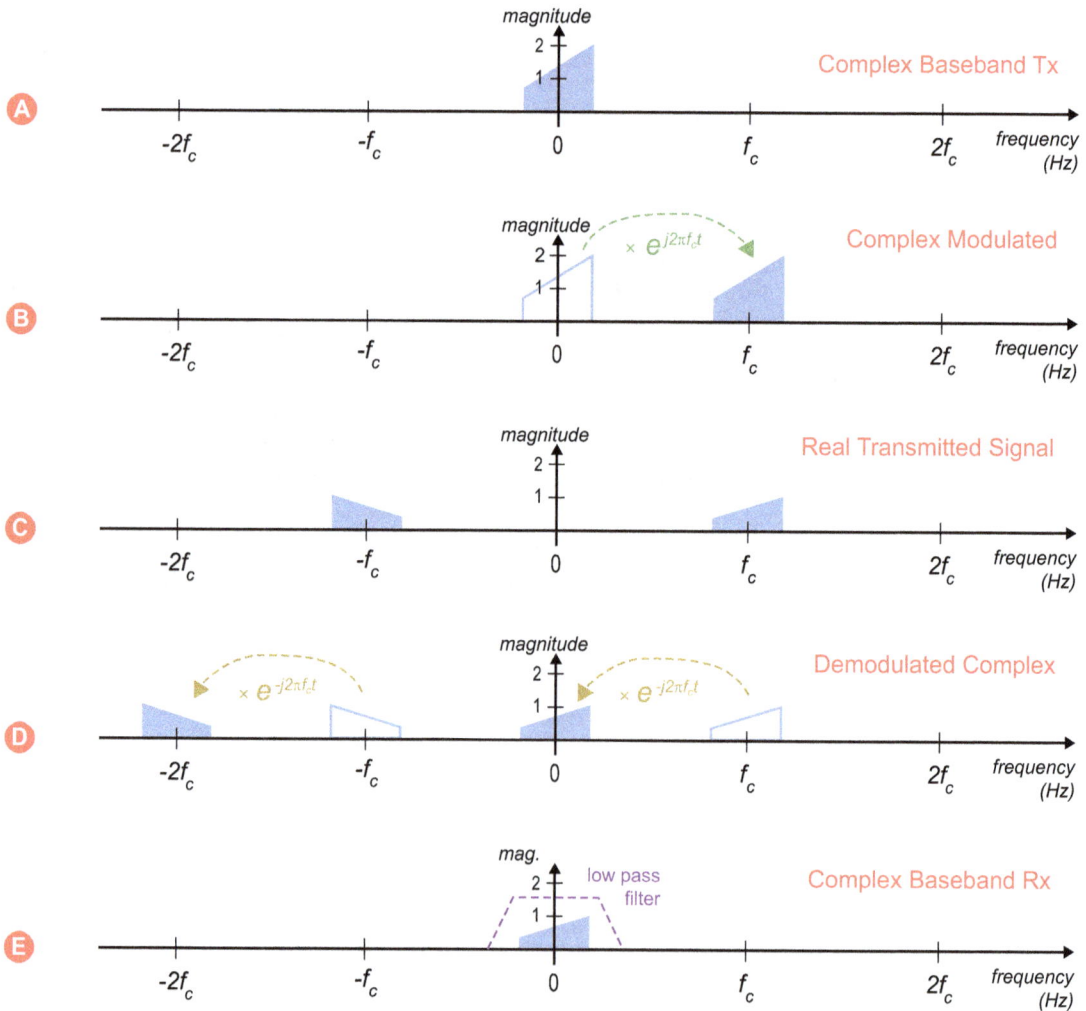

Figure 7.21: Complex signal spectra for quadrature modulation and demodulation using complex notation.

7.5.1. Simple Mathematical Example of Passband to Complex Baseband

A numerical example will now be presented to confirm the concepts of complex demodulation covered so far. In this example, we consider a bandpass signal in the range 800 Hz to 1,220 Hz, which consists of four tones (cosine waves) at frequencies of 800 Hz, 900 Hz, 1,080 Hz, and 1,220 Hz. The signal $y(t)$ is given by

$$y(t) = 8\cos(2\pi 800t) + 6\cos(2\pi 900t) + 4\cos(2\pi 1080t) + 2\cos(2\pi 1220t). \tag{7.51}$$

and it is shown in the upper part of Figure 7.22. To generate this signal, we assume that a baseband signal has previously been generated and modulated onto a carrier at 1,000 Hz (note these are all artificially low numbers to make the example easy to follow). The signal is real, and therefore has a symmetric spectrum when plotted on complex frequency axes. Note that, because no sine terms are present, therefore there are no non-zero components in the imaginary spectra and we do not plot it.

Recalling that $\cos(\omega) = \frac{1}{2}(e^{j\omega} + e^{-j\omega})$, we can express $y(t)$ in complex exponential form as

$$y(t) = \frac{8}{2}(e^{j2\pi 800t} + e^{-j2\pi 800t}) + \frac{6}{2}(e^{j2\pi 900t} + e^{-j2\pi 900t})$$
$$+ \frac{4}{2}(e^{j2\pi 1080t} + e^{-j2\pi 1080t}) + \frac{2}{2}(e^{j2\pi 1220t} + e^{-j2\pi 1220t}) \tag{7.52}$$
$$= 4e^{j2\pi 800t} + 3e^{j2\pi 900t} + 2e^{j2\pi 1080t} + e^{j2\pi 1220t}$$
$$+ 4e^{-j2\pi 800t} + 3e^{-j2\pi 900t} + 2e^{-j2\pi 1080t} + e^{-j2\pi 1220t}$$

and this can be considered as the transmitted signal shown at point **C** in Figure 7.21.

Next, the signal from (7.52) can be demodulated using a complex exponential, to form

$$y(t)e^{-j2\pi 1000t} = 4e^{j2\pi(800-1000)t} + 3e^{j2\pi(900-1000)t} + 2e^{j2\pi(1080-1000)t} + e^{j2\pi(1220-1000)t}$$
$$+ 4e^{-j2\pi(800+1000)t} + 3e^{-j2\pi(900+1000)t} + 2e^{-j2\pi(1080+1000)t} + e^{-j2\pi(1220+1000)t} \tag{7.53}$$
$$= 4e^{-j2\pi 200t} + 3e^{-j2\pi 100t} + 2e^{j2\pi 80t} + e^{j2\pi 220t}$$
$$+ 4e^{-j2\pi 1800t} + 3e^{-j2\pi 1900t} + 2e^{-j2\pi 2080t} + e^{-j2\pi 2220t}$$

which is depicted in the middle spectrum of Figure 7.22, corresponds to the signal at point **D** in Figure 7.21. Finally, the signal (both real and imaginary parts) is passed through a low pass filter to obtain the complex baseband signal, which is shown in the final spectrum in Figure 7.22 and relates to point **E** in Figure 7.21.

$$z(t) = \text{LPF}\{y(t)(e^{-j2\pi 1000t})\} = 4e^{-j2\pi 200t} + 3e^{-j2\pi 100t} + 2e^{j2\pi 80t} + e^{j2\pi 220t}$$
$$+ 4e^{-j2\pi 1800t} + 3e^{-j2\pi 1900t} + 2e^{-j2\pi 2080t} + e^{-j2\pi 2220t} \tag{7.54}$$

low pass filtered terms

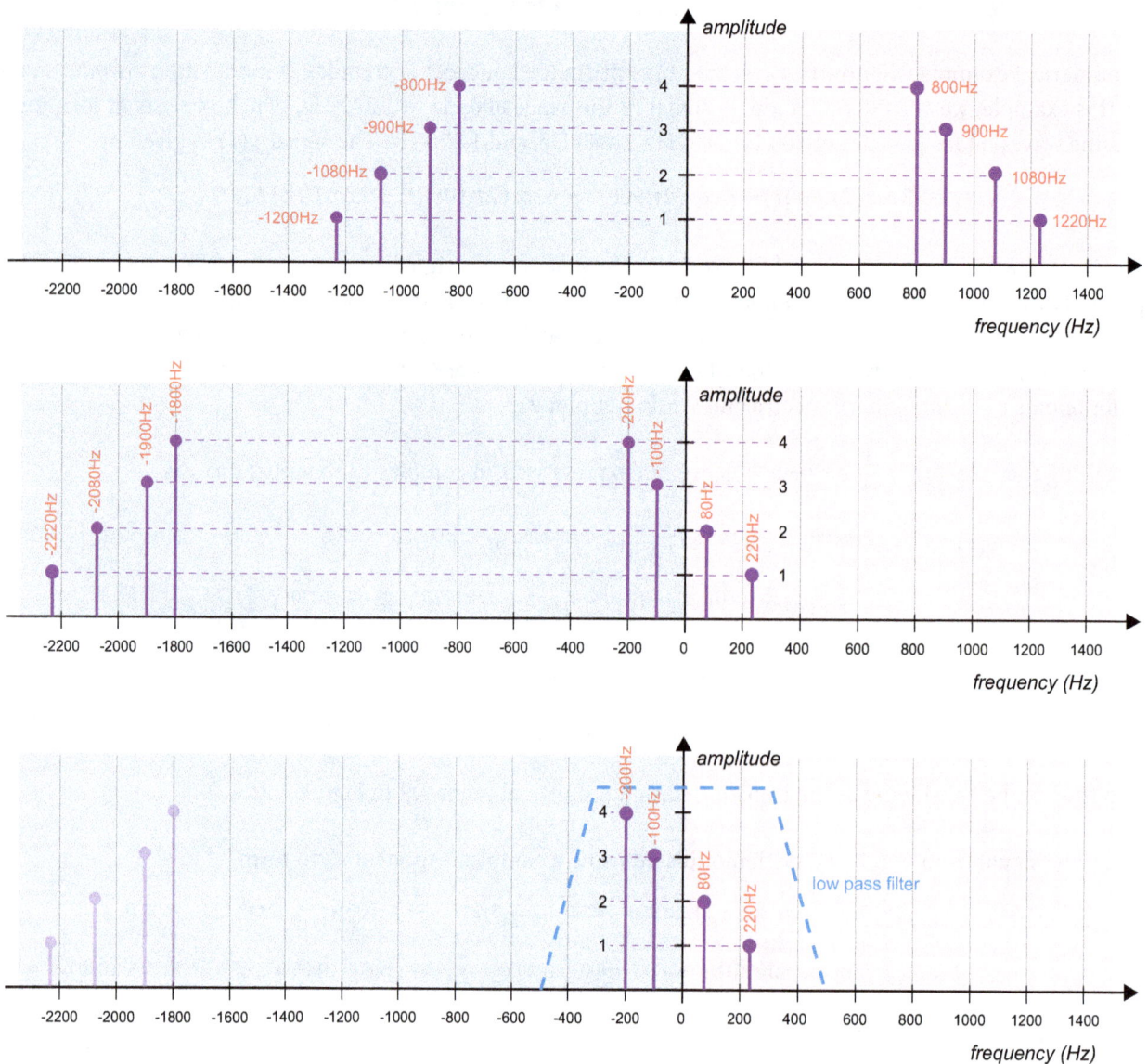

Figure 7.22: Numerical example of complex demodulation:
(upper) bandpass signal centred around 1,000 Hz;
(middle) complex demodulated by a negative 1,000 Hz complex exponential, creating
(bottom) demodulated components centred around baseband (0 Hz), and at -2,000 Hz (filtered out).

Later, in Chapter 10, we will consider the requirement to reverse a baseband signal in terms of frequency (in other words, to 'flip' it). From a complex exponential perspective, this is easy to do by simply mixing with the positive exponential $e^{j2\pi 1000t}$ (instead of the negative exponential as used in Eq. (7.53)) which shifts the signal in a positive frequency direction. The negative component of the two-sided spectrum in Figure 7.22 translates to around 0 Hz at baseband. The spectrum at baseband is now flipped, and slopes in the opposite direction.

7.6. Receiver Frequency Offset Error and Correction

Returning back to Figure 7.20, with a transmit carrier frequency of f_c, the receiver local oscillator frequency should be set to f_c also. However, there will be a small error, f_Δ, which represents the deviation in frequency between the frequencies that are *actually* synthesised in the transmit and receive local oscillators (as a result of component tolerances, etc.). The carrier frequency used to demodulate the signal is therefore represented as $f_c + f_\Delta$ Hz, rather than the expected value of f_c Hz.

Depending on the technology, and the actual carrier frequency (100's of kHz or MHz or GHz... and so on), f_Δ could be a few Hz, or a very small fractional value of, say, 0.0001 Hz. But f_Δ will not be zero! So, to some degree, the receive carrier will deviate from the transmit carrier, and frequency locking is required.

If we consider the frequency deviation f_Δ as a continuously varying phase, $\theta(t)$ of the carrier f_c, the carrier frequency at the receiver is instantaneously $f_c + f_\Delta = f_c + \theta(t)$. Consequently, for small frequency errors, a received symbol constellation will appear to rotate or spin (as illustrated earlier in Figure 7.17(b)). Therefore, if we see a constellation spinning, then this indicates that frequency locking has not been implemented, or is not working.

The resulting demodulation to baseband in Figure 7.23 from input **P** mixed (multiplied) with the complex exponential carrier to create output **Q**, can be described mathematically as:

$$z(t) = y(t)e^{-j2\pi(f_c + f_\Delta)t}$$
$$= \left[g_1(t)\cos(2\pi f_c t) - g_2(t)\sin(2\pi f_c t)\right] e^{-j2\pi f_c t} \cdot e^{-j2\pi f_\Delta t} \tag{7.55}$$

which, compared to the ideal demodulation from (7.41), has as frequency error. The lowpass filter then band-limits the signal at **R**. The effect of this frequency error when viewed in the complex or two-sided spectrum is simply to shift the demodulated spectrum an extra f_Δ (left if positive and right if negative) with respect to 0 Hz, as shown in Figure 7.23. Therefore as shown in Figure 7.23, the mixing to baseband was intended to be $e^{-j2\pi f_c t}$ if there was perfect frequency locking (and shifting left by f_c Hz (as in Figure 7.21 above); however, the f_Δ frequency error in the receiver carrier of $e^{-j2\pi(f_c + \Delta f)t}$ has shifted the signal of interest by an extra f_Δ Hz. Therefore, unlike in Figure 7.21 where the signal of interest is centred on 0 Hz, in Figure 7.23 its position is offset by f_Δ from the 0 Hz centre.

This type of effect, where there is a difference between the frequencies used to modulate and demodulate the signal, is an almost inevitable occurrence in practical wireless communications systems. There are various reasons for this, including that the oscillators used to generate the transmit and receive carriers will not be perfectly matched, and therefore will be subject to component tolerances. Another possibility is that the transmitter and receiver are moving relative to each other in space, and therefore a Doppler shift is generated, which appears as a frequency shift. The good news is that such errors can be corrected in the receiver using synchronisation techniques. In the RFSoC RF-ADC there are many strategies to support this.

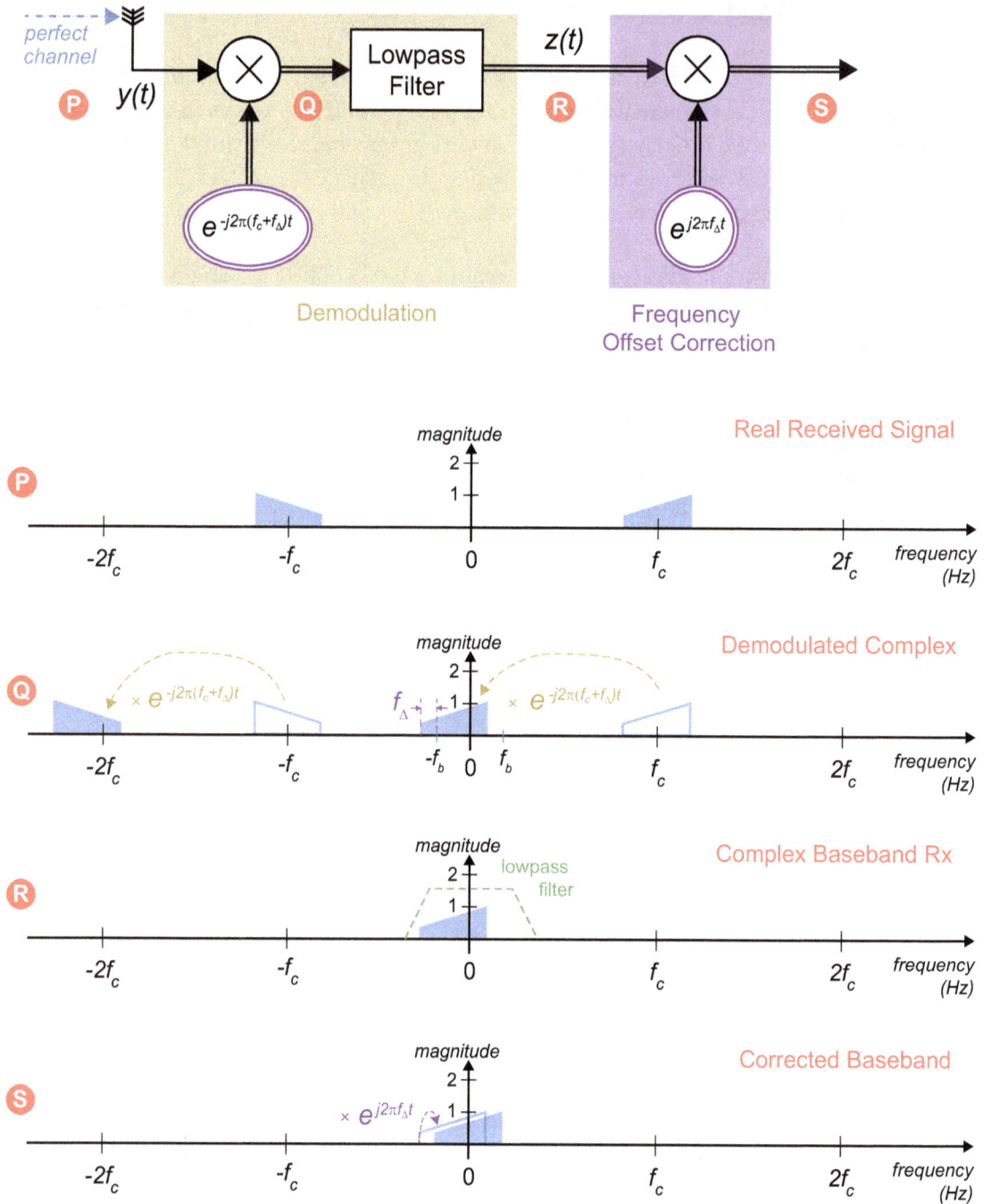

Figure 7.23: Effects of frequency error in a complex receiver — shifting and correction of demodulated signal

When the complex signal is received using the carrier frequency at $f_c + f_\Delta$, a *frequency correction* can be applied by multiplying the incoming spectrum **R** by:

$$e^{j2\pi\Delta ft} \tag{7.56}$$

to obtain the output at **S** as shown in Figure 7.23. This is simply a complex multiplication if f_Δ is fixed, and a changing value if the value of f_Δ varies (as is likely), which must be tracked by synchronisation methods.

Whether fixed or varying, the offset frequency f_Δ must be calculated in some way — and that is part of the synchronisation that is undertaken in a radio receiver and is a part of the RFSoC RF-ADC. In this chapter in our simplified analysis, we assume that the frequency error is already known. However, the important point, as this example demonstrates, is that an error in the frequency used to demodulate a signal can thereafter be compensated by a further multiplication with a complex exponential term (which again is just a single complex number).

Noting the equivalence of the trigonometric and complex models shown in Figure 7.20, and in particular that $z(t)$ is the same in both models, this method of frequency correction is equally valid in both cases.

7.7. Equivalence of Quadrature and Complex Modulator

As has been established in this chapter, working with the complex exponential representation of a quadrature modulator makes mathematical manipulation and frequency domain design easier and more tractable. Key to successfully applying these principles is being aware of the difference between a one-sided spectrum for a real signal, and a two-sided spectrum for a real spectrum (which will be symmetric), and a two-sided spectrum for a complex signal (which will be asymmetric).

In Figure 7.24 we present the equivalence of the quadrature mixer and the complex exponential mixer, including the RF-ADCs and RF-DACs from the RFSoC (more on this in Chapter 10) in order to present the digital quadrature modulators and demodulators. Shown in Figure 7.24(c) is a *re-drawn* version of the quadrature mixer, where the sine modulator part is simply folded along a horizontal axis; we present this here because, in some of the SDR architectures we will present and derive in Chapter 8, we use the folded version to reduce space requirements in some SDR architecture figures (for example, Figures 8.1 to 8.7).

7.7.1. Complex and Real I/O using RFSoC RF Data Converters

In this chapter, we have observed the usefulness of complex exponential notation for the design and implementation stages of quadrature modulators and demodulators. The RFSoC architecture makes specific provision for this via the RFDC blocks, which can be configured to handle both complex and real inputs and outputs.

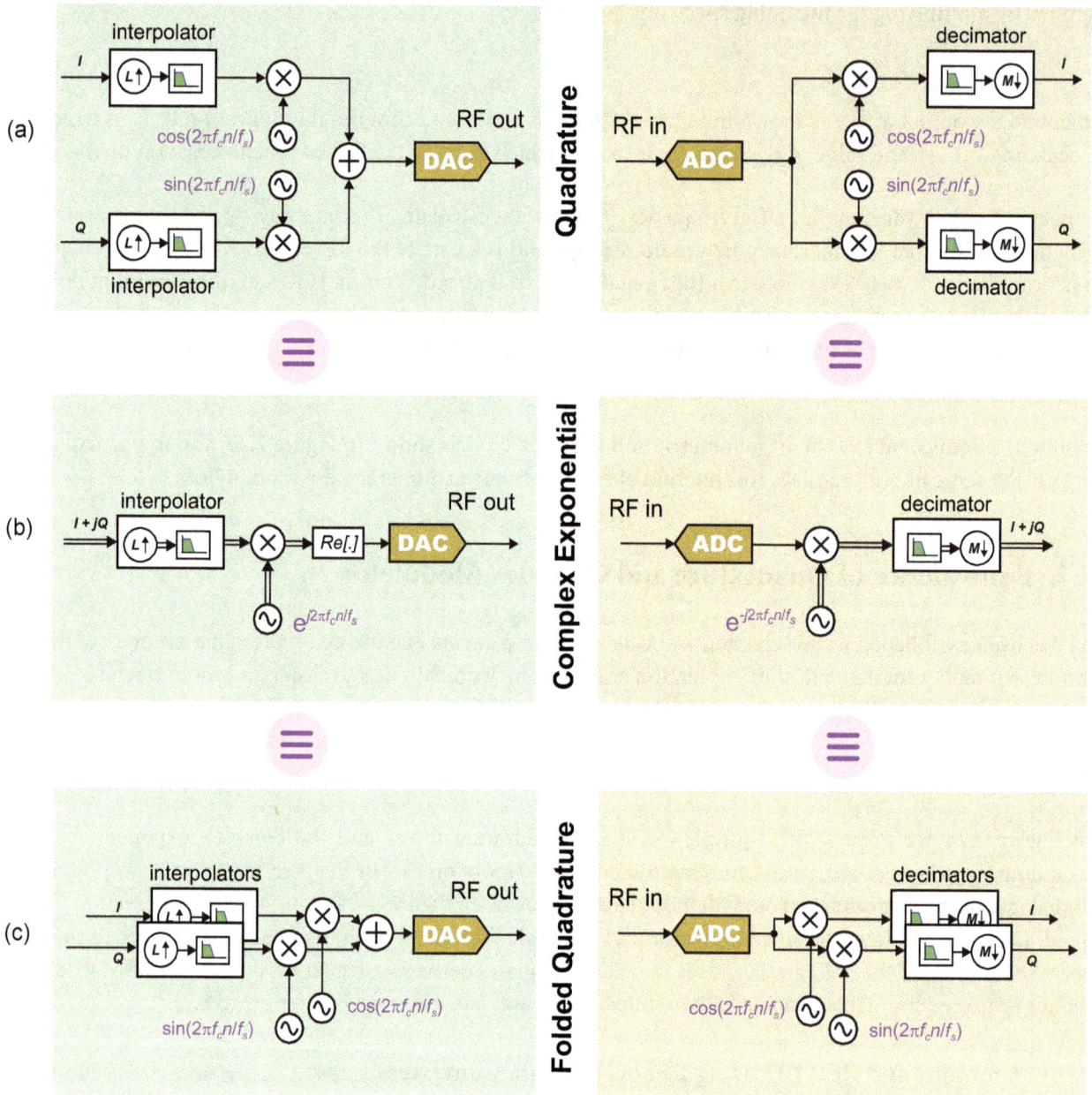

Figure 7.24: (b) The standard quadrature modulator/demodulator, (b) the 'folded' quadrature modulator/demodulator, and (c) the equivalent complex representation of the modulator/demodulator for the RFSoC RF-ADCs and RF-DACs.

The RF-ADCs can of course be used singly for real inputs, and in pairs for quadrature or 'complex' (I and Q) inputs. For instance, an RFSoC device with 8 RF-ADCs can support up to 8 real input channels, or 4 quadrature or 'complex' channels). Similarly, the RF-DACs can be used individually to generate real output signals, or in pairs to generate quadrature or complex (I and Q) signals.

The selection of either real or complex external ports depends on the type of radio architecture being implemented, and will be explored further in Chapter 10. Additionally, complex inputs and outputs may be useful in instrumentation applications, e.g. for spectral analysis (see Chapter 5), or to generate test waveforms.

7.8. Chapter Summary

This chapter has reviewed quadrature modulation and demodulation, and presented the mathematics describing these processes in both trigonometric and complex exponential form. Important underpinning concepts of the complex spectrum and the link with Euler's equation were established, and examples presented. The major conclusion of this chapter is that complex notation can greatly simplify the analysis of modulation and demodulation, hence its wide adoption. Support for complex architectures and designs is available in the RFSoC device architecture and associated design tools. In Chapter 8 we will look at a few structured SDR transmit/receive architectures, and in Chapter 10 we will review the parameters and settings for using the RF-ADC quadrature mixer for receiving RF signals.

Notebook Set E

Complex Frequency Domain

Following on from Chapter 7, you can now explore a series of notebooks stepping through three modulation and demodulation schemes. These notebooks use Python, along with the NumPy package for computation [281], and the MatplotLib package for plotting [257]. The example signals are plotted in both the time and frequency domains at each stage in the pipeline, which gives visual support to the underlying mathematics.

There are three notebooks as part of this chapter, each representing a different modulation scheme. These can be accessed through Jupyter Labs, as in previous chapters, and none of them require access to an RFSoC development board. The notebooks and their relative locations are:

ALL	01_amplitude_modulation.ipynb — *rfsoc_book/notebook_E/01_amplitude_modulation.ipynb*
ALL	02_qam_modulation.ipynb — *rfsoc_book/notebook_E/02_qam_modulation.ipynb*
ALL	03_complex_qam.ipynb — *rfsoc_book/notebook_E/03_complex_qam.ipynb*

These notebooks step through the process of modulating and then demodulating a baseband signal of interest, beginning with a simple amplitude modulation methodology, before introducing a QAM scheme from a trigonometric perspective. An example of a complex QAM scheme is also presented, which confirms that this approach is fundamentally identical to the traditional QAM method.

Before exploring these examples in depth, recall that carrier modulation is the process of multiplying together (or 'mixing') two signals, causing the output signal to be 'shifted' in frequency. We can take advantage of this property to multiply a baseband information signal with a high frequency carrier signal, to produce a modulated signal ready for transmission. In the context of RFSoC devices and implementations, this carrier waveform has a frequency in the range of hundreds of MHz to several GHz.

E.1. Amplitude Modulation

The first notebook in this set, ***01_amplitude_modulation.ipynb***, introduces a basic amplitude modulation and demodulation approach, as depicted in Figure E.1. Throughout these notebooks, a simple tone is used as the "information signal", for demonstration purposes. A more generic baseband signal can be modulated using the same concepts as those featured here.

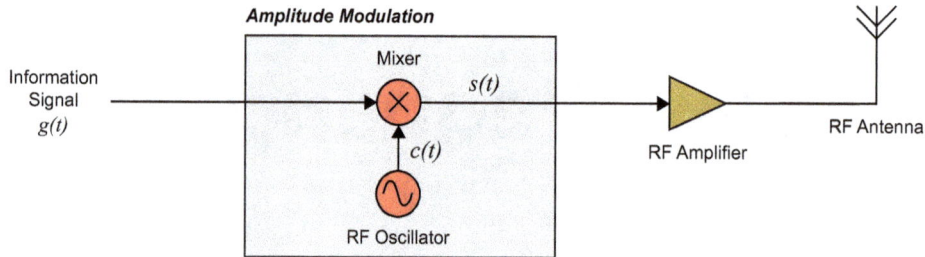

Figure E.1: Functional block diagram of amplitude modulation.

The information signal, $g(t)$, the carrier waveform, $c(t)$, and the modulated signal, $s(t)$, are all plotted as presented in Figure E.2. Inspecting the modulated waveform in the frequency domain also confirms that modulation generates frequency components at $f_c - f_{baseband}$ and $f_c + f_{baseband}$.

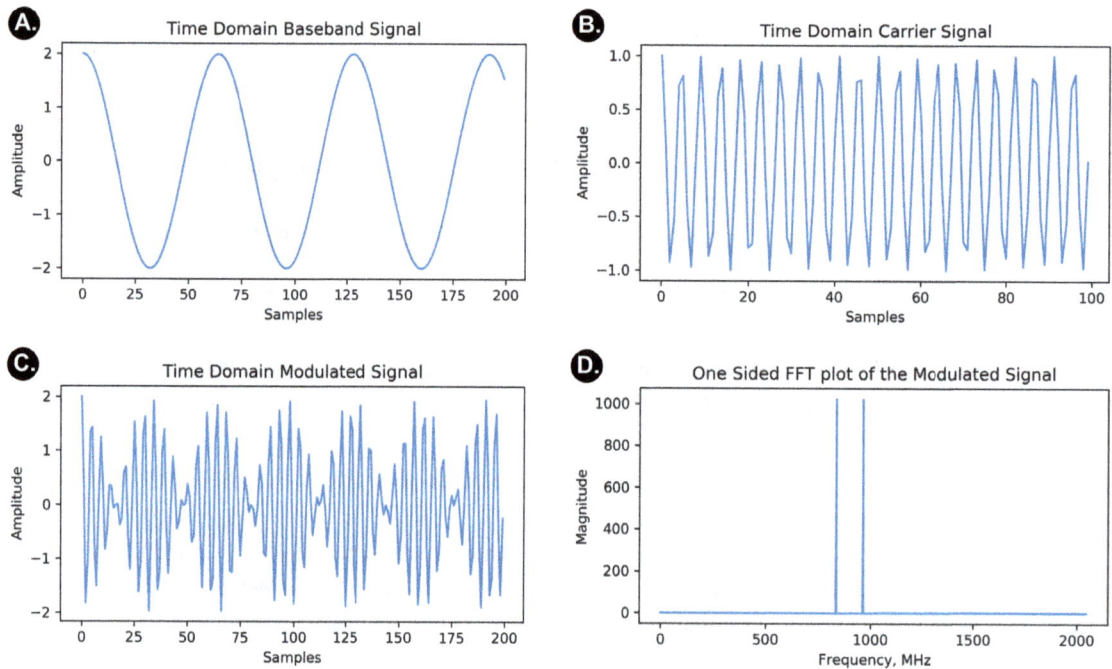

Figure E.2: The time domain representation of the baseband signal (a), carrier waveform (b), modulated waveform (c), and the frequency domain representation of the modulation waveform (d).

This signal is then demodulated by mixing $s(t)$ with the known carrier signal, producing an signal which we can confirm contains the desired baseband signal, but also high frequency components at $2f_c - f_{baseband}$ and $2f_c + f_{baseband}$. By applying a low pass filter we remove these high frequency components, leaving a reconstruction of the original information signal, with half the original amplitude. In addition, the notebook explores what happens when the demodulation takes place with a phase error in the local RF oscillator.

E.2. Quadrature Amplitude Modulation

The second notebook, ***02_qam_modulation.ipynb***, explores the QAM scheme, which uses orthogonal carriers to transmit twice as much data in a given bandwidth, compared to a single carrier (refer back to Section 7.3 if a recap is needed). This transmitter design can be seen in Figure E.3.

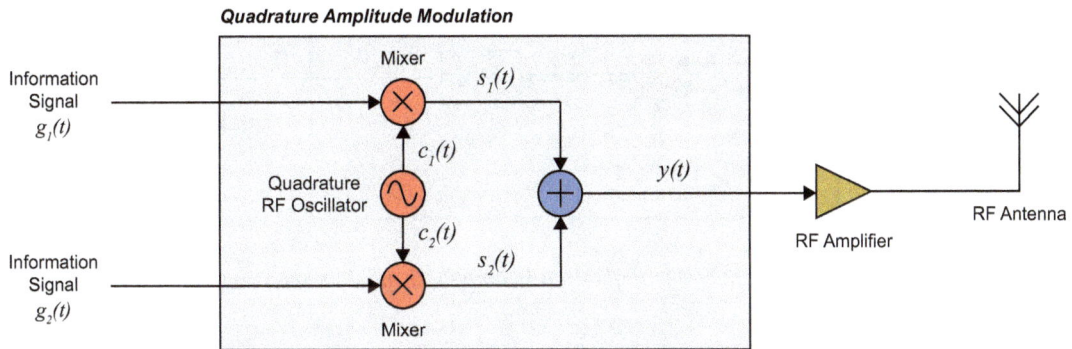

Figure E.3: Functional block diagram of QAM modulation.

This notebook follows the same format as the first — the information signals, $g_1(t)$ and $g_2(t)$, the modulated signals, $s_1(t)$ and $s_2(t)$, and the summed signal $y(t)$, are plotted for inspection. An example plot from the notebook can be inspected in Figure E.4. Here the modulated signals, $s_1(t)$ and $s_2(t)$, are presented in the time and frequency domain (represented as I and Q). These signals are simply summed to derive the output signal $y(t)$.

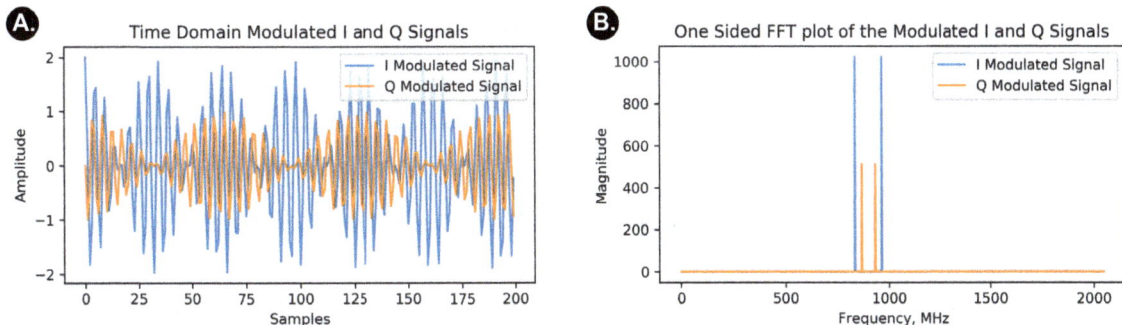

Figure E.4: The time domain representation (a) and frequency domain representation (b) of the modulated signals .

To demodulate this signal, $y(t)$ is mixed with sine and cosine oscillators at the same frequency as the transmit side, before being low pass filtered to remove high frequency terms. This produces two separate reconstructed signals which match $g_1(t)$ and $g_2(t)$, but with half the amplitude. We also examine the effects of phase error on the QAM demodulation process.

E.3. Complex Quadrature Amplitude Modulation

QAM seen previously in Figure E.3 uses real signals. A complex representation is also possible, and this is often used to simplify the mathematics. The third notebook, *03_complex_qam.ipynb*, demonstrates this complex representation. The transmitter for the complex QAM scheme is given in Figure E.5.

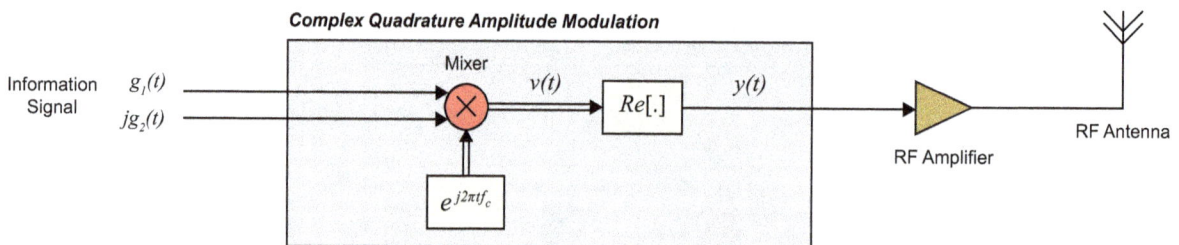

Figure E.5: Functional block diagram of quadrature amplitude modulation using complex notation.

Here we a see a complex input, $g(t) = g_1(t) + jg_2(t)$, where $g_1(t)$ and $g_2(t)$ are equivalent to the input signals in the previous notebook. This complex information signal can be modulated by mixing with a complex exponential at a 'frequency' of f_c Hz. Only the real (in-phase) part of the signal is selected to derive the output $y(t)$, which is presented using time and frequency domain plots in Figure E.6.

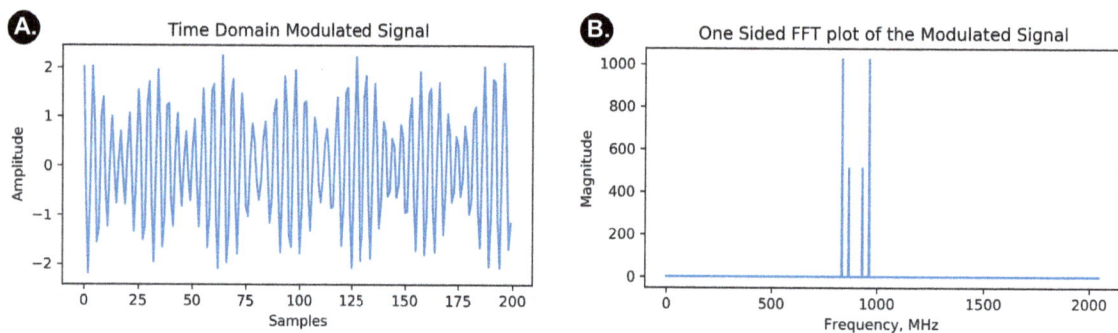

Figure E.6: The time domain representation (a) and frequency domain representation (b) of the output signal, $y(t)$.

By observing the plot of this modulated signal, we confirm that it is equivalent to the modulated signal, $y(t)$, seen in the previous notebook. Similarly, we confirm that the complex representation of demodulation is equivalent to the original trigonometric model.

Chapter 8

SDR Architectures

Kenny Barlee

As introduced previously, Software Defined Radio is a generic term that refers to radio systems in which some or all of the PHY components traditionally implemented with dedicated hardware (e.g. mixers, filters, modulators, demodulators) are instead implemented using DSP algorithms in software, or on programmable hardware. They are an extremely flexible type of radio, as they can be used to transmit and receive many different types of waveform and enable the front end of 5G and soon to be 6G radio implementations.

8.1. SDR Architectures Through the Ages

Since the 'Software Radio' term was first coined by Mitola in 1995 [264], there have been a number of evolutions in DAC/ADC sampling technology. In turn, this has resulted in a number of distinct architectural 'generations' of SDR. Early SDRs used a Digital Signal Processor IC for kHz-bandwidth baseband signal generation, and all of the modulation to RF (and subsequent RF filtering and RF amplification etc.) was carried out using discrete analogue components. The RFSoC platform, being perhaps the most advanced modern SDR platform, can generate GHz-bandwidth signals using FPGA programmable logic cores, meaning that upconversion, filtering, Digital Pre-Distortion (DPD) and even modulation onto the RF carrier can all be performed digitally. In such a Direct-RF SDRs, the only remaining analogue stages are RF filtering and RF amplification.

8.1.1. Baseband Sampling / Analogue IF Baseband Sampling SDRs (1990s)

Back in the late 1990s, A/D converters running at 100 kSps (kilo Samples per second) with a 16-bit resolution was the latest (and rather expensive) technology. As illustrated in Figure 8.1 and 8.2 in these first generation 'digital radios', the analogue section downconverted signals from the RF carrier in either one or two stages

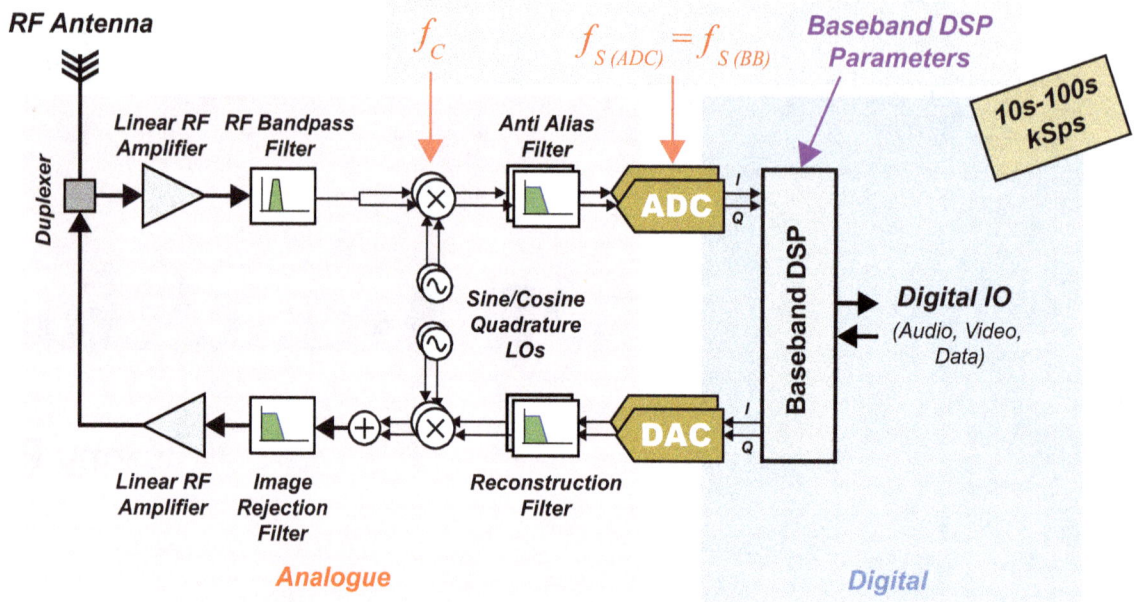

Figure 8.1: *High level architecture of a Baseband-Sampling Software Defined Radio.*

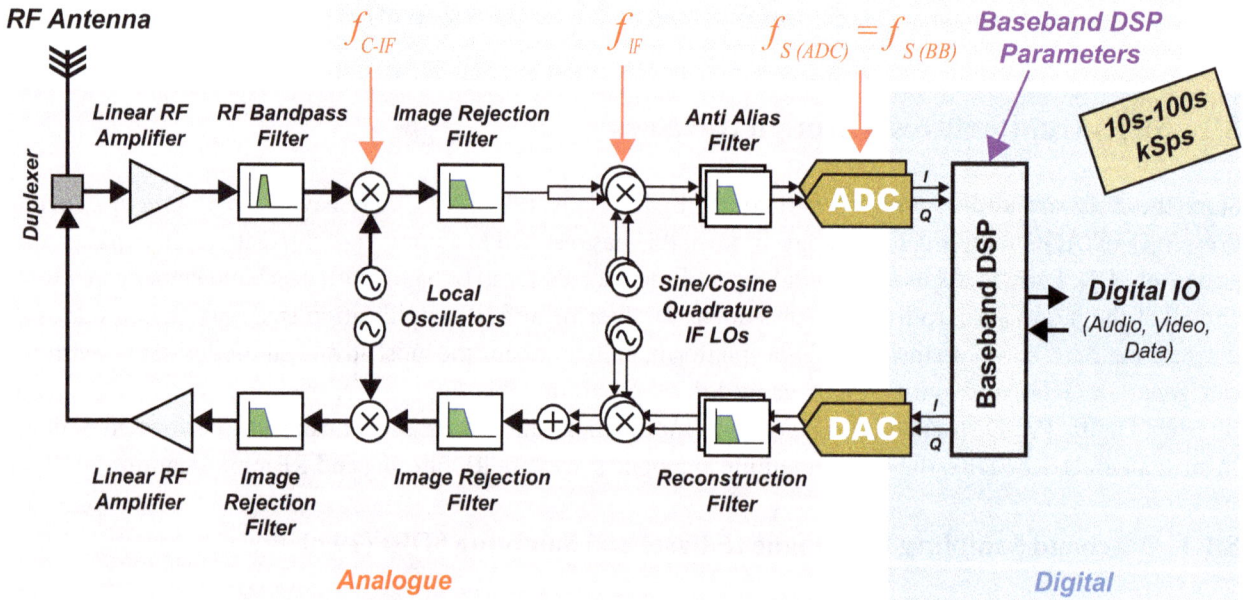

Figure 8.2: *High level architecture of an Analogue IF Baseband-Sampling Software Defined Radio.*

using an analogue Local Oscillator (LO). (In these figures and others in this chapter, we are illustrating using the 'folded' quadrature modulator signal flow graph, as reviewed in Figure 7.24 on page 268). The two-stage version, as demonstrated in Figure 8.2, features an Intermediate Frequency (IF) stage, and uses a second analogue LO to further downconvert the IF signal to baseband (i.e. a superheterodyne architecture, as developed by Edwin Armstrong for FM Radio). Historically, these approaches were used where, due to the limitations of DAC and ADC technology (and in particular the achievable sampling rates), this was the only viable position for the A/D interface.

The baseband signal is then sampled and digitised using an ADC, and DSP operations are used to perform the final processing stages to recover the transmitted information. The second generation mobile phones of the 1990s — those that received GSM signals — were likely to have used this architecture.

8.1.2. Digital IF Sampling SDRs (2000s)

With the sampling speeds of A/D interfaces gradually increasing to the MSps range, the next generation of digital radios saw the sampling and digitisation processes being performed at an IF, rather than baseband. IFs of around 40 MHz (for example), could be supported by an ADC that sampled at, say, 125 MHz. The first DSP stage of this architecture involved using a Direct Digital Downconverter (DDC) to shift IF signals to baseband using demodulation and decimation filtering, as shown in Figure 8.3. Further DSP processing was then performed once the signal was at baseband. In this architecture, more functionality was implemented in the digital domain, giving greater flexibility for the SDR.

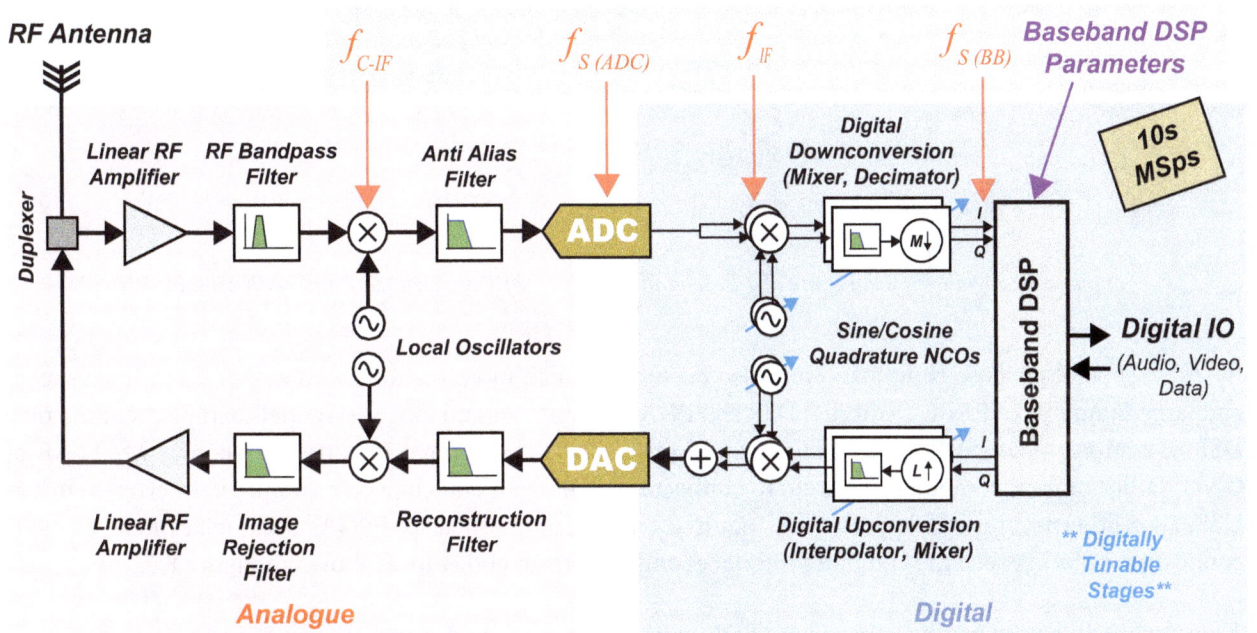

Figure 8.3: *High level architecture of a Digital IF Sampling Software Defined Radio.*

8.1.3. Baseband Sampling SDR with Tunable RF (2010s)

As SDRs were becoming more widely used, IC manufacturers began to develop single chip SDR front ends that combined some of the analogue and digital stages together; bringing digital/software tunability to the analogue oscillator, filtering and amplifier stages too, as highlighted in Figure 8.4. This made the SDRs far more flexible, and able to operate across a wide frequency range for the first time. A/D interfaces were now able to reach sample rates of 100's of MSps, increasing achievable baseband signal bandwidths to 10's of MHz; meaning that SDRs could be used for prototyping and implementing the popular radio standards that we use day-to-day, such as Wi-Fi and LTE.

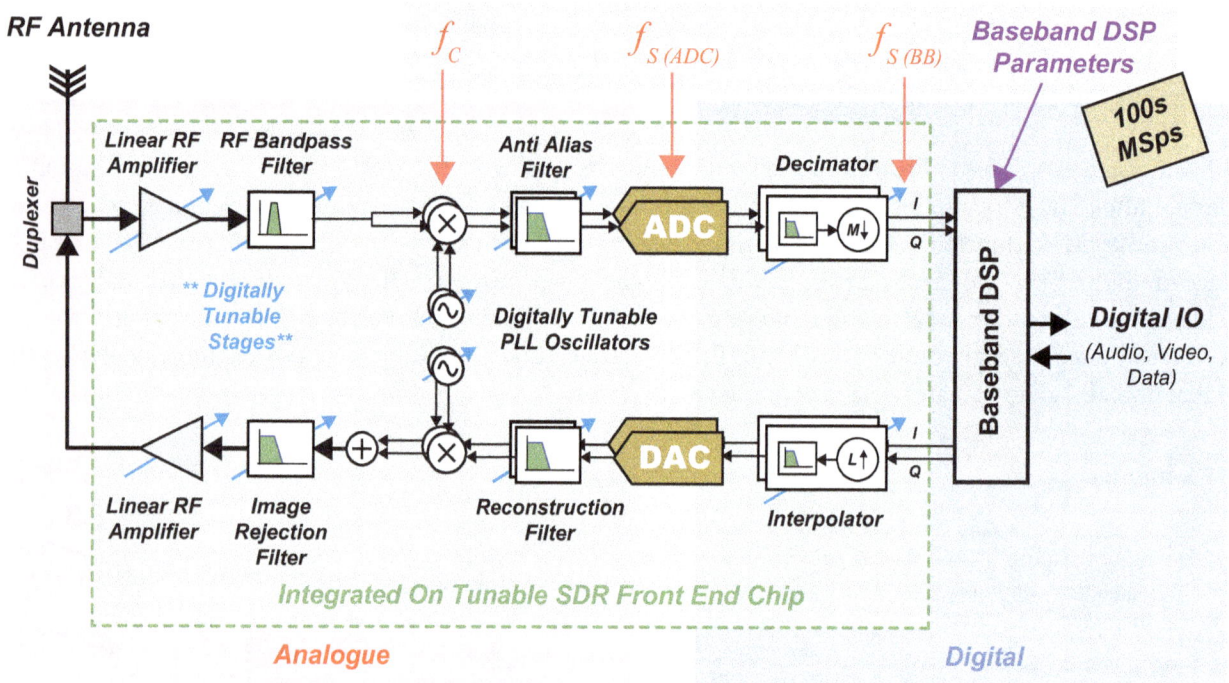

Figure 8.4: High level architecture of a Baseband Sampling Software Defined Radio with Tunable RF.

By 2010, general purpose computers were also becoming much more capable, with higher CPU clock speeds and larger amounts of RAM. Dedicated DSP or FPGA hardware was no longer essential to implement the final DSP operations of the SDR system. These could instead be implemented in software (such as MATLAB or GNU Radio) on host computers. In order to connect the SDR front end chips to a computer, interfaces such as USB, Gbit Ethernet and PCI Express were required on SDR motherboards. FPGAs became key here, as interconnection hubs between the computer interfaces and SDR front end chip, as shown in Figure 8.5.

These two major steps forward opened the SDR market to a much wider community — hobbyists, prosumers and student researchers — for the first time, as previously SDR solutions were really only used in advanced research and military applications. Economies of scale reduced the cost of this SDR transceiver hardware to

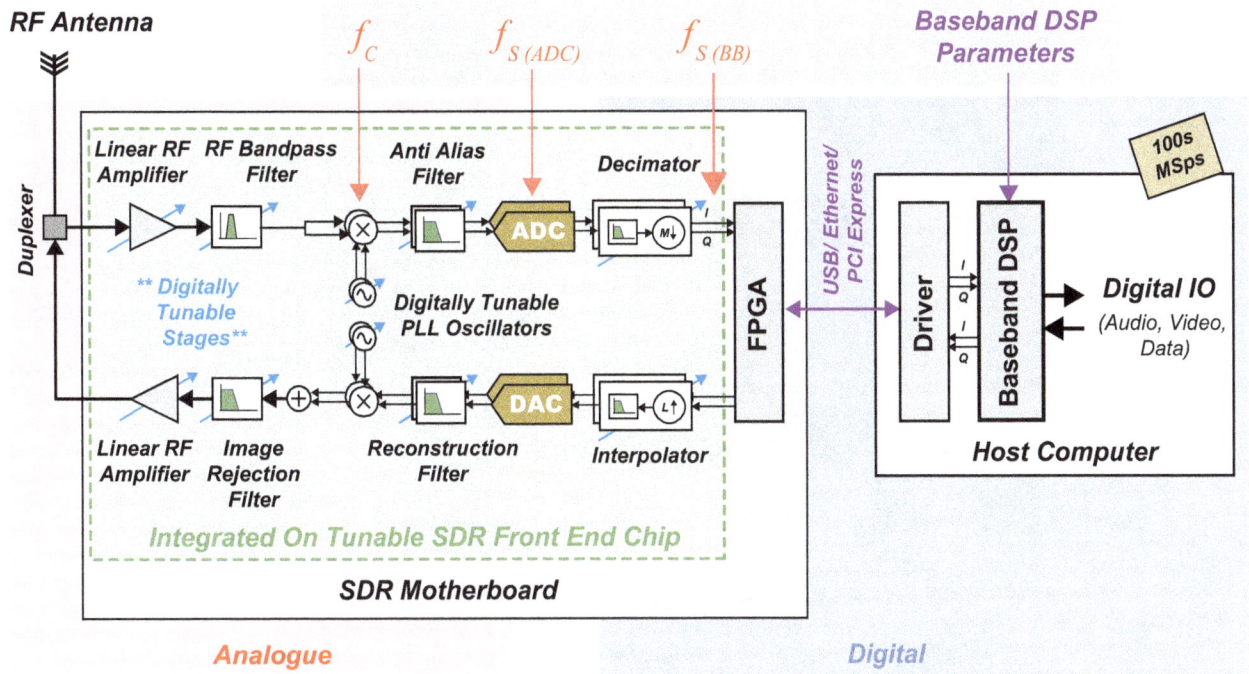

Figure 8.5: Example system level implementation of Baseband Sampling Software Defined Radio with Tunable RF, connected to remote host computer which carries out baseband DSP operations in software.

under \$1,000, and the SDR revolution really took off. Examples of these types of radio include the USRP™ B210 (equivalent to the "SDR Motherboard" shown in Figure 8.5), and the Zynq SDR (a combination of an AMD Zynq -based development board such as the ZedBoard™, and a third party SDR front end) [104].

8.1.4. Direct-RF SDRs (Almost-All-Digital) (2020s)

We have now reached the stage where it is possible to run A/D converters at multi-GSps rates, meaning that (almost!) all-digital radios can be implemented for an increasing number of bands. Ultimately, the move has been made to sample RF signals directly, as illustrated in Figure 8.6, and to subsequently downconvert them from RF frequencies to baseband in a single stage, using DSP. This architecture was first realised in a fully integrated form in 2018, with the creation of the Gen 1 RFSoC. The Gen 3, released in 2022, is able to sample at rates of up to 10 GSps.

An almost-all-digital Direct-RF SDR requires very little analogue processing — mostly the front end RF filters, and RF amplifiers. From an SDR perspective, the fact that almost all functionality is implemented digitally is highly significant — it means that the operation of the radio can be controlled and even dynamically updated at runtime using software, as shown in Figure 8.6. While many (but not yet all) 5G networks are using direct-RF SDR front ends, we can expect for future 6G implementations that all radios will be of this form.

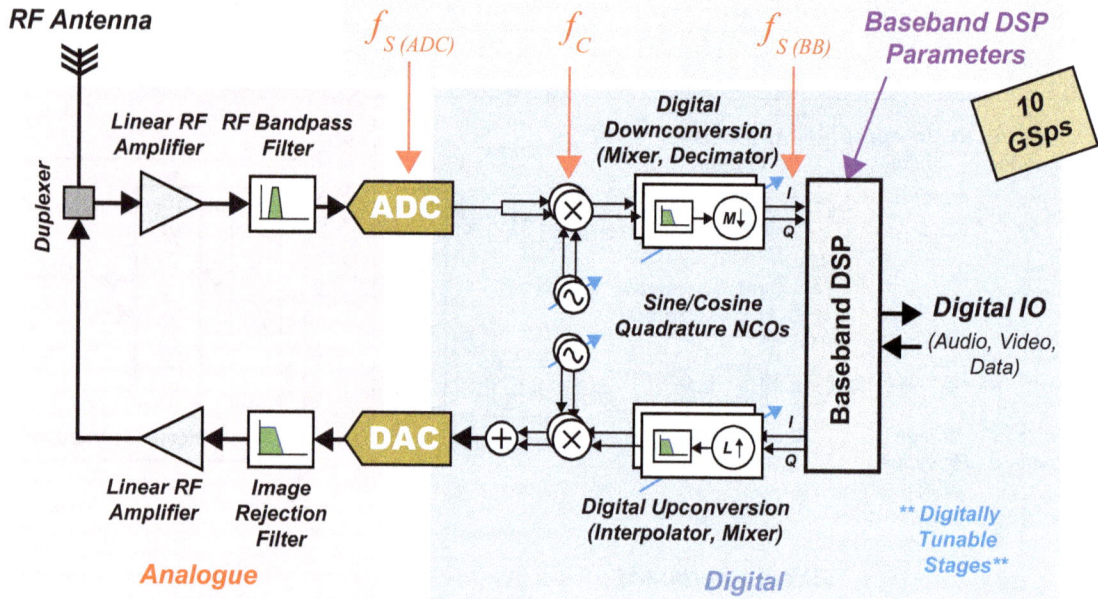

Figure 8.6: High level architecture of a Direct-RF Software Defined Radio.

8.2. Multiple Input, Multiple Output (MIMO)

All of the SDR architectures presented in the previous section are depicted as Single Input Single Output (SISO) radios. Commonly, SDRs support at least 2 Transmit (2Tx) and 2 Receive (2Rx) paths. This is known as a 2Tx 2Rx or 2x2 Multiple Input Multiple Output (MIMO) SDR. Using MIMO technology can increase the throughput, and helps to increase the signal strength and resilience by sending the signal on multiple paths from the transmitter to the receiver.

SDRs with multiple antennas feature duplicate copies of the SDR front end for each of the Tx and Rx paths. For the 2x2 example, this means that the SDR will feature a total of four ADCs and four DACs (2 complex sampling pairs of each), as shown in Figure 8.7.

The RFSoC family of SDRs are all able to support MIMO. There are a number of devices available, which range between 2x2 and 16x16 Transmit and Receive (TRx) interfaces.

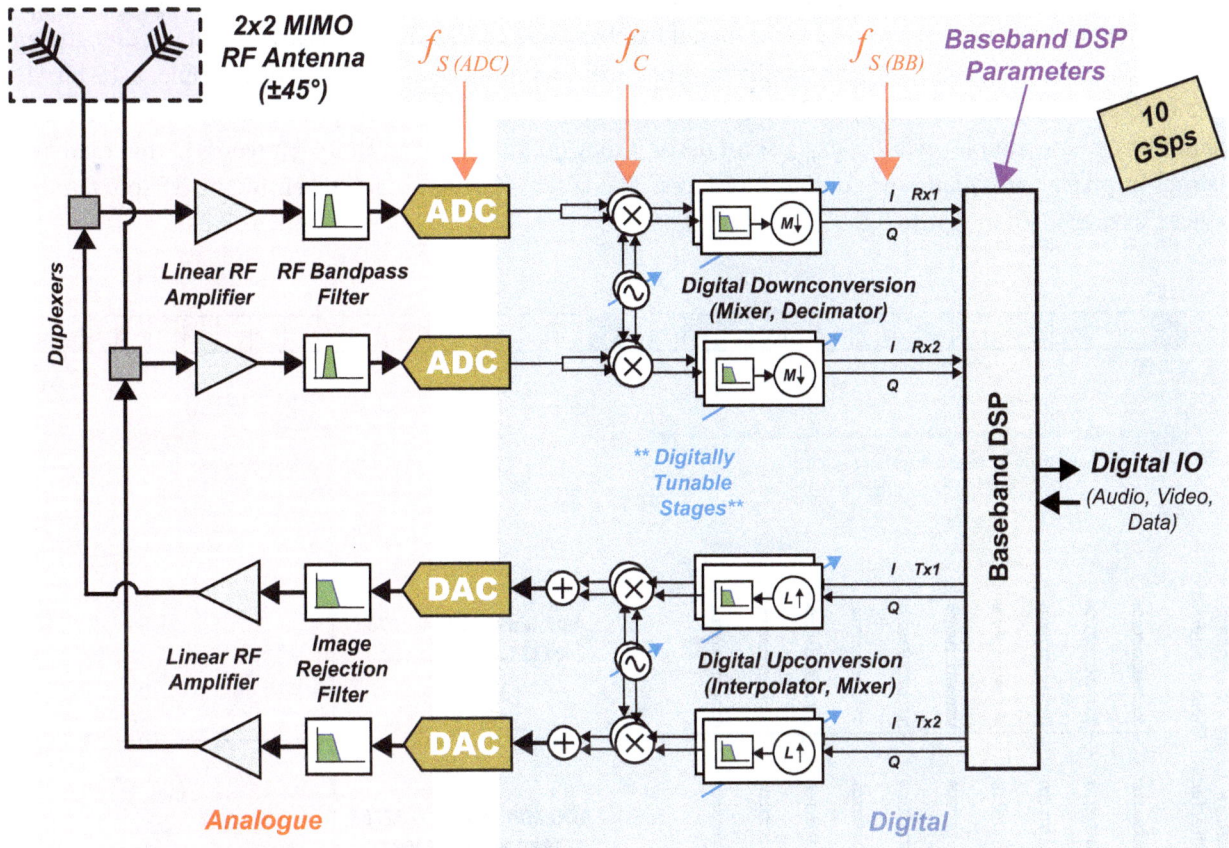

Figure 8.7: 2x2 MIMO front end of a Direct-RF Software Defined Radio.

8.3. The Digital Baseband Stage

All of the SDR architectures outlined in Section 8.1 feature a *digital baseband* processing stage. This stage can be implemented on various programmable hardware, such as a DSP IC or FPGA, or as software running on general computing equipment.

8.3.1. Receive Path

On the receive side, baseband samples of radio signals are input as complex digital words. This means that they are composed of In phase and Quadrature components, and represent a window of the radio spectrum, from

$$\left(f_c - \frac{f_s}{2} \right) \text{ to } \left(f_c + \frac{f_s}{2} \right) ,$$

downconverted to baseband (i.e. centred around 0 Hz); where fc is the SDR centre frequency, and fs_{BB} is equal to the sampling rate at the input to the digital baseband stage. The samples have a digital wordlength, and the method of packing this data into the digital word will vary from SDR to SDR. For example, a 16-bit SDR with complex sampling may output 32-bit binary words (16 bits for the In phase sample, 16 bits for the Quadrature phase sample) every $1/fs_{BB}$ seconds; or it may output two separate 16-bit words at that rate, using distinct In phase and Quadrature output interfaces. The latter is the approach taken by the RFSoC family of devices, as presented in Figure 8.8.

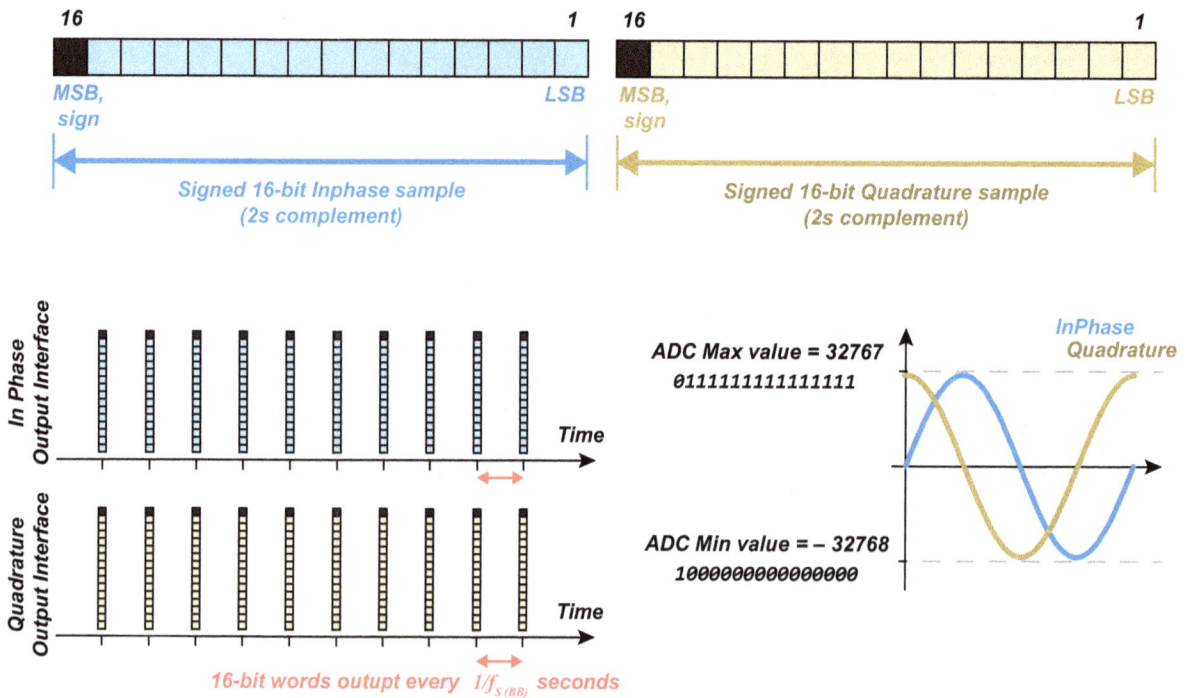

Figure 8.8: *With this example, two 16-bit digital words are received every $1/fs_{BB}$ seconds. One word represents the signed 16-bit In Phase sample, and the other, the signed 16-bit Quadrature sample.*

Samples are represented using 2's complement notation. This means the absolute maximum positive value that can be represented by each of the ADCs on the receive path is 0111111111111111 in binary (32767 when represented in decimal), and the minimum is 1000000000000000 (minus 32768 when represented in decimal).

One of the functional blocks in the SDR receive path is Active Gain Control (AGC). The purpose of AGC is to increase or decrease the amplitude of the received signal, by dynamically adjusting its gain to ensure that the signal occupies the ADC range as fully as possible, whilst not saturating the ADCs (which results in corrupted data). Figure 8.9 provides an illustration of AGC in action (part (b) in the lower two-thirds of the diagram), and compares it to an alternative configuration with no AGC present (part (a), in the upper third).

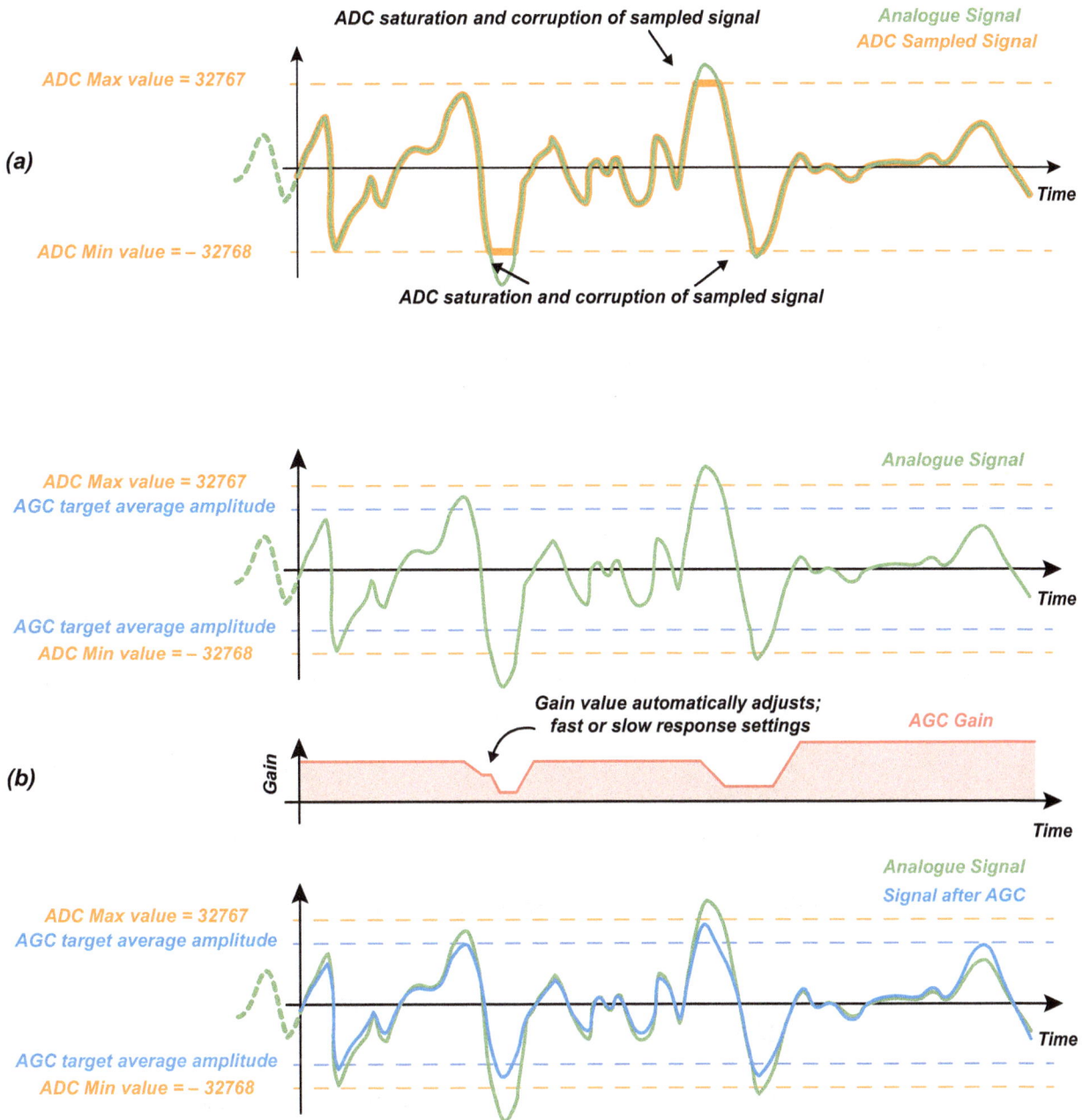

ADC saturation and corruption of sampled signal

Analogue Signal
ADC Sampled Signal

ADC Max value = 32767

(a)

Time

ADC Min value = – 32768

ADC saturation and corruption of sampled signal

Analogue Signal

ADC Max value = 32767
AGC target average amplitude

Time

AGC target average amplitude
ADC Min value = – 32768

Gain value automatically adjusts;
fast or slow response settings

AGC Gain

Gain

(b)

Time

Analogue Signal
Signal after AGC

ADC Max value = 32767
AGC target average amplitude

Time

AGC target average amplitude
ADC Min value = – 32768

Figure 8.9: The impact of Active Gain Control: (a) ADC saturation will cause corruption of the sampled signal; (b) Active Gain Control can be used in the SDR front end to automatically reduce the signal amplitude, to mitigate the risk of saturation.

Using as many bits as possible from the ADC's operating range, as is achieved by AGC scaling, maximises the dynamic range of the digital signal, which can be approximated by

$$\approx 6.02 \times N \tag{8.1}$$

where N is the target number of bits representing the signal; or

$$\approx 6.02 \times log_2(\alpha) \tag{8.2}$$

where α is the target signal amplitude in decimal representation.

Generally, therefore, the AGC will aim to have the average sample amplitude represented in these digital words around 80% of the maximum value. For this example, we will assume the AGC aims to have the ADCs produce samples with values around ±26214. According to Eq (8.2), this means that the dynamic range of the received signal is around 88dB.

The value of the dynamic range is highly significant, as this essentially determines the represented noise floor of the ADC. An example of this is shown in Figure 8.10.

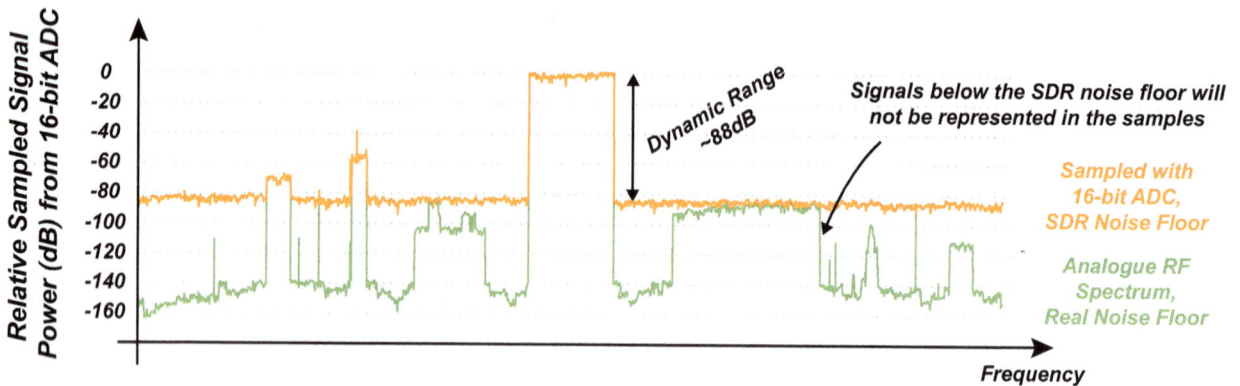

Figure 8.10: *Increasing the dynamic range reduces the SDR noise floor, allowing more signals to be detected.*

As a result of the AGC stage (as well as gains from the antenna and other RF amplifiers), the sampled signal power is <u>not</u> equal to the true RF signal power. Rather, it is relative to the properties of the ADC. This effectively means that the SDR receiver is uncalibrated. In order to use an SDR for an accurate signal measurement (for example, use as a spectrum analyser), calibration is required in order to adjust the gain of the sampled signal power so that it matches the true RF signal power. With knowledge of the SDR front end antenna gain, RF amplifier gain, instantaneous gain of the AGC, and the ADC characteristics, it is possible to implement a gain compensation stage, to calibrate the SDR.

(For an interactive AGC PYNQ Notebook, please check out https://github.com/strath-sdr/pynq_agc).

To recap, we receive a 32-bit digital word every $1/fs_{BB}$ seconds, and the digital word contains complex samples of from the I and Q ADCs. Each of these are 16-bit, and the average amplitude of the sample values is around ±26214. These complex samples represent a window of the frequency spectrum, downconverted to baseband (i.e. centred around 0 Hz). The received signal will still be in its modulated form.

"Great, now what?!", you might ask. The final stages of the receiver must be implemented.

Over the next few pages, we explore two different types of receivers at opposite ends of the complexity scale.

FM Radio Receiver

If the signal being received was an analogue FM Radio signal, we would now have an FM-modulated waveform at baseband, and require to demodulate it using DSP techniques. First, a digital filter would be used to isolate the 200 kHz bandwidth of the FM station of interest from the downconverted FM band (which will probably contain numerous stations). This filtering operation is depicted in Figure 8.11, and it appears as the first stage of the block diagram shown in Figure 8.12.

Next, this FM modulated signal is passed through an FM demodulator, such as a complex discriminator. Then we would demultiplex the stereo FM multiplex, and recover the *left* and *right* audio channels. During these processes, if fixed point representation is being used (rather than floating point), it is likely that the wordlength will grow throughout the demodulation process, due to the arithmetic effects of filtering and other DSP processing stages. *To give an example of this, a multiplier block—which is a fundamental component of many DSP operations—multiplying two n-bit numbers together should be configured with a greater number of bits on the output, in order to ensure saturation and data corruption does not occur. Taking our 16-bit samples and considering the maximum values they can represent, the output might need to be as large as 32-bits.*

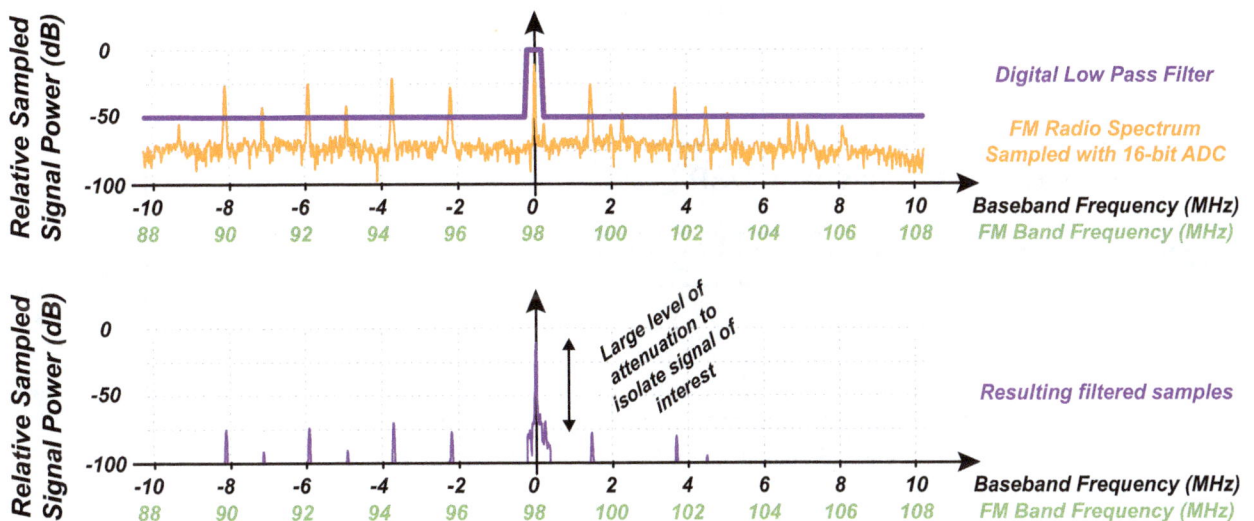

Figure 8.11: Filtering to isolate the baseband FM Radio signal of interest.

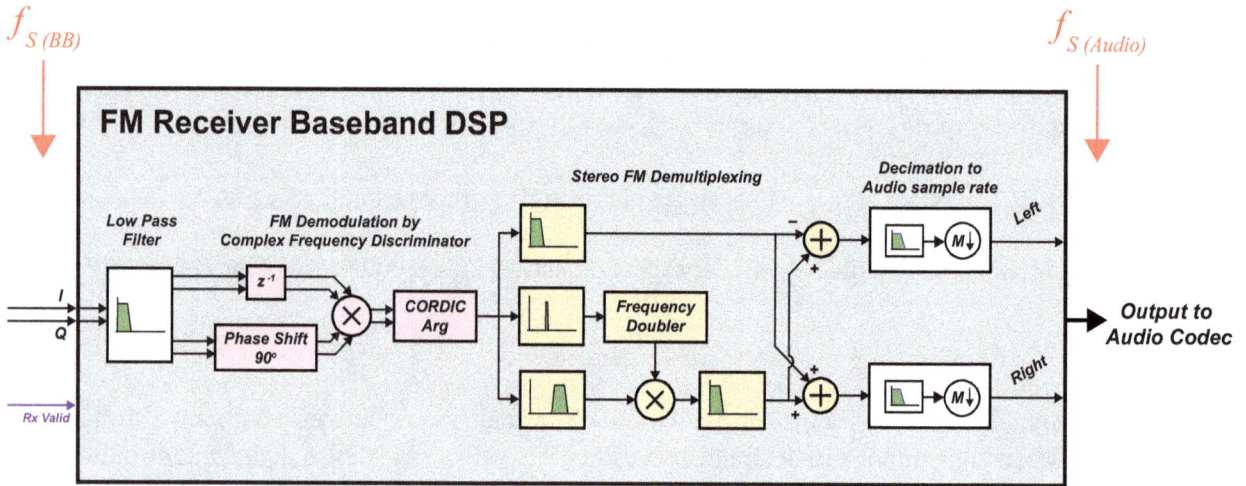

Figure 8.12: *Baseband Stereo FM Radio receiver block diagram, with demodulator and stereo FM demultiplexer.*

The sampling rate could then be reduced to a more conventional audio rate (e.g. $f_{s\,(Audio)}$ = 44.1 kHz) using decimation, and the fixed point wordlength reduced to a value compatible with the output audio codec, through truncation of LSBs. At this stage the SDR has completed its task, and recovered the transmitted stereo audio signal.

OFDM Receiver

As a more complex example, let's consider an OFDM receiver (as would be required to receive a Wi-Fi signal). A block diagram of this radio is provided in Figure 8.13.

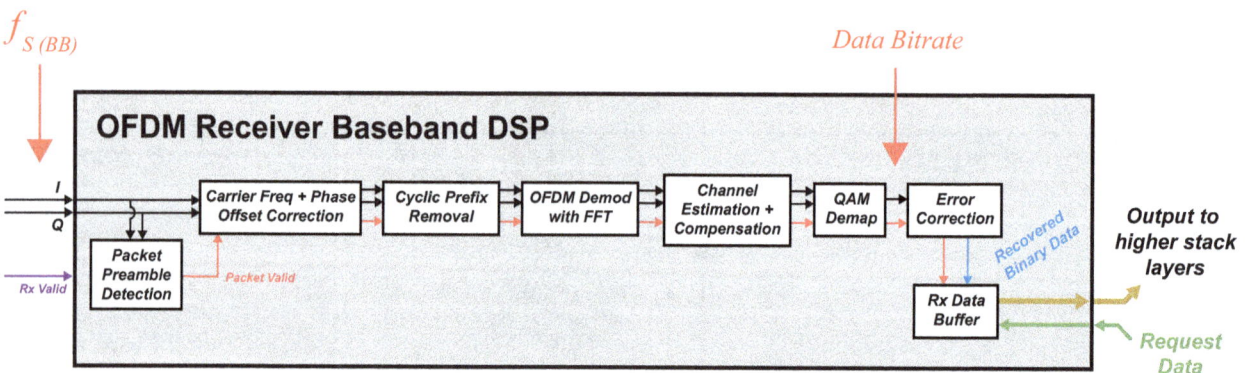

Figure 8.13: *Baseband OFDM demodulation block diagram.*

After digital filtering to isolate the signal of interest, preamble detection is carried out to detect the start of an OFDM packet. Following this, frequency and phase synchronisation are performed to compensate for any frequency tuning offsets, and sampling offsets in the ADCs (which are guaranteed to exist, as all oscillators have tolerances). The next stages are cyclic prefix removal and OFDM demodulation using an FFT. Channel estimation and channel compensation systems correct offsets across the bandwidth of the baseband signal that are introduced by the RF environment and radio spectrum channel, and finally, QAM demapping is performed to recover the binary data stream, and an error correction stage may be used to correct bit errors. (Much more detail on OFDM is provided in Chapter 16, when each of these stages are described fully).

At this point, the PHY layer demodulation of the OFDM signal has been completed, and a binary data stream has been recovered. However, this has not yet resulted in useful data for an end user application. The communication stack receive protocols need to be implemented to convert the binary data stream into the web page, document, music or video stream (etc.) that was transmitted.

While the PHY digital baseband stage is normally implemented on a FPGA, because of the parallel processing capabilities of FPGAs, a partner CPU will likely be used to process some of the higher software layers of the stack. An SoC that couples an FPGA with a CPU, such as the RFSoC, is an ideal platform — it would be possible to implement the entire receiver communications stack on the single chip, with some of the functionality running on the processor.

Having reviewed the receivers for both FM and OFDM, we now consider the opposite side of the link — the corresponding FM and OFDM transmitters.

8.3.2. Transmit Path

The digital baseband stage of an SDR transmitter is tasked with producing samples of a baseband signal that can be upconverted to an RF carrier and transmitted by the SDR front end. Working backwards, the SDR front end / digital baseband interface has exactly the same requirements as the receive path. Complex digital words containing I and Q samples with a fixed wordlength need to be output from the digital baseband stage every 1/fs_{BB} seconds. Continuing with our 16-bit SDR example, this corresponds to 32-bit digital words. Again, as we want to maximise the dynamic range of the DAC, and this means that the target amplitude of the generated samples should be around 80% of the wordlength, i.e. ± 26214.

FM Radio Transmitter

To build an analogue FM Radio transmitter using an SDR, the digital baseband stage must process a source audio stream to create a stereo FM multiplex, perform baseband FM modulation, and finally adjust the sample rate, sample format and sample amplitude to reach the target 16-bit wordlength with average sample amplitude of ± 26214, at the rate fs_{BB}. Such a transmitter is illustrated in Figure 8.14. Where the source audio stream is supplied in fixed point, it is important to ensure that no loss of data occurs during the multiplexing and modulation process. It is common practice to allow the sample wordlength to increase throughout these arith-

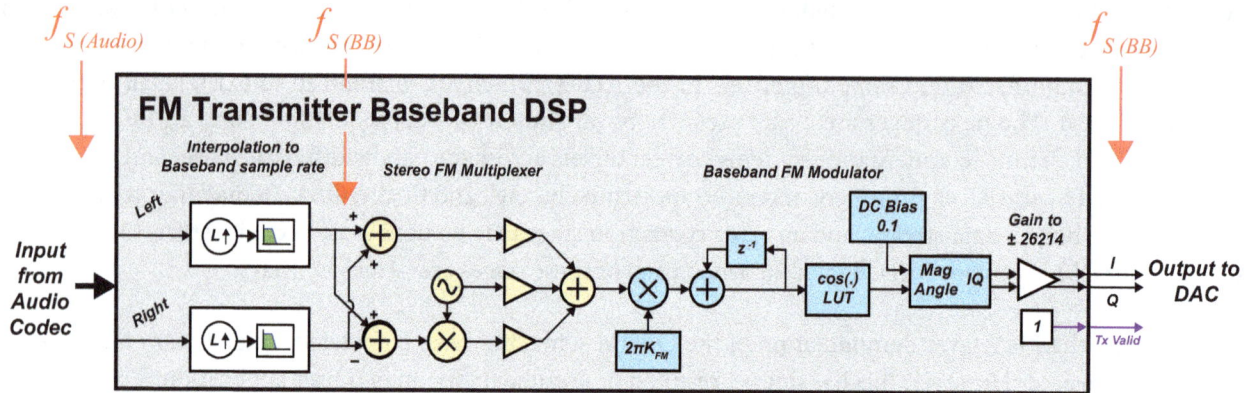

Figure 8.14: Baseband stereo FM modulation block diagram.

metic processing stages, before reducing it as the final stage in the process, in order to optimise the signal quality and dynamic range.

With knowledge of the DAC characteristics, it should be possible to accurately calculate the resulting analogue signal power, then configure the SDR front end and external RF amplifiers, and select antennae according to the RF Effective Isotropic Radiated Power (EIRP) permitted in the licence issued by the spectrum regulator. More on this in Section 8.5.

OFDM Transmitter

The initial stages of an OFDM transmitter will likely be implemented in software, with the various levels of the communication stack receiving instructions to send data packets over the air interface. By the time the data reaches the PHY layer in the SDR's digital baseband processing stage, this should be in the form of streaming binary data (1's and 0's). Data arriving to the PHY layer from upper layers is shown at the left hand side of Figure 8.15. Where error correction coding is used, this data initially passes through an FEC encoding stage.

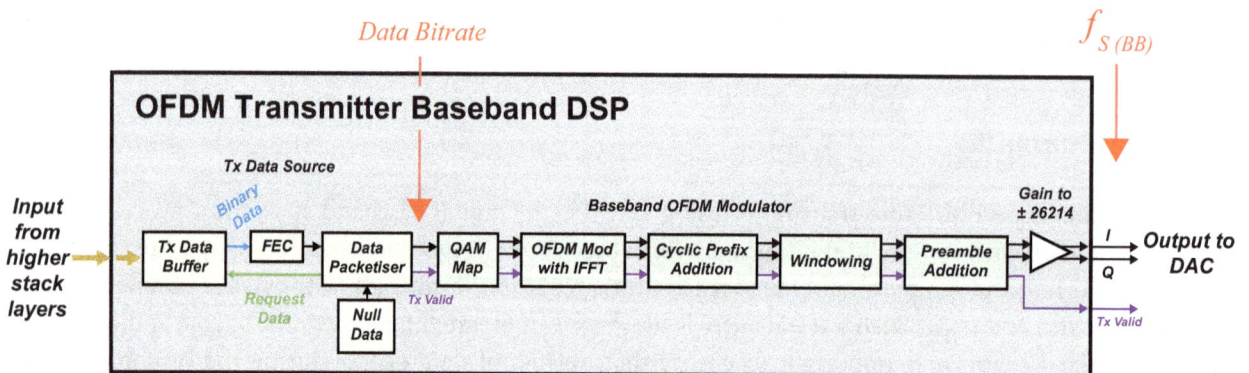

Figure 8.15: Baseband OFDM modulation block diagram.

When the transmitter is designed to operate in a 'bursty' fashion, rather than by transmitting a continuous data stream, the digital baseband stage must insert (and then process and transmit) 'null' bits that result in no output being sent to the SDR front end.

The OFDM modulator (as will be explained in greater detail in Chapter 16) takes the binary/null data, and performs QAM mapping, thus creating QAM symbols. Next, these pass through the IFFT to create OFDM symbols. A cyclic prefix is added, and windowing is performed to smooth discontinuities. Windowing also helps reduce the Peak to Average Power Ratio (PAPR), i.e. it reduces large amplitude spikes, which can help maintain a constant average sample amplitude (useful for optimising the dynamic range). The preamble, used for OFDM packet detection in the receiver, is then added. Again, the wordlength should be allowed to grow significantly during the modulation process. Finally, the sample format and amplitude are adjusted to reach the target 16-bit wordlength with average sample amplitude of ± 26214.

8.4. Digital Up- and Downconversion

As described in Section 8.1, most SDR architectures (including those implemented using the RFSoC) require Digital Upconversion and Downconversion stages. These conversions sit between the A/D and digital baseband stages, and comprise a frequency translation of the signal (between baseband and the modulated carrier frequency, and vice versa), and a change in sampling rate.

In the remainder of this section, we provide an overview of the generic signal processing stages involved in each, particularly from the perspective of sampling rates and computational requirements. This generic background precedes a more detailed discussion of the hardened DDCs and DUCs present within the RFSoC, which follows in Chapters 9 and Chapters 11, respectively.

We begin with the DDC, which is arguably the more intuitive, followed by a briefer review of the DUC.

8.4.1. Digital Downconverter (DDC)

Digital Downconverter (DDC) forms part of the receiver, and is the first processing stage following the ADC. The architecture of the DDC is presented in Figure 8.16. The DDC first shifts the incoming modulated signal from a carrier frequency to baseband, by mixing it with the output of a Numerically Controlled Oscillator (NCO). The NCO is normally implemented in the FPGA/ PL, with the output values coming either from a pre-computed dictionary of samples stored in a Lookup Table (LUT), or by calculating the output dynamically using a Co-Ordinate Rotation DIgital Computer (CORDIC) processor [354].

In the examples presented here, we consider that the incoming signal is modulated onto a carrier at fc Hz, which could be the RF carrier in the case of a Direct-RF radio, or an IF carrier if the radio has an analogue conversion between RF and IF. This means that the NCO used to demodulate the signal should generate sine and cosine outputs at fc Hz. The mixing process shifts the sampled signal to baseband, and cyclically shifts all other sampled signals captured by the ADC to other frequencies.

During the demodulation stage, the sampling rate of the system is equal to that of the ADC, fs_{ADC}, which can be extremely high (e.g. 4 GHz or even higher, in the case of RFSoC). Since demodulation shifts the signal of interest to 0 Hz, the sampling rate can be substantially reduced. This results in a sampling rate at the output of the DDC that is much closer to (but still higher than) the baseband signal bandwidth. We will denote the resulting baseband sample rate as fs_{BB}.

In a DDC, sampling rate reduction is normally accomplished using a cascade of decimators (rather than a single decimator), as indicated in Figure 8.16. Notice that we require two sets of these cascades, one each for the *In Phase* and *Quadrature* branches.

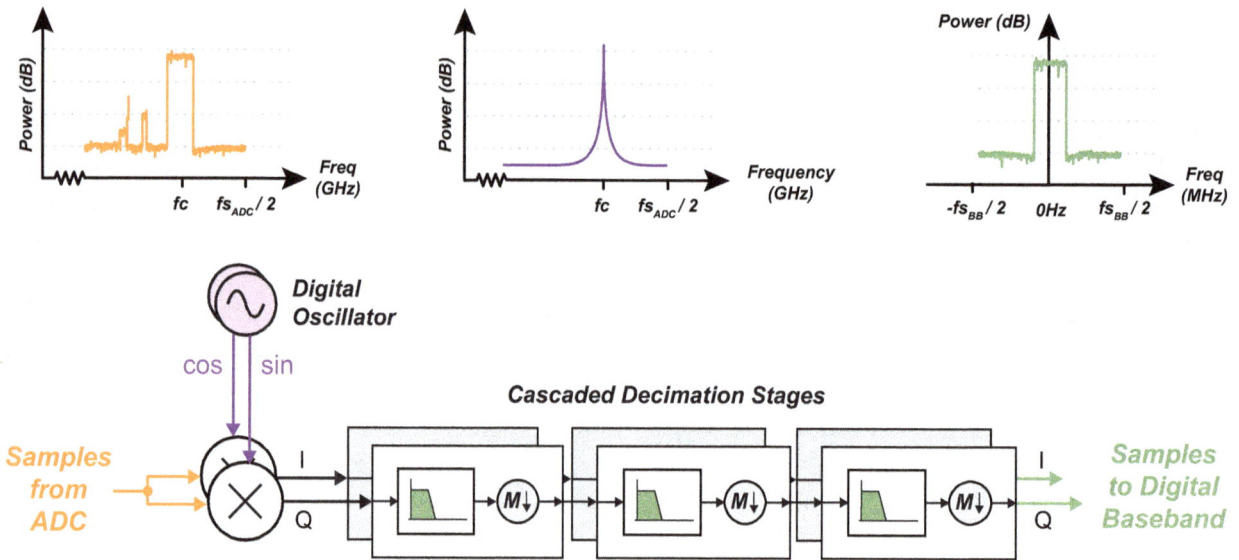

Figure 8.16: Digital Downconversion.

The reason for selecting a decimation chain, as opposed to a single-stage decimator, is computational efficiency — the number of MAC operations that need to be performed per second can be reduced by opting for a multi-stage design. We will now explore this issue with an example, which builds on the simple two-stage decimator previously discussed in Section 4.7.6.

Let us suppose that a DDC has the following parameters:

Table 8.1: Parameters for DDC design example

Description	Symbol	Value
ADC sampling rate	f_{s_A}	4 GHz
Carrier frequency	f_c	1.6 GHz
Signal bandwidth	B	200 MHz
Baseband sampling rate	f_{s_B}	250 MHz
Decimation ratio[1]	M	16

> 1. The decimation ratio, M, is the ratio between the input and output sampling rates, i.e. 4 GHz / 250 MHz = 16.

Before and after the signal is demodulated to baseband, the remainder of the spectrum is likely to contain energy across other frequency bands. When receiving signals from the RF spectrum Over the Air (OTA), this energy will comprise AWGN, transmissions by other users of the radio spectrum, and so on. We can model the unwanted components generically as noise, as shown in Figure 8.17. Note that the received signal is **real** immediately after the ADC, and **complex** after demodulation (hence the spectrum becomes two-sided).

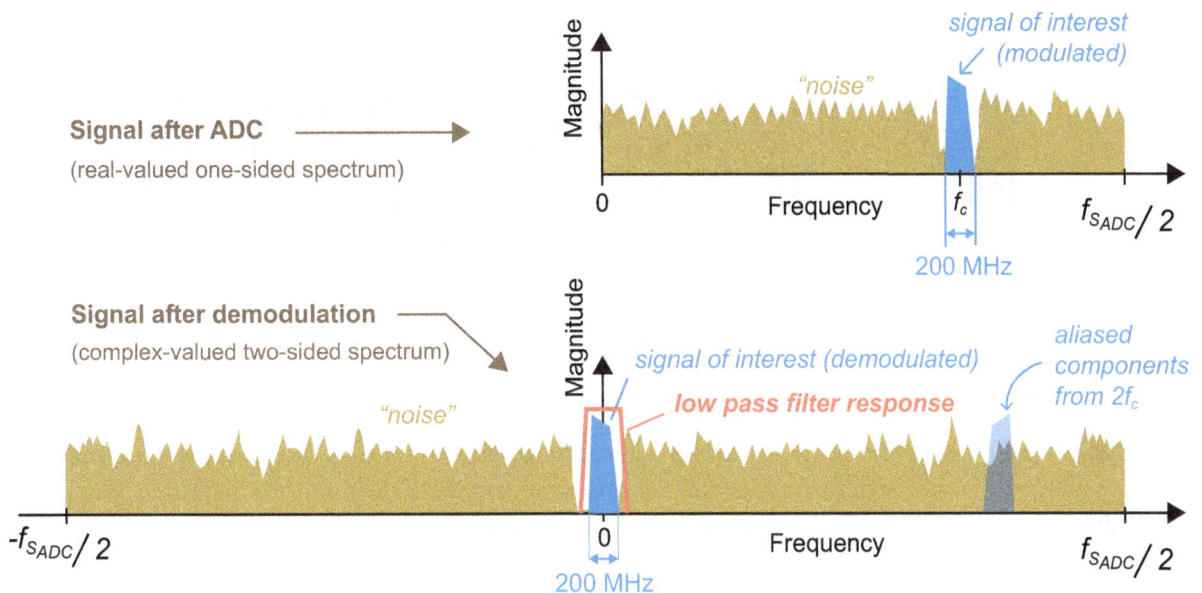

Figure 8.17: Demodulation of signal in presence of noise, with decimator low pass filter requirement indicated.

Therefore, we consider that each of the In Phase and Quadrature components are processed by a separate decimation chain, composed of filters with real-valued coefficients. (An alternative would be to depict the complex signal being processed by a cascade of filters with real-valued coefficients.)

We will now consider the design of one of these decimation cascades, i.e. for an individual I or Q branch (the same design would be replicated for both branches).

If the noise were *not* removed prior to downsampling, then it would alias into the band containing the signal of interest. Therefore, low pass filtering is required; the filter must preserve the signal bandwidth while attenuating the noise, and avoiding aliasing. A sharp transition band is often needed, which translates into a long filter, with many weight multiplications to be computed. A filter design for the example scenario is presented in Figure 8.18 (i.e. a single stage solution). In this 'equiripple' design, the passband ripple is 0.1dB, stopband attenuation is 60dB, and the passband and stopband edges are set at 100 MHz and 150 MHz, respectively.

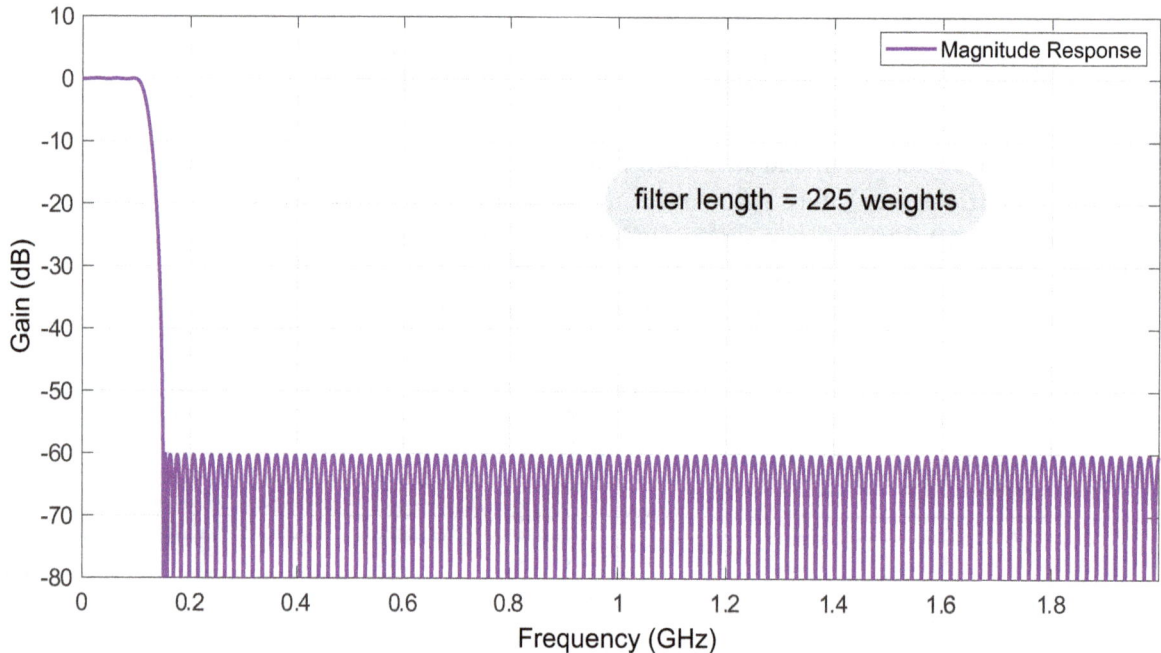

Figure 8.18: Filter response for single stage decimator (decimating by 16).

Given that the designed filter has $W = 225$ weights, and if we assume that the polyphase implementation method is adopted (as was reviewed in Section 4.7, meaning that the filter performs arithmetic processing at the output rate of 250 MHz), then the computation rate for the single stage decimator design can be expressed as:

$$R_{ss} = W \times f_{s_{BB}} = 225 \times 250 \text{ MHz} = 56.25 \text{ GMACs/s}. \tag{8.3}$$

An alternative approach is to partition the overall ratio of 16 into smaller stages of decimation, an approach that can be taken for any required ratio, other than a prime number. The viable partitionings for the current example of 16 are set out in Table 8.2 (including the original, single stage design).

There are two main advantages of partitioning the filtering task into smaller stages.

Firstly, with smaller decimation ratios, the properties of halfband (and more generally, L-band) filters can be exploited, wherein approximately $1/L$ of the filter weights are exactly zero and do not need to be computed. This leads to a reduction in computation rate; halfband filters are especially attractive for this reason.

Secondly, and perhaps less obviously, the design of the filters can be relaxed. Each of the filters *only needs to preserve the passband[1], and apply a stopband across the region that will alias into the passband*. This means that the transition band can be extremely wide in some cases, leading to relatively inexpensive filters.

Table 8.2: Candidate partitioned decimation ratios for the x16 decimator

Description	Stage 1	Stage 2	Stage 3	Stage 4	Total
Single stage decimator	16	-	-	-	16
Two-stage decimator (a)	8	2	-	-	16
Two-stage decimator (b)	2	8	-	-	16
Two-stage decimator (c)	4	4	-	-	16
Three-stage decimator (a)	4	2	2	-	16
Three-stage decimator (b)	2	4	2	-	16
Three-stage decimator (c)	2	2	4	-	16
Four-stage decimator	2	2	2	2	16

This 'staged' approach best illustrated with an example. We will base a cascaded design on the Three-stage decimator (c) design, which has decimation ratios of 2, 2, and 4 for the first, second, and third stages, respectively. (Note: the selection does not mean to indicate that this is the most efficient design!)

Stage 1: Decimate by 2

At the first stage, the filter is designed to pass the region between 0 Hz and half of the sampling rate at the output of the cascaded decimator, i.e. 125 MHz in this example. The range of frequencies that would alias into this band, after downsampling by a factor of 2, extends from 1.875 GHz to 2 GHz, and therefore the stopband

---

1. In the context of this decimation example, we regard the passband as the region from 0 Hz to half of the baseband (i.e. output) sampling rate. Any further filtering to extract the signal of interest can take place subsequently.

should be placed here. The transition band may extend across the entire band between 125 MHz and 1.875 GHz, as any remnant energy in this frequency band can be removed by a subsequent filter in the cascade. In this first decimation stage, the sampling rate is reduced by a factor of 2, from 4 GHz to 2 GHz. All filtering computation can be performed at the (lower) output sampling rate using polyphase techniques.

Stage 2: Decimate by 2

The sampling rate at the input to Stage 2 is 2 GHz. Once again, the low pass filter is designed to pass the frequency band from 0 Hz to 125 MHz, and to stop the band that would alias into that region, i.e. 875 MHz to 1 GHz. The frequencies between 125 MHz represent the transition band, and will experience a sliding degree of attenuation across the band. The filter computation runs at the output rate of 1 GHz.

Stage 3: Decimate by 4

The final filter in the cascade decimates by a factor of four, and therefore there are three bands that would alias into the passband region of 0 to 125 MHz. As a result, Stage 3 requires a filter that cuts off at a lower frequency and has a sharper transition than the previous ones.

With the defined passband region extending from 0 to 125 MHz, the three alias regions exist from 125 to 250 MHz, 250 MHz to 375 MHz, and 375 to 500 MHz. The low pass filter should therefore cut off at around 125 MHz. Recall that a 'brick wall' filter is unrealisable, i.e. there must be some non-zero transition bandwidth between the passband and stopband. This transition bandwidth can be set with knowledge of the signal of interest. In this case, we know that the signal occupies the band between 0 Hz and 100 MHz, and therefore a filter can be designed with a transition band between 100 MHz and 150 MHz, while avoiding any aliasing into the signal band of interest.

Once again, the filter computation can run at the output, decimated rate, if implemented in polyphase form. Therefore, the filter operates at a rate of 250 MHz in this case.

Three-stage Decimator Summary

The three filter stages outlined above can be summarised graphically, as shown in Figure 8.19.

The implementation of this cascaded approach can be analysed in a similar manner as the single stage decimator. The metrics for each of the individual stages are set out in Table 8.3. In all cases, the 'equiripple' method was used to design the filters, with passband ripple set to 0.1dB, and a stopband attenuation of 60dB.

Therefore, the total computation rate of the cascaded decimator design is the sum of the individual stages, i.e.

$$R_{CASC} = 12 + 7 + 14.25 = 33.25 \text{ GMACs/s} \tag{8.4}$$

and this would be replicated for each of the I and Q branches.

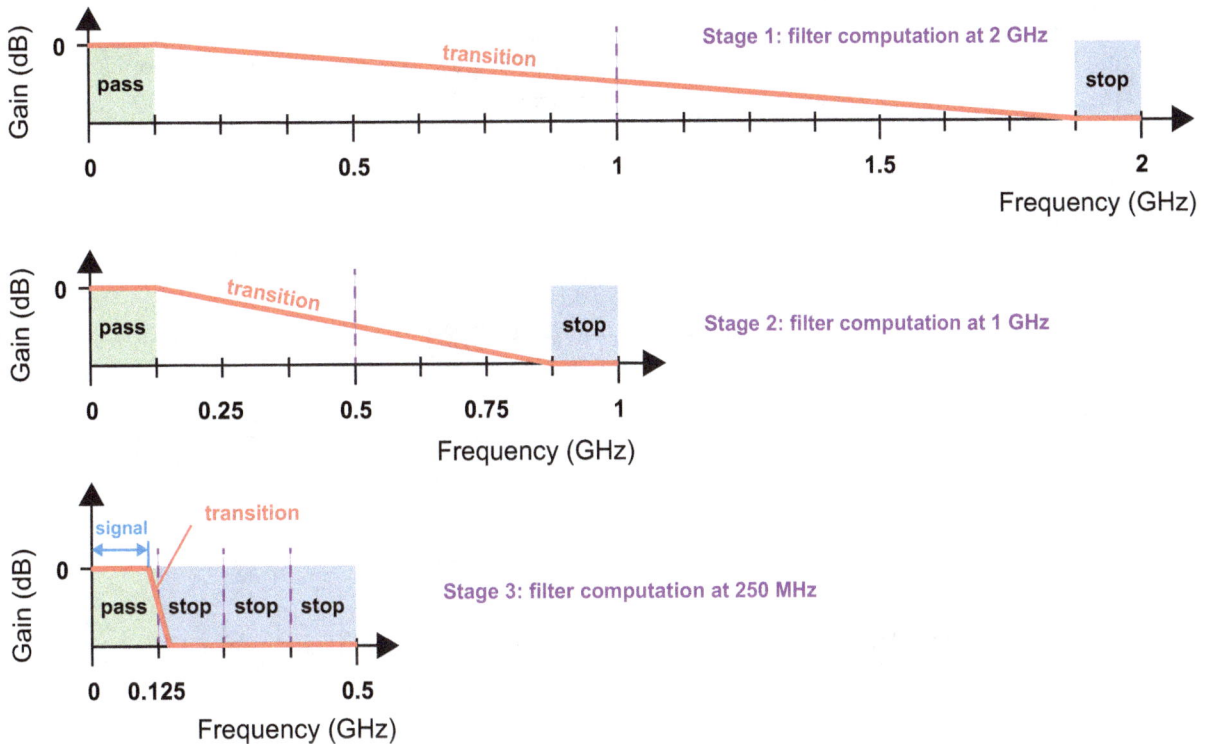

Figure 8.19: Details of the cascaded decimator stages.

Table 8.3: Parameters for DDC design example

Stage	Decimation Ratio	Filter Length (No. Weights)	Filter Sample Rate	Computation Rate
1	2	6	2 GHz	6×2 GHz $=$ 12 GMACs/s
2	2	7	1 GHz	7×1 GHz $=$ 7 GMACs/s
3	4	57	250 MHz	57×250 MHz $=$ 14.25 GMACs/s

The multi-stage approach is therefore much less computationally expensive than the single stage decimator: at 33.25 GMACs/s, as opposed to 56.25 GMACs/s, i.e. a saving of over 40%. Some of the other candidates outlined in Table 8.2 may produce even more efficient decimator implementations.

The hardened DDC in the RFSoC uses a similar, multi-stage approach as outlined here, with either three or four stage cascaded decimators (depending on the device generation). A hardened, LUT-based NCO is included within the same blocks.

8.4.2. Digital Upconverter (DUC)

The Digital Upconverter (DUC), which is part of the transmitter, performs a similar but mirrored set of operations to the DDC. A block diagram representation of a typical DUC architecture is provided in Figure 8.20.

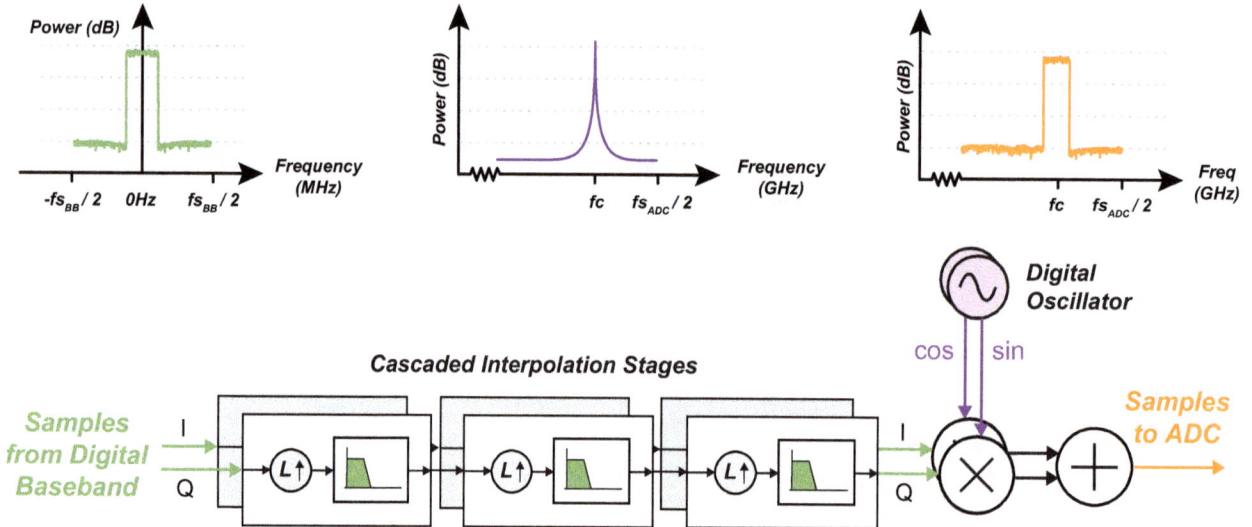

Figure 8.20: Digital Upconversion.

First, the low sampling rate of the digital baseband stage, fs_{BB}, is increased to the much higher rate used by the DAC, fs_{DAC}, by an interpolator.

The interpolated signal is then modulated by mixing it with sine and cosine signals generated by an NCO, at the desired carrier frequency, fc. Modulation has the effect of shifting the signal up in frequency, such that it is centred at fc, which could be equivalent to the eventual RF transmit frequency (if using the Direct-RF SDR architecture), or an IF frequency, if an external IF-to-RF modulation stage is used, i.e. a superheterodyne approach.

The DUC interpolator can be implemented as a single stage that changes the sampling rate by the full interpolation ratio, or by cascading several smaller interpolators together, which collectively perform the total rate change. For instance, an interpolation by 16 could be partitioned into an equivalent set of stages as was outlined in Table 8.2 for the DDC decimator. Like the decimators considered earlier, the interpolators incorporate low pass filters (in this case, to remove the spectral images generated by upsampling, as previously outlined in Section 4.7), and they can be implemented in polyphase form such that all filter processing takes place at the input rate, i.e. the lower of the two sampling rates.

An equivalent analysis could be performed to evaluate the computational cost of candidate cascaded interpolator designs. If a three-stage interpolator with factors of 4, 2, and 2 was chosen (i.e. the mirror image of the decimator design examined in Section 8.4.1), the filtering operations indicated in Figure 8.19 for the decimator would be performed in reverse. For instance, Stage 1 would interpolate the input signal from a sampling rate of 250 MHz to 1 GHz. This would be composed of an upsampler, raising the sampling rate by a factor of four, in the process creating three spectral images (one in each of the 'stop' regions indicated in Figure 8.19), which would then be removed using a low pass filter with a relatively sharp cut-off. Stage 2 would then upsample that signal by a factor of two, from 1 GHz to 2 GHz, creating one spectral image in the 'stop' region, which would be removed using a much more relaxed low pass filter. Stage 3 would also interpolate by a factor of 2, from 2 GHz to 4 GHz, in a similar manner to Stage 2.

8.4.3. Other Filter Types

In our treatment of the decimator and interpolator within the DDC and DUC, respectively, we have assumed that FIR filters have been used, implemented in the polyphase form (which is the most efficient method of realising them). Given the RFSoC focus of this book, this is the most relevant approach, as the RFSoC's integrated DDC and DUC both use this architecture.

It should be acknowledged, however, that there are other options for implementing DDC decimators and DUC interpolators, beyond those discussed so far. Alternative methods are also relevant to RFSoC development (even if not the most direct option!) given that designers can create their own custom designs, bypassing the hardened DUC and DDC blocks.

A popular approach is to use a CIC filter (previously introduced in Section 4.7.7) to perform a final stage of interpolation in the DUC, or a first stage of decimation in the DDC. The CIC is an especially efficient type of multirate filter that does not require any multiplications; it has the disadvantage of introducing a characteristic 'droop' in the passband, but this is readily corrected with a CIC compensation filter. Another option is to use almost-linear-phase multirate IIR filters.

8.5. Front End Analogue Signal Conditioning, and Antennae

When a radio spectrum regulator issues a licence for a transmitter, they will normally define a permitted maximum signal power, and a spectral mask that the radio must abide by.

Radio waves propagate from an antenna (connected to the SDR) through the air to the receiver. Signals emitted from the antenna will have an electrical power value, known as the EIRP (Effective Isotropic Radiated Power). This can be calculated as follows:

$$EIRP_{dBm} = P_{SDR\ output\ (dBm)} + G_{RF\ Amplifier\ (dBm)} + G_{Antenna\ (dBi)} \qquad (8.5)$$

where $P_{SDR\ output\ (dBm)}$ is the electrical power output from the SDR, $G_{RF\ Amplifier\ (dBm)}$ is the gain from any RF amplifiers used, and $G_{Antenna\ (dBi)}$ is the isotropic gain of the antenna.

8.5.1. Signal Conditioning

The power output from the SDR will likely be extremely low, in the range of 5mW/7dBm (for comparison, a Wi-Fi router in the USA is permitted to transmit at up to 4W/36dBm in the 5GHz channel; x800 higher power). Therefore, it is almost certain that external RF Power Amplifiers (PAs) will be required in order increase the power and allow the signal to transmit more than a few meters. RF PAs will amplify a wide band of frequencies (sometimes hundreds of MHz wide), often wider than the bandwidth of the SDR. In turn, analogue RF Bandpass Filters are required in order to ensure the PAs do not transmit unwanted energy at other frequencies.

Most SDRs, if you examine the output ports, have separate connectors for Tx and Rx antennae. Normally in radio systems, Tx and Rx are combined into a single antenna port, TRx, and this is then connected to an antenna. This can be achieved using an RF Duplexer. Duplexers are components which enable bi-directional communication (Tx, Rx) over a single RF path and antenna element. This is especially important in FDD radios, which operate Tx and Rx paths simultaneously.

A more in depth explanation of this signal conditioning for a cellular basestation use case is presented in Section 17.3.1 of *RFSoC Applications in Cellular Networks*.

8.5.2. Antennae

An antenna is a passive object that radiates (transmits) and receives the electromagnetic waves that enable wireless communication. The radiation pattern and performance depends on the antenna size, shape, and band optimisations. The type of antenna chosen varies upon the RF band in use and the use-cases.

Antennae have a nominal antenna gain. This gain is entirely passive, as there are no actual amplifiers in an antenna. It is measured in dBi, which stands for *decibels relative to the isotrope*. The isotrope is a hypothetical antenna that radiates uniformly in all directions (i.e. it has spherical radiation from a central point) - which is a very poor design. All real-world antennae perform better than this imaginary isotropic antenna, and therefore have a nominal dBi gain relative to it.

There are three main categories of antenna you are likely to use in your network, as illustrated in Figure 8.21. These are:

- **Omni-directional:** An antenna that radiates uniformly in 360 degrees around it. These are normally cylindrically shaped and installed on the top of a mast or building. They are ideal for providing coverage around a flag pole mast in the middle of a town square, for example. They generally have lower performance than sector antennae.

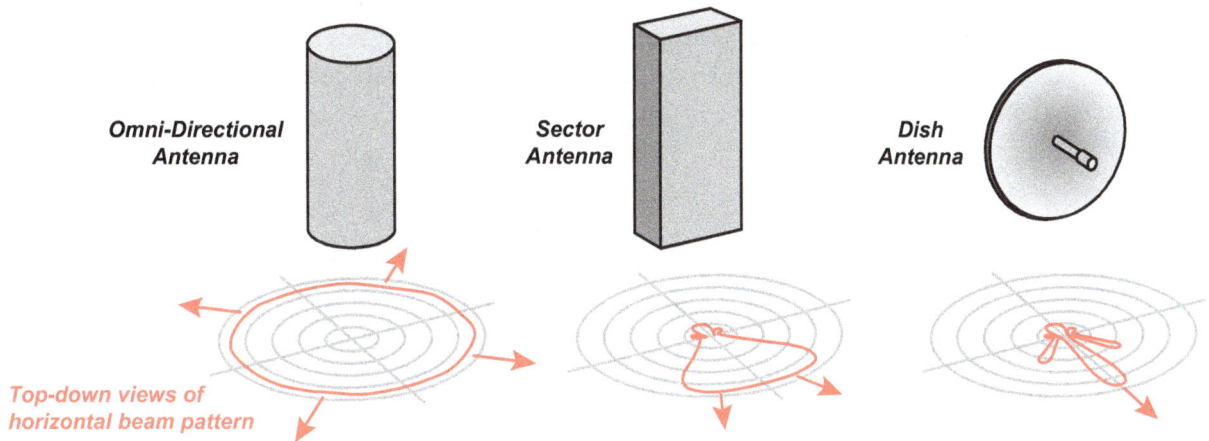

Figure 8.21: Various antenna formats.

- **Sector:** This type of antenna is directional. Commonly you will see 60°, 90° and 120° sector antennae used in 4G/5G mobile and Wi-Fi FWA networks. A 60° sector will produce a beam that has most of its energy within a 60 degree horizontal wedge, if viewed from the top down. They generally perform better than omni-directional antennae, as they are tuned to only radiate in one direction. If a receiver is located outside of the transmitter antenna beam, it will struggle to connect to the base station.

- **Dish:** Dish antennae are curved parabolic objects (like a TV satellite dish), that are used for point to point transmissions. They are used in microwave backhaul radios, and in LEO satellite systems.

MIMO antennae are required when multiple transmit and receive paths are used. A 2x2 MIMO antenna will feature two antenna ports, and two physical elements inside the antenna. These will likely be offset by 90°, giving one transmit/receive path on a vertical plane, while the other is on a horizontal plane. 4x4 MIMO antennae will feature four ports, and so on.

8.6. Chapter Summary

In this chapter, we have reviewed the impact that DAC and ADC sampling rate has had on SDR architectures over the last few decades, focusing in particular on the Direct-RF variant enabled by the multi-GHz RFSoC. Two examples of SDR architectures were presented, with very different levels of complexity: at the lower end of the scale, an FM transmitter and receiver; and as a more sophisticated example, a full OFDM transmitter and receiver. More detail on the OFDM architectures outlined here is provided in Chapter 16.

Two other common elements of SDRs were discussed. Firstly, the digital up- and down-conversion operations, which are parts of the digital transmitter and receiver, respectively. We saw that, within the DUC, the transmit signal is interpolated and modulated onto the carrier; while the DUC performs demodulation followed by decimation. We also reviewed some of the more detailed considerations for implementing the decimator of a DDC (which are equally applicable for DUC interpolators). Secondly, some of the important front end analogue hardware was introduced. The main categories of antennae were reviewed, with the most relevant use-cases noted for each antenna type.

Chapter 9

RF Data Converters: Analogue to Digital

Lewis Brown

When discussing modern SDR hardware platforms, the most crucial attribute is the ability to directly sample key Radio Frequency (GHz signals) without the need for an intermediate stage. This chapter provides an overview of the Radio Frequency Analogue to Digital Converters (RF-ADCs) present in the RFSoC and their functionality for sampling radio signals. We also discuss RF-ADCs as part of the RF Data Converter (RFDC) blocks, and explore the stages of the RF-ADC pipeline required to process an RF signal, before concluding with a review of how to program the RF Data Converter RF-ADC in Vivado. This chapter is supported by the AMD RFDC User Guide, PG269 [90], which makes excellent follow-up reading to this chapter and the following Chapter 10.

9.1. Analogue to Digital Conversion

Before exploring the world of RF-ADCs, it is worthwhile recapping on the general process of converting a signal from the analogue domain to the digital domain, which entails sampling and quantising an analogue signal. Sampling takes place at fixed time intervals, known as the sampling period, T_s, which is the reciprocal of the sampling frequency, f_s.

Quantisation maps the amplitudes of these samples to a set of discrete values. In an ideal ADC, any measured analogue voltage would be perfectly mapped to an equivalent digital value, correct to an infinite number of binary places. Of course, in practice this is not possible and the numbers of bits used in the quantisation process is limited; Gen 1 and 2 RFSoC RF-ADCs have 12-bit precision and Gen 3 and DFE devices have 14-bit precision. The measured analogue voltage must be fitted to the closest available quantisation level from the set of 2^n levels, where n is the number of bits in the ADC. The gap between quantisation levels, otherwise known as the resolution, depends on the smallest value that the ADC can resolve based on the maximum input voltage [258], [323]. An n-bit ADC has a resolution of

$$\Delta_v = \frac{V_{max} - V_{min}}{2^n},$$

(9.1)

where V_{max} and V_{max} are the maximum and minimum input voltages, respectively.

Example: *A 12-bit ADC with a maximum input voltage range of $\pm 1\,V$ has a resolution of:*

$$\Delta_v = \frac{1\,V - (-1\,V)}{2^{12}} = \frac{2\,V}{4096} = 0.000448V = 0.488\,mV.$$

(9.2)

As a result of the quantisation process, differences in amplitude exist between the real, analogue values of the sampled signal, and the digital values resulting from quantisation. This *quantisation error* of a single sample is determined by the resolution of the ADC [108]. Assuming that samples are rounded to the closest quantisation level, the maximum quantisation error is $\Delta_v/2$ (for note, in earlier chapters Δ_v also denoted as 'q'), so for a 12-bit ADC the worst case error would be ± 0.244 mV. To optimise the accuracy achieved by an ADC, manufacturer calibrations should be undertaken to minimise other sources of error, such as gain and offset errors. If desired, please refer back to Chapter 4 for further review of the sampling and quantisation processes.

9.2. ADCs for RF Frequencies

Many common mobile and wireless communications signals are located in the low and mid-band RF spectrum, with frequencies ranging up to several Gigahertz (GHz), probably 6 GHz, albeit the top of the mid-band varies a little in the literature. To sample such a signal, a specialised high-speed ADC is required. Modern RF-capable ADCs, or RF-ADCs, can achieve multi-Gigasample per second, GSps, (or GHz) sampling rates, enabling the direct sampling of signals in this low and mid band RF GHz signal range [173], [338].

9.2.1. The 1st Nyquist Zone

Recall the Nyquist Sampling Theorem as discussed in Chapter 4. We noted that the sampling rate should be set to greater than twice the maximum frequency component of the sampled analogue signal, in order to avoid aliasing. Therefore, the range of frequencies which can be directly sampled by any given ADC can be represented as 0 Hz to $f_s/2$ Hz. This is known as the 1st Nyquist Zone.

Given the very high sampling rates of the RFSoC's RF-ADCs, the 1st Nyquist Zone can span multiple GHz. As a result, the RF frequencies of many frequency bands used in wireless communications can be directly digitised without the need for an intermediate mixing or demodulation stage from high frequency to baseband.

Example: With a sampling rate of $f_s = 4$ Gsps, analogue signals present in the range 0 to 2 GHz (i.e. the 1st Nyquist Zone of the RF-ADC) can be directly digitised in the conventional manner. However, there may be energy (unwanted frequencies) present in the spectrum above the 1st Nyquist Zone, and as such an analogue anti-aliasing low pass filter should be included prior to the ADC, to remove all of those frequency components above $f_s/2$, as illustrated in Figure 9.1.

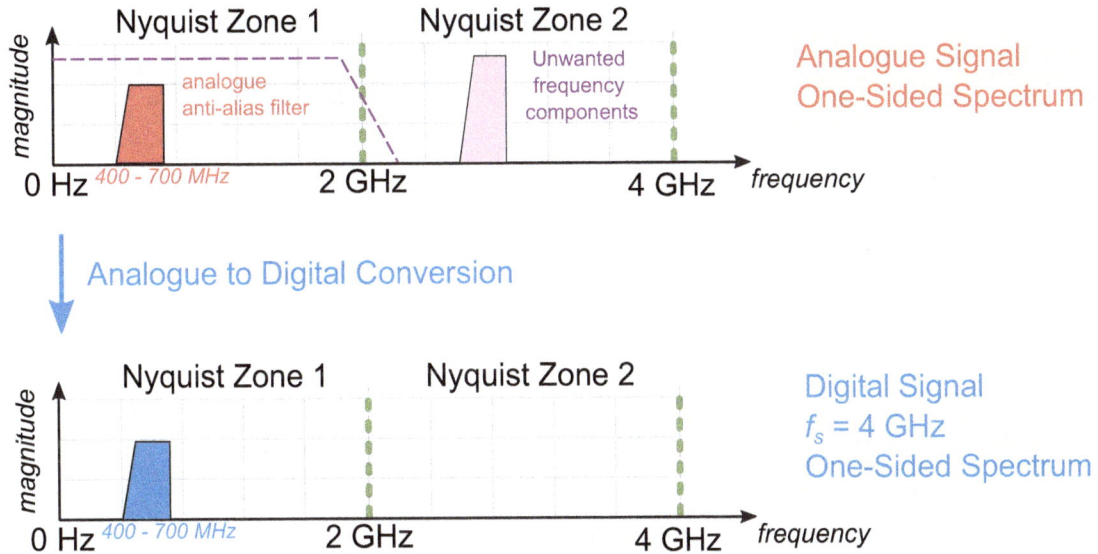

Figure 9.1: 1st Nyquist Zone signal acquisition with analogue lowpass filter.

Given the example sampling rate of $f_s = 4$ GSps, several common communications signals can be captured in the 1st Nyquist Zone [285]. This includes:

- Broadcast television (~470 to 700 MHz in the UK)

- Global Navigation Satellite Systems (several bands from 1.164 GHz to 1.610 GHz)

- Low range Internet of Things (IoT) standards (867 to 869 MHz, 902 to 928 MHz)

- Several cellular bands (800 MHz, 900 MHz, 1.4 GHz, 1.8 GHz in the UK).

Note: Many of these bands differ globally.

9.2.2. The 2nd Nyquist Zone and Above

It is usually desirable to avoid aliasing, however in some cases it can be exploited to deliberately 'fold' a signal down into the 1st Nyquist Zone, allowing these signals to be digitised by the ADC. This technique is particularly relevant for RF-ADCs, as it provides access to signals present in the 2nd Nyquist Zone and above, i.e. frequencies greater than $f_s/2$.

By sampling an aliased version of the 0 Hz to $f_s/2$ frequency range, an RF-ADC can be used to directly sample the 2nd Nyquist Zone without an IF demodulation stage. It is assumed that an appropriate bandpass filter precedes the RF-ADC, to remove any components present at other frequencies that are not of interest (particularly within the 1st Nyquist Zone, which is the band being folded into!). This process of applying a bandpass filter to a 2nd Nyquist Zone signal, and aliasing it into the 1st Nyquist Zone, can be seen in Figure 9.2.

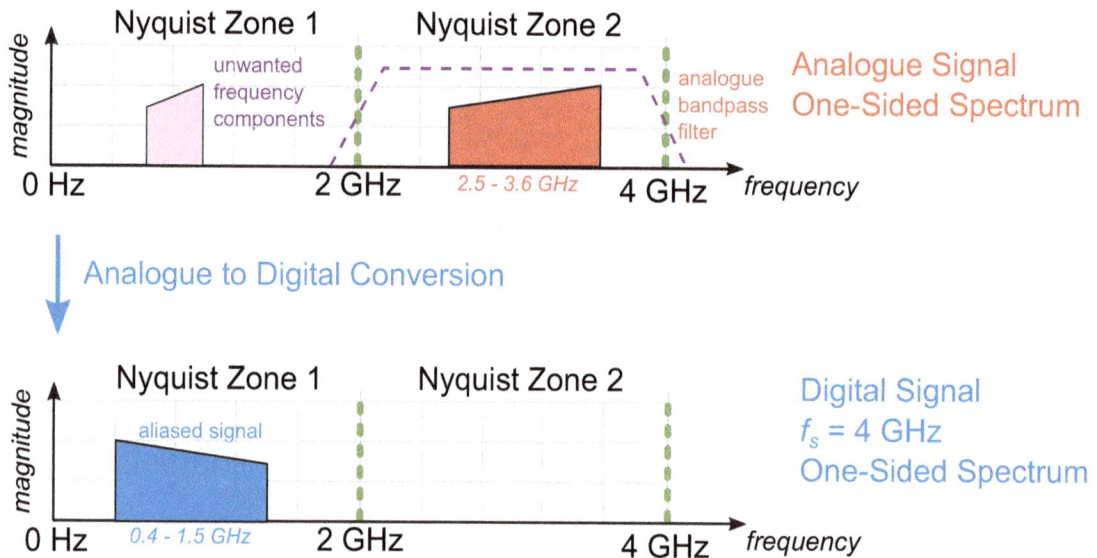

Figure 9.2: 2nd Nyquist Zone signal acquisition with analogue bandpass filter.

Notably, the aliased signal shown in Figure 9.2 has been 'flipped' left-to-right in its orientation; we will later show how to compensate for this effect and flip back as may be required.

Taking the same example as before, with f_s = 4 GHz, and extending to the 2nd Nyquist Zone (2 to 4 GHz), additional higher frequency mobile and wireless signals that can be received by our RF-ADC include:

- 2.4 GHz Wi-Fi Band (2.4 GHz to 2.5 GHz)

- Bluetooth (2.45 GHz)

- Citizens Broadband Radio Service (CBRS) in the USA (3.55 GHz to 3.7 GHz)

- Cellular Bands for 4G LTE (2.1 GHz, 2.3 GHz, 2.6 GHz) [8]

- Cellular Bands for Frequency Range 1 5G (3.4 GHz, 3.6 to 4 GHz) [8]

The 3rd Nyquist Zone would therefore bring further possibilities again, including the 5 GHz Wi-Fi band from 5.15 GHz to 5.725 GHz. There are some practical limitations of exploiting aliasing in this way, however, as RF-ADCs have a limited operating range (signal components experience increasing attenuation above the stated maximum input frequency for the device). The maximum input frequencies for Gen 1, Gen 2, Gen 3, and DFE devices are listed as 4 GHz, 5 GHz, 6 GHz, and 7.125 GHz, respectively [88].

9.2.3. Analogue Bandpass Filter

As reviewed in the previous sections, many RF applications involve fixed frequency bands, and therefore a specific frequency range can be targeted with the RF-ADC. When a system has been designed for a particular standard or application, an RF bandpass filter can be used to isolate that signal before the signal reaches the ADC and thus provide good out of band rejection and improve SNR at the device. An example of such an RF filter is sketched in Figure 9.3.

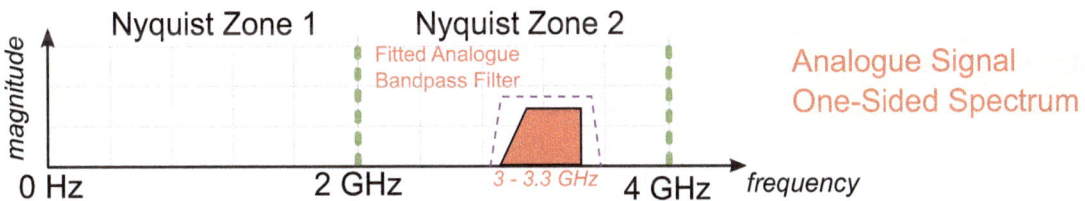

Figure 9.3: Tightly fitted analogue bandpass filter in the 2nd Nyquist Zone.

In Figure 9.3, the analogue bandpass filter can be designed to fit around the signal of interest, rather than passing the entire Nyquist Zone. In a production system, if the frequency band is known and fixed, then the best performance could be achieved by designing an analogue filter to fit relatively tightly around that band, thus removing most unwanted frequency components before they reach the RF-ADC. This approach has the added benefit of not only removing frequencies from other Nyquist Zones, but also unwanted signals within the target Nyquist Zone, resulting in better noise suppression.

On the other hand, as discussed in earlier chapters, a key benefit of an SDR is the flexibility of design, allowing for operation over a wide range of frequencies. The 'tight bandpass' approach is not suitable when an SDR is required to be maximally flexible, or to receive multiple signals that may reside at different frequency bands within the same Nyquist Zone. In these cases, it is preferable to apply a filter that passes the entire Nyquist Zone containing the signal(s) of interest. This may also involve a cheaper analogue filter.

To receive signals from the entire 1st Nyquist Zone, a low pass filter with cut-off frequency around $f_s/2$ would be optimal in most circumstances. This would prevent frequencies in the 2nd Nyquist Zone and above being aliased into Nyquist Zone 1, where the signal resides. For Nyquist Zone 2, a bandpass filter would be required. In this case, recall that the contents of Nyquist Zone 2 will be folded into Nyquist Zone 1 (i.e. intentional aliasing). The bandpass filter clears Nyquist Zone 1 of other frequency components, which would otherwise be superimposed on the desired aliased signal from Nyquist Zone 2 whilst also removing higher frequency components from Nyquist Zone 3 and above, that would otherwise (unintentionally) alias into Nyquist Zone 1.

9.3. RF-ADCs on RFSoC

RF-ADCs are advantageous in the sense that they can directly digitise many modern communication signals. Additionally, the high sampling rates of RF-ADCs enables instrumentation applications such as RF spectrum analysis for low and mid band mobile/wireless spectra, which was explored in depth in Chapter 5. To understand more about the operation of the RF-ADCs, we will now consider their architecture more closely.

9.3.1. The RFSoC Device Family

As reviewed in Chapter 3, the defining feature of the RFSoC is its hardened RFDC blocks, which allow the device to function as an RF transmitter and receiver (with the addition of analogue circuitry and antenna(e)). Most devices also include SD-FEC blocks for implementing error correction schemes. These resources, alongside the PL and PS, which are equivalent to the MPSoC devices, make the RFSoC devices extremely flexible, and therefore very suitable as an SDR implementation platform.

As of the time of writing, three generations of RFSoC have been released, with Gens 1 and 3 each comprising several devices. In addition to the three generations of standard RFSoC devices, there is also a *RFSoC DFE* device family which is optimised for 5G New Radio applications and features a targeted set of hardened processing blocks. The naming convention for these sets of devices is as follows:

- Gen 1: ZU2xDR

- Gen 2: ZU39DR

- Gen 3: ZU4xDR

- RFSoC DFE: ZC6xDR

RFSoC devices can be compared in terms of the number of channels, resolution (number of bits used for the quantisation process), and the maximum sampling rate supported. Sampling rate dictates the range of signal frequencies that can be received, and the resolution defines the noise floor due to quantisation noise. Table 9.1 highlights some examples from the set of available devices.

Table 9.1: RF-ADC comparison across RFSoC devices [89].

Generation ('Gen')	Device	Number of ADCs	ADC Resolution (number of bits)	Maximum Sampling Rate (GSps)
1	ZU28DR	8	12	4.096
	ZU29DR	16	12	2.058
2	ZU39DR	16	12	2.220
3	ZU43DR	4	14	5.0
	ZU46DR[a]	8	14	2.5
		4	14	5.0
	ZU48DR	8	14	5.0
	ZU49DR	16	14	2.5
DFE	ZU65DR	6	14	5.9
	ZU67DR[b]	8	14	2.95
		2	14	5.9

a. The rows are additive, i.e. the ZU46DR contains a total of 12 RF-ADCs.
b. Similar to the above, the ZU67DR contains a total of 10 RF-ADCs.

Notably Gen 1 and 2 RFSoC devices both feature 12-bit RF-ADCs, while Gen 3 and DFE devices have 14-bit RF-ADCs, meaning that they have increased resolution and hence lower quantisation noise. The maximum sampling rate also gradually increases as we progress through the generations. To better understand the differences within a generation we must first investigate the structure, or hierarchy, of the RF Data Converters.

9.3.2. The RF-ADC Tile Hierarchy

The RFDC contains the RF-ADCs and RF-DACs of the RFSoC device, structured in a hierarchy of tiles and blocks. Each RF-ADC and RF-DAC is contained within a block and one, two or four blocks make up a tile, depending on the device. These are referred to as Single, Dual and Quad tiles respectively. High-level illustrations of Quad and Dual tiles are provided in Figures 9.4 and 9.5, respectively.

Gen 1 RFSoC devices can contain either Dual or Quad tiles, but each device contains only one kind of tile. For example, the ZU28DR device has four Dual tiles, each with two blocks, for a total of 8 RF-ADCs. Gen 2 comprises a single RFSoC device, the ZU39DR, which contains four Quad tiles, giving a total of 16 RF-ADCs.

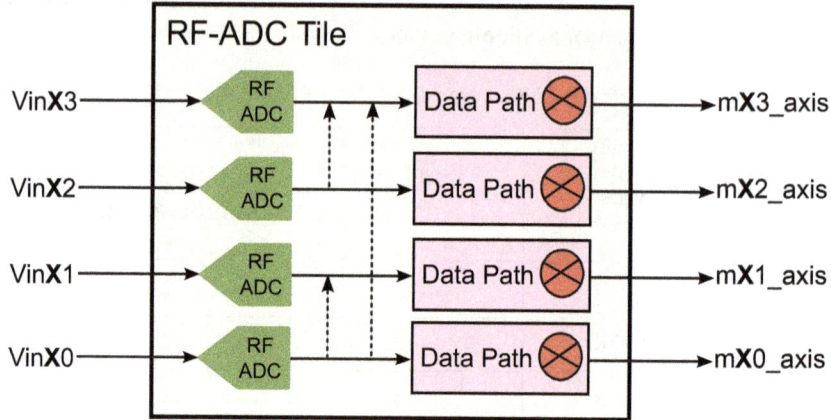

Figure 9.4: Quad RF-ADC tile.

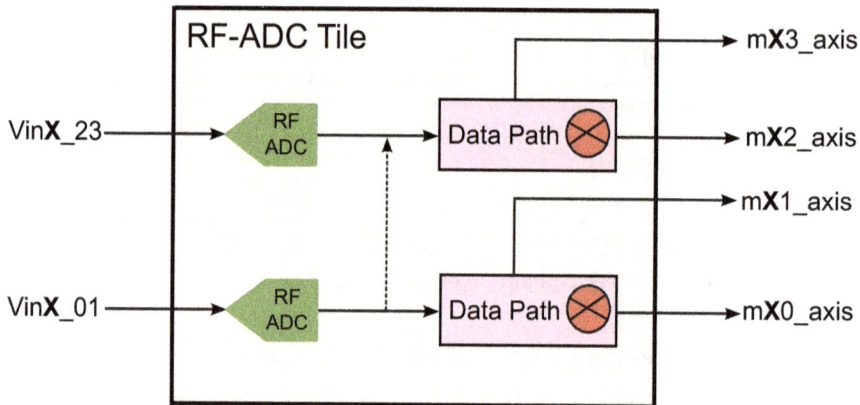

Figure 9.5: Dual RF-ADC tile.

Gen 3 devices feature either Quad or Dual tiles exclusively, or a combination of the two, depending on the device. The ZU48DR and ZU49DR devices are similar to the two devices previously discussed, consisting of four Dual and four Quad tiles, respectively, for a total of 8 and 16 RF-ADCs. The ZU46DR device has a mixed configuration and contains 12 RF-ADCs in total, consisting of two Dual tiles and two Quad tiles. The ZU43DR device contains four Single tiles, each with one RF-ADC, and therefore has a total of 4 RF-ADCs. At the time of writing this is the only device with Single tiles.

DFE devices are similar in structure to Gen 3 devices. At the time of writing, two DFE devices are available: the ZU65DR and the ZU67DR. The former contains three Dual tiles, for a total of 6 RF-ADCs, and the ZU67DR also contains three tiles: two Quad tiles and one Dual tile, and thus a total of 10 RF-ADCs.

9.3.3. Interleaving Factor

Each RF-ADC within the RFSoC device is composed of multiple sub-ADCs which are interleaved together to improve the maximum sampling rate. By using ADC interleaving, an input signal is sampled simultaneously by each sub-ADC (which share a common clock relationship). Therefore, by using m sub-ADCs, the effective sampling rate is increased by a factor of m compared to a single ADC. This is known as the interleaving factor.

For successful interleaving, the clock phase relationship between ADCs is crucial. This is defined as

$$\Theta_n = 2\pi\left(\frac{n-1}{m}\right), \tag{9.3}$$

where Θ_n is the sampling phase for the nth sub-ADC, and $n = 1\ldots m$.

In a Dual tile, each RF-ADC consists of 8 interleaved sub-ADCs whilst a Quad tile RF-ADC consists of 4 sub-ADCs. As a result, within an RFSoC generation, a Dual tile has twice the sampling rate of a Quad tile. For example, referring to Table 9.1 we can see that the ZU46DR contains Quad tiles with a maximum sampling rate of 2.5 GSps and Dual tiles with a maximum sampling rate of 5.0 GSps. Given the varied options available, a suitable target device can be chosen to meet the requirements of the application.

9.3.4. RF-ADC Tile Composition

Regardless of tile configuration, each RF-ADC block within a tile contains a high-performance input buffer alongside a pipeline of components, including the RF-ADC itself; a Quadrature Modulation Correction (QMC) unit, which can correct for any imbalances in the external (analogue) signal paths of a quadrature system; complex mixers for demodulation; and decimation filters for reducing the sampling rate. This RF-ADC processing pipeline is optimised for direct conversion from RF signal frequencies.

Gen 1 Tiles

Both Quad and Dual tiles are structured similarly, and the clock circuitry includes a Phase Lock Loop (PLL) that is driven by an external reference clock. A designer can choose to either enable this PLL, or bypass it with an external sampling clock. All RF-ADCs within the tile share the same clock source and infrastructure.

Each RF-ADC within a tile has an associated Digital Down Converter (DDC), which contains a digital complex mixer, and a programmable decimator. The decimator contains a chain of halfband filters that can be programmed to decimate by an overall factor of 1, 2, 4, or 8. Each RF-ADC within the tile has its own dedicated mixer and decimator.

The Gen 1 Quad tile, shown in Figure 9.6, contains four RF-ADCs arranged as two pairs. Each RF-ADC can be configured either in isolation, or as part of a pair, for real and complex signals, respectively.

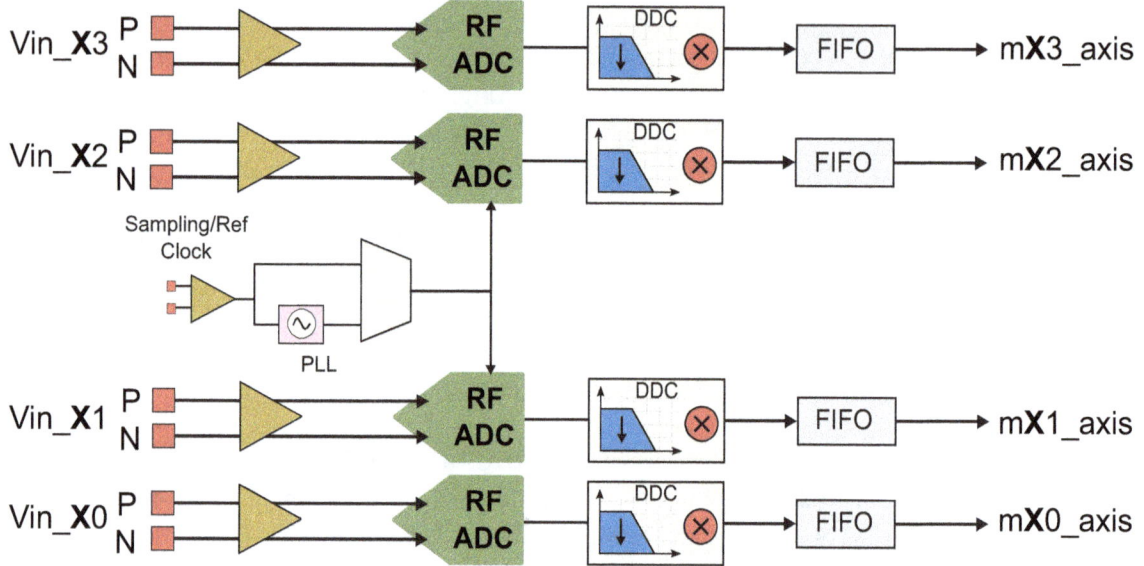

Figure 9.6: Gen 1 and 2 Quad tile.

Figure 9.7 shows a Gen 1 and 2 Dual Tile configuration. This is very similar to the Quad tile, however only two RF-ADCs are present. These can be operated either individually or as a pair, enabling complex signals to be received.

Reception of complex signals can only be achieved when the RF-ADCs within a tile are operated as a pair. Here, the even numbered RF-ADCs are used for I data, and the odd numbered RF-ADCs are used for Q data.

We can now simplify the operation of each RF-ADC block within a tile, and its associated DDC to a single linear pipeline, as shown in Figure 9.8.

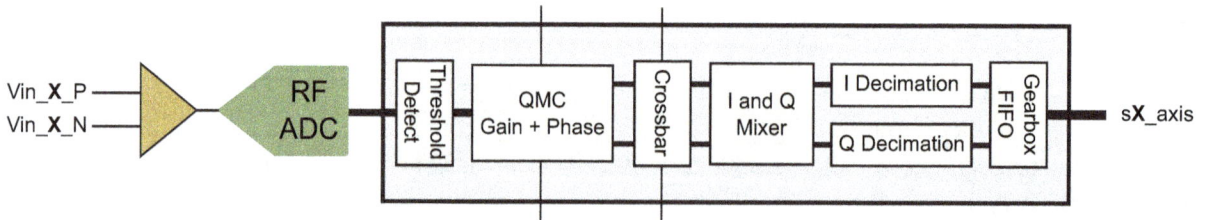

Figure 9.8: Gen 1 and 2 High Level Block Diagram of the RF-ADC Pipeline.

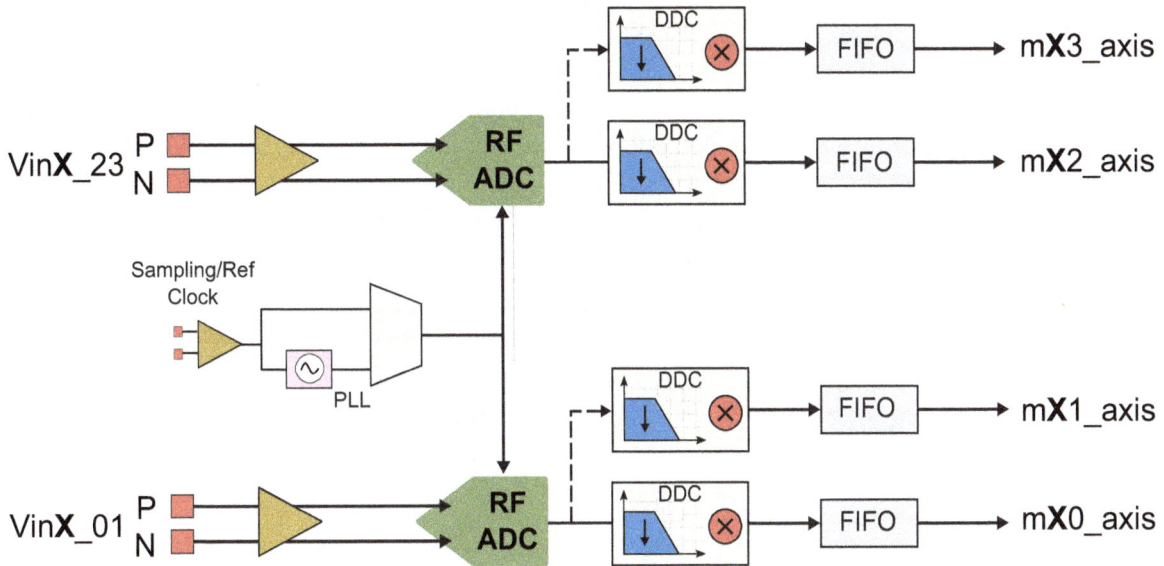

Figure 9.7: Gen 1 Dual tile.

The steps of this pipeline are listed below. The RF-ADCs are highly controllable via the RFDC interface, and as such the user has a significant amount of control at each stage.

1. Following an input buffer to control the rate of signal acquisition, the analogue input signal is sampled by the RF-ADC to convert it to a digital signal.

2. Next, a threshold detector can be employed to detect and record the amplitude levels of the input.

3. If the received signal is complex, the QMC block can be used to compensate for any imbalance between the I and Q signals.

4. The complex mixer then modulates the input signal to baseband to ease processing.

5. The I and Q decimators are capable of decimating the signal, before interfacing with the PL via the gearbox FIFO.

Note: A Quad tile contains four RF-ADC pipelines, whilst a Dual tile contains two.

Gen 3 Tiles

Gen 3 tiles operate in much the same way as Gen 1 and 2 tiles, and their high level functionality is similar to that shown in Figure 9.8. Additionally, Gen 3 tiles feature a Digital Step Attenuator (DSA) that precedes the

RF-ADC. Another key difference is that a more advanced decimation filter chain is provided, to enable decimation by factors of 1x, 2x, 4x, 6x, 8x, 10x, 12x, 20x, 24x and 40x.

The DSA is used in situations where the analogue signal amplitude or power varies over time, allowing the amplitude of the variable signal to be adjusted to optimal values for the RF-ADC. This variable power could be due to a variance within the received signal strength, or interference signals, for instance. Traditionally, the DSA would be an external component (such as a variable gain amplifier), however Gen 3 devices combine this functionality with the internal input buffer, as shown in Figure 9.9.

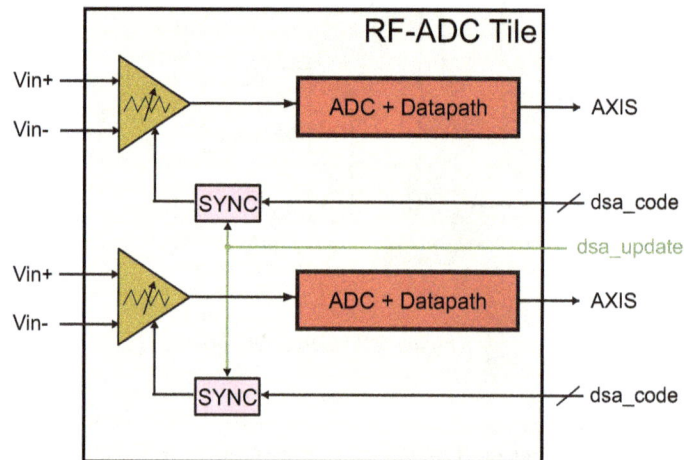

Figure 9.9: Gen 3 DSA, showing that the input buffer is combined with the DSA.

The DSA is also used for the automatic prevention of over-voltage states, where the input signal is too large for the RF-ADC to handle. This comes in two forms: *Over Amplitude* and *Outside Common-Mode Range*. For *Over Amplitude*, where the amplitude of the signal in the input buffer is too large for the RF-ADC, the DSA is triggered by a flag in the buffer and is automatically set. *Outside Common-Mode Range*, as implied, is triggered when the value at the input is either over or under the reliable common-mode range. This can be resolved by disabling the input buffer for its own protection. The location of these over voltage protections in relation to the DSA is shown in Figure 9.10.

Gen 3 RF-ADCs also allow for clock distribution in the RF-ADCs and RF-DACs, as shown in Figure 9.11. In normal operation, each tile operates from an independent tile clock. As discussed for Gen 1, this clock can originate from an external source or be generated by the on-chip PLL. Clock distribution, as the term implies, allows a tile to distribute its tile clock to other adjacent tiles in a chain. Note however that only RF-ADCs 1 and 2 can operate as the source for high frequency clock distribution (with RF-ADC 1 being recommended).

This clock can be forwarded through any number of adjacent tiles, so long as a different clock does not interrupt the chain. In the example provided in Figure 9.11, RF-ADC1 can distribute to RF-ADC0, as well as RF-ADC2 and then RF-ADC3. RF-ADC3 can only be distributed to if RF-ADC2 is also distributed to.

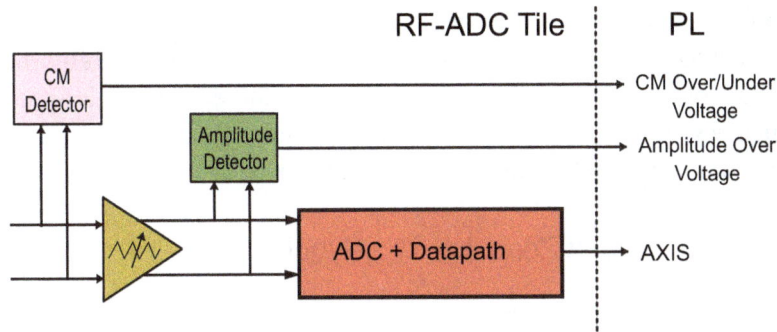

Figure 9.10: Detection of Over Voltage states.

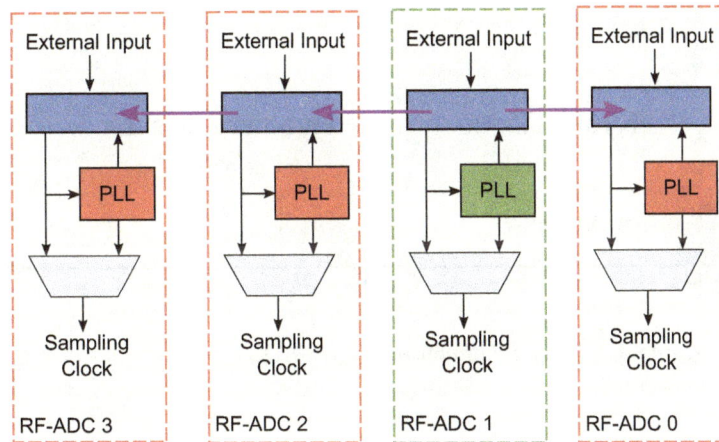

Figure 9.11: Recommended clock distribution configuration (RF-ADC1 distributes to all other RF-ADC tiles).

9.4. RF-ADC Processing Stages: The Digital Complex Mixer

We will now explore two of the key components of the RF-ADC chain in more detail, namely the complex mixer and decimator, to better understand the reconfigurable nature of the RFDCs.

As previously discussed, the RF-ADC is capable of receiving signals at RF frequencies up to several GHz. Once digitised, the signal is demodulated (shifted to baseband), such that it is then centred around 0 Hz. Key to this operation is the **Digital Complex Mixer** found in each RF-ADC. The complex mixer multiplies the incoming signal with sine and cosine waves generated from a Numerically Controlled Oscillator (NCO). This has the effect of shifting the input signal up or down in frequency, with the frequency generated by the NCO determining the direction and extent of this shift.

The Complex Mixer functions on the concept of heterodyning — e.g. when two sine wave signals are mixed (equivalent to multiplication), we obtain two components; one at $f_1 - f_2$ and another at $f_1 + f_2$. The same principle applies when a bandpass signal is demodulated. Generally, one of the resulting signals is useful and falls within the bandwidth of the mixer, whilst the other signal is filtered out by a lowpass filter. Within the RF-ADC, the NCO frequency should ideally match the carrier frequency and phase of the received signal, f_c, so that the differential signal, $f_1 - f_2$ can be found at baseband (centred at exactly 0 Hz), as shown in Figure 9.12.

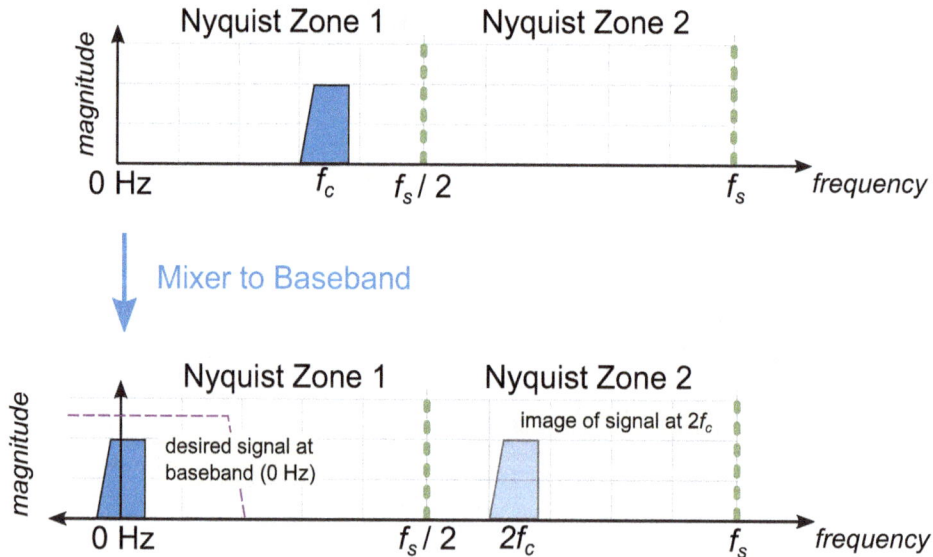

Figure 9.12: Demodulation using the RF-ADC digital complex mixer.

Note: The magnitude of both output signals is half that of the input (modulated) signal at f_c! For details of the mathematical background, please refer to Chapter 7.

The I/Q mixer can be operated in three modes: Coarse, Fine and Bypass. These modes can be selected by the user at run time. The architecture of the mixer is highlighted in Figure 9.13. Note that the bypass mode can also be accessed through the coarse mixer block.

9.4.1. The Coarse Mixer and Bypass Mode

The Coarse Mixer component implements both the Coarse mode and the Bypass mode. The Coarse mixer only allows for a very restricted set of frequencies: $f_s/4$, $f_s/2$, and $-f_s/4$; however it can operate with much lower power than *Fine* mode. At these frequencies, a sine or cosine wave can be represented with a minimal number of samples, e.g. a sine wave of frequency $f_s/4$ can be represented with only 4 samples per cycle, {0, 1, 0, -1}. It is trivial to generate these values in a repeating sequence, and to demodulate an input signal using this signal, obviating the need to store an extensive set of samples using a Lookup Table (LUT), as in *Fine* mode.

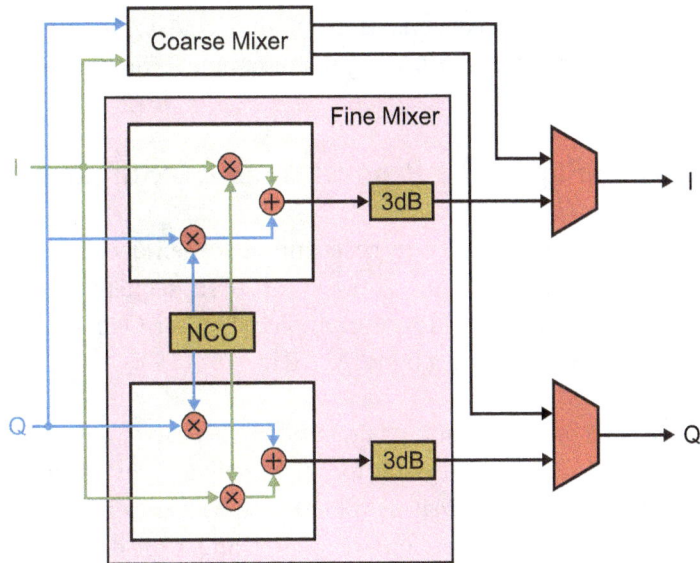

Figure 9.13: Complex mixer architecture.

The restricted set of frequencies produced in *Coarse* mode is used for demodulation due to some unique traits. For example, $f_s/2$ is a rather special mixer frequency, as it has the effect of 'flipping' a spectrum. This can be used to correct the inversion of the spectrum when an even Nyquist Zone has aliased to baseband.

The Coarse mixer component can also be used to implement *Bypass* mode, where the signal path simply bypasses the I/Q Mixing stage. This means that the received signal does not undergo any demodulation, directly passing the high frequency signal to the PL.

9.4.2. The Fine Mixer

In *Fine* mixer mode, the mixer uses an NCO, which can generate any arbitrary frequency between $-f_s/2$ and $f_s/2$. This is done by setting:

- A step size input to the phase accumulator (automatically calculated in the design tools);

- The size of the LUT;

- The desired frequency.

Optional additional parameters can be set, including a phase adjustment and the addition of dithering.

When a signal present in an even Nyquist Zone is aliased to the 1st Nyquist Zone by the RF-ADC, its spectrum is inverted (flipped left-to-right). This can be corrected by setting a negative frequency in the NCO when the desired signal is located in an even Nyquist Zone, thus shifting the negative image to 0 Hz. Similarly, a negative NCO frequency can be used when the desired signal is in an odd Nyquist Zone, to shift the positive image to 0 Hz. Examples of both are provided in Chapter 10, along with example RFSoC receiver configurations.

9.5. RF-ADC Processing Stages: the Programmable Decimator

Once the received signal has been demodulated to baseband, it is then decimated to a lower sampling rate, which is advantageous because it reduces the computational cost of subsequent processing of the signal. For RF decimation, this can mean reducing a bandwidth of multiple GHz (for Gen 1 RFSoCs, up to $f_s/2$ = 2.048 GHz) down to baseband frequencies of around a hundred MHz.

The decimator is implemented as a programmable chain of filters that each implement a sample rate reduction. For Gen 1 and 2 RFSoC devices, the decimator can perform rate reduction by a factor of 1x, 2x, 4x or 8x (where reduction by 1 is a complete bypass of the decimating filters). Gen 3 extends the decimation options to a larger set of integer factors. In the remainder of this section, we first introduce the decimator architecture from Gen 1 and 2 devices, which is simpler and thus easier to understand, before moving on to review the Gen 3 decimator architecture.

9.5.1. Gen 1 and 2 Decimation

Gen 1 and 2 decimation is achieved by a set of halfband filters: FIR0, FIR1 and FIR2. These lowpass filters each decimate by a factor of 2, and are cascaded together to form the selection of decimated outputs. The block diagram of a programmable decimator for a Gen 1 RF-ADC is shown in Figure 9.14. The magnitude responses of the individual cascaded filters (FIR2, FIR1, and FIR0) are shown in Figure 9.15.

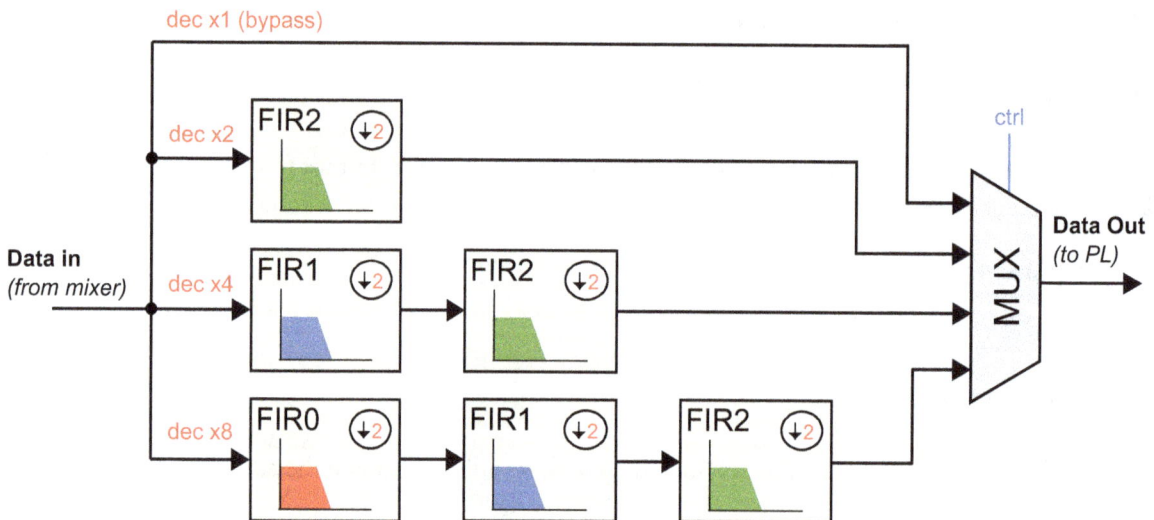

Figure 9.14: Gen 1 and 2 decimation filter chain.

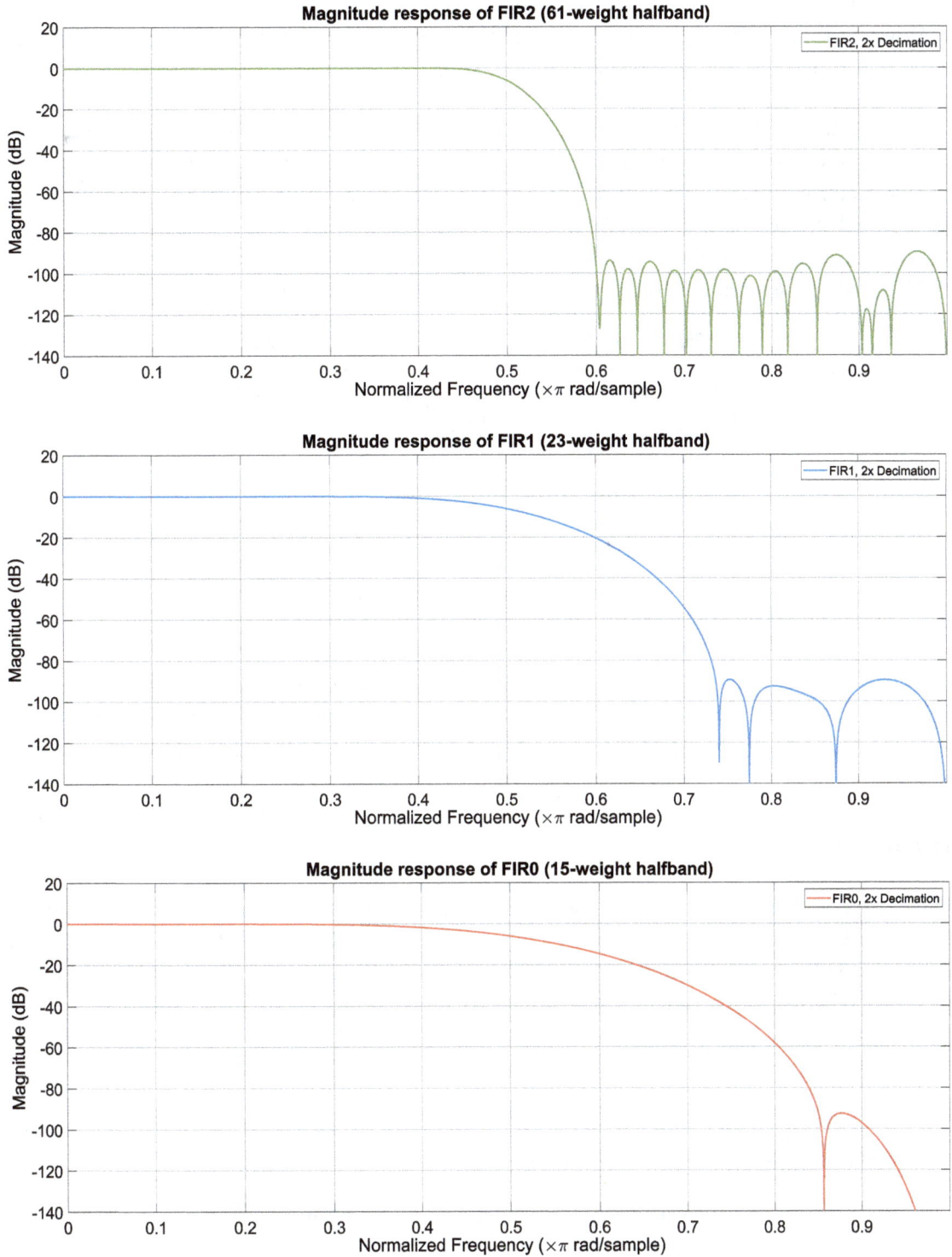

Figure 9.15: Fixed magnitude responses of FIR2 (green), FIR1 (blue), and FIR0 (red).

The design of the FIR2, FIR1 and FIR0 filters is fixed in hardware, and the filter weights are not user-customisable. However, these halfband responses provide the low pass filtering needed to remove frequency components that would otherwise alias upon sample rate reduction.

Notice that FIR2 has the sharpest cut-off. If decimation by 2 is required, only this FIR is used, and as a result the magnitude response of the decimator is equivalent to that of FIR2. When decimation by 4 is required, FIR1 is included subsequent to FIR2; while for decimation by 8, all three filters are used in cascade. FIR1 and FIR0 can have more relaxed responses than FIR2 (requiring fewer coefficients) because the cascaded response still satisfies the requirements, as will be demonstrated by the filter magnitude responses shown over the next few pages. Table 9.2 gives the number of filter weights for each of the halfband filters.

Table 9.2: Filter lengths for Gen 1 and 2 decimation halfband filters.

Filter	Number of Weights[a]
FIR2	61
FIR1	23
FIR0	15

a. As halfband filters, around 50% of the weights are zero-valued!

Figure 9.14 previously confirmed that three different decimation ratios can be achieved (a factor of 2, 4, or 8), by using either a single halfband filter (for x2 decimation), or a cascade of two or three filters, for x4 or x8 decimation, respectively. The remainder of this subsection demonstrates how these decimator responses are composed, beginning with the simplest case.

Decimation by 2

When the decimator is set to decimate by a factor of 2, then only the FIR2 filter is required. This response is shown in Figure 9.16, where the RF-ADC sampling rate is 4 GSps, and therefore f_s = 4 GHz. Notice that this response corresponds to the FIR2 response shown in Figure 9.15. FIR2 passes 80% of the Nyquist bandwidth, i.e. all frequencies between 0 Hz and $0.2f_s$, where f_s is the input sampling rate.

Example:

For our previous example input sampling frequency, f_s = 4 GHz, we achieve a (decimated) output sampling rate from the filter of f_d = 2 GHz, with a corresponding Nyquist bandwidth of $f_d/2$ = 1 GHz. Therefore;

$$80\% \text{ of Nyquist bandwidth} = 80\% \text{ of 1 GHz} = 0 \text{ to 800 MHz} ,$$

or,

$$0.2 f_s = 0.2 \times 4 \text{ GHz} = 800 \text{ MHz},$$

$$40\% \text{ of the output } f_d = 0.4 \times 2 \text{ GHz} = 800 \text{ MHz}.$$

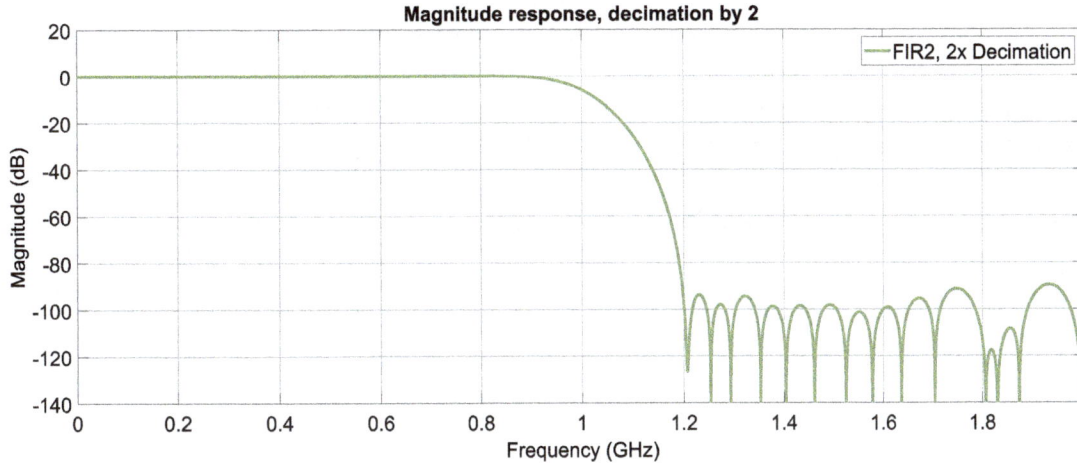

Figure 9.16: Magnitude response for decimation by 2, f_s = 4 GSps.

Decimation by 4

If the FIR1 and FIR2 filters are cascaded to decimate by a total factor of 4, what does the overall filter response look like? First, we need to look at both individual filter responses referenced to the input sampling rate, f_s, which is shown in Figure 9.17. Note that an image of the FIR2 filter response appears above $0.25 f_s$.

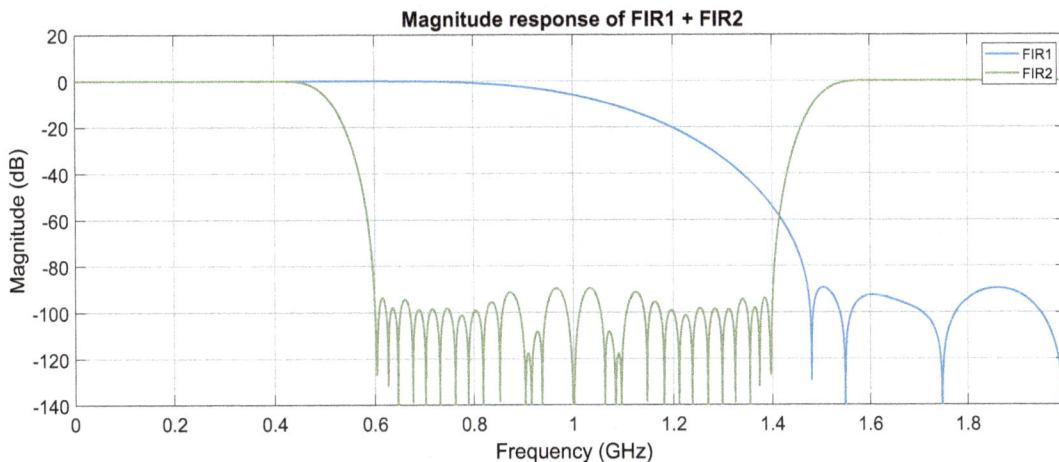

Figure 9.17: Decimation by 4 individual filter responses (FIR2 and FIR1), referenced to input sampling rate.

Notice that the response of FIR1 is more relaxed than that of FIR2. This is because FIR1 is only required to attenuate signals that fall within the image of the FIR2 response, between $0.375f_s$ and $0.5f_s$. Such frequencies would fold into the area of interest covered by the baseband FIR2 response, between 0 Hz to $0.125f_s$.

The response of the two filters in cascade is the superposition of the two individual responses from Figure 9.17, and is shown in Figure 9.18. Notice that there is greatest attenuation between about $0.3f_s$ and $0.35f_s$ (i.e. 1.2 to 1.4 GHz) where both the FIR and FIR2 filters attenuate by at least -20dB. Where both filters have gains of 0dB (in the baseband region), the cascaded response also has a gain of 0dB, as is required to preserve the signal of interest; all other frequencies are attenuated by at least ~90dB.

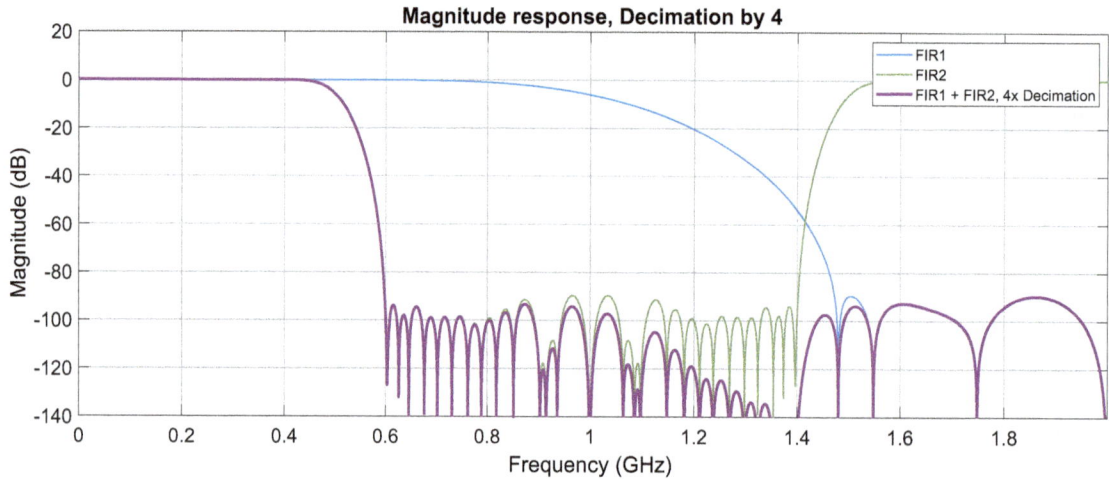

Figure 9.18: Decimation by 4 cascaded response (emboldened).

Decimation by 8

For a decimation factor of 8, the FIR0, FIR1 and FIR2 filters are all cascaded together. All three individual filter responses are plotted in Figure 9.19, referenced to the input sampling rate, f_s. An image of the FIR1 filter response appears above $0.25f_s$ (1 GHz) and similarly, there are also multiple images of the FIR2 filter. Therefore, the FIR0 filter is used to attenuate frequencies which fall within these images, once again preventing folding into the baseband area of interest.

In this case, the three filter responses combine to form the cascaded response shown in Figure 9.20. As with the decimation by 4 design, the stopband attenuation is at least ~90dB. The most prominent lobes in the stopband occur at those frequencies where at least one of the filters (or their images) has a gain of 0dB. For instance, the region between about $0.2f_s$ and $0.3f_s$ (i.e. between 0.8 and 1.2 GHz), where an FIR2 image can be seen, exhibits higher gain than some other parts of the stopband of the filter response.

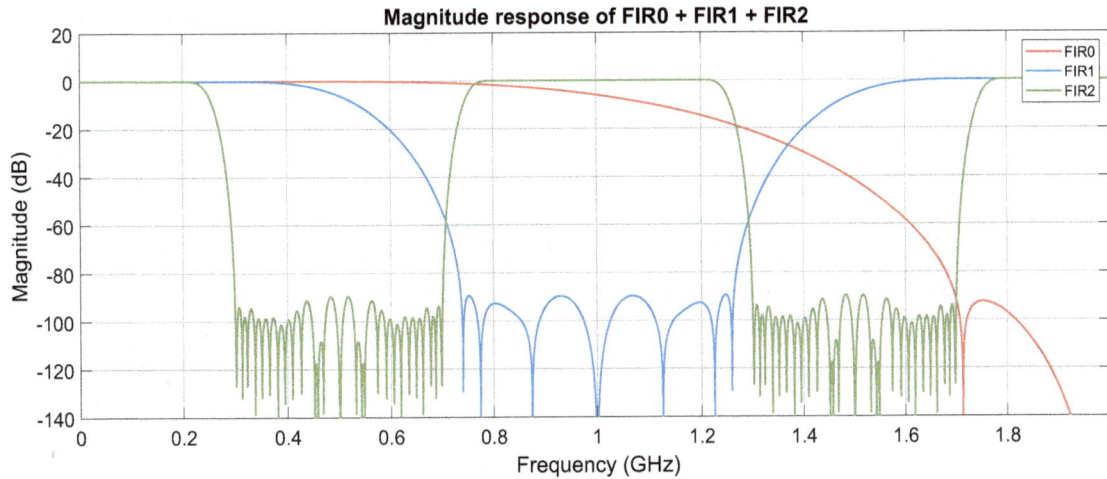

Figure 9.19: Decimation by 8 individual filter responses (FIR2, FIR1 and FIR0), referenced to input sampling rate.

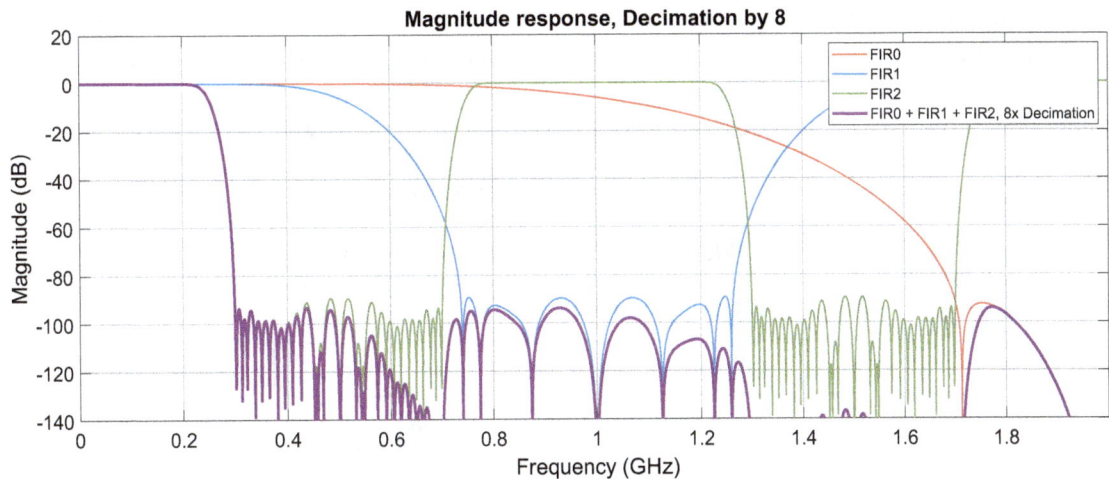

Figure 9.20: Decimation by 8 cascaded response (emboldened).

9.5.2. Gen 3 Decimation

Gen 3 RFSoCs expand upon the available decimation rates in the DDC. On these devices, decimation by a factor of 1x (bypass), 2x, 3x, 4x, 5x, 6x, 8x, 10x, 12x, 16x, 20x, 24x and 40x is possible, meaning that a more sophisticated filter chain is required than with a Gen 1 or 2 device. The block diagram for this decimation filter chain is shown as Figure 9.21.

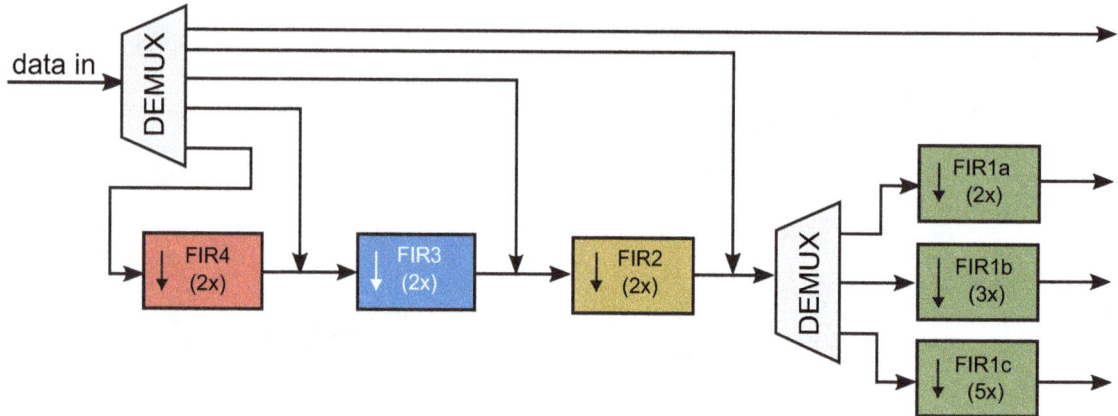

Figure 9.21: Gen 3 decimation filter chain.

Here, we see four stages of decimation filters cascaded together, with each able to be bypassed independently. Notably, the final stage of the chain, FIR1, contains three decimation filters — FIR1a (2x), FIR1b (3x) and FIR1c (5x) — and only one of these can be enabled for any specific configuration. The FIR2, FIR3 and FIR4 blocks each have a decimation factor of 2x, corresponding to the FIR filters of the Gen 1 and 2 RFSoCs.

The number of weights in each filter are as given in Table 9.3. Note that FIR3 and FIR4 are identical filters.

Table 9.3: Filter lengths for Gen 3 decimation filters

Filter	Number of Weights
FIR1a	59
FIR1b	89
FIR1c	143
FIR2	23
FIR3	15
FIR4	15

As in the simpler Gen 1 and 2 decimators, each of the decimate-by-2 filters is a halfband filter, meaning that approximately half of the weights of filters FIR 1a, FIR2, FIR3, and FIR4 are exactly zero-valued. We can see that the selectable FIR1 filters have a notably larger number of weights than the preceding filters, giving a much tighter response. Where appropriate, a single FIR1 filter is used to provide the entire decimation factor, meaning that the overall decimation magnitude response is equivalent to the FIR1 filter. To provide other decimation factors in addition to those available with a single FIR1 filter (2x, 3x, 5x), one or more of FIR2, FIR3 and FIR4 are enabled in a filter chain.

Example:

To illustrate how different decimation ratios are achieved, the configurations for several different decimation factors are provided in Table 9.4. Note that this is not an exhaustive list.

Table 9.4: Examples of FIR filters required for selected decimation factors.

Decimation Factor	Enabled Filters	Decimation Chain
2x	FIR1a	2
4x	FIR2 + FIR1a	2 x 2 = 4
6x	FIR2 + FIR1b	2 x 3 = 6
10x	FIR2 + FIR1c	2 x 5 = 10
20x	FIR3 + FIR2 + FIR1c	2 x 2 x 5 = 20
24x	FIR4 + FIR3 + FIR2 + FIR1b	2 x 2 x 2 x 3 = 24

We will now examine the magnitude response of a Gen 3 decimation factor that is not supported by Gen 1 devices. Figure 9.22 shows the magnitude response of a decimation by 24x, using FIR4, FIR3, FIR2 and FIR1b in a filter cascade. As before, an input RF-ADC sampling rate of 4 GSps is chosen for this decimation example.

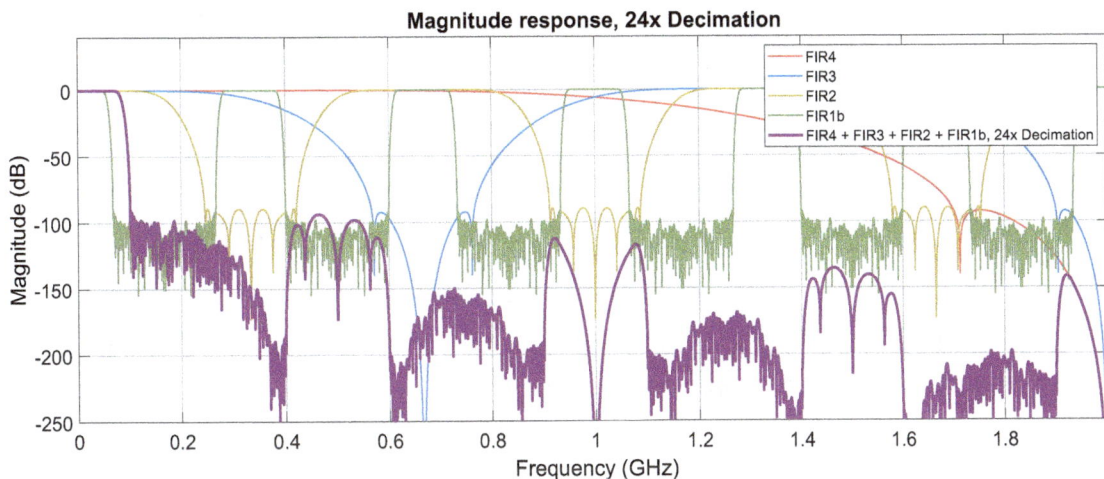

Figure 9.22: Gen 3 decimation by 24 magnitude response of cascaded filters.

Similar to previous examples, it is possible to recognise images of the filter responses that operate at lower sampling rates. Due to knowledge of the positions of these images, the filters that operate at the highest rates can be designed with relaxed specifications, thus optimising the amount of filter computation required.

9.6. Principles of RF-ADC Operation

The RFSoC is a powerful device, able to serve several different RF receiver architectures due to its support for both real and complex RF-ADC inputs. We will now examine architectures for both cases in more detail.

9.6.1. Real Signal Architectures

As discussed previously in Section 9.2, suitable amplification and analogue filtering (lowpass or bandpass, for Nyquist Zones 1 and 2, respectively) are required prior to the RF-ADC.

If the received RF signal arrives directly from the antenna via analogue signal conditioning (filters, amplifiers, etc.) on a single wire then it can be considered a *real* input signal. Additionally, if the received RF signal is mixed down to IF using a real analogue oscillator, then it remains real at the point of reaching the RF-ADC. Figure 9.23 illustrates the Direct-RF architecture, followed by an RF-to-IF intermediate stage architecture.

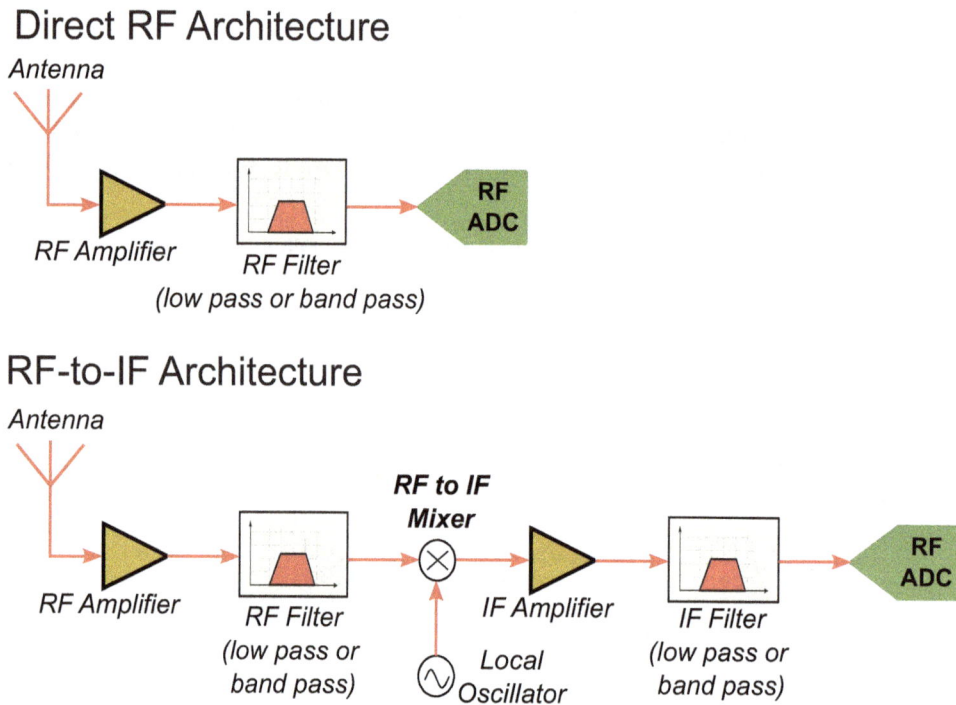

Direct RF Architecture

Antenna

RF Amplifier   RF Filter
               (low pass or band pass)

RF ADC

RF-to-IF Architecture

Antenna

RF Amplifier   RF Filter   **RF to IF Mixer**   IF Amplifier   IF Filter
               (low pass or   *(low pass or*
               band pass)   Local Oscillator   *band pass)*

RF ADC

Figure 9.23: Two real signal receiver architectures: (upper) direct architecture, and (lower) intermediate stage architecture.

In the first scenario, the RF signal is directly digitised by the RF-ADC, meaning that it does not undergo demodulation to either baseband, or to an IF frequency, prior to digitisation. All of the demodulation can therefore take place digitally. By implication, the RF bandpass signal must reside within the RF input range of the RF-ADC (i.e. within either Nyquist Zone 1 or 2). For example, in the case of a Gen 1 RFSoC RF-ADC, RF signals with centre frequencies up to almost 4 GHz could be received.

If the received RF signal is above the 2nd Nyquist Zone, then it may be desirable to first demodulate it to a suitable IF frequency. This can be achieved by an RF-to-IF mixer, as seen in the lower diagram within Figure 9.23. If required, an additional IF mixer stage could also be used, first from RF to a 'high IF' frequency, and secondly from the 'high IF' to a 'low IF' frequency. Regardless of whether a single or a two-stage architecture is used, the eventual IF frequency must be within the operating range of the RF-ADC (again, 0 to 4 GHz for a Gen 1 RFSoC device). Note that analogue bandpass filtering is required after the RF-IF mixing stage, as the mixing process will also generate unwanted spectral components at higher frequencies.

Example:

Let's provide two simple scenarios involving Wi-Fi signals, where the two architectures may be used:

- The direct-RF architecture could support a 2.4 GHz Wi-Fi signal, as it falls within the operating bandwidth of the RF-ADC; however, it could not receive a 5.8 GHz Wi-Fi signal as this carrier frequency is too high.

- The 5.8 GHz Wi-Fi signal could instead be received by the second model. An IF-mixing stage would be used to shift the signal to a frequency within the operating range of the RF-ADC (for instance, centred at 1 GHz) before digitising using the RF-ADC.

9.6.2. Complex Signal Architecture

If a received RF signal is demodulated using an analogue I/Q (complex) mixer, it becomes a complex analogue signal, i.e. we obtain two signal components, one demodulated with a cosine (I) and the other with a sine (Q). Together these form a complex signal, and therefore a pair of RF-ADCs is required. The complex format permits higher bandwidth signals to be received. The resulting architecture can be seen in Figure 9.24.

In this architecture, the RF signal is demodulated to baseband using a complex mixer (the local oscillator generates two outputs separated by 90 degrees), which creates two signals, denoted as I and Q (*In Phase* and *Quadrature*). The I and Q phases then separately pass through amplifiers and low pass filters (which remove unwanted high frequency components prior to digitalisation). There are however a number of issues to be aware of when considering the I/Q mixing stage. These are:

- The local oscillator may not produce outputs that are separated by exactly 90 degrees. The mathematics of demodulation relies upon orthogonality between the I and Q phases. This means that a small portion of the I phase may creep into the Q phase, and vice versa.

- The gain applied to the I and Q phases may differ slightly. This effect can be mitigated by adopting coupled amplifiers or a dual amplifier, rather than two completely separate amplifiers.

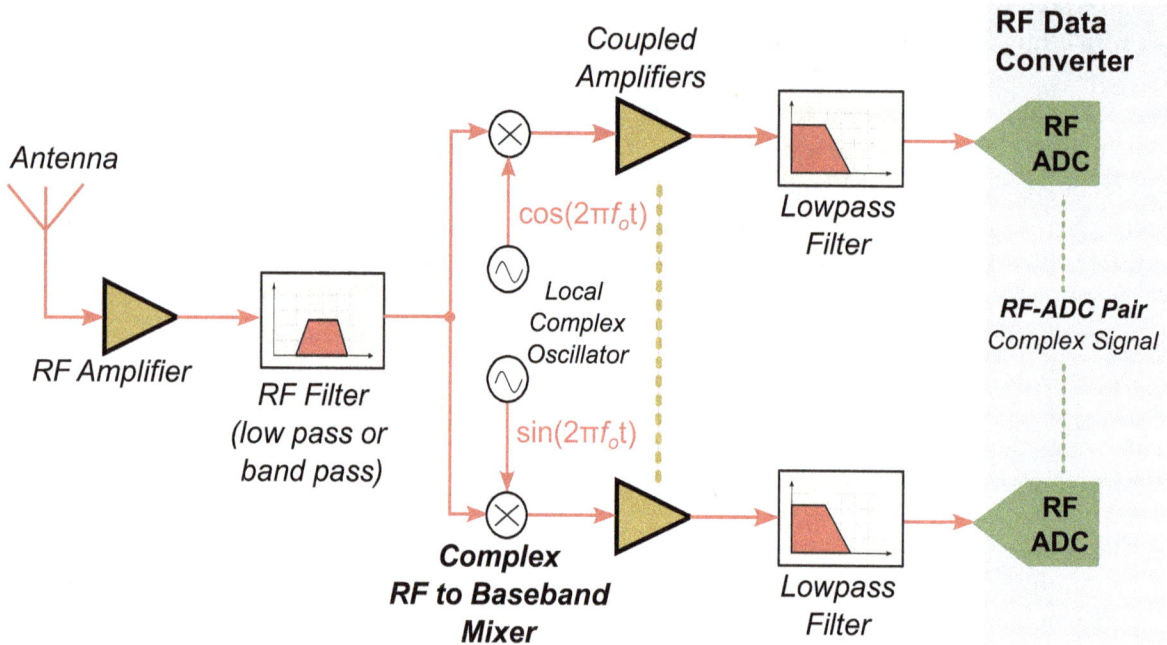

Figure 9.24: Complex receiver architecture.

- The responses of the low pass filters applied to the I and Q phases may differ slightly, due to component tolerances, as well as environmental and ageing effects.

- The signals in the I and Q paths may experience different voltage offsets.

Each RF-ADC includes a QMC block that can be used to correct for the above effects, which would otherwise be detrimental to signal integrity. The designer must incorporate their own logic to detect errors and generate the correction signals.

Given the difficulties laid out above, why choose the I/Q mixing stage architecture? One key reason is that the I/Q complex format can represent signals with twice the bandwidth of a real input. Therefore, a signal with a bandwidth between $f_s/2$ and f_s can be directly digitised. Using our earlier example, for a Gen 1 RFSoC, $f_s = 4$ GHz, we could directly digitise a signal with a bandwidth of 3 GHz.

This architecture would therefore be useful for next generation communications standards (such as 5G New Radio and O-RAN, discussed in a later Chapter 17) supporting extremely wide bandwidth signals such as in the mmWave bands (~30 GHz to 300 GHz), or for radar applications — for instance the Short-Range Radar band which offers up to 4 GHz of bandwidth. Another possible application is in instrumentation, for example spectrum analysis.

9.6.3. Operating Modes

To support these differing radio architectures, the RF-ADC tiles can be operated in two different modes: Real-to-Complex, and Complex-to-Complex.

In Real-to-Complex (R2C) mode, an RF channel is input to the RFDC as a real signal, and is then mixed with the output of a complex NCO to form a complex I/Q output. For the Dual RF-ADC tile, the upper and lower halves of the tile can each handle one input RF signal in this manner. Similarly, each RF-ADC in the Quad tile can support one input signal, giving four channels in total.

In Complex-to-Complex (C2C) mode, an RF channel is input to the RF Data Converter as a complex signal (I/Q), and is then mixed using a complex oscillator to form a complex output (also I/Q). In this configuration, two RF channels are required for an RF signal, one for the analogue I input and another for the analogue Q input. Therefore, a Dual RF-ADC tile can only support one RF signal, and a Quad RF-ADC can support two.

9.6.4. Reception of Multiple Bands

The primary application area for the RFSoC device is wireless communications, some examples of which are discussed in detail in later chapters. In some circumstances, it may be desirable to combine the receive channels for more than one standard, such that they can share an input from a single RF-ADC. The RF-ADCs support this use case through Multi-Band reception.

The RF-ADC can be operated in 'multi-band mode', where the input analogue signal consists of signals mixed to different carrier frequencies. A single RF-ADC tile can support 2 or 4 frequency bands in a multi-band configuration (for Dual and Quad tiles respectively). The support for multi-band operation within a single tile recognises that the first stage (RF-ADC sampling) is common for all bands. As a result, one signal with a wide bandwidth, containing the multiple signals of interest, can be digitised. Following this, distinct stages of I/Q mixing and decimation are required for each of the bands, according to their centre frequencies and bandwidths. Therefore, separate I/Q mixers and decimators are needed for each band. If required, different decimation ratios can be applied for each band, and samples passed to the PL at different rates.

Example:

A Quad RF-ADC tile can be used to receive four different bands within the same Nyquist Zone. The single input signal contains signals mixed to four carrier frequencies within the 1st Nyquist Zone: 450 MHz, 750 MHz, 1.5 GHz and 1.7 GHz. The spectrum of this input signal is illustrated in Figure 9.25.

Once digitised by the RF-ADC, if Multi-Band mode is enabled, the signal is routed to all four digital complex mixers in the tile. These mixers can be configured independently, allowing each to extract a different band from the combined input signal. For this example, the four mixers would be set to match the initial bands at 450 MHz, 750 MHz, 1.5 GHz and 1.7 GHz respectively. Once demodulated and decimated, the bands can be

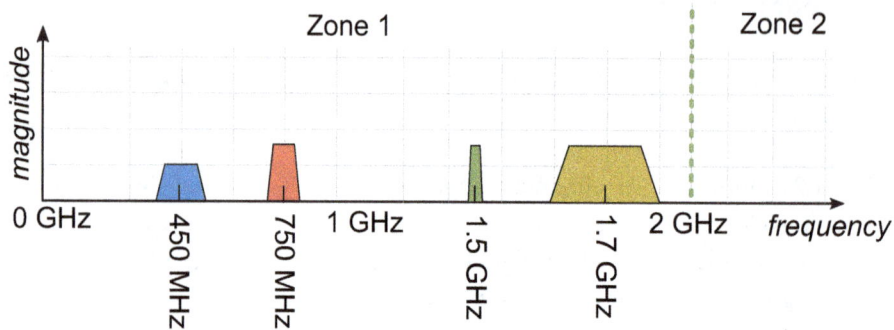

Figure 9.25: Four bands contained within a single input signal.

forwarded separately to the PL for further processing. The operation of a single RF-ADC tile in multi-band mode is shown in Figure 9.26.

As previously discussed, with a real architecture, one RF-ADC in the tile would be used. As in this example, with a complex architecture two RF-ADCs are used: one for the I signal and the other for the Q signal. Each band can be decimated independently following the mixer stage, allowing each band to be passed to the PL at different clock rates.

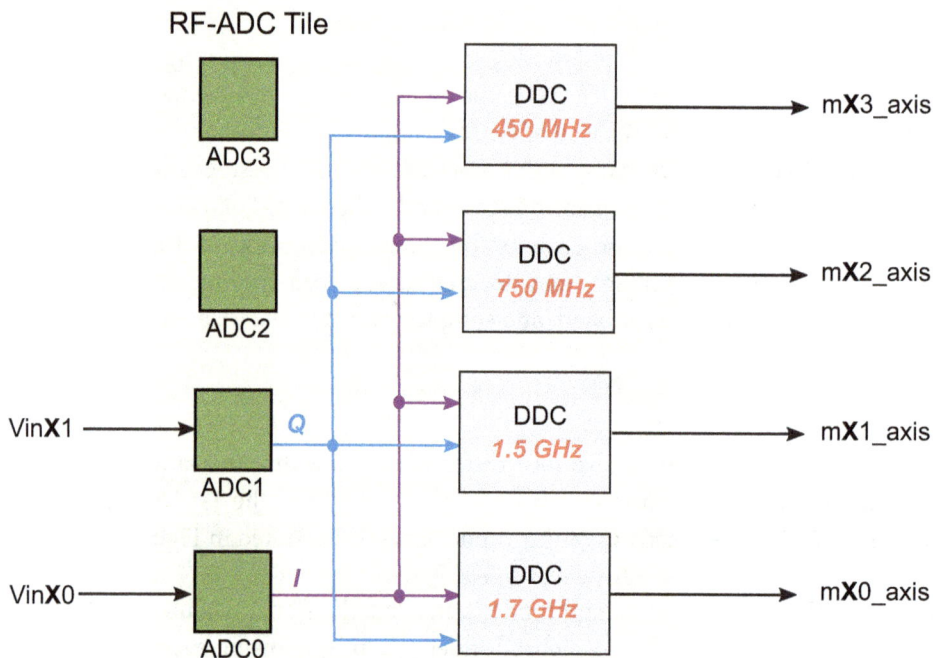

Figure 9.26: Quad tile configured to extract four signal components from a multiband input signal.

9.6.5. Multi-Tile Synchronisation

The RFDC contains flexible clocking and data interfaces to enable a wide variety of applications. Natively, each tile is clocked independently, and the latency within a tile is consistent. However, for some applications, multiple tiles or even multiple RFSoC devices are required, and therefore it is highly desirable to match latencies across tiles. This can be enabled through the multi-tile synchronisation option, distributing a SYSREF clock across the desired tiles. ADC Tile 0 is treated as the 'manager tile' to which all others are synchronised, and as such it must be enabled for multi-tile synchronisation.

9.6.6. Nyquist Zones in Real World Scenarios

As previously discussed, an ideal SDR should be designed to be maximally flexible, and as such it should offer software programmable operation over the widest possible bandwidth. This would imply applying an analogue filter across one entire Nyquist Zone, with all band selection being accomplished digitally. An ideal analogue filter would have a 'brick wall' response, similar to that shown in Figure 4.18 on page 92 in the context of digital filters. with immediate transitions between passband and stopband located exactly at the boundaries of the Nyquist Zones. However, in practical terms, this is impossible to achieve.

In practice, filters have one or more transition bands, i.e. a range of frequencies with sliding gain between the passband and stopband. The aim is therefore to limit the maximum signal bandwidth to slightly less than one Nyquist Zone. Consequently, the effects of attenuation and aliasing are particularly pronounced at the edges of each Nyquist Zone, as shown in Figure 9.27 for the RF-ADC sampling rate of f_s = 4 GSps. For this reason, signals located very close to the Nyquist Zone boundaries should be avoided via frequency planning, a technique which is discussed in detail in Chapter 12.

To give a simple example, let's consider that the received signal falls within the edge region of a Nyquist Zone. This is undesirable, but can be readily addressed by adjusting the RF-ADC sampling frequency. In the example depicted in Figure 9.27, the band from 1.9 - 2.0 GHz falls within the edge region of the 1st Nyquist Zone, but if the RF-ADC sampling frequency was reduced from 4 GSps to 3 GSps, the signal would instead be located in the central part of the 2nd Nyquist Zone. .

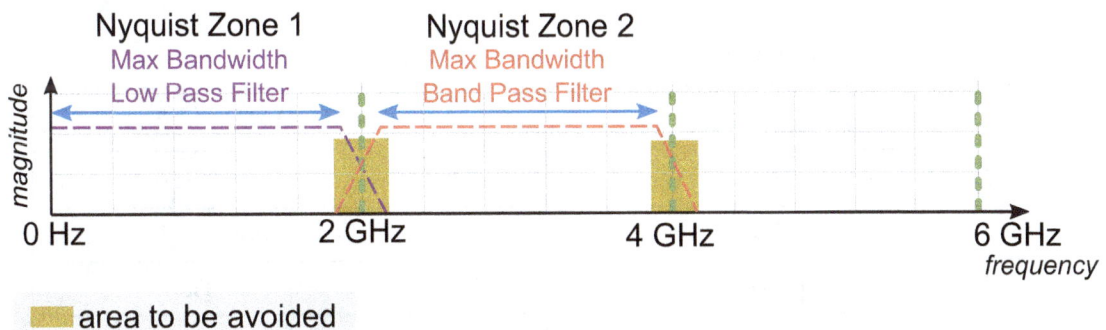

Figure 9.27: Real world Nyquist Zone filter design (f_s = 4 GSps).

The analogue low pass or bandpass filter prior to the RF-ADC is an external component, and therefore is an independent design choice linked to the application or use case, and signal frequency band(s). Development boards may include some pre-packaged analogue filter support that complements the RF-ADCs on the RFSoC board.

9.7. Design Tool Support

The hardened RF-ADCs can be added to any RFSoC Vivado IP Integrator [61], [65] project by including the **RF Data Converter** IP core, which can be seen in Figure 9.28. This core acts as a configurable wrapper around both the RF-ADCs and RF-DACs, which can be customised per tile.

Each RF-ADC and RF-DAC can be enabled and configured individually. The pins available on the GUI will update to match the enabled RF-ADCs and RF-DACs. In addition, the customisation options differ depending on the chosen RFSoC device. The available tiles and blocks for instance will depend on whether the device has Dual or Quad tiles.

For the IP core, the RFDC tiles are named following their schematic bank allocation. For programming convenience, through the software driver documentation they are also referred to as Tiles 0 to 4. As Vivado uses both of these terminologies, Table 9.5 associates a Tile with both a bank allocation name and a software driver name.

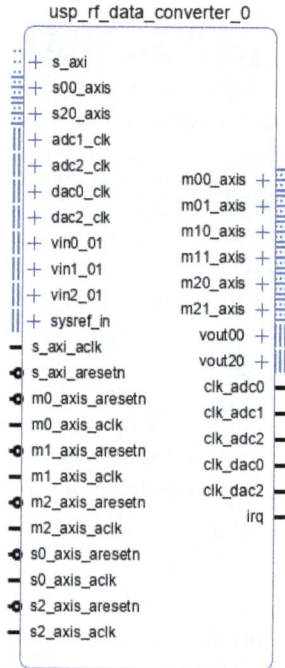

Figure 9.28: RF Data Converter IP core (seen in IP Integrator).

Table 9.5: RF Data Converter Tile Naming Convention.

RF-ADC		RF-DAC	
Tile 224	ADC Tile 0	Tile 228	DAC Tile 0
Tile 225	ADC Tile 1	Tile 229	DAC Tile 1
Tile 226	ADC Tile 2	Tile 230	DAC Tile 2
Tile 227	ADC Tile 3	Tile 231	DAC Tile 3

Example:

In Figure 9.28, the RFDC is configured for the ZCU208 RFSoC board, using the ZU48DR device part. Here we have three RF-ADCs enabled (Tile 224 ADC 0, Tile 225 ADC 0 and Tile 226 ADC 0). Due to clock sharing, clocks are available from the IP Core for tiles 225 and 226, as *adc1_clk* and *adc2_clk* respectively. Here we use the bank allocation tile names within the GUI to refer to the enabled RF-ADCs and the software driver names used for the clocks. Similarly, we have enabled two RF-DAC tiles, Tile 228 DAC 0 and Tile 230 DAC 0, which are also represented on the GUI by their clock pins *dac0_clk* and *dac2_clk*.

9.7.1. RF-ADC Tile Configuration

When configuring an RF-ADC in IP Integrator, a number of options are available to configure the tile as a whole and to enable and configure individual RF-ADCs.

Note: The majority of the configuration options below can be initially set through IP Integrator and then dynamically reconfigured by the data converter Application Programming Interface (API), either in C or using the RFDC PYNQ Python Driver [39]. It is important that any desired tiles are first enabled and connected properly in the IP Integrator design.

Common RF-ADC Tile Configuration Options

This first group of options is found towards the top of the tile configuration page, and contains parameters common to every RF-ADC within a tile. These options can be seen in Figure 9.29.

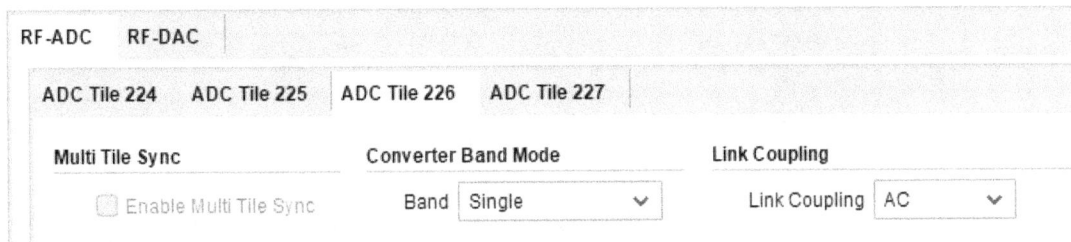

Figure 9.29: RFDC IP core tile configuration GUI, common RF-ADC tile configuration options.

There are three options of note here:

- **Multi Tile Sync** — synchronises the clocks of grouped RF-ADCs across tiles.

- **Link Coupling** — determines if the tile is AC or DC coupled. Most applications will be AC coupled.

- **Converter Band Mode** — determines whether the tile is operating in single or multi-band mode.

Individual RF-ADC Configuration Options

The second group of options are configurable individually for each RF-ADC block within the tile. These options are further broken down into four categories: generic options, data settings, mixer settings and analogue settings. The RF Data Converter GUI for these options is shown in Figure 9.30.

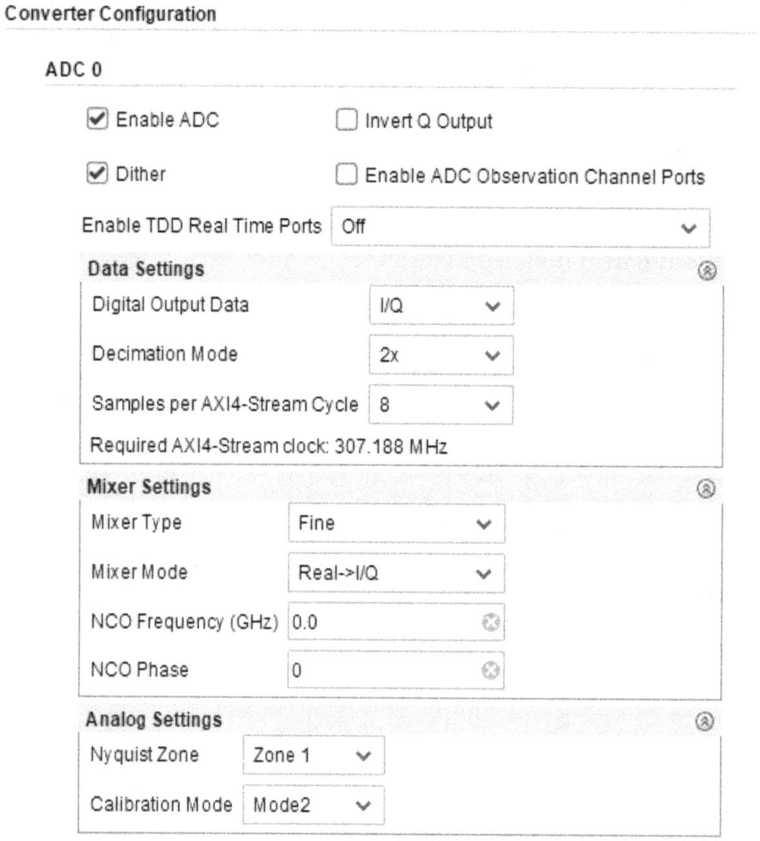

Figure 9.30: RFDC IP core tile configuration GUI, individual RF-ADC configuration options.

General Settings:

- **Enable ADC** — Each RF-ADC can be enabled individually.

- **Invert Q Output** — Configurable only when I/Q output data is selected and the fine mixer is enabled. Inverts the output of the Q channel.

- **Dither** — Adds a small amount of noise to the signal to improve spectral purity. Should be enabled unless the sample is under 0.75x the RF-ADC maximum sample rate.

- **Bypass Background Calibration (Gen 1 and 2)** — If enabled, the background calibration logic is implemented in the IP core. Only available in Real-to-Real mode.

- **Enable TDD Real Time Ports (Gen 3)** — Adds the *tdd_mode* port to the IP, which provides power savings by powering down sections of the RF-ADC.

- **Enable ADC Observation Channel Ports (Gen 3)** — Adds the *tdd_obs* port to the IP. Provides an observation port for the specific RF-ADC block.

Data Settings:

- **Digital Output Data** — Configures the data of the RF-ADC as Real or Complex. For complex data, paired RF-ADCs must be enabled.

- **Decimation Mode** — Sets the decimation value of the DDC within the RF-ADC. Valid values are:

  - *Gen 1 and 2:* 1x (bypass), 2x, 4x, 8x.

  - *Gen 3:* 1x (bypass), 2x, 3x, 4x, 5x, 6x, 8x, 10x, 12x, 16x, 20x, 24x and 40x.

- **Samples per AXI4-Stream Word** — Configures the number of samples to be output per AXI-stream sample (1 to 12).

Mixer Settings:

- **Mixer Type** — choose which mixer type to use: bypass, coarse or fine. The available options are dependent on which Digital Output Data option is selected.

- **Mixer Mode** — Select between real-to-real, real-to-complex or complex-to-complex output.

- **Coarse Mixer Frequency** — Sets the frequency of the Coarse Mixer { $f_s/2$, $f_s/4$, $-f_s/4$ }.

- **Fine Mixer Frequency** — Sets the frequency of the Fine Mixer between -10 and 10 GHz.

- **Fine Mixer Phase** — Sets the phase of the Fine Mixer, between -180 and 180 degrees.

Analogue Settings:

- **Nyquist Zone** — Select between even or odd Nyquist zone operation.

- **Calibration Mode** — Selects between calibration optimization strategies. Mode 1 is best for signals within 10% of the Nyquist Zone boundaries i.e. $f_s/2$ ±10%, f_s ±10%, $3f_s/2$ ±10% etc. Mode 2 is optimal for signals outside of this range. Gen 3 introduces 'Autocal' mode which is suitable for all input frequencies.

9.7.2. RF-ADC Clocking Configuration

The system clocking tab within the RF Data Converter GUI enables clocking settings for each tile to be configured individually. Enabled tiles are configurable, with non-enabled tiles greyed out. Figure 9.31 shows this clocking tab, with configuration for the ZCU208 RFSoC board.

AXI4-Lite Interface Configuration

AXI4-Lite Clock (MHz) 100.0

Tile Clocking Settings

Tile	Sampling Rate (G SPS)	Max Fs (G SPS)	PLL	Reference Clock (MHz)	PLL Ref Clock (MHz)	Ref Clock Divider	Fabric Clock (MHz)	Clock Out (MHz)	Clock Source	Distribute Clock
ADC 224	4.915	5.000	☑	491.500 ▾	491.5	1 ▾	307.188	307.188 ▾	ADC225 ▾	Off ▾
ADC 225	4.915	5.000	☑	491.500 ▾	491.5	1 ▾	307.188	307.188 ▾	ADC225 ▾	Input Refclk ▾
ADC 226	4.915	5.000	☑	491.500 ▾	491.5	1 ▾	307.188	307.188 ▾	ADC226 ▾	Off ▾
ADC 227	2.0	5.000		2000.000 ▾	-	1 ▾	0.0	15.625 ▾	ADC227 ▾	Off ▾

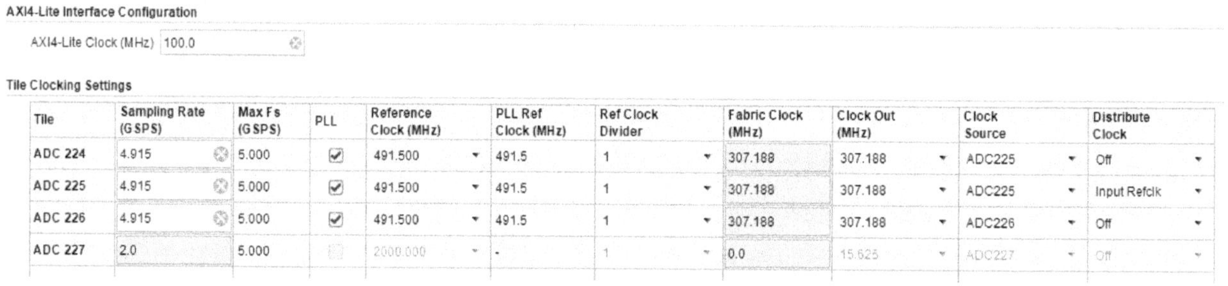

Figure 9.31: RF-ADC Clocking Configuration, set for the ZCU208 RFSoC board.

The available settings can be summarised as:

- **Sampling rate** — The individual tile sampling rate in GHz, maximum value is device dependent.

- **Max Sampling Rate** (f_s) — Indicates the maximum sampling rate of the chosen device. Cannot be changed by the designer.

- **PLL Enable** — Check box to enable the PLL of the tile. If disabled, the tile PLL is bypassed. If enabled, the tile generates its own clock based on a provided reference clock.

- **Reference Clock** — Set the input clock to the tile. Must be an integer multiple of the sampling rate. The drop-down offers options based on the selected sampling rate. This clock drives the tile PLL or the data converters directly based on PLL enable settings.

- **PLL Ref Clock** — Cannot be changed by the designer; displays the reference clock frequency in the PLL. Affected by reference clock value and reference clock divider. Only given if PLL is enabled.

- **Ref Clock Divider** — If PLL is enabled, sets division of the reference clock. For general applications should not be changed.

- **Fabric Clock** — Displays the minimum required clock to drive the Data Converter.

- **Clock Out** — Optional output clock from the tile that can be used to drive the AXI Stream (AXIS) clock.

- **Clock Source (Gen 3)** — Configure which tile clock will drive each tile. Clocks can only be distributed to and from adjacent tiles.

- **Distribute Clock (Gen 3)** — Configure if the selected tile will distribute its clock. Options are off (no distribution), input reference clock (forward the tile reference clock) and output PLL clock (sampling clock generated by on tile PLL). For Gen 1 and 2 devices, clock distribution options are not available.

9.8. Chapter Summary

This chapter has introduced the RF-ADC in detail as a crucial element of the RFSoC device, exploring in depth its key components and operating principles, including the decimation chains and complex mixer components that are included alongside each RF-ADC. We also summarised the available RFSoC devices, examining the number of RF-ADCs available on each device, maximum sample rates and ADC bit depth. RFSoC tile layout was then reviewed, introducing Quad and Dual tile configurations for Gen 1, 2 and 3 RFSoCs, noting the additional features of Gen 3 tiles. Finally, we examined the RF Data Converter IP Core within Vivado IP Integrator, and noted how to use and customise the RF-ADC tiles.

Although the current chapter has covered the RF-ADCs available on an RFSoC device in great detail, it only tells half the story of the RFDCs. Chapter 11 will review the RF-DACs, which enable the transmission of RF signals, and have only been touched on up to this point. In the next chapter will be present some scenarios of using the RF-ADC to receive signals in various bands (Nyquist Zones 1 and 2, and RF frequencies outside of Zones 1 and 2 requiring initial stages of IF mixing prior to the RF-ADC).

Chapter 10

RF Data Converters: Example Receiver Architectures

Lewis Brown

Following the review of the structure and operation of the RF-ADCs in Chapter 9, next we explore a set of example receiver architectures for different RF frequency receiver range scenarios. With the options to operate across Nyquist Zones 1 and 2, to add external IF circuitry to cater for higher frequencies, and to handle complex inputs (i.e. two channels, real and imaginary, or I and Q), four primary configurations will be presented. We will also show how the designer has the flexibility to bypass the internal DDC of the RF-ADC entirely, in favour of implementing a custom solution in the PL.

For the purposes of presenting these example receivers, we will use an RF-ADC sampling rate of f_s = 4 Gsps, in order to keep the numbers simple and the explanations intuitive. It should be noted, however, that each of the RFSoC generations can achieve higher RF-ADC sampling rates, and the designer can of course choose other lower RF-ADC sampling rates than the maximum to suit a specific application requirements.

In the series of examples in this section, **analogue signals** (which are real!) will be presented as real or **single sided frequency spectra** showing only positive frequencies from 0 Hz upwards and with red shading ▨ for showing the frequency domain spectral energy (see an example at Figure 10.2(a), page 340). **Digital signals** obtained after sampling with the RF-ADC will use blue shading ▨ to represent the spectra and will be shown as **two-sided frequency spectra** (i.e. both the 'positive' and 'negative' frequencies, see an example in Figure 10.2(b), page 340 — refer back to Section 7.1.2, on page 234 to review complex spectra). If the complex spectra is symmetric about the 0 Hz line, then this is from a **real time domain signal** and if asymmetric, then this is from a **complex time domain signal**, i.e. with I/Q channels representing the real and imaginary parts.

Analogue signal paths will be shown as red lines ⟶ and real channel digital signals as blue lines, ⟶ and a **complex digital signal** with I and Q, or real and imaginary components, as a double blue line arrow ⟹ . Signals of interest in the frequency domain will be given an asymmetrically shaped spectrum and sloped side on the left) as in Figure 10.2(a), page 340. This is so that we can easily and visually keep track when the signal spectrum gets flipped (or reversed), during the various mixing and demodulation stages. (In some applications reversing the spectrum may be the desired operation and will show how to achieve this later in this Chapter.)

The examples that follow will use some of the knowledge of the complex frequency domain from Chapter 7, and present the notations and block diagrams for both real path receivers (with separate I and Q paths and quadrature cosine and sine mixers/oscillators), and the equivalent complex (real and imaginary) oscillators / mixer block diagrams where the more compact complex notation is used, and the receiver mixes down with a complex exponentials, i.e. $e^{j\omega} = \cos\omega + j\sin\omega$. Please refer back to Chapters 5 and 7 for background on these topics if appropriate.

10.1. Example 1: Nyquist Zone 1, Direct-RF

For this first example we examine the simplest case, and most straightforward RFSoC RF signal receiver, referred to as the Direct RF architecture. In this example the analogue RF signal occupies spectrum in Nyquist Zone 1 and therefore it can be directly digitised by the RF-ADC with no aliasing process required (from higher Nyquist Zones), and no stage(s) of analogue IF (intermediate frequency) required. The signal frequency band in this presented example occupies the frequency range from 400 to 700 MHz (300 MHz of bandwidth), as shown in Figure 10.2(a), and thus has a centre frequency of $f_{centre} = f_c = 550$ MHz.

With the RF-ADC $f = 4$ Gsps, the Nyquist Zone 1 then extends from 0 - 2 GHz ($f_s/2$), and therefore the signal range of interest, 400 to 700 MHz, falls within this Nyquist Zone[1]. Therefore a Direct-RF architecture, which was previously shown in the upper part of Figure 9.23 on page 324, can be used to receive this signal.

10.1.1. Equivalent Quadrature Mixer and Complex Mixer Architectures

A block diagram representing this Direct RF receiver architecture is shown in Figure 10.1(a) in its real signal quadrature mixer form, and in Figure 10.1(b) as the complex mixer form of the same Direct RF receiver. The first stage of the receiver in both the quadrature and equivalent complex versions of the Direct RF receiver in Figure 10.1 is the RF analogue signal conditioning, including some RF gain and an anti-alias filter, with both components being external to the RFSoC device. The low pass (anti-alias) filter passes signal energy in the Nyquist Zone 1 and attenuates the unwanted frequencies above 2 GHz in Nyquist Zone 2 (else the energy in this band would otherwise alias down into Nyquist Zone 1). The dotted line in Figure 10.2(a) shows the

---

1. Note that we will usually try to state the sampling frequency of an ADC in samples per second (sps), i.e. the RF-ADC here is $f_s = 4$ Gsps, which is essentially the same as $f_s = 4$ GHz . However after the ADC/analogue stage we will use Hz to denote sample rates as the signal traverses through various DSP components, filters, and other signal processing stages.

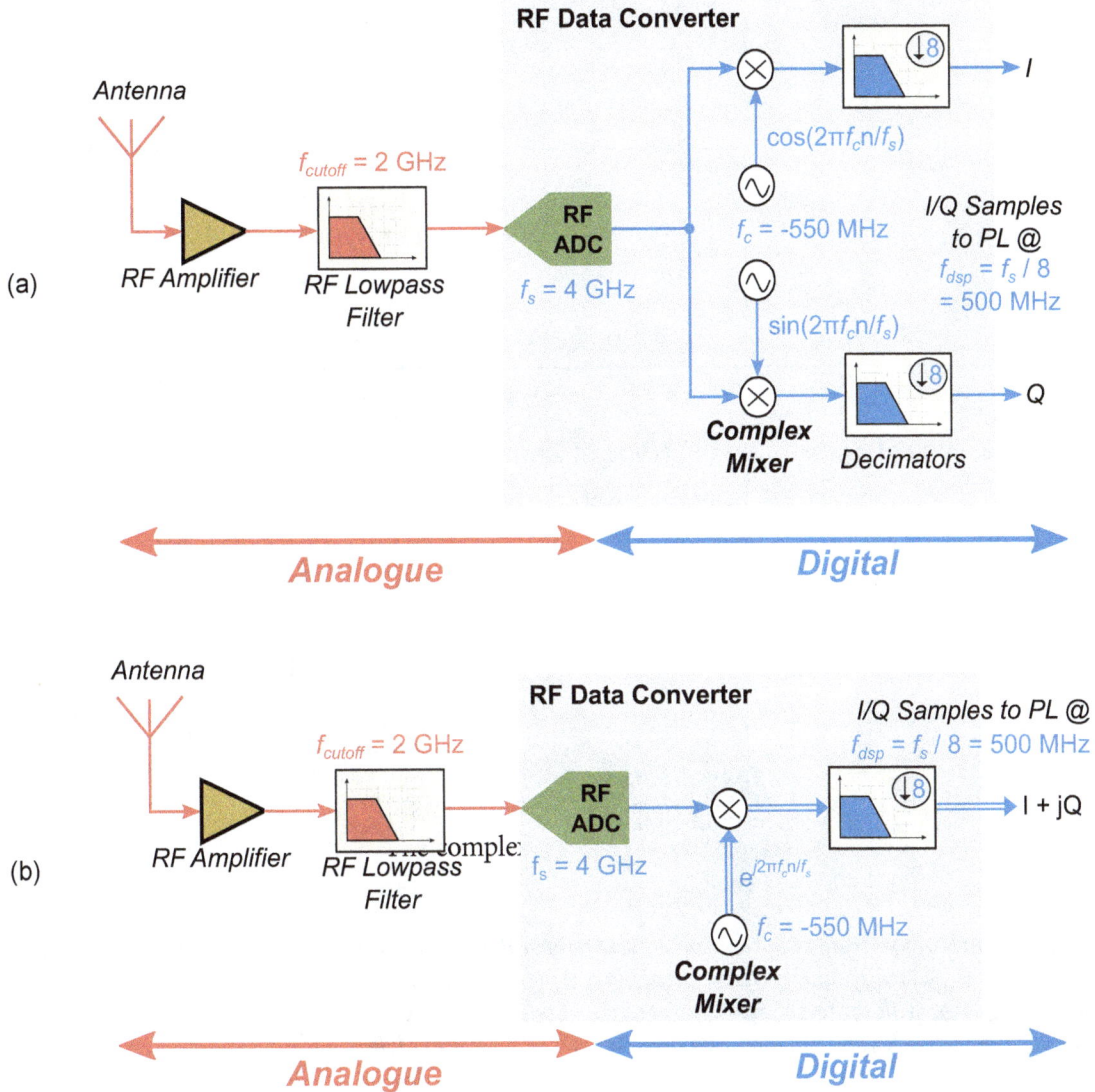

Figure 10.1: Example 1: (a) Direct-RF Quadrature Mixer receiver to receive a 400 - 700 MHz (i.e. 300 MHz bandwidth-signal of interest in Nyquist Zone 1. (b) Complex mixer equivalent of the architecture in (a).

magnitude response of the low pass filter that passes the Nyquist Zone 1 (i.e. ideally implemented as flat passband *Gain* = 1 (i.e. 0dB, recalling that $(20\log_{10}(Gain) = 0$ dB) from 0 to 2 GHz.

The output of the low pass anti-alias filter is input to the RF-ADC sampling at f_s = 4 Gsps, and outputs the real digital signal (and therefore *symmetric* two sided spectrum) shown in Figure 10.2(b). This signal is then input to the mixers shown as two real multipliers in the real path architecture of Figure 10.1(a), and as a multi-

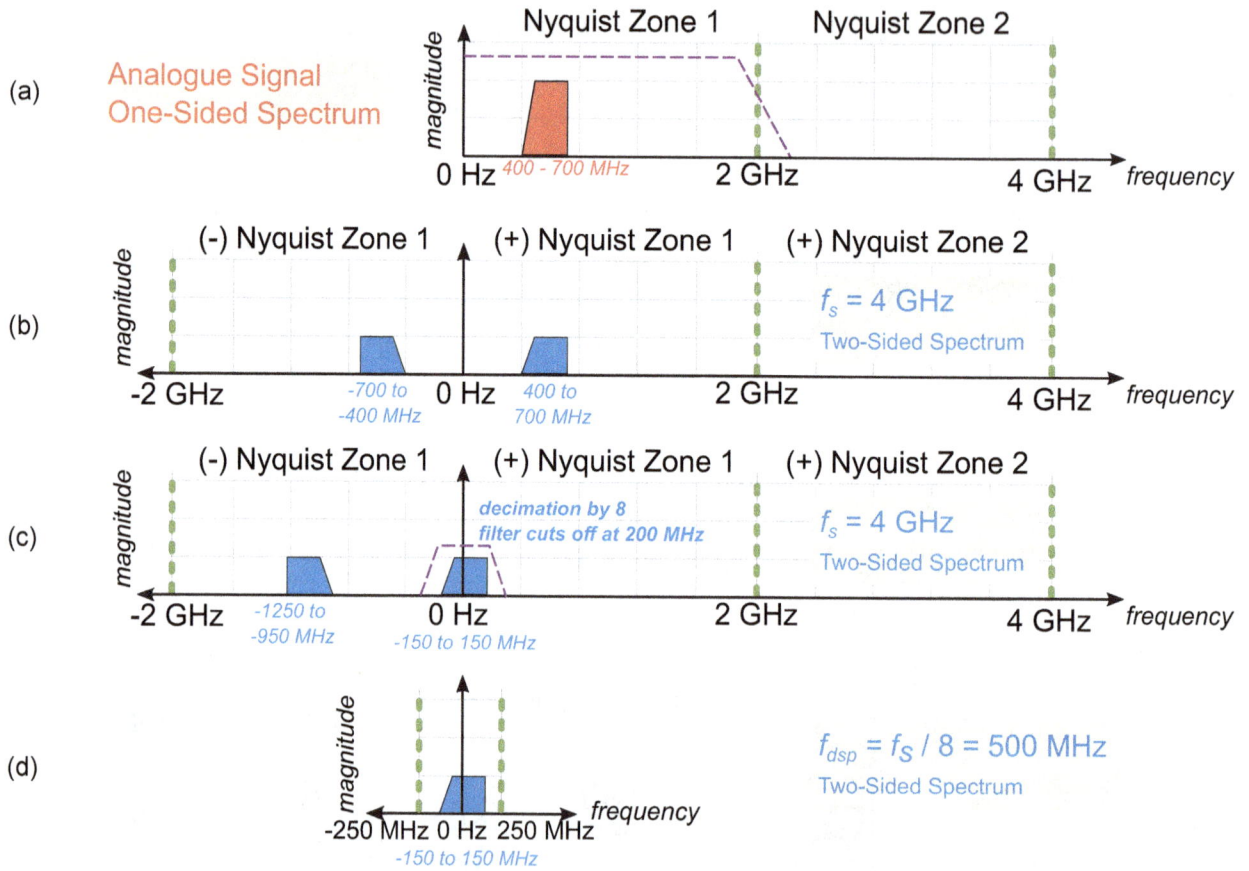

Figure 10.2: Example 1: Spectrum of the analogue input signal (400 - 700 MHz) and anti-alias filter low pass filter characteristic. (b) Two-sided symmetric spectrum of the real digitised signal following RF-ADC sampling at 4 Gsps. (c) After mixing stage, with the target signal now present at baseband (-150 to 150 MHz) and negative frqeuency band. (d) Final demodulated and decimated baseband signal (-150 to 150 MHz), now sampled at $f_{\text{dsp}} = 500$ MHz).

plier with one real input and one complex input for the equivalent complex path architecture of Figure 10.1(b). In Figure 10.1(a) the RF-ADC output is multiplied by the '-550 MHz' cosine and sine oscillators, recalling that, $\cos(-\theta) = \cos(\theta)$ and $\sin(-\theta) = -\sin(\theta)$, i.e.

$$\cos(2\pi(-550e6)n/f_s) = \cos(2\pi(550e6)n/f_s) \quad \text{and} \quad \sin(2\pi(-550e6)n/f_s) = -\sin(2\pi(550e6)n/f_s)$$

In the complex path of Figure 10.1(b), the mixer or multiplier has a real input and a complex input, and we can see that the positive frequency band is shifted to baseband and the negative frequency band is shifted to be centred at 2×-550 MHz $= -1100$ MHz. The complex NCO frequency of -550 MHz, i.e. $e^{-j2\pi(550e6)n/f_s}$ as shown in Figure 10.1(b) relates to the quadrature oscillators as:

CHAPTER 10: RF Data Converters: Example Receiver Architectures

$$e^{-j2\pi(550e6)n/f_s} = \cos(-2\pi(550e6)n/f_s) + j\sin(-2\pi(550e6)n/f_s)$$

$$= \cos(2\pi(550e6)n/f_s) - j\sin(2\pi(550e6)n/f_s)$$

In the equivalent real path valued architecture, Figure 10.1(a) as written above, the cosine and sine NCOs set to 550 Hz and -550 Hz respectively which is precisely equivalent to the complex path architecture in Figure 10.1(b). The mixer in Figure 10.1(a) is operating in Real-to-Complex mode, generating separate I and Q baseband signals which are shown as the real and imaginary components respectively at the output as shown in Figure 10.1(b) - again as reviewed here and in Chapter 7 the two architectures are identical, and the use of the complex notation is for mathematical convenience in Figure 10.1(b) as compared to Figure 10.1.

10.1.2. Mixing Target RF Signal to Complex Baseband

In Figure 10.2(b), we show the two-sided complex spectrum of the real digitised signal, immediately after sampling by the RF-ADC. As the signal is real at this stage, its spectrum is symmetrical around 0 Hz and we can denote the positive complex frequency band as $f_{positive}$ and the negative complex frequency band as $f_{negative}$.

The complex digital mixer at -550 MHz can then be used to shift the signal $f_{positive}$ band down to baseband and centred around 0 Hz. Using the fine mixer NCO in the RFSoC (see again Section 9.4.2, page 315), and selecting a frequency of $f_c = -550$MHz (i.e. the centre of the 400-700 MHz band), the pair of spectra are shifted leftwards in the frequency spectrum plot by 550 MHz. As a result, the complex mixed versions of the signal are generated at:

$$f_{positive} - f_c = (400 \text{ to } 700) \text{ MHz} - 550 \text{ MHz} = (-150 \text{ to } 150) \text{ MHz}, \quad \text{and}$$

$$f_{negative} - f_c = (-700 \text{ to } 400) \text{ MHz} - 550 \text{ MHz} = (-1250 \text{ to } -950) \text{ MHz}.$$

The result of this demodulation can be seen in Figure 10.2(c), with one copy of the signal centred at baseband (0 Hz), and another at centre frequency of -1100 MHz. The signal is now *asymmetrical* with respect to 0 Hz, and therefore complex. Note that the real architecture in Figure 10.1(a) does *exactly the same*, but more difficult to interpret and sketch when using separate real spectra versions of I and Q, as opposed to considering the complex signal $I + jQ$ of Figure 10.1(b).

10.1.3. Decimation of Mixer outputs for PL (Programmable Logic) Stage

The final low pass decimation filter chain then low pass filters the signal to 250 MHz, and downsamples by a factor of 8, from 4 GHz to a signal sampling frequency of $f_{dsp} = 500$MHz $(4000/8)$ giving the spectra in Figure 10.2(d) with $f_{dsp}/2 = 250$ MHz shown as the green dotted lines. The downsampled rate of $f_{dsp} = 500$MHz is still high enough to represent the complex bandwidth of 300 MHz bandwidth signal from

341

-150 MHz to 150 MHz. Following this Direct RF receiver, the separate I and Q signals are then passed to the PL for the next stage of processing.

Note that a lower f_{dsp} rate (down to 300 MHz) could have been chosen and still represent the signal. (Or indeed you might choose a higher f_{dsp} rate). The engineering decision on the f_{dsp} sample rate is for the designer to make, and will be based on various factors such as available clocks, or interaction with other DSP signal paths where, for example, might perhaps looking to match to the sample rates for those other paths.

10.1.4. Flipping or Reversing the Received Frequency Band

Notice the shape of the final baseband spectrum now centred at 0 Hz in Figure 10.2 — it slopes on the left hand side and same as original RF signal in Figure 10.2(a). If we had wanted to shift the $f_{negative}$ *image* spectrum to baseband (i.e. the band initially centred at -1100 MHz, with the slope on the right), then we would instead have configured the NCO to generate a frequency of +550 MHz for the mixing process (i.e. a complex oscillator of $e^{j2\pi(550e6)n/f_s}$). This would have shifted the magnitude spectra of $f_{negative}$ and $f_{positive}$ and from Figure 10.2(c) to the right (rather than the left), and generated bands at 0 Hz and +1.1 GHz. This baseband version would have the slope on the right, and therefore the original real spectrum from Figure 10.2(a) has been flipped or reversed in frequency (not the desired operation though in this example).

10.2. Example 2: Nyquist Zone 2, Direct-RF

For this second example, we consider a signal of interest between 3 and 3.5 GHz. We can use the same, $f_s = 4$ Gsps RF-ADC sample rate as previously, this signal falls in Nyquist Zone 2, i.e. between $f_s/2 = 2$ GHz and $f_s = 4$ GHz. This signal can still be directly digitised by the RF-ADC, and as such the Direct-RF architecture can be used, as in Section 10.1 for Example 1 above. A block diagram of this receiver is provided in Figure 10.3 which shows both the real signal path quadrature mixer and the complex mixer and the various filter stages and decimation stages.

10.2.1. Selecting Nyquist Zone 2 with a Bandpass Filter

First, an analogue RF bandpass filter passes Nyquist Zone 2, thus attenuating frequencies in Nyquist Zone 1, as well as frequencies in Nyquist Zone 3 and above (they all alias down!) as shown in Figure 10.3. The signal is then sampled by the RF-ADC operating at $f_s = 4$ Gsps, and the signals from Nyquist Zone 2 will alias to Nyquist Zone 1 as a result of the sampling process (recall the fundamentals of aliasing — for a reminder see Section 4.1.4, page 77). Hence the aliased spectrum will be flipped or reversed compared to the original signal in the 3 GHz to 3.5 GHz band as shown in Figure 10.4(b) which shows the real (as its symmetric) two sided signal spectrum at the output of the RF-ADC.

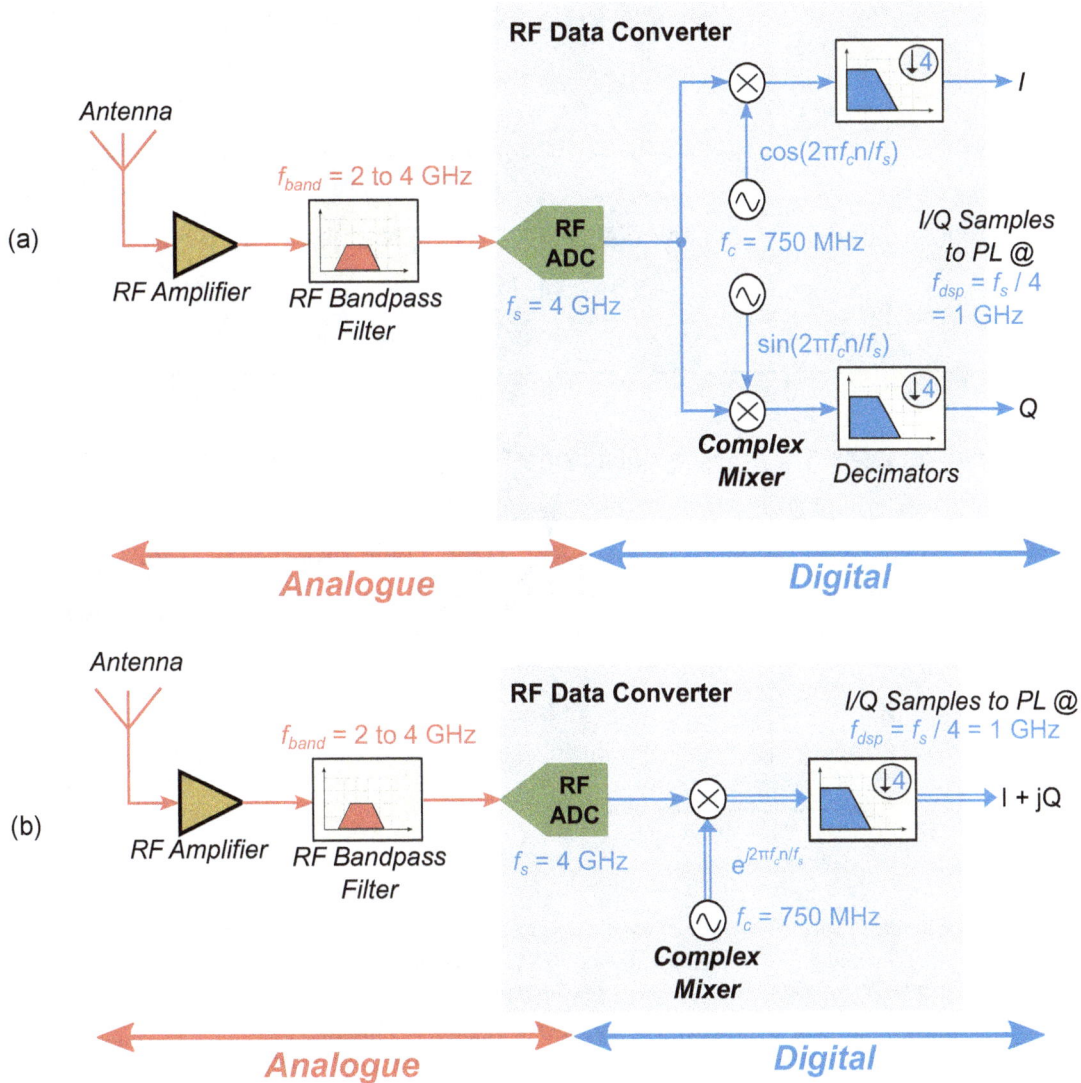

Figure 10.3: Example 2: (a) Nyquist Zone 2 Direct-RF Quadrature Mixer receiver to receive a 3 GHz to 3.5 GHz (i.e. 500 MHz bandwidth) signal of interest in Nyquist Zone 2. (b) Complex mixer equivalent of the architecture in (a).

10.2.2. Mixing Target RF Signal to Complex Baseband

The signal of interest can now be mixed to baseband by mixing with a '+750' MHz NCO signal in the quadrature mixer and in the complex mixer as show in Figure 10.3(c). Note that different to Example 1, we now select to shift the f_{negative} band from (-1000 MHz to -500 MHz and centred at -750 MHz) to be mixed to

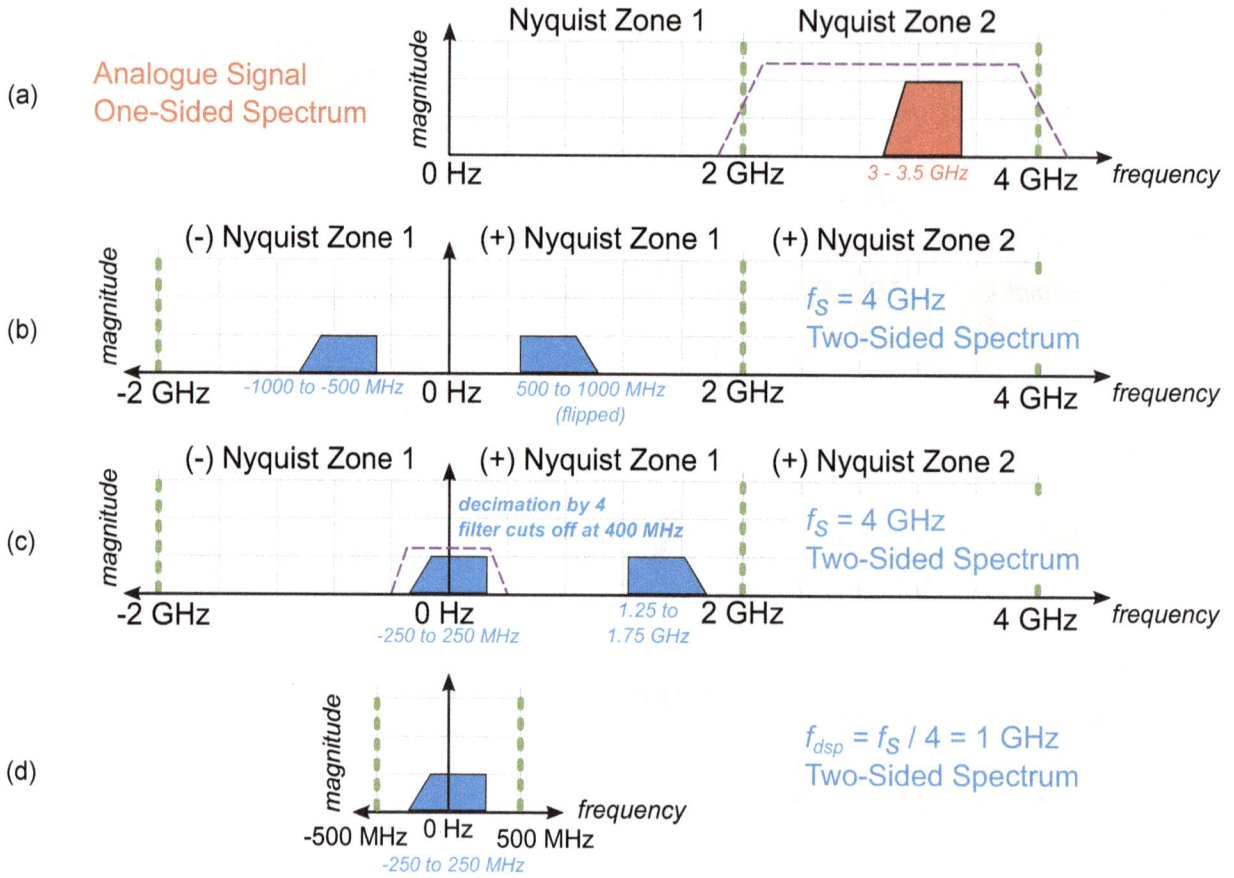

Figure 10.4: *(a) Example 2: Spectrum of the analogue input signal (3 - 3.5 GHz) and band pass filter. (b) Two-sided spectrum of the real digitised signal following RF-ADC sampling at 4 Gsps. (c) After mixing the desired signal is now at baseband (-250 to 250 MHz). (d) Final demodulated and decimated baseband signal (-250 to 250 MHz), now sampled at f_{dsp} = 500 Msps).*

0 Hz in order to get the same spectrum at baseband (i.e. not flipped in frequency which would happen if we mixed by '-750' MHz which would mix down the positive frequency components at 500 MHz to 1000 MHz).

So again, mixing the negative spectrum, f_{negative} to be centred at 0 Hz and baseband is accomplished by mixing with an NCO frequency of f_c = 750 MHz, i.e. the centre frequency of the aliased band, this produces two signals from the mixing up of the spectra at f_{positive} and f_{negative}:

$$f_{\text{positive}} + f_c = (500 \text{ to } 1000) \text{ MHz} + 750 \text{ MHz} = (1250 \text{ to } 1750) \text{ MHz}, \quad \text{and}$$

$$f_{\text{negative}} + f_c = (-1000 \text{ to } -500) \text{ MHz} + 750 \text{ MHz} = (-250 \text{ to } 250) \text{ MHz}.$$

The signal is now centred at 0 Hz (baseband), and an image of the signal is centred at 1.5 GHz, as shown in Figure 10.3(c).

10.2.3. Decimation of Mixer outputs for PL (Programmable Logic) Stage

In this example, the signal bandwidth is wider than in Example 1 in Section 10.1 (500 MHz, as opposed to 300 MHz). Therefore a higher sampling rate is required and a lower downsampling factor of 4 is selected, meaning that the I/Q samples are transferred to the PL for further processing at the higher rate of $f_{\text{dsp}} = 1\,\text{GHz}$, compared to 500 MHz in the previous example.

As a result of the low pass filtering of the decimator stage to 500 MHz cut-off, the image centred at 1.25 GHz is filtered out, resulting in the baseband signal shown in Figure 10.4(d).

10.3. Example 3: Nyquist Zone 1, Analogue IF Stage

In the third example, we wish to receive a QAM generated signal which is 100 MHz wide and across the range from 5.8 to 5.9 GHz. An RF frequency in this range cannot be directly digitised by an RF-ADC with sampling rate $f_s = 4$ GSps, and hence a prior stage comprising of an analogue local oscillator (LO) mixer from RF to IF is required, which is external to the RFSoC.

10.3.1. Analogue Mixing to an IF (Intermediate Frequency) in the RF-ADC Range

Figure 10.5 shows an architecture for digitising this signal in 5.8 to 5.9 GHz range via an IF stage. In this example, the received analogue 100 MHz signal occupies the band from 5.8 - 5.9 GHz (we will denote the bottom of the band as $f_1 = 5.8\,\text{GHz}$ and the top of the band as $f_2 = 5.9\,\text{GHz}$). This range is too high in frequency to be digitised directly, even in Nyquist Zone 2, which spans from 2 to 4 GHz.

Figure 10.6(a) illustrates the RF signal of interest, at $f_{\text{centre}} = 5.85$ GHz, which is first isolated with a suitable RF analogue band pass filter to attenuate frequencies that will exist above and below this spectrum. Therefore we first mix down to an IF frequency using an RF-to-IF analogue local oscillator (cosine wave!) operating at a frequency of $f_0 = 5.55\,\text{GHz}$ as shown in Figure 10.5. Of course other f_0 local oscillator frequencies could be used — the low band output from the mixer just needs to land in the 0 to 2 GHz of Nyquist Zone 1 (albeit in Nyquist Zone 2 will work too, but usually most straightforward to work in Nyquist Zone 1).

Analogue mixing with the 5.55 GHz local oscillator at frequency f_0 with the signal from f_1 to f_2 creates both lower and upper bands. Recalling the identity: $\cos A \cos B = 0.5 \cos(A + B) + 0.5 \cos(A - B)$, then

$$\cos(f_1 \text{to} f_2) \cos(f_{LO}) = \underbrace{\cos((f_1 \text{to} f_2) + f_{LO})}_{\text{lower band}} + \underbrace{\cos((f_1 \text{to} f_2) - f_{LO})}_{\text{upper band}}$$

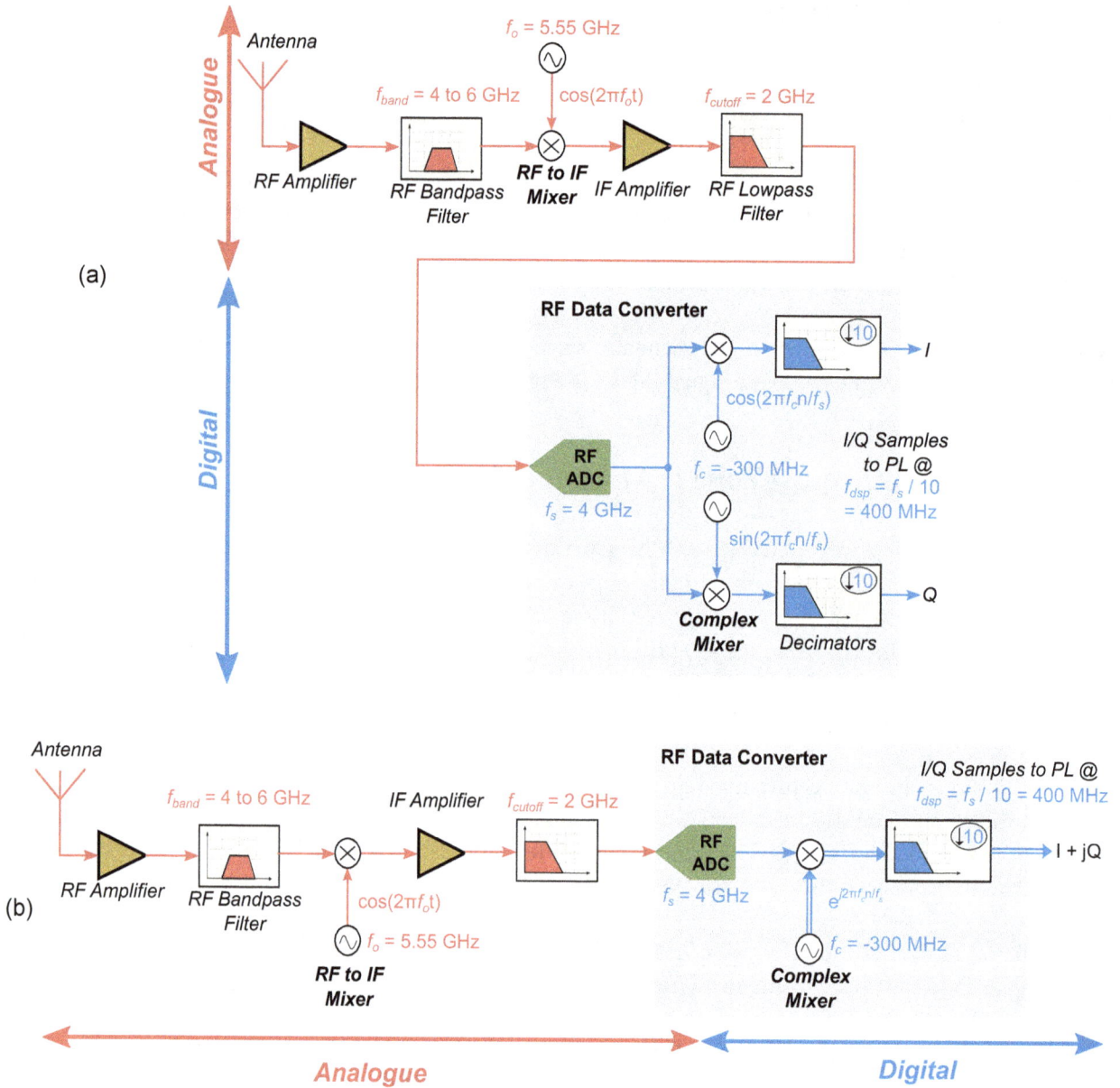

Figure 10.5: Example 3: (a) IF receiver for a 100 MHz signal, centred at 5.85 GHz and mixed to an IF in the RF-ADC Nyquist Zone 1. (b) Complex equivalent of the architecture in (a).

and therefore the centre frequency, f_{IF}, of the band appearing in the RF-ADC Nyquist Zone 1 (0 to 2 GHz) is at frequency:

$$f_{IF} = f_{centre} - f_o = 5.85 - 5.55 = 0.3 \text{ GHz} = 300 \text{ MHz}$$

Noting the signal is 100 MHz wide, then the frequency range f_1 to f_2 (5.8 GHz to 5.9 GHz) is translated to the baseband frequencies from f_{1b} to f_{2b}:

$$f_{1b} = |f_1 - 5.5| = |5.8 - 5.55| = 250 \text{ MHz} \quad \text{and} \quad f_{2b} = |f_2 - 5.5| = |5.9 - 5.55| = 350 \text{ MHz}$$

therefore creating an (analogue) IF frequency range of 250 to 350 MHz as shown in Figure 10.6(b). Note that as well as the 300 MHz centred 'lower' band there will also be components created in the range 5.55 + 5.85 GHz to 5.55 + 5.95 GHz, i.e. 11.4 GHz to 11.5 GHz. This 'upper' band can be filtered by other stages, but will also be filtered by the RF low pass filter stage shown in Figure 10.5 and which is also functioning as the 2 GHz cut-off frequency anti-alias filter for the RF-ADC sampling at 4 Gsps (the low pass anti-alias frequency response is again illustrated with the dotted line in Figure 10.6(b)).

10.3.2. Sampling the IF centred Signal of Interest in Nyquist Zone 1

Once demodulated to IF in Nyquist Zone 1 as in Figure 10.6(b), and the signal of interest centred at 300 MHz, the 2 GHz anti-alias low pass filter removes frequencies above $f_s/2$, and also the image band high frequency components. The signal is then digitised by the RF-ADC sampling at 4 Gsps and in Figure 10.6(c) we show the two sided symmetric spectrum (i.e. real) for the digital signal centred at 300 MHz. The subsequent steps are similar to those of Example 1 in Section 10.1 above and we shift the spectrum to the left to mix the upper spectrum to baseband (with a complex mixer/oscillator frequency in Figure 10.5(b) at -300 MHz) creating the complex spectrum of Figure 10.6(d).

Demodulation from IF to baseband has shifted the entire two sided spectrum down in frequency by 300 MHz, such that the target signal occupies the frequency range from -50 to 50 MHz, and the negative image occupies the frequency range from -650 MHz to -550 MHz, as seen in Figure 10.6(d).

10.3.3. Decimation of Mixer outputs for PL (Programmable Logic) Stage

Following the complex mixer, the signal is decimated by low pass filtering and downsampling by 10, producing the final baseband signal as in Figure 10.6(e) and with a sampling rate of f_{dsp} = 400MHz (= 4GHz / 10) and passes to the PL stage for further DSP and processing. The low pass response of the decimator removes the unwanted image of the complex signal centred at -600 MHz. The signal output from this final processing stage can be seen in Figure 10.6(e). (Note, the signal decimation by 10 to achieve a 400 MHz sampling rate requires that a Gen 3 RFSoC device is used for rate changes of 10 and above; Gen 1 and 2 cater for a maximum decimation ratio of 8).

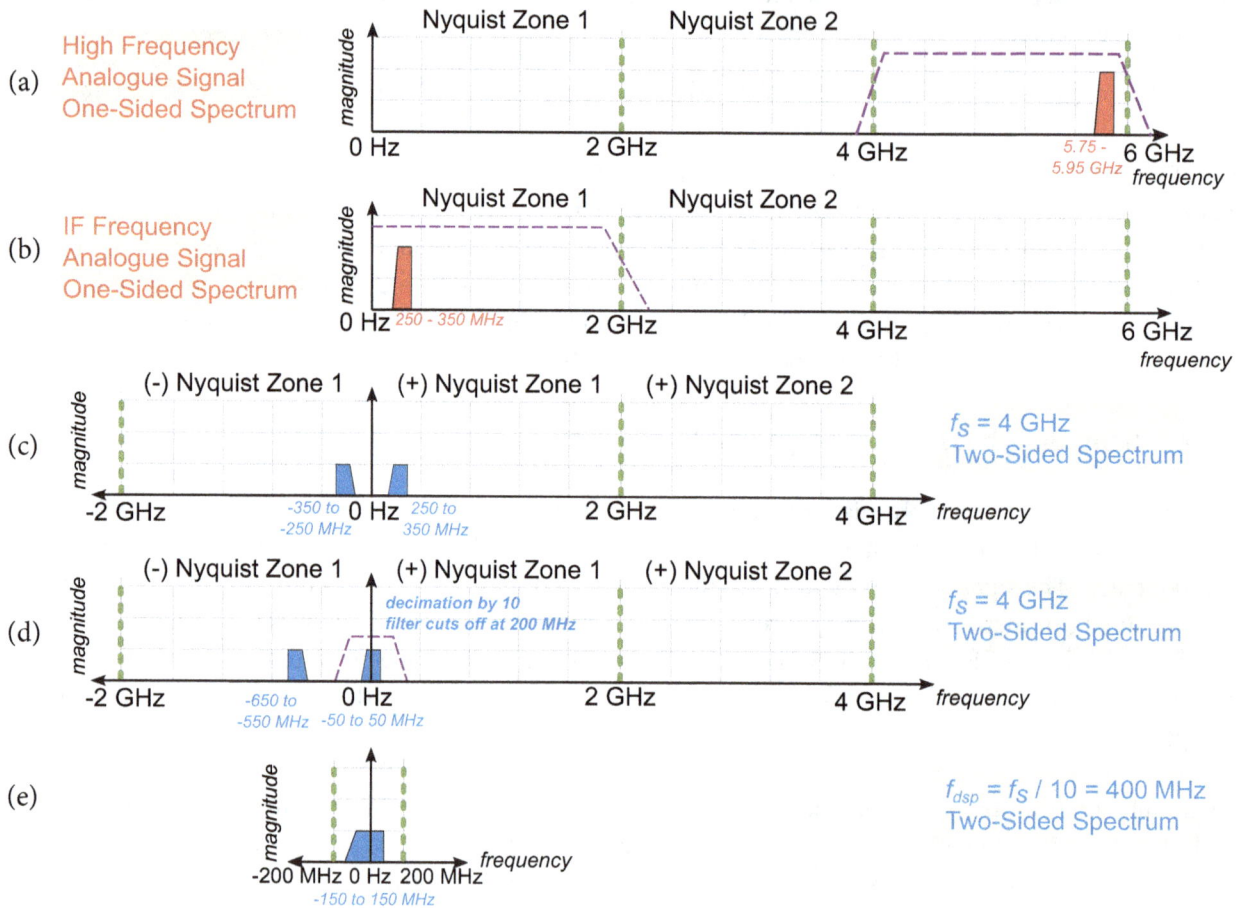

Figure 10.6: *Example 3: Spectrum showing the (a) analogue target 100 MHz wide RF signal of interest centred at f_{centre} = 5.85 GHz, (b) the target mixed down to an IF frequency at 300 MHz with an oscillator at f_o = 5.55GHz, (c) the two sided complex spectrum of the real signal input to the RF-ADC, (d) frequency shifting by complex mixing with 300 MHz complex oscillator, and (e) final decimated signal sampled at f_{dsp} = 400 MHz sent to the PL.*

10.4. Example 4: I/Q Mixer to Analogue Baseband and Sampling

For the fourth example, we will use the analogue quadrature mixer receiver of Figure 10.7, to receive the signal interest centred at 27 GHz, with a bandwidth of 3 GHz (from 25.5 GHz to 28.5 GHz), and shown in Figure 10.8(a). As we have selected an RF-ADC sampling rate of f_s = 4 GHz then while we can digitise from 0 to 4 GHz with Nyquist Zones 1 and 2, clearly the 3 GHz signal centred at 27 GHz greatly exceeds $f_s/2$, and as such, cannot be directly converted from analogue to digital.

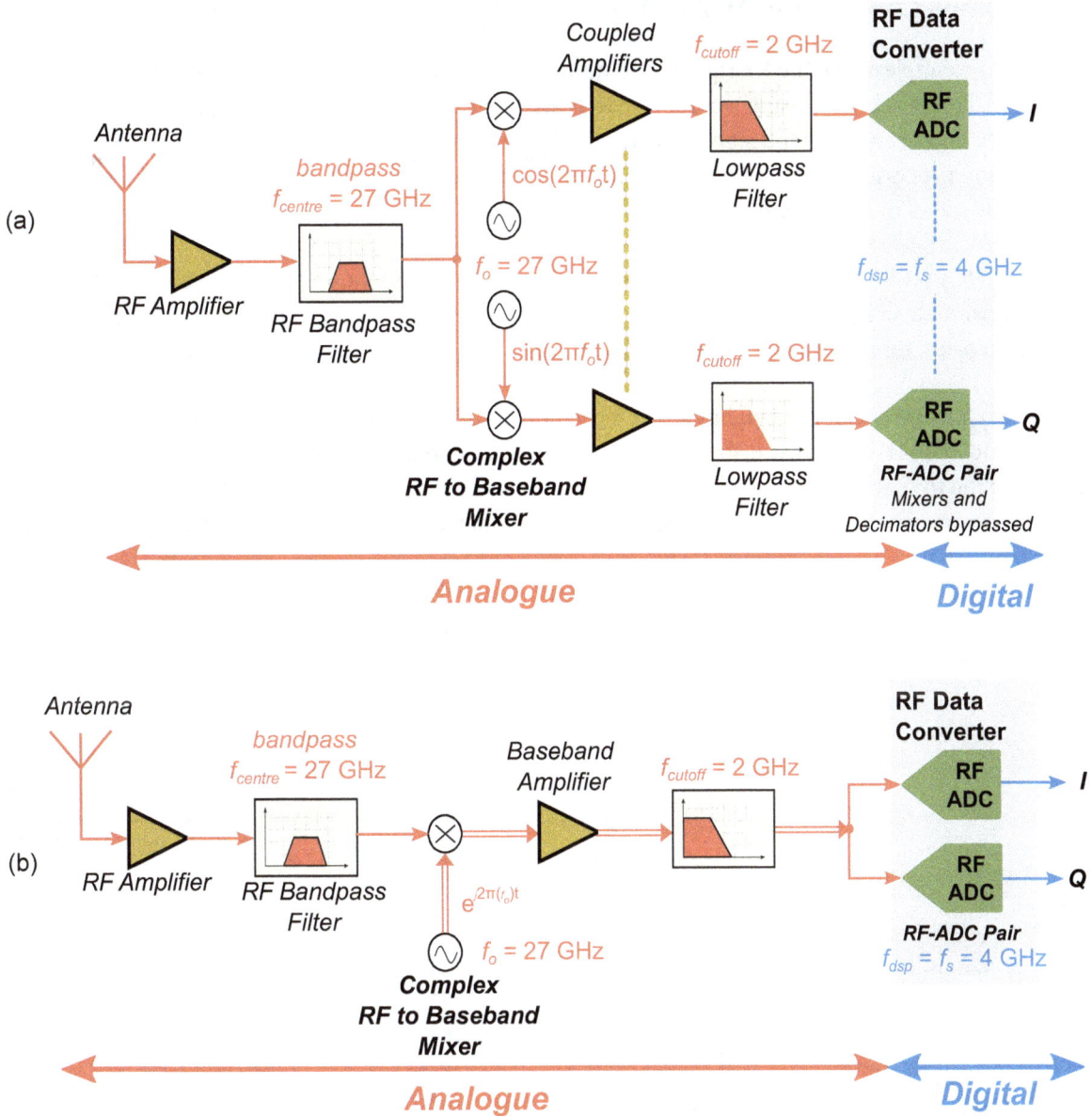

Figure 10.7: Example 4: Analogue quadrature mixer receiver architecture used for receiving a 3 GHz wide signal from 27 GHz carrier frequency. (b) Complex equivalent of the architecture in (a).

10.4.1. Quadrature Analogue Mixing to Analogue IF and Baseband Sampling

Unlike previous examples, here the RF signal is demodulated to baseband in the analogue domain, prior to the RF-ADC stage. This is achieved using a complex (I/Q) analogue mixer with centre frequency $f_o = -27$ GHz, which generates a complex baseband signal Figure 10.8(b) that is supplied to the input of the RF-ADC pair. (As above the mixing processing will also create components at the 'upper' sideband of 2 x 27 GHz, but these are easily filtered by stages of analogue filtering.)

With the RF-ADCs set to $f_s = 4$ Gsps, the complex format permits up to a 4 GHz signal bandwidth, whereas the maximum with a real signal would be 2 GHz (in practice, the usable bandwidth might be slightly lower in both cases, to account for realistic analogue filter responses). This is often referred to as quadrature sampling.

Note for this architecture, as the signal has already been demodulated to baseband, externally to the RF-ADC, then there is no requirement to use the RF-ADC quadrature/complex mixer, and it is therefore bypassed entirely for mixing to baseband purposes. However the complex mixer still may be used as part of a synchronisation scheme to correct for any carrier frequency offset that may exist (recall the carrier offset review back in Section 7.6).

The spectra of the target signal and the demodulated baseband analogue signal, and the complex baseband (signal is asymmetric) can be seen in Figure 10.8(a) and (b).

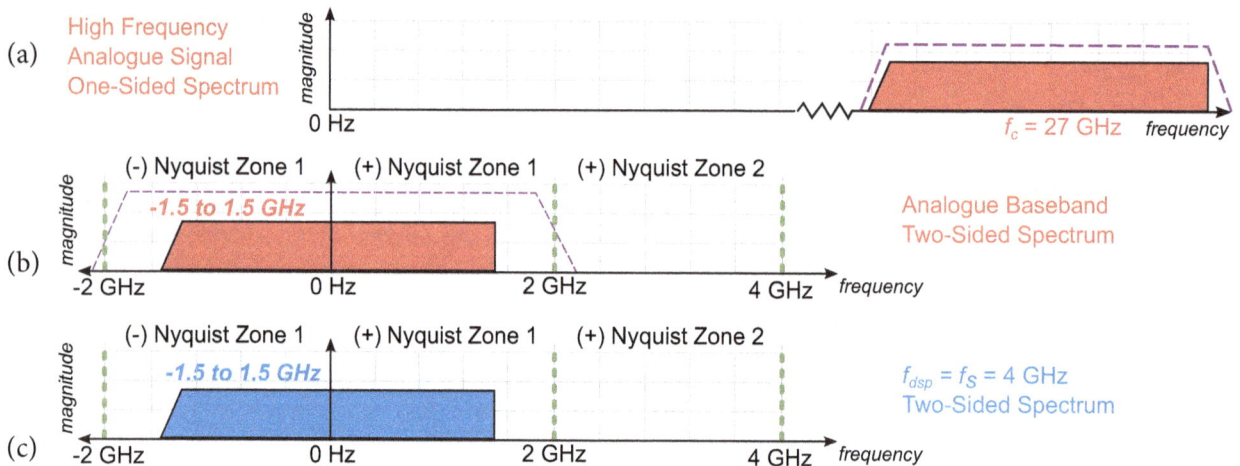

Figure 10.8: (a) Example 4: Spectrum showing the analogue target signal centred at 27 GHz. (b) After analogue complex mixing with NCO at 27 MHz down to complex baseband, and (c) after sampling with two RF-ADCs (I and Q) at f_s = 4 Gsps.

10.4.2. Mismatch and Analogue Correction Options

A challenge inherent to this approach is that the two analogue branches may have small analogue mismatches and differ slightly in terms of gain, phase, and DC offset, due to component tolerances, I/Q mixer imbalance, and other real-world effects, hence causing some issues with the I/Q signal balance. However these mismatches can be addressed by the RFSoC's Quadrature Modulator Correction (QMC) block can be used to compensate and thus correct for these analogue impairments.

Given the very wide signal bandwidth in this example, there is no scope for decimation, and the I/Q samples are transferred from the RF-ADC tile to the PL at the rate of 4 Gsps. The digital signal passed from the RF-ADC tile to the PL is shown in Figure 10.8(c). Given that this sampling rate exceeds that of the PL clock frequency, a super-sampling architecture is required for subsequent stages, i.e. a parallelised structure capable of processing multiple samples per clock cycle.

10.5. Example 5: Custom Architectures

The previous four examples have explored typical use-cases and architectures for using the RF-ADCs on the RFSoC, but this is not exhaustive! Use of the quadrature and complex mixers and decimators on the RF-ADC tile is optional — designers can choose to bypass either or both of these, and instead build their own custom architectures in the PL, if desired. The custom architecture shown in Figure 10.9 is provided as an example.

Here we assume that the received signal has a 100 MHz bandwidth and is centred at 850 MHz (in Nyquist Zone 1), and is sampled by the RF-ADC at a rate of 3.6 GHz. The complex mixer in the RF Data Converter block is used to demodulate the signal to baseband using a complex NCO operating at -850 MHz.

However in this example, the integrated programmable decimators (see again Figure 10.1) are bypassed in favour of custom decimators created on the PL, and with the RF-ADC samples are transferred to PL at a rate of 3.6 Gsps (unchanged from the input rate, as no decimation has taken place). In the PL and for the purposes of this example, the designer implements their own proprietary decimator to reduce the sampling rate by a custom factor of 9, prior to any further processing at this rate.

10.5.1. RF-ADC Programmable Decimators vs. Custom Decimators

The RF-ADC tile's programmable decimators offer a discrete set of decimation ratios, which varies by generation. In each case, the responses of the decimation filters are fixed and cannot be customised by the designer. For most applications, it would seem preferable to leverage these hardened facilities within the RF Data Converter blocks, as they provide a good range of options and operate at lower power than equivalent functionality implemented in the PL. However, occasionally it may be necessary to design a custom architecture to cater for specific design requirements; for instance, where extremely low latency (and low group

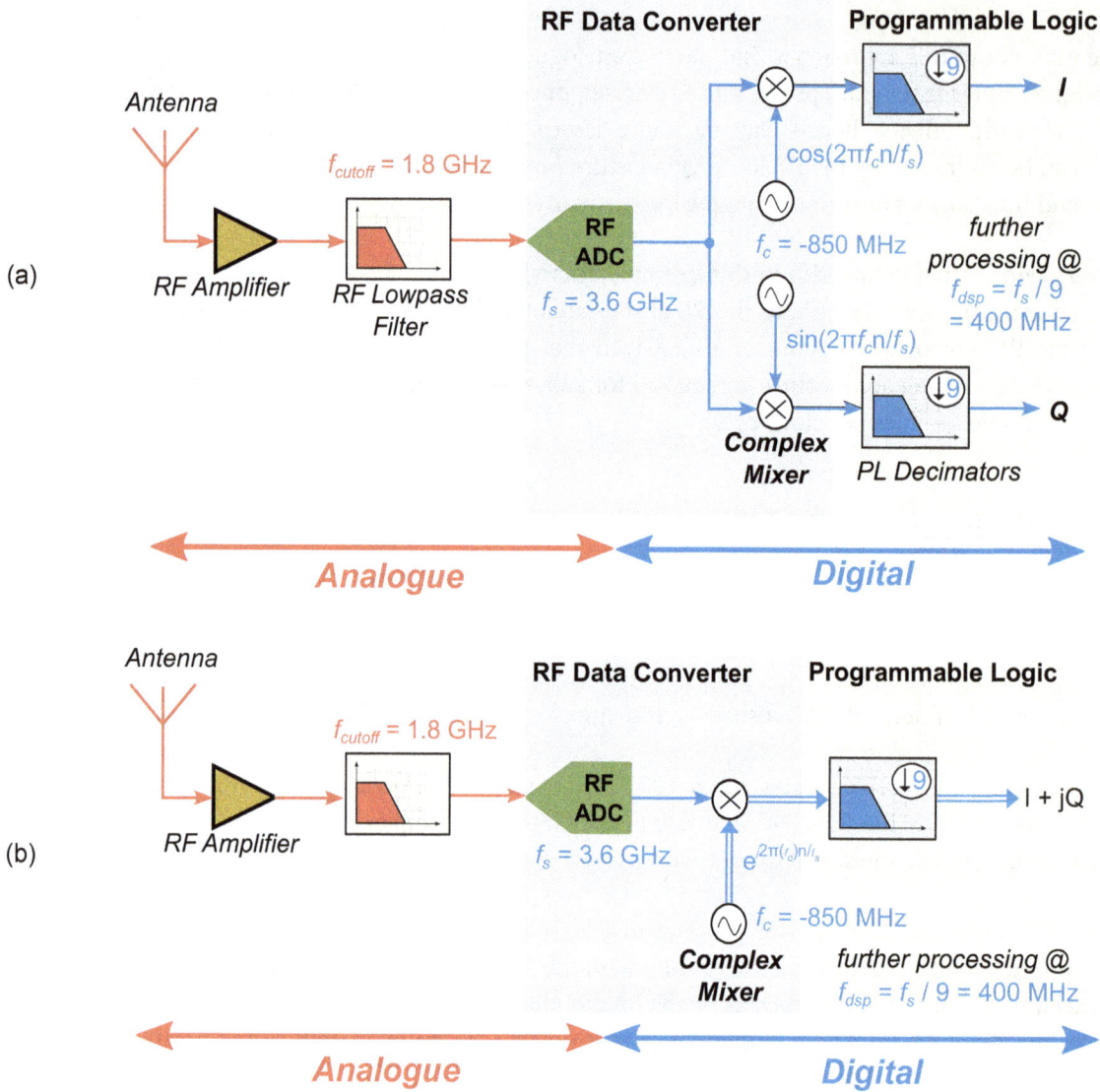

Figure 10.9: Example 5: Custom architecture, using the PL to implement decimators instead of using the RF-ADC decimators.

delay in a filter) is required between the signal first being received and reaching algorithms operating on the PL. As reviewed in Chapter 9, the ability to individually bypass the complex mixer, and/or the programmable decimator within the RF-ADC tile, provides this flexibility.

10.6. Chapter Summary

This chapter has expanded upon the RF-ADC architectures introduced in Chapter 9 by presenting and discussing a set of design examples. First, we explored a Nyquist Zone 1 signal which could be directly digitised using a real architecture using a single RF-ADC. A second example was then presented for a Nyquist Zone 2 signal. It was shown that the complex mixer can be configured to compensate for the spectrum 'flipping' introduced through aliasing of Nyquist Zone 2 signals.

Thirdly, an architecture with an analogue IF stage was developed. it was noted that an analogue demodulation stage is necessary for receiving signals containing components above the RF-ADC sampling frequency. By using such an intermediate stage, the signal could be demodulated to Nyquist Zone 1, and then sampled by the RF-ADC.

The fourth example considered a signal of bandwidth exceeding $f_s/2$ with respect to the RF-ADC sampling frequency. This necessitated a complex analogue architecture, as well as a pair of RF-ADCs to receive the I and Q components of the signal. Finally, custom architectures were explored, highlighting the flexibility to bypass the hardened functionality in the RF-ADC tiles, and implement specialised mixer and decimation functionality in the PL if desired.

Chapter 11

RF Data Converters: Digital to Analogue

Josh Goldsmith

As with the RF analogue to digital converters discussed in the previous two chapters, the RF Digital to Analogue Converters (RF-DACs) are another crucial component of the RFSoC. Due to the multi-GHz sampling rates that they support, in many cases the RF-DACs enable the transmission of radio signals directly at radio frequencies, without the need for an IF stage. In this chapter we provide a brief overview of the digital to analogue conversion process and detail the structure and components of the RF-DACs on RFSoC devices.

11.1. Digital to Analogue Conversion Recap

Before delving into the details of the RFSoC's RF-DACs, it is worth briefly reviewing the operation of DACs in general. For additional detail, please refer back to Section 4.6.2.

The role of a DAC is to convert discrete-time digital samples into continuous-time analogue signals. It achieves this by converting the digital samples to a continuous signal, and then applying a low pass filter to remove any high frequency components that have been generated.

To convert the discrete signal into a continuous signal, DACs commonly use a *Zero-Order Hold* (ZOH) technique, which can be considered a counterpart to the sampling technique used in ADCs. The DAC holds the amplitude of samples between clock ticks, creating a staircase effect as illustrated in Figure 11.1.

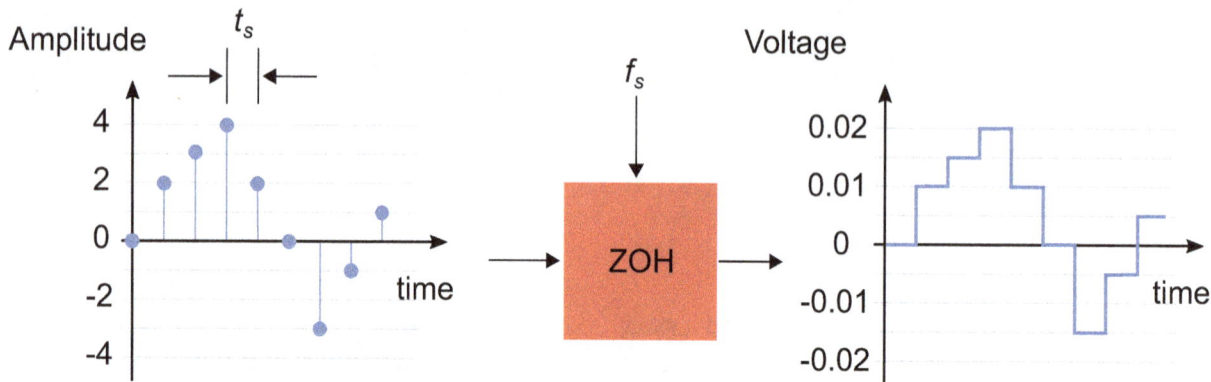

Figure 11.1: Staircase response after the zero-order hold sampling technique.

Each sample in the signal has now effectively been converted to a rectangular function with a height equal to the amplitude, and a width equal to the sample period, as shown in Figure 11.2.

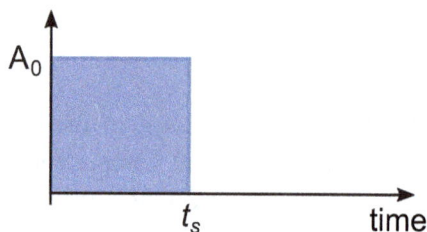

Figure 11.2: Impulse response of the zero-order hold rectangular function.

In the frequency domain this rectangular function can be described by the sinc function

$$H(f) = \frac{sin(\pi f/f_s)}{\pi f/f_s},$$ (11.1)

where f_s is the sampling frequency of the DAC.

The effect of the zero-order hold sampling process is two-fold. The "steppiness" of the resultant signal creates high frequency components, not present in the original signal, that manifest as spectral images across the entire frequency spectrum. Additionally, the sinc-like frequency response introduces non-linear gain across the frequency band between 0 Hz and $f_s/2$, potentially distorting the signal being reconstructed by the DAC. In Figure 11.3 we can see both of these effects in action, with the original signal and its images shown in red, and the sinc response of the DAC as the blue line.

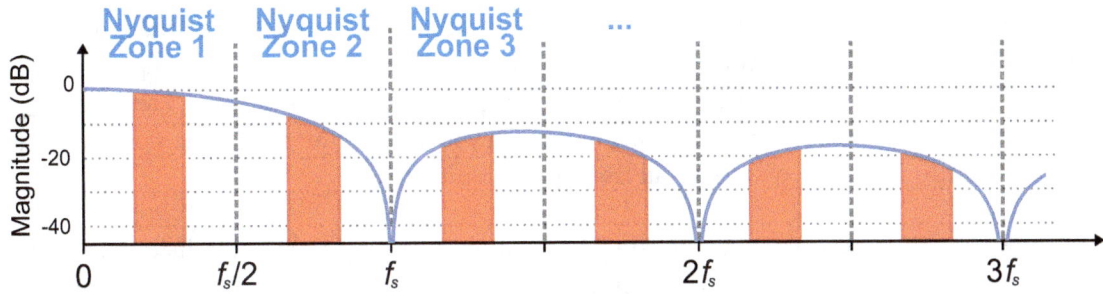

Figure 11.3: Frequency spectrum after zero-order hold process.

The spectral images resulting from digital to analogue conversion are generally undesirable, particularly in radio applications where these high frequency components could interfere with adjacent radio channels. As can be seen from the sinc response shown in Figure 11.3, the images in the upper Nyquist zones are attenuated to an increasing degree, however, this is typically inadequate for most applications, and thus a low pass filter is required to further attenuate them. This filter is normally referred to as a **reconstruction filter** (also an *anti-imaging* or *image rejection* filter), and it has a cut-off frequency equal to the DAC Nyquist rate ($f_s/2$) which removes the spectral images and smooths out the time domain waveform, as shown in Figure 11.4.

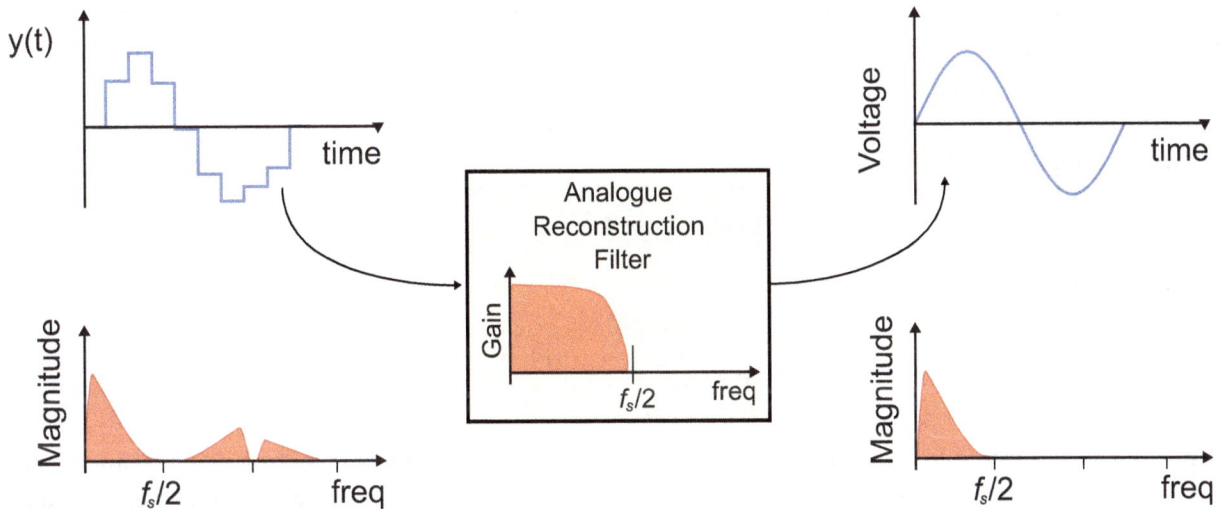

Figure 11.4: Operation of the reconstruction filter, showing how the filter 'smooths' (actually low pass filters!) out the time domain waveform by removing the high frequency signal components.

The characteristic sinc response of the DAC causes a steady drop in amplitude across the first Nyquist zone (and subsequent Nyquist zones) with a difference of around 4dB between DC and $f_s/2$ — equating to more than a 50% drop in power! There are two options available for dealing with this effect. One is to use a sample rate much greater than the highest frequency component of the signal (known as oversampling). In this way, the signal occupies only the region of the DAC response with approximately linear gain. The other is to use an *inverse sinc* filter that compensates for this loss in amplitude by increasing the gain of the signal at the affected frequencies. We discuss this filter in more detail later, in Section 11.2.2.

Two other important characteristics of a DAC are its resolution, expressed in bits, and its minimum sampling period, measured in time. Equivalently, the latter could be expressed as a maximum sampling frequency in Hertz, which is the reciprocal of the minimum sampling period.

The resolution of a DAC describes the number of possible values with which it can represent the amplitude of a signal. With reference to the staircase waveform from Figure 11.1, the resolution of the DAC determines the height of each step. For example, a 10-bit DAC can represent $2^{10} = 1024$ different values. Internally within a DSP system, a higher resolution may be used (i.e. a greater number of bits may be used to represent the signal, than are used at the point of conversion to analogue). Upon digital-to-analogue conversion, each sample is quantised to the closest value that the DAC can reproduce, either above or below it; this adds a source of error to the signal and is known as quantisation noise. The number of bits used in the DAC therefore has a direct effect on the SNR — the higher the resolution of the DAC, the better the SNR will be.

If resolution is considered as the height of the steps in the staircase from Figure 11.1, then the minimum sample period describes the width of each step. This is determined by the maximum rate at which the DAC can operate. Because time is inversely related to frequency, such that $f = 1/t$, the smaller the sample period of the DAC, the higher the maximum frequency signal component that it can reproduce. Following Nyquist theory, this maximum frequency component is just less than half of the DAC sampling rate. For example, if the minimum sample period of a DAC is 10μ s, then it can faithfully reproduce signals in the range

$$0\,\text{Hz} < f_{DAC} < 50\,\text{kHz} \tag{11.2}$$

where the upper limit is determined as

$$\frac{1}{2} \times \frac{1}{10\mu s} = 50\,\text{kHz}. \tag{11.3}$$

DACs can be optimised for a wide range of applications, from low frequency control signals with sampling rates in the tens of Hertz, right up to high frequency RF signals in the Gigahertz range. In the next section we look at the latter of these, in the form of the radio frequency DACs (RF-DACs) on the RFSoC family of devices.

11.2. DACs for RF Frequencies

In many cases, RF-DACs allow the RFSoC to transmit signals directly onto the target RF bands, without the need for IF circuitry. In Chapter 9, we discussed how the RF-ADCs are able to sample signals over the entire spectrum from 0 Hz to f_s (i.e. Nyquist Zones 1 and 2) by taking advantage of aliasing constructively. The RF-DACs are also capable of using both Nyquist zones for transmission, albeit using a different method.

11.2.1. Normal Mode and Mix-Mode

As discussed in the previous section, the zero-order hold sampling process produces images of the original signal in the upper Nyquist bands. Although these images are undesirable in the normal operation of a DAC, they can also be exploited to transmit signals with frequency components above the Nyquist rate of the DAC. The location of the images on the spectrum is related to the frequency components of the original signal, and to the DAC sampling rate. After digital-to-analogue conversion, spectral images will be present at $N \times f_s \pm f$, where N is an integer (1, 2, 3, ...), in theory extending along the spectrum to infinity, albeit with increasing attenuation due to the sinc response observed in Figure 11.3. Because the location of the images is calculable, we can exploit this effect to transmit a signal at any desired position on the spectrum.

Let's illustrate the concept with a simple example, with reference to the general response of the DAC presented in Figure 11.3. If we wanted to transmit an 8 MHz sine wave using a DAC with a maximum sample rate of 10 MHz, this would appear not to be possible, as the highest frequency the DAC can represent according to Nyquist sampling theorem is just less than 5 MHz, i.e much lower than the 8 MHz tone of interest. However, as any signal generated by the DAC within the 0-5 MHz range will also produce a spectral image at $f_s - f$, we can instead generate a sine wave at 2 MHz, thus creating an image at 8 MHz. An appropriate filter is required to pass the 8 MHz sine wave, and reject the original 2 MHz tone, as well as any subsequent images located in the upper Nyquist zones. Rather than using a low pass reconstruction filter, we instead use a bandpass filter tuned to suppress frequencies outside the $f_s/2$ to f_s frequency band (i.e. Nyquist Zone 2).

When used to generate signals in the second Nyquist Zone, the sinc-shaped frequency response of the DAC (arising from the zero-order hold sampling process) compromises signal quality. As we saw in Figure 11.3, the sinc response causes the amplitude of the DAC output to decrease at higher frequencies, and notably it is also less flat in the second Nyquist Zone than in the first Nyquist zone — this causes distortion, particularly affecting wide-bandwidth signals. This issue can be overcome by changing the zero-order hold sampling technique (which we hereafter refer to as *Normal Mode*) to one that is more optimised for generating signals in the second Nyquist Zone. In the context of the RFSoC, this technique is referred to as *Mix-Mode* [90], while in some other literature it may also be known as *RF mode*.

In Mix-Mode the rectangular pulses used in the zero-order hold process are modified by inverting the amplitude halfway through the sample period, as shown in Figure 11.5.

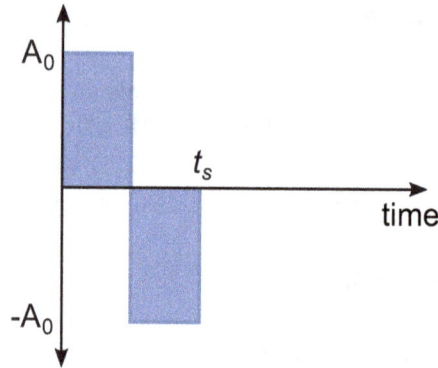

Figure 11.5: Mix-mode impulse response.

In the frequency domain this new function can be described by the equation,

$$H(f) = \frac{sin^2(\pi f / 2f_s)}{\pi f / 2f_s} . \tag{11.4}$$

This change in sampling technique to *Mix-Mode* has a direct effect on the output frequency response, such that it reshapes the energy output by the DAC. This is due to the difference impulse responses and transfer functions of normal mode and mix-mode: compare Figures 11.2 and 11.5, and Equations (11.1) and (11.4). As can be seen in Figure 11.6, the output frequency response of the DAC in Mix-Mode exhibits higher and flatter gain across the second Nyquist Zone, making it much better suited for operation in this frequency band.

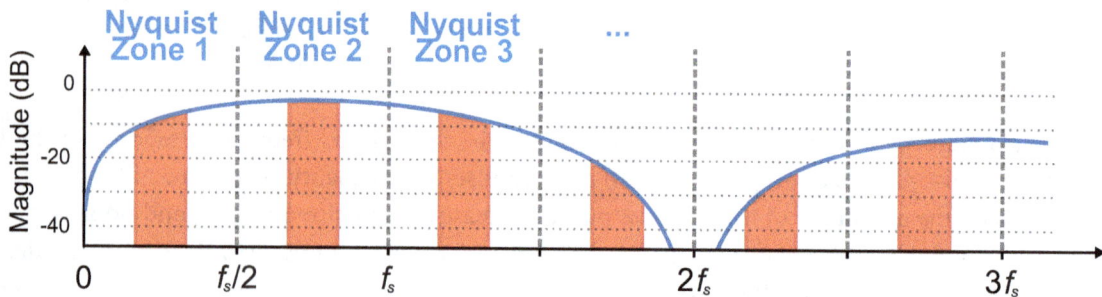

Figure 11.6: Frequency spectrum after mix-mode process.

Figure 11.7 compares the frequency responses of the original, Normal Mode operation of the DAC, and Mix-Mode. Clearly, Normal Mode is preferable in Nyquist Zone 1, and Mix Mode in Nyquist Zone 2. Even so, we observe that there is still a drop in gain at the upper edge of Nyquist Zone 1 (in Normal Mode), and both the upper and lower edges of Nyquist Zone 2 (in Mix-Mode). To compensate for these effects, the signal can be passed through a digital filter (prior to the DAC) that applies a correction to the frequency response.

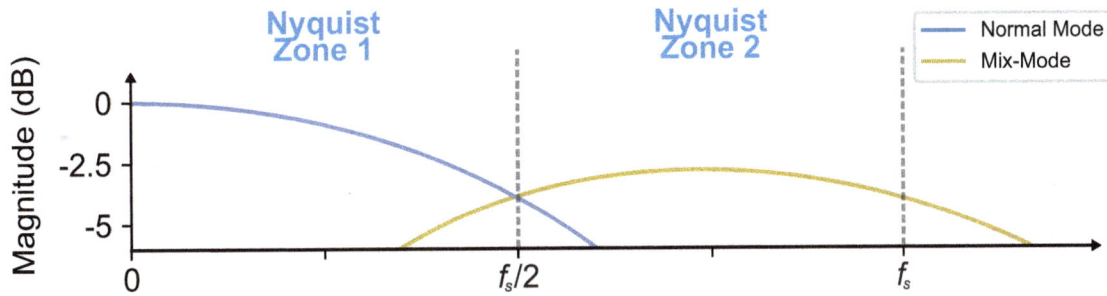

Figure 11.7: DAC output frequency responses for both normal mode and mix-mode.

11.2.2. The Inverse Sinc Filter

As noted over the last few pages, the frequency response of a DAC is not flat, particularly around the edges of the Nyquist Zones, meaning that signals within these regions will be attenuated and often distorted. To combat this, an inverse sinc filter can be applied to correct for the non-linear magnitude response across the target Nyquist Zone [90]. By applying this correction filter, an approximately flat response can be achieved over 90% of Nyquist Zone 1, as shown in Figure 11.8.

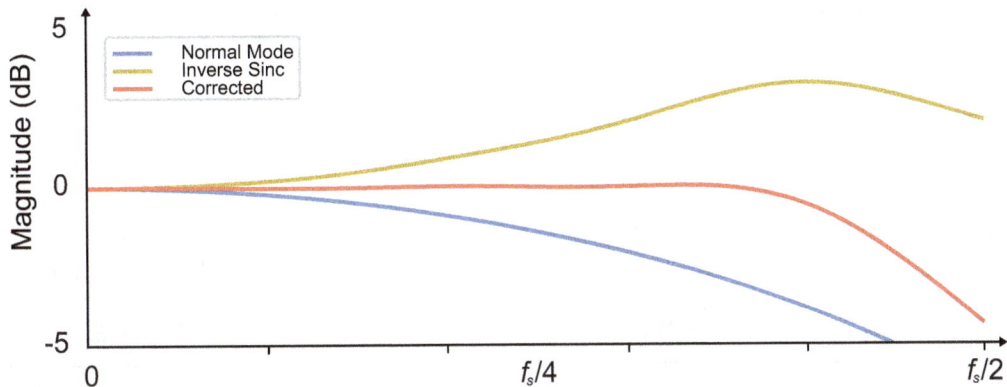

Figure 11.8: Frequency response correction using an inverse sinc filter (Normal Mode).

The frequency response in Mix-Mode can be optimised in a similar way, by applying an inverse sinc filter to correct the non-linear gain of the mix-mode response across Nyquist Zone 2. The result of applying inverse sinc correction is a flatter response over around 80% of Nyquist Zone 2, as shown in Figure 11.9.

Care must be taken when applying inverse sinc filters as they inherently increase the gain of the signal and can cause overflow or saturation. Therefore, it is usually necessary to attenuate the signal slightly before the filter is applied.

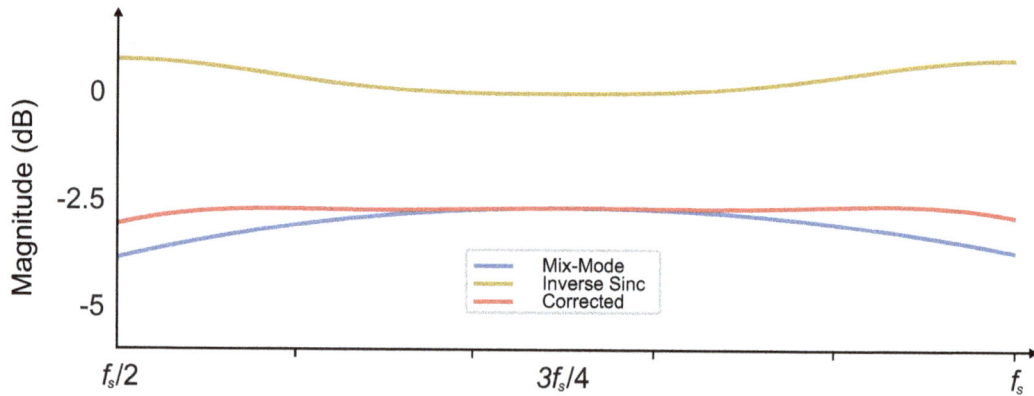

Figure 11.9: Frequency response correction using an inverse sinc filter (Mix Mode).

The RF-DACs on RFSoC devices include a digital inverse sinc filter that can be applied (optionally). It can be enabled by configuring the Vivado RFDC IP at design time, or at run-time with use of the software API. The frequency responses plotted in Figures 11.8 and 11.9 reflect the actual filters implemented as part of the RF-DAC architecture (the coefficients are defined in[90]).

In the following sections we review the RF-DAC architecture in general, and explore some of its other components.

11.3. RF-DACs on RFSoC

The quantity, configuration, structure, and capabilities of the RF-DACs vary between RFSoC devices and generations. Table 11.1 highlights the differences in configuration of the RF-DACs of a selection of devices.

Table 11.1: Comparison of DAC configurations across a selection of RFSoC devices [89].

Generation ('Gen')	Device	Number of DACs	DAC Resolution (number of bits)	Maximum Sampling Rate (Gsps)
1	ZCU28DR	8	14	6.554
	ZCU29DR	16	14	6.554
2	ZCU39DR	16	14	6.554
3	ZCU48DR	8	14	9.85
	ZCU49DR	16	14	9.85
DFE	ZCU65DR	6	14	10.0

An obvious difference between the RF-DACs and RF-ADCs is that the RF-DAC resolution is 14-bits across all devices, compared to the mixture of 12-bit and 14-bit configurations available for the RF-ADCs. The maximum sample rate of the RF-DAC is also significantly higher than the RF-ADC. Over the next few sections, we will go on to discuss the differences (and similarities) between the structure of the RF-DACs and RF-ADCs on the RFSoC.

11.3.1. The RF-DAC Tile Hierarchy

The RFSoC is structured into a hierarchy of tiles and blocks, with each individual RF-DAC and RF-ADC contained within a block, and one (single), two (dual), or four (quad) blocks making up a tile, depending on the device. The number of tiles available also varies according to the device.

All Gen 1 and Gen 2 RFSoC devices include Quad RF-DAC tiles, whereas Gen 3 devices can be composed of Dual tiles, Quad tiles, or a mixture of Dual and Quad tiles. For example, the Gen 1 ZCU28DR device has two Quad DAC tiles, making a total of 8 RF-DACs available; whereas the Gen 3 ZCU46DR has two Quad tiles and two Dual tiles, making a total of 12 RF-DACs. It is also worth noting that one device, the ZCU43DR, contains four Single tiles, with one RF-DAC output per tile, making a total of 4 RF-DACs. As of the time of writing, this is the only device that contains Single tiles.

Quad tiles can be configured either as four individual real signal outputs, or as a pair of complex I/Q signal outputs. The Dual tiles are similar, insofar as they can be configured for either real or complex outputs, however the number of individual outputs is halved.

11.3.2. RF-DAC Tile Composition

Each block in an RF-DAC tile contains a pipeline of components before reaching the actual digital to analogue converter, including a gearbox FIFO, interpolation chain, complex mixer, QMC block, coarse delay, and digital filtering. Many of the pipeline components are identical in function between RFSoC device generations, and at least some of the components are identical to those in the RF-ADCs. A detailed explanation of the pipeline and its components is given in later sections, but first, it is worth reviewing the architectural differences between the tile and block structure of the Dual and Quad tiles over the various device generations.

Gen 1 and 2 Tiles

The RF-DAC tiles within Gen 1 and 2 RFSoC devices are configured to have four (Quad) RF-DACs and one PLL per tile. Each RF-DAC has 14-bit resolution with a highly configurable 16-bit digital signal processing pipeline. As with the RF-ADCs, an external reference clock drives the PLL, which can optionally be bypassed to allow a direct sampling clock to feed into each block on the tile, if preferred. To suit different analogue front-end designs, the output power of an RF-DAC tile is also configurable. In the case of Gen 1 and Gen 2 devices, two modes are available: 20 mA and 32 mA.

The Quad tiles organise the four RF-DAC blocks into two pairs. Each pair can be configured separately as two individual real outputs, or as a pair of separate *In Phase* (I) and *Quadrature* (Q) channels. The RF-DAC that is used within the tile is dependent on the mode of operation. For example, if the tile is configured to output separate I and Q channels, even-numbered DACs are used for I and odd-numbered DACs are used for Q.

Figure 11.10 shows the structure of the Gen 1 and 2 Quad tiles. Each RF-DAC block can be simplified into a single linear pipeline containing multiple signal processing components, as shown in Figure 11.11.

The RF-DAC block pipeline allows for a high degree of reconfigurability, both at design-time and run-time. The input to the block first passes through a gearbox FIFO, which enables conversion between the different data rates of the PL and the RF-DAC, and separates the concatenated I and Q channels (if both are used).

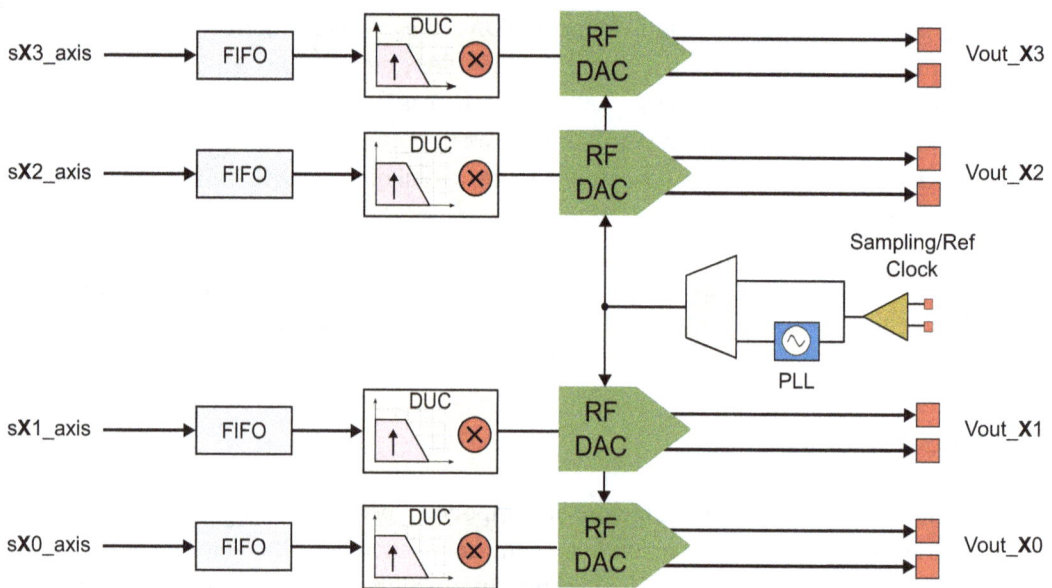

Figure 11.10: *Block diagram of the Gen 1/Gen 2 RF-DAC tile structure.*

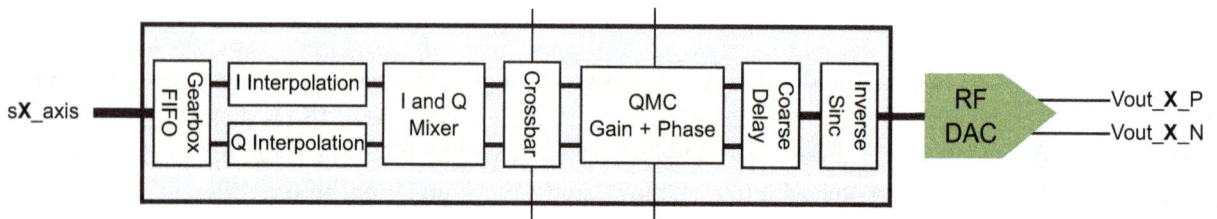

Figure 11.11: *Simplified RF-DAC Gen 1/Gen 2 block pipeline.*

The I and Q paths are then passed through separate interpolation chains, implementing a rate change of up to 8x. The complex mixer can then be used to modulate the separate I/Q paths to a higher frequency carrier. Together, the interpolation chain and the complex mixer make up the Digital Up Converter (DUC), which can be bypassed if interpolation and/or modulation are not required.

If the DUC is enabled, and if a complex signal is being used, the QMC block is able to correct any gain or phase offsets that are experienced between the I and Q channels along the signal path. Further, any timing offsets that exist between channels can be compensated by the coarse delay block. The final stage of the pipeline before digital-to-analogue conversion is the inverse sinc filter. Unlike the previous stages, the inverse sinc filter does not require the DUC to be enabled.

Gen 3 Tiles

The RF-DAC tiles on Gen 3 RFSoC devices are configured to have either one (Single), two (Dual), or four (Quad) RF-DACs and one PLL per tile. The Quad tiles work in much the same manner as their Gen 1 and 2 counterparts, with the main differences being in the RF-DAC blocks. As with Gen 3 RF-ADCs, the RF-DAC tiles also allow for clock distribution between adjacent tiles.

The Dual tiles contain two RF-DAC outputs per tile, as shown in Figure 11.12. However, each Dual RF-DAC tile still contains four DUCs and related pipeline components — the purposes of which we will discuss over the next few sections. As with the Quad tiles, the Dual tiles can also be configured to output separate I and Q channels, although as they only have two RF-DAC outputs, only one I/Q instance is available per tile.

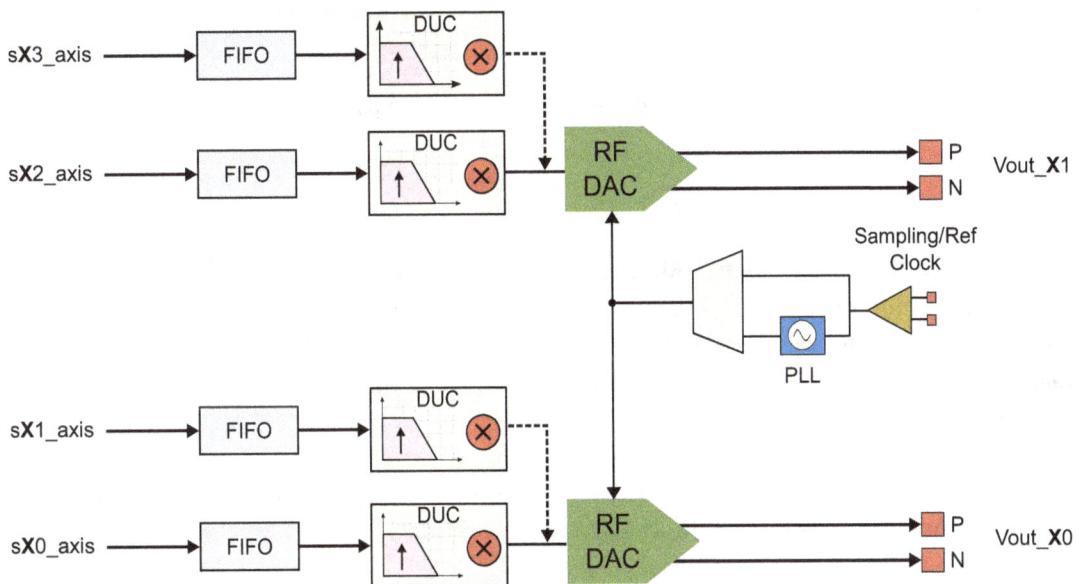

Figure 11.12: Block diagram of the Gen 3 RF-DAC Dual tile structure.

Single RF-DAC tiles follow the same structure as the Dual tiles, in the sense that they still contain four DUCs, but only provide one RF-DAC output instead of two. Because of this, the Single RF-DAC tiles cannot be configured for separate I and Q outputs.

Although the DSP pipeline of the Gen 3 RF-DAC blocks is similar to that of Gen 1 and 2, there are some key differences worth detailing. Figure 11.13 shows the RF-DAC block pipeline for the Gen 3 devices.

The first change to the RF-DAC blocks is the addition of Variable Output Power (VOP) to the RF-DAC itself. This feature allows much finer control of the RF-DAC output power, with an enhanced 10-bit resolution providing 1024 different levels, and is backwards compatible with the 20/32 mA modes available on Gen 1 and 2 devices. The VOP can be controlled either from the PL via IP interface ports, or from the PS via the software API.

The programmable interpolation chain has also been updated to enable 13 different interpolation factors, including a trivial 1x interpolation. These are: 1x (bypass), 2x, 3x, 4x, 5x, 6x, 8x, 10x, 12x, 16x, 20x, 24x, and 40x. The addition of higher rate changes and, generally, more rate change options, greatly reduces the need for any supplemental interpolation on the PL, freeing up FPGA space for additional DSP logic.

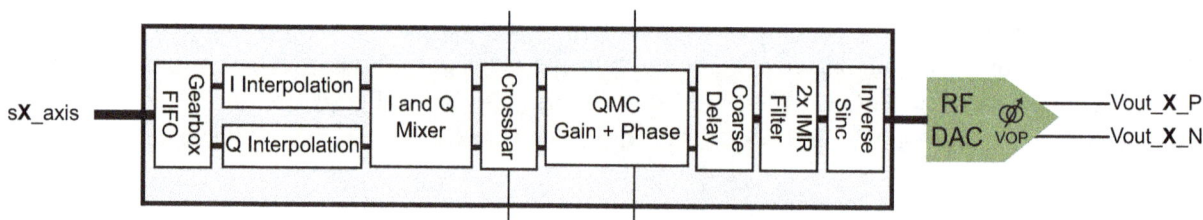

Figure 11.13: Simplified RF-DAC Gen 3 block pipeline.

Another change to the pipeline is the addition of the IMage Rejection (IMR) filter. The IMR filter can be configured as either low or high pass, which aids in suppressing images in the second and first Nyquist zones, respectively. The IMR filter can only be used in combination with the DUC, and introduces an additional 2x interpolation to the pipeline — this means that, if the IMR is enabled, the maximum interpolation rate is increased to 80x.

Finally, in Gen 3 devices, the inverse sinc filter has been updated to include correction for Mix-Mode (i.e. Nyquist Zone 2) as well as Normal Mode, as discussed in Section 11.2.

11.4. RF-DAC Processing Stages: The Programmable Interpolator

After data is passed through the gearbox FIFO, shown on the left hand side of Figure 11.13, the first processing stage increases the sampling rate, such that it is closer to (or equal to) the DAC sampling rate used for transmission. As described in Chapter 4, interpolation is the process of increasing the sample rate of a signal by inserting zeros in between the original samples (known as *upsampling* or *zero-stuffing*), then filtering the resultant signal to remove the spectral images that are generated.

On the RFSoC, the interpolation stage is part of the DUC and it can operate in either real or I/Q (complex) modes, depending on the RF-DAC configuration. First and second generation devices have a different interpolation chain structure than the third generation devices, as outlined next.

11.4.1. The Gen 1 / Gen 2 Interpolation Chain

Interpolation on the first and second generation devices is achieved by three cascaded stages of upsamplers and low pass FIR filters, each performing a rate change of 2x, denoted FIR2, FIR1, and FIR0. Each filter stage can be bypassed, and the output of each stage can be routed to the final output of the interpolation chain, as shown in Figure 11.14.

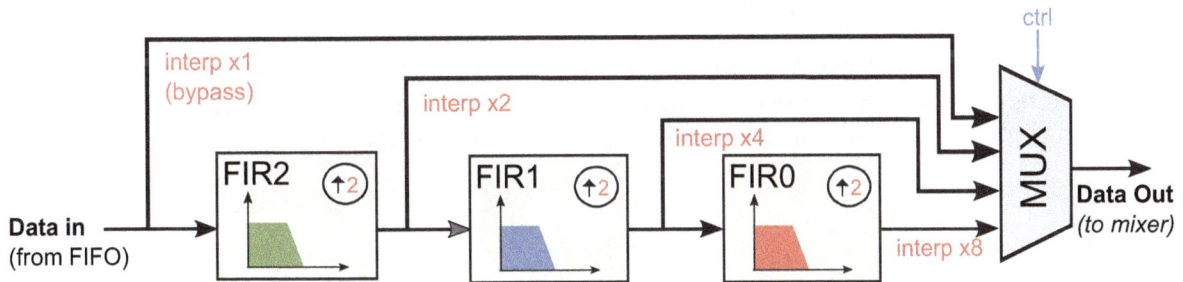

Figure 11.14: Gen 1 / Gen 2 interpolation chain.

Although the amount of interpolation is programmable by the user, the coefficients used for each of the FIR filter stages is fixed. The frequency responses of each of the three FIR filter stages are shown in Figure 11.15.

As with the RF-ADC decimation chain, the first FIR filter stage has the sharpest cut-off, with the subsequent filters having progressively more relaxed responses due to the increasing gap between the signal of interest, and the position of the spectral image to be attenuated. Each of the three interpolation stages can either be bypassed, or used in cascade with the others, which produces four different interpolation options: 1x (bypass), 2x, 4x, and 8x.

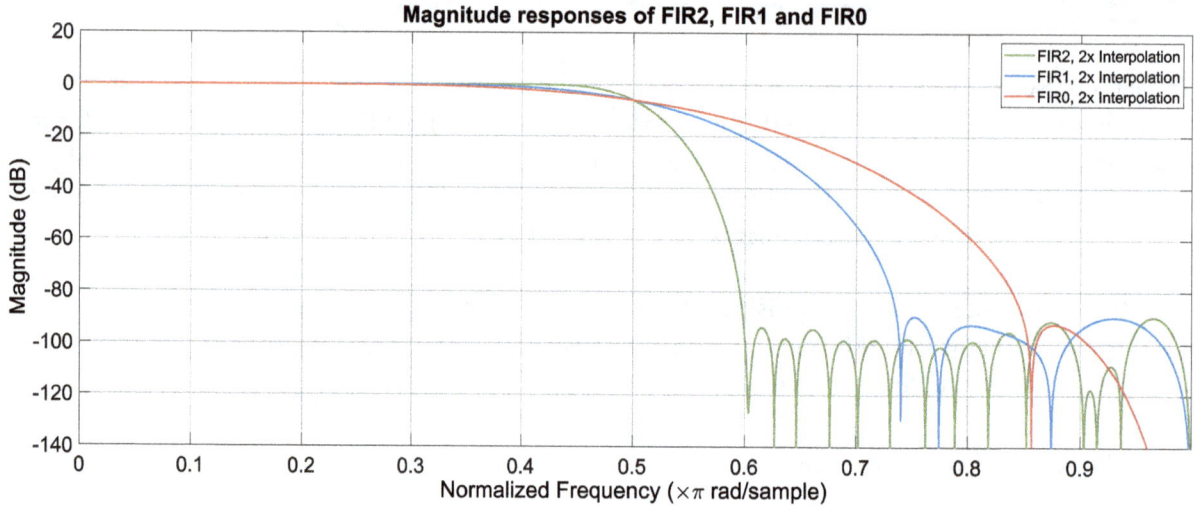

Figure 11.15: Magnitude response of the three FIR filters in the Gen 1/Gen 2 interpolation chain.

11.4.2. The Gen 3 Interpolation Chain

The interpolator in Gen 3 devices offers a significant increase in rate change options compared to Gen 1 and 2 devices. The interpolation chain consists of four cascaded stages and, as with the previous generations, each stage can optionally be bypassed. Figure 11.16 shows a block diagram of the Gen 3 interpolation chain.

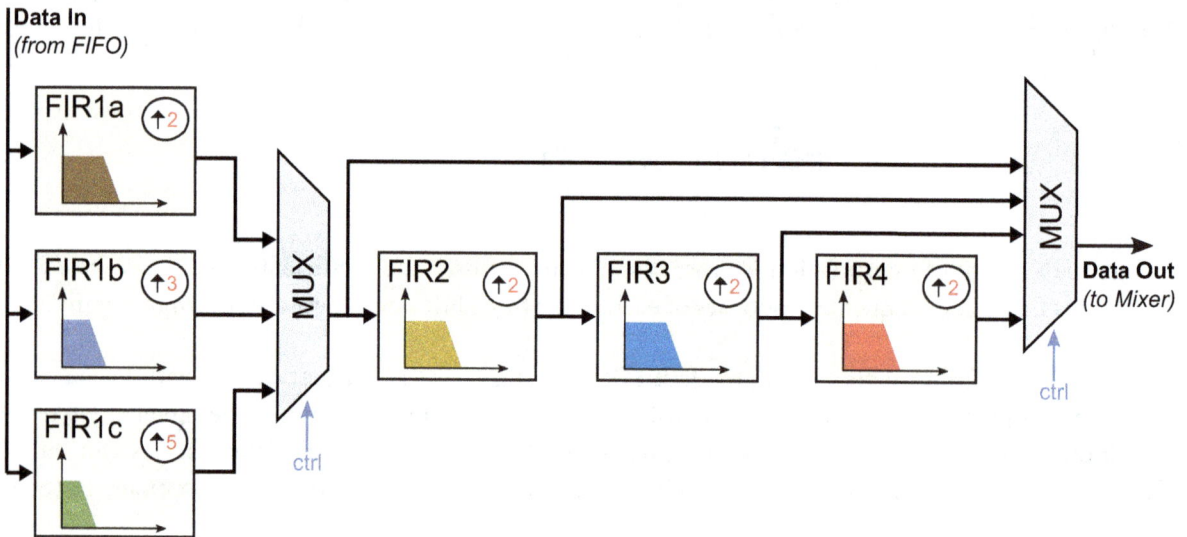

Figure 11.16: Gen 3 interpolation chain.

Using a combination of these filter stages, the Gen 3 interpolation chain can achieve the set of interpolation ratios outlined earlier, i.e.: 1, 2, 3, 4, 5, 6, 8, 10, 12, 16, 20, 24, and 40. A few examples of how these filters can be cascaded together to produce each possible interpolation factor are provided in Table 11.2.

Table 11.2: Examples of interpolation factors supported by Gen 3 RFSoC devices.

Interpolation Factor	Enabled Filters	Interpolation Chain
2	FIR1a	2
4	FIR1a + FIR2	2 x 2 = 4
6	FIR1b + FIR2	3 x 2 = 6
10	FIR1c + FIR2	5 x 2 = 10
20	FIR1c + FIR2 + FIR3	5 x 2 x 2 = 20
24	FIR1b + FIR2 + FIR3 + FIR4	3 x 2 x 2 x 2 = 24

The first stage of the interpolation chain comprises of three multiplexed upsamplers and low pass FIR filters, denoted as FIR1a, FIR1b, and FIR1c; allowing 2x, 3x, or 5x interpolation, respectively. Only one filter in this group can be used at a time. As these filters collectively represent the first stage of the chain, they all have steep transition bands, compared to the subsequent stages. The frequency responses for each of the filters in the first stage are shown in Figure 11.17.

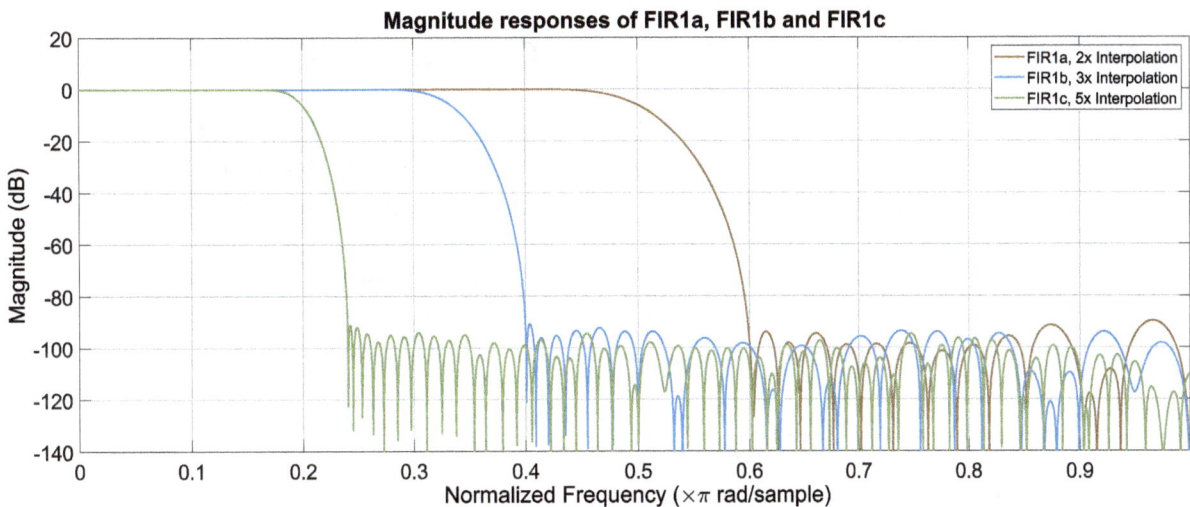

Figure 11.17: Magnitude responses of FIR1a, FIR1b, and FIR1c in the Gen 3 interpolation chain.

The second, third, and fourth stages of the interpolation chain all have a 2x interpolation factor, and each can be bypassed individually. As with the interpolation chain in the previous generation devices, each subsequent filter has a more relaxed response, with FIR3 and FIR4 both having identical responses. The frequency responses of stages 2, 3, and 4 are shown in Figure 11.18; these have been generated based on the filter specifications published in [90].

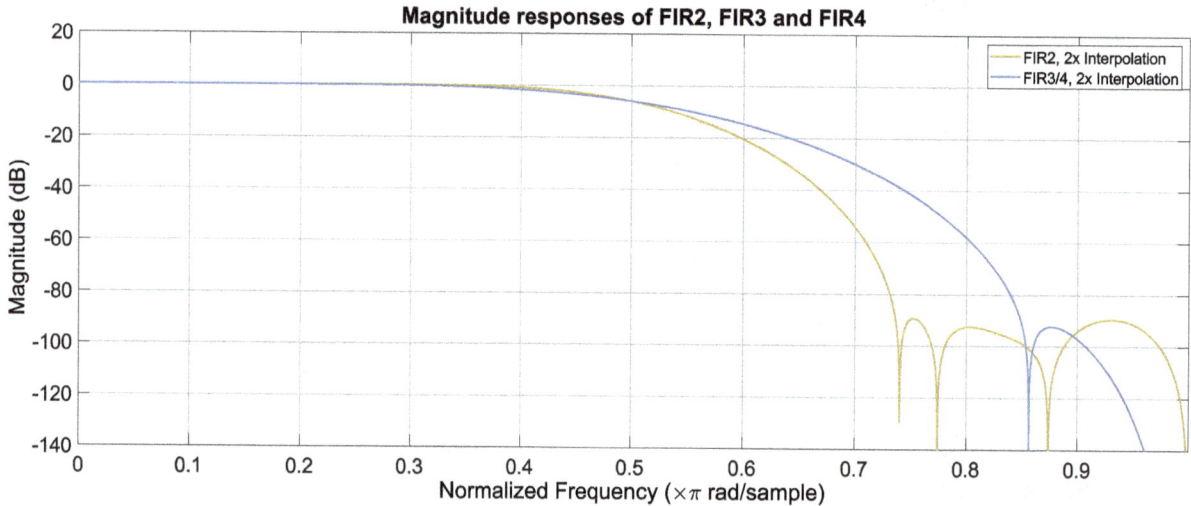

Figure 11.18: Magnitude responses of FIR2, and FIR3, / FIR4 in the Gen 3 interpolation chain.

11.5. RF-DAC Processing Stages: The Digital Complex Mixer

Functionally, the complex mixer within the RF-DAC is identical to that from the RF-ADC; the difference being that its role in the RF-DAC is to modulate the data, rather than demodulate it. The function of the mixer is to shift the signal into a target frequency band by mixing the input signal with a higher frequency carrier.

The RF-DAC mixer includes a 48-bit digital NCO fine mixer, enabling modulation with arbitrary frequency carriers, and a coarse mixer for mixing with one of a small set of preset carriers. The mixer supports I/Q to real, and I/Q to I/Q modes. A block diagram of the complex mixer is shown in Figure 11.19.

11.5.1. The Coarse Mixer and Bypass

The coarse mixer is the simpler of the two mixer modes and only allows mixing with a restricted set of carriers, i.e. $f_s/2$, $f_s/4$, or $-f_s/4$. However, the coarse mixer can operate at much lower power than the fine mixer, as its architecture is simpler, and in particular, it does not involve storing and accessing an extensive lookup table of sine wave amplitudes.

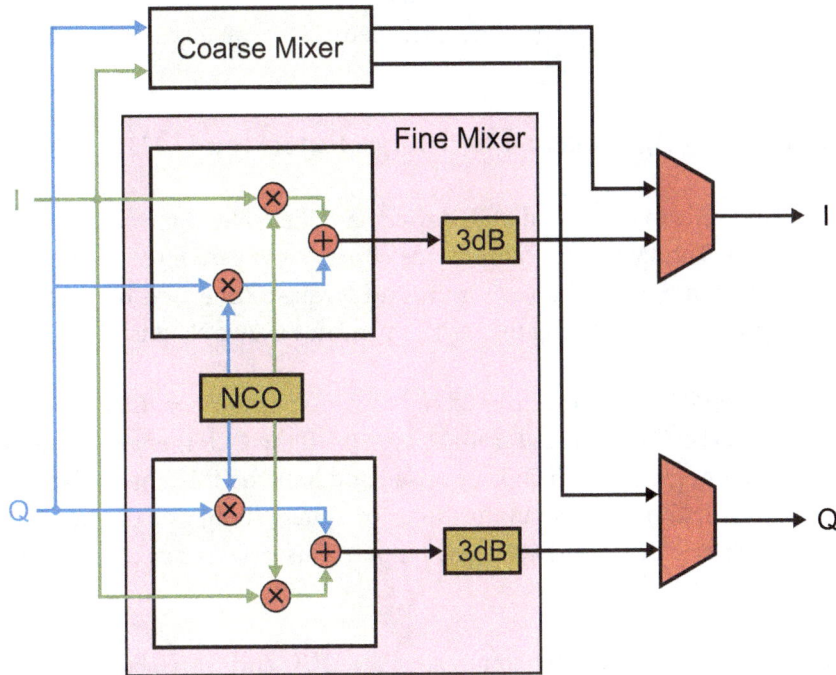

Figure 11.19: Block diagram of the RF-DAC mixer.

The coarse mixer can also be used to bypass the entire mixing component, whereby the signal is simply passed through the mixer unmodulated. In this mode, only real input and real output signals are permitted.

11.5.2. The Fine Mixer

The fine mixer consists of an NCO and set of modulators (multiplication and addition/subtraction operators) and is capable of shifting an information signal by an arbitrary frequency, including control over the phase. The NCO provides fine control over the generated frequency, in addition to 18-bit phase adjustment, both of which are programmable by the user. If using multi-tile synchronisation, the phase of the NCO can also be synchronised across tiles.

As can be seen from Figure 11.19, the output of the fine mixer includes a 3dB attenuation block that helps to avoid potential overflow in the mixing process (functionally, the attenuation block multiplies the output signal amplitude by 0.707, which corresponds to a halving of signal power, i.e. 3dB attenuation). The attenuation block can be enabled or disabled as required.

After mixing, signals within even-numbered Nyquist zones are inverted (flipped left-to-right in frequency), while signals within odd-numbered Nyquist zones are not. Any inversion can be reversed by using a negative frequency in the NCO.

As the IMR filter in Gen devices introduces an additional 2x interpolation, the sample rate at the mixer is equal to half of the DAC output sample rate (when IMR is enabled). Next, we discuss the IMR filter in more detail.

11.6. RF-DAC Processing Stages: The Image Rejection (IMR) Filter

The IMR filter is available only in Gen 3 and DFE devices and it enables removal of unwanted spectral images after mixing has occurred. The IMR filter can provide either a low pass filter, retaining the signal in the 1st Nyquist Zone and attenuating the signal image in the 2nd Nyquist Zone; or a high pass filter, attenuating the signal in the 1st Nyquist Zone and retaining the signal image in the 2nd Nyquist Zone.

In addition to filtering, the IMR also introduces 2x interpolation to the signal. Because of this, the maximum interpolation rate of the RF-DAC is increased from 40x to 80x. If the IMR is enabled, and only 2x interpolation is required, the pre-mixer interpolation chain is bypassed and only the IMR interpolator is used. The 2x interpolation introduced by the IMR filter has ramifications for system design, as the sample rate at the output of the complex mixer will be half that of the RF-DAC output. This issue is discussed in more detail in Section 11.7.5.

In the RF-DAC data pipeline, the IMR filter comes after the DUC and QMC, but before the inverse sinc filter. While the IMR can be enabled and disabled by the user at run-time, it can only be used when the DUC is enabled, meaning that the IMR cannot be used when the DUC is in bypass mode. Figure 11.20 shows the RF-DAC data path with the various routing options for the IMR filter.

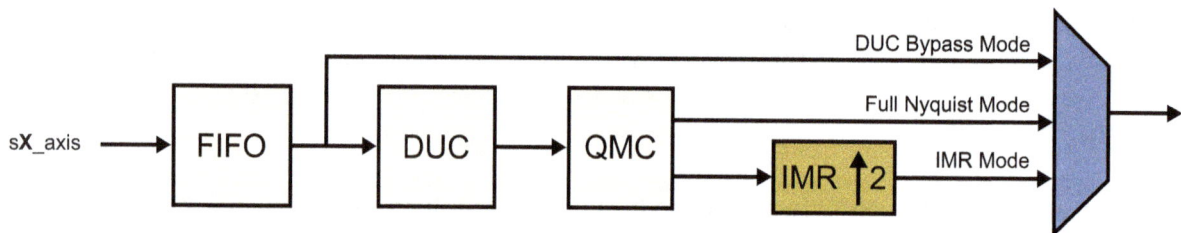

Figure 11.20: RF-DAC IMR data path.

The high and low pass filters of the IMR filter are symmetric, both providing 60dB stopband attenuation with reference to the carrier (60 dBc). It is important to note that the cut-off frequencies of the Nyquist filters are in relation to the sample rate *before* interpolation, meaning that the $f_s/2$ cut-off frequency relates to the sample rate at the output of the DUC, rather than at the output of the IMR.

As an example, consider that the sample rates after the DUC and IMR are 1 Gsps and 2 Gsps, respectively. In this case, the IMR filter cut-off frequency would be 500 MHz, rather than 1 GHz. Figure 11.21 shows the frequency responses of both IMR filters.

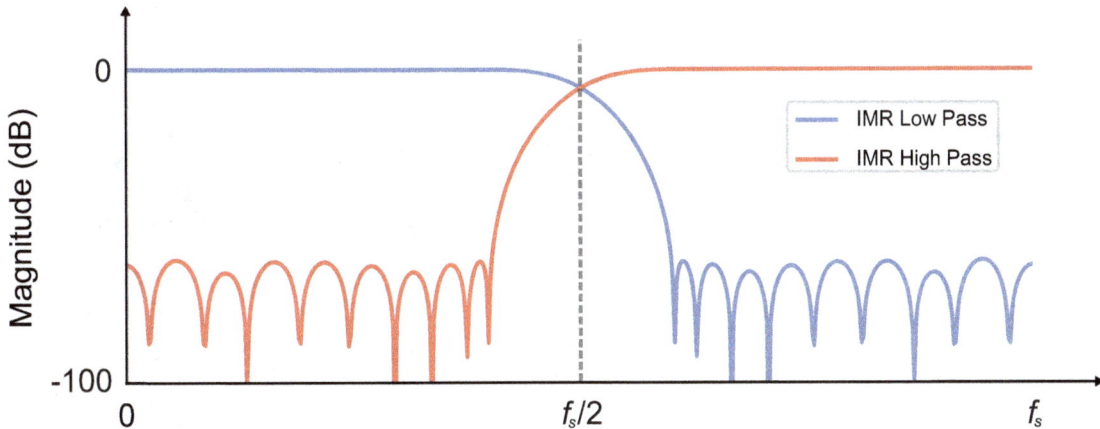

Figure 11.21: IMR low and high pass filter frequency responses.

11.7. Principles of RF-DAC Operation

In earlier chapters, we discussed various receiver architectures available on the RFSoC using the RF-ADCs. The RFSoC is similarly capable of serving several different transmitter architectures due to its ability to work with both real and complex RF-DAC outputs. Much of the discussion regarding both real and complex RF receiver architectures has been covered in Chapter 9, with many of the processing stages corresponding closely to the equivalent stage of the transmitter architecture (but in reverse order). For this reason, we will only briefly discuss transmitter architectures, and quickly move on to the details that are more specific to RF-DACs.

11.7.1. Real Signal Architecture

For real outputs there are two main options for the transmitter architecture: direct conversion to RF, or using an IF stage.

In *direct-RF*, the signal output by the RF-DAC is passed to the antenna via the analogue front-end (i.e. filtering and amplification in the analogue domain). If an IF stage is used, the output of the RF-DAC additionally undergoes a stage of analogue mixing that modulates the signal to a frequency above baseband, but lower than the eventual RF carrier frequency. The option to use an IF stage is usually chosen when the RF frequency is higher than can be generated in the digital domain, given the sampling rate of the RF-DAC.

In both cases, the signal can either be purely real, or I/Q modulated, the main point being is that the signal at the output of the RF-DAC is a single composite analogue output. Figure 11.22 depicts both of these architectures.

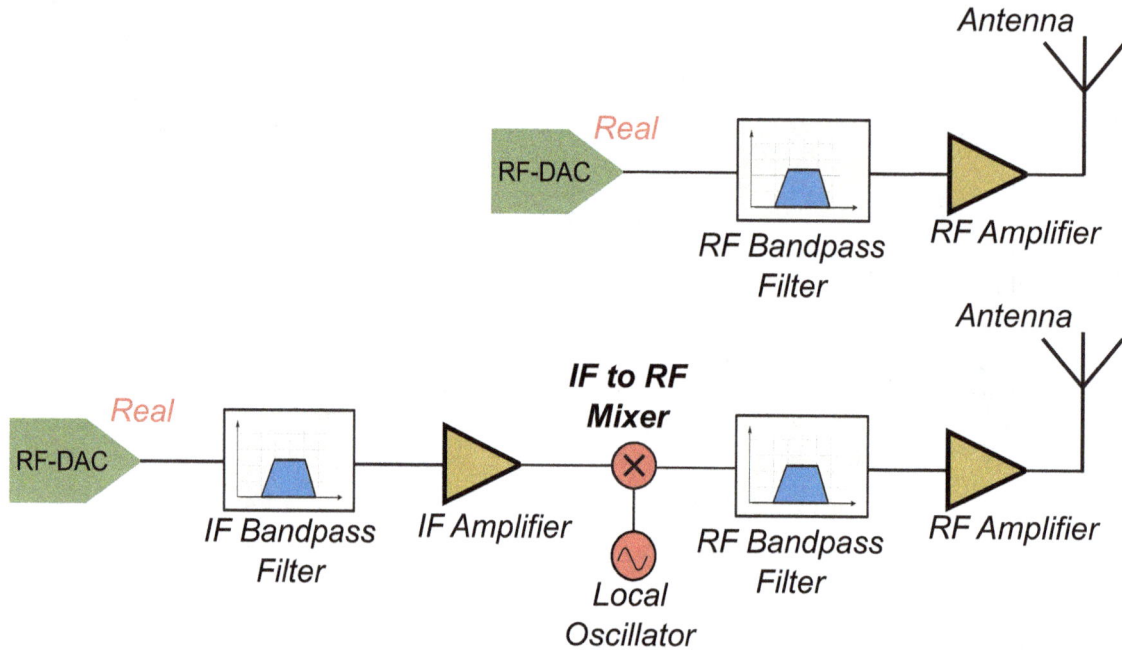

Figure 11.22: Example of two real signal transmitter architectures. Direct RF (top), and IF (bottom).

11.7.2. Complex Signal Architecture

In a complex signal architecture, the I and Q components have separate internal signal paths, and leave the RF-DACs as two separate analogue signals, offset in phase by 90 degrees. In this scenario the combination of the I and Q channels takes place in the analogue domain, along with any necessary correction of phase or gain offsets.

Figure 11.23 shows an example of this complex signal architecture. As this transmitter modulates directly from baseband to RF, low pass filters are included to retain only the baseband signal components and attenuate spectral images from the 2nd Nyquist Zone and above. This contrasts with the real transmitter designs from Figure 11.22, where the signal is modulated to IF or RF in the digital domain, and which instead require bandpass filters around the signal bands of interest.

11.7.3. Operating Modes

To facilitate these architectures, the RF-DACs can be operated in two different modes: *Complex-to-Real*, or *Complex-to-Complex*.

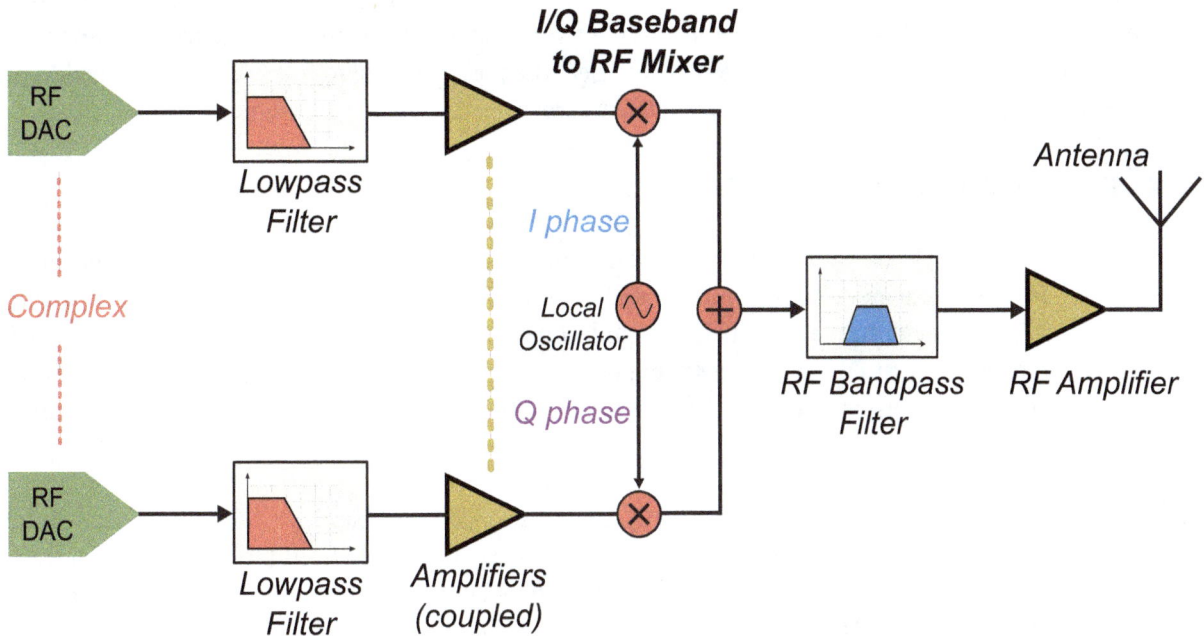

Figure 11.23: Example of a complex transmitter architecture.

In Complex-to-Real (C2R) mode, the I/Q outputs of the DUC are summed together to create a single channel prior to the RF-DAC analogue output. In this mode, Quad RF-DAC tiles can output four separate channels on a single tile, while Dual RF-DAC tiles can output two.

In Complex-to-Complex (C2C) mode, the I/Q outputs of the DUC are kept separate throughout the entire data path. At the point of the crossbar (shown in Figures 11.11 and 11.13), the Q signal is routed to an odd-numbered RF-DAC, while the I signal continues along an even-numbered RF-DAC. The two signals are then passed to two separate RF-DAC analogue outputs. In this mode, Quad RF-DAC tiles can output two separate C2C channels, while Dual RF-DAC tiles can only output a single C2C channel.

The DUC is required for both C2R and C2C modes. The input to the RF-DAC tile must be a single AXI4-Stream signal with I and Q channels concatenated together. The gearbox FIFO is responsible for separating the complex signal into its I and Q components.

A further option exists wherein a real-only signal is passed from the RF-DAC input to the RF-DAC analogue output. There are two possible configurations for this mode. The first is that the real-only signal is passed through the RF-DAC input directly to the RF-DAC analogue output, bypassing the DUC entirely, removing the options for mixing or interpolation. The second configuration, if mixing and interpolation are required, is to use the coarse mixer. However, the use of the coarse mixer greatly restricts the possible frequencies that the signal can be modulated onto. If the fine mixer is required, then it is possible to use C2R mode by making the imaginary part of the concatenated input signal entirely zero-valued.

11.7.4. Transmission of Multiple Bands

A single RF-DAC is capable of simultaneously transmitting multiple signals that have been modulated onto different carriers. This is achieved using multi-band mode, wherein several signals are combined into a composite RF-DAC analogue output. Multi-band mode involves combining the outputs of multiple DUCs within a single tile, where each DUC processes a single signal. As the use of DUCs is required, they cannot be used in bypass mode. When using multi-band mode, latencies between the individual DUCs are automatically synchronised across the tile.

Each tile can be configured for real dual-band, I/Q dual-band, real quad-band, or I/Q quad-band. In dual-band mode, two DUCs are used, while in quad-band mode four DUCs are used. All DUCs must be configured in the same way. For example, if I/Q quad-band is used, all four DUCs must be configured for I/Q output. In real mode a single analogue RF-DAC output is used, whereas in I/Q mode, two RF-DAC analogue outputs are used. As with C2C mode, odd-numbered RF-DACs are used for real outputs, whereas even-numbered tiles are used for the imaginary outputs.

Although the Dual tiles have only two RF-DAC analogue outputs, four DUCs are present in the tile. This allows both dual-band and quad-bands, and both real and I/Q outputs in the Dual RF-DAC tile devices. As Single RF-DAC tiles also contain four DUCs, multi-band operation is also supported. However, given that they only contain one RF-DAC analogue output, transmission of separate I/Q channels is not possible.

As multiple signals are summed together in multi-band mode, it is possible that overflow can occur. To avoid this possibility, the signals are automatically attenuated at the output of each DUC before they are summed together. For dual-band, 6dB of attenuation is applied per channel, while in quad-band, 12dB of attenuation per channel is used [90]. If required, the attenuation stage can be bypassed by means of the software API.

Example:

A Quad RF-DAC tile can be configured to transmit four separate signals within the same Nyquist zone. Four separate inputs are required from the PL to the RF-DAC tile. Each signal is passed through its respective DUC and mixed to an individual carrier frequency. After mixing, the outputs of the DUCs are summed together to create a single composite signal along the DAC0 data path. If used, the composite signal can then be passed through the IMR and inverse sinc filter before being converted to analogue at the RF-DAC output. Figure 11.24 shows the RF-DAC data path for this example.

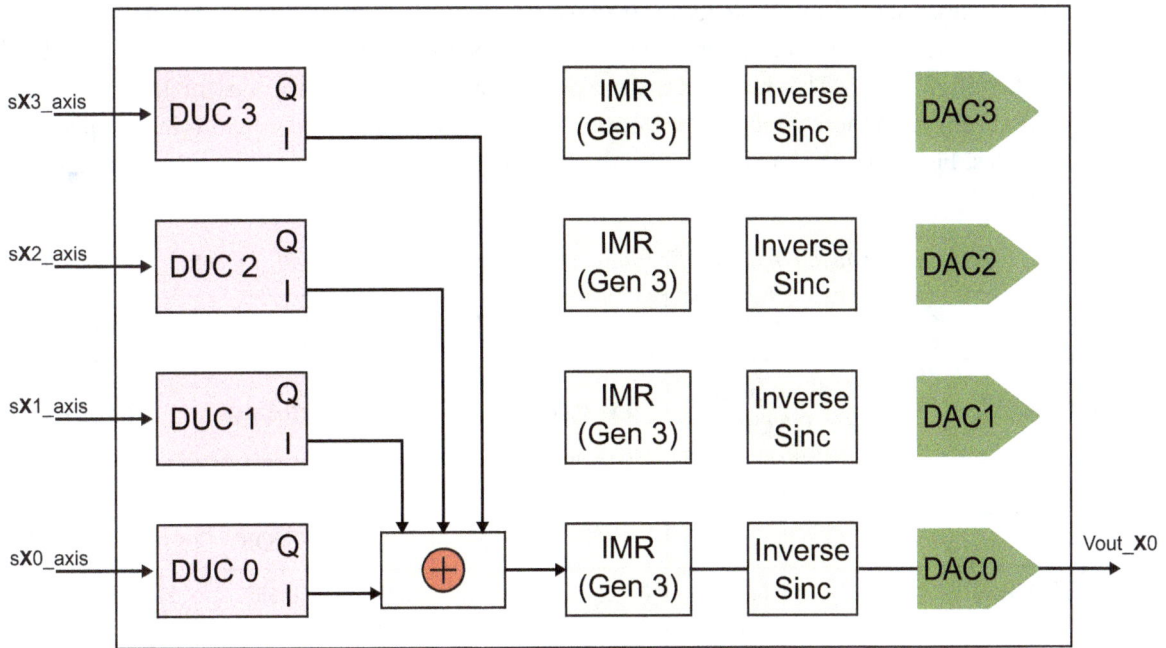

Figure 11.24: high level block diagram of a real quad-band configuration for a Quad RF-DAC tile.

11.7.5. Nyquist Zones in Real World Scenarios

Earlier in the chapter, we discussed the theory of digital to analogue conversion and the role of filtering in removing unwanted images from the signal. Typically, at the output of the DAC, an analogue reconstruction filter would be used to remove these images — a low pass filter for the first Nyquist zone, or a bandpass filter for the second Nyquist zone, each with a bandwidth equal to the size of a single Nyquist zone.

Ideally, this filter would have a brick wall response, removing any images outside the filter's passband entirely. In practice, however, there will be a transition band between the cut-off frequency and the desired stopband, where the amplitude of the frequency response drops off more gradually. This has implications when transmitting signals, as any signal that is close the edge of the Nyquist zone will have an image that may not be sufficiently suppressed by the reconstruction filter. For this reason it is recommended to avoid transmitting IF or RF modulated signals within the edge regions of the Nyquist zones, typically ±5% of the sample rate.

There are additional considerations to be made for designs that use the IMR filter in Gen 3 devices. As seen in the frequency response of the low and high pass IMR filters in Figure 11.21, there is likely to be insufficient suppression of images that fall within the edges of the Nyquist zone. As the IMR filter introduces 2x interpolation after the mixer, this further reduces the bands that the input signal should be modulated onto by the DUC. Areas around the edges of the Nyquist zones should be avoided (noting that, when using the IMR, Nyquist Zones are defined in relation to the sample rate at the IMR input, rather than at its output).

To address these issues, Gen 3 RFSoC devices introduce the concept of the RF-DAC **datapath mode**. Four separate modes act as a guide for designers when using the various applicable combinations of RF-DAC components and configurations. This includes the use of Normal Mode or Mix-Mode, the IMR filter, and the DUC. Table 11.3 summarises these configuration options, as well as the usable bandwidth in the spectrum for each of these modes. Figure 11.25 visually clarifies the bands that should be avoided when using the various datapath modes.

Table 11.3: Datapath mode configuration options (reproduced from [90], Table 63)

Mode	Mode 1		Mode 2		Mode 3		Mode 4	
Short Name	**Full Nyquist DUC**		**IMR Low-pass**		**IMR High-pass**		**DUC_Bypass**	
IMR x2	OFF		ON		ON		OFF	
Mix-Mode	OFF	ON	OFF	ON	OFF	ON	OFF	ON
Useable bandwidth (Fs)	0 - 0.45	0.55 - 0.95	0 - 0.2	0.8 - 0.95	0.3 - 0.45	0.55 - 0.7	0 - 0.45	0.55 - 0.95

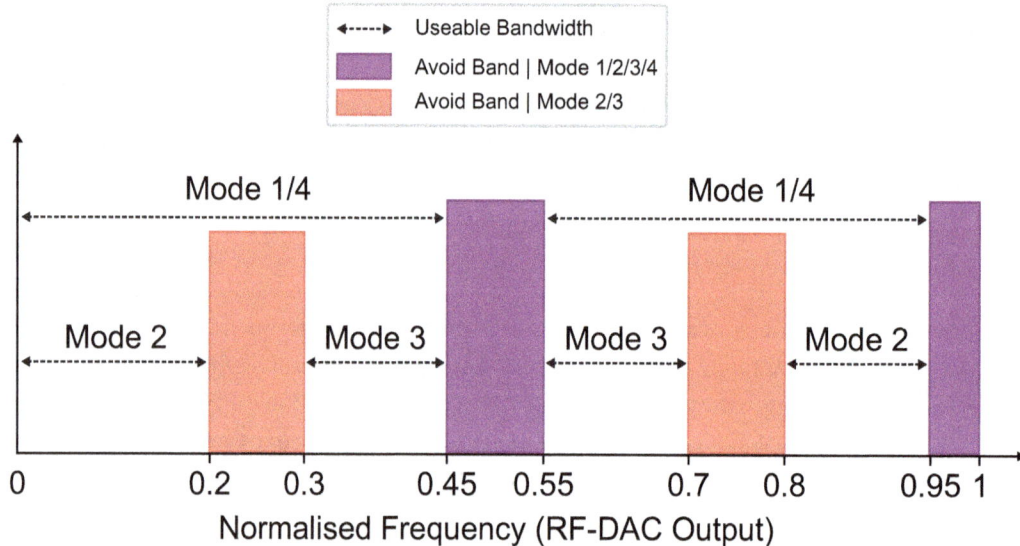

Figure 11.25: Frequency band restrictions for the RF-DAC.

11.8. Design Tool Support

As with the RF-ADCs, the RF-DACs can be added to any Vivado project with the inclusion of the **RF Data Converter** IP core. This IP core allows the setup and configuration of the RF-DAC tiles and blocks within the IP core settings. Many of the details of the IP core configuration are covered in Chapter 9, therefore only the details relevant to the RF-DAC are discussed here.

11.8.1. RF-DAC Tile Configuration Overview

In the GUI for the main RFDC IP core, the designer can configure the RF-DACs by selecting the 'RF-DAC' tab, indicated in Figure 11.26, within which various options are available for configuration of the tile as a whole ('common settings').

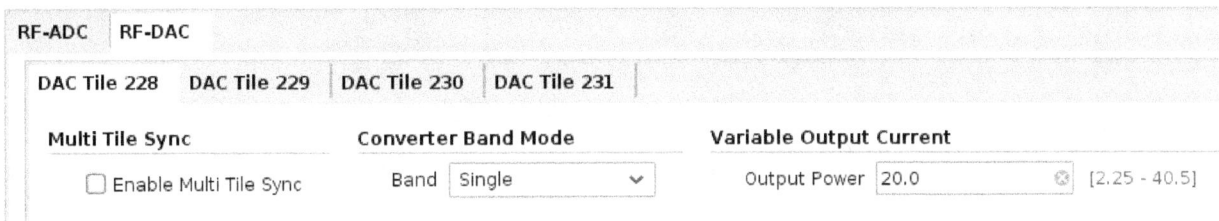

Figure 11.26: RFDC IP core tile configuration GUI, common RF-DAC tile configuration options.

Additionally, various parameters can be controlled independently for each individual RF-ADC.

A summary of both categories of configuration options is provided over the next few pages.

11.8.2. Common RF-DAC Tile Configuration Options

There are three configuration options common to each RF-DAC tile, as seen in Figure 11.26. These can be summarised as follows:

- **Multi Tile Sync** — Similar to the RF-ADC option, multi-tile synchronisation can be enabled to align RF-DAC latency across tiles.

- **Converter Band Mode** — This allows the use of multi-band mode within the RF-DAC tile. Several options are available, dependent on device and tile type.

- **Variable Output Current (Gen 3/DFE)** — Allows the output current of the RF-DAC to be controlled. This option is persistent across all RF-DACs within the tile, and cannot be set for individual RF-DACs.

11.8.3. Individual RF-DAC Configuration Options

Each individual RF-DAC can be configured with its own individual settings. These fall into four categories, as seen in the RFDC IP core tile configuration GUI shown in Figure 11.27, and summarised thereafter.

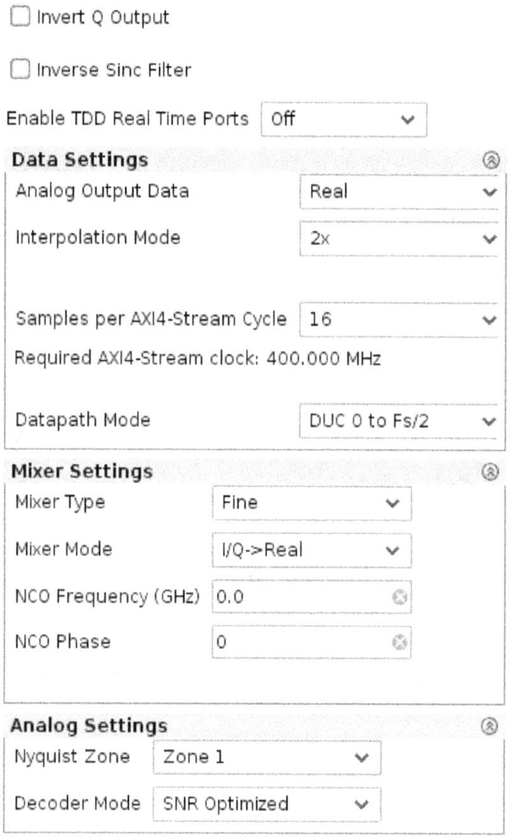

Figure 11.27: RFDC IP core tile configuration GUI, individual RF-DAC tile configuration options.

Generic Settings

- **Invert Q output** — Configurable only when I/Q output data is selected and the fine mixer is enabled. If selected, this option inverts the Q channel.

- **Inverse Sinc Filter** — This option allows the designer to enable or disable the inverse sinc filter. On Gen 3 devices, the filter selection is determined by the Nyquist zone in use.

- **Enable TDD Real Time Ports (Gen 3/DFE)** — Provides output ports to the IP that allow power savings by enabling the powering down of RF-DACs or RF-ADCs. This option is useful when a Time Division

Duplex (TDD) communications protocol is used, i.e. two terminals communicate with each other by transmitting / receiving on the same frequency band, using a time-sharing scheme. In this case, there is potential for the transmitter to be powered down some of the time.

Data Settings

- **Analog Output Data** — Configures the data of the RF-DAC as real or complex. For complex mode, both pairs of RF-DACs must be enabled.

- **Interpolation Mode** — Sets the interpolation rate of the DUC within the RF-DAC.

- **Sample per AXI4-Stream Cycle** — Configures the number of samples the input to the RF-DAC uses (1 to 16).

- **Datapath Mode (Gen 3/DFE)** — Allows the selection of the datapath mode (see Section 11.7.5).

Mixer Settings

- **Mixer Type** — Configures which mixer the DUC uses (coarse, fine, or bypass). In Gen 3/DFE devices, the bypass option is removed and is instead selected within the Datapath Mode setting.

- **Mixer Mode** — Configures which mode the mixer operates in (I/Q-to-Real, or Real-to-Real).

- **Coarse Mixer Frequency** — Sets the frequency of the coarse mixer to one of the available options, i.e. $f_s/2$, $f_s/4$, or $-f_s/4$.

- **NCO Frequency** — Specifies the frequency of the fine mixer, between -10 and 10 GHz.

- **Fine Mixer Phase** — Sets the phase of the fine mixer, between -180 and 180 degrees.

Analogue Settings

- **Nyquist Zone** — Selects between even and odd Nyquist zone operation.

- **Decoder Mode** — Select between RF-DAC optimisation modes. The RF-DAC can be configured as optimised for either low SNR, or high linearity.

11.9. Chapter Summary

This chapter has detailed the RF-DAC as a component of the RFSoC. After a brief recap of analogue to digital conversion in general, we reviewed the operation of RF-DACs, and their ability to transmit efficiently in higher Nyquist zones, in particular using *Mix Mode* to target the 2nd Nyquist Zone. The issue of the non-ideal sinc frequency response was identified, and it was demonstrated that an inverse sinc filter can mitigate its effects.

The range of RFSoC devices was reviewed, discussing the differences between generations, and the architectural features of the RF-DAC tiles. This included the interpolation chains, the coarse and fine mixers to modulate signals onto carriers, and the role of the IMR filter in Gen 3 devices. Following on, we examined the principles of RF-DAC operation, and the different configurations of real and complex architectures that the RF-DACs can enable. We also examined how composite multi-band signals can be composed using several DUCs, and transmitted from a single RF-DAC output, and how to use the Nyquist zones effectively when operating in different datapath modes. Finally, the use of the RFDC IP core to implement an RF-DAC was briefly reviewed.

Chapter 12

RF Data Converters: Figures of Merit and Frequency Planning

Josh Goldsmith

With RF-sampling devices pushing more of the radio components into the digital space, quantifying the noise and spurious emissions (spurs) originating in the devices themselves has become an important aspect of radio system analysis. In this chapter we explore the various sources of noise and spurs that can occur when using RF-sampling data converters, and look at a variety of methods of quantifying and measuring their effects. We then go on to discuss the concepts of frequency planning and how it can be used to reduce the effects that noise and spurs have on signal integrity. Finally, we look at some of the characterisation measurements that are most relevant to RF-sampling data converters, such as the RFSoC.

12.1. Noise and Spurious Emissions

Ideally, any analogue signal that passes through an ADC is perfectly reproduced in the digital domain. Similarly, in ideal conditions, a DAC converts any digital signal representation into a perfect reproduction in the analogue domain. In practice, both noise and spurious emissions are created by the data conversion process, which can distort and interfere with the original signal, affecting the overall performance of the system. In this section we explore the various sources of noise and spurs that may be encountered.

12.1.1. Data Converter Noise

Noise in a system can come from a variety of different sources but, within the data converter itself, two main sources are usually discussed: quantisation noise and thermal noise.

As reviewed in Section 4.2, quantisation refers to the conversion of an analogue signal's amplitude (where any arbitrary amplitude can be represented) to a set of discrete amplitude values. Resulting from this conversion process, there is an error between the original and discretised signals, known as the quantisation error. Figure 12.1 demonstrates the quantisation error (shown in red, with a sawtooth-like shape), for the case of a coarsely quantised sine wave.

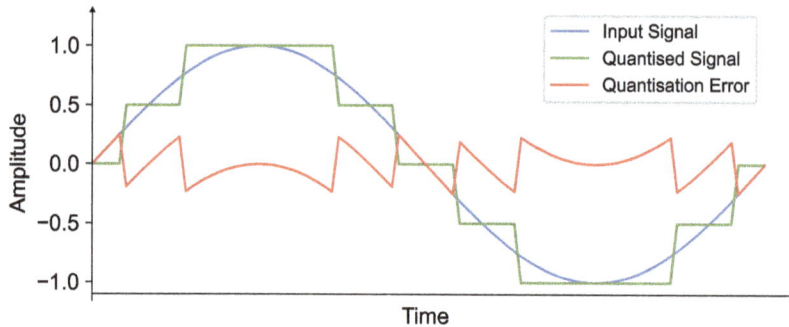

Figure 12.1: Quantisation errors arising from the sampling process

The errors arising from quantisation can be modelled as the addition of a noise source, of power $q^2/12$, where q is the quantisation interval. This assumes that the analogue signal input to the ADC is sufficiently random that the quantisation errors are themselves uniformly distributed [13].

The resolution of a data converter (i.e. the number of bits) has a direct effect on the power of the quantisation noise present in a signal. The higher the resolution, the smaller the quantisation interval, q, and therefore the lower the quantisation noise will be. It is also important to ensure that the signal driving the data converter utilises the full amplitude range in order to maximise the Signal to Quantisation Noise Ratio (SQNR).

While quantisation noise is dependent on the input signal, and the data converter resolution, thermal noise is independent of these factors and will always be present in the system. Thermal noise is present in all electrical components and is caused by quantum effects in the data converter circuitry. As thermal noise is a product of the data converter design itself, the user cannot exert control over this effect, although it can be easily measured by testing the output of a data converter with a DC-only input signal.

In lower resolution data converters, quantisation errors are typically the primary source of noise. However, as resolution is increased, thermal noise becomes more prominent. Other sources of noise can also interfere with data converters, such as clock noise and amplifier noise. However, as these noise sources come from components separate to the data converter, their effect can be reduced by good circuit design.

It is worth noting that the principle source of noise for RF-ADCs is clock jitter, which is the variation of the clock period between clock cycles. Due to the importance of this issue, RFSoC boards use dedicated clocking infrastructure that contain multiple clock conditioning chains to reduce the jitter as much as possible. For example, the ZCU208 and ZCU216 RFSoC development boards use a separate clocking daughter board, the CLK104, that uses two reference clocks, a jitter cleaner, and clock synthesizer to create the necessary clock signals for the data converters [22].

Noise is usually measured in relation to the power of a full-scale input signal (in this case, one that uses the full range of quantisation values, typically set at 0 dBFS), which is normally a sinusoid. The overall ratio between signal power and noise power is the Signal to Noise Ratio (SNR), which we discuss in more detail later in this chapter.

12.1.2. Harmonic Distortion

Harmonic distortion is caused by non-linearities in the data converter signal path (including components external to the data converter itself, such as amplifiers). These non-linearities result in additional frequency components being generated, known as harmonics, which are mathematically related to the input signal by the relationship

$$HD_n = nf_{in}, \tag{12.1}$$

where n is the order of the harmonic (2, 3, 4, etc.) and f_{in} is the frequency of the input signal. In other words, harmonic components appear at integer multiples of the input signal frequency, across the entire spectrum.

Figure 12.2 shows the positions of the first four harmonics on a frequency plot, for the example of a 10 MHz input sine wave. The first four harmonics (HD2, HD3, HD4, and HD5) are located at 20 MHz, 30 MHz, 40 MHz, and 50 MHz, respectively. The magnitude of the harmonics is dependent on the linearity of the data converter signal path, and may also be affected by the frequency of the input signal.

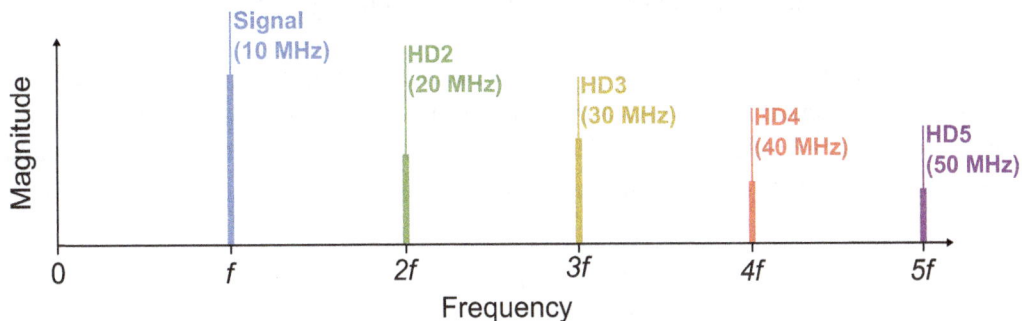

Figure 12.2: An input signal and its first four harmonic components

In considering a data converter in isolation, i.e. omitting external component effects, designers can refer to manufacturer data sheets for information on harmonic performance. For example, the relevant RFSoC data sheet states that the HD2 component of RFSoC Gen 1 Quad tile RF-ADCs has a magnitude of -85 dBc for a -1 dBFS input signal at 240 MHz, and a magnitude of -65 dBc at 3.5 GHz [87]. Typically, only the first two harmonics are stated in data converter specifications, as these tend to dominate.

A common measurement used to describe data converters is Total Harmonic Distortion (THD). THD is defined as the ratio between the power of the sum of all harmonic components, and the power of the input signal. Typically only the first 6 harmonics are considered and the DC component is excluded [12].

12.1.3. ADC Interleaving Spurs

Previously, in Chapter 9, it was noted that the RFSoC's RF-ADCs use an interleaving technique to enable much higher sample rates than would be achievable with a single ADC. Rather than using one ADC with a sample rate of f_s, M ADCs are used together to achieve a sampling rate of Mf_s. For example, an interleaved set of four 500 Msps ADCs has an effective sampling rate of $4 \times 500 = 2,000$ Msps, or 2 Gsps.

Ideally, each of the ADCs has identical characteristics but, in reality, this is extremely difficult to achieve, if not impossible. The mismatches between the ADCs manifest as interleaving spurs, of which there are three main types: DC offset, gain, and phase/time [233], [353].

DC offset spurs occur when the DC value between the interleaved ADCs are mismatched. This causes the DC value to switch back and forth between these values as each ADC is sampled, causing a spur that is related to the sampling rate and the number of interleaved ADCs. The kth DC offset spur is located at

$$f_{DC_k} = \frac{k}{M} f_s \qquad (12.2)$$

where M is the number of interleaved ADCs, f_s is the sample rate, and k is an integer (0, 1, 2,..., M-1). For a 2-interleaved ADC, spurs are present at both DC and $f_s/2$, while a 4-interleaved ADC will have spurs present at DC, $f_s/4$, and $f_s/2$. The magnitude of the spurs increases with the degree of mismatch between DC offsets. An example of DC offset spurs for a 2-interleaved ADC is shown in Figure 12.3.

Gain and phase mismatches in the interleaved ADCs relate to the frequency components of the input signal and, therefore, the positions of these spurs are relative to the signal frequency components. Gain mismatch refers to differences in amplitude of the interleaved ADCs, while phase offset arises from timing differences between them. Both gain and phase mismatches result in spurs at

$$f_{gp_k} = \frac{k}{M} f_s \pm f_{in} \qquad (12.3)$$

where f_{in} is the input signal frequency. Figure 12.4 illustrates how gain and phase mismatches occur, for the example of a 2-interleaved ADC [241].

To avoid gain and phase spurs RF-ADC vendors seek to properly match each interleaved ADCs and ensure signal path latencies are equal. However, even if these measures are taken at the design stage, at least some level of interleaving mismatch should be expected. It is also possible to calibrate the ADCs during operation to reduce the impact of these spurs [353]. For example, the RF-ADCs on RFSoC devices support two calibration processes that attempt to compensate for any residual gain, phase, and DC offset mismatches. The *foreground* calibration process is run during the initialisation stages of the tile, while the *background* calibration process operates throughout ADC operation, providing real-time adjustments [90].

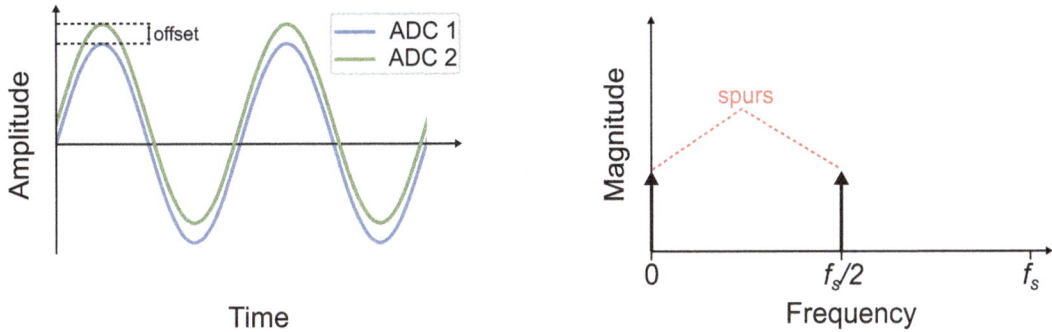

Figure 12.3: *DC offset mismatch for a 2-interleaved ADC*

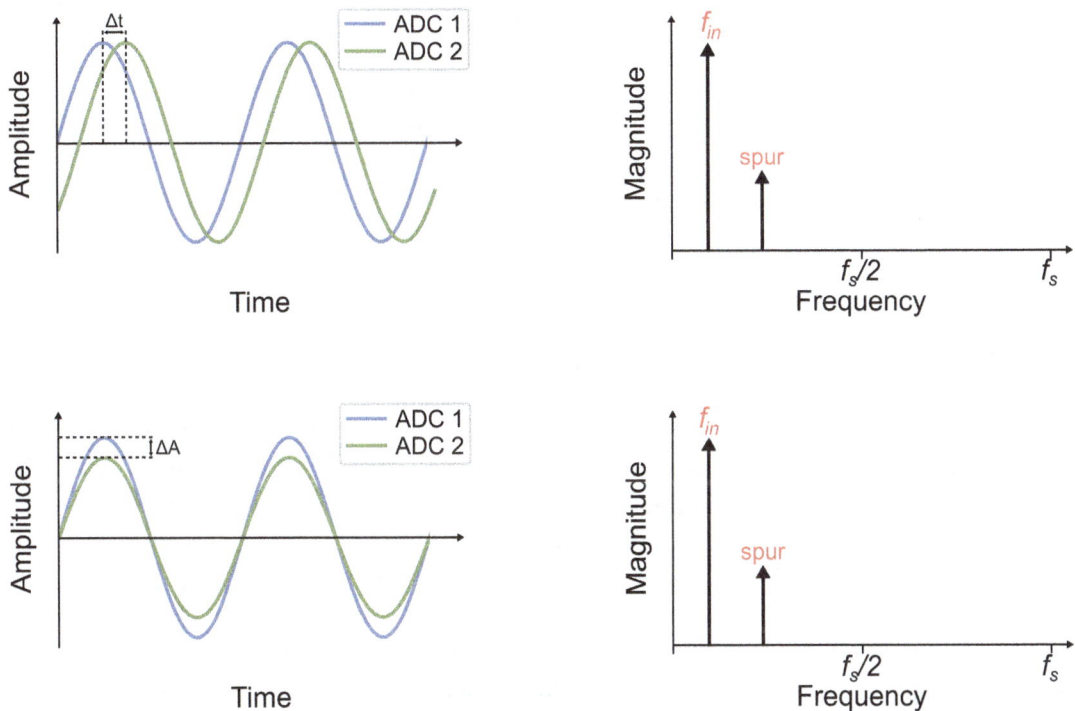

Figure 12.4: *Phase (top) and gain (bottom) mismatches in a 2-interleaved ADC*

Additional interleaving spurs are also present, as a result of lower order harmonics mixing with internal ADC clock frequencies. These spurs will be located at frequencies given by

$$f_{hmix_k} = \frac{k}{M} f_s \pm HD_n,$$
(12.4)

where HD_n is the frequency of the nth order harmonic spur. For example, in a 4-interleaved ADC, the second harmonic component results in spurs located at $f_s/2 - HD_2$ and $f_s/4 \pm HD_2$ [336]. The power of these spurs is affected by both the power of the harmonic and the amount of gain and phase mismatch between the interleaved ADCs. However, due to the calibration process on the RFSoC minimising this mismatch, the impact of these spurs is typically negligible.

12.1.4. Clock Spurs

The clocks that drive the data converters can also contribute to spurs during the conversion process. One important factor is the relationship between the clock frequency and the input signal frequency. Generally, it is best to avoid an integer ratio between these two values, as it can have a direct effect on quantisation noise. If f_{clk}/f_{in} is an integer, quantisation noise tends to be concentrated around the harmonics of the input frequency, instead of being uniformly spread across the Nyquist band. This can lead to a non-uniform noise pattern in the spectrum, which can decrease the data converter's dynamic range.

Data converter clocks can also introduce their own spurs, although these are difficult to quantify, and can be as much down to circuit design as they are to the quality of the clock. One type of spur that can be easily calculated is the set of PLL mixing products. These spurs occur because the clock signal effectively mixes with the input signal, similar to the heterodyning process used to modulate a signal. This effect causes spurs to occur at

$$f_{pllmix} = f_{in} \pm f_{clk}$$
(12.5)

where f_{clk} is the frequency of the PLL clock.

12.1.5. Spurious Emissions of Bandlimited Signals

So far we have only considered the frequency content of spurs in relation to single-frequency input signals. Typically, communication signals contain a number of frequencies that comprise the signal bandwidth, up to tens or hundreds of MHz wide (i.e. bandlimited). As a result, any spurs related to the input signal will also have a relative bandwidth, increasing the opportunity for them to cause interference and distortion. Moreover, the greater the signal bandwidth, the higher the likelihood that a spur will overlap and interfere with it.

In the case of a bandlimited communications signal, harmonic distortion results in harmonic bands rather than single tones. In general terms, harmonics are a product of the input signal (i.e. nf_{in}); as f_{in} is now a range of frequencies, so too are the resulting harmonic bands. For instance, if a communications signal occupied the bandwidth from 400 MHz to 500 MHz, then the lower end of the HD2 harmonic band would be given by 2×400 MHz, while the upper end would be 2×500 MHz, resulting in a 200 MHz harmonic bandwidth between 800 MHz and 1 GHz.

Therefore, the general relationship is that the nth harmonic band has a bandwidth n times greater than the original signal. A second order harmonic is twice the bandwidth, while the third order harmonic is three times the bandwidth, and so on, as shown in Figure 12.5. Gain and time interleaving spurs are a little more forgiving, as their bandwidth is always equal to the bandwidth of the input signal. However, there are generally more interleaving spurs to contend with, especially in ADCs that have a large interleaving factor.

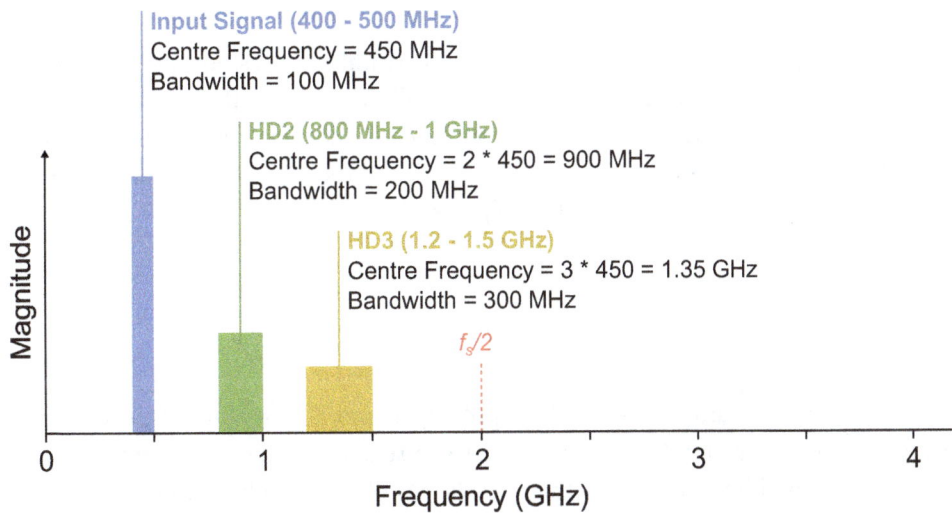

Figure 12.5: Harmonic bands resulting from a bandlimited signal

Collectively, the bandlimited spurious components have direct consequences for system design, relating to both DACs and ADCs. In the case of DACs, the spurious bands occupy a significant portion of the spectrum, and must be filtered out before reaching the antenna, to avoid generating interference for other radio signals. For both ADCs and (somewhat unintuitively) DACs, spurs with frequency components above the Nyquist rate will alias into the first Nyquist Zone, and potentially interfere with or distort the signal of interest

The benefit of RF-sampling data converters is the ability to use oversampling as a means of reducing the above mentioned effects. We discuss these concepts in more detail later in this chapter.

12.2. Traditional Data Converter Characteristics

As we have seen from the previous section, noise and spurs can adversely affect data converter performance by impacting dynamic range and interfering with the signal of interest. Therefore, the effects of noise and spurious signals must be accounted for when describing the characteristics of a data converter. To quantify these effects, there are a variety of measurements available that give a fuller understanding of the data converter's ability, each describing a particular aspect in relation to both noise and spurious content. In this section we explore the traditional methods of characterising DACs and ADCs in relation to these effects.

12.2.1. Spurious Free Dynamic Range (SFDR)

Spurious-Free Dynamic Range (SFDR) is probably the most readily understood performance characteristic, and is an important metric in communication applications. SFDR measures the useable dynamic range of the data converter with respect to the various spurious components that are present in the Nyquist band. It is defined as the ratio between the RMS value of the input signal, and the RMS value of the largest spur present in the bandwidth of interest (which is typically the first Nyquist Zone), i.e.

$$\text{SFDR} = 20 log_{10}\left(\frac{A_{input}(RMS)}{A_{spur}(RMS)}\right) \text{dBc} \tag{12.6}$$

where A_{input} and A_{spur} are the amplitudes of the input signal and largest spur, respectively, and dBc is the units in decibels relative to the carrier. In this case, the carrier refers to a sinusoidal input signal, and therefore the harmonic components will be tones rather than bands. SFDR can also be defined in relation to full-scale input, rather than to the carrier, given in dBFS.

A simpler method to measure SFDR is to find the difference in the magnitudes of the two frequency components. Figure 12.6 shows an example of how SFDR is measured in the frequency domain.

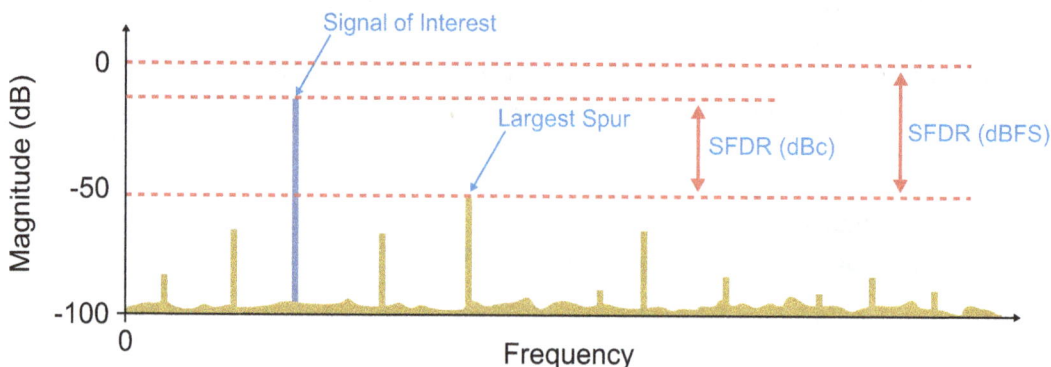

Figure 12.6: Measuring SFDR

When the input signal is at full-scale, the harmonic components tend to be the largest spurs. However, as the input signal drops in amplitude, other spurs can become more prominent. The SFDR metric takes all spurious components into account, and it is also common to see SFDR measurements that exclude the harmonics. In addition to this, as the amplitude of spurs can depend on the input signal frequency, SFDR measurements are often given for various input frequencies. For example, the datasheet for Gen 1 RFSoC devices provides SFDR measurements for 240 MHz, 1.9 GHz, 2.4 GHz, and 3.5 GHz, excluding the second and third order harmonics [12],[87].

12.2.2. Signal-to-Noise Ratio

Signal-to-noise ratio (SNR) is one of the most widely known measurements for quantifying data converter performance, used to measure noise relative to an input signal. SNR is defined as the ratio between the power of the input signal and the total noise power, excluding any harmonics or other spurious components,

$$\text{SNR} = 10 log_{10}\left(\frac{P_{input}}{P_{noise}}\right) \text{ dB}. \tag{12.7}$$

We can see from (12.7) that, as the power of the noise decreases, the SNR value increases, and vice versa. If the signal and noise have equal power, then the SNR is 0 dB. Another expression of the SNR of a data converter is sometimes given as

$$\text{SNR} = 6.02N + 1.76 \text{ dB} \tag{12.8}$$

where N is the resolution of the data converter, in bits. It is clear from (12.8) that the SNR increases with the resolution of the data converter. For example, a 12-bit data converter will have an SNR of 74 dB.

However, care should be taken when using this equation. Firstly, the SNR value given by (12.8) is an ideal case and only accounts for the quantisation noise of a data converter, not any other sources of noise. As a result, this equation is sometimes referred to as the signal-to-quantisation-noise ratio (SQNR) instead. Secondly, this equation assumes the RMS value of the input signal is at full-scale amplitude. If the input signal is less than full-scale an additional term must be added.

$$\text{SNR} = 6.02N + 1.76 + 20 log_{10}\left(\frac{A_{input}}{A_{FS}}\right) \text{ dB} \tag{12.9}$$

where A_{input}/A_{FS} is the ratio between the amplitude of the input signal and the full-scale amplitude of the data converter. Hence if the amplitude of the input signal is equal to the full-scale amplitude, this term will disappear. However, if the amplitude of the input signal is less than the full-scale amplitude, then the result will be negative, decreasing the SNR.

Finally, (12.8) and (12.9) also assume that the bandwidth of the signal is equal to the Nyquist rate, such that $f_s = 2B$. If the sampling rate is greater than twice the bandwidth, another term must be included, also known as the *processing gain* [12].

$$ SNR = 6.02N + 1.76 + 20log_{10}\left(\frac{A_{input}}{A_{FS}}\right) + 10log_{10}\left(\frac{f_s}{2B}\right) \ dB \quad (12.10) $$

Similar to the reduction to (12.9), we can see that if $f_s = 2B$, then the final term in (12.10) disappears. We can also observe that, as the sample rate increases in relation to the input signal bandwidth, the SNR increases proportionately. This has positive consequences for devices like RFSoC, as high oversampling rates can greatly increase the SNR. For example, if the sample rate is four times the input signal bandwidth, i.e. ($f_s = 4B$), the SNR increases by 3 dB. In fact, every doubling of the sample rate results in an additional 3 dB of SNR. The SNR of the data converter continues to increase until the quantisation noise power falls below other sources of noise, such as thermal noise. If we take our previous example of a 12-bit data converter, but use a sample rate 4 times greater than input signal bandwidth, the ideal SNR increases from 74 dB to 77 dB.

It is important to recognise that, when increasing the sampling rate, the total energy of the noise remains the same. However, as the noise is spread over a wider bandwidth, its density decreases, as observed in Figure 12.7 (notice the reduction in quantisation noise). With that said, because the bandwidth of the oversampled signal now occupies less of the Nyquist band, it is possible to filter and/or decimate to reduce the total noise energy.

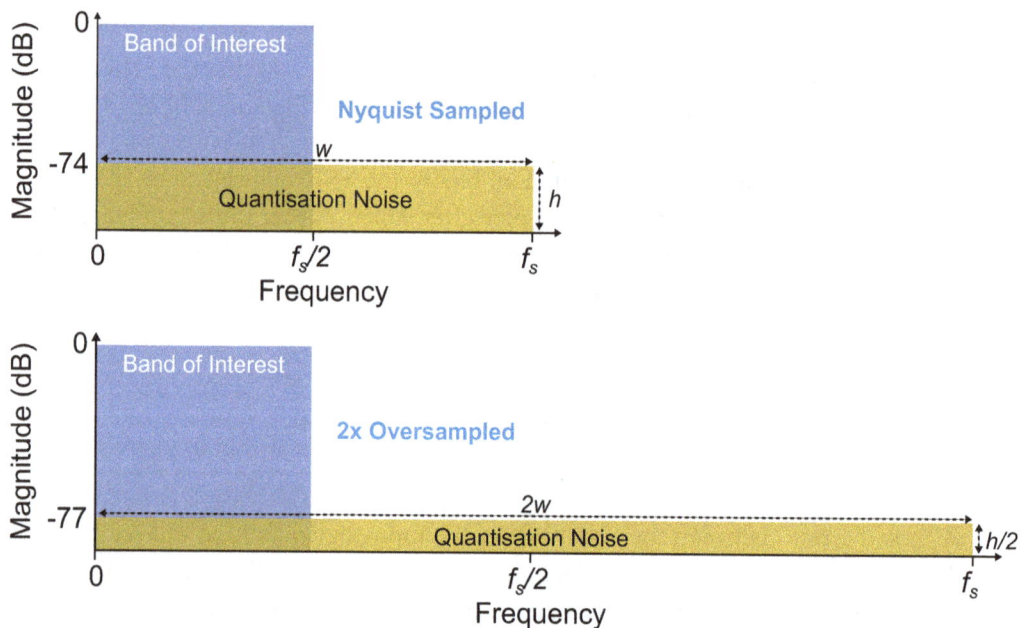

Figure 12.7: The effect of oversampling on quantisation noise density

12.2.3. Signal to Noise and Distortion Ratio

Signal to Noise and Distortion Ratio, usually shortened to SINAD or SNDR, is an important measurement in communication systems. It not only includes all noise components (quantisation noise, thermal noise etc.), but also all other spectral components such as harmonics, interleaving spurs, and so on. It is defined as

$$\text{SINAD} = 10 log_{10}\left(\frac{P_{input}}{P_{noise} + P_{distortion}}\right) \text{ dB},\qquad(12.11)$$

where P_{input} is the power of the input signal, P_{noise} is the noise power, and $P_{distortion}$ is the Total Harmonic Distortion (THD), excluding DC. Thus, SINAD gives an overall measurement of the dynamic performance of the data converter.

12.2.4. Effective Number of Bits

Effective Number Of Bits (ENOB) is a measurement of the overall dynamic range a data converter is capable of, over the full Nyquist band. The best scenario is that a data converter's resolution (in bits) and its ENOB are equal, although in practice this is usually not the case. For example, a data converter may be advertised as having a resolution of 10-bits but, due to the noise and spectral components discussed in this chapter, its actual resolution may be only 8.5-bits.

A lower ENOB value reduces the dynamic range, as the minimum and maximum values the data converter is able to produce are limited. Because ENOB is affected by both the noise and spectral components, it is usually expressed in relation to SINAD using (12.8). The equation is solved for N, with SNR being replaced with SINAD, i.e.

$$\text{ENOB} = \frac{\text{SINAD} - 1.76}{6.02}.\qquad(12.12)$$

12.2.5. Limitations of Traditional Metrics

As the traditional characteristics we have discussed in this section assume bandwidths close to the Nyquist rate, they are less relevant to data converters that operate at RF frequencies and have high oversampling rates.

Typically a separate, but related, set of measurements are used to characterise the performance of RF data converters that better reflect the high oversampling rates and digital up/down conversion employed: noise spectral density, intermodulation distortion, and adjacent channel leakage [53]. We go on to discuss these terms in more detail later in this chapter but, first, we will explore the use of frequency planning to avoid poor performance in terms of the metrics reviewed so far.

12.3. Frequency Planning

Noise and spurs are present even in the best designed and manufactured data converters, and so strategies must be employed to mitigate their effects. Frequency planning is one such strategy. It exploits the fact that many spurs are deterministic, and their frequency content computable. As it is relatively simple to predict where spurs will occur in the spectrum based on a known set of parameters (sample rate, input frequency etc.), frequency planning involves tuning these parameters to avoid overlaps between spurious components and the signal of interest, and positions spurs at frequencies that can readily be removed by filtering.

While it is possible to compute spur frequencies, there are a number of additional factors to consider when frequency planning in RF-sampling data converters. Digital complex mixers can shift spur frequencies in unexpected (but deterministic) ways, harmonics can become problematic when transmitting in upper Nyquist zones and, in RF-ADCs, aliasing can fold spurs into the band of interest, which may even overlap with the signal.

In this section we explore these concepts and provide examples of good frequency planning practices. The theoretical review is complemented by a small set of Jupyter notebooks (see Notebook F on page 411), which interactively explore the spectral artefacts reviewed in this chapter, and present a Jupyter-based frequency planning tool. The latter is based on a similar tool released by AMD [43].

12.3.1. Effects of Aliasing

In our discussion so far, we have only calculated the frequencies of spurs in direct terms, and omitted any discussion of aliasing effects in the case of ADCs. As a result of the ADC sampling process, any spurs located in the upper Nyquist Zones will be aliased back into the first Nyquist Zone, increasing the chances that a spur will interfere with the signal of interest. Therefore, when frequency planning for receiver architectures, aliasing must be taken into account; in other words, the frequencies of spurs *after sampling* must be calculated. While we have already discussed aliasing in detail in previous chapters (in particular, Chapter 4), it is worth briefly revisiting the concept here to frame it within the context of frequency planning.

The frequency of a spur after aliasing depends on its original Nyquist Zone. Of course, any spur located in Nyquist Zone 1 will not be subject to aliasing and so no further analysis is required. A spur occurring at a frequency of f_{spur} in Nyquist Zone 2, however, will fold into Nyquist Zone 1 at a frequency of $f_s - f_{spur}$. Any spurs occurring in higher Nyquist Zones will also experience aliasing according to the pattern depicted in Figure 4.5 on page 78. For instance, a spur appearing in Nyquist Zone 4 would be aliased to Nyquist Zone 1, with a frequency of $2f_s - f_{spur}$.

Noting this repeating pattern, it is trivial to calculate the location of spurs in the upper Nyquist Zones after aliasing. A simple algorithm is to take the frequency of the spur (f_{spur}) modulo the sample rate (f_s), and then subtract the result from f_s if it is above the Nyquist rate. An example implementation of this algorithm is provided next, written in the Python language.

```
f_alias = f_spur % fs
if f_alias > fs / 2 :
    f_alias = fs - f_alias
```

In Figure 12.8, we observe how the second and third harmonics (HD2, HD3) alias down from Nyquist Zone 2 when sampled by an ADC operating at 4 GHz. As the input centre frequency is at 1.15 GHz, HD2 and HD3 are centred around 2.3 GHz and 3.45 GHz, respectively, placing them both in Nyquist Zone 2. Since the input signal bandwidth is 100 MHz, HD2 has a bandwidth of 200 MHz, and HD3 has a bandwidth of 300 MHz. After aliasing, HD2 is centred around 4 - 2.3 = 1.7 GHz, and HD3 is centred around 4 - 3.45 = 0.55 GHz.

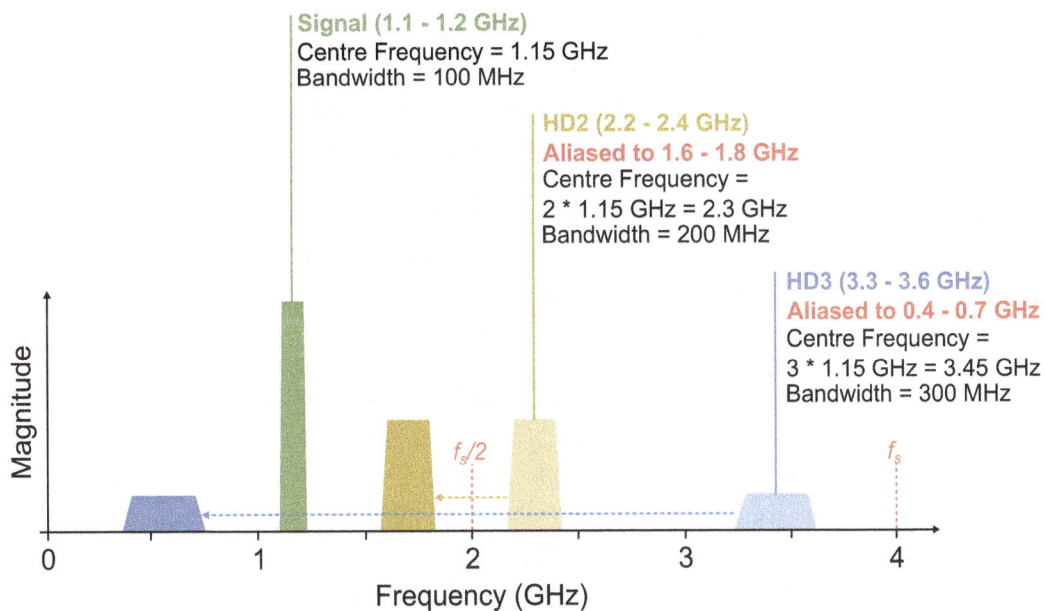

Figure 12.8: Example of aliased spurs

It can be seen from this example that HD2 and HD3 are now positioned on either side of the signal of interest, meaning that a bandpass filter must be used to sufficiently suppress the spurs, instead of the usual low pass filter. This scenario leaves a frequency gap of around 400 MHz between the signal and the spurs (10% of the sample rate) at either side, allowing the bandpass filter to have relatively relaxed transition bands. In the case of wider signal bandwidths, however, spurs are likely to be closer to the signal of interest, requiring filters with stricter characteristics. We discuss filtering in more detail later in this section.

As mentioned earlier in this chapter, DACs are also affected by aliasing. Due to non-linearities within the DAC architecture, harmonics will appear at multiples of the input frequency across the entire spectrum. Because the harmonics caused by these non-linearities can occur *before* the sampling process, each harmonic within one Nyquist Zone will have an associated image in all the other Nyquist Zones.

If we take the first and second Nyquist Zones as an example, any harmonics present in the first Nyquist Zone will have an image at $f_s - HD_n$ in the second Nyquist Zone, while any harmonics in the second Nyquist Zone will have an *aliased* image in the first Nyquist Zone, also at $f_s - HD_n$. While these images can make frequency planning more complicated, the locations of the aliased harmonic images are trivial to calculate. This may be done using the same equations discussed for the ADC case.

12.3.2. Effects of Digital Downconverters (DDCs)

DDCs, which contain digital mixers and decimation chains, also have an effect on the location of spurs. The mixer is used to move the received RF signal to baseband, or an IF closer to baseband. However, along with the signal of interest, the mixer also shifts any spurs that are present. Furthermore, because aliasing occurs before the mixing process, it is the aliased version of the spurs that are shifted. It can be difficult to visualise this process, so we will use a simple example to better explain the concept.

Let us imagine an RF-sampling receiver with a 100 MHz ADC, an input signal with a frequency of 40 MHz, and a complex mixer with an NCO frequency of 35 MHz. For simplicity we will assume that the input signal is a real sine wave, and only the first harmonic spur, HD2, will be considered. We denote the frequency band between $-f_s/2$ and 0 Hz as Nyquist Zone -1.

After sampling, HD2 aliases from 80 MHz to 20 MHz. Because the input signal is real, it also includes negative images at -40 MHz (corresponding to the sine wave signal of interest), and at -20 MHz (the aliased version of HD2) in Nyquist Zone -1. When the input signal is mixed with the NCO, all frequency components (including the images in Nyquist Zone -1) are shifted down in frequency by 35 MHz, i.e. the frequency generated by the NCO. Recall that the frequency domain is periodic based on the sampling rate, f_s, such that any component shifted left of $-f_s/2$ (i.e. below Nyquist Zone -1) appears somewhere within Nyquist Zone 1, while any positive frequency component that shifts above $f_s/2$ appears in Nyquist Zone -1. With this 'frequency wrapping' in mind, we list the result of the mixing process below, and illustrate the effects in Figure 12.9.

- The input sine wave shifts from 40 MHz to 5 MHz.

- The negative image of the input sine wave shifts from -40 MHz and wraps around to 25 MHz.

- HD2 is shifted from 20 MHz to -15 MHz.

- The negative image of HD2 is shifted from -20 MHz and wraps around to 45 MHz.

In this example, we observe that the complex mixing process has shifted the frequencies of the input sine wave, its HD2 spur, *and also* the complex images of these components. In fact, there are now three components within Nyquist Zone 1, i.e. the region between 0 Hz and $f_s/2$.

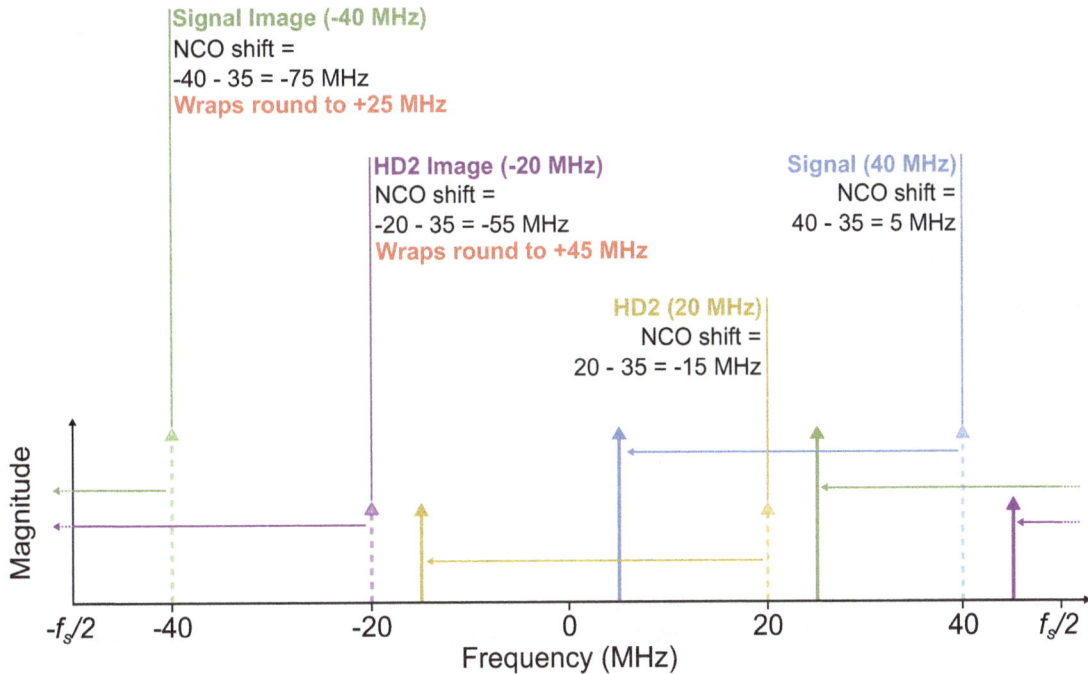

Figure 12.9: Example of how the NCO shifts frequency components and their images in the complex spectrum (dashed lines — original frequencies; solid lines — new frequencies after NCO shift)

More generally, when considering higher-order harmonics and other spurs, the spectrum can become congested after the mixing process — this increases the likelihood that a spur will interfere with the signal of interest. One way to reduce these effects is simply to increase the sample rate. This can reduce the amount of frequency wrapping that occurs, helping to keep the spurs sufficiently separated from the signal of interest to enable their easy removal by filtering.

Like mixing, decimation can also change the frequency content of spurs. Recall that the process of decimation by an integer value, R_M, involves anti-alias filtering to suppress any signals above half of the new sample rate, followed by a downsampling stage that discards $R_M - 1$ out of every R_M samples to achieve the new rate.

As a result of decreasing the sample rate, any spurs below $-f_s/2$ or above $f_s/2$ (with reference to the new sampling rate) are subject to aliasing. These components should have already been attenuated by the anti-aliasing filter; however, realistic filters do not have an ideal 'brick-wall' response, and therefore any spurs located close to the Nyquist rate may not be sufficiently suppressed. This must be taken into consideration when frequency planning as, by reducing the sample rate, remnants of spurs may move closer to the signal of interest, making them more difficult to subsequently filter out. We examine the previously described example in Figure 12.10, for the case of $2 \times$ decimation.

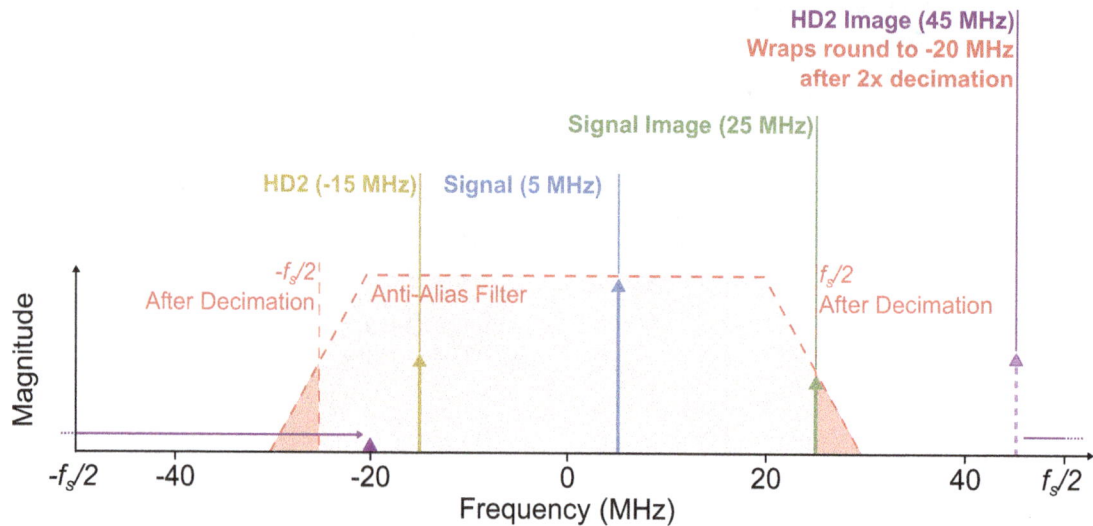

Figure 12.10: Example of how decimation can affect the frequency and amplitude of spurs
(dashed lines — frequency before decimation; solid lines — frequency after decimation, which includes cases of no change)

The outcomes arising from decimation can be summarised as follows:

- The 2 × decimation reduces the Nyquist rate from 50 MHz to 25 MHz.

- Having previously been demodulated, the input sine wave and HD2 are within the new Nyquist Zone 1 (between 0 Hz and half of the new sample rate, i.e. 25 MHz), and therefore remain unchanged.

- The image of the input signal now resides at exactly at the new Nyquist rate of 25 MHz, and is not fully attenuated by the anti-alias filter.

- The HD2 image is now outside the new Nyquist Zone 1, and is wrapped from 45 MHz to -20 MHz. However, the anti-alias filter should suppress this spur sufficiently, and therefore it can be disregarded.

In the example just presented, demodulation precedes decimation. This is typical in receiver architectures, and the RFSoC's DDCs are designed to support that order of processing. Any receiver architecture with an initial decimation stage would require dedicated analysis, as the frequencies of the spurs would be affected differently.

It is worth noting that, as the interpolation and mixing processes in DUCs occur before digital to analogue conversion, the frequencies of spurs occurring in RF-DACs are much simpler to calculate. Because spurs are not present in the digital signal path, only the input frequencies and sample rate entering the DAC are required for analysis. However, any spurs that do occur after conversion will be shifted by any mixing process performed in the analogue domain, such as in the IF sampling radio systems discussed in Chapter 8.

12.3.3. Filtering to Eliminate Spurs

It is important to note that the spurs discussed in this chapter result directly from the data conversion process. Good frequency planning can provide adequate protection around the signal of interest, allowing most spurs to be eliminated by filtering. The direction of the conversion process (i.e. digital to analogue, or analogue to digital) determines the type of filters to be employed.

Any spurs generated by an ADC will be present in the signal it provides to the digital section of the receiver. Consequently, a digital filter is required to remove these spurious components. Digital filters implemented in the PL of the RFSoC provide a high degree of reconfigurability, and the designed response can be modified to cater for a variety of frequencies and bandwidths. Digital filtering also enables SDR applications, where filters can be dynamically configured from software, allowing a wide range of radio standards to be supported by a single device. In receiver architectures that directly down-convert to baseband in the digital domain, effective frequency planning allows much of the required filtering to be achieved by the decimation chain.

Continuing the example from Figure 12.10, in Figure 12.11 we show how the introduction of an additional 2 × decimation stage can adequately suppress the spurs that are present.

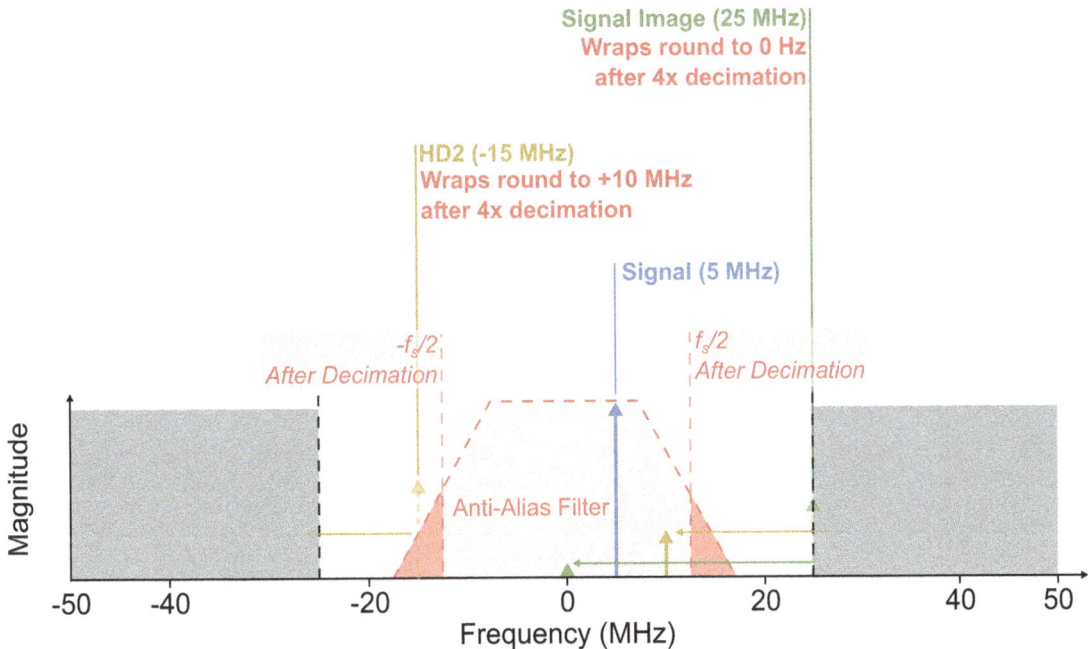

Figure 12.11: Example of decimation filtering to suppress spurs
(dashed lines — frequency before decimation; solid lines — frequency after decimation)

The result of incorporating an additional stage of decimation can be summarised as follows:

- The additional $2\times$ decimation reduces the Nyquist rate from 25 MHz to 12.5 MHz.

- The input signal is located within the new Nyquist Zone 1 and remains unchanged.

- HD2 is now outside the new Nyquist Zone -1, and therefore it experiences frequency wrapping from -15 MHz to 5 MHz. As the spur is positioned at -15 MHz at the time of filtering, close to the new Nyquist rate, it will not be completely suppressed by the filter.

- At 25 MHz, the signal image is also outside the new Nyquist Zone 1 and, because its frequency is equal to f_s at the decimated rate, it will wrap to DC. However, the anti-alias filter should sufficiently suppress this image, and it can be disregarded.

In digital to analogue conversion, the spurs created by the DAC are present in the analogue domain, meaning that an analogue filter is required to remove any spurious signal content. Because channel leakage is so important in transmitter designs, care must be taken to ensure that any spurs are well enough suppressed before reaching the antenna to avoid interfering with adjacent radio channels. Image rejection filters should suppress any spurs outside the Nyquist Zone of operation. However, for oversampled systems where the sample rate is many times higher than the band of interest, or where an upper Nyquist Zone is used for transmission, any spurs located within this Nyquist Zone will be transmitted along with the original signal. In this case, additional filtering will be required. Figure 12.12 shows how an oversampled DAC with a reconstruction filter still passes spurs in Nyquist Zone 1, requiring the subsequent application of an analogue bandpass filter.

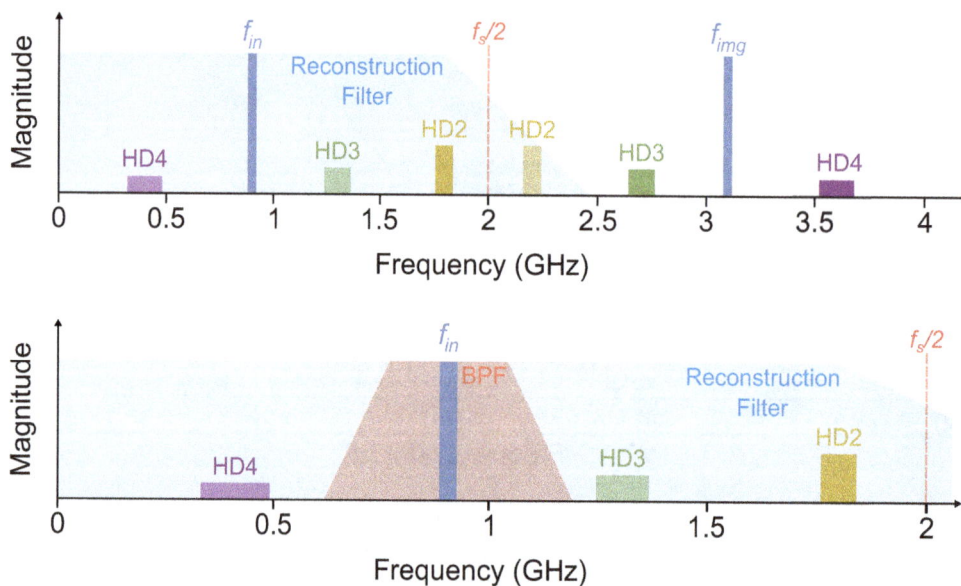

Figure 12.12: Example of analogue filtering to suppress spurs
(faded frequency components denote aliases or images of the harmonics)

Analogue filters tend to be less configurable than digital ones, reducing the amount of bands and frequencies that a single filter can serve. With that said, it is possible to employ multiple filters at the analogue front-end to serve different frequency bands. In this configuration the output of the DAC can be routed to the appropriate filter(s) under software control. The use of multiple analogue filters can however increase design complexity and BOM (Bill of Materials) costs, due to the additional components required, and the need to minimise cross-talk, which may be unsuitable for some applications.

The requirements for a filter's transition band and stopband attenuation are determined by the frequencies of spurs in relation to the input signal, and their relative amplitude. The closer the spurs are to the signal of interest, the tighter the transition band needs to be. Also, the more significant the amplitudes of the spurs, the greater the stopband attenuation that is required. A good frequency plan will avoid spurs occurring close to the signal of interest, lessening the demands on the filtering process.

12.3.4. Planning the Spectrum

Having examined how and where spurs occur in the spectrum, and the requirement of filters to sufficiently suppress them, we can now explore the strategies involved in creating a good frequency plan.

As reviewed in previous sections, the location of a spur is dependent on the frequency of the input signal and the data converter sample rate, while the width of the spur is dependent on the input signal bandwidth. The role of frequency planning is to adjust these parameters to achieve the 'cleanest' possible spectrum around the signal of interest, by avoiding the generation of spurs at adjacent frequencies. However, most radio applications have standardised signal characteristics including fixed centre frequencies and bandwidths — therefore, when creating a frequency plan for such scenarios, only the sample rate can be changed.

In this section we look at two simple examples for a transmitter and receiver based on the characteristics of RFSoC devices as stated in the data sheet [87]. These examples demonstrate how different configurations can be used to make a good frequency plan.

The data sheet states values for RF-DAC and RF-ADC second and third order harmonics [87]. In the case of the RF-ADCs, DC offset and gain/time interleaving spurs are additionally specified. The following examples therefore consider these two sets of spurs.

It should be noted that the magnitude of the spurs given in the following examples are measured in relation to the input signal (or carrier), given in units of dBc. For the RF-DAC examples the input signal has a magnitude of 0 dBFS, while the receiver examples use an input signal magnitude of -1 dBFS.

Example 1: Transmitter

In this example, a signal with a centre frequency of 3.6 GHz and a 100 MHz bandwidth is to be transmitted on a third generation Quad tile RF-DAC. The RF-DAC is set for low noise mode, and to operate with a 32 mA output current.

Setting the sample rate of the RF-DAC to 5.5 GHz, and operating in mix-mode, we can use a centre frequency of 1.9 GHz to transmit a signal at 3.6 GHz in Nyquist Zone 2. Based on the equations presented in previous sections, it can be determined that HD2 has a centre frequency of 3.8 GHz with a 200 MHz bandwidth, and that HD3 has a centre frequency of 5.7 GHz with a bandwidth of 300 MHz. Due to aliasing, HD3 will have components in both the first and second Nyquist zones with centre frequencies of 200 MHz and 5.3 GHz, respectively. Similarly, HD2 will have an aliased version in the first Nyquist zone at 1.7 GHz. Referring to the data sheet, we can expect HD2 and HD3 to have magnitudes of around -60 dBc and -68 dBc, respectively.

Figure 12.13 illustrates that this configuration places HD2 very close to the signal image intended for transmission, requiring a filter transition band of around 50 MHz, which equates to less than 1% of the sample rate. Such a filter would be very expensive to implement, and therefore it is desirable to seek an alternative solution. Note that the amplitude of the fundamental signal at 1.9 GHz will be slightly lower than depicted here, due to the sinc-shaped roll-off of the DAC (see Chapter 11).

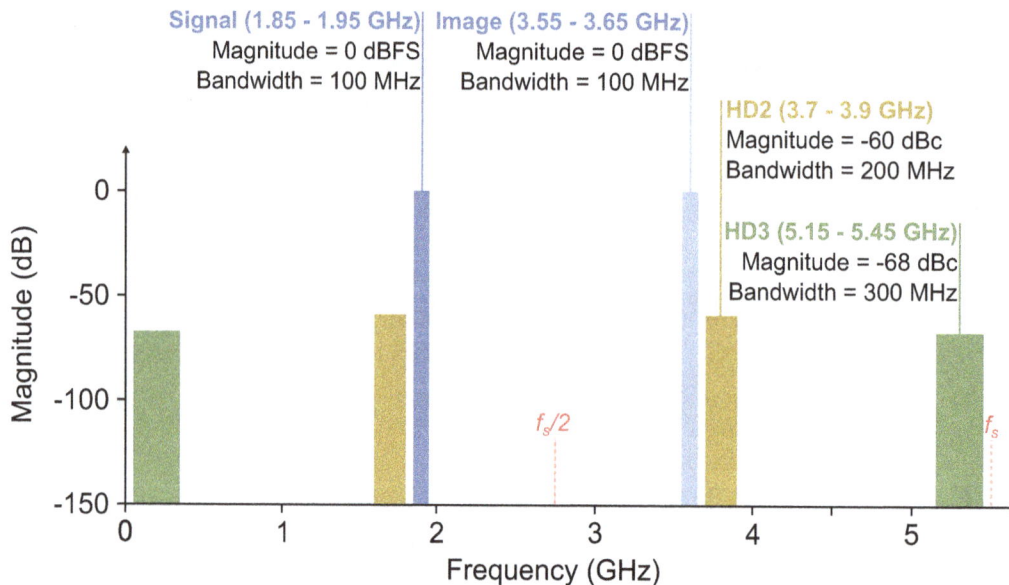

Figure 12.13: Transmitter configuration which contains interfering spurs

To reduce the requirements of the filter needed to suppress the spurs, an alternative approach is taken to increase the sample rate and transmit at 3.6 GHz in Nyquist Zone 1.

As shown in Figure 12.14, using a sample rate of 9 GHz, HD2 and HD3 will be located in the second Nyquist Zone at 7.2 GHz, while their aliases will be in the first Nyquist Zone at 1.8 GHz. This leaves ample separation between the signal image and the spurs, and therefore the filter transition band can be substantially relaxed. With the signal transmitted directly at 3.6 GHz, the data sheet indicates that the magnitudes of HD2 and HD3 increase to around -59 dBc and -60 dBc, respectively (these values are not exact for this specific configuration, but they give a good indication of expected performance).

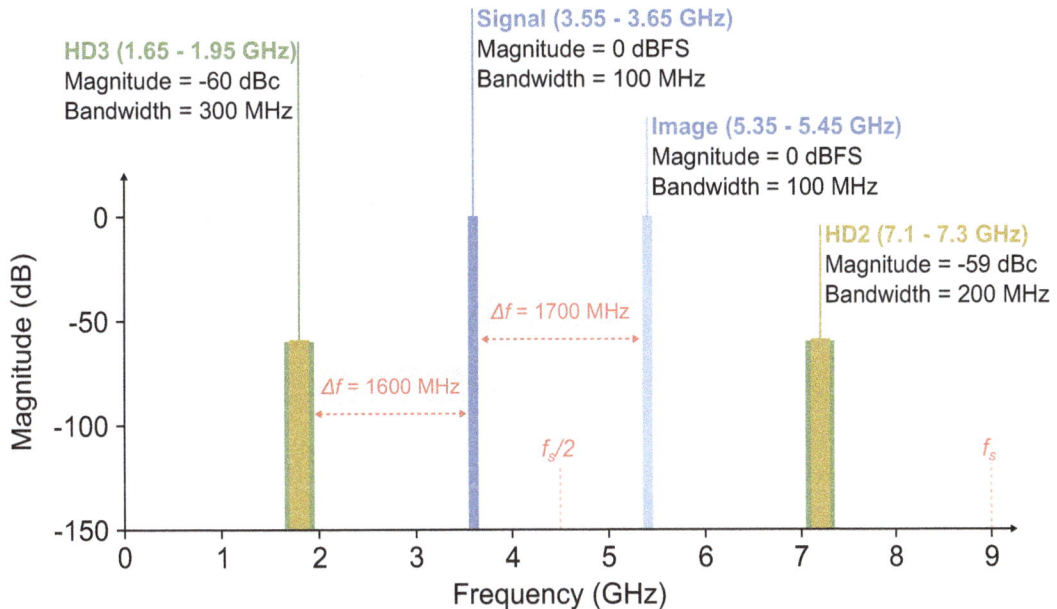

Figure 12.14: Transmitter configuration which avoids interfering spurs

Example 2: Receiver

In this example, a signal with a centre frequency of 240 MHz and a 40 MHz bandwidth is received on a Gen 3 Dual RF-ADC tile. The RF-ADC is set with 0 dB DSA attenuation. With these parameters, the data sheet indicates expected HD2 and HD3 values of around -87 dBc and -77 dBc, respectively, with the gain/time interleaving spurs at approximately -92 dBc, and the DC offset interleaving spurs around -86 dBc. Using a sample rate of 1 GHz, we can calculate the frequencies of these spurs after aliasing, as shown in Figure 12.15.

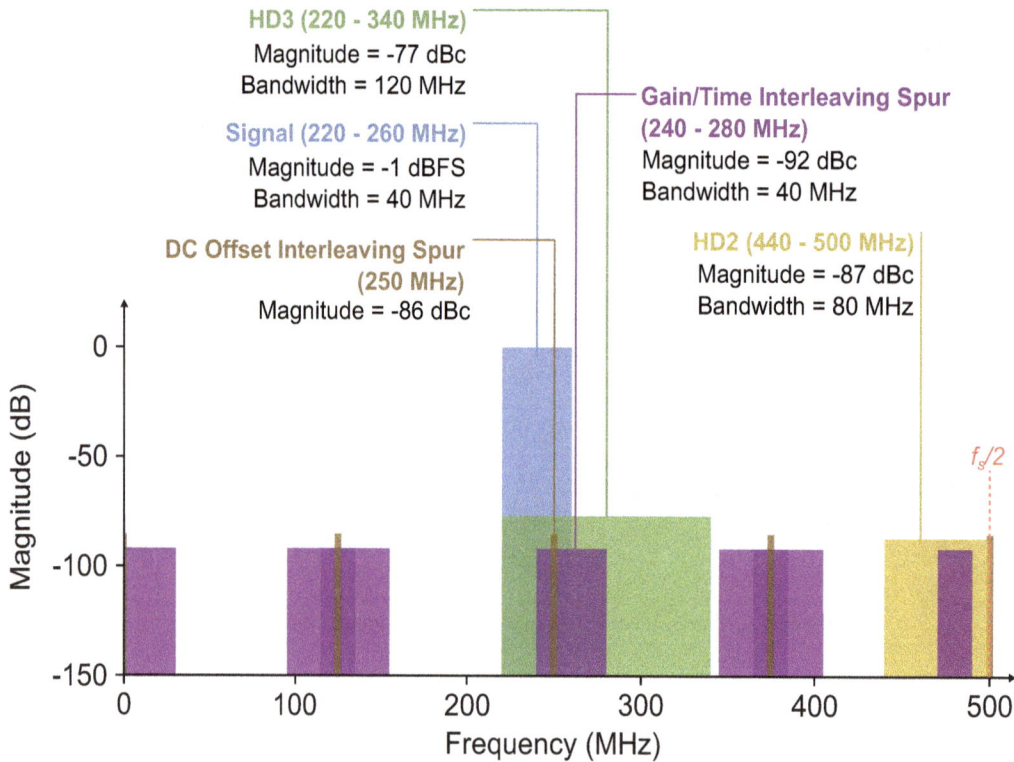

Figure 12.15: Receiver configuration which contains interfering spurs

It is evident from Figure 12.15 that HD3 and two interleaving spurs directly overlap the signal of interest, making it impossible to remove these spurs with a filter. This represents a poor frequency plan!

Instead, an alternative sample rate of 2.8 GHz is used and the spur frequencies are recalculated, as shown in Figure 12.16. We can confirm that, using this configuration, no spurs directly overlap with the signal of interest, and that there is a separation between the signal and closet spurs of at least 90 MHz (around 3% of the sample rate) on either side. This represents a favourable frequency plan, and permits the design of a bandpass filter of reasonable cost to remove the spurs.

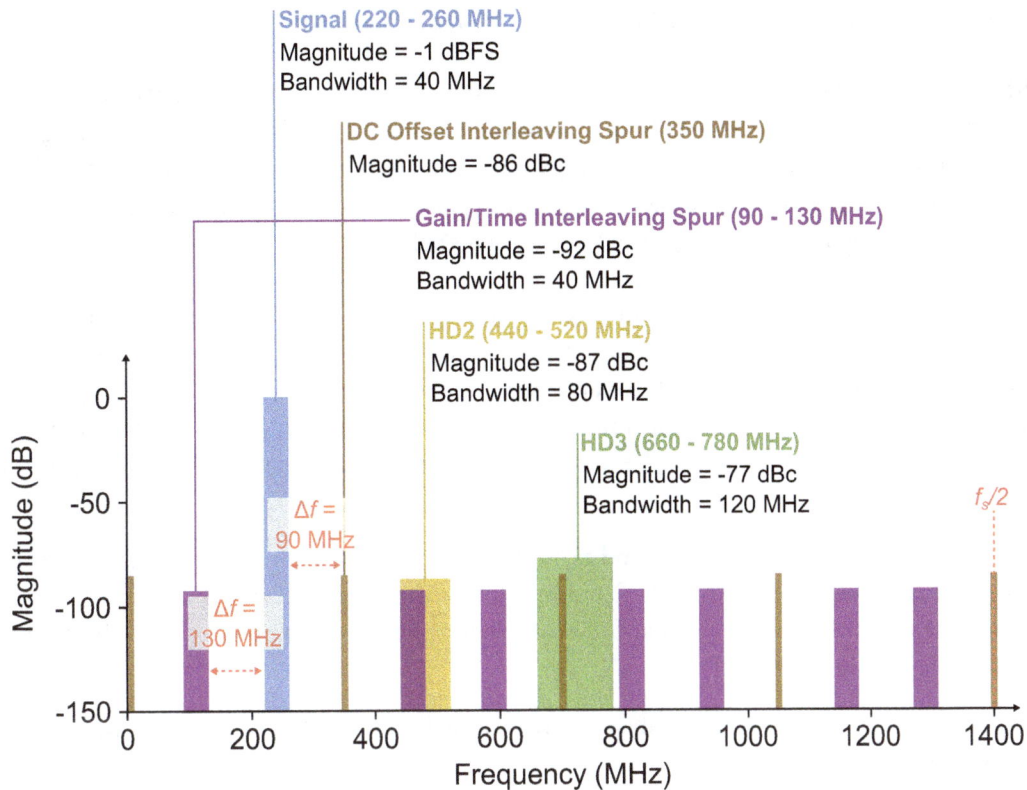

Figure 12.16: Receiver configuration which avoids interfering spurs

12.4. RF-Sampling Data Converter Characteristics

When discussing the traditional characteristic measurements for data converters, we noted that they are of limited relevance to RF-sampling devices, due to the high oversampling rates used. Instead, data converter manufacturers have generally agreed upon a different set of characteristics that better describe the capabilities of RF-sampling data converters. These are: *noise spectral density, intermodulation distortion*, and *adjacent channel leakage*, each of which we discuss in this section.

12.4.1. Noise Spectral Density

While SNR gives a good understanding of noise power in relation to an input signal, it is measured across the entire Nyquist band. In oversampled systems, SNR becomes less relevant as the noise is spread over a wider bandwidth, reducing its density and amplitude. With every doubling of the sample rate, the noise density decreases by 3 dB. Additionally, as it is now spread over a larger bandwidth, noise outside the band of interest can be easily filtered out, further increasing SNR within the signal bandwidth of interest.

It can be difficult to compare oversampled data converters based on their SNR values, because the bandwidth of the signal, the sample rate, and the additional filtering used for a given application all affect the SNR. Instead, manufacturers use Noise Spectral Density (NSD) as an alternative way of measuring and expressing the noise performance of RF-sampling devices. Furthermore, NSD gives a truer picture of the sensitivity of the data converter, determining its ability to capture low power signals

NSD measures the noise power per 1 Hz bandwidth, making it simpler to calculate the total noise power within a particular band of interest. NSD is measured in relation to the full-scale input and is usually given in dBFS/Hz, where FS denotes 'Full-Scale' [53]. An approximation of NSD can be calculated using the ideal SNR of an N-bit data converter, and the theoretical noise floor of an FFT with N_{FFT} points (in some calculations NSD is also calculated using SINAD). As we have seen earlier in this chapter, the SNR of an ideal data converter can be calculated by (12.8).

Since SNR is the ratio between signal power and noise power across the entire Nyquist band (0 to $f_s/2$), the total noise power across the band can be calculated by subtracting the signal power from the full-scale input power. If we assume that the input signal is at full-scale, then the total noise power across the band is simply the negative of the SNR value in dBFS [12].

The FFT can be thought of as a spectrum analyser with a bandwidth of f_s/N_{FFT} per FFT bin, and so the noise power per bin can be calculated as

$$P_{\text{noise/bin}} = 10\log_{10}\left(\frac{N_{FFT}}{2}\right) \text{ dB .} \tag{12.13}$$

To calculate the noise power per 1 Hz bandwidth, the number of FFT bins must equal the sample rate, becoming

$$P_{\text{noise/Hz}} = 10\log_{10}\left(\frac{f_s}{2}\right) \text{ dB .} \tag{12.14}$$

The NSD can then be calculated by subtracting the noise power over the entire Nyquist band and the noise power per 1 Hz FFT bin

$$NSD = -SNR - P_{\text{noise/Hz}} \quad \text{dBFS/Hz} \tag{12.15}$$

$$NSD = -(6.02N - 1.76) - 10\log_{10}\left(\frac{f_s}{2}\right) \quad \text{dBFS/Hz .} \tag{12.16}$$

For example, the NSD of an ideal 4 Gsps 14-bit ADC can be calculated as

$$NSD = -(6.02 \times 14 - 1.76) - 10 log_{10}\left(\frac{4 \times 10^9}{2}\right) = -175.53 \quad \text{dBFS/Hz}. \tag{12.17}$$

The true value of the NSD will differ from this ideal case because other sources of noise will be present, decreasing the SNR, and thus the NSD. Input signal frequency can also affect NSD. Therefore, the actual value must be obtained from direct measurement. Most manufacturers supply NSD values for various input frequencies, which gives a good characterisation of the overall performance of the data converter. For example, for first generation RF-ADC Quad tiles, typical values of NSD are given as -150 dBFS/Hz and -146 dBFS/Hz for input frequencies of 240 MHz and 2.4 GHz, respectively.

12.4.2. Two Tone Intermodulation Distortion

Two Tone Intermodulation Distortion (IMD) is measured by generating two sine waves that are close to each other in frequency (we denote the frequencies of these test tones as f_1 and f_2). Any non-linearity in the system causes the two input signals to intermodulate, producing spurs at predictable frequencies. The linearity of the data converter can therefore be characterised by measuring the power in these spurs.

Usually only second order (IM2) and third order (IM3) products are measured, with IM3 products being the most important as they tend to be closer in frequency to the input signal (away from this test scenario, the implication is that IM3 spurs are more difficult to filter out).

IM2 and IM3 are measured in relation to the carrier (i.e. input signal), and usually stated in dBc. IM2 products are located at $f_2 \pm f_1$, while IM3 products are located at $2f_1 \pm f_2$ and $2f_2 \pm f_1$. Figure 12.17 shows an example of the second and third order intermodulation products (IM2 and IM3) produced from two input test tones at f_1 and f_2 [12].

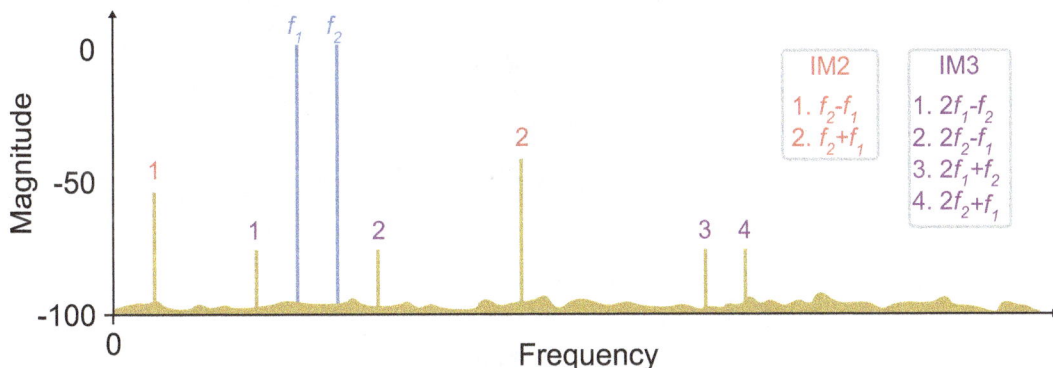

Figure 12.17: Second and third order intermodulation products produced from two input signals, f_1 and f_2.

As with spurs caused by harmonic distortion, the amplitude of the intermodulation products is dependent on the linearity of the device. The more linear the device, the lower the amplitude of the IMD products. Manufacturers' data sheets typically only state third-order products, and provide the input frequencies and amplitudes used for measurement. For example, Gen 3 RFSoC RF-ADC Quad tiles have a typical IM3 measurement of -75 dBc for a 1.9 GHz input frequency, and the two input test tones are separated by 20 MHz [87].

As bandlimited communication signals are composed of multiple frequencies, intermodulation products are also bandlimited. This can cause issues in radio systems because IM3 products can 'leak' into adjacent channels, potentially interfering with other spectrum users. While many components on radio systems are non-linear, producing their own IM3 distortion products, IM3 is an important metric in RF-sampling devices as it characterises the ability of a data converter to meet radio design specifications.

12.4.3. Adjacent Channel Leakage Ratio

In modern radio communications, spectrum is considered a scarce resource, and its usage is controlled by a national or regional regulator. Users transmitting on allocated frequency bands must do so in a way that avoids, or minimises, the interference caused to other users of the spectrum on nearby frequencies. It is therefore important to quantify how much a transmitted signal 'leaks' into adjacent channels. The standard measurement of this is the Adjacent Channel Leakage Ratio (ACLR).

ACLR is a standardised spectrum measurement that is used in radio standards such as 4G LTE and 5G NR. It is a measurement of the amount of power that leaks into adjacent channels (mainly caused by IM3 intermodulation products) relative to the power of the signal within a target channel, and usually stated in dBc [53]. Figure 12.18 shows a measurement of ACLR for a channel directly adjacent to the target channel. An equivalent measurement can also be used to quantify leakage in 'next adjacent' channels, i.e. those further separated in frequency from the reference channel.

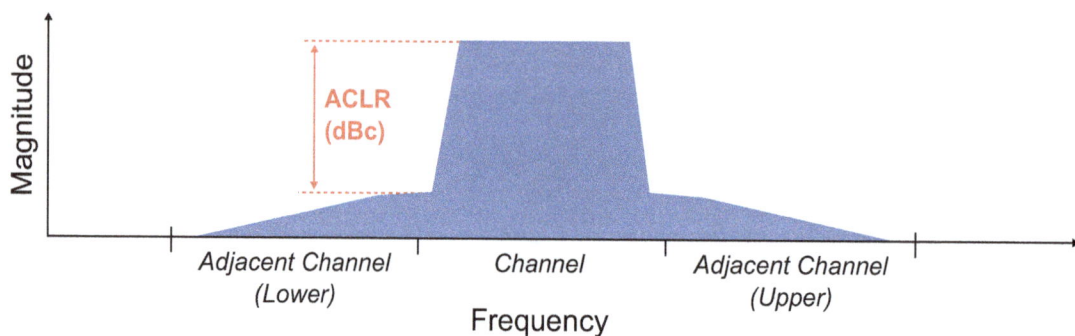

Figure 12.18: Measuring the Adjacent Channel Leakage Ratio (ACLR)

12.5. Chapter Summary

In this chapter we have explored some of the concepts behind frequency planning and the reasons why it is an important aspect of radio application design in the era of RF-sampling data converters. The chapter began by looking at the various types of noise and spurious emissions that occur as a result of the data conversion process. We saw that in traditional Nyquist-rate data converters, quantisation is the primary source of noise, and that the impact of quantisation noise can be reduced through oversampling. The origins of harmonics and ADC interleaving spurs were reviewed, as well as how to calculate the frequencies at which they appear.

The traditional methods of characterising data converters using measurements such as SFDR, SNR, and ENOB were reviewed, and it was noted that various types of noise and spurs can affect these values. We discovered that these measurements are less suited for RF-sampling data converters because they assume Nyquist-rate sampling, and noted their values can be improved with good frequency planning.

Following on from there, the chapter explored the concepts of frequency planning; in particular, investigating how aliasing and mixing can affect spur location, and how filters can be used to suppress spurs that are located outside the band of interest. Finally, we looked at alternative data converter characteristics that are more suited to RF-sampling devices, such as NSD, IM3, and ACLR.

As a practical extension, you are invited to try out some frequency planning for yourself! Please see the introduction immediately following this chapter, which outlines the interactive notebooks accompanying the book.

Notebook Set F

Frequency Planning

Chapter 12: *Figures of Merit and Frequency Planning* looked at the various sources of spurious emissions (spurs) and how to mitigate against their effects with the use of frequency planning. Two examples were provided, showing the use of frequency planning to design radio applications for a transmitter and receiver. The chapter also walked through what constitutes a good or bad frequency plan. Before using the next set of notebooks, it is recommended that you read Chapter 12 (if you have not already done so) as it reviews the underlying concepts. Some of these key details are briefly revisited here.

In this chapter, we introduce a frequency planning tool (based on a similar tool distributed by AMD), for the RFSoC range of devices. The tool is written entirely in Python and can be run within a Jupyter notebook, directly from a browser hosted on a computer or RFSoC board. It allows the spur locations for both the RF-ADCs and RF-DACs to be quickly and easily determined. Spurs are calculated using parameters set by the user, such as the sample rate and signal frequency (which determines the Nyquist Zone of operation). The signal and spurs are displayed on a plot, which automatically updates when any parameters are changed.

There are two notebooks on frequency planning. Their names and relative locations are listed below.

> `ALL`   01_frequency_planning.ipynb — *rfsoc_book/notebook_F/01_frequency_planning.ipynb*

> `ALL`   02_rfsoc_frequency_planner.ipynb — *rfsoc_book/notebook_F/02_rfsoc_frequency_planner.ipynb*

F.1. Introduction to Frequency Planning

Frequency planning is a radio system design technique used to mitigate against the interfering spurs that are generated during digital-to-analogue and analogue-to-digital conversion. The most prominent spurs are typically the harmonics, located at integer multiples of the input frequency, which are caused by non-linearities in the data converter and external components. Phase-Locked Loop (PLL) mixing spurs may also be present, caused by the sampling clock mixing with the input signal, in a process similar to heterodyne mixing. Additionally, interleaved ADCs generate spurs due to DC offset, gain, and phase mismatches between the sub-ADCs, and their frequencies are related to the input frequency and/or sample rate of the data converter.

The first notebook, ***01_frequency_planning.ipynb***, explores some of the underlying concepts behind frequency planning. We begin by exploring the effects of non-linear devices and how non-linearity can create harmonic components not present in the original signal. We discuss the mathematics behind non-linear responses and allow the user to change parameters and compare results. For instance, a simulation of a non-linear amplifier is performed and directly compared to a linear amplifier. Results conclude that the non-linear amplifier introduces harmonics at integer multiples of the fundamental frequency, while the linear amplifier does not introduce harmonics.

Spurs are then investigated in more detail. In particular, we explore spurs that are caused by interleaving ADCs. We look at DC Offset Interleaving Spurs (OIS), Gain/Time Interleaving Spurs (GTIS) and harmonic interleaving spurs, and calculate their frequencies. We also consider the effects of aliasing, and how this can change the frequency properties of a spur in a predictable way, based on Nyquist zones.

A simulation of a non-linear interleaved ADC is then performed, and many of the different spurs that occur are generated. At the end of the simulation, we present a plot showing the location of the spurs in the frequency domain. This plot is given in Figure F.1, where it is possible to see the OIS, GTIS, harmonics and interleaving spurs alongside the ADC output.

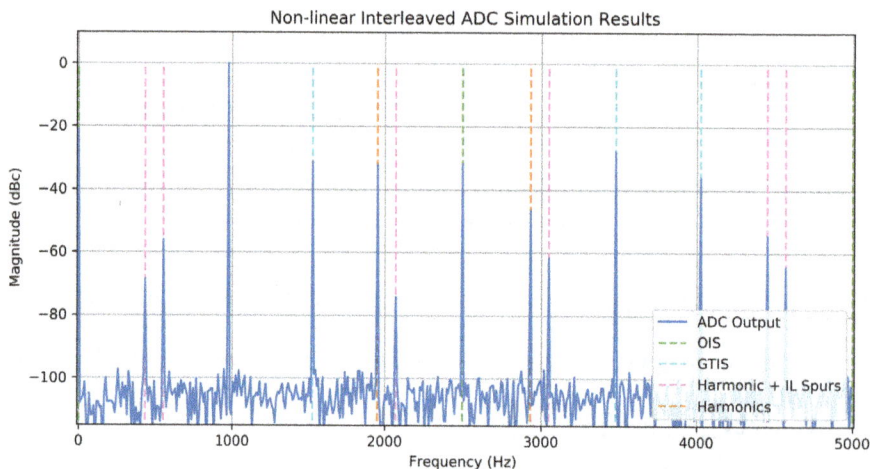

Figure F.1: An output plot of a non-linear interleaved ADC simulation. The plot presents the OIS, GTIS, harmonics and interleaving spurs with the ADC output.

Finally, the notebook concludes by considering the design of filters, and how they can be used to suppress spurs around the signal of interest.

F.2. RFSoC Frequency Planner

The second notebook, titled ***02_rfsoc_frequency_planner.ipynb***, contains the RFSoC Frequency Planner, which enables users to quickly find the location of spurs given a set of configurable input parameters. The frequency planner is split into four separate tools:

- RF-ADC

- RF-DAC

- Digital Down-Converter (DDC)

- Digital Up-Converter (DUC)

An example of the RFSoC frequency planning tool is presented in Figure F.2. In particular, the DDC tool is shown. Several sections of the frequency planner are highlighted: the tool selection tab (1); the parameter control (2); and the frequency plan plot (3). The tool selection tab allows the user to navigate between the ADC, DAC, DDC, and DUC plans. The parameter control contains several widgets that tune the frequency planner's properties. Lastly, the frequency plan plot in the middle of Figure F.2 updates in real-time as the user tunes the frequency plan parameters.

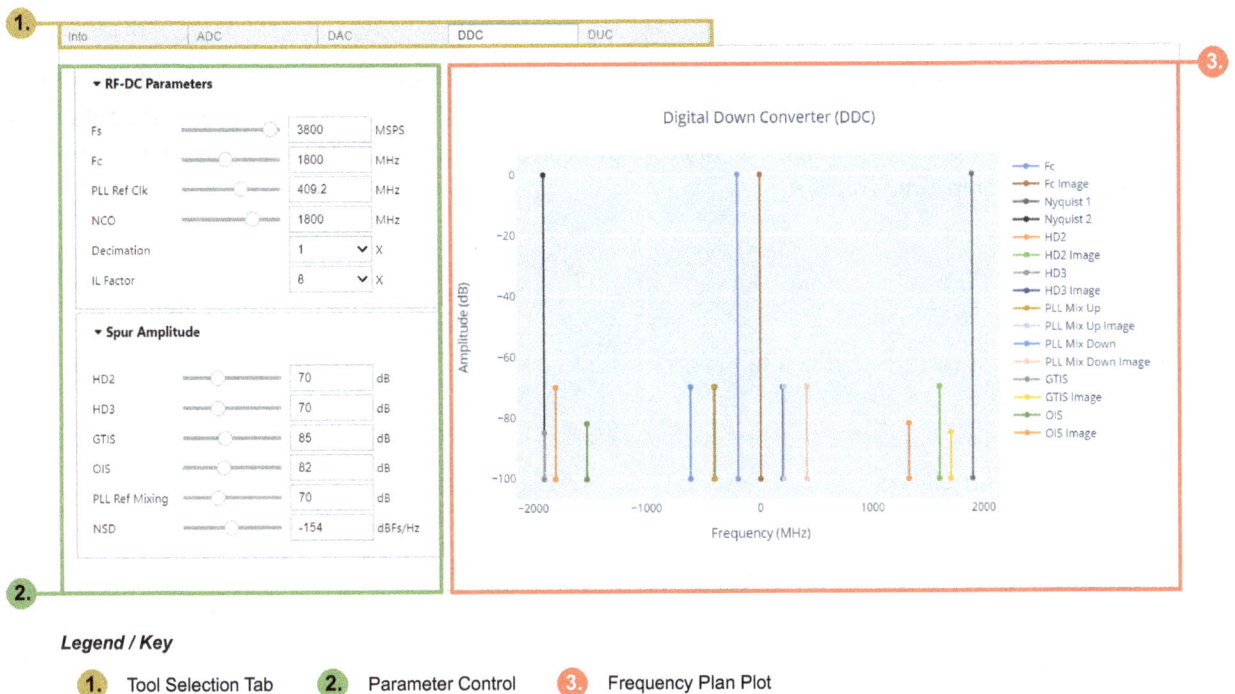

Figure F.2: The RFSoC frequency planning tool with annotations.

The goal of frequency planning is to determine a set of parameters that minimise the interference affecting the signal of interest. The frequency plan should also leave sufficient spur-free frequency bands above and below the signal to allow the spurs generated elsewhere in the spectrum to be suppressed with filtering. The frequency planning tool makes this task easier by allowing the user to quickly tune different parameters and see the results before committing to a specific configuration. We will now describe each of the frequency planning tools, starting with the RF-ADC.

The RF-ADC tool has five parameters that the user can change: the sampling rate, the centre frequency, the bandwidth of the input signal, the PLL reference clock, and the ADC interleaving factor. Changing each parameter will update an adjacent plot, displaying the location of the spurs in relation to the input signal after aliasing. The user can tune the parameters until a favourable configuration is found. Note that the interleaving factor is dependent on the target device. Quad RF-ADC tiles have an interleaving factor of 4, while Dual RF-ADC tiles have an interleaving factor of 8. Users can refer to [88] to find the appropriate value for their specific device. Additionally, this tool also calculates the *Calibration Mode* the user must select for a given configuration. Mode 1 is optimised for signals located between $0.4f_s$ and $f_s/2$, while mode 2 is optimised for signals between DC and $0.4f_s$.

The RF-DAC tool offers the same configurable parameters as the RF-ADC tool, with the exception of the interleaving factor, which is not relevant to the RF-DAC. Additionally, this tool calculates the sampling mode the RF-DAC must use, depending on the given input parameters. Note that, operating within Nyquist Zone 2 requires the *Mix-Mode* configuration of the fine mixer to be enabled.

The DDC and DUC tools allow spurs to be located after the mixing and rate change stages (decimation or interpolation, respectively). In both of these tools, signals and spurs are limited to single tones rather than frequency bands. Additionally, the amplitude of each spur can be controlled by the user. Associated amplitude values of spurs can be retrieved from [87].

Chapter 13

Design Tools and Workflows for RFSoC SDR

Josh Goldsmith, Lewis Brown,

Marius Šiaučiulis, and Graeme Fitzpatrick

Many of the earlier chapters have reviewed the features of the RFSoC device, and the theoretical basis of SDR system design and implementation. The Notebook Sets provide an opportunity to run, learn, and experiment with examples run on an RFSoC device (or simply in a notebook, for theory-based examples).

This chapter discusses the components, tools, and processes necessary to design RFSoC SDR systems, from a practical perspective. There are several possible design flows, and the opportunity to use your preferred style of working — whether that involves AMD and/or third party development tools, low or high levels of design abstraction, and different design entry and integration methods. As well as setting out the main options for hardware, software, and combined hardware-software co-design, we also share some of our own experiences of developing SDR designs and demonstrators based on RFSoC platforms and PYNQ.

13.1. High Level Design Process

As discussed earlier in this book, the RFSoC device is split into two main customisable parts, the Processing System (PS) and the Programmable Logic (PL). When designing systems for the RFSoC it is worth first considering how the functionality of the design will be split between these two elements. For example, will data be generated or processed on the PS, or will this be performed solely on the PL? While the FPGA can be used to efficiently accelerate DSP algorithms, other processes may be better suited to the PS. Deciding how and where this split happens is an important first step in the design process.

Additionally, while the hardened RFDCs and SD-FEC Intellectual Property (IP) cores are integrated with the PL, they can be controlled from both the PL, by means of interface ports, and the PS, with the use of software drivers (this may also be the case for any custom or licensed IPs included in a PL design). For some designs, configuring these IP cores from PL may be desirable if latency is a key issue. However, for SDR designs where the radio may need regular reconfiguration, it may be preferable to use software drivers on the PS.

As a high-level preview of a design featured later in the book (the OFDM transceiver from Chapter 16 and Notebook Set I), we consider how its functionality is divided between the PS and PL. The transceiver contains both transmitter and receiver processing, as well as additional functionality for demonstration purposes. The transmitter generates random data, and then converts this data into an OFDM signal, before sending it to the RF-DAC. The receiver performs the opposite process, accepting data from the RF-ADC, undertaking OFDM receive processing, and recovering the data. Both the transmitter and receiver logic also require control inputs that allow configuration changes. Further, data is tapped off at various points on the transmit and receive paths for inspection.

Figure 13.1 shows a simplified block diagram of the OFDM transceiver model, displaying how these processes are split between the PS and PL. As can be seen from the figure, the data and signal processing is primarily performed on the PL, while the PS is mainly used for control. Additionally, the data inspection process is split between both PS and PL. The decision to split these functions between the PS and PL in this way was as much to do with the software being used to develop it as its intended purpose as an educational tool.

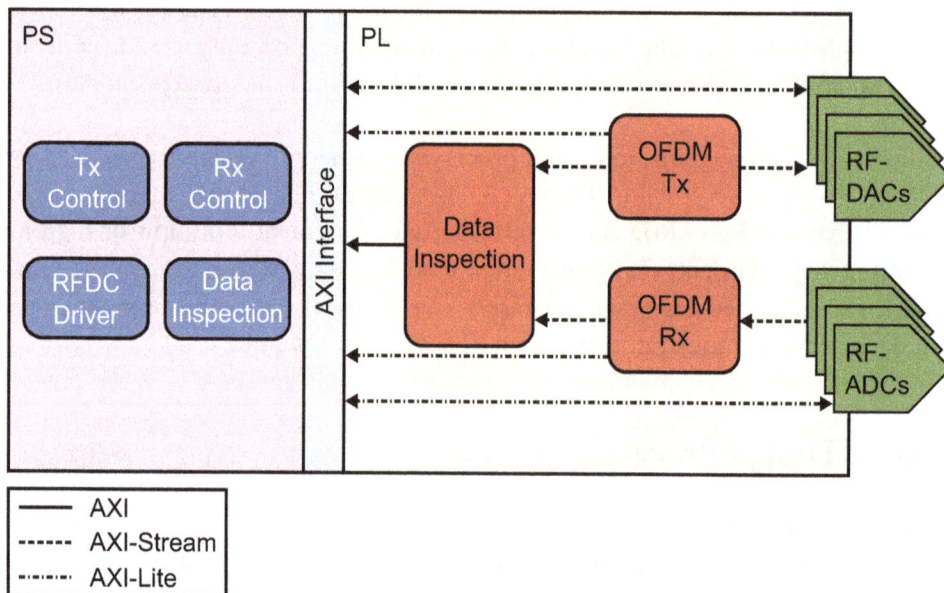

Figure 13.1: Simplified block diagram of the OFDM transceiver example from Chapter 16 and Notebook Set I.

The rest of this chapter looks at the various software tools and design methods available for developing systems for both the PL and PS. We also detail the design flow process that we adopted to develop many of the designs featured in this book.

13.2. Programmable Logic (PL) Design

All Gen 1, 2, and 3 devices in the RFSoC family have a set of standardised FPGA resources, which include CLBs and DSP48E2 slices, as well as memory in the form of Block RAMs and Ultra RAMs. These resources are fully reconfigurable by the user, enabling a wide variety of applications and use cases — for SDR design, the PL is particularly powerful for implementing signal processing algorithms.

Alongside logic fabric, RFSoC devices also feature hardened resources such as SD-FEC blocks [45] and RF Data Converters [90] which have been covered in detail in previous chapters[1]. These hardened blocks provide optimised implementations of computationally intensive functions, and the parameters that control their operation can be reprogrammed to suit the target application. For the RFDCs, this includes setting sampling rates, decimation and interpolation rates (and more), as covered in Chapters 9 through 11. These resources can be directly connected to user-configured PL logic fabric to build complex designs such as SDR transceivers.

When designing RFSoC systems, external interfaces should be considered, especially for evaluation boards such as the ZCU208 [77] or RFSoC4x2 [42]. As previously discussed, external RF signals can be transmitted and received through the RFDCs, via baluns and SMAs. Additional IO may also be desirable, for instance to support an external display, high speed QSFP data offload, or Pmod interfaces. These external IO interfaces are enabled through reconfigurable connectors within the PL, and offer great potential to integrate the RFSoC as part of a larger system (e.g. including SDR functionality along with other features).

The remainder of this section reviews the design tools and processes used to develop the hardware design for an RFSoC-based SDR system. A top-down approach is taken: we first consider the system-level design, before reviewing methods of creating the IP blocks that form lower levels of the design hierarchy.

13.2.1. Vivado IP Integrator

The most efficient method to merge custom hardware logic, pre-existing IP Cores, and IO together is through *IP Integrator (IPI)* [25], a tool that forms part of AMD's Vivado Integrated Development Environment (IDE). Pre-generated IP cores can be instantiated and connected to form a block design, either using the IPI GUI, as shown in Figure 13.2, or programmatically via Tool Command Language (Tcl) scripting [335].

IPI designs are generally constructed by combining instances of IP cores through *interfaces*, i.e. groups of signals that share a common function. The main benefit of designing with interfaces is a reduction in

---

1. The RFSoC DFE architecture provides additional hardened blocks optimised for mobile infrastructure deployments, as highlighted in Chapter 3.

complexity — a single graphical connection in the tool can represent a (potentially very large!) group of signals and buses. If each of these signals and buses were represented visually and had to be connected independently, the design complexity, effort, and time, would be much greater. Through the use of interfaces, a single connection made in the IPI GUI, or equivalently a single Tcl command, can create the complete connection between an IP core and another component in the system design[2].

Designing with interfaces also enables use of IPI Design Rule Checks (DRCs), which are aware of the required signals for specific interface types, and thus can provide validation that signals are connected correctly. These checks occur in real time as the design is assembled, and can be compared to syntax error messages in a programming language.

If a Vivado project is created to target a specific evaluation platform (e.g. the ZCU208 RFSoC development board [77]), then IP Integrator has knowledge of the applicable external FPGA pins. This allows the IPI *Connection Automation* feature to be leveraged to tie IO ports of the design to external pins of the board, thus

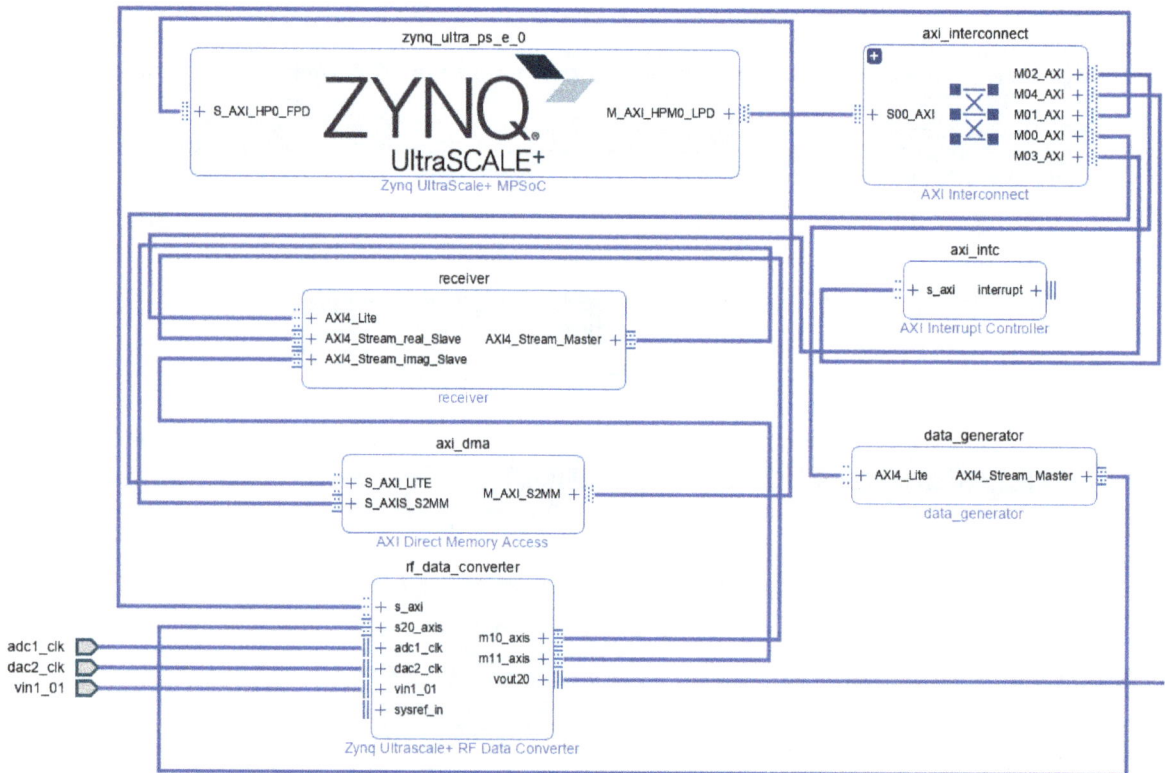

Figure 13.2: Example of an IP Integrator hardware system design.

---

2. Connections can also be made by manipulating individual signals (also known as ports) to give a greater degree of control. An IPI design may involve a combination of both interfaces and port manipulation, depending on the IP cores used.

saving designer effort. The appropriate physical pin and clocking constraints are created automatically through this process. IP Integrator also automates the creation of a top level wrapper file, which describes the interface of the complete system in Hardware Description Language (HDL) code.

Example:

A conceptual IP Integrator design is shown in Figure 13.3. The design represents an RFSoC loopback test configuration (where the outputs of the RF-DACs are connected, via SMA cable and analogue filters, directly to the RF-ADC inputs of the same device).

Data is generated through a custom 'Data Generation' IP Core, which represents the transmitter in the circuit. The IP core generates a test signal, which it supplies to an RF-DAC tile. Through loopback, this data re-enters the PL through the RF-ADC, before being supplied to a second, 'Receiver' custom IP core, which uses DSP techniques to process the received signal. A DMA IP core is used to transfer the processed signal to the PS, where it can be observed for testing and/or demonstration purposes.

This design highlights a key feature of an RFSoC IP Integrator design — the Processing System IP Core — which represents the PS part of the device. A number of configurable options are available within this IP core, including: IO options such as SPI and I2C; various clocking options including peripheral clocks and PL fabric clocks; and configurable memory such as DDR4.

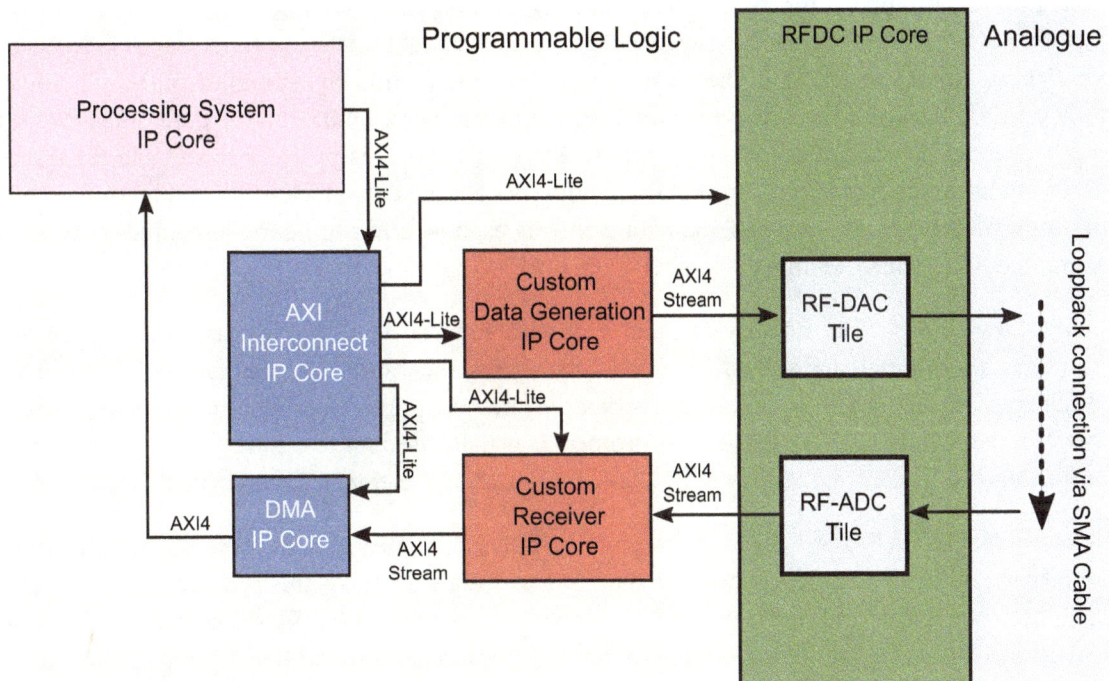

Figure 13.3: High level overview of a simple IP Integrator SDR design.

13.2.2. AXI Interfaces for IP Integrator

One of the key interface types when designing using IP Integrator is the AXI protocol, which is used to connect processing elements and IP Cores which require high bandwidth, low latency communication.

Recall from Chapter 3 that RFSoC devices implement the fourth iteration of this standard (AXI4) in both PL and PS, and that there are three key types of AXI4 buses available for RFSoC design: AXI4, AXI4-Lite (both for memory-mapped interfaces [95]) and AXI4-Stream (for point-to-point data transfer [94]). IP Cores are connected by AXI4 buses through slave and master interfaces. Each of the independent channels within the AXI4 bus consists of a set of information signals, as well as *tvalid* and *tready* signals which provide a two-way handshake mechanism. For information transfer, a transaction is initiated when the master interface provides a *tvalid* signal. A slave interface then confirms that it is able to accept information with a *tready* signal.

We can see the usage of the AXI4 protocol in Figure 13.3. Specifically, there are: AXI4-Stream interfaces between IP cores; AXI4-Lite interfaces connecting from PS to PL IP cores, via an AXI Interconnect IP core; and an AXI DMA IP core for passing data from PL to PS at high throughput rates. When designing custom IP cores in external software, it is possible to specify ports as any of the three AXI4 bus types, and to specify an IO port as a slave or master interface.

The typical use of each AXI4 type is described in more detail:

- **AXI4** — This protocol is for connections requiring memory-mapped links between processing elements (the RFSoC PS) and IP Cores. It is capable of either single-beat transfers or burst transfers with up to 256 data beats per transfer, and data can be moved bidirectionally simultaneously. There are five independent channels for AXI4: *read address*, *read data*, *write address*, *write data* and *write response*. Each channel has dedicated resources for transferring data. This protocol is suitable for transferring large quantities of data to and from main memory — a large data block in PS can be transferred to PL through multiple bursts of up to 256 data elements each — and can be implemented in an IPI design using the AXI4 Direct Memory Access IP Core [18].

- **AXI4-Lite** — A simple memory-mapped link with reduced handshaking signals, resulting in lower resource allocation than full AXI4. AXI4-Lite shares the same five channels as AXI4. The main difference between AXI4 and AXI4-Lite is that the latter supports only single-beat transfers, and not the burst transfers supported by AXI4. This protocol is usually used for low bandwidth communication, for instance with the control registers of IP Cores and processing elements, often from a PS-based IP driver.

- **AXI4-Stream** — AXI4-Stream supports point-to-point data streaming. It provides burst transfers of an unrestricted size. No address channel is required, as the use-case for this protocol is the direct flow of data between source and destination IP Cores within a device. The AXI4-Stream protocol is therefore particularly useful for signal processing in video, communications and networking applications, and is the method used to transferr data to and from the RFDCs of the RFSoC device.

13.2.3. Intellectual Property (IP) Cores and Libraries

Vivado enables the use of IP cores to aid the development of a hardware design. An IP Core is a self-contained hardware block, equivalent to an HDL description, that can be integrated into system designs. IP Cores are usually highly parameterisable, and stored in an IP repository, from where they can be deployed in different designs as required. IP cores are available from AMD for the RFDCs and SD-FEC blocks, along with a rich library of high value cores for DSP and communications that accelerate SDR hardware design, as well as many more IPs for a range of applications [31], [62]. The availability of pre-made, pre-verified blocks for radio applications is a notable attribute of the RFSoC design ecosystem and has the potential to greatly accelerate SDR system design.

As well as domain-specific IPs like those mentioned above, the availability of the wider IP catalogue means that a designer can reuse common functionality such as Direct Memory Access (DMA) [18] and peripheral interfaces, thus avoiding the overhead of developing and testing them. This allows the designer to focus on the creation of the specific application design, drastically reducing development time.

To give an example, Figure 13.4 shows an AXI Interconnect IP core from the Vivado GUI (specifically, from the *IP Integrator* tool, which was introduced in Section 13.2.1). The IO interfaces which enable connections to the wider PL design appear on the IP core representation in IPI: this core has two input interfaces, *S00_AXI* and *S01_AXI*, and one output interface, *M00_AXI*. Clock ports are provided such that each of these interfaces can be clocked independently, if required.

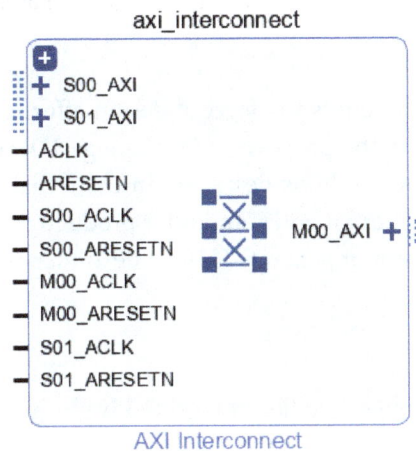

Figure 13.4: AXI Interconnect IP from the Vivado IP Catalogue (shown in Vivado GUI).

Designers also have the option to design and package their own IP Cores, and to create their own repositories. This is useful when custom hardware blocks require to be developed as part of a system design, and even more so when these IP Cores have reuse potential across multiple projects. We will now review the various methods available to create custom IP cores.

13.2.4. Tools and Methods for Custom IP Creation

As a designer, the ability to efficiently create custom hardware functionality for a design or application is crucial. Designers can create their own repositories of custom, user-designed IP cores for importing into Vivado and integration as part of an IPI system design.

Several different tools and methods can be used to create IP Cores. Design methods that use high level software languages such as C and C++, or which enable graphical methods of block-based design, offer a layer of abstraction over traditional HDL design entry. The major advantage of higher-level methods is an acceleration of the design process, which can lead to reduced costs and faster time-to-market.

The remainder of this section will review a number of options for IP core creation, including both AMD tools and third-party offerings. Some of the third party tools also provide workflows that integrate with Vivado as part of their background operations.

HDL Code Development

The traditional method of hardware design is to develop HDL code using one of the two major HDLs, VHDL and Verilog, or SystemVerilog (a superset of Verilog that also includes verification features). The HDL description specifies the interface and functionality of the design, and an accompanying testbench is normally developed for verification purposes. HDL IP cores can be developed and simulated in the Vivado IDE. HDLs support hierarchy, making it easy to develop, test, and integrate hardware modules, and to reuse existing modules.

The major drawback of HDL development is the level of design effort and time involved — although a high degree of designer control is possible, the process of developing HDL is labour-intensive. Specialist coding skills are also required. The tools and methods described in the remaining subsections all raise the level of design abstraction compared to HDL design entry, and offer productivity gains as a result[3]. It is useful to note that these higher-level tools all represent or generate HDL code 'under-the-hood'.

Vitis High Level Synthesis (HLS)

Vitis HLS [67] is an AMD tool that can be installed alongside Vivado, and enables the development of IP Cores from C and C++ functions. The tool automates code modifications, through the use of *pragmas*[4], to optimise C based functions to achieve hardware design goals (for instance, to maximise data throughput, or minimise latency). Therefore, functions can be fully compiled, simulated and debugged as software algorithms before being synthesised into IP Cores. Due to the tight integration between Vitis and Vivado, the designer can readily create an IP core repository from within Vitis HLS that is compatible with their Vivado version.

---

3. There are certain situations where writing in HDL is still the preferred approach, particularly when aiming to meet tight timing constraints, or to minimise resource usage.
4. Pragmas are instructions that are included within the HLS source code, to influence the synthesis of the code.

HDL Coder and SoC Blockset

MathWorks *HDL Coder* [249] is an alternative tool for hardware design, based in the MATLAB and Simulink environment [251]. Using HDL Coder, HDL code can be generated from compatible MATLAB functions and block-based Simulink models [253], to create IP Cores suitable for use in Vivado designs. A simple example HDL Coder design is seen later in the chapter, in Figure 13.17.

When HDL Coder is configured at the start of a design session, MATLAB and Simulink are linked with a supported version of Vivado, which provides tool integration and allows design flow automation from within the MathWorks environment. The main user interface is the *HDL Workflow Advisor*, a wizard that assists in the generation of HDL from a top-level model. This method also allows the IO of the model to be easily mapped to the AXI4 Input and Output interfaces of the generated IP Core.

The functionality of HDL Coder can be expanded with the *SoC Blockset* [255], which adds Simulink blocks for FPGA memory, external IO, and PS scheduling, as well as for hardened RFSoC resources such as the RFDCs. These blocks can also be generated into HDL code, enabling complete Vivado block designs to be generated from Simulink. The SoC Blockset is closely tied to the *SoC Builder* tool, which will be explored in Section 13.4. When using the SoC Blockset, PS and PL elements can be composed as a single Simulink model — together with SoC Builder, this facilitates hardware-software co-design for creating entire embedded systems for RFSoC devices.

Model Composer

AMD offers its own model-based HDL support within the Simulink environment, in the form of *Vitis Model Composer*[5] [68]. This tool can optionally be installed alongside Vivado, and provides more than 200 blocks maintained by AMD, from which HDL can be generated. A designer is able to import custom blocks, including those written (by hand) in VHDL or Verilog, or developed in Vitis HLS. In this way, IP cores can be simulated and exported to hardware comprising a mixture of library blocks and user-developed ones.

Model Composer provides a set of HDL-capable blocks specially designed and optimised for DSP applications, including blocks for Finite Impulse Response (FIR) filters, Fast Fourier Transforms (FFTs), and trigonometric operations based on the Co-Ordinate Rotation DIgital Computer (CORDIC) algorithm [354]. These blocks have been optimised for hardware deployment, and include implementation options that allow the designer to target specific PL architecture resources like DSP48E2 slices and Block RAMs, depending on the block type.

---

5. The *System Generator* block-based design tool was incorporated into *Model Composer* from Vivado version 2021.1.

Open-Source HDL Generation Options

Several alternative HDL and IP Core generation options are available from the open-source community. Clash is a high-level HDL that borrows the rules and syntax of the Haskell programming language, and is mapped to VHDL and/or Verilog at compile time [296]. MyHDL [274] and Amaranth [15] are both Python-based projects that take a similar approach, by converting to VHDL or Verilog at compile time. Chisel [122] is comparable to Vitis HLS, adding HDL-capable pragmas to the Scala programming language (rather than C/C++). Chisel allows for reusable HDL while offering register level coding, raising the abstraction level from HDL whilst maintaining low level control. It should be noted that, as community-supported projects, these tools may provide less guidance for IP core generation than those featured earlier.

13.2.5. Vivado Hardware Design

Vivado enables the creation of hardware designs for programming onto the RFSoC and other AMD programmable devices [65]. The entire hardware design workflow can be undertaken within Vivado — from the initial design description through to programming the target device. A designer can choose to operate the Vivado software through either a GUI, or programmatically via scripting languages such as Tcl.

At the lowest level, Vivado uses HDL to describe the design that will be implemented on the PL, and it supports the VHDL, Verilog, and SystemVerilog[6] languages. All PL-based components of a design must be included in the HDL description, including external IO, memory, any hardened resources used (such as RFDCs or SD-FEC blocks), and interfaces to the PS. It is important to highlight that the steps leading to a complete hardware description in HDL can be completed using other tools and design methods, i.e. the designer can work at a higher level of abstraction. As discussed earlier in this chapter, IP Integrator is recommended for system design, and tools such as Vitis HLS, Model Composer or HDL Coder can be used for IP creation.

Once the HDL description of the PL design is completed, a series of further steps are required to generate a programming file with which to configure the PL with the designed circuit. This is known as the hardware design flow, and is outlined next.

Hardware Design Flow

The Vivado hardware design flow is illustrated in Figure 13.5 and can be summarised as follows:

- **Hardware design entry** — A hardware design must first be described using an HDL. Simulation and testing are undertaken at this stage to verify the intended functional operation. As noted earlier, higher level design tools can be used to generate HDL and thus accelerate this hardware modelling stage.

---

6. VHDL and Verilog are standardised by IEEE and subject to revisions. At the timing of writing, the current version of Vivado supports VHDL-1993 [210]and parts of VHDL-2008 [211], and the 2001 revision of Verilog [209], as well as the 2009 revision of SystemVerilog [208].

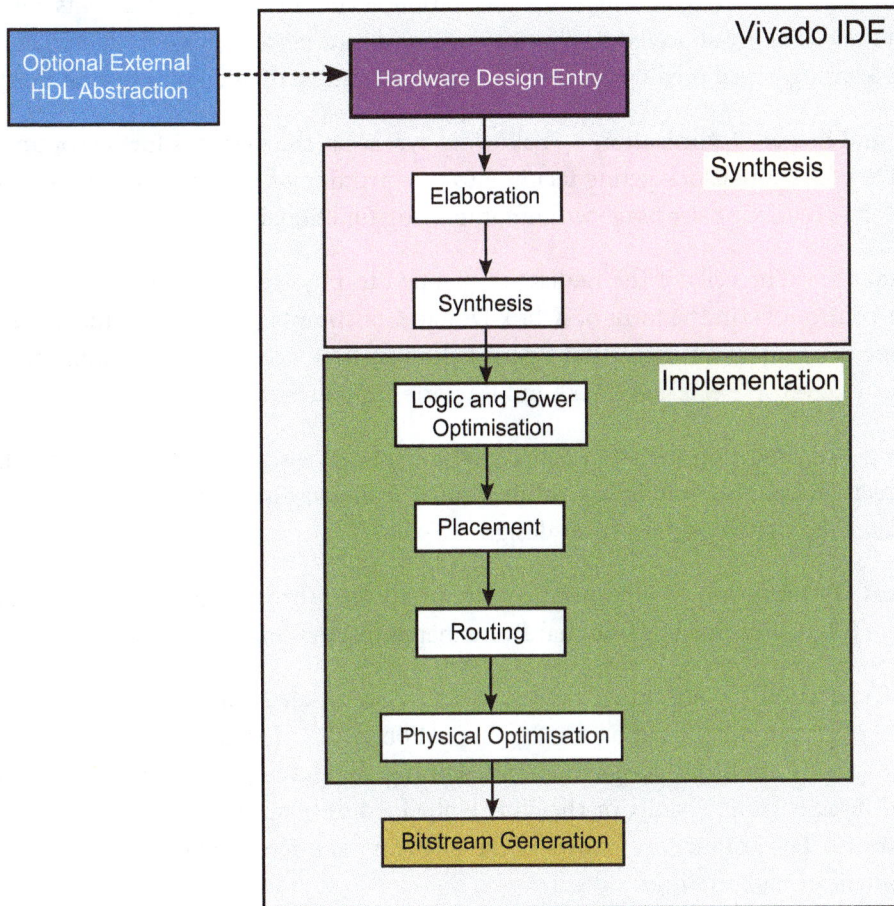

Figure 13.5: Overview of the Vivado design flow.

- **Elaboration** — During this first stage of the synthesis process, the HDL code is parsed to locate descriptions of standard logic elements such as flip flops and multiplexers. This produces an elaboration netlist, i.e. a file listing the components and connections in the circuit, as interpreted from the HDL code. The elaboration netlist is technology independent — it is not based on knowledge of the specific PL components that the design can be mapped to.

- **Synthesis** — This process step configures the HDL code to a technology-specific netlist — mapping the circuit described by the elaborated netlist to the PL resources (logic gates, LUTs, DSP48E2 slices, etc.) of the target device. At this stage no placement has occurred — the netlist lacks information about *where* on the device the circuit should be placed. The definition of timing constraints, IO planning and hardware floorplanning can be performed at this stage. Timing estimates are shown following synthesis, indicating the expected critical path of the design and whether it should meet timing requirements (more on critical path and timing in Section 13.6.3).

- **Implementation** — After synthesis, implementation is performed. Components and connections specified in the netlist are allocated to physical logic and routing resources on the target device, guided by machine learning-based predictions [64]. This process can be divided into several substeps:

  - **Logic and Power Optimisation** — Following synthesis, the netlist is further optimised to reduce resource utilisation. Clock gating techniques (the pruning of clocks which are not in use) are also deployed to reduce power usage without impacting functionality.

  - **Placement** — The cells of the netlist are mapped to physical device resources according to any design constraints (in the form of Xilinx Design Constraints (XDC) files) that have been input to the process [135]. If an evaluation board is known to Vivado, key constraints are automatically applied, i.e. available external clocks and IO ports are specified.

  - **Routing** — Connections between netlist components are assigned to physical routing resources on the target device. The routing algorithms prioritise global resources such as IO and clocking, and then route the signals from the user design.

  - **Physical Optimisation** — Optimisations are performed throughout placement and routing using accurate timing information from hardware mapping, resulting in a reduced critical path.

- **Bitstream Generation** — The implemented design is translated into a bitstream (i.e. a binary file containing configuration information for the PL portion of the RFSoC), that can be downloaded to the device to program it with the designed circuit. During the development phase, the bitstream can be downloaded directly from Vivado, or the Vitis embedded development environment, or using PYNQ runtime software. For production hardware, the bitstream is often stored in non-volatile memory and loaded automatically at boot time.

Each of these processes can be driven from within the Vivado IDE, or programmatically. Various options are available at each stage; these can be user-specified to guide the achieved results. For instance, there is an option to define a routing strategy that explores the design space more thoroughly than with the default settings.

13.2.6. IP Cores for Hardened Resources

To conclude this section on PL design, we discuss the use of IP cores corresponding to the two specialised, hardened blocks on Gen 1, 2, and 3 RFSoC devices: namely the RFDCs, and SD-FEC blocks. Additional hardened blocks are available on RFSoC DFE devices, as previously outlined in Chapter 3, although we will not consider those specifically here.

RF Data Converter IP Core

The RF Data Converter IP allows the RF-DAC and RF-ADC blocks to be included in an IP integrator design. Figure 13.6 shows a high-level overview of the RFDC IP Core, highlighting the AXI4 Interfaces connecting the RFDC to the PL and PS, as well as the external interfaces to transmit and receive analogue signals.

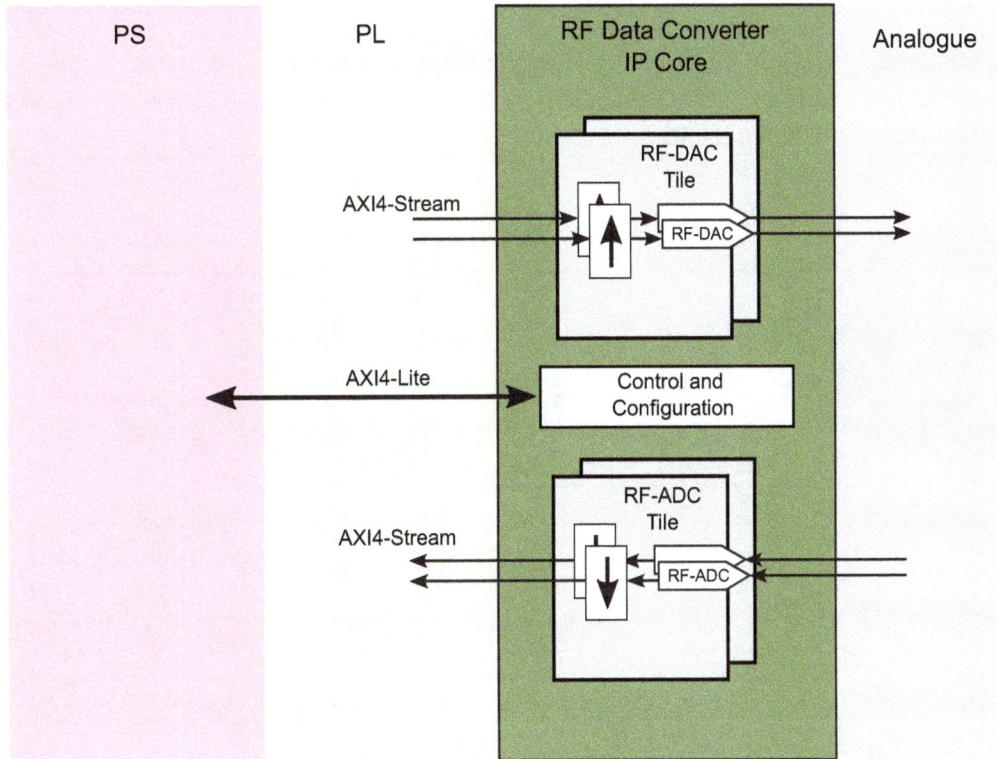

Figure 13.6: High Level overview of RFDC IP Core showing AXI4 Interfaces.

Each RFDC tile within the IP Core is tied to an AXI4-Stream interface, with each RF-ADC tile connected to a master interface and each RF-DAC tile to a slave interface. Therefore, large amounts of data can be transferred to and from the RFDCs. From IP Integrator the RFDC IP Core can be configured (either programmatically through Tcl, or with a GUI as with other standard IP Cores from the IP catalogue). Options are available to enable tiles and blocks, set sample rates, configure the mixer and decimators, among many more options. For more details of the RFDC, see Chapters 9 through 11.

The RFDC IP Core also contains an AXI4-Lite interface, which connects to a processor in the design (usually the PS). The connector can enable a software-based driver to configure the RF Data Converters at run-time, without needing to rebuild a bitstream. For timing sensitive applications, it may be sensible to configure the design at the PL level, as this can reduce latency compared to configuring from a PS-based software driver. A comprehensive Python driver exists within PYNQ, enabling straightforward use of the RFDCs.

SD-FEC IP Core

The SD-FEC IP Core is similar to the RFDC IP Core, in the sense that AXI4 interfaces are used to connect the hardened resource to a wider PL and PS design. Figure 13.7 provides an overview of the SD-FEC core, and shows its connections to and from the PL.

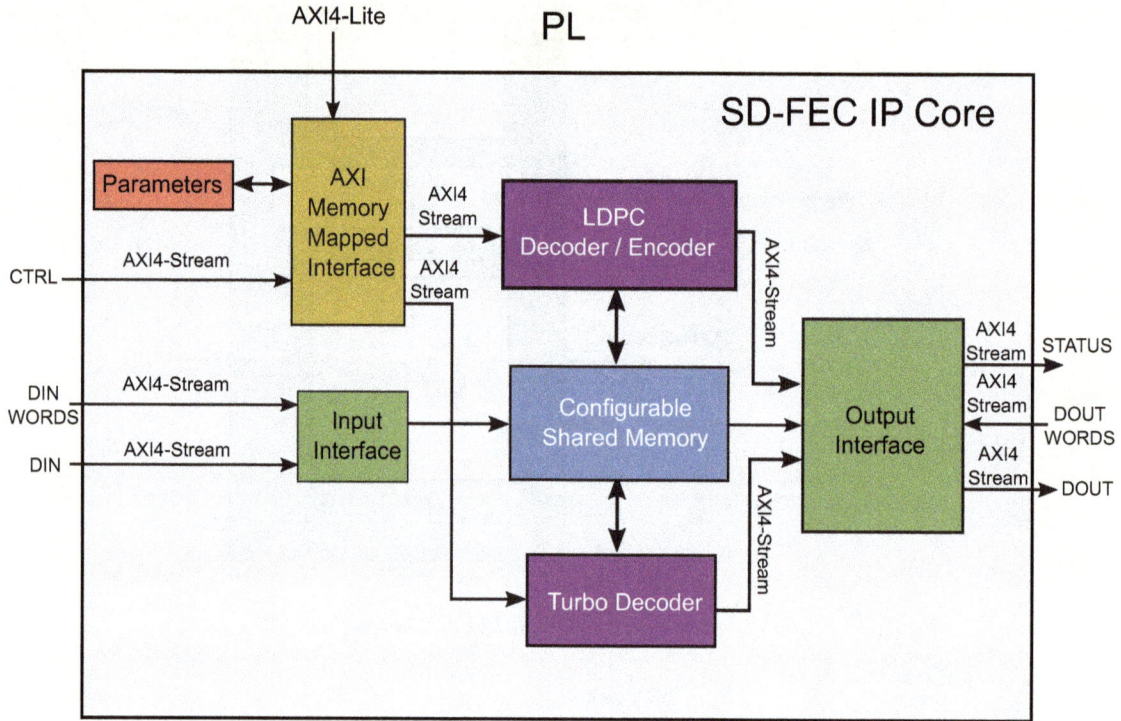

Figure 13.7: High level overview of SD-FEC IP Core showing AXI4 Interfaces.

An AXI4-Lite bus connects to the PS and is used for supplying parameters such as LDPC code definitions, which can be controlled via a software driver. Multiple code types can be loaded using the AXI-Lite bus, and then at run-time, a specific code can be chosen from the loaded types (on a per code block basis) using the AXI4-Stream interface. AXI4-Stream interfaces are also used for input and output data, the dimensions of which are parameterisable. The last of these can be abstracted from the design by setting a fixed size in the GUI.

Practical design using the SD-FEC IP will be covered in extended detail in Chapter 15.

13.3. Processing System (PS) Design

As reviewed in Chapter 3, RFSoC devices incorporate a Processing System (PS) alongside the PL and hardened IP resources. The PS includes Arm cores capable of executing software programs. The RFSoC PS subsystem design is closely related to the MPSoC CG device family, and contains up to two Real-time Processing Units (RPUs) and up to four Application Processing Units (APUs). The Arm Cortex-R architecture based RPU cores are optimised for real-time and safety-critical applications and thus can be programmed in bare metal or using a Real Time Operating System (RTOS). The APU cores are based on substantially more powerful Arm Cortex-A53 architecture, and additionally support full operating systems such as Linux or Android™.

The following section of the chapter will give a high-level overview of different software stacks and workflows available when designing for RFSoC PS. For more in-depth information on the available resources and features of the PS, please refer to Chapter 3, and also to our earlier book on the Zynq UltraScale+ MPSoC, which as noted above, features a very similar PS [131].

13.3.1. Software Stacks

When designing for RFSoC PS, a suitable software stack has to be chosen to meet the design requirements. A software stack is a set of base software that a developer can build upon by adding their own custom software to suit the intended application. Depending on whether the defined software system has two or more concurrent functions, requires real-time operations to be performed, or has rich user interface requirements, a well chosen stack can reduce the development time and increase user satisfaction.

Different software stacks also offer distinct abstraction layers. In lower abstraction layers, developers need to manage more system resources such as memory, I/O and peripherals. Higher abstraction layers provide unified Application Programming Interfaces (APIs) to handle system resources, but introduce memory and latency costs.

The remainder of this section reviews some of the candidate software stack options for RFSoC SDR systems. For further discussion of software stacks, please also refer to [131].

Bare-Metal

Bare-metal refers to the program instructions being executed directly on hardware, with no intervening OS. It is the lowest level of abstraction, and all system resources need to be managed by the programmer. The software is executed in a single thread with optional interrupts handling high-priority and urgent processes.

AMD supports bare metal application development for their SoCs containing APU, RPU and MicroBlaze processors through the Vitis Unified Software Development Environment [86]. Key libraries for PS and PL peripherals, standard C libraries (*libc* and *libm*) as well as additional middleware libraries for file system, encryption and networking are supplied within the standalone board support packages (BSPs) [70]. The BSP provides a simple, single threaded environment to base your application upon.

Bare-metal is best for designs that have simple requirements: for instance, a software task that involves only a single action loop; or a routine which can be driven by interrupts. Bare-metal is also suitable when fine-grained control is required on the part of the developer, e.g. where low level optimisation of the software implementation is needed to achieve a performance metric, such as low latency or power usage. For applications with greater complexity, more concurrent tasks, or where low-level hand-optimisation is unnecessary, other software stacks are preferred.

Real Time Operating System (RTOS)

A Real Time Operating System (RTOS) is a step above bare-metal. It is a relatively simple, OS-like layer that defines a unified way of creating, scheduling and managing tasks as well as system resources. The RTOS offers deterministic process scheduling, and guarantees that designated critical system tasks will execute in response to events and interrupts within a predefined time frame.

Unlike a full-blown OS such as Linux, RTOS libraries are linked with the application code, which produces a single executable that is run when the board boots. Most RTOSs do not support loading or unloading of code dynamically, i.e. code cannot be changed at runtime.

AMD provides an official open-source port of FreeRTOS [29] for APU, RPU and MicroBlaze processors implemented on AMD SoCs. Additionally, some third party RTOSs such as Microsoft Azure RTOS [259] or The Zephyr Project [369] have deployment examples for selected AMD SoC-based boards.

Linux OS

Linux is a free and open source operating system kernel originally authored in 1991 by Linus Torvalds [341]. The kernel is one of the first programs loaded on system startup, and provides a unified, architecture-independent API to control and communicate with system hardware resources such as memory, processes, network interfaces and device storage. The GNU/Linux distribution (commonly referred to as Linux OS) is a general-purpose operating system that includes the Linux kernel [342] as well as supporting system software and libraries, many of which come from another Free/Libre and Open Source Software (FLOSS) project called GNU [166]. It can run on many different processing architectures, including x86 as found in standard desktop PCs and servers, as well as the Arm processors found in RFSoC.

Linux provides the highest abstraction level out of the discussed software stacks, as most of the user applications run in the Linux userspace, and communicate with the kernel (and in turn with hardware components) via System Call Interface [231]. Normally developers do not need to be concerned with the System Calls directly, as most applications are written at an even higher level, using libraries such as GNU C library (*glibc*) [177] or *musl* [273]. This separation of user and kernel space ensures that applications can access shared resources in a safe manner using the higher level interface [220]. Access to a library of OS service APIs for most common system resources including file systems, networking, and inter-process communication means that developers can focus on their application instead of system management.

By default, Linux provides a Completely Fair Scheduler (CFS) which allocates CPU time fairly to all applications depending on the total number of runnable processes and the time they have already been active. Real-time-like capabilities can be achieved using other inbuilt scheduling algorithms or switching to an alternative, real-time optimised kernel. Although real-time Linux can be a great choice for many applications, it cannot match the deterministic and hard real-time scheduling performance of dedicated bare-metal or RTOS implementations, due to the significant resource overhead required to run the operating system itself.

13.3.2. Synchronous Multi Processing (SMP) and Asynchronous Multi Processing (AMP)

Complex heterogeneous processing systems, such as the PS of RFSoC devices, support advanced deployment architectures for the previously mentioned software stacks. The PS of the RFSoC can be configured and used in two primary modes: *Synchronous Multi Processing (SMP)* and *Asynchronous Multi Processing (AMP)* [86]. These two modes can be outlined as follows.

Synchronous Multi Processing

In SMP mode multiple processors are governed by a single instance of an operating system. The OS handles most of the complexity of distributing and managing system resources such as processing cores, caches, load balancing and peripheral interrupts. The SMP mode is only supported by the Arm Cortex-A53 APUs of the RFSoC PS.

Asynchronous Multi Processing

An AMP system has multiple CPUs of the same or different architectures running independent programs. Each core has its own address space, and may-or-may-not run an OS. Depending on the design, the system can incorporate a communications facility between the cores, or they can run completely independently of each other. AMP allows designers to run a mix of Bare-metal, RTOS and Linux OS on the same processing system, with each of the PS cores running an independent software stack. For example, a design can use the Arm Cortex-R running bare-metal or a RTOS stack for mission-critical software, whilst other cores in the system (the Arm Cortex-A application cores) can run user-facing software.

13.3.3. Tools

A number of software development tools can be used to generate the PS software elements of an RFSoC SDR design. In this section, we highlight selected tools that are particularly useful for developing elements of a software stack for RFSoC-based systems.

PetaLinux

PetaLinux Tools [35], [36] is a free-to-use embedded Linux Software Development Kit (SDK) for AMD SoCs and MicroBlaze [34] soft processors. It is based on the open-source Yocto Project® [366], but provides a higher abstraction layer focused on development for AMD silicon devices.

PetaLinux is not a Linux distribution in itself, but provides all the necessary tools to customise, build and deploy embedded Linux images tailored to your application. This includes creating and updating Board Support Packages (BSPs)[7], as well as building the various components of a Linux System, e.g. the kernel, bootloader, root file system and so on. Embedded Linux images built with PetaLinux can be deployed on the board over network or JTAG interfaces. Simulation is also possible using the bundled QEMU full-system emulator when hardware is not available [297].

PetaLinux also integrates with other AMD tools. For instance, hardware platform developed using Vivado can be used to configure a PetaLinux project. Vitis can use a PetaLinux project as a base to develop software applications.

Vitis Software Development Workflows

The Vitis Unified Software Platform is a new AMD tool that combines all SoC development aspects into a single unified environment [57],[58]. It includes the features from now-deprecated Xilinx SDK, SDSoC™ and SDAccel™ tools, as well as the previously mentioned Vitis HLS, and extends them with the next-generation application acceleration development flow. Vitis is the official AMD tool for creating hardware-accelerated designs on heterogeneous multiprocessing systems, and supports AMD Alveo™, Versal™ Adaptive Compute Acceleration Platform (ACAP), and Zynq-based SoCs.

Vitis Embedded Software Development flow is part of Vitis Unified Software Platform focused on embedded software development [60], and provides the following features:

- Workflow to partition the PS resources of the SoC for different SMP and AMP configurations.

- Software development stack and cross-compilers for bare-metal, RTOS and Linux applications targeting the APU, RPU and MicroBlaze cores.

- Simulators and debuggers

- Program analysers for profiling application performance.

---

7. A Board Support Package is a hardware-specific collection of libraries, drivers and information about the hardware [55]. It forms the lowest layer of the application software stack that other applications bind to, or run on top of. A BSP is necessary for an embedded application or operating system to boot and function in a given hardware environment.

- Xilinx Runtime (XRT) – API and low-level drivers for applications to connect with the target hardware platform.

- Vitis accelerated libraries [56] – open-source, performance-optimised hardware-accelerated libraries for common maths, DSP and statistics functions, as well as domain-specific workflows like Machine Learning (ML), image processing and more.

Vitis supports two different embedded design flows, which are summarised as follows [59]:

- The ***traditional embedded software or fixed-platform design flow***, which relies on a completed hardware design built using the block-based Vivado IP integrator or HDL system description, exported as a fixed platform. In addition to the bitstream, the fixed platform file includes the processor domain (AMP/SMP) configuration, OS, boot files and software drivers. This flow is best suited for projects where the hardware design has been completed and only the software application part of the design is remaining.

- The ***extensible platform-based flow***, illustrated in Figure 13.8. In this flow, the embedded processor platform contains the extensible hardware design as well as the project's software components. The

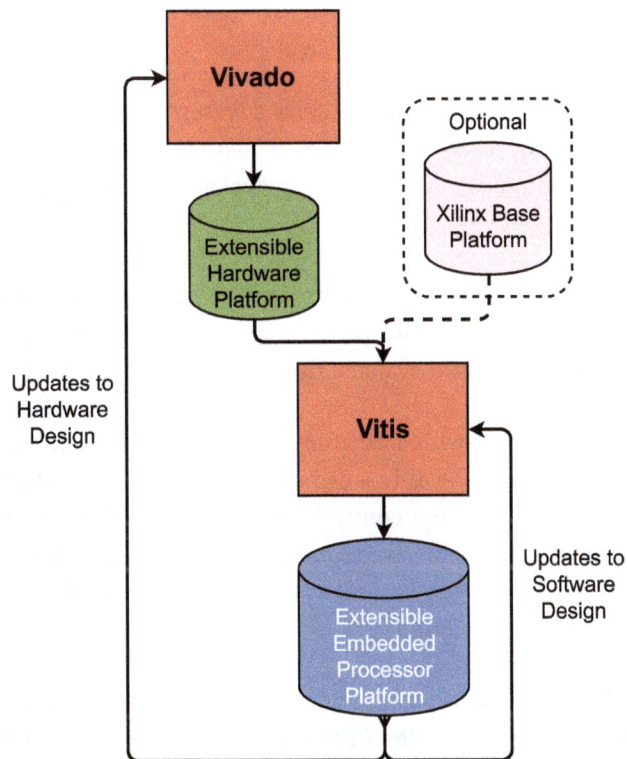

Figure 13.8: High level overview of the Vitis extensible platform-based flow.

433

creation of an extensible platform starts with a hardware platform (hardware specification file) that can either be provided by AMD as a base platform, or a Vivado hardware design created using the traditional Vivado IP Integrator or HDL design methods. The hardware platform can be in either a *pre-synthesis* form, which does not contain a bitstream but is quicker to build, or a *post-implementation* form, which contains an implemented design together with a final bitstream. Whichever of these hardware specification options adopted, the hardware platform is imported into Vitis, and acts as the foundation of the hardware design for the Vitis linker to add the software component configuration to (i.e. processor domains, DeviceTree, OS), and thus create the custom extensible embedded processor platform.

The availability of an AMD base platform is an important enabler of concurrent development for heterogeneous systems, as the software design can be started without having the finished hardware bitstream. This enables teams consisting of multiple engineers to work on their respective parts of the design in parallel. Once the extensible embedded processor platform is generated, it can be updated by Vivado (for hardware architecture changes) and Vitis (for PS software). This enables an iterative design flow without having to rebuild the platform from scratch, thus saving development time.

Further information on the integration of AMD Vivado and Vitis workflows is provided in Section 13.4 on Hardware-Software Co-Design.

PYNQ

PYNQ is an open-source project from AMD that provides a feature-rich development environment for AMD platforms, including ZYNQ, MPSoC, RFSoC, and Alveo. PYNQ is mainly distributed as a Ubuntu-based OS image, and it comes with a pre-configured, web-based JupyterLab IDE, along with Python libraries that allow convenient interaction with the PL.

PYNQ allows developers to use the high-level Python programming language and its libraries for data analysis, scientific computing, data visualisation, image processing, machine learning and other applications. In particular, it supports designers to create and evaluate their own custom hardware-accelerated designs.

At the core of PYNQ is the Overlay class, which exposes the PL design and individual IP registers as an accessible Python object. PYNQ provides out-of-the-box drivers for some of the most popular hardware IPs, such as AXI DMA, AXI GPIO, AX IIIC, AXI interrupt controller and others. Additionally, MicroBlaze soft processors are also supported and can be programmed directly from the Python environment. The provided libraries work well as a starting point for most designs, and can be further extended by writing your own custom libraries for other IPs included in your PL design.

PYNQ features extensively throughout this book, and as part of the accompanying example notebooks. An introduction to the PYNQ framework is provided in Notebook Set A (see page 7). In many of these notebooks, we adopt RFSoC-PYNQ, an extension of PYNQ that incorporates support for RFSoC platforms.

MathWorks Tools

As discussed in Section 13.2.4, MathWorks provide IP generation capabilities for AMD FPGAs and SoCs through their *HDL Coder* toolbox. MathWorks also develops add-on extensions to target the Arm-based PS of the RFSoC and other AMD SoC platforms.

The *Embedded Coder* toolbox [247] is an extension of *MATLAB Coder* [252] and *Simulink Coder* [254] that allows developers to generate industry standard-compliant C/C++ code from MATLAB and Simulink projects. Embedded Coder can be further extended using hardware support packages to enable greater optimisation and extra features. The *Xilinx Zynq Support Package* (which also has support for MPSoC & RFSoC) can be used in conjunction with the HDL Coder toolbox to automatically generate software interfaces for IP cores developed using MathWorks tools.

ARM Cortex A Support for Embedded Coder [246] further extends the Embedded Coder toolbox by providing Arm Neon [97] Single Instruction Multiple Data (SIMD) accelerator support for select signal processing MATLAB System objects and Simulink blocks. This allows compute-intense algorithms such as FFT/IFFT, multirate FIR decimation/interpolation, and other filters to be run on optimised and independent execution hardware that is capable of multiple operations at the same time.

GNU Radio

GNU Radio is a Free and Open-Source toolkit for SDR design development [178]. It is designed to run on a CPU-based processing system, but can be extended with custom *Out Of Tree* (OOT) modules to utilise GPU, DSP, or FPGA acceleration. GNU Radio Companion provides a block-based Graphical User Interface (GUI) for GNU Radio that is designed to enable rapid creation of SDR designs, without the need to reinvent fundamental components [180]. Figure 13.9 provides an example of the GNU Radio Companion interface: blocks are configured and connected together to form a flowgraph that models the functionality of a radio.

At the time of writing, the only RFSoC-based radio with official GNU Radio support is the Ettus Universal Software Radio Peripheral (USRP) X410 model [152]. The USRP Hardware Driver (UHD) allows the board to work in a stand-alone mode, with GNU Radio running on the PS or in tandem on an external PC.

On other development boards, a custom OOT module can be written to enable the RFSoC's features in GNU Radio. An embedded python block can be used in conjunction with PYNQ libraries to control the RFSoC device [179].

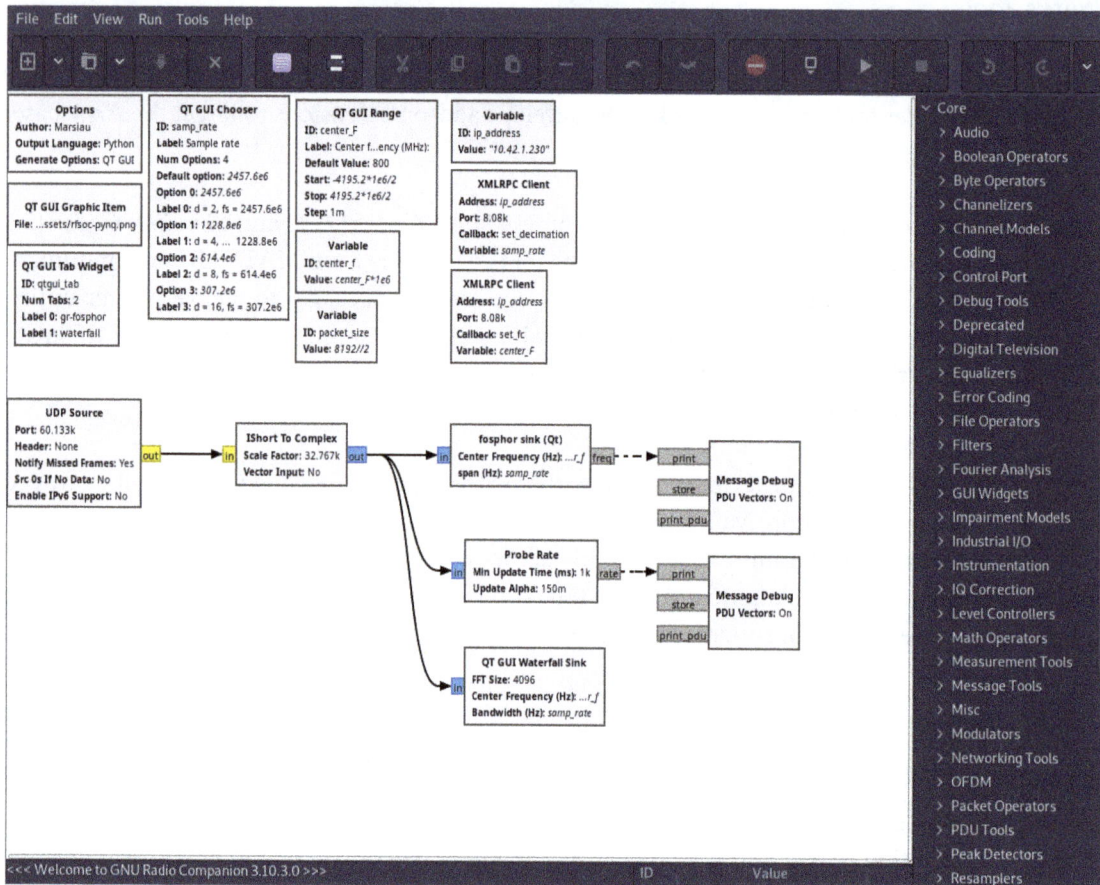

Figure 13.9: A screenshot from the GNU Radio Companion environment, showing an example radio flowgraph.

13.4. Hardware-Software Co-Design

Now that we understand the tools available for hardware and software RFSoC design, it is possible to examine a design methodology that considers both aspects throughout the design process. Hardware-Software Co-Design (or sometimes simply 'co-design') is an umbrella term for techniques that consider embedded systems development as an integrated, concurrent process of hardware and software development, rather than two individual design processes that eventually come together as a single output [133]. The term originates in the 1990s, and was first used to describe the partitioning of operations in a multi-core system to target the most appropriate microprocessor or application-specific core for the required resource, as in [193]. Such a design process therefore reduces power and resource requirements by efficiently allocating tasks.

The concept of co-design may seem second nature to modern sensibilities, and it is readily applicable to RFSoC development. For example, logic fabric is an ideal platform for deterministic, parallel algorithms, whereas processors are better for implementing sequential algorithms with decision-making and branching. Tasks can be appropriately partitioned across the PL and PS of the RFSoC device using a co-design approach.

13.4.1. Vivado Workflow as Hardware-Software Co-Design

The co-design methodology can be highlighted by re-examining the Vivado workflow considered previously. Figure 13.10 illustrates this integrated design approach within the Vivado and Vitis ecosystem, which encompasses partitioned but highly dependent PL and PS design, implementation and hardware verification.

In such a model, PL and PS elements of a design are partitioned appropriately based on specific tasks, just as they were in the 1990s co-design description, with both elements considered throughout the design process. The PL design aspect of the co-design process incorporates the hardware design elements discussed in Section 13.2 — abstraction methodologies such as Vitis Model Composer and HLS are typically used to design the required custom IP cores, which are configured alongside library IP cores from AMD (and third party tools, if applicable) as an IP Integrator block design.

As discussed previously in the context of the IP Integrator workflow, aspects of PS functionality can be directly configured, such as the peripheral interfaces and PS-PL connections in use. The PS configuration aspect is

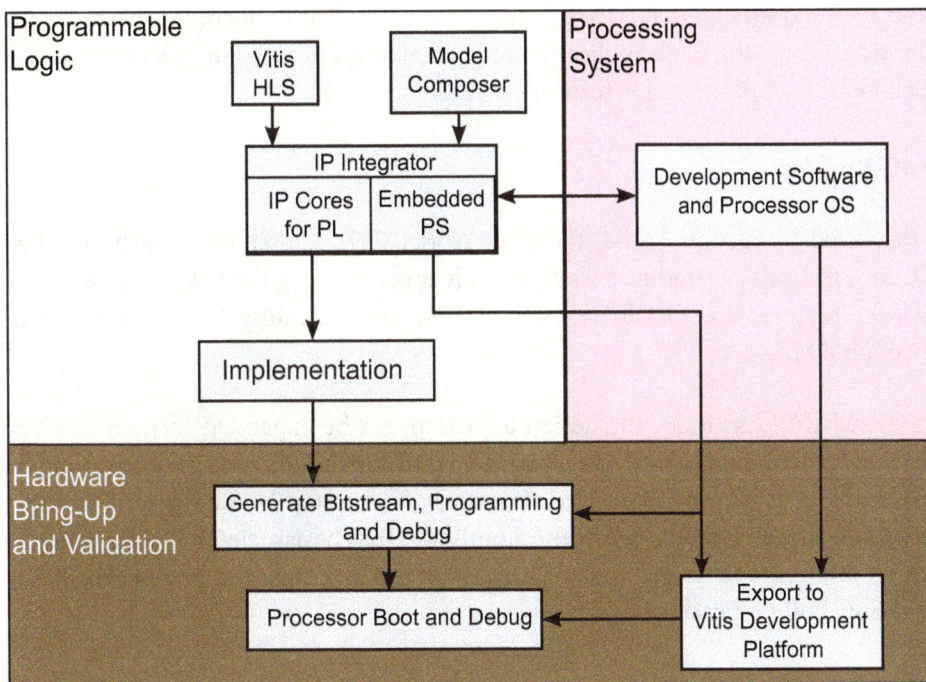

Figure 13.10: Hardware-Software Co-Design Overview for the Vivado and Vitis workflow.

accompanied by software development for the design, whether in Bare-Metal C through Vitis, as part of a Linux-based system like PYNQ.

The Vivado IDE is then used for Synthesis, Implementation and Bitstream generation to program the device. Dynamic Function eXchange (DFX) elements can also be incorporated here (coming up in Section 13.5).

The hardware and software elements of a design are usually tested together, as a PS-based driver (for example) can only be fully tested if the hardware design is successfully programmed. Equally, functionality in the hardware is often configured, enabled and operated via AXI4-Lite interfaces controlled by the PS. Therefore, the debugging and error solving process is part of the hardware-software co-design methodology — error fixing in the hardware design requires testing in the PS, whilst changes in the PS require to be tested using the hardware design.

Another key part of the co-design process is to ensure that the system as a whole boots correctly and reliably, from power-on to bringing up the OS (if used), initialising the various subsystems, and being ready for operation.

13.4.2. Alternative Co-Design Tools

Some third party tools offer alternative approaches to hardware-software co-design than the workflow shown in Figure 13.10. These methods typically allow PL and PS design within a single development environment, and have a link to a Vivado version that is used "behind the scenes" to perform Synthesis, Implementation and Bitstream Generation. Often, the same application offers a method to program the target hardware device and control the PS, allowing for full hardware-software co-design.

MathWorks SoC Builder

As previously discussed in Section 13.2.4, the MathWorks HDL Coder tool can be used as a method of abstracting HDL and IP Core generation. By additionally incorporating the SoC Blockset, a model design can be created that fully represents a Vivado IP Integrator design, including RFDCs, PS control and memory management through DMAs.

Utilising additional MathWorks tools, the design approach can be expanded further: if a Vivado version is linked, the *SoC Builder* tool can be used to create an HDL-capable design, generate a bitstream, and then download and execute it on a target SoC [256]. The tool offers a step-by-step wizard which guides the user through the Vivado workflow, without leaving the Simulink IDE. The associated Vivado version operates in the background of this process, and as such, an IP Integrator block design can be opened in Vivado following completion of the SoC Builder process.

Assuming that the target evaluation board is configured correctly, and connected to the host PC running SoC Builder, a connection can be made from the wizard via ethernet. The generated bitstream can be loaded to the device and the application run directly from the IDE.

RFNoC for GNU Radio

The RF Network on Chip (RFNoC) project is an open source, third-party software tool from the SDR manufacturer, Ettus Research™, which aims to simplify the integration of IP Cores into the signal processing chain of its Universal Software Radio Peripheral (USRP) SDR units [150],[151]. RFNoC can be used as an add-on package to the *GNU Radio* software, which is primarily a development environment for the PS elements of a system design.

The RFNoC package enables hardware design alongside GNU Radio for hardware-software co-design. The package provides selected 'RFNoC blocks', which add DSP functionality such as FIR filters and FFTs. These blocks can be added to a GNU Radio design alongside Vivado IP catalogue IP cores, as well as any custom IP cores that use AXI4 interfaces. As with the SoC Builder tool, RFNoC requires knowledge of an installed Vivado version, which is used in the background to perform synthesis, implementation and bitstream generation. Uniquely, all IP Cores within an RFNoC design are connected via an AXI4 Crossbar, which means that the order of blocks in the processing chain can be altered at runtime, without changing the bitstream. This adds a great deal of flexibility to a design. The drawback is increased latency, as samples have to cross this AXI4 Crossbar between each processing block, meaning that timing requirements become more of an issue.

13.5. Dynamic Function eXchange (DFX)

Dynamic Function eXchange (DFX) is the AMD design flow for dynamic partial reconfiguration of FPGA logic fabric. This technology allows an application to swap between different functional blocks implemented in hardware, without reprogramming the entire device [63]. The hardware design of the FPGA is split into a static portion, and a number of reconfigurable partitions known as DFX Regions. In this way, a bitstream is generated to implement the static portion of the FPGA, while several separate 'partial' bitstreams are created to fill the DFX region with sub-designs known as Reconfigurable Modules (RMs).

As the PL portion of the RFSoC is equivalent to an FPGA, DFX methods can be applied to SDR systems design on RFSoC. In combination, RFSoC and DFX offer the potential for extremely flexible radio implementations.

13.5.1. DFX Operation

The concept of DFX regions is illustrated in Figure 13.11.

The yellow region represents the static logic in the PL design — this is the portion that remains functional and is unaffected by the reconfiguration of the DFX region by loading partial bitstreams.

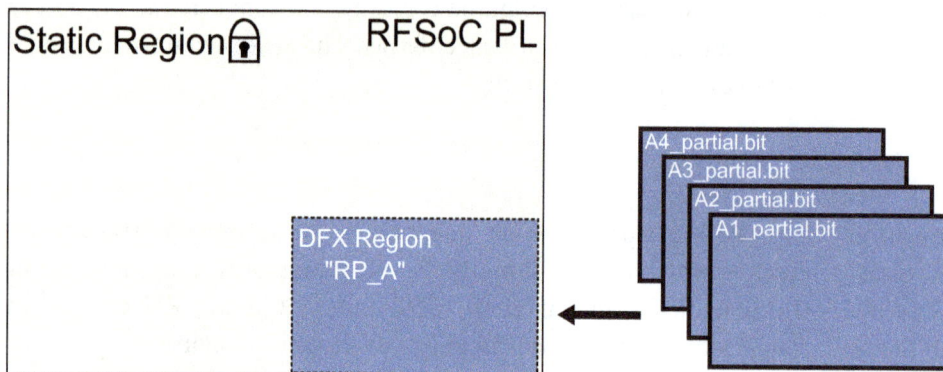

Figure 13.11: Conceptual overview of DFX.

The DFX region of the PL (bounded by the blue box within the PL) is the reconfigurable partition. This region is reprogrammed with the contents of the RMs, which can be loaded into the design via partial bitstreams.

The primary bitstream implements the static design, including placing and routing resources around the outside of the DFX region. Each partial bitstream can be loaded into the design separately to reconfigure the PL within the DFX region while the design is running. As the design surrounding the DFX region is static, the interfaces at the boundary between the static design and the DFX region must be kept consistent for all configurations.

By replacing subsets of PL at runtime, designs can be realised more efficiently in terms of resources and power budget. For instance, if any two independent sets of functionality do not require to be used at the same time, they can be loaded into the design only when needed. This can provide significant savings in power consumption, and the hardware resources required to implement the system as a whole.

13.5.2. Security and IP Protection

DFX can also be used to bolster the security of a proprietary design. In particular, DFX enables the use of Asymmetric Key Encryption (AKE) [63]. The technique uses public and private key pairs, where the static portion of the RFSoC houses the private key in the PL. New public keys can be generated at any time that can pair with the private key, but the private key is protected. If an attacker intercepts a transmission and obtains the public key, they cannot decrypt the message without knowing the private key.

Using the AKE scheme for PL device configuration, the majority of the design is held within an encrypted partial bitstream, combining cryptography with DFX regions to protect the configuration file. The static region contains only logic for decryption, and protects the rest of the proprietary design. This system ensures that even if there is a hardware attack on the RFSoC while it is powered, the private key is extremely difficult to access because it resides in the PL. Further, when the device is not powered, the private key does not exist and therefore cannot be compromised. This use-case differs from that described earlier for swapping the functionality implemented by DFX regions in a design, in the sense that there may only be a single partial bitstream for the DFX region, and the DFX region may encompass most of the PL.

DFX techniques also aid in the protection of proprietary IP. In Ultrascale+ devices, including RFSoC, designs for RMs can be made while only knowing the details of the DFX region itself — access to the static design is not required. This method is known as the Abstract Shell design flow for DFX [63]. It allows the details of a proprietary static design to be hidden in accordance with commercial legal agreements, for instance when a company subcontracts the design of a subsystem.

13.5.3. Advantages of DFX

As well as the benefits of flexibility, and resource and power efficiency outlined earlier, DFX offers benefits for ongoing system support. The ability to generate new RMs to partner with an existing static design means that updates can be delivered to systems that are already deployed, using relatively small bitstream files. This could include new features and bug fixes, as well as more comprehensive new functionality.

As of Vivado 2021.1, DFX regions can be introduced within an IP Integrator block design using block design containers. By applying this method, reconfigurable modules can be contained as sub-block designs within a system. Therefore, SoC hardware designs (including RFSoC devices) can include DFX functionality without deviating from the Vivado co-design methodology introduced in Section 13.4.

SDR systems can support a larger set of DSP signal chains, communications protocols, and hardware-accelerated algorithms, by exploiting DFX techniques to swap functionality as required, thus greatly improving the flexibility of the radio design. This helps to make RFSoC a uniquely powerful platform for SDR applications.

13.6. Clocks, Timing, and Sample Rates

Clocking is a crucial stage of hardware IP Integrator design, with multiple clocks often required within a design to drive its various interfaces. This section will explore AXI4 protocol clocks in particular, set in the context of the RFDC IP Core, demonstrating how clocks can be configured in the IP Integrator environment. Relationships between clock rates and sampling rates will also be discussed.

13.6.1. AXI4 Protocol Clocking

As discussed earlier in this chapter, the majority of IP Cores within IPI designs are connected through AXI4 protocol interfaces [94],[95]. The three types of AXI4 bus (AXI4, AXI4-Lite and AXI4-Stream) often require different types of clocking. If an IP Core that utilises these interface types is included in an IPI design, clock ports for each slave and master interface will be available on the IP block in the IPI block diagram.

Each of these clock ports requires a corresponding reset signal, driven from an instance of the Processor System Reset IP Core (with a clock provided from the same source as that driving the respective AXI4 link).

13.6.2. RF Data Converter Clocking and Sample Rates

Recall that the RFDCs are hierarchical in nature, and that most RFSoC devices feature a number of RF-ADC and RF-DAC tiles. Each tile contains one, two or four RF-ADCs or RF-DACs, with individual RF-ADCs and RF-DACs referred to as *blocks*.

A high-level overview of the RF Data Converter IP Core is shown in Figure 13.12 (for complete details of the RFDC IP Core, see [90]). Note in particular the AXI4-Lite and AXI-Stream clocks on the IP Core; these are the subject of discussion later in this section.

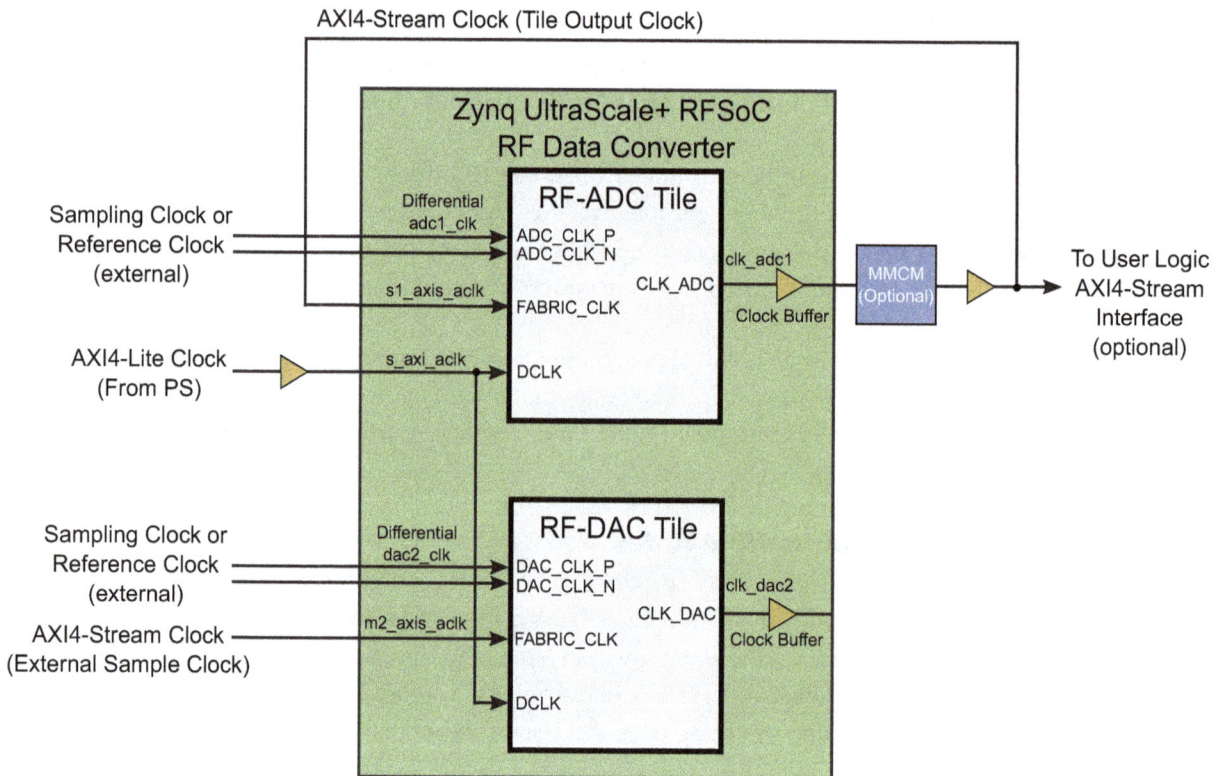

Figure 13.12: Instance of the RF Data Converter IP Core with clocking highlighted.

RFDC AXI4-Stream Interface Clocks

AXI4-Stream interfaces are typically used to transfer data between processing blocks in a 'dataflow' style hardware design, where data is transferred continuously through a number of processing stages, and between the output of one IP core block, and the input to the next. Consequently, it is often desirable to match the frequency of the clock driving this bus to the desired sample rate of the data passing through the IP Cores.

For RFSoC designs featuring an RFDC, the AXI4-Stream interface clocks can be provided from an external fabric clock or from the output clock of enabled RF-ADC and RF-DAC tiles[8]. The frequency required for the input AXI4-Stream input is determined by the sample rate of the tile, which has a minimum value around 1GHz and a device-specific maximum value. Use of the mixers within the tile, along with the decimation or interpolation factor and SSR options, can reduce the required fabric clock frequency.

The output clock frequency of a tile can be chosen from selected 2^n divisions of the sampling rate. This frequency can be further customised within the PL fabric through the use of the Clocking Wizard IP Core [23], which can implement a Mixed-Mode Clock Manager (MMCM) module. In this way, the IP Core can generate multiple output clocks with defined phase and frequency relationships to the input clock. This can be particularly useful if the tile output clock is unable to produce the frequency required for the tile AXI4-Stream input interface. In such a case, it may be necessary to scale the clock within the FPGA fabric.

Example:

Two AXI4-Stream clocks can be seen within Figure 13.12. The enabled RF-ADC tile provides an output clock which drives the AXI4-Stream input to the same tile, as well as AXI4-Stream interfaces of other IPs within the design that operate at the same frequency. The enabled RF-DAC tile, on the other hand, has an AXI4-Stream clock provided from an external source. It may be assumed that any signal processing IP Cores operating on the data prior to RF-DAC transmission would also be clocked by this external source.

RFDC AXI4 and AXI4-Lite Interface Clocks

Figure 13.12 also features an AXI4-Lite clock, *s_axi_aclk*, which is input to all enabled tiles within the RFDC and allows a PS-based driver to reconfigure the RFDCs. Memory-mapped interfaces such as AXI4 and AXI4-Lite provide connections between the PL and PS. Therefore, these interface types are typically clocked from PL-attached pins.

Tile Reference Clock

As shown in Figure 13.12, each tile within the RFDC has its own differential input clock which can be provided either from a sample clock (where the clock rate equals the sample rate), or from a lower frequency reference clock, which can be upscaled to the sample rate using an internal tile PLL. As the latter reference clock

---

8. Note that, in order for a configuration to be valid, the required input frequency for each block within a tile must match!

approach includes an additional PLL stage, it provides reduced clock jitter, creating a 'smoother' clock. This is therefore often the preferred method of clocking the RFDC tiles.

Frequency planning, as discussed in Chapter 12, should be considered prior to designing with the RFDCs, as the frequency plan will inform the selection of a suitable sampling rate and PLL reference clock frequency. The reference clock can be selected as any integer division of the sample rate and is provided by the reference PLL clock (LMK) and RF PLL clock (LMX), both of which were introduced in Chapter 3.

The LMK and LMX clocks must be programmed using registers prior to operation. This can be done manually using the clock programming connector of the board, or controlled from the PS through C or the PYNQ Python driver [74]. The necessary register programming configurations for the LMK and LMX clocks can be exported from the *TICS Pro* software [337]. The programmed LMX output frequency should match the RFDC tile reference frequency. For certain supported evaluation boards, including the ZCU111 [75] and RFSoC4x2 [42], select LMX frequencies are provided with the PYNQ RFDC clock driver to reduce the required design effort. Within the IP Integrator design, the RFDC is connected to the LMX reference clock through external IO. Two such interfaces can be seen in Figure 13.12, labelled as the differential *adc1_clk* and *dac2_clk*.

Certain evaluation boards, particularly those featuring Gen 3 devices, only provide this external LMX clocking to certain tiles, and require the clock distribution feature to use additional tiles. For example, the LMK and LMX clocks for the ZCU208 evaluation board [77] are provided via the CLK104 add-on card[9] [22]. The ZU48DR device on this board contains 4 RF-DAC and 4 RF-ADC tiles, however the LMX clocks are output via SMA cable to DAC Tile 2 and ADC Tile 1 only. To use other DACs and/or ADCs on the board, these tiles must be enabled within the design, and the clock distributed to other desired tiles. This clock distribution method is further illustrated in Figure 13.13.

Synchronisation Clock

The RFDC IP Core can also optionally take a system reference clock *sysref* as an input. This input signal is common to all tiles in the RFDC, and provides the clock for multi-tile and multi-device synchronisation (if it is required within the designed system). For multi-tile designs, the *sysref* signal is an external input that connects to a master tile within the RFDC. This tile then acts as the reference to which other tiles are synchronised.

13.6.3. Maximum PL Clock Rates

When developing hardware designs in the PL part of an RFSoC device (or in FPGA logic generally), there are two maximum clock rates that the designer should be aware of: (i) the potential maximum clock rate of the device; and (ii) the (generally somewhat lower) maximum clock rate arising from the user design.

---

9. See Chapter 3 (Section 3.4.2) for more details on the CLK104 RF Clock Add-On Card.

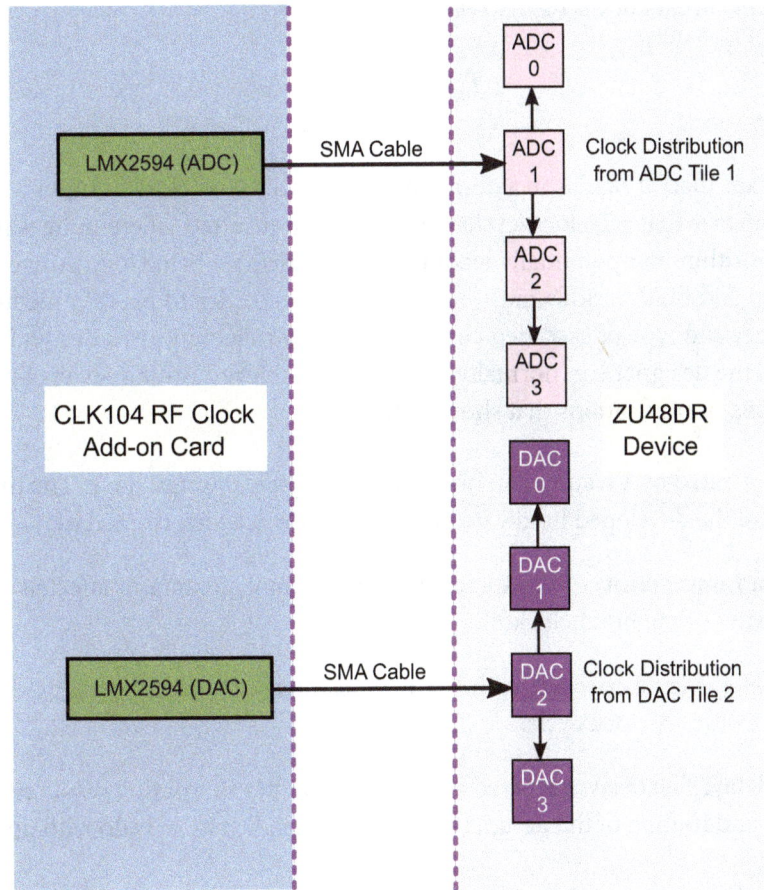

Figure 13.13: Conceptual Diagram of CLK104 Add-On Card Connected to a Single RF-ADC and RF-DAC Source Tile. Clock Distribution is required to provide a reference clock to other RFDC Tiles.

The first of these is a physical constraint of the chip, and is usually published in the product data sheet (there can also be speed grade variants of the same chip, each with a different maximum supported clock rates). The latter depends on the developed hardware, and the choices that have been made by the designer; in DSP systems, factors such as arithmetic wordlength specifications and the use of pipelining can influence the achievable clock rate, as well as the percentage utilisation of the PL (the more fully the PL is used, the more challenging the task of routing all of the signals, and the lower the maximum clock frequency is likely to be).

Designs will rarely achieve a clock rate approaching the device maximum, and as such, design-specific timing metrics are generally a greater concern to the designer. At a simple level, the maximum clock frequency is limited by the *critical path delay*, τ_{CPD}, which is the time taken for a signal to propagate along the longest combinatorial logic path between two clocked elements in a design (known as the *critical path*). The design-specific maximum clock frequency is *approximately*[10]

$$f_{max} \approx \frac{1}{\tau_{CPD}}.$$

(13.1)

This relationship means that all signals in a design must reach the next clocked input within one clock cycle — otherwise, it would appear that a 1-clock-cycle delay had been inserted where none was desired, thus altering the implemented algorithm and potentially leading to unpredictable behaviour. Any changes to signal values within flip-flop setup and hold periods must be avoided too, in order to prevent metastability issues. Recognising that the number and type of cascaded combinatorial logic elements in a design has a major effect on its timing performance, the designer can normally improve the achieved timing results by identifying the critical path and making design modifications that shorten it.

A timing report is generated by Vivado as an output of the *Implementation* stage. The timing report details the timing performance of the developed hardware on the target device, determined by:

- **Clock skew and uncertainty:** The efficiency of clock implementation, affected by clocking constraints and the properties of the input clock(s).

- **Logic delay:** The delay associated with the logic elements (i.e. combinatorial hardware) that a signal passes through within a clock cycle.

- **Net or route delay:** The delays incurred as a signal traverses PL routing resources. This is dependent on the placement and routing of the design (a process undertaken by Vivado with optional user directives).

One or more of these factors can be optimised when the timing performance of a design is crucial. For more information, and candidate approaches for optimising results, see [47].

After any attempts have been made to optimise timing performance, a continued failure to "meet timing" means that the desired clock rate is higher than the PL can support, for a particular design. This implies that the clock rate must be reduced.

In cases where the desired sample rate is very high, e.g. close to the interface between the PL and RFDCs, this may prompt a design change, e.g. an increase in the amount of decimation or interpolation undertaken in the RFDCs, to permit a lower sample rate to be used in the PL. Where the data rate must remain high, an SSR interface may be used to increase the data width. In this way, the clock rate can be reduced whilst maintaining the desired data rate.

---

10. Other factors such as flip-flop setup and hold times also contribute.

13.6.4. Super Sample Rate (SSR)

The Super Sample Rate (SSR) technique involves blocks of consecutive samples being processed in parallel using replicated hardware, as opposed to conventional sample-by-sample processing. When SSR is adopted, the required sampling rate can be supported at a lower clock frequency. For example, an SSR value of 8 means that the clock frequency is reduced by a factor of 8 (compared to a conventional sample-by-sample architecture), without reducing the overall sampling rate. As a consequence, the resource cost of implementing the hardware increases by approximately a factor of 8.

Example:

An RFSoC design features an RF-ADC tile to PL AXI4-Stream interface. It is desired to pass data through a particular custom IP core at a rate of 500 MHz, however the maximum achievable clock rate is 410 MHz. To meet these design requirements, an SSR 2 implementation can be adopted. By passing two data samples in parallel every clock cycle, the required clock rate can be halved from 500 MHz to 250 MHz, which comfortably meets the applicable timing constraints. These two alternative designs are illustrated in Figure 13.14, and SSR is further discussed in the next section.

As introduced in Chapter 3, an SSR interface makes the transition between regular and SSR sections of a system. SSR is perhaps most likely to be used in an RFSoC at the interface between the RFDCs and PL. Here, it can be introduced into a design through the RFDC IP Core: by increasing the 'samples per AXI4-Stream Cycle' parameter to an integer value greater than 1, an SSR interface is created between the RFDC and the PL.

Many commonly required cores, including the FIR Compiler IP core, provide support for SSR. In addition, the Model Composer library contains an entire SSR Blockset (through System Generator DSP support) allowing for custom IP Core generation with SSR interfaces [69]. Through careful design, the AXI4-Stream interfaces throughout a design can be integrated with SSR implementations, greatly reducing the required clock rate.

13.7. Antennas, Filters, and Amplifiers

External radio components are often required when prototyping with the RFSoC. For instance, an antenna may be necessary to improve signal acquisition, or an external filter is required to suppress spectral aliasing. In this section, we will briefly explore a few Commercial Off-The-Shelf (COTS) options that may be useful when designing a receiver front-end for the RFSoC. In particular, we will discuss antennas, amplifiers, and external filters. Our discussion will be around simple, easy to build, front-end designs. We will not cover specialised design of external equipment. Lastly, we will only focus on signal acquisition equipment. Please seek professional support if you require external equipment for a radio transmitter.

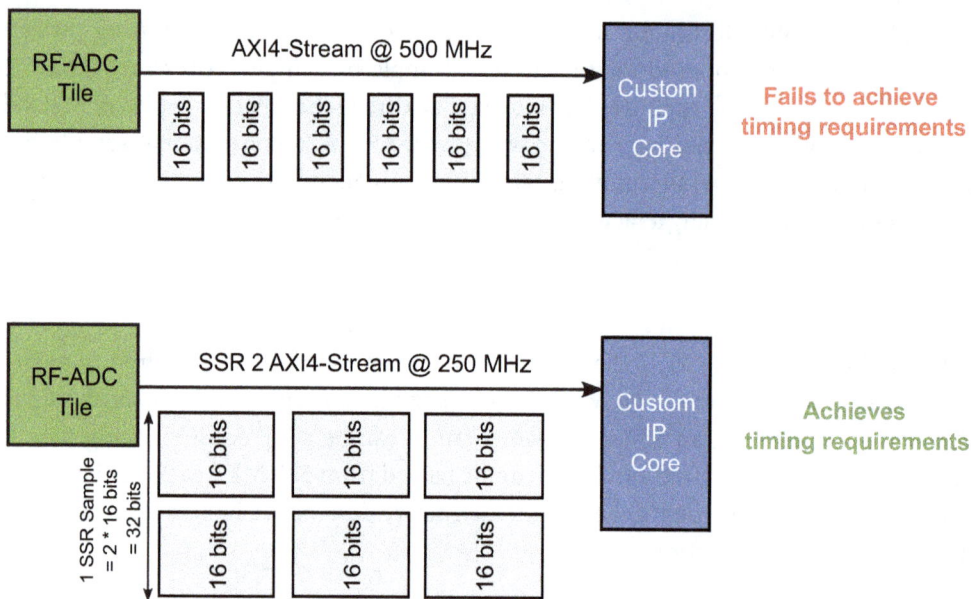

Figure 13.14: *RF-ADC to custom IP core design required to pass data at 500 MHz.*
An SSR of 2 has been adopted to meet timing requirements.

Figure 13.15 contains a diagram illustrating a simple front-end configuration for an RF-ADC. The antenna is used to acquire a signal, the external filter suppresses undesirable frequency bands, and the amplifier improves signal acquisition by increasing the power of the signal to a desirable level.

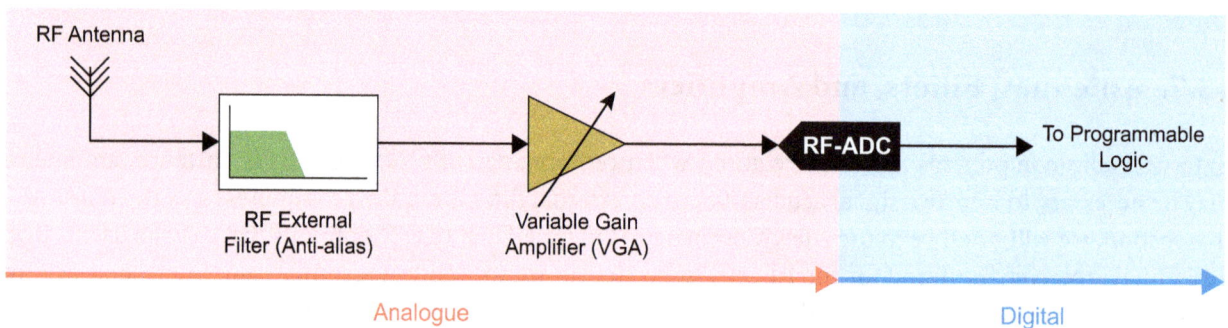

Figure 13.15: *A common receiver front-end configuration for an RF-ADC channel input.*

Antennas

Adding an antenna to the input of the RF-ADC will significantly improve signal acquisition. There are many commercial off-the-shelf (COTS) antennas that are useful for general exploration of the radio spectrum. We recommend using the following antennas:

- The wide-band antenna bundle known as the NooElec™ NaTLSnake [278] has several antennas that operate across several frequencies between 100MHz and 1800MHz.

- There is an application specific antenna known as the ANT-2.4-LCW-SMA from Linx Technologies™ [232], which is suitable for Wi-Fi, Bluetooth, and ZigBee communication standards.

Passive and active antennas are available: passive antennas do not amplify the acquired signal, whereas active antennas include an integrated amplifier that adds power to the signal. Both of the above highlighted antennas are passive. Care should be taken over antenna selection, particularly where active antennas are involved, and ensuring operation within the rated values to avoid causing damage to the RFSoC [87].

Amplifiers

It may be necessary to add a small amount of amplification at the front-end of your RFSoC platform to acquire radio signals at an appropriate power. Adding an amplifier is not completely necessary when prototyping radio designs. We recommend only adding an amplifier if absolutely required. The following generic amplifiers are useful:

- The NooElec VeGA Barebones - Ultra Low-Noise Variable Gain Amplifier (VGA) module [280].

- The NooElec LaNA - Wideband Ultra Low-Noise Amplifier (LNA) Module [277].

Please be mindful that excessive signal amplification may damage your RFSoC platform. Care should always be taken to ensure that signal power is in the appropriate range to be acquired by the RF-ADC input. Attenuators, such as those given in [279], may be useful to safely prevent over-amplification.

External Filtering

There are a range of external filters that can be applied to the input of the RF ADC channel to improve signal acquisition and to suppress spectral aliasing. Filtering requirements are determined based on the frequency bands that you would like to inspect. We recommend the following:

- The Mini-Circuits® VLF-1800+ low pass filter [262] to inspect frequencies up to 1800MHz.

- There are many filters that are available for particular applications and wireless standards. For instance, the Mini-Circuits VBF-2435+ bandpass filter [263] is suitable for acquiring Wi-Fi signals.

Similarly to antennas, active filters can increase the signal voltage, whereas passive filters do not. Therefore, care should be taken if connecting the output of an active filter to the RFSoC. Both of the example filters highlighted here are of the passive type.

13.8. SDR Design Flow for PYNQ on RFSoC

So far, this chapter has explored the various methods and software tools available for developing designs for the PS and PL portions of RFSoC devices. In this section, we take the reader through the design flow process that we (the authors) have used to make the various designs that are featured within the Notebook chapters of this book. This includes both the software packages and the design methodology. This is not to say that the methods detailed in this section represent the "standard" design approach for the RFSoC, nor is it necessarily the best way to do things. We simply present a methodology that has worked for us during our experience designing systems for RFSoC. The reader is free to use all, parts, or none of the information we detail here, although we hope that at least some of this shared experience is useful!

The design process detailed in this section can be separated into six separate steps, which are shown in Figure 13.16 and discussed over the next few pages.

13.8.1. Initial Design Process

The first step, and arguably the most important, is the initial design process. It is here where the system designer has to consider a number of factors, such as the range of frequencies to be received and/or transmitted by the RFDCs, and the bandwidth(s) of these signals. This step is also where frequency planning should be used in order to determine if the desired configuration will result in any in-band interference. These factors will determine the required sampling rates and clock rates to be used within the design. For example, if the signal bandwidth is much less than the required sample rate for the RFDCs, then interpolation or decimation stages can be used to reduce the clock rate on the PL. However, if large bandwidths are required then the use of SSR may be necessary to meet PL timing constraints.

Another factor to consider is the communication between IPs on the FPGA, as well as communication between the PL and PS. Many IPs provided within Vivado are AXI-compliant (as reviewed in Section 13.2.2, the three AXI protocols represent ideal interfaces for passing data between IPs), and therefore it tends to be beneficial for any custom IPs to also adopt AXI-compliant interfaces. If a large amount of data is to be passed between PS and PL, AXI-Stream DMAs will likely be required, whereas for lower-bandwidth signals, such as control signals, AXI-Lite interfaces are usually sufficient.

It is important to consider at this stage how the PS software will be structured and how it will interact with the PL hardware. AMD provides a number of drivers for its IPs that can be controlled from the PS, such as the AXI DMA. If custom IPs are to be developed, then it may also be necessary to develop custom drivers for them.

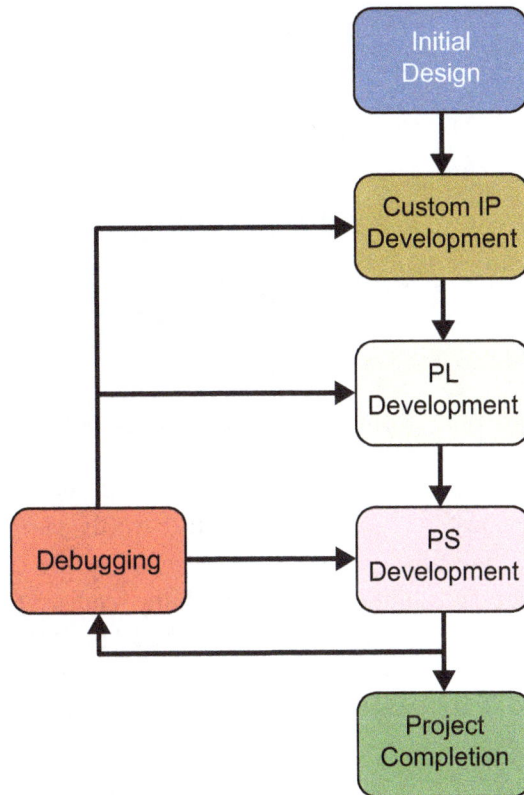

Figure 13.16: The design flow for creating SDR designs for RFSoC, using PYNQ.

Additionally, if the design requires the use of AXI-Lite control signals, it is important that the PL hardware conforms to AXI-Lite standards and handle them appropriately..

13.8.2. Development of Custom IPs

The next step is the development and testing of the custom IPs to run on the FPGA. The authors' preferred tools for this stage are Model Composer and HDL Coder, both of which run within the MATLAB/Simulink environment. The use of Simulink allows system designers to quickly simulate and test IP functionality before integrating it into the Vivado design, e.g. Simulink Source blocks can be leveraged to provide appropriate stimulus for simulation, and Sink blocks can be used to visualise results in the time and frequency domains. Both tools are able to incorporate AXI-compliant ports in the IPs that they generate.

An example HDL Coder design is shown in Figure 13.17, highlighting the HDL model that forms the IP core, as well as the generation of input stimulus, and inspection of output signals.

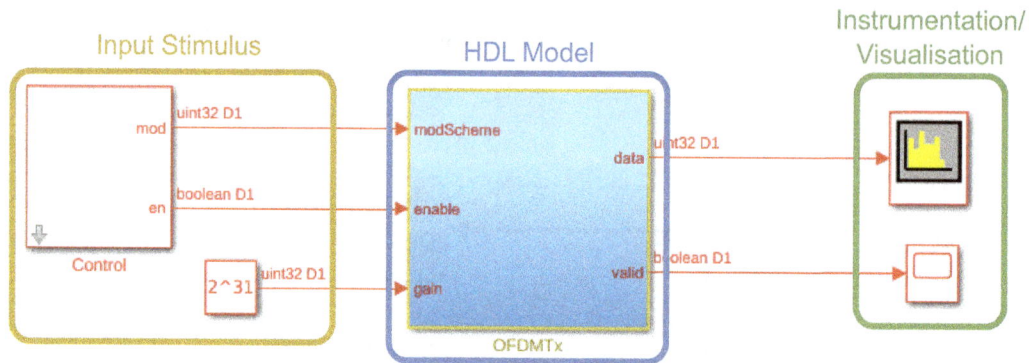

Figure 13.17: An example HDL Coder block design.

Clocking within Simulink uses the concept of *sample time* (in other words, sample period), which is the reciprocal of the sample rate. As such, the *sample time* parameter refers to the clock rate required for the IP. Generally, a single clock drives all interfaces of the IP, set to match the required sample rate. If any rate changes are required within a custom IP, then it is important to note that the IP must be driven by a clock operating at the highest sample rate. The tools then generate other clock(s) within the IP, to drive blocks that operate at the lower sampling rate(s); required sample rates (and hence clock rates) are normally related by integer factors.

When designing custom IPs, it is important to consider the inputs and outputs, and how they will interact with the wider design. As mentioned previously, the use of the AXI standard is worthy of strong consideration, and will allow these IPs to work well with other AXI-compliant IPs within Vivado. AXI-Stream can be used to pass RF data through processing chains, while AXI-Lite is useful for control signals, such as enables and resets.

Simulink provides both signal scope and spectrum analyser tools that allow the system designer to test and verify the functionality of the model before converting it to an HDL IP. The spectrum analyser is very useful for determining that signal data is being processed correctly, while the signal scope can be used to identify problems with AXI-Stream interface signals.

13.8.3. Integrating Custom IPs into Vivado, and Hardware System Design

Once the custom IPs have been developed and tested, they can then be imported into Vivado and integrated into a full PL design. This step can be the most critical, as it requires the system designer to have a good understanding of the underlying hardware and its capabilities. It is within Vivado where the system designer connects the interfaces between the custom IPs designed in MATLAB/Simulink, and the other IPs such as the RF data converters; as well as any communication between the PL and PS.

Additionally, this is where clock rates and sample rates are set within the design. Care must be taken to ensure that each IP is being driven by the correct clock, as unexpected behaviour can arise otherwise, which can be difficult to debug. Most designs featured in this book are run from three independent clock sources: (i) the PL clock, which is used to drive AXI-Lite interfaces; (ii) the DAC clock, which drives the IPs associated with the transmit path; and (iii) the ADC clock, which drives the IPs on the receive path. In some cases, it may be necessary to use clock converters to generate the frequencies that the IPs require.

The most direct method of creating the PL design, and the one that the authors have adopted, is to use Vivado's built-in IP Integrator tool. IP Integrator allows the system designer to use a visual, block-based method of connecting the IPs together; this is similar to the Simulink environment, and the Model Composer and HDL Coder development methods described in Section 13.8.2. Unlike Simulink, however, simulating and testing the entire design is non-trivial, and requires the system designer to create suitable stimulus and testbenches for all interfaces to verify the operation of the IPs.

An issue that hardware system designers will eventually come across is timing closure, i.e. ensuring that the design can support the desired clock rate. This can be especially challenging for designs that use a significant amount of hardware resources, and where a high sample rate is required. While the subject of timing closure could fill a chapter on its own (if not an entire book!), there are a couple of simple tricks that can make it easier for a design to 'meet timing'. The first is, if possible, to reduce clock rates, especially between long chains of IPs. For SDR designs, this may involve decimating the signal earlier in the signal processing chain of a receiver, or, conversely, interpolating later in the signal processing chain of a transmitter. Another option is to change the *Implementation Strategy* in Vivado, which defines the approach that is taken to placement and routing. Vivado defines a number of different strategies, including some that focus primarily on timing closure (often with extended run times, but the potential to achieve better results). With that said, in larger designs this may be insufficient, and additional tactics may be required. More information on timing closure can be found in [47].

13.8.4. Development of PS Software

After the PL design is complete, the software for the PS can be developed. As will be apparent from the various designs featured within this book, PYNQ is the authors' preferred framework for developing software on the PS. There are many benefits to using PYNQ, including good driver support for standard IPs, as well as simple methods to create drivers for custom IPs. The use of Python also enables code to be written quickly and efficiently, making it relatively quick to identify any issues on both the PL and PS. Furthermore, the Python ecosystem contains a myriad of established libraries, such as *Plotly* and *SciPy*, which make it simple to develop code used for visualisation and analysis of signals, directly on the PS.

Due to its ease-of-use, PYNQ provides a quick method of confirming that a PL design performs as expected. AXI-Lite registers can be read from and written to using the MMIO library, the RFDC driver can be used to configure the RF data converters, and the DMA driver makes it simple to pass data between the PS and PL via shared memory. If data is only passed between the PL and RFDCs, Integrated Logic Analysers (ILAs) run on the PL, or external equipment such as spectrum analysers and/or signal generators, may be required to verify the functionality of the transmit and receive paths.

13.8.5. Debugging

Debugging is a frustrating process, even for the most experienced engineers, and it can often be difficult to determine exactly where a fault lies. Using the design method we have detailed in this section, it is generally easier to first check whether the PS software is at fault, as the iteration time for making a change and checking whether it fixes the bug is usually much faster than for the PL.

As is common in the debugging process, the solution is usually simple, but determining where the fault is can be very time consuming. A useful tool to aid this process is the Python debugger, pdb, which allows the user to set breakpoints and methodically step through the PS code. Once the fault is found and fixed, the debugger code can be removed from the project.

If the PS software is confirmed to be operating as expected, it may be necessary to revisit the PL design stages to solve the problem. Depending on the size of the PL design, it can take a long time to generate a new bitstream and, thus, PL debugging can be a time-consuming process, especially if it is necessary to regenerate the bitstream multiple times, to attempt different solutions. For this reason, it is beneficial to extensively simulate and test the PL hardware before moving to the software development stage.

While there can be any number of different faults within a design, from our experience there are two common problems that can occur when developing systems for the RFSoC. As discussed earlier in this section, an incorrect clock frequency can result in unpredictable behaviour. The RFDCs are typically driven by external clocking infrastructure, and therefore it is important to ensure the correct programming of the clocks, and that they are delivering the expected frequency.

Another common problem arises from how AXI-Stream signals are generated and handled within custom IPs. Along with the *tdata* signal, AXI-Stream IPs can also contain various other interface signals, including *tvalid*, *tready*, and the end-of-packet signal, *tlast*. If these signals are not synchronised properly with the *tdata* signal, then IPs further along the signal path may not operate correctly. For example, the PYNQ DMA driver requires the packet size to be of a specific length when setting up the data transfer. If the *tlast* signal is pulsed too early or too late, indicating the end of the data packet, this will cause the DMA to hang. Similar issues can occur with the *tready* signal, if not handled correctly. More information on the AXI-Stream protocol can be found in [20].

13.8.6. Project Completion

Once the PS and PL have been verified to work as expected, the software can then be formalised. This may include making drivers for the custom IPs, or (if appropriate to the intended use case) creating stimulus or visualisation tools, such as the many designs featured in this book. Another possibility that is used extensively in these designs, is to abstract away much of the functionality into classes and functions, and then import these into the notebook. This allows the code in the Jupyter notebooks to be more streamlined and easier to follow.

13.8.7. Further Opportunities

As mentioned at the start of this section, the method we describe here is only one way of designing systems for the RFSoC. There are many combinations of software available for both PL and PS to make your own design, but this is the method we used to create the various designs featured within this book. Additionally, the source code and design files for all these featured designs is available under a permissive license, and may be used as is, or as a template to create your own designs.

13.9. Chapter Summary

This chapter has presented an overview of the design flow for RFSoC, and outlined the design of the hardware (PL-based) and software (PS-based) aspects of an RFSoC system. Issues around clocks and timing, and the applicability of the flexible DFX method to SDR, were given a special mention. The idea of software-hardware co-design, i.e. considering the system design in an integrated manner throughout the process, was highlighted.

The latter part of the chapter provided some practical guidance on SDR design components that exist outside of the RFSoC itself, such as RF connectors, amplifiers, analogue filters, and antennae; as well as discussion of the PYNQ-based RFSoC design flow that has been adopted by the authors for system development.

456

Notebook Set G

RFSoC Radio Demonstrator

So far we have used PYNQ for a variety of interesting RFSoC applications, which include the RFSoC spectrum analyser and the frequency planner, and we will later investigate an evaluation notebook for the SD-FEC blocks. PYNQ is an excellent framework for achieving dynamic user control over RFSoC radio designs. FPGA signals and data paths can also be observed and visualised using plots and graphs. In this series of notebooks, we will investigate a fully functioning radio system on the RFSoC platform. This includes exploring how PYNQ can be used to perform real-time inspection and visualisation of an RFSoC radio design by observing different stages of a radio pipeline.

There are three notebooks to investigate throughout this chapter using Jupyter Labs on your RFSoC platform. The notebooks and their relative locations are listed as follows:

`RFSoC` 01_rfsoc_radio_system.ipynb — ***rfsoc_book/notebook_G/01_rfsoc_radio_system.ipynb***

`RFSoC` 02_rfsoc_radio_observe.ipynb — ***rfsoc_book/notebook_G/02_rfsoc_radio_observe.ipynb***

`RFSoC` 03_rfsoc_radio_helloworld.ipynb — ***rfsoc_book/notebook_G/03_rfsoc_radio_helloworld.ipynb***

G.1. The Radio System

The aim of the ***01_rfsoc_radio_system.ipynb*** notebook is to provide users with an overview of the underlying radio architecture by exploring the transmitter and receiver radio pipelines. This includes investigating the *xrfdc* Python package, which is responsible for configuring the RFSoC's data converters. We will also use another important package, *xrfclk*, which configures the RFSoC's clocking network. Users will also be introduced to the radio dashboard, which is a graphical user interface that controls the system. A diagram illustrating the radio system architecture can be seen in Figure G.1.

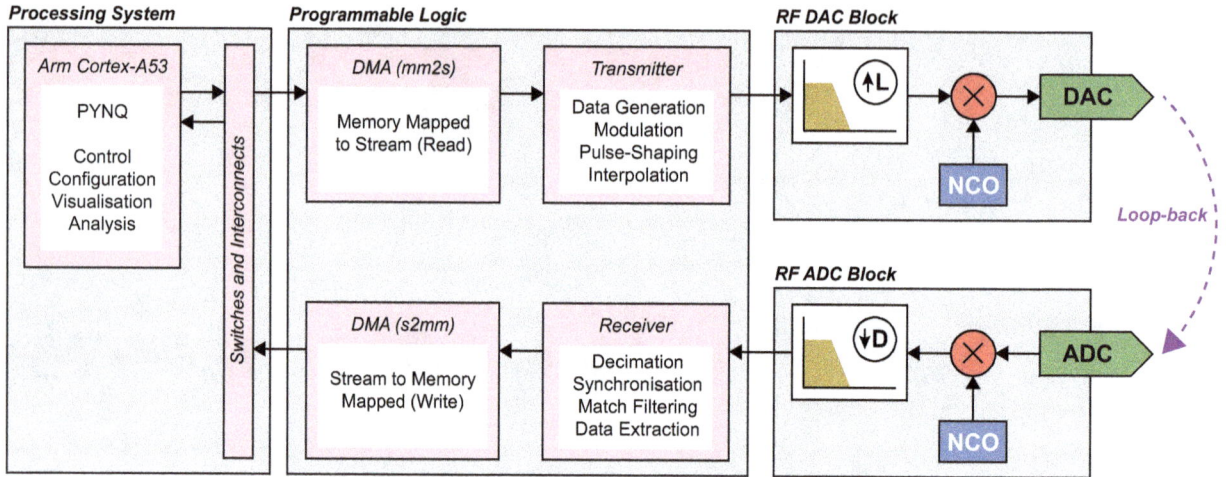

Figure G.1: Architecture overview of the RFSoC demonstration system.

This notebook introduces the structure of the data frames used by the transmitter to send data to the receiver, which acts as a simple communications protocol. Figure G.2 shows the data frame structure used by the RFSoC radio demonstrator. Notice that the data frame has three main sections: the preamble, the header, and the payload. Usually the transmitter sends a known sequence of bits (the preamble) that is detected by the receiver using synchronisation techniques. Following this, additional information such as the frame number, data flags, and frame length are included in a header (these are later extracted and used by the receiver to interpret the frame). The payload is the data that is carried by the frame. .

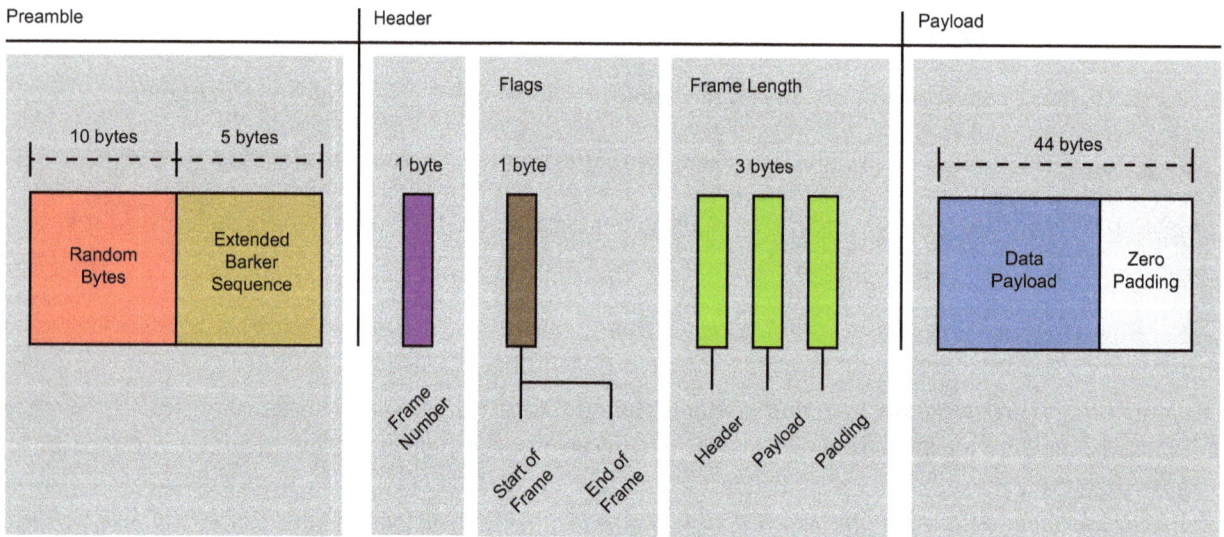

Figure G.2: The data frame structure used by the RFSoC radio demonstration system.

G.2. Observing the Radio Pipeline

After introducing the RFSoC radio design, we will begin observing the transmit and receive radio pipelines while the system is operating. This includes performing real-time introspection of the radio system by 'tapping-out' from various stages of the transmitter and receiver pipelines, i.e. drawing out copies of internal signals for analysis, without affecting the processing pipelines they originate from. These signals are then used to generate waveforms for visualisation (an example plot is shown in Figure G.3). You can begin observing the radio pipeline by running the ***02_rfsoc_radio_observe.ipynb*** notebook.

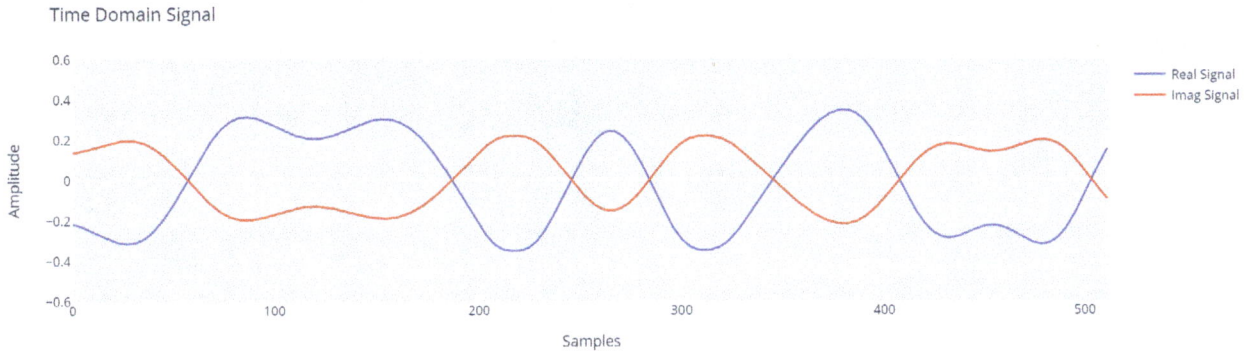

Figure G.3: Example plot of complex time domain data from the RFSoC radio demonstration system.

We will also investigate the use of two digital modulation schemes: Binary Phase Shift Keying (BPSK), and Quadrature Phase Shift Keying (QPSK). You can swap between BPSK and QPSK during system operation and plot the received constellation diagrams, which ideally should resemble those seen in Figure G.4.

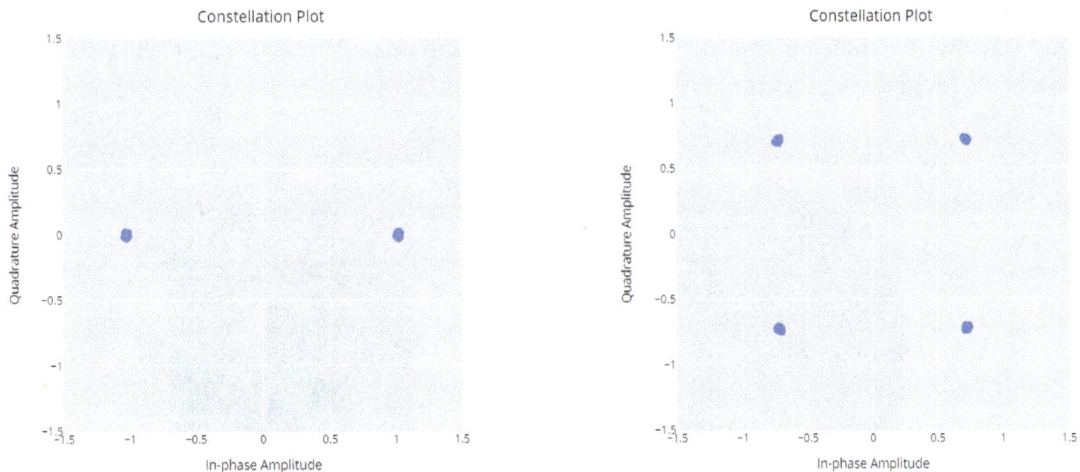

Figure G.4: Example plots of BPSK constellation diagram (left) and QPSK constellation diagram (right).

G.3. Transmit and Receive: "Hello World!"

At this stage, you will be familiar with the RFSoC radio demonstration system and the digital modulation schemes used by the transmitter and receiver. We can now begin sending and receiving more meaningful data on our radio platform. To begin, open and run the notebook named ***03_rfsoc_radio_helloworld.ipynb***.

This example will use the RFSoC radio system to send a "Hello World!" message between the transmitter and receiver. We will send this message using three different techniques. The first method uses a simple function call to send "Hello World!". The second method uses Python widgets to allow you to easily enter and transmit messages using buttons and a text window (an example of this application can be seen in Figure G.5). The third method allows you to broadcast the "Hello World!" message repeatedly using the Python threading library.

Figure G.5: A collection of Python widgets that allows the user to transmit a message (left) and receive a message (right) using the RFSoC radio demonstration system.

The last part of this notebook series allows you to transmit and receive an image using the RFSoC radio demonstration system. The radio system transmits an image repeatedly and rotates the image by 90° after every successful transmission. For this demonstration, you can choose between the BPSK and QPSK modulation schemes for transmission. An example output of the image transmit and receive RFSoC demonstration system is presented in Figure G.6.

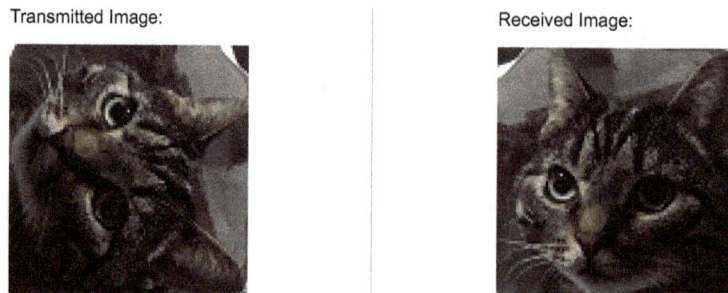

Figure G.6: Transmitting and receiving an image using the RFSoC radio demonstration system.

Chapter 14

Forward Error Correction

Douglas Allan

This chapter introduces the fundamental theory of Forward Error Correction (FEC). FEC is used to detect and correct errors and thus increase the robustness and reliability of a digital communication system. The term forward error correction refers to the fact that errors can be corrected without explicit need for re-transmissions. The ability to detect and correct errors is one of the key advantages of digital over traditional analogue communications systems. The use of adaptive Modulation and Coding schemes (MCS) enables robust performance across a range of Signal to Noise Ratios (SNRs) at the expense of reduced spectral efficiency, either through loss of data rate or a corresponding increase in bandwidth. The implementation of FEC schemes also implies increased resource usage and power consumption, due to the additional processing required at both transmitter and receiver. As such, practical codes must provide good error correcting performance while also ensuring that processing complexity is minimised as much as possible.

Shannon's theorem states the maximum possible spectral efficiency that can be achieved by a digital communications system with error correcting coding for a given E_b/N_0, whilst maintaining an arbitrarily small probability of bit error (P_{be}). This maximum spectral efficiency is known as the Shannon limit, beyond which error free communication is not possible. The theory implies the existence of an error correcting code or codes that can reach the Shannon limit, but does not specify how these codes would be constructed. In recent years, several practical coding schemes have been developed which achieve close to the Shannon limit, including Turbo Codes and Low Density Parity Check (LDPC) codes. Different varieties of FEC are found in all modern wireless standards including 4G Long Term Evolution (LTE), 5G New Radio (NR) and Wi-Fi 6 (IEEE 802.11ax), to name a few.

A thorough treatment of FEC would likely require multiple chapters or even an entire book in itself. Since this is not the main purpose of the book, we aim to provide the reader with enough detail to appreciate the fundamental role of FEC in digital communications, and its design and implementation in current wireless standards. In particular, Hamming codes, Convolutional codes, Turbo codes and LDPC codes are discussed. In addition, the design and configuration of the AMD SD-FEC IP Core is described.

14.1. Motivation

Error detection and correction schemes play a fundamental role in ensuring the reliable communication and storage of digital information. In real world environments, communications and storage systems are subject to unwanted noise and other sources of degradation, that contribute to an increase in the error rate and a reduction in performance and reliability. Both error detection and correction schemes have been developed to detect and attempt to correct errors caused by environmental and other factors.

Due to the utility of FEC schemes, they have become ubiquitous in all modern digital communications protocols such as Wi-Fi (IEEE 802.11) and Ethernet which form the backbone of the internet; mobile broadband networks (4G/5G); terrestrial and satellite networks for broadcasting of digital radio and television; emergency services networks, and deep space communications, to name a few prominent examples. They are also widely used in digital storage and media systems such as CDs, DVDs, and both volatile and non-volatile computer memory. As an example, error correcting schemes are used in CDs to correct errors resulting from scratches and other damage.

In this book, we are principally interested in error detection and correction schemes used in the context of wireless digital communications systems, such as Wi-Fi and 4G LTE / 5G NR. In particular, focus will be placed on convolutional codes, Turbo codes and LDPC codes. Hamming codes will also be described as a means of introducing the reader to the fundamental principles of error correction coding.

14.2. Principles of Forward Error Correction (FEC)

In a wireless digital communication system, information is transmitted in the form of bits, i.e. 0's and 1's. The bits are grouped into discrete symbols that modulate an RF carrier to transmit information across the wireless channel. The discrete nature of a digital communications system is advantageous because the receiver must only distinguish between a finite number of possible symbol values, making it easier to successfully decode the information in the presence of noise. In contrast, in analogue systems, the receiver must distinguish between a potentially infinite number of signal values, making it more difficult to decode information reliably in noise.

Although digital communications systems do offer better noise immunity than their analogue counterparts, it is still possible to have bit errors. The number of bits received in error is proportional to the SNR of the received signal. In the context of digital communications, the SNR is defined as,

$$SNR = \frac{E_b}{N_0}\frac{R_b}{B}, \tag{14.1}$$

where R_b is the bit rate, B is the bandwidth, E_b is the energy per bit and N_0 is the noise power per 1Hz of bandwidth. The ratio E_b/N_0 is colloquially known as the "SNR per bit" and can be used to compare the error rate performance of different modulations schemes, since it is independent of bit rate and bandwidth. For each modulation scheme, there is a defined mathematical relationship between the Probability of Bit Error, P_{be}, and E_b/N_0.

Figure 14.1 shows the theoretical P_{be} vs. E_b/N_0 for Binary Phase Shift Keying (BPSK), Quaternary Phase Shift Keying (QPSK) and 16-Quadrature Amplitude Modulation (QAM) modulation schemes in an AWGN channel. Note, E_b/N_0 is expressed in *dB*, and the BPSK and QPSK curves actually sit on top of each other.

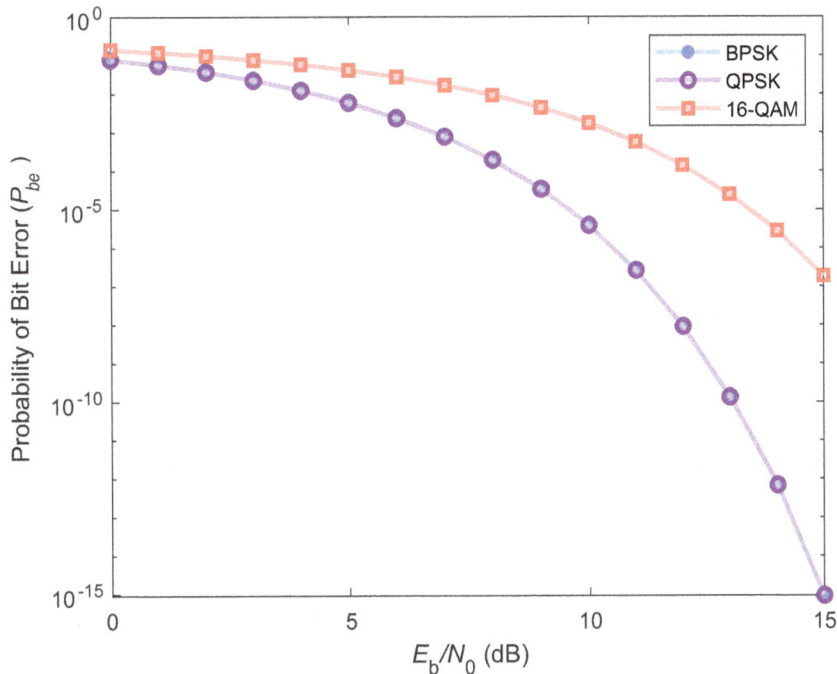

Figure 14.1: Theoretical P_{be} vs. E_b/N_0 for BPSK, QPSK and 16-QAM in an AWGN channel.

The theoretical relationship for each modulation scheme places a lower bound on the achievable error rate for a given E_b/N_0. In other words, for a particular E_b/N_0, the error rate on average will not be lower than a certain value. For example, considering BPSK at an E_b/N_0 of 5dB, P_{be}, is approximately 0.006, which means there will be on average approximately 60 errors in every 10,000 bits. In the case of 16-QAM, there are approximately 400 errors per 10,000 bits. The 16-QAM error rate performance is worse than BPSK / QPSK because there is a larger symbol alphabet, which makes it more difficult for the receiver to distinguish between symbols in the presence of noise.

FEC schemes involve generating n coded bits from k data bits, where the group of n bits is called a codeword. The coding scheme is designed to ensure that the minimum distance between n bit codewords is greater than would be the case for the original k data bits. This means that when errors occur, the likelihood of one codeword being confused for another is reduced, allowing the receiver to determine the most likely codeword that was transmitted and therefore correct the errors. The ratio $r = k/n$ is known as the code rate, and it decreases as more redundancy is added. The $n - k$ redundant bits are known as 'parity' bits in reference to the most basic error detecting code, which adds a single parity bit to a block of k bits to indicate the evenness or oddness of the number 1's in a data block.

The relative increase in performance afforded by FEC is known as the coding gain, which is defined as the amount E_b/N_0 can be reduced while still maintaining a given P_{be}, measured in *dB* [176]. In order to illustrate coding gain, Figure 14.2 compares P_{be} vs. E_b/N_0 for uncoded BPSK, and Hamming (7,4) coded BPSK with Hard Decision Decoding (HDD). The mechanics of Hamming (7,4) encoding and HDD will be described in further detail later on. Looking at the purple curve for the Hamming-coded BPSK example, it can be observed that for a P_{be} of less than $\sim 10^{-3}$ the coding gain is positive, i.e. the Hamming coded BPSK signal achieves the same P_{be} as the uncoded BPSK signal at a lower E_b/N_0. However, for error rates above this threshold, the coding gain is actually negative, i.e. a higher E_b/N_0 is required for the coded signal to achieve the same error rate as the uncoded signal.

The coding gain is not uniform because coding leads to a loss in E_b/N_0. As has been established, the process of coding involves the addition of redundant bits to the transmitted signal. Although this enables errors to be corrected in the receiver and a coding gain to be achieved, it contributes to a loss of E_b/N_0 because the bit energy is spread across multiple redundant bits [176]. The loss of E_b/N_0 is proportional to the code rate, with a lower code rate implying a greater loss in E_b/N_0 due to the addition of more redundancy.

Due to the finite error correcting capability of the code, at lower SNRs and higher error rates, it becomes impossible to compensate for the loss in E_b/N_0 and still achieve a positive coding gain, compared to the uncoded case. Improving the coding gain at higher error rates is not very important, since the communication is inherently unreliable. However, improving coding at lower values of P_{be} e.g. 10^{-5}, ensures that reliable and robust communication can be achieved at progressively lower values of E_b/N_0.

Shannon's theorem shows that achieving error free communication (or close to it) at lower and lower values of E_b/N_0 means sacrificing spectral efficiency. In other words, more redundancy has to be added to ensure adequate performance at low values of E_b/N_0. The loss of spectral efficiency manifests as a decrease in data rate or a corresponding increase in bandwidth. To illustrate this, let us assume that the bit rate allowed by a particular channel is R_b bits per second (bps). In an uncoded system, the maximum data rate would equal R_b, since all bits correspond to data. However, when coding is included, the maximum data rate drops to rR_b, i.e. it drops by a factor of k/n in order to accommodate the redundant bits. Alternatively, we could retain the

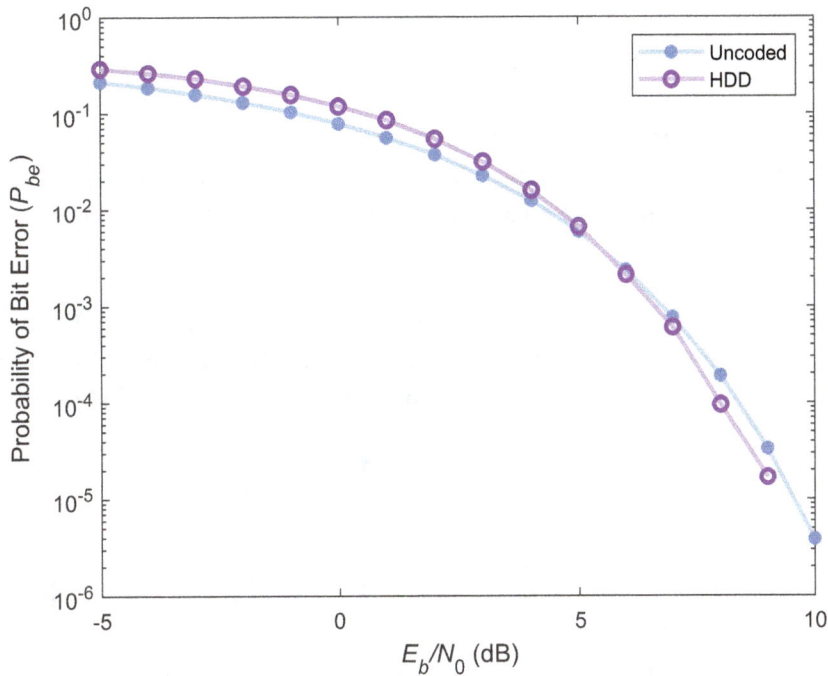

Figure 14.2: Illustration of Coding gain.

same data rate, but this would mean increasing the overall bit rate from R_b to R_b/r *bps* which means increasing the required bandwidth [176]. It is usually preferable to accept a reduced data rate since bandwidth is an expensive resource.

It is sometimes possible to jointly design coding and modulation to achieve a coding gain, without having to reduce the data rate or increase the required bandwidth [176]. The process of designing both the code and the modulation jointly is known as coded modulation, with the most prominent example of this being trellis coded modulation, developed in the early 1980s [164], [176].

Having introduced the fundamental ideas underpinning FEC, we will now move on to discuss the details of the encoding and decoding processes for several prominent codes, including Hamming Codes, Convolutional codes, Turbo codes and LDPC codes.

14.3. Hamming Codes

Hamming codes are examples of linear block codes. In linear block coding, a linear transformation is applied to blocks of k bits in succession to produce a series of n bit codewords. The transformation is designed to retain a total of 2^k codewords from each of the 2^n possible codewords. Each of the 2^k codewords corresponds to a specific k bit input block. Through the coding process, the minimum distance between codewords has been increased to allow one or more errors to be corrected. The block code is typically denoted as an (n, k) block code.

For binary codes such as Hamming codes, the minimum distance is measured using *Hamming distance*, which is defined as the total number of places where two codewords are different. Figure 14.3 illustrates a Hamming distance of 3 between two vectors, where the places with different values are highlighted in red.

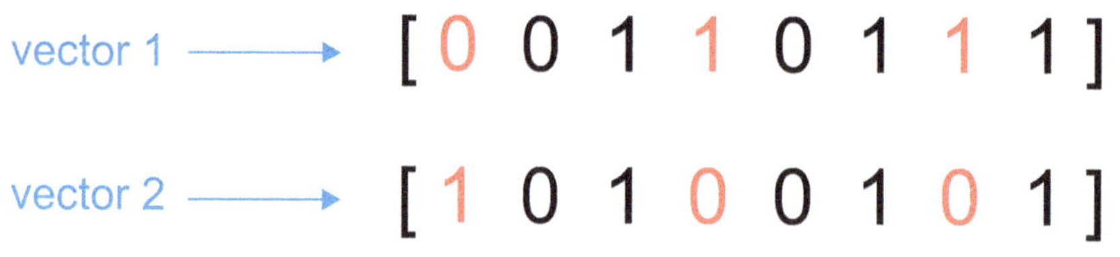

$$\text{vector 1} \longrightarrow \begin{bmatrix} 0 & 0 & 1 & 1 & 0 & 1 & 1 & 1 \end{bmatrix}$$

$$\text{vector 2} \longrightarrow \begin{bmatrix} 1 & 0 & 1 & 0 & 0 & 1 & 0 & 1 \end{bmatrix}$$

Figure 14.3: Hamming distance of 3 between two binary vectors.

In the case of block codes, the number of errors that can be corrected by the code is related to its minimum Hamming distance, $d_{min} = n - k = 3$, which is defined as the minimum Hamming distance between any two codewords in the code. The error correcting capability of the code is [176],

$$t = \left\lfloor \frac{d_{min} - 1}{2} \right\rfloor, \tag{14.2}$$

where t is the number of errors that can be corrected. This relationship is intuitive because the larger d_{min} is, the more errors that would be necessary for one codeword to be confused with another. A Hamming (7,4) code has $d_{min} = 3$, so it can correct one error in a codeword.

The goal of the decoder is to find the most likely transmitted codeword, given the received codeword. In a mathematical sense, this corresponds to the codeword which maximises the Conditional *Probability Density Function* (PDF),

$$P(C_i|R) \qquad i = 0, 1, \ldots 2^{k-1}, \tag{14.3}$$

where P denotes the PDF, R is the received and possibly corrupted codeword, and C_i is the ith possible codeword. The receiver would compute 2^k conditional PDFs, and the one with the maximum value would correspond to the most likely transmitted codeword. Having found the most probable transmitted codeword, it would then be possible to determine the most likely k-bit data block.

The maximum likelihood codeword is the one closest in distance to the received codeword [176]. In HDD decoding, this corresponds to the codeword with the minimum Hamming distance from the received codeword. Therefore, the receiver does not need to compute a series of conditional PDFs to determine the most likely codeword, but rather a series of hamming distances. In SDD, the raw received signal is used, so *Euclidean distance* or *Log Likelihood Ratio (LLR)* metrics are more common. All of these decoding metrics will be discussed in more detail later in this section.

The linear operation used to generate codewords in Hamming codes is a matrix multiplication, and the matrix used to generate the code is called the Generator matrix, denoted as G. The methods used to derive generator matrices are beyond the scope of this chapter, however an example of a generator matrix for a Hamming (7,4) code is

$$G = \begin{bmatrix} 1 & 1 & 1 & 0 & 0 & 0 & 0 \\ 1 & 0 & 0 & 1 & 1 & 0 & 0 \\ 0 & 1 & 0 & 1 & 0 & 1 & 0 \\ 1 & 1 & 0 & 1 & 0 & 0 & 1 \end{bmatrix}. \tag{14.4}$$

In this example, G has $k = 4$ rows and $n = 7$ columns. The codeword is generated as,

$$C = mod(UG, 2), \tag{14.5}$$

where C is the 1×7 codeword, U is the 1×4 data block, and $mod(\ , 2)$ denotes a modulo 2 operation. Therefore, the code has added 3 parity bits.

The Hamming code above is an example of a non-systematic code, because the original data do not appear explicitly in the generated codeword, except in the case where the input block is all zeros. This contrasts with systematic codes which include the original bits in the coded output. Hamming codes can be either systematic or non-systematic.

The parity check matrix, H, for the above Hamming (7,4) code is,

$$H = \begin{bmatrix} 1 & 0 & 1 & 0 & 1 & 0 & 1 \\ 0 & 1 & 1 & 0 & 0 & 1 & 1 \\ 0 & 0 & 0 & 1 & 1 & 1 & 1 \end{bmatrix}. \tag{14.6}$$

The parity check matrix has dimensions $(n - k, n)$, and each row represents a different parity check equation. Each of these equations must be satisfied for a codeword to belong to a code. If a codeword belongs to the code, then the following equality is true,

$$mod(CH^T, 2) = 0, \tag{14.7}$$

where 0 denotes an $(n-k) \times 1$ vector with all entries equal to zero, and ^{T} denotes transposition. The quantity cH^T is referred to as the syndrome and is useful in syndrome decoding schemes. In syndrome decoding, the decoder computes,

$$x = mod(\hat{C}H^T, 2),$$ (14.8)

where x is the syndrome with dimensions $(n-k) \times 1$ and $\hat{C}$ is the received and potentially corrupted codeword. If x is an all-zero vector, then the receiver determines that the codeword is error free. However, if it is non-zero, the block contains errors. This decoding method has equivalent performance to finding the codeword in the code with the minimum Hamming distance from the received codeword.

As mentioned earlier, HDD involves deciding whether a received symbol corresponds to a 0 or a 1, prior to the decoder. This involves passing the received symbols through a decision device, which usually compares the received symbols to a threshold to decide if the transmitted bit was a 0 or a 1. This process is illustrated in Figure 14.4 for a generic BPSK transceiver with Hamming coding.

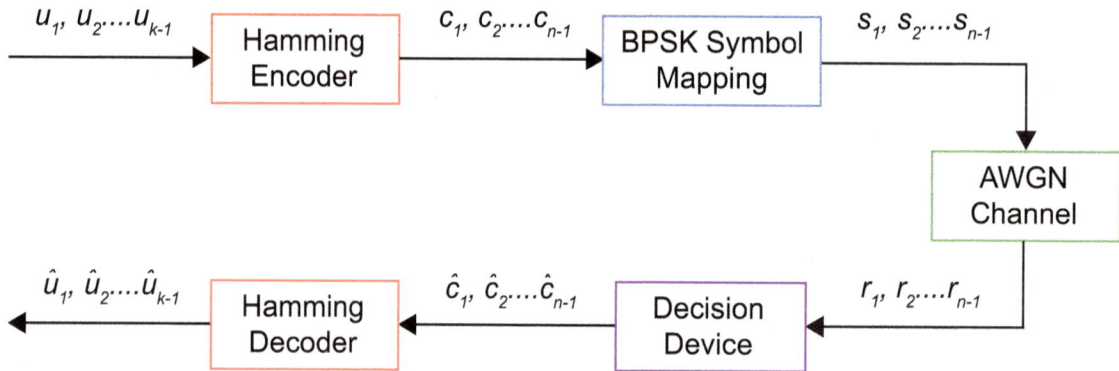

Figure 14.4: High level illustration of hard decision decoding process for BPSK modulation.

In many cases, the decision device will not make the correct decision and, hence, additional errors will be introduced through the HDD process. However, in SDD schemes, no hard decision is made prior to the decoder, thus removing this unwanted source of error.

The simplest SDD scheme relies on computing the Euclidean distance between the received noisy codeword and the possible transmitted codewords. The codeword with the minimum Euclidean distance from the received codeword is chosen as the most likely transmitted codeword. The Euclidean distance is defined as,

$$d_e = \sqrt{\sum_{i=1}^{L} (r_i - s_i)^2},$$ (14.9)

where r_i is the ith received symbol and s_i is the ith transmitted symbol and L is the number of symbols corresponding to a single codeword. In general, it is necessary to design the coding scheme in conjunction with the modulation to use the Euclidean distance metric.

An alternative SDD metric is the LLR,

$$LLR = ln\left(\frac{P(b = 1|r_i)}{P(b = 0|r_i)}\right),$$

(14.10)

where $P(b = 1|r_i)$ is the probability that the transmitted bit was a 1, given symbol r_i, and $P(b = 0|r_i)$ is the probability that the transmitted bit was a 0, given received symbol r_i and $ln(\ )$ is the natural logarithm. In the case of BPSK, only one LLR is calculated due to the fact the each symbol represents one bit. For higher level modulation schemes, multiple LLRs care calculated for each symbol since each symbol represents multiple bits.

In the case of BPSK in an AWGN channel, the symbols are normally distributed with different mean values, depending on whether the original symbol represents a 0 or a 1. This can be understood more clearly by observing a noisy BPSK constellation, as in Figure 14.5.

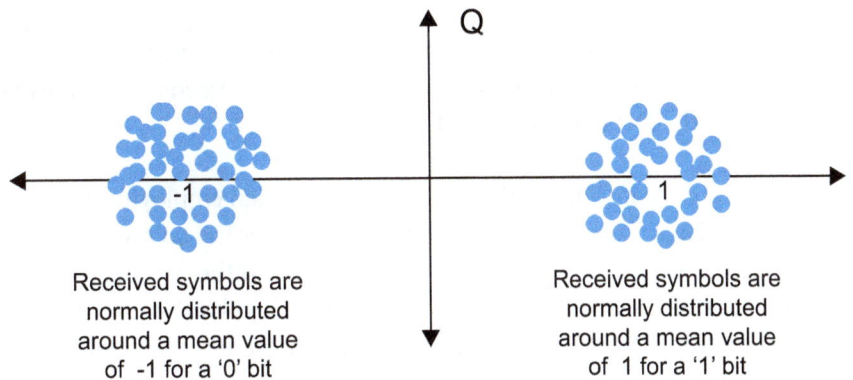

Received symbols are
normally distributed
around a mean value
of -1 for a '0' bit

Received symbols are
normally distributed
around a mean value
of 1 for a '1' bit

Figure 14.5: Example of a noisy BPSK constellation.

It is clear that symbols representing 1 are centred around a mean value of 1 and symbols representing a 0 are centred around -1. By computing the LLR, the receiver attempts to determine which distribution a given symbol is most likely to belong to, i.e. the distribution with a mean of 1, or the distribution with a mean of -1. In both cases, the PDFs have the following form,

$$P(x) = \frac{1}{\sqrt{2\pi\sigma}}e^{\frac{-1}{2}\left(\frac{x-\mu}{\sigma}\right)^2},$$

(14.11)

where x is the random variable (representing individual received symbols in this case), μ is the mean of the distribution, σ is the standard deviation, and e is the base of the natural logarithm. As mentioned previously, μ will either be -1 or 1, depending on which possibility is being tested.

The likelihood ratio in (14.10) reduces to the form e^x, which is inconvenient to compute directly in the receiver. Therefore, a natural logarithm is applied to the expression, which eliminates the e term and just leaves the exponent — hence the name 'log likelihood ratio'. It can be shown that the LLR expression for BPSK reduces to [198],

$$L(\hat{c}_i) = \frac{2r_i}{\sigma^2},$$

(14.12)

where $L(\hat{c}_i)$ is the LLR for the ith received coded bit, $\hat{c}_i$, and σ^2 is an estimate of the noise variance. LLR expressions for higher level modulations schemes such as 16-QAM and 64-QAM can be found in [311] and [343]. If the received symbol is more likely to represent a 1, the likelihood ratio will be greater than 1 and, due to the fact that the natural logarithm of any value greater than 1 yields a positive result, the LLR will be positive. Conversely, if the received symbol is more likely to represent a 0, the LLR will be negative. The magnitude of the LLR gives an indication of the level of confidence that a received symbol represents either a 1 or a 0.

Figure 14.6 compares P_{be} vs. E_b/N_0 for four different cases: (i) uncoded BPSK; (ii) Hamming (7,4) coded BPSK with HDD; (iii) Hamming (7,4) coded BPSK with SDD decoding based on Euclidean distance; and (iv) Hamming (7,4) coded BPSK with SDD based on the LLR.

It is clear that a much more significant coding gain can be achieved when employing soft decisions rather than hard decisions in the receiver. Thus is because SDD techniques consider the relative likelihood that a given bit is a 0 or a 1 rather than making a sometimes incorrect decision prior to the decoder. As such, they are able to determine the correct transmitted codeword with greater accuracy than hard decision decoding, albeit at the expense of increased computational cost. There is no performance difference between the LLR and Euclidean distance methods, so the LLR approach is preferred — this method is also used heavily in decoding of convolutional codes, Turbo codes and LDPC codes.

In fading channels, bursts of errors can occur due to deep fading, leading to a significant loss in performance in terms of coding gain, even at high SNRs. Therefore, techniques such as interleaving and concatenated coding have been developed in order to mitigate the effects of burst errors on error correction performance [343].

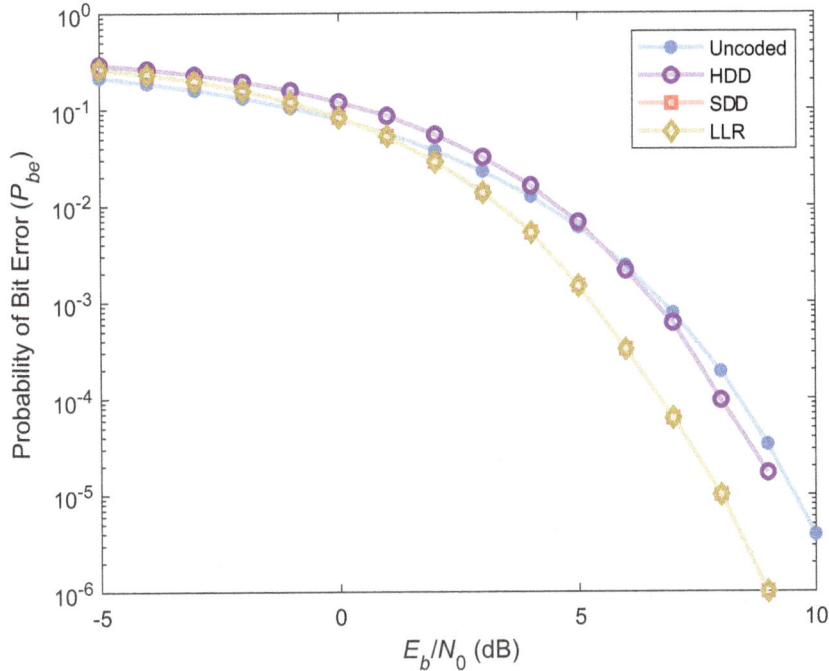

Figure 14.6: P_{be} *vs.* E_b/N_0 *for Hamming(7,4) codes with HDD and SDD.*

14.4. Convolutional Codes

The next important class of linear error correcting codes are convolutional codes. As the name suggests, convolutional codes generate codewords through convolution, i.e. the modulo-2 sum of the current bit and a finite number of previous input bits. Convolutional codes were first proposed in the 1950s and, in 1967, Andrew Viterbi proved that they could be maximum likelihood decoded with moderate complexity, leading to their widespread use in wireless communication systems [147],[352].

Since their earliest use in deep space communications, convolutional codes have been adopted in many wireless standards such as Digital Video Broadcasting – Terrestrial (DVB-T), 4G, and the Wi-Fi family of standards. They are also the foundation of more advanced coding schemes such as Turbo codes.

For convolutional codes, the particular design of the filter structure determines the error correcting performance of the code. Unlike block codes, convolutional codes do not use a defined block length, and therefore the length of the input data stream can potentially be infinite. In practice though, the input data stream will have a defined length, e.g. in packet-based communications protocols such as Wi-Fi. Figure 14.7 illustrates an example of a convolutional coder [310].

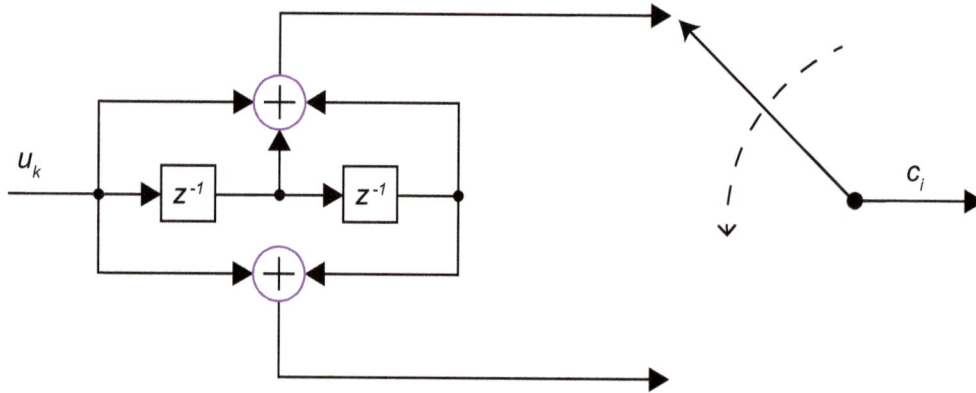

Figure 14.7: Example convolutional code r=1/2, K=3 and [7,5]$_8$ [310].

The first thing to notice about the convolutional coder is the *shift register*, whose length determines the number of past bits which contribute to the current output. In this case, the convolutional code has $r = 1/2$, meaning that two output bits are produced for every one input bit. The code is non-systematic so that the input bit does not appear directly in the coded output.

The output bits are generated by modulo-2 addition (or equivalently, *XOR*ing) of current and past inputs. The convolutional code is parametrised by its constraint length, K, which is equal to the length of the shift register plus one, and its generator polynomials, which define how the different bits in the shift register are combined to produce the output. In this case, the constraint length is $K = 3$ and the generator polynomials are

$$G_1 = x^2 + x + 1, \tag{14.13}$$

and,

$$G_2 = x^2 + 1. \tag{14.14}$$

The generator polynomials can be represented as the binary vectors [1 1 1] and [1 0 1], which are 7 and 5 in Octal format, respectively. The Octal format is typically used as a shorthand method of describing the generator polynomial. Therefore, the convolutional code from Figure 14.7 is parametrised as $K = 3$, $[7, 5]_8$. The terms u_k and c_i represent the kth data bit and i^{th} coded bits respectively.

The error correcting performance of a convolutional code is related to a concept called *free distance*. This term will be described in more detail in Section 14.7, which explores the performance of convolutional codes. Since convolutional codes are based on a shift register, they have memory or 'state'. Hence, the code can be described using a state diagram. The number of states is 2^{K-1}, which in this case is 4, because there are 2 bits in the shift register. As a result, the possible states are 00, 01, 10, and 11. The state diagram is shown in Figure 14.8.

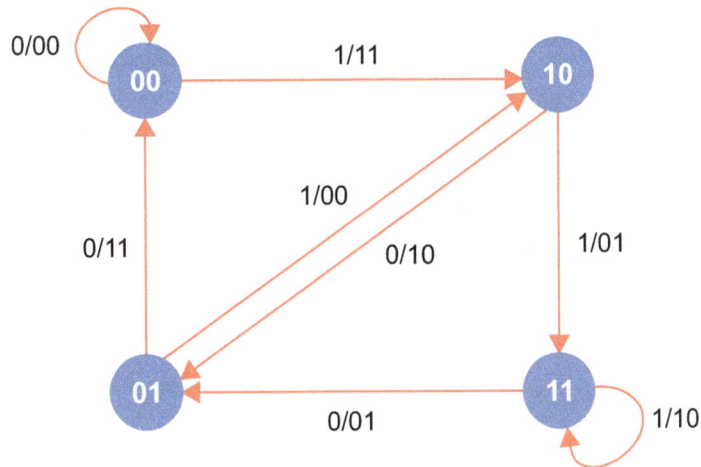

Figure 14.8: State diagram for example Convolutional code [310].

The state diagram shows the encoder behaviour at a single time instant. For example, Figure 14.8 shows that if the encoder is in state 00 and the input is a 0, the encoder will remain in state 00 and the output will be 00. Conversely, if the input is a 1, the encoder will transition from state 00 to state 10 and the output will be 11. The possible state transitions and outputs described in the state diagram are time invariant, i.e. they will be the same for any future time instant.

The state diagram does not show how the state transitions evolve with time in response to a given input sequence. In order to achieve this, a *trellis diagram* is employed. The trellis diagram is very important as it forms the basis of the decoding method, i.e. the Viterbi decoder and in the MAP decoders described in Section 14.6.

Figure 14.9 illustrates the Trellis diagram for the convolutional code described in this section, with a 6-bit input sequence, $U = [1\,0\,0\,1\,1\,1]$.

As can be observed in Figure 14.9, the encoder states are denoted by circles, and the circles are replicated for each time step k, to reflect the fact that the encoder could be in any one of the states at each time instant. The 2-bit values along the top of the trellis are the coded bits for each time step, k. Together, these bits form the transmitted coded sequence, C, corresponding to input sequence, U. For this example, the output codeword is $C = [1\,1\,1\,0\,1\,1\,1\,1\,0\,1\,1\,0]$.

In this case, it is assumed that the encoder begins in the 00 state. Therefore, given an input of 1, the next state is 10 and the encoder output is 11. The transition from state 00 to state 10 at $k = 1$ is drawn as a red line connecting the two states. This line is referred to as a *branch* of the trellis. In this example, there are a total of 6 branches corresponding to the transitions at each time step. The individual branches combine to form a *path*

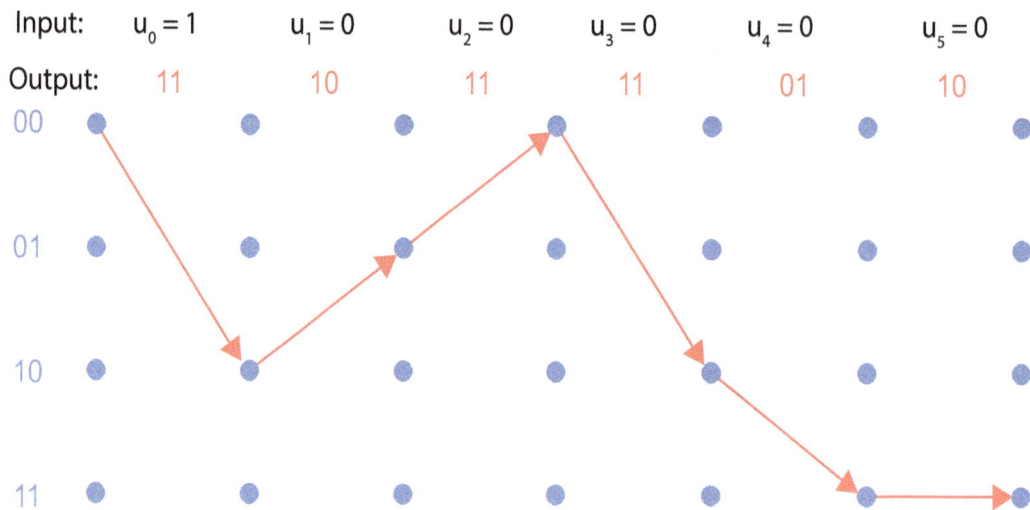

Figure 14.9: Trellis diagram for example convolutional code with input u=[1 0 0 1 1 1].

through the trellis. Each possible input sequence, U, leads to a unique path through the trellis and the goal of the decoding algorithm is to find the most likely path, given knowledge of the encoder structure and the received and corrupted coded sequence, $\hat{C}$. As mentioned, there is technically no limit on the length of U, which means that there is a potentially infinite number of possible paths. It is not feasible for a decoder to consider all of the possible paths through the trellis. However, the Viterbi decoder addresses this problem by retaining only the most likely paths at each time step, and eliminating less likely paths from consideration.

14.5. Viterbi Decoder

In order to understand how the Viterbi decoder avoids having to consider all possible paths, let us examine all of the possible transitions or branches of the trellis and corresponding outputs, at a single time step. These are illustrated in Figure 14.10 for the example Convolutional encoder, introduced in Section 14.7.

After the initial transient or warm up period, the encoder can be in any one of the four states at time step k, in the diagram shown in Figure 14.10. After the transient period, the encoder is said to be in steady state. Up until this point, some encoder states are not possible, due to the fact that the encoder can only be in one state at $k = 0$, and it takes some minimum time for all possible states to be reached. The encoder attains steady state at time K, i.e. after the transient period of $K - 1$.

As mentioned earlier, the possible transitions or branches from one state to another are invariant, i.e. they do not change. For example, if the encoder is in state 00, it can only ever transition to state 00 if the input is a 0, or

Time step *k*

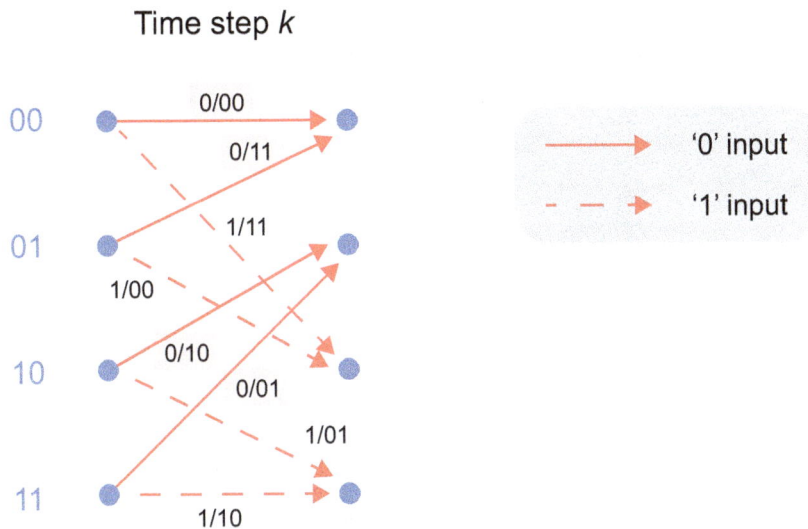

Figure 14.10: All possible branches of trellis at a single time step for convolutional encoder.

state 10 if the input is a 1. This means that there are 8 identical branches to consider at each time step (with two emanating from each state). In the general case, there are $2 \times 2^{K-1}$ branches at each time step (assuming an $r = 1/2$ encoder). Note, this assumption is valid because in practice it is easier to use a convolutional encoder with a base rate of 1/2 and employ *puncturing* to achieve higher code rates [176].

At this stage, you may be asking the question, "How does the Viterbi decoder choose the most likely branches at each stage?" Firstly, a *path metric* is assigned to each state on the left-hand side of Figure 14.7. Having a path metric for each state is sufficient, since all paths must originate from either of the four (or 2^{K-1} in the general case) possible states. At each time step, the Viterbi decoder then computes *branch metrics for* each of the branches. The particular branch metric used differs between decoding methods, but we will assume that it is based on Hamming distance. Therefore, the Hamming distance is computed between the received coded bits and the outputs for each possible branch. Let us assume that the received bits at time step k are *01*. In this case, the Hamming distances for each branch are illustrated in Figure 14.11.

It is clear from the diagram that, for each state, there are two possible branches from current states to next states. For example, to arrive at state 00, the branch could have originated from either state 00 (corresponding to an input of 0) or from state 01 (also corresponding to an input of 0). We can also observe that for each next state, there are two converging branches from current states.

The Viterbi decoder exploits this fact to eliminate paths by selecting only the most likely of the converging paths. To do this, it first adds the branch metrics for the converging paths to their corresponding path metrics. For example, for state 00, the branch metric for the path originating from current state 00 is added to its path metric. Equally, the branch metric for the path originating from state 01 is added to its path metric up to time

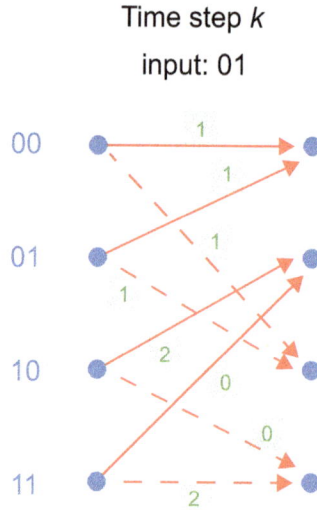

Figure 14.11: Hamming distances for all branches at time step k, assuming input of 01.

k. The two resulting path metrics are then compared, and whichever has the minimum value (assuming the Hamming distance metric), corresponds to the most likely converging path. The other path is then eliminated from further consideration. The path metric for the path in question is then updated with its new value. The process described here is known as *Add Compare Select (ACS)* and is expressed mathematically as [245],

$$PM[n_s] = min(PM[c_{s1}] + BM[c_{s1} \to n_s], PM[c_{s2}] + BM[c_{s2} \to n_s]), \qquad (14.15)$$

where n_s denotes the next state, c_{s1} denotes the first current state, c_{s2} denotes the second current state, BM denotes branch metric and PM denotes path metric. Note, ACS equation can also use the *max* operator depending on the branch metric that is employed. Intuitively, the converging path that is selected corresponds to a coded sequence that is closest in Hamming distance to the received codeword, up to time k.

Let us assume for argument's sake that by time k the path metrics are 5, 7, 6 and 8 for the paths emanating from states 00, 01, 10, and 11 respectively. Applying (14.15) to the example, four of the converging paths are retained (one for each state) and the other four paths are discarded, as illustrated in Figure 14.12.

As can be observed, the converging paths with the lowest updated path metric were selected as the surviving paths. However, for state 01, both converging paths had an accumulated path metric of 8. In the case of a tie, the survivor path is chosen randomly, like flipping a coin.

The remaining paths after ACS are called the *survivor paths* and these are stored in a matrix. The matrix has dimensions $[2^{K-1}, N]$ where N is the final time step being considered. Therefore, at each time step, the 2^{K-1} (4 in this case), survivor paths are stored in one column of the survivor path matrix. The survivor paths are recorded as the states from which the surviving paths originated at time, k. After the full coded sequence

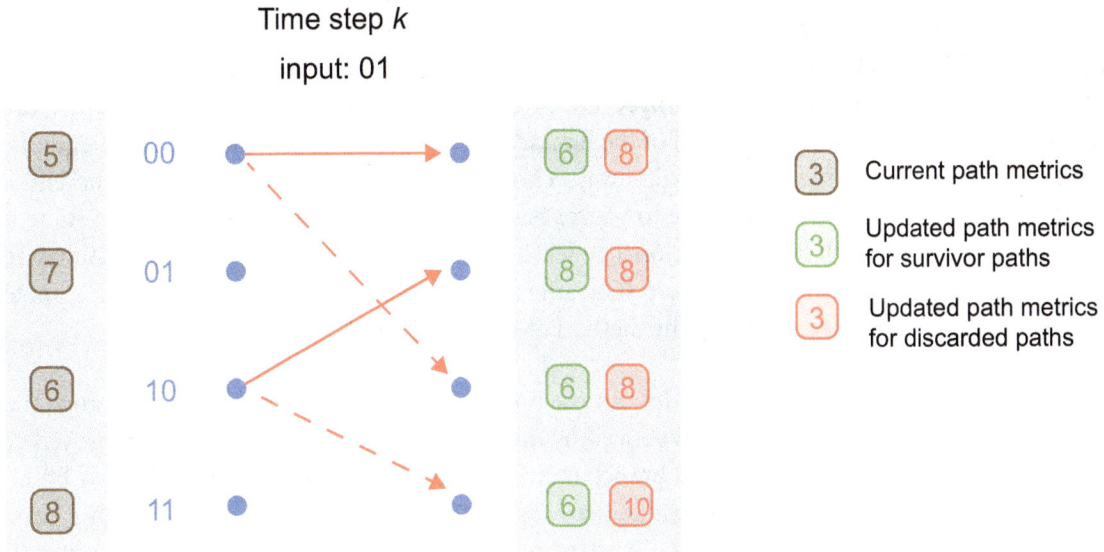

Figure 14.12: Survivor paths after performing ACS procedure.

has been observed at time N, the Viterbi decoder simply selects the surviving path with the minimum overall path metric (again assuming Hamming distance). This path is judged by the decoder to correspond to the coded sequence that is closest in Hamming distance to the received codeword, and therefore the maximum likelihood sequence. For SDD techniques such as Euclidean distance and LLR, the metrics are different but the same principle applies.

The final stage of the Viterbi decoder is to estimate the original input data sequence. In order to do this, the decoder uses a method called *traceback*. This process starts by indexing the survivor path matrix at the final column (corresponding to time N) and the row corresponding to the state where the most likely path ends. For example, let us assume that the final state is determined to be 00, corresponding to row 1 of the survivor path matrix. Let us then assume that for entry $(1, N)$, the value is 01, which tells the decoder that the path came from state 01. Since the decoder is aware of the structure of the encoder, it knows that this transition must have been triggered by an input of 1. Therefore, it estimates that the last bit in the input sequence is a 1. The decoder then proceeds through every column of the survivor path matrix until it has estimated the entire input sequence.

It should be noted that the value of N does not necessarily need to be the equal to the length of the original input sequence. This is because for large values of N, the survivor path matrix can be come very large, which can lead to prohibitive costs in terms of memory. Instead, a *traceback depth* that is smaller than the length of the input sequence is specified. As a rule of thumb, For $r = 1/2$ convolutional coders, a traceback depth of $5K$ provides a good trade-off between decoding performance and memory consumption.

Determining the state where the most likely path ends requires finding the maximum of all 2^{K-1} path metrics. which can add latency for large values of K. Moreover, there is a chance that two paths may have the same final path metric, which can result in the Viterbi decoder making an incorrect decision on the original input sequence. These problems can be resolved by a technique called *zero termination* which involves inputting a sequence to the encoder at the end of the data sequence, which ensures that the encoder ends in an all-zero state. Hence, it means the decoder will always choose the path ending in state 00 at the end of the decoding process. This technique is applicable to packet based communications systems where a finite length input sequence is passed through the convolutional coder. Zero termination involves adding additional redundancy to make the encoder return to the all-zero state after the input sequence has ended, which is not always acceptable; this issue can be addressed using the method of *tail biting* [267].

As with the Hamming codes, the Viterbi decoder can use both hard decision and soft decision branch metrics. The use of soft decisions avoids the errors introduced by hard decision decoding, leading to improved coding gain. Assuming an $r = 1/2$ encoder, the LLR branch metrics are calculated by correlating the two-bit LLRs at each time step with a length 2 vector whose value depends on the coded output associated with a particular branch. As such, the LLR approach finds the path whose corresponding output correlates best with the received LLR sequence.

In practice, using raw LLR values in the Viterbi decoder is computationally intensive and costly, because they need to be represented by large wordlengths. Therefore, it is typical for the LLRs to be quantised through an 'analogue to digital' conversion process, to 3 or 4 bit values, before being passed into the decoder. This makes the decoder much more computationally and resource efficient [250]. The performance of Viterbi decoding with both hard and soft decision metrics will be explored further in Section 14.7. Before we reach that discussion, the MAP/BCJR, Log MAP and Max Log MAP decoders for convolutional codes will be introduced. These decoders are important as they form the basis of Turbo coding, which is reviewed in Section 14.9.

14.6. The BCJR, Log MAP and Max Log MAP Algorithms

Although the Viterbi decoder is the most common decoding algorithm for Convolutional codes, there is another competing approach which applies the concept of *a posteriori* probabilities, known as the *BCJR* algorithm [10],[102]. The BCJR algorithm was first proposed in 1976 and is named after its four inventors: Bahl, Cocke, Jelinek and Raviv. It is also known more generically as the *Maximum a Posteriori (MAP)* decoding algorithm [10].

Due to its computational complexity and lack of performance advantage compared to the Viterbi decoder, the BCJR algorithm is usually not chosen for decoding convolutional codes. However, it does have one clear advantage over the Viterbi decoder, namely that it produces soft decision estimates for the original sequence, i.e. a series of a posteriori LLRs for each input bit. Since it accepts a soft input (i.e. a series of LLRs for the received coded sequence) and produces a soft output (i.e. a series of LLRs for the original bit sequence), it is known as a *Soft Input Soft Output (SISO)* algorithm. In contrast, the traditional Viterbi decoder accepts a soft input and produces a hard output, which means it is a *Soft Input Hard Output* (SIHO) algorithm. A SISO variation of the Viterbi algorithm called the Soft Output Viterbi Algorithm (SOVA) was developed by Hagenauer in 1989 [186]. In addition, simplified versions of the BCJR algorithm known as the Log MAP and Max Log MAP algorithms were proposed in the 1990s [10].

The iterative decoding of Turbo codes leads to near "capacity achieving" performance [107]. More details on Turbo codes will be given in Section 14.9.

14.6.1. The BCJR Algorithm

In all cases, the BCJR and its variants compute a series of N a posteriori LLRs for each input bit, u_k, noting that there is a total of N bits in the original input sequence. This contrasts with the input LLRs (introduced earlier) which are computed on the received coded bits, $\hat{c}_i$. Therefore, the SISO algorithm accepts LLR inputs and produces LLR outputs. The *a posteriori* LLRs are denoted as [10],

$$L(u_k|\hat{C}) = ln\left(\frac{P(u_k = 1|\hat{C})}{P(u_k = 0|\hat{C})} \right). \tag{14.16}$$

It is known as an "a posteriori" LLR because it measures the probability that a given bit, u_k, is a 1 or a 0 given observation of the full received coded sequence. In decoding of a convolutional code, the signs of the a posteriori LLRs are used to determine the original bit sequence.

In order to gain a level of insight into how this algorithm works, let us first consider how to estimate the conditional probability, $P(u_k = 1|C)$ The conditional probability can be re-written as,

$$P(u_k = 1|\hat{C}) = \frac{P(u_k = 1, \hat{C})}{P(\hat{C})}, \tag{14.17}$$

where $P(u_k = 1, C)$ is the joint probability of $u_k = 1$ and $\hat{C}$, i.e. the probability that $u_k = 1$ and the received coded sequence is $\hat{C}$. The same approach can be applied to $P(u_k = 0, \hat{C})$ and, therefore, $L(u_k|C)$ can be expressed in terms of joint probabilities,

$$L(u_k|\hat{C}) = ln\left(\frac{P(u_k = 1, \hat{C})}{P(u_k = 0, \hat{C})} \right). \tag{14.18}$$

Note, the $P(\hat{C})$ terms cancel when formulating the ratio in (14.18). To calculate the joint probability $P(u_k = 1, \hat{C})$, we need to consider all branches of the trellis corresponding to an input of 1. These are illustrated in Figure 14.13 for the example encoder, introduced in Section 14.4.

Time step k

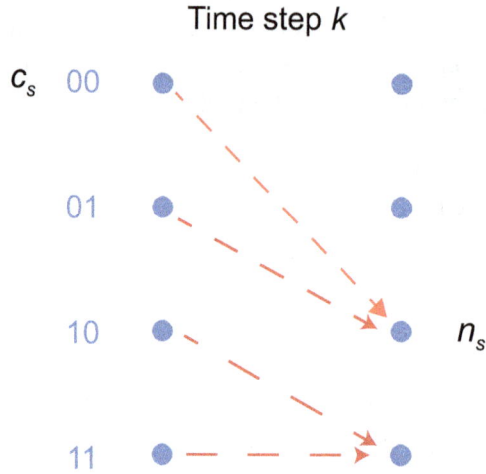

Figure 14.13: Branches of trellis corresponding to an input of 1 at time step k.

As we know, each branch corresponds to a transition from one state to another. Since there are four possible transitions corresponding to an input of 1 (2^{K-1} in the general case) and the probability of any of them occurring is simply the sum of the joint probabilities for each sate transition [10],

$$P(u_k = 1, \hat{C}) = \sum_i P(c_s, n_s, \hat{C}),$$ (14.19)

where $P(c_s, n_s, \hat{C})$ is the joint probability that the encoder transitions from state c_s to n_s, $\hat{C}$ is the received sequence, and i indexes the possible transitions associated with an of input 1. An identical expression can also be derived for $P(u_k = 0, \hat{C})$.

Having derived the expression in (14.19), the question remains as to how to estimate the $P(c_s, n_s, \hat{C})$ terms for each possible branch at each time step k. The essential idea underpinning this is to consider the received sequence, $\hat{C}$, to be composed of three sub-sequences: one representing the past components of $\hat{C}$, one representing the present value of $\hat{C}$ and, finally, one representing the future values of $\hat{C}$ [10]. As such, $P(c_s, n_s, \hat{C})$ is further factorised as,

$$P(c_s, n_s, \hat{C}) = P(c_s, n_s, \hat{C}_p, \hat{C}_k, \hat{C}_f),$$ (14.20)

where $\hat{C}_p$, $\hat{C}_k$ and $\hat{C}_f$ are the past, present and future values of $\hat{C}$ respectively. Through further derivation, this expression is reduced to the product of three sub-terms [10],

$$P(c_s, n_s, \hat{C}) = \alpha_k(c_s)\zeta_k(c_s, n_s)\beta_{k+1}(n_s),$$ (14.21)

where the α, ζ and β terms now relate to the past, present and future sub-sequences of $\hat{C}$ respectively. Firstly, $\alpha_k(c_s)$ is expressed mathematically as [10],

$$\alpha_k(c_s) = P(c_s, \hat{C}_p), \tag{14.22}$$

which is the joint probability that the encoder is in current state, c_s, at time k and the past values of $\hat{C}$ are $\hat{C}_p$. Hence, there are 2^{K-1} α terms at each time step (one for each possible state).

Secondly, the $\zeta_k(c_s, n_s)$ term is,

$$\zeta_k(c_s, n_s) = P(\hat{C}_k, n_s | c_s). \tag{14.23}$$

This is the conditional joint probability that value of $\hat{C}$ at time k is $\hat{C}_k$ and the encoder transitions to state n_s given that it is in current state c_s. Note, the value $\hat{C}_k$ represents two bits because there are two coded bits for every time step k. Finally, the term $\beta_{k+1}(n_s)$ is,

$$\beta_{k+1}(n_s) = P(\hat{C}_f | n_s), \tag{14.24}$$

This is the probability that the sequence, $\hat{C}$, has the future values, C_f, given the encoder transitions to state n_s, at time $k+1$. By definition, measuring this probability requires knowledge of what happens at time $k+1$ from the perspective of time k. This is clearly not possible and hence the algorithm must wait until all elements of $\hat{C}$ have been received before calculating $\beta_{k+1}(n_s)$. Therefore, α and ζ are calculated on the forward pass of the trellis and β is calculated on the backward pass, i.e. once the entire sequence has been received. As such, the BCJR algorithm is sometimes known as the *forward-backward* algorithm. The quantity $P(u_k = 1, \hat{C})$ can now be written in terms of α, ζ and β as,

$$P(u_k = 1, \hat{C}) = \sum_i \alpha_k(c_s)\zeta_k(c_s, n_s)\beta_{k+1}(n_s). \tag{14.25}$$

An identical expression can be derived for $P(u_k = 0, \hat{C})$ except the index j goes through all branches associated with an input of 0,

$$P(u_k = 0, \hat{C}) = \sum_j \alpha_k(c_s)\zeta_k(c_s, n_s)\beta_{k+1}(n_s). \tag{14.26}$$

Therefore, the LLR expression for each bit u_k is,

$$L(u_k|\hat{C}) = \ln\left[\frac{\sum\limits_{i}\alpha_k(c_s)\zeta_k(c_s, n_s)\beta_{k+1}(n_s)}{\sum\limits_{j}\alpha_k(c_s)\zeta_k(c_s, n_s)\beta_{k+1}(n_s)}\right]. \tag{14.27}$$

Now that we have an expression for the a posteriori LLRs, the final step is to derive final expressions for α, ζ and β. For the sake of brevity, we do not attempt to derive these in this book, but the reader is referred to [10] for a detailed treatment of the BCJR algorithm. In the case of an AWGN channel, $\zeta_k(c_s,n_s)$ reduces to [10],

$$\zeta_k(c_s, n_s) = e^{\displaystyle u_k\frac{L(u_k)}{2} + \frac{L_c}{2}\sum_{q=1}^{2}C_{kq}\hat{C}_{kq}}, \tag{14.28}$$

where $L(u_k)$ is the a priori LLR for bit u_k, C_{kq} is the qth expected output bit associated with the branch in question at time k, C_{kq} is the qth received coded bit at time k, and L_c is called the channel reliability value. Note, the $L(u_k)$ term becomes important in Turbo decoding, where each decoder passes improved estimates of $L(u_k)$ to the other decoder over several iterations. It can be observed that the ζ terms have a form similar to the normal probability density function, which is to be expected for an AWGN channel. The correlation on the right hand side is identical to the correlation used to calculate the branch metrics in the Viterbi decoder for an LLR input. Since there are $2 \times 2^{K-1}$ $P(c_s,n_s,\hat{c})$ terms, i.e. one per possible branch, there are $2 \times 2^{K-1}$ $\zeta_k(c_s,n_s)$ terms to calculate at each time step k.

The α values are calculated using the following recursive formula [10],

$$\alpha_k(n_s) = \sum_{i=1}^{2}\alpha_k(c_{si})\zeta_k(c_{si}, n_s). \tag{14.29}$$

The process of calculating $\alpha_k(n_s)$ for the example encoder with $n_s = 00$ is illustrated in Figure 14.14.

As can be observed, the α terms are calculated in the forward direction, i.e. $\alpha_k(n_s)$ is calculated from the previous values $\alpha_k(c_{s1})$ and $\alpha_k(c_{s2})$. It is clear that the branches converging to state 00 are mutually exclusive (only one can occur at a given time), so $\alpha_k(n_s)$ is given by the sum of the two products $\alpha_k(c_{s1})\zeta_k(c_{s1},n_s)$ and $\alpha_k(c_{s2})\zeta_k(c_{s2},n_s)$. Since there are a total of 2^{K-1} possible states, there are 2^{K-1} $\alpha_k(n_s)$ values calculated at each time step, k. Similarly to α, the β values are calculated as,

$$\beta_k(c_s) = \sum_{j=1}^{2}\beta_k(n_{si})\zeta_k(c_s, n_{si}). \tag{14.30}$$

Time step *k*

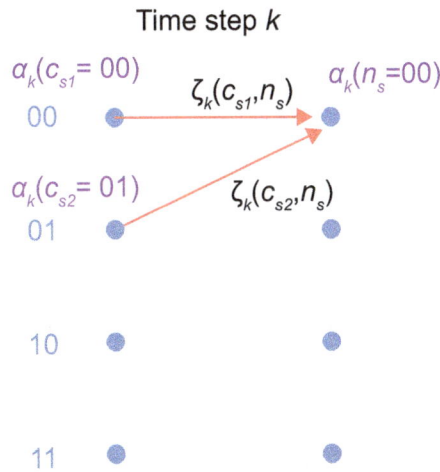

Figure 14.14: Alpha calculation at time step k in the BCJR algorithm.

As described earlier, the β values cannot be calculated until the entire sequence, $\hat{C}$, has been received. Therefore, they are calculated on the backward pass through the trellis, with the β values from the current time step being used to calculate the β values at the previous time step. To initialise β, the final state of the encoder must be known, and this may be achieved using zero termination. Once all of the α, β and ζ values have been calculated through the forward and backward passes of the trellis, the a posteriori LLRs can be calculated using (14.27) and the BCJR algorithm is complete. The α and β terms are typically normalised to address numerical stability issues [10].

14.6.2. The Log MAP and Max Log MAP Algorithms

From the previous section, it is clear that the BCJR algorithm is very computationally complex, requiring a large number of multiplications and to calculate the α, β and ζ values for each time step. In order to reduce this complexity and make the algorithm more suitable for software and hardware implementation, the Log MAP and Max Log MAP variants were proposed. In implementations of Turbo decoders, these algorithms are generally preferred over the BCJR algorithm.

Both of these simplified algorithms are based on the fact that multiplication and division are equivalent to addition and subtraction in the logarithm domain. The α, β and ζ values are replaced by A, B and Γ. The natural logarithm also eliminates the e term in (14.28) which simplifies the calculation of the Γ terms. The Γ term is computed by taking the natural logarithm of both sides of (14.28) [10],

$$\Gamma_k(c_s, n_s) = \frac{u_k L(u_k)}{2} + \frac{L_c}{2} \sum_{q=1}^{n} C_{kq} \hat{C}_{kq}. \tag{14.31}$$

In a similar fashion, to get the A terms we take natural logarithms of both sides of (14.29),

$$A_k(n_s) = ln\left[\sum_{i=1}^{2} \alpha_k(c_{si})\zeta_k(c_{si}, n_s)\right]. \tag{14.32}$$

If we use the fact that $ab = e^{lna + lnb}$, (14.32) can be re-written as,

$$A_k(n_s) = ln\left[\sum_{i=1}^{2} e^{A_k(c_{si}) + \Gamma_k(c_{si}, n_s)}\right]. \tag{14.33}$$

It can be observed that the expression in (14.33) has the form $ln(e^a + e^b)$, which is equivalent to,

$$ln(e^a + e^b) = max(a, b) + ln(1 + e^{|a-b|}), \tag{14.34}$$

where a and b are just generic exponents. In the Log MAP algorithm, the expression in (14.34) is used directly with $a = A_k(c_{s1}) + \Gamma_k(c_{s1}, n_s)$ and $b = A_k(c_{s2}) + \Gamma_k(c_{s2}, n_s)$, where c_{s1} and c_{s2} are the first and second states that can transition to state n_s at time k. A similar approach can be used to derive the B terms, that are computed on the backward pass through the trellis. In the Max Log MAP algorithm, the $ln(1 + e^{|a-b|})$ correction term is dropped, which simplifies the computation since only a *max* operation is performed. The final LLR computation for the Log MAP and Max log MAP algorithms is,

$$L(u_k|\hat{c}) = max_{S_1}[A_k(c_s) + \Gamma_k(c_s, n_s) + B_{k+1}(n_s)] - max_{S_0}[A_k(c_s) + \Gamma_k(c_s, n_s) + B_{k+1}(n_s)], \tag{14.35}$$

where S_1 represents all transitions or branches corresponding to an input of 1 and S_0 represents all transitions corresponding to an input of 0. As before, the *max* operator differs between the Log MAP and Max Log MAP decoders. In the case of the Log MAP, the *max* operation involves more than two variables and therefore it cannot be computed directly. However, it can be estimated through recursion (See appendix of [10]). It is clear that all of the multiplications and divisions associated with the BCJR algorithm have been replaced with additions and subtractions and therefore the computational complexity has been reduced significantly, which is the purpose of these algorithms.

Since the Max Log MAP uses an approximation without the correction term, it has slightly poorer performance than the Log MAP algorithm in Turbo decoding because it leads to biased extrinsic information. The log MAP and BCJR algorithms are indistinguishable in terms of decoding performance. In practical implementations of Turbo decoding, the Max Log MAP is usually the preferred approach.

14.7. Performance of Convolutional Codes

In this section, the performance of convolutional codes and their various decoding approaches will be assessed. As mentioned previously, the error correcting capability of a convolutional code is determined by a quantity known as *free distance*. The number of errors that can be corrected is given by,

$$t = \left\lfloor \frac{fd-1}{2} \right\rfloor, \tag{14.36}$$

where *fd* denotes free distance. The free distance is defined as the minimum Hamming distance between the all-zero codeword, and codewords corresponding to paths beginning and arriving in a future all-zero state [245]. By definition, the paths that depart and then arrive back at the all-zero state at a future time instant are those closest to the path corresponding to the all-zero codeword. The one whose output codeword is closest in Hamming distance to the all-zero codeword requires the smallest number of errors to be confused with the all-zero codeword, and thus defines the error correcting capability of the code. Note, the path whose codeword has the minimum Hamming distance from the all-zero codeword does not always correspond to the shortest path starting and arriving at a future time in the all-zero state. Figure 14.15 illustrates the free distance for the example encoder in Section 14.4.

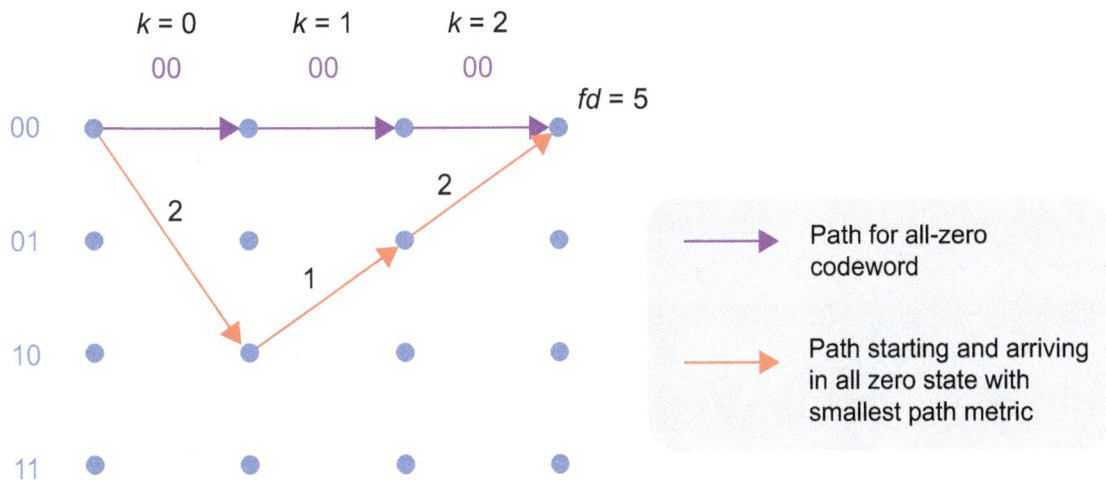

Figure 14.15: Free distance for an example convolutional code.

It can be observed that the path goes $00 \rightarrow 10 \rightarrow 01 \rightarrow 00$, i.e. the path has three branches corresponding to 6 coded output bits. In this case, the path with the minimum Hamming distance is the shortest possible path leaving and then arriving back in the all-zero state. The Hamming distance from the all-zero codeword, or alternatively the path metric for this path (assuming HDD), is 5, and therefore, *fd* = 5. Substituting into (14.36), the convolutional code can fix *t* = 2 errors in every 6 coded bits. As long as there are no more than 2 errors in every 6 coded bits, the convolutional code can correct all of the errors in a coded sequence [176],[245]. As such, convolutional codes do not tend to cope well with bursts of errors.

The process of designing a convolutional code usually begins by defining a maximum value for K (which defines the amount of resources and computation required for encoder and decoder). Assuming $r = 1/2$ and a particular value for K, there are a limited number of possible values for polynomials G_1 and G_2. Therefore, the designer of the code will search for the values of G_1 and G_2 that maximise the free distance and hence the error correcting capability of the code [245].

Having described the factors defining the performance of a convolutional code, we will now assess the P_{be} vs. E_b/N_0 performance for the example encoder, with different decoding metrics and algorithms. Specifically, the Viterbi decoder with Hamming distance and LLR metrics, and the BCJR, Log MAP and Max Log MAP and decoders. As before, BPSK modulation is assumed in all cases. The results are shown in Figure 14.16.

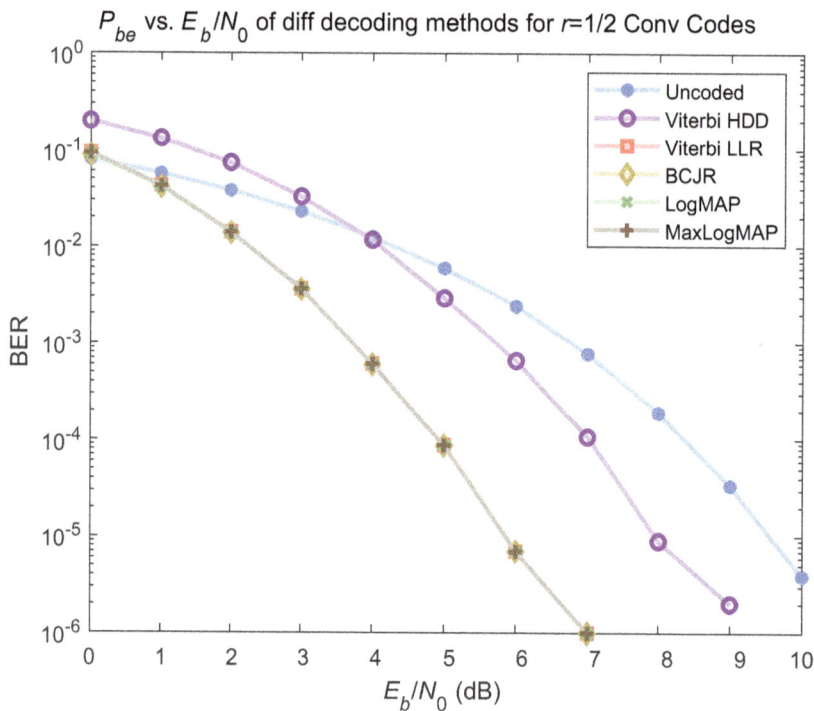

Figure 14.16: Performance comparison of different decoding methods for Convolutional codes.

Firstly, for the case of the Viterbi decoder with HDD, a positive coding gain is achieved for E_b/N_0 values greater than 4 *dB*, where it reaches a maximum of just over 1 *dB*. In contrast, the Viterbi decoder with LLR begins to exhibit a positive coding gain above an E_b/N_0 of 1 *dB* and about 2 *dB* better than the HDD decoder. This clearly demonstrates the performance benefit inherent in using soft decision decoding.

In addition, it can be observed that the BCJR, Log MAP and Max Log MAP decoders achieve identical performance to the Viterbi decoder, based on LLRs. This is consistent with earlier statements that the BCJR and its variants do not offer any performance advantage over the Viterbi decoder for convolutional codes. The performance difference between Log MAP and Max Log MAP only becomes evident in Turbo decoding.

14.8. FEC for Fading Channels

In most cases, codes are designed to deal with a specific number of errors within a given period of time. This approach is well suited to AWGN channels and higher SNR conditions, where the probability of a particular bit being in error is independent of any previous bits, and thus it is less likely for errors to occur in bursts. In multipath fading channels, deep fades occurring due to destructive interference of signal paths can lead to long bursts of errors, which leads to a significant loss of performance. This is especially true in slow fading channels where the deep fade can last for a long period of time [176]. The resilience of a coding scheme to deep fades can be improved through techniques such as interleaving and concatenated coding.

The basic idea behind interleaving is to spread coded bits across time, such that when a burst of errors occurs, it affects bits that were not adjacent to each other in the original bit stream. As such, after de-interleaving in the receiver, the error bits are well separated in time, and the decoder is able to correct them more effectively. The simplest form of an interleaver is a block interleaver, designed for *(n,k)* block codes. The block interleaver is effectively a matrix with n columns and d rows, where d is known as the *depth* of the interleaver. Figure 14.17 illustrates a block interleaver for a Hamming (7,4) code with *d=4*.

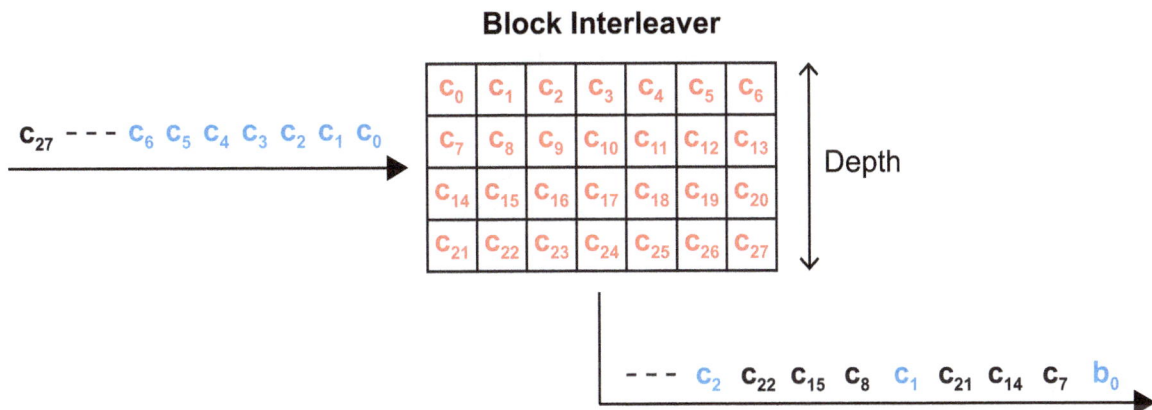

Figure 14.17: Block interleaver for a Hamming (7,4) code with d = 4.

At the left-hand side of the diagram, the coded bits are produced by the Hamming (7,4) code. As we know, a single codeword constitutes 7 bits, as highlighted in blue. Once the block interleaver is full, the bits are read out column-wise instead of row-wise. As such, there is a gap of *d - 1 = 3* bits between each bit in a codeword. Therefore, assuming an error burst of d bit periods, only 1 bit in the codeword is affected rather than 4 bits without interleaving. This leaves only one error for the code to correct rather than 4, which would be impossible. In general, if the interleaver is designed such that $dT_s > T_c$ where T_s is the bit period and T_c is the channel coherence time, then each individual bit in a codeword will experience independent fading channels, leading to much improved performance in the presence of deep fades. An interleaver that is designed to achieve $dT_s > T_c$ is known as a *deep interleaver* [176].

Another approach that provides resilience in fading environments is concatenated coding. In a concatenated code, the encoder consists of two constituent encoders — an inner code and outer code. These are usually separated by an interleaver to provide additional protection against burst errors. A generic architecture for a concatenated code is shown in Figure 14.18.

Generic Concatenated Code

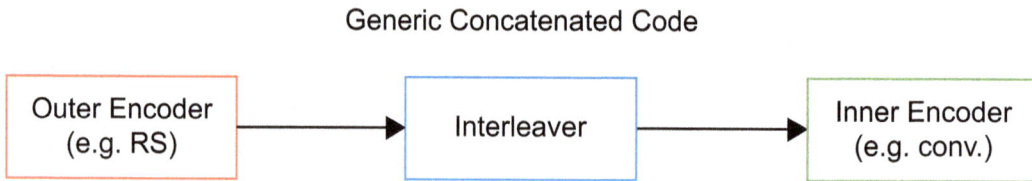

Figure 14.18: Concatenated coding scheme.

One of the earliest examples of a concatenated code was used in a deep space application for the Voyager unmanned spacecraft [235]. This code was chosen because it gave a better coding gain than the alternative unconcatenated scheme at the error rate 10^{-6}, which was deemed necessary to support image compression algorithms. It was found that the decompression algorithm was very sensitive to bit errors and especially burst errors. In the original system, a convolutional encoder was chosen with $K = 7$ and $r = 1/2$. Despite the good performance of convolutional codes, they still make decoding errors and these tend to occur in bursts. Therefore, an RS code was chosen as the outer code due to its burst error correcting capability [235]. In fading channels, the outer code and interleaving process can be used to compensate for the sensitivity of the inner coder to burst errors, and thus increase the encoder resilience in deep fading channels.

14.9. Turbo Codes

Having laid the necessary groundwork it the previous sections, it is now possible to introduce *Turbo Codes*. Turbo codes were first proposed by C. Berrou et al in their landmark 1993 paper "Near Shannon Limit Error-Correcting Coding and Decoding: Turbo Codes" [107]. It was shown that the parallel concatenated convolutional coding scheme and iterative decoding procedure could achieve a P_{be} of 10^{-5} at an E_b/N_0 of 0.7dB, which is 0.7dB from the Shannon limit (which specifies a P_{be} of 10^{-5} at an E_b/N_0 of 0dB for a binary modulation scheme with $r = 1/2$) [107].

As mentioned previously, a Turbo encoder usually consists of two parallel convolutional encoders which operate on a data sequence and an interleaved version, respectively. This approach contrasts with the serial concatenation scheme described in Section 14.8, where the output of the first encoder was interleaved and then passed to the second encoder.

The constituent encoders of a Turbo encoder are typically identical (in which case the Turbo code is symmetric) and are Recursive Systematic Convolutional (RSC) codes [107]. Recall that a systematic encoder uses the data bits directly in the coded output, and that a recursive encoder has an IIR structure. RSC encoders generally have a greater error correcting capability than their non-recursive and non-systematic counterparts [107]. Figure 14.19 shows a high-level architecture for a generic Turbo encoder [10].

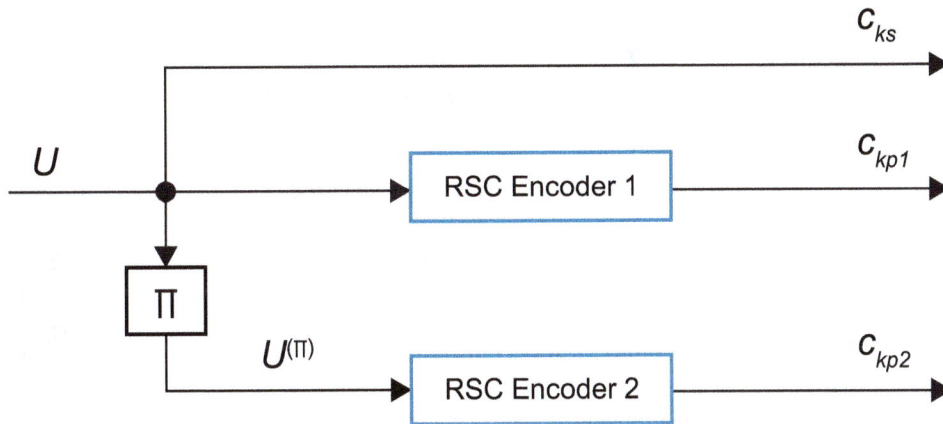

Figure 14.19: Turbo encoder.

On the left hand-side, the input stream, U, enters the first RSC encoder which is assumed to have a code rate of $r = 1/2$. Unlike a conventional convolutional encoder, the Turbo code operates on a block of bits with a defined length. Before entering the second encoder, the input block, U, is passed through an interleaver denoted by the symbol, Π.

As was established earlier, an $r = 1/2$ convolutional encoder produces 2 bits for every 1 input bit. In a systematic architecture, the first bit is simply the original input bit and the second bit is a parity bit. Therefore, each constituent encoder in the Turbo encoder produces a systematic bit and a parity bit, leading to a total of four output bits for every input bit. However, the systematic bit of the second encoder is discarded and only the parity bit is retained. The second systematic bit is discarded because it can be derived from the first systematic bit using the interleaver at the receiver (the systematic bits at the output of the second encoder are just an interleaved version of the original bit sequence).

Having discarded the second systematic bit, the three outputs of the Turbo encoder are the systematic bit, c_{ks}, the 1st parity bit, c_{kp1}, and the 2nd parity bit c_{kp2} at time k. Therefore, the base rate of the encoder is $r = 1/3$. As with standard convolutional codes, the output can be punctured to achieve higher code rates [10] (the details of this process are beyond the scope of this chapter).

As usual, it is assumed that both encoders begin in the all-zero state. The encoders are also zero terminated to ensure that they end in the all-zero state, as required by the BCJR algorithm and its variants. Unlike the non-recursive convolutional encoder, it is not possible to zero terminate the code by passing $K-1$ zeros to the input because of the presence of feedback. In [137], a method was derived to zero terminate a recursive convolutional code, which is best illustrated with an example, as shown in Figure 14.20 [201].

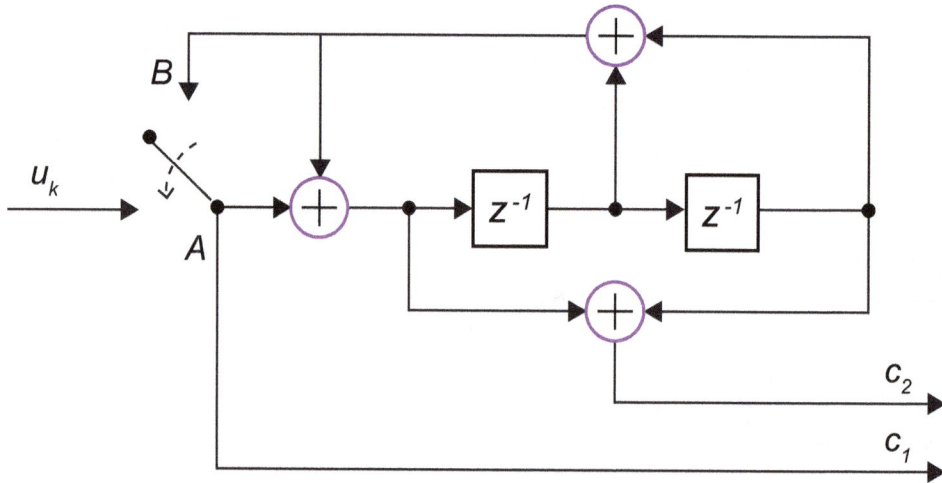

Figure 14.20: Zero termination of recursive convolutional code.

When the input data is being passed into the encoder, the switch remains in position A. After all data bits have been processed, the switch moves to position B. Therefore, the 'inputs' are in fact the final $K-1$ bits stored in the shift register. The XOR operation of a bit with itself on the left-hand side results in an output of 0, which eventually leads to each element of the shift register returning to zero (i.e. the encoder terminating in an all-zero state). The $K-1$ termination bits lead to an additional $2(K-1)$ bits appended at the end of each encoder output, hence a total of $4(K-1)$ additional bits at the output of the Turbo encoder. Once the final output is formed for the input block, U, the bits can be serialised and prepared for transmission over the channel.

Due to the fact that each encoder is systematic and $u_k = c_{ks}$, it can be shown that [10],[107],

$$L(u_k|\hat{C}) = L(u_k) + L(\hat{c}_{ks}) + L_e(u_k), \tag{14.37}$$

where $L(u_k)$ is the a priori LLR for bit u_k, $L(\hat{c}_{ks})$ is the LLR for the received systematic coded bit, and $L_e(u_k)$ is the 'extrinsic' information about bit u_k, which is gained through the decoding process. The extrinsic information is additional knowledge derived from the parity bits.

It is clear that $L(u_k|\hat{C})$ (the output of the MAP decoder), is the sum of two of its inputs ($L(u_k)$ and $L(\hat{c}_{ks})$) and the extrinsic information, $L_e(u_k)$. This means that the only 'new' information contributed to the estimate of

$L(u_k|\hat{C})$ by the decoding process is the extrinsic information. Therefore, because it represents additional information, it is $L_e(u_k)$ that is passed between the constituent decoders in the iterative decoding process. More specifically, the $L_e(u_k)$ term at the output of each decoder is used as a more accurate estimate of $L(u_k)$ at the input of the other decoder [10].

The underlying concept is that by passing the additional information gained about $L(u_k|\hat{C})$ at the output of one decoder to the input of the other decoder, this will lead to a progressive improvement of the estimate of $L(u_k|\hat{C})$ over several iterations. However, eventually the algorithm will converge to a solution and the information passed between decoders will no longer improve the estimate of $L(u_k|\hat{C})$. The Turbo decoding process is usually stopped after a fixed number of iterations, or when a convergence criterion is satisfied. Note, it can sometimes take a large number of iterations to converge to a solution, so the former approach is typically preferred to ensure that the decoding time is deterministic. The process of passing extrinsic information between decoders means that a Turbo decoder is sometimes called a *message passing* algorithm. Figure 14.21 illustrates the Turbo decoding architecture [10].

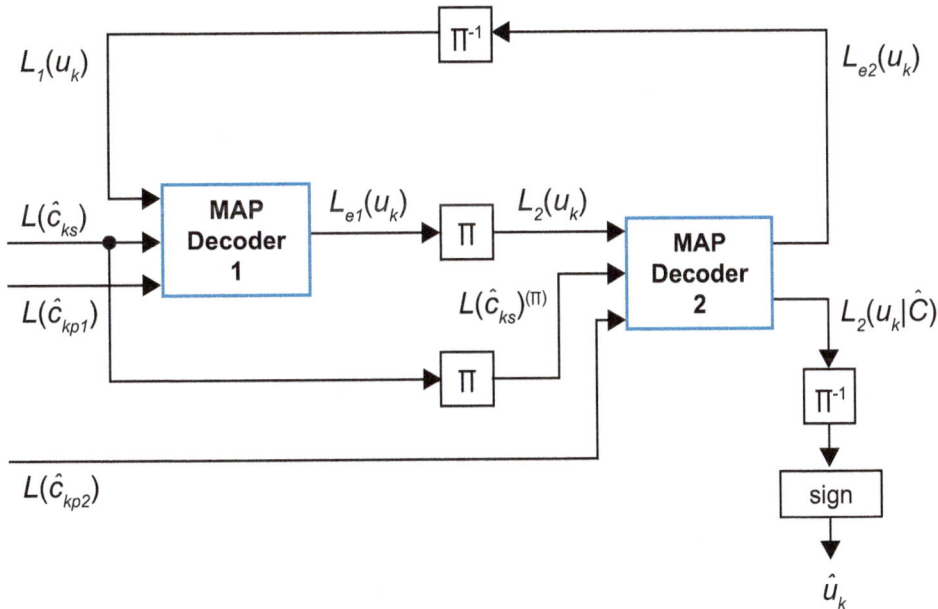

Figure 14.21: Turbo decoder.

From the perspective of the first decoder, it can be observed that the inputs are the LLR for the systematic bit, $L(\hat{c}_{ks})$, the LLR for the first parity bit, $L(\hat{c}_{kp1})$, and $L_1(u_k)$, which is the improved estimate of $L(u_k)$ derived as a de-interleaved version of the extrinsic information $L_{e2}(u_k)$, from the second decoder.

The extrinsic information from Decoder 1 is estimated as,

$$L_{e1}(u_k) = L_1(u_k|\hat{C}) - L(\hat{c}_{ks}) - L_1(u_k), \tag{14.38}$$

where $L_1(u_k|\hat{C})$ is the estimate of $L(u_k|\hat{C})$ at the output of the first decoder. A similar equation can be derived for the second decoder.

It can be observed that $L_{e1}(u_k)$ is interleaved before being passed into the second decoder as $L_2(u_k)$. This is because the input to the second decoder is an interleaved version of the input to the first decoder. In the event that the first decoder makes a burst of errors, the interleaver is able to spread out the errors, thus minimising their impact on the second decoder.

The interleaving process also ensures that the inputs to each decoder are sufficiently uncorrelated to maximise the usefulness of the information that each decoder can pass to the other [306]. If each decoder had to use the same input information, they would produce the same output, and thus would not be able to exchange any new or useful information with each other. The choice of an appropriate interleaver is the most important aspect in the design of Turbo codes but does incur significant latency [306]. After all iterations are completed, the output LLRs of the second decoder, $L_2(u_k|\hat{C})$, are de-interleaved to form the final estimates of $L(u_k|\hat{C})$ and a series of hard decisions are made to produce the final estimate of the data bits, u_k.

In order to assess the performance of Turbo codes, we will consider the Turbo coder used in the 4G LTE standard. The Turbo coder is used for coding of several transport channels in the LTE standard including the Downlink – Shared Channel (DL-SCH), Uplink – Shared Channel (UL–SCH) and the Paging Control Channel (PCH) [4]. Both of the constituent RSC encoders have constraint length $K = 4$ and the overall transfer function is [4],

$$G(D) = \left[1, \frac{g_1(D)}{g0(D)}\right].\tag{14.39}$$

The '1' in (14.39) represents the systematic bit. The generator polynomials $g_0(D)$ and $g_1(D)$ are given by,

$$g_0(D) = D^3 + D^2 + 1,\tag{14.40}$$

and $g_1(D) = D^3 + D + 1,\tag{14.41}$

where $g_0(D)$ is the feedback polynomial and $g_1(D)$ is the feed forward polynomial. The code is expressed in Octal format as $[13\ 11]_8$. The turbo coder is defined for a specific set of input block lengths, K, with the maximum size equal to 6,144 bits. The interleaving function is given by,

$$\pi(i) = mod(f_1 i + f_2 i^2, K),\tag{14.42}$$

where f_1 and f_2 are determined by the block length K as defined in Table 5.1.3-3 of [4], and i is the bit index, ranging from 0 to K-1. Figure 14.22 shows the achieved P_{be} vs. E_b/N_0 for the LTE Turbo coder with BPSK modulation and an input block length of $K = 6,144$ bits. The Max Log MAP algorithm is employed for each constituent decoder.

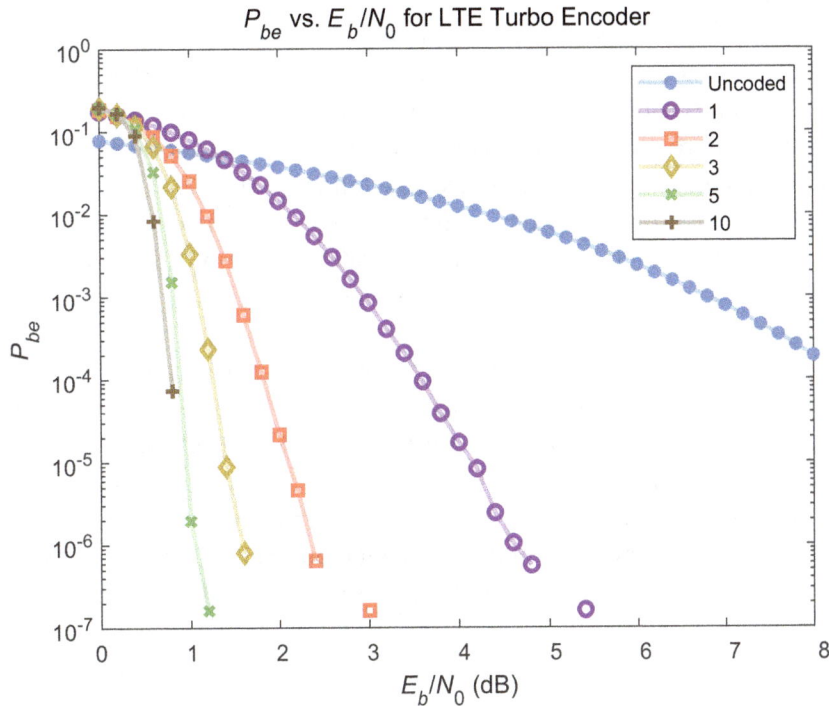

Figure 14.22: Performance of LTE Turbo Coder with K = 6,144 block size.

The plot shows the P_{be} vs. E_b/N_0 curves for 1, 2, 3, 5 and 10 iterations of the Turbo decoder, respectively. It can be observed that the error correction performance improves as the number of iterations increases. However, the performance improvement between 5 and 10 iterations is much smaller than between 1 and 5 iterations. In general, there will be a negligible improvement in the estimate of $L(u_k|\hat{C})$ after a certain number of iterations, at which point the decoding algorithm can be stopped. This maximum number of iterations is usually set to a value between 10 and 20. In the case of 10 iterations, after an initial negative coding gain, P_{be} drops off very sharply, and a significant coding gain is achieved compared to the uncoded case. This demonstrates the excellent performance that can be achieved using Turbo coding.

14.10. LDPC Codes

The final important class of codes that will be described in this chapter are Low Density Parity Check (LDPC) codes. In recent years, they have found application in several wireless communications standards including most recently the 5G NR standard. In similar fashion to Turbo codes, LDPC codes exhibit excellent error correcting performance (as close as 0.0045dB from the Shannon limit has been demonstrated [123]) and can be near optimally decoded using iterative graph based algorithms, with moderate computational complexity. LDPC codes generally outperform Turbo codes at higher code rates and hence are better suited to high throughput applications.

LDPC codes were discovered by Robert G. Gallager and were first published in his 1960 PhD thesis, "*Low Density Parity Check Codes*" [169]. As such, they are sometimes called Gallager codes. Due to the limited computational power available at the time, it was not possible to demonstrate their near capacity achieving performance, which lessened their impact. Furthermore, the belief that the concatenated RS and convolutional code was the most practical approach for error correction, led to LDPC codes being largely forgotten for the next 30 years [111]. One notable exception to this rule was Tanner [111],[333], who first showed that LDPC codes could be represented using bipartite graphs and invented the min-sum and sum-product decoding algorithms. As such, the bipartite graphs describing LDPC codes are called "Tanner Graphs". The AMD SD-FEC IP core employs the normalised min-sum algorithm for decoding [45].

In the early 1990s, Turbo coding showed the excellent performance that can be achieved using iterative decoding algorithms, which combined with the fact that Turbo codes were patented, created renewed interest in LDPC codes [111]. One of the early proponents were Mackay and Neal, who demonstrated the advantages of codes with low density parity check matrices and showed that pseudo-randomly generated LDPC codes performed within $1.2dB$ of the Shannon limit [111],[236]. Since then there has been a large amount of research and development into the design, construction and implementation of LDPC codes and they have been implemented in numerous wireless standards, with one of the most recent being the 5G NR standard [6].

14.10.1. Encoding

LDPC codes are linear block codes, that employ a parity check matrix that has a low density of 1's – hence the name Low Density Parity Check. In more formal terms, the parity check matrix is designed to be a *sparse* matrix, i.e. one where there is a small number of 1's compared to 0's. The sparseness property of the parity check matrix has a number of advantages including generation of codes with very good distance properties and enabling encoding and decoding algorithms with reasonable complexity.

Before introducing LDPC codes, let us briefly re-introduce the concept of a parity check matrix. Recall that the parity check matrix can be used to determine if a codeword belongs to a specific code. This is expressed mathematically as,

$$mod(cH^T, 2) = 0, \tag{14.43}$$

where H is the *(n-k, n)* parity check matrix and c is the $1 \times n$ codeword. If the above equality is satisfied, then the codeword is a member of the code. In decoding, this relationship can be used to determine if a codeword has been received with errors. The receiver calculates the syndrome and if the result is non-zero, it can detect and then go on to correct errors. In addition, the parity check matrix can be used to derive a generator matrix for the code.

Each row of the parity check matrix corresponds to a different parity check equation, and each equation represents a different combination of bits which must XOR to zero, if the codeword belongs to the code. To illustrate this, let us consider the following parity check matrix for a (8,4) code [228],

$$H = \begin{bmatrix} 0 & 1 & 0 & 1 & 1 & 0 & 0 & 1 \\ 1 & 1 & 1 & 0 & 0 & 1 & 0 & 0 \\ 0 & 0 & 1 & 0 & 0 & 1 & 1 & 1 \\ 1 & 0 & 0 & 1 & 1 & 0 & 1 & 0 \end{bmatrix}. \tag{14.44}$$

For this matrix, the parity check equations are,

$$c_1 \oplus c_3 \oplus c_4 \oplus c_7 = 0, \tag{14.45}$$

$$c_0 \oplus c_1 \oplus c_2 \oplus c_5 = 0, \tag{14.46}$$

$$c_2 \oplus c_5 \oplus c_6 \oplus c_7 = 0, \tag{14.47}$$

$$\text{and } c_0 \oplus c_3 \oplus c_4 \oplus c_6 = 0. \tag{14.48}$$

The first equation states that for codeword $(c_0, c_1, c_2, c_3, c_4, c_5, c_6, c_7)$, the bits c_1, c_3, c_4 and c_7 should XOR to 0 for it to belong to the code. The remaining two equations specify different combinations of bits that must also XOR to 0. For example, the codeword (1,1,0,0,0,0,1,1) is a member of the code because it satisfies all of the parity check equations, i.e (14.45) - (14.48).

LDPC codes can be divided into two classes; regular LDPC codes and irregular LDPC codes. In a regular LDPC code, the number of 1's per column, w_c, is the same for each column of H. Equally, the number of 1's per row, w_r, is also constant for every row. Conversely, for an irregular LDPC code, the number of 1's per row and column vary, so w_c and w_r are dependent on the row or column index, which an be denoted as $w_r(i)$ and $w_c(j)$, where i and j are the row and column indices respectively. Irregular codes generally exhibit better performance than their regular counterparts albeit at the expense of increased encoding complexity [237].

In order to satisfy the sparseness criterion, it is necessary to have a large parity check matrix. However, this is not an issue, because the fundamental idea underpinning LDPC codes is the use of large input block lengths, since this leads to codes with a large minimum distance and, therefore, improved error correcting capabilities [170]. In Gallager's initial formulation, he proposed regular LDPC codes and showed that for $w_c >= 3$ and $w_r >= 3$, the LDPC code had excellent distance properties [169].

There are various methods of encoding LDPC codes, but broadly speaking they either involve deriving a generator matrix from the parity check matrix or encoding from the parity check matrix directly, such as in [302]. The generator matrix approach involves finding a generator matrix, G, that satisfies the equality,

$$GH^T = 0. \tag{14.49}$$

A common approach to solving this problem is to make the code systematic, by performing a series of Elementary Row Operations (EROs) on H (row permutations, modulo-2 additions of rows and column permutations if necessary [298]), such that it has the form,

$$H = [P|I] , \qquad (14.50)$$

where I denotes an n-k identity matrix and P is a $(n-k) \times k$ matrix. Having reduced the parity check matrix to this form, G, can be put in the form [298],

$$G = [I|P^T], \qquad (14.51)$$

where I is a k identity matrix. Recall that the codeword, c, is then generated through a vector matrix multiplication, i.e. $c = uG$. The fact that the first part of G is a $k \times k$ identity matrix, leads to a systematic code where the first k bits in the codeword are simply the information bits and the final n-k bits are the parity bits. The complexity of performing the necessary EROs to reduce H to the form in (14.50) is $O(n^3)$ and, since P is generally no longer sparse, the complexity of generating the codeword is similar to matrix multiplication which is $O(n^2)$ [298].

In [302], the authors describe a method of encoding using H directly, i.e. without deriving a generator matrix G. The resulting code is systematic and involves dividing H into two parts: a systematic part and a parity part. The systematic part, H_1, has dimensions $(n-k) \times k$ and the parity part, H_2, has dimensions $(n-k) \times (n-k)$. Since the code is systematic, the codeword c is expressed as,

$$c = [u, p] , \qquad (14.52)$$

where u is the original $1 \times k$ information bit vector and p is the $1 \times (n-k)$ parity bit vector. Therefore, the task of encoding involves finding the parity bits, p. Given H has been split into H_1 and H_2 and that c is expressed as in (14.52), the syndrome can be written as,

$$uH_1^T + pH_2^T = 0 . \qquad (14.53)$$

Note, the additions are all modulo 2 as before. Taking into account the fact that modulo 2 addition and subtraction are the same, (14.53) can be re-written as,

$$pH_2^T = uH_1^T . \qquad (14.54)$$

To solve for p, Gaussian elimination is performed, to put H_2 into lower triangular form, as illustrated in Figure 14.23.

After this the n-k parity bits p can be computed using *forward substitution*. In a similar fashion to the generator matrix, the pre-processing step of reducing H to the form shown in Figure 14.23 through Gaussian elimi-

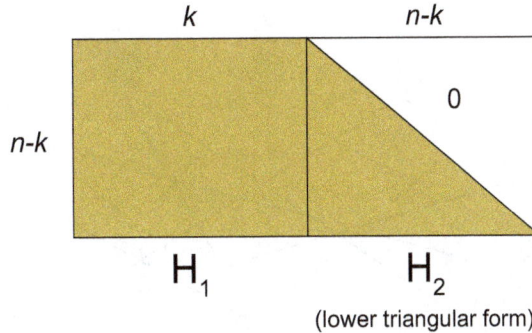

Figure 14.23: H matrix, reduced to lower triangular form.

nation, has complexity $O(n^3)$ [302]. Again, the matrix H is no longer sparse and so encoding has complexity $O(n^2)$ [302]. The authors then propose an algorithm that places H into an "approximately lower triangular" form, which reduces the encoding complexity to $O(n)$, i.e. the encoding complexity scales linearly with the block length. In order to avoid the preprocessing step, it is possible to design codes such that H has the approximate lower triangular form, as required by the algorithm.

14.10.2. Decoding

As mentioned previously, one of the main advantages of the sparseness property of the parity check matrix is that it allows the LDPC codes to be decoded using iterative algorithms, with reasonable complexity. As was the case for convolutional and Turbo codes, the decoding process is implemented using a graphical approach. However, instead of a trellis, the graph in this case is a Tanner graph.

The Tanner graph is an example of a bipartite graph and consists of three components; *variable nodes, check nodes* and *edges*. The variable nodes are more commonly known as v-nodes and each represent a different bit in the codeword. Similarly, the check nodes are known as c-nodes and each represents a different parity check equation. The v-nodes and c-nodes exchange information or messages via the edges. Figure 14.24 illustrates the Tanner graph for the parity check matrix in (14.44).

It can be observed that there are 8 variable nodes and 4 check nodes, corresponding to the 8 coded bits and 4 parity check equations respectively. Firstly, four edges are connected to the check node f_0, corresponding to v-nodes v_1, v_3, v_4 and v_7 which map to bits c_1, c_3, c_4 and c_7. Therefore, this represents the first parity check equation in (14.45). Secondly, four edges are connected to the second check node, f_1, corresponding to v-nodes v_0, v_1, v_2 and v_5 which map to bits c_0, c_1, c_2 and c_5. This represents the second parity check equation. The pattern repeats for the remaining c-nodes.

In the decoding process, both v-nodes and c-nodes perform local calculations based on the exchange of extrinsic information (similarly to Turbo codes) between one another, i.e. information that is not already known to a given node. This repeats iteratively until the best estimate of the original codeword is determined.

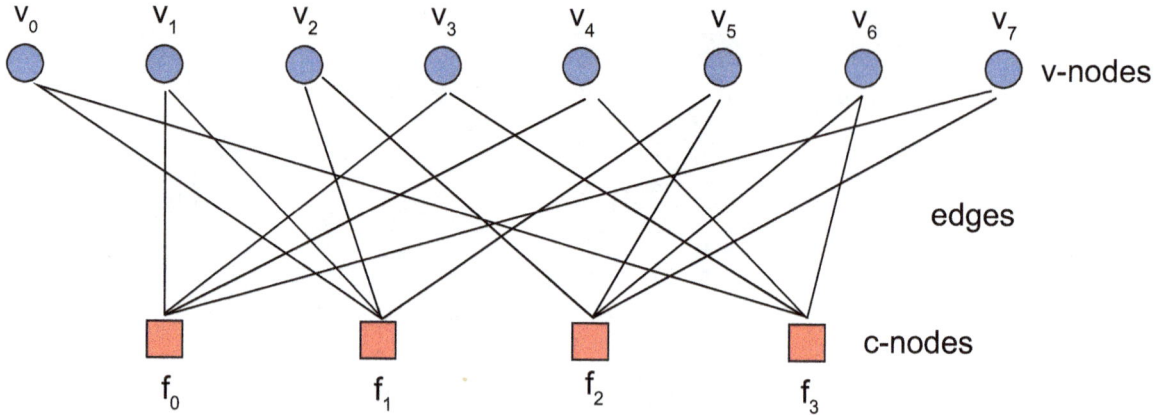

Figure 14.24: Tanner graph for LDPC code.

The decoder can use either hard decision or soft decision decoding methods but, as was the case with previous coding schemes, the latter is preferred due to its superior performance. However, in order to illustrate the decoding process, we will start by considering a hard decision method.

On the first iteration, the only information available to v-node, v_i, is $\hat{c}_i$, i.e. the ith received coded bit. To understand what the check node does, let's start by considering the first one, f_0. As mentioned, this is connected to v-nodes v_1, v_3, v_4 and v_7. The v-nodes in question each send their extrinsic information (which is just comprised of the bits $\hat{c}_1$, $\hat{c}_3$, $\hat{c}_4$ and $\hat{c}_7$ in the first iteration) to f_0. For each bit, f_0 computes the value it should have to satisfy the first parity check equation, assuming all other contributing bits are correct [170],[228]. The same process occurs for the other c-nodes.

To illustrate this, let us consider that the codeword (1,0,0,1,0,1,0,1) was transmitted and there was an error at bit, $\dot{c}_1$, leading to a received codeword (1,1,0,1,0,1,0,1) [228]. At the start of the decoding process, the bits in the received codeword are sent to the relevant check nodes. The check-nodes then perform local calculations that are sent back to the appropriate v-nodes as extrinsic information. Table 14.1 shows the received and sent messages between the v-nodes and c-nodes for the first iteration [228].

Table 14.1: Messages passed between v-nodes and c-nodes.

C-node	Received	Sent
f_0	$v_1 \to 1, v_3 \to 1, v_4 \to 0, v_7 \to 1$	$0 \to v_1, 0 \to v_3, 1 \to v_4, 0 \to v_7$
f_1	$v_0 \to 1, v_1 \to 1, v_2 \to 0, v_5 \to 1$	$0 \to v_0, 0 \to v_1, 1 \to v_2, 0 \to v_5$
f_2	$v_2 \to 0, v_5 \to 1, v_6 \to 0, v_7 \to 1$	$0 \to v_2, 1 \to v_5, 0 \to v_6, 1 \to v_7$
f_3	$v_0 \to 1, v_3 \to 1, v_4 \to 0, v_6 \to 0$	$1 \to v_0, 1 \to v_3, 0 \to v_4, 0 \to v_6$

The messages sent to the v-nodes from each c-node are then used to perform a majority vote and decide the most likely transmitted bit, as shown in Table 14.2 [228].

Table 14.2: Majority Voting scheme for bit decisions.

v-node	Received Bit (c_i)	Messages from c-nodes	Decision
v_0	1	$f_1 \rightarrow 0, f_3 \rightarrow 1$	1
v_1	1	$f_0 \rightarrow 0, f_1 \rightarrow 0$	0
v_2	0	$f_1 \rightarrow 1, f_2 \rightarrow 0$	0
v_3	1	$f_0 \rightarrow 0, f_3 \rightarrow 1$	1
v_4	0	$f_0 \rightarrow 1, f_3 \rightarrow 0$	0
v_5	1	$f_1 \rightarrow 0, f_2 \rightarrow 1$	1
v_6	0	$f_2 \rightarrow 0, f_3 \rightarrow 0$	0
v_7	1	$f_0 \rightarrow 1, f_2 \rightarrow 1$	1

As can be observed, the v-node v_0 receives the input bit c_0 and the extrinsic information 0 and 1 from c-nodes f_1 and f_3. This can be understood by inspecting the Tanner graph, which shows that v_0 is connected to c-nodes f_1 and f_3. Therefore, it follows that v_0 will receive its extrinsic information from c-nodes f_1 and f_3. The v-node v_0 receives [1 0 1] (including message bit and extrinsic information), which allows it to decide that 1 is the most likely value for the bit c_0 through a majority vote. The same process is repeated for all other v-nodes.

After the v-nodes have made their decisions, the syndrome is calculated and if the result is zero, the decoder determines that it has corrected all errors in the received codeword. If not, the decoder continues to iterate until the parity check equations are satisfied or a maximum number of iterations is reached. Again, it is usually the case that a fixed number of iterations is employed to ensure decoding time is deterministic. Table 14.2 shows that the bit decisions after the first iteration are (1,0,0,1,0,1,0,1), meaning that the error in position c_1 is corrected. Therefore, in this case, the decoder terminates after one iteration [228]. By applying the sparseness constraint on the parity check matrix, a sparse Tanner graph is produced, i.e. one where there are few edges connecting v-nodes to c-nodes. This means that the decoding process can be computationally efficient even for very large block sizes (which are necessary to achieve error correction capabilities near the Shannon limit).

Although describing the hard-decoding process is useful for illustrative purposes, it is not implemented in practice for the same reasons as other coding schemes, i.e. that hard decisions introduce an additional source of errors. As with the previous coding schemes, LDPC codes can also be soft-decision decoded. One of the most prominent soft-decision decoding methods is *sum-product algorithm.*

In the sum-product algorithm, extrinsic information is passed from v-nodes to c-nodes and c-nodes to v-nodes in the form of probabilities rather than binary values. More specifically, the messages are sent as LLRs because processing in the logarithmic domain simplifies the update equations for both v-nodes and c-nodes (multiplications can be replaced by additions). The messages sent between v-node i and c-node j are calculated as,

$$L(q_{ij}) = L(\hat{c}_i) + \sum_{j' \in C_{i\backslash j}} L(r_{j'i}),$$
(14.55)

where $L(q_{ij})$ is the extrinsic information exchanged between the ith v-node and j^{th} c-node, $L(\hat{c}_i)$ is the LLR for received coded bit $\hat{c}_i$ and $L(r_{ji})$ is the extrinsic information exchanged between the jth c-node and i^{th} v-node. Note, the summation in (14.55) includes contributions from all c-nodes connected to v-node i except c-node j, which is the c-node receiving information. In other words, it only includes extrinsic information which is unknown to the c-node in question. The set of all c-nodes connected to v-node i but excluding c-node j is denoted by $C_{i\backslash j}$ and this set is indexed by the variable j'.

The messages sent between c-node j and v-node i are of the form,

$$L(r_{ji}) = \left(\prod_{i' \in V_{j\backslash i}} \alpha_{i'j} \right) \cdot \phi\left(\sum_{i' \in V_{j\backslash i}} \phi(\beta_{i'j}) \right).$$
(14.56)

Similarly to before, the c-node computations only include extrinsic information, i.e. information from v-nodes other than the v-node in question. The set of all v-nodes connected to c-node j excluding v-node i is defined as $V_{j\backslash i}$ and this is indexed by the variable i'. The terms ϕ, α and β are given by,

$$\phi(x) = -log\left(tanh\left(\frac{x}{2}\right) \right),$$
(14.57)

$$\alpha_{ij} = sign(L(q_{ij})) \text{ and}$$
(14.58)

$$\beta_{ij} = |L(q_{ij})|.$$
(14.59)

At each iteration of the algorithm, it computes an estimate of the final LLR for each coded bit denoted as $L(Q_i)$ and given by,

$$L(Q_i) = L(\hat{c}_i) + \sum_{j \in C_i} L(r_{ji}).$$
(14.60)

In the final decision, the inputs from all c-nodes connected to v-node i are considered to form an estimate of the final LLR for each received coded bit. This is effectively the same majority voting system that was used in the hard decision method, except that the bit have been replaced by LLRs. The algorithm is essentially improving the estimate of $L(\hat{c}_i)$ over several iterations. After each iteration, a hard decision is performed on $L(Q_i)$, i.e.

$$\hat{c}_i = L(Q_i) > 0. \tag{14.61}$$

If after this decision the syndrome resolves to zero, the most likely codeword has been found and the algorithm stops. Otherwise, it will continue until a maximum number of iterations is reached. The name *sum-product* algorithm comes from the fact that the v-node update equation in (14.55) involves summations and the c-node update equation in (14.56) involves products.

As can be observed from (14.57), the c-node operation in the sum-product algorithm involves the calculation of hyperbolic tan and log functions. This incurs a significant computational burden which is overcome by the min-sum algorithm, which utilises a simplified c-node update equation. It is derived based on the following approximation [118],

$$\phi\left(\sum_{i' \in V_{j\backslash i}} \phi(\beta_{i'j})\right) \approx min_{i' \in V_{j\backslash i}}(\beta_{i'j}). \tag{14.62}$$

In essence, the various summations and products are replaced by the *min* operator, which is significantly less computationally intensive. Similarly to the Max Log MAP algorithm used in Turbo decoding, the min-sum algorithm tends to produce biased extrinsic information, which reduces decoding performance. The performance is improved in the normalised min-sum algorithm, which applies a scaling factor to the extrinsic information between v-nodes and c-nodes at each iteration [194]. The normalised min-sum algorithm is used in the AMD SD-FEC IP core.

14.10.3. LDPC Codes in the 5G NR Standard

In the 5G NR standard, LDPC codes are used for coding of the DL-SCH, UL-SCH and PCH. Other channels such as the Broadcast Control Channel (BCH) and the Downlink / Uplink Control Information (DCI/UCI) use polar codes, which are not covered in this chapter. Therefore, the use of LDPC and polar encoding represents a departure from the Turbo and convolutional codes used in the 4G LTE standard [194].

The LDPC codes used in the 5G NR standard are known as *Quasi Cyclic LDPC (QC-LDPC)* codes. For these codes, the parity check matrix is comprised of an array of sub-matrices which are either sparse Circulant Permutation Matrices (CPMs) or all zero matrices. In CPMS, each row is equal to the previous row cyclically shifted one place to the right, and the first row is a cyclically shifted version of the last row [229]. QC-LDPC codes have the property that the codewords are cyclic shifts of each other, which can be exploited for efficient encoding [229].

The dimensions of each CPM is known as the *lifting size* of the code and the full parity check matrix is defined using a *Base Graph Matrix (BGM)*. The BGM, H_{BG}, stores the CPMs as integer values which define their structure and the parity check matrix H is found by replacing each integer element with an appropriately structured $Z_c \times Z_c$ matrix, where Z_c is the lifting size. The use of a base graph matrix reduces the memory requirements for storing the parity check matrix.

The LDPC code in the 5G NR standard encodes a block of bits B, where $B>0$. If the block to be encoded is larger than a specified maximum code block size, K_{cb}, the block of B bits is segmented into smaller code blocks, which are coded individually. The standard specifies two possible LDPC base graphs; LDPC base graph 1 and LDPC base graph 2 [6]. In general, base graph 1 is used for larger input block sizes and hence higher code rates and base graph 2 is used for smaller block sizes and lower code rates.

The maximum input block sizes for base graphs 1 and 2 are K_{cb} = 8,448 and K_{cb} = 3,840 respectively, after which the blocks are segmented and coded individually. The code block segmentation process ensures that potentially very large transport blocks from the MAC layer do not have to be re-transmitted as part of the HARQ process in the event of errors being detected at the receiver; rather only the relevant blocks where the errors occur need to be re-transmitted. Each segmented code block will contain data bits, a 24 bit Cyclic Redundancy Check (CRC) for error detection in the receiver and filler bits to ensure the block size is $K = 22Z_c$ for base graph 1 and $K = 10Z_c$ for base graph 2. The numbers 22 and 10 relate to the no. of columns of the BGM corresponding to systematic bits, for base graphs 1 and 2 respectively.

The value of Z_c is chosen by indexing a LUT to find the minimum value of all possible lifting sizes that satisfies,

$$K_b \times Z_c \geq \overline{K}, \tag{14.63}$$

where K_b = 22 for base graph 1, K_b = 10, 9, 8 or 6 for base graph 2 depending on the block size and $\overline{K}$ is the length of the code block including data bits and CRC. The BGM matrix has dimensions *46 × 68* for base graph 1, which leads to an H with dimensions *17,664 × 26,112* with Z_c = 384 Recall that the parity check matrix has dimensions *(n-k, n)* where k is the number of data bits per block and n is the number of coded bits. Therefore, for an input block size k = 8,448, the resulting coded block size is n = 26,112, which means that 17,664 parity bits are added. The code is usually punctured to achieved a specific code rate, e.g. r = 1/3 is achieved by puncturing to a codeword size of n = 25,344. The subsequent rate matching process further modifies codewords to achieve higher or indeed lower code rates.

Figure 14.25 shows the P_{be} vs. E_b/N_0 curve for base graph 1 with a block size of K_{cb} = 8,448 and Z_c = 384, which is punctured to a rate of r = 1/3. As before, BPSK modulation is assumed and decoding is performed using the normalised min-sum algorithm with a scaling factor of α = 0.75. The maximum number of iterations is 25. It can be observed that, as was the case for Turbo codes, after the initial negative coding gain, there is a sharp drop off in P_{be}. For example, at P_{be} of 10^{-3}, the coding gain is approximately *5dB* over uncoded BPSK. This demonstrates the excellent performance that can be achieved with LDPC codes.

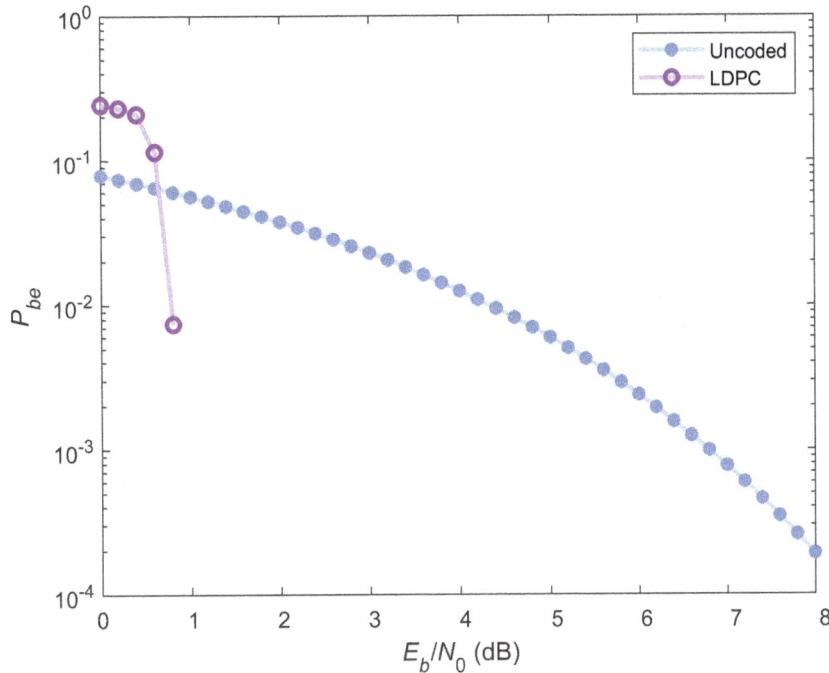

Figure 14.25: P_{be} vs. E_b/N_0 for 5G NR QC-LDPC Code with K_{cb} = 8448 and r=1/3.

14.11. Chapter Summary

This chapter has introduced the fundamental theory of error correction coding in digital communications and storage systems. Error detection and correction schemes are employed in a vast array of communications protocols. They are also widely used in other non-IP digital transmission systems including DAB, DVB-T and DVB-S, for terrestrial and satellite broadcasting of content such as news, sports and general television programming. Furthermore, they are used for reliable storage of digital information in both volatile and non-volatile computer memory.

Error correction schemes are designed to reduce the P_{be} and thus increase the reliability and robustness of digital communications and storage systems in the presence of noise and other forms of degradation. In FEC schemes, codewords are formed by adding redundancy (in the form of additional bits) to the original data, which has the effect of increasing the distance between codewords and allows a receiver to detect and ultimately correct bit errors. FEC schemes are designed to correct a finite number of errors within a data block or stream and are characterised by their coding gain, which measures the amount SNR can drop while still maintaining a given P_{be} compared to an uncoded system. There is an inherent trade-off between the coding gain provided by a code and the loss of spectral efficiency and SNR incurred by the addition of redundant bits.

Therefore, a careful balance must be struck between coding gain and redundancy, when designing FEC schemes for practical systems.

A detailed overview of four linear FEC schemes and their associated decoders has been presented: Hamming codes, Convolutional Codes, Turbo Codes and LDPC codes. Both Turbo codes and LDPC codes are examples of so-called "capacity achieving" codes. This is because for certain configurations, they have been shown to achieve a P_{be} within a fraction of a *dB* from the Shannon limit. The Shannon limit specifies a lower bound, beyond which error free communication is considered to be impossible. In general, achieving close to the Shannon limit requires increasingly complex encoding and decoding schemes. However, both Turbo codes and LDPC codes can be encoded and decoded with reasonable complexity, which makes them suitable for practical implementations.

Turbo codes are a form of concatenated code that uses two parallel RSC encoders to generate a coded bit stream from a block of data bits and an interleaved version. In the decoder, an iterative decoding process is used where "extrinsic" soft information is exchanged between two constituent MAP decoders (for the original and interleaved data blocks respectively). The exchange of extrinsic information allows a progressive improvement of the estimate of a set of a-posteriori LLRs, each corresponding to a different input bit, u_k. In general, increasing the number of iterations improves the estimate of the LLRs and therefore leads to a larger coding gain [107].

LDPC codes are linear block codes which are based on the observation that increasing the input block size leads to a larger minimum codeword distance and therefore improved error correction capability [170]. The codes are based on the design of a sparse parity check matrix (one with a low density of 1's), which enables large block sizes to be encoded and decoded with reasonable complexity. In a similar fashion to Turbo codes, an iterative graphical decoding method is used, based on the Tanner graph. The Tanner graph is a graphical representation of the parity check matrix and consist of v-nodes (one for each coded bit) an c-nodes (which represent the parity check equations). Extrinsic soft information is passed between v-nodes and c-nodes to improve the estimate of a set of LLRs for each coded bit and thus allow errors to be corrected. Due to the sparseness of the parity check matrix, the Tanner graph is also sparse allowing the code to be encoded and decoded with moderate complexity even for large block sizes. Due to their various benefits, both LDPC and Turbo codes have found application in several modern wireless standards including 4G LTE, 5G NR and Wi-Fi 6.

Chapter 15

Practical SD-FEC Design

Lewis McLaughlin

Following on from the previous chapter, which provided a theoretical review of FEC techniques, here we turn the focus to practical design with the hardened SD-FEC blocks that are available on RFSoC devices. An example design is presented to demonstrate how to integrate the SD-FEC blocks into a larger system running on the RFSoC.

15.1. SD-FEC Blocks and IP Core

Selected Zynq Ultrascale+ RFSoC devices contain Soft Decision Forward Error Correction (SD-FEC) integrated blocks. Table 15.1 details the number of SD-FEC blocks that are included in each of the available RFSoC devices (as at the time of writing).

Table 15.1: RFSoC devices and available SD-FEC blocks.

Device (ZUXXDR)	Gen. 1					Gen. 2	Gen. 3						DFE	
	21	25	27	28	29	39	42	43	46	47	48	49	65	67
SD-FEC	8	0	0	8	0	0	0	0	8	0	8	0	0	0

These SD-FEC blocks are implemented in dedicated silicon within the Programmable Logic (PL), and are optimised for performing forward error correction. They can perform both LDPC and Turbo decoding, in addition to LDPC encoding. The Turbo decoding is fixed and employs the standard for decoding LTE. The LDPC encoding and decoding is flexible, and the block supports both standardised and custom LDPC codes.

An IP core is available in Vivado, shown in Figure 15.1, that enables the use of these integrated blocks. The IP core can operate in one of two modes: 5G New Radio (NR) and non-5G NR. More information regarding the IP core and its operation can be found in PG256 [45]. We will instead focus more on the practicalities of incorporating this IP core into a larger system.

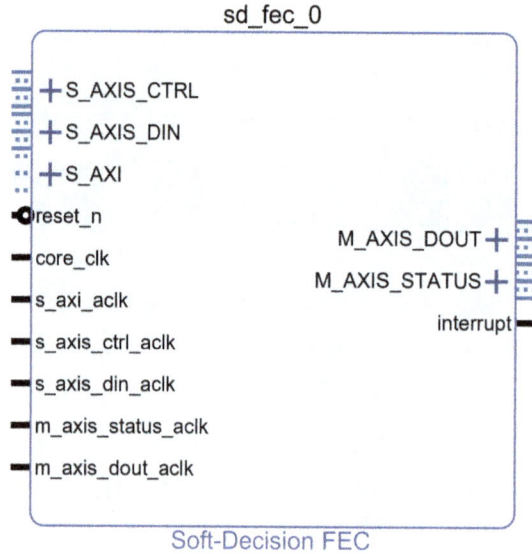

Figure 15.1: SD-FEC IP core as seen in Vivado IP Integrator.

15.2. Hardware Design

The remainder of this chapter will discuss a practical example design of a system which implements LDPC encoding and decoding on an RFSoC, using PYNQ to interact with the design at runtime. Figure 15.2 provides a high-level block diagram illustrating a generic radio pipeline which employs soft decision forward error correction. Section numbers are provided in the blocks to indicate where more information regarding that stage can be found.

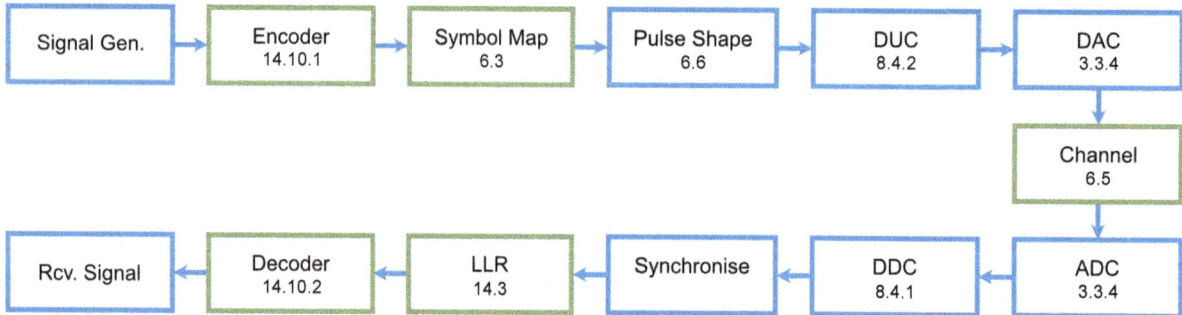

Figure 15.2: Block diagram of generic radio pipeline which employs SD-FEC.

The example hardware design explored in this section will concentrate on the blocks highlighted in green as these are specific to performing SD-FEC on RFSoC and the minimum required to demonstrate the operation of the SD-FEC blocks. The whole system will be confined to the PL and Processing System (PS) of the RFSoC, as including loopback through the data converters would require pulse shaping to allow for transmission, and introduce an unknown latency or phase shift in the channel that would require a synchronisation stage.

Figure 15.3 illustrates the proposed example design. Green blocks will target the PS and blue blocks, the PL. Red highlights the two hardened FEC blocks within the PL that will be used in the system to perform the encoding and decoding.

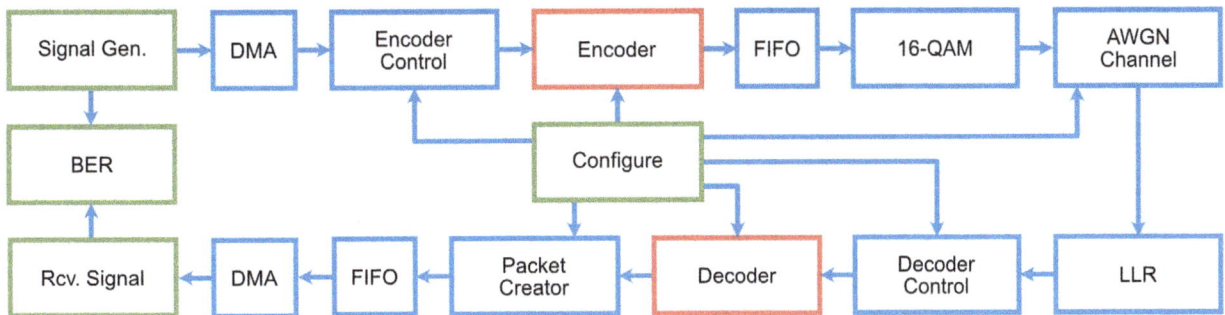

Figure 15.3: Block diagram of example RFSoC design.

AXI Direct Memory Access (DMA) IP cores are used to move data between the PS and PL. A Tx buffer containing data blocks of random bits is generated in Python on the PS and input to a custom FEC controller, by way of a DMA, which is responsible for transmitting the necessary control word to the SD-FEC encoder for each data block. An SD-FEC block is configured as an encoder using the Data-Over-Cable Service Interface Specifications (DOCSIS) 3.1 standard available for selection within the IP core. The encoded data blocks are then baseband modulated using the 16-QAM scheme and sent through an Additive White Gaussian Noise (AWGN) channel. The resulting noisy signal is baseband demodulated into soft bits (log likelihood ratios) which is accepted by the SD-FEC block configured as a decoder. Decoded data is then packaged into a data packet which is equal to the size of the Rx buffer in the PS where it is received. The Tx and Rx buffers are compared and the Bit Error Rate (BER) is calculated in the PS to assess the performance of a given LDPC code.

The following subsections will discuss the design and configuration of each stage of the system outlined in Figure 15.3. Each stage targeting the PL will be packaged as an IP core so as to be incorporated with existing IP cores (DMA, FIFO and SD-FEC) in Vivado IPI. Some hardware design block diagrams will indicate the fixed-point data type on signals. The notation employed is *<sign:wordlength:fractional>* where a 0 or 1 at *sign* signifies unsigned and signed respectively. For example, a signed number with 10 integer bits and 5 fractional bits would be represented as <1:15:5>. AXI4-Lite register address offsets are specified in square brackets. A final section (Section 15.3) will cover the run-time configuration of the IP cores in addition to the BER analysis using PYNQ.

15.2.1. LDPC Codes

The first design step is to determine the LDPC code(s) that will be used in the system. The SD-FEC core supports both standardised and custom LDPC codes. Supported standards include: 5G, LTE, Wi-Fi IEEE 802.11, and DOCSIS 3.1. These standards can be selected directly when configuring the IP core in Vivado. While this example design will use the supported standard DOCSIS 3.1, it is beneficial to understand how to create the code definition file for a custom LDPC code.

Custom codes can be used by instead selecting *custom* from the drop down in the IP core's configuration window and providing a path to a code definition file, which is a YAML formatted file with a *.txt extension. The code definition file must contain: the encoded block size, n; the input block size which contains the information bits, k; the sub-matrix size, p; and the base matrix definition, *sm_array*, which itself must contain a list of row, column, and shift values. For a full list of the accepted code definition parameters, refer to the SD-FEC Integrated Block product guide [45] in the *LDPC Code Definition Parameters* table.

These code definition parameters can be derived from a parity check matrix. To illustrate this, the parity check matrix for the DOCSIS 3.1 Initial Ranging LDPC code is used as an example, shown in Figure 15.4.

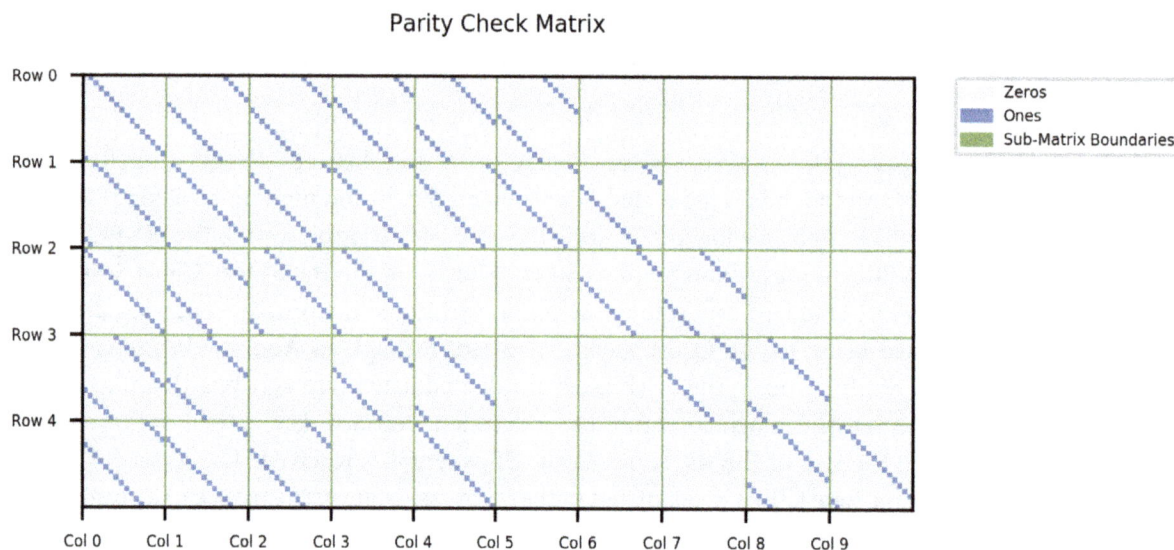

Figure 15.4: Parity check matrix for DOCSIS 3.1 Initial Ranging code.

The parity check matrix is a $k \times n$ matrix which comprises a number of $p \times p$ sub-matrices. These sub-matrices can either be an all-zero matrix or a cyclically right-shifted identity matrix. In this case, the parity check matrix for DOCSIS 3.1 Initial Ranging is an 80×160 matrix with 50 16×16 sub-matrices. These parameters can be taken directly to begin constructing the YAML file.

Figure 15.5 shows the YAML file which specifies the code definition parameters. A name is provided in this example, however this in not a requirement if only one code is specified in the file. Multiple codes do require that each is given a name, however. The block size (n), number of information bits (k) and the sub-matrix size (p) are all supplied. The base matrix definition (*sm_array*) is assembled by taking the sub-matrices which contain identity matrices, and then providing their row and column numbers along with the number of right shifts undergone by the identity matrix.

```yaml
docsis_init_ranging:                                    - {row: 2, col: 0, shift: 0}        YAML
  n: 160                                                - {row: 2, col: 1, shift: 9}
  k: 80                                                 - {row: 2, col: 2, shift: 3}
  p: 16                                                 - {row: 2, col: 3, shift: 2}
  sm_array:                                             - {row: 2, col: 6, shift: 11}
  - {row: 0, col: 0, shift: 1}                          - {row: 2, col: 7, shift: 7}
  - {row: 0, col: 1, shift: 11}                         - {row: 3, col: 0, shift: 6}
  - {row: 0, col: 2, shift: 10}                         - {row: 3, col: 1, shift: 8}
  - {row: 0, col: 3, shift: 12}                         - {row: 3, col: 3, shift: 10}
  - {row: 0, col: 4, shift: 7}                          - {row: 3, col: 4, shift: 3}
  - {row: 0, col: 5, shift: 9}                          - {row: 3, col: 7, shift: 10}
  - {row: 1, col: 0, shift: 2}                          - {row: 3, col: 8, shift: 4}
  - {row: 1, col: 1, shift: 1}                          - {row: 4, col: 0, shift: 12}
  - {row: 1, col: 2, shift: 14}                         - {row: 4, col: 1, shift: 13}
  - {row: 1, col: 3, shift: 15}                         - {row: 4, col: 2, shift: 11}
  - {row: 1, col: 4, shift: 14}                         - {row: 4, col: 4, shift: 0}
  - {row: 1, col: 5, shift: 14}                         - {row: 4, col: 8, shift: 5}
  - {row: 1, col: 6, shift: 12}                         - {row: 4, col: 9, shift: 2}
```

Figure 15.5: YAML formatted code definition file of DOCSIS 3.1 Initial Ranging parameters.

The YAML file is then given a *.txt extension and is available to be sourced during configuration of the SD-FEC IP core.

It is important to consider both the number of information bits in a data block, k, in addition to the number of bits in an encoded block, n. These numbers can influence how the system is designed. For instance, if we again look at DOCSIS 3.1 Initial Ranging parameters, the number of parity bits, $n - k$, is equal to the number of information bits, k. This is said to have a code rate of a half, meaning that the output of the encoder will produce double the number of bits that are input to it. Figure 15.6 again illustrates the example design, but this time highlights the varying amount of information at particular stages of the pipeline when using this LDPC code and a 16-QAM scheme.

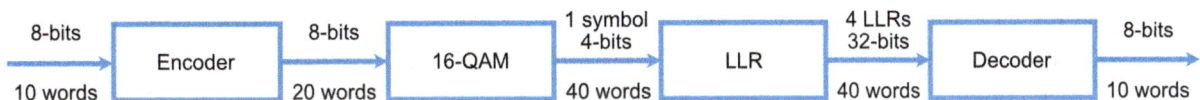

Figure 15.6: Information growth in system.

What does this mean for clock rates or wordlengths upstream? For systems that are clocked at a higher rate than the sample rate, word growth may not cause much concern as there could be enough slack in the system to cope with more data. However, if we are passing large frames of multiple data blocks into the system at the same sample rate as the clock rate, such as in our example of moving multiple data blocks of random bits from the PS into the PL using a DMA, we must consider more carefully how the system is designed.

The complication arises from the fact we would like to perform processing (symbol mapping, adding noise, soft demodulation) on groups of bits within the wordlength input to the system, as opposed processing the wordlength as a whole. We can retain the same input wordlength and sample rate if we perform parallel processing. However, this will result in higher resource utilisation. We can separate the wordlength into smaller groups and time interleave these at a higher rate to avoid using more programmable logic resources. Doing this would require increasing the clock rate and as a result parts of the design may not meet timing closure. These are both valid options with their own trade-offs which should be considered for a given application. Another option is to insert FIFOs at points to buffer data and then apply back pressure using AXI4-Stream between stages. Back pressure in AXI4-Stream is when the slave interface indicates that it is not ready to accept data from the master interface connected to it and so the master interface pauses further transmission until the slave indicates that it is again ready to receive. Although this might not result in the best throughput or indeed resource consumption, it is the simplest to implement and is the methodology adopted in this example.

We will be moving data from the PS to the PL with a wordlength of 8 bits. The SD-FEC IP core supports the DOCSIS 3.1 standard which supplies five LDPC codes: Short, Medium, Long, Initial Ranging, and Fine Ranging. Table 15.2 details the encoded block sizes as well as the number of information bits for each LDPC code. These sizes, k and n, are also divided by our wordlength of 8 bits so we can determine its suitability.

Table 15.2: Sizes of information and encoded blocks in bits and bytes for five DOCSIS 3.1 LDPC codes.

LDPC Code	Information Bits (k)	Encoded Block Size in Bits (n)	$k/8$	$n/8$
DOCSIS Short	840	1120	105	140
DOCSIS Medium	5040	5940	630	742.5
DOCSIS Long	14400	16200	1800	2025
DOCSIS Initial Ranging	80	160	10	20
DOCSIS Fine Ranging	288	480	36	60

As evidenced in the table, there is only one code which cannot be divided evenly by our wordlength of 8 bits: DOCSIS Medium. What does this mean for our system? If the encoder outputs data with a width of 8 bits, the first 742 words would be valid, however the 743rd word would only contain valid data on the four least significant bits. To accommodate this, we must add some control logic to part of the design which will ignore the bits that are not valid. The fact that we are using a 16-QAM scheme which has a symbol size of 4 bits makes the symbol mapping stage a suitable point to integrate the control logic. However, depending on the system, this control logic could be placed elsewhere.

15.2.2. Encoder Configuration

Having decided on an LDPC coding standard, we can configure our encoder. Table 15.3 details the Non-5G configuration used in this design.

Table 15.3: SD-FEC IP core configuration for LDPC encoding.

Function		Interface		Runtime Loading	
Configuration		**Parameters (S AXI)**		Physical Utilization	N/A
Standard	DOCSIS 3.1	**Interface**	Runtime-Configured	Throughput Utilization	100
Turbo Decode Code Parameters		**Code Parameters**	N/A		
Turbo Decode	N/A	**DIN**			
Algorithm	N/A	**Interface**	Pre-Configured		
Scale	N/A	**Lanes**	1		
LDPC Decode Code Parameters		**Word Configuration**	Fixed		
LDPC Decode	No	**Number of Words**	1		
Support W>1	N/A	**DOUT**			
Code Definition	N/A	**Interface**	Pre-Configured		
Overrides	N/A	**Lanes**	1		
LDPC Encode Code Parameters		**Word Configuration**	Fixed		
LDPC Encode	Yes	**Number of Words**	1		
Code Definition	N/A				

As mentioned, we are using the DOCSIS 3.1 standard supported by the SD-FEC IP core and have configured this IP core as an encoder. In this example design, one 8-bit word is transferred per AXI4-Stream transaction on both the data interfaces, *DIN* and *DOUT*. Only one lane is used on each interface. A lane has a width of 128 bits and can accommodate up to 16 8-bit words. Although *s_axis_din_tdata* and *m_axis_dout_tdata* have widths of 128, only the least significant byte (bits 7:0) is used on each, in this configuration.

While the SD-FEC IP core can be configured to have 4 lanes, giving a total width of 512 bits, it is important to note that only one lane is processed per clock cycle of the core. A small Clock-Domain-Crossing (CDC) FIFO exists at the AXI4-Lite and AXI4-Stream interfaces of the SD-FEC block. This converts the input and output widths of the *DIN* and *DOUT* interfaces between 128 bits for processing inside the block and either 1, 2 or 4 times this width externally. For high-throughout applications this can allow for large wordlengths of data to be processed with a high degree of parallelism outside of and at a lower clock speed than the SD-FEC core, which can be advantageous when trying to meet timing closure.

The SD-FEC block is highly flexible and the number of words input and output of the data interfaces can also be varied at run time. This is achieved by selecting the option *Unconfigured* from the drop-down associated with the *Interface* parameter under the *DIN* and *DOUT* headings on the *Interface* tab of the IP core configuration window. Doing so will expose the interfaces *DIN_WORDS* and *DOUT_WORDS* which can be used to configure the number of bytes transferred on each cycle. This example opts to have fixed word configurations.

15.2.3. SD-FEC Controller

Every data block input to the SD-FEC core through *DIN* requires a control word input on the *CTRL* interface. A status word is also output for each processed block. This is illustrated in Figure 15.7 which depicts the *DIN*, *DOUT*, *CTRL* and *STATUS* interfaces when an SD-FEC block is encoding data. Decoding is similar however the lengths of *DIN* and *DOUT* would differ.

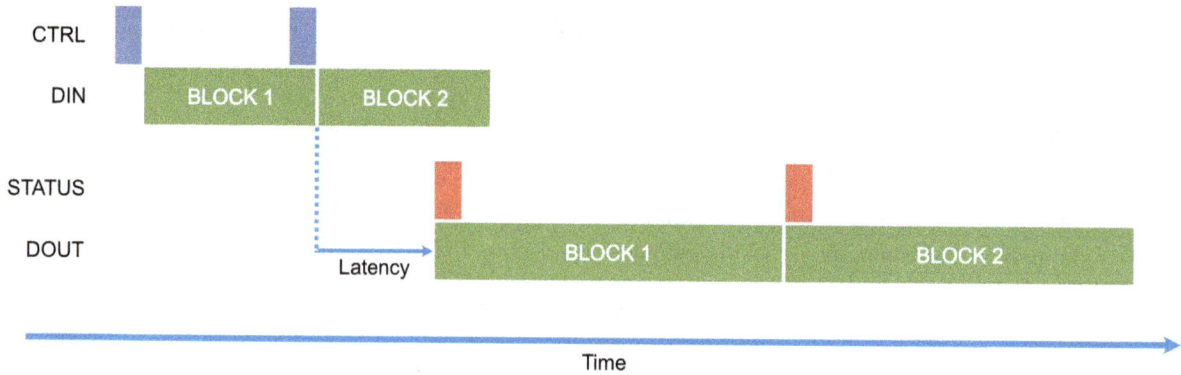

Figure 15.7: SD-FEC core interfaces when encoding data.

The control word comprises a number of fields within a 32-bit word, when operating in non-5G NR mode, shown in Table 15.4. The SD-FEC core only requires certain fields depending on whether it is configured as an encoder or decoder. The rows highlighted in red indicate fields used by both configurations. The remaining rows are exclusive to the decoder configuration.

Table 15.4: Control register for encoding/decoding in Non-5G NR mode.

Field (bits)	Type
External Block ID (31:24)	uint8
Maximum Number of Iterations (23:18)	uint6
Terminate on No Change (17)	bit1
Terminate on Pass (16)	bit1
Include parity Output (15)	bit1
Hard Output (14)	bit1
Reserved (13:7)	uint7
Code Number (6:0)	uint7

The external block identifier can be any integer between 0 and 255. It is passed directly through to the status register and helps identify which encoded or decoded block the reported status belongs to. The code number relates to the code parameters that have been loaded into the SD-FEC block's internal memory allowing for easy switching between loaded codes. The maximum number of iterations can be a value between 1 and 63. This refers to the number of iterations performed when decoding. A higher number of iterations incurs larger latency. These are 'maximum' iterations because the process can terminate earlier if no change in hard bits is detected for the whole block between iterations (*Terminate on No Change* = 1) or if the parity check passes (*Terminate on Pass* = 1). The decoder can also return a hard output (*Hard Output* = 1) or soft output (*Hard Output* = 0) or include the parity in the output (*Include Parity Output* = 1). These options will effect the amount of data output from the decoder.

If a control word is not supplied, the SD-FEC core will not process data and prevent further data from being input. Similarly, the status word must be read from the *STATUS* interface or this will also cause the SD-FEC core to stall. The CDC FIFOs which exist at the interfaces will allow for some blocks to be output, but once the *STATUS* CDC FIFO fills, the core will stall.

We can exploit the fact that our application uses the same settings for processing many blocks of data and design a controller that can be updated intermittently over AXI4-Lite, and output a control word using AXI4-Stream at the rate required by the LDPC code being used. Figure 15.8 shows a suitable controller. Two memory mapped registers, *aximm_ctrl* and *aximm_length*, provide the control word and the number of words in a block, respectively. A counter counts up to the number of words in a block, after which it is reset and begins counting from zero again. Every time the count is equal to zero, *m_axis_ctrl_tvalid* and *m_axis_ctrl_tdata* are driven high if there is also valid data coming in (*s_axis_tvalid* = 1). Doing this produces a valid, single sample AXI4-Stream transaction containing the value of *aximm_ctrl* on the *tdata* signal at the start of each block.

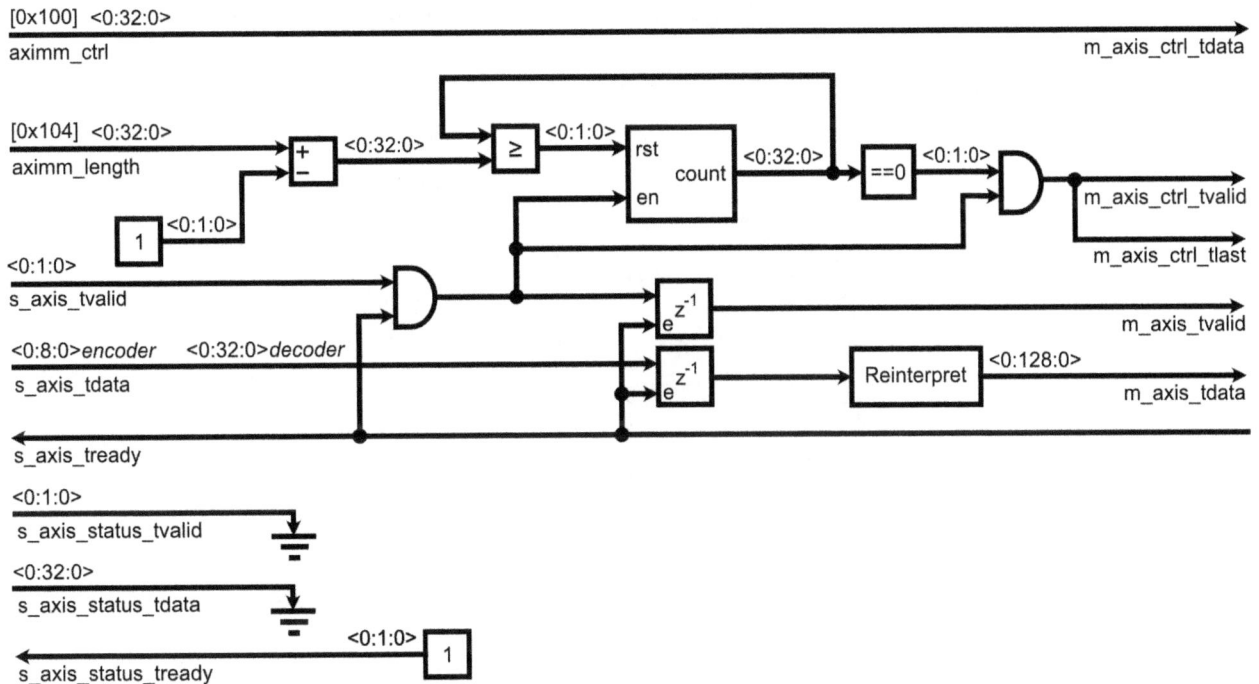

Figure 15.8: Block diagram of SD-FEC controller core.

We do not use the status value output from the SD-FEC core in this application, but to avoid stalling the SD-FEC block, the controller keeps *s_axis_status_tready* high, meaning that status words are accepted by the controller but immediately terminated. The *s_axis_tvalid* and *s_axis_tdata* signals are passed through the controller, delayed and output on the master interface as *m_axis_tvalid* and *m_axis_tdata*. The delay ensures that the control words arrive at the controller just before the data to be processed, as illustrated in Figure 15.7. By reinterpreting the input data from 8 or 32 bits on the slave interface to 128 bits on the master interface, we avoid any critical warnings regarding interface mismatches between our controller IP core and the SD-FEC IP core in Vivado. Our data byte(s) will be the least-significant byte(s) on the master interface of the SD-FEC controller. The two widths come from the fact that the encoder will be accepting wordlengths of 8, output from the PS, whereas the decoder will be accepting wordlengths of 32, output from the IP core demodulating the data into soft bits which is covered in Section 15.2.6.

Finally, it is important to incorporate a *tready* signal. The SD-FEC block may not be ready to process new data at points and we must ensure that this information is relayed back to the DMA so that it can pause further transactions. Otherwise, we would drop samples and this could result in the whole system stalling and certainly skew any BER analysis. The *tready* signal should also prevent the counter from counting further when low and hold any data in the two delay registers when low so as not to lose information.

15.2.4. Symbol Mapping (Baseband Modulation)

The symbol mapping employed in the system uses the 16-QAM scheme and Gray coding, where adjacent symbols differ by only 1 bit to improve the BER. Figure 15.9 depicts the symbol mapping used.

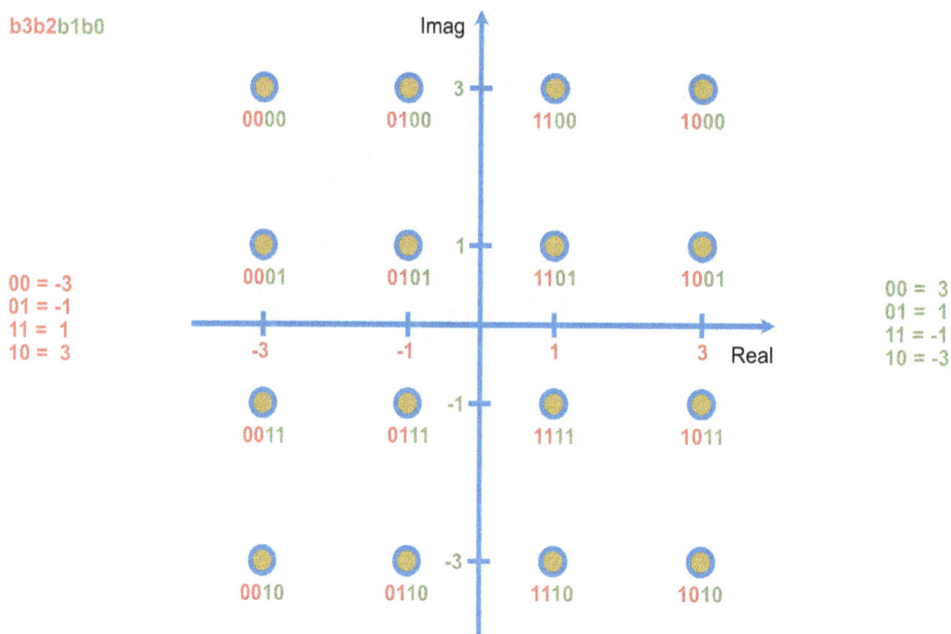

Figure 15.9: Gray coded symbol mapping for 16-QAM.

This mapping can be easily realised by taking the two most significant bits to obtain the real value, and the two least significant bits to obtain the imaginary value, of a symbol. Figure 15.9 highlights this as the real values are shown in red and the imaginary values in green. The block diagram shown in Figure 15.10 demonstrates how this translates into hardware. Two multiplexers output symbols depending on the two bits input to the select port. The output of the multiplexers are then concatenated together into a byte, where the least significant bits contain the real part and the most significant, the imaginary part. When using AXI4-Stream it is important for the *tdata* signal to be an integer multiple of 8 bits, so that it is not removed in the synthesis process.

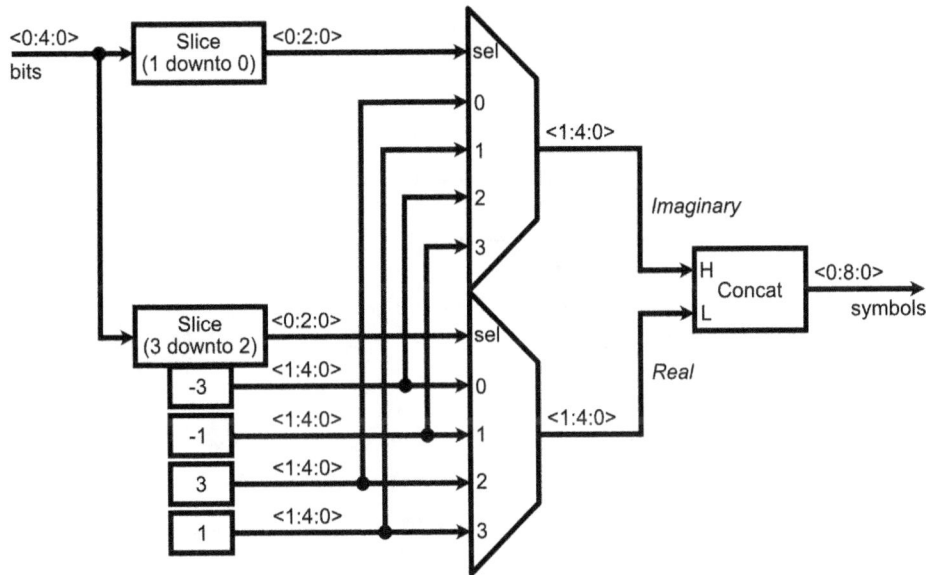

Figure 15.10: Gray coded 16-QAM symbol mapper.

The wordlength output from the encoder is 8 bits and one 16-QAM symbol is mapped to four of these bits. We must therefore reconcile this difference in wordlengths. One option would be to have two symbol mappers, each taking a slice of 4 bits and processing in parallel. This is a good option if throughput is a concern as the input and output sample rate would be the same. The following stages would have to be designed to accommodate two symbols. Additionally, outputting two symbols would no longer make this stage a suitable point to introduce control logic to ignore symbols that are not valid, as in the case of the last byte of a block encoded using DOCSIS Medium. Instead, this system opts to use one symbol mapper, meaning that the input rate must be halved to allow sufficient time for the mapper to generated two symbols for every word input.

Figure 15.3 shows how this is achieved. The 16-QAM symbol mapper from Figure 15.10 can be seen in the bottom right corner. Back pressure must be applied to halve the data input rate. This is implemented with a 1-bit counter that toggles *s_axis_tready* at the rate the IP core is clocked at. A FIFO between the output of the encoder and the input to this IP core ensures no samples are dropped. This FIFO will hold *tvalid* high on its master interface when there is valid data ready to be output from it, therefore *s_axis_tvalid* and *s_axis_tready* are input to an AND gate to create a new *valid_in* signal that can be used within the symbol mapping IP core to indicate when new valid data has entered the core.

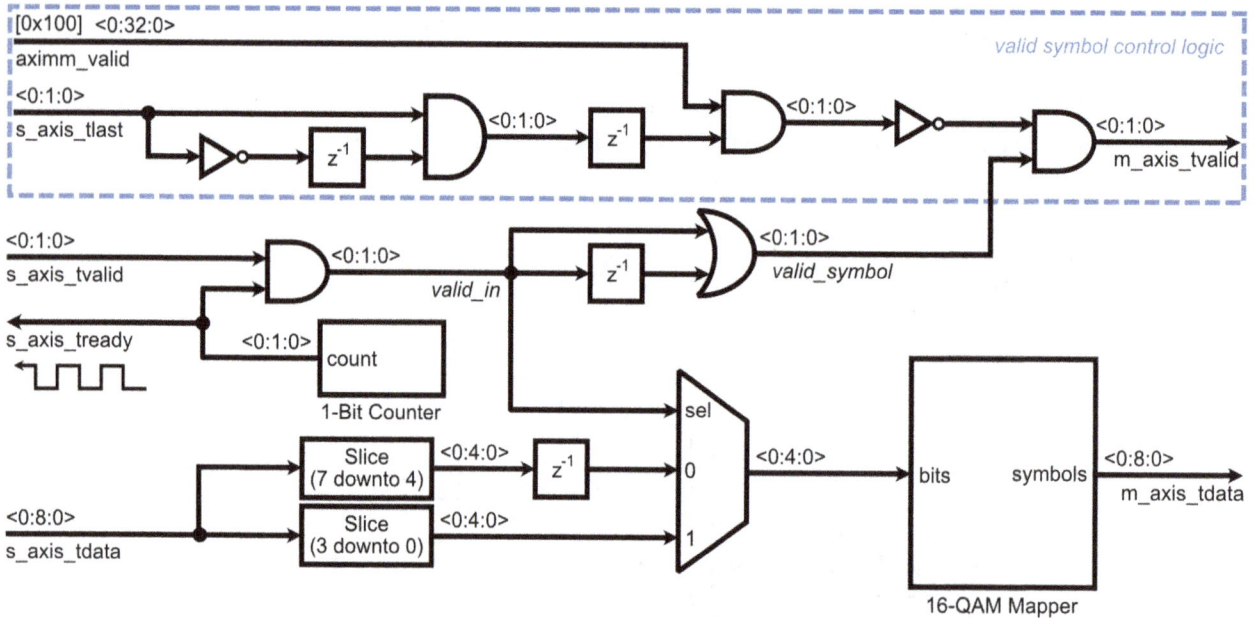

Figure 15.11: Top-level block diagram of the symbol mapping IP core.

The input to the symbol mapper block is created by first splitting *s_axis_tdata* into two 4-bit words, and then time-interleaving these into one signal using a multiplexer. The selection line of the multiplexer is driven by the newly created *valid_in* signal. The multiplexer now outputs two valid 4-bit samples each time the select line transitions from one to zero. To reflect this, a further valid signal named *valid_symbol* is created using a delay block and an OR gate. This holds *valid_in* high for two samples.

The final component of the symbol mapping IP core is the valid symbol control logic that was discussed when considering LDPC codes in Section 15.2.1. The purpose of this control logic is to drive *m_axis_tvalid* low when bits do not correspond to a valid symbol. As the number of valid symbols is dependant on the LDPC code, an AXI4-Lite memory-mapped register (*aximm_valid*) allows the control logic to be updated at run-time from the PS, depending on the LDPC code that is in use. We know that only one code, DOCSIS Medium, requires this logic. Therefore, for all other codes we can set *aximm_valid* to zero, meaning that the output of the rightmost AND gate is high when *valid_symbol* is high. When DOCSIS Medium is being used, *aximm_valid* is set to one. We then utilise the knowledge that only the last 4 bits in an encoded block are not valid and use the *s_axis_tlast* signal to drive *m_axis_tvalid* low. The SD-FEC block outputs a *tlast* signal to indicate the last word of an encoded or decoded block. As we are applying back pressure and halving the input rate, *s_axis_tlast* will be two samples long. Therefore, we use a rising edge detector — made from a NOT gate, delay block and AND gate — to reduce the high width of *s_axis_tlast* to one sample and delay this so that it aligns with the very last symbol in an encoded block.

15.2.5. AWGN Channel

The next stage in the radio pipeline is our AWGN channel. Figure 15.12 shows the top level schematic of this IP core. Complex white Gaussian noise is generated and added to the real and imaginary part of each symbol, which have been obtained by separating *s_axis_tdata* into its least and most significant 4 bits. The real and imaginary components of the now noisy symbol are concatenated again into a 32-bit word before being output on *m_axis_tdata*.

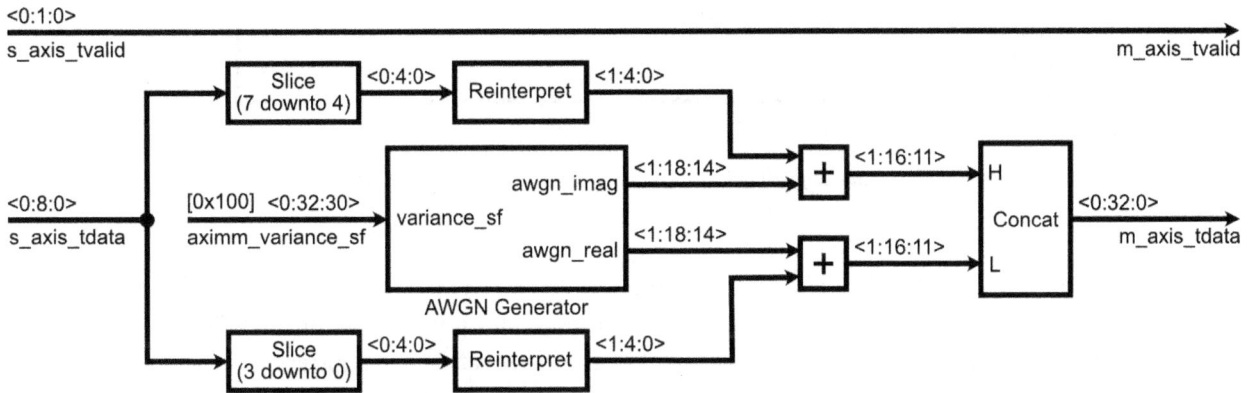

Figure 15.12: AWGN generator top level.

The white Gaussian noise is generated from two uniformly distributed random numbers using the Box-Muller transform [113]. This transform takes two sources of uniformly distributed random numbers, U_1 and U_2, of the interval [0, 1] and generates pairs of normally distributed random numbers, z_1 and z_2, with mean, $\mu = 0$, and variance, $\sigma^2 = 1$. Equations (15.1) and (15.2) illustrate how the transform is performed.

$$z_0 = \sqrt{-2 \ln U_0} \, cos(2\pi U_1) \qquad\qquad (15.1)$$

$$z_1 = \sqrt{-2 \ln U_0} \, sin(2\pi U_1) \qquad\qquad (15.2)$$

Figure 15.13 shows a hardware implementation of the Box-Muller transform.

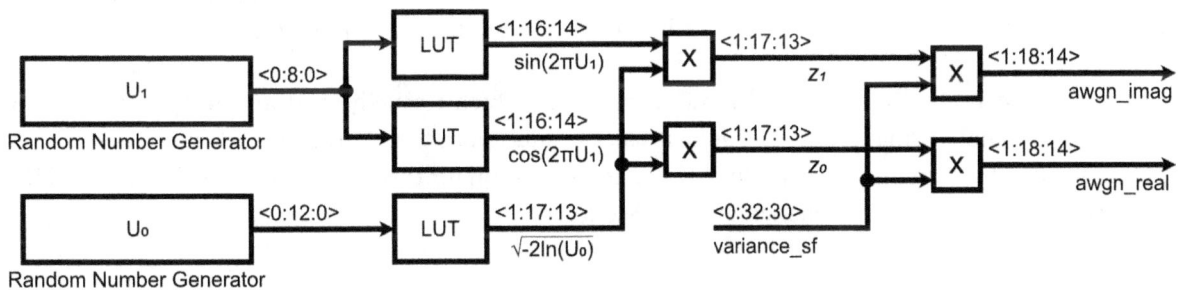

Figure 15.13: AWGN generator using Box-Muller transform.

U_0 and U_1 are generated by concatenating the 1-bit output of multiple Linear Feedback Shift Registers (LFSRs) together into larger wordlengths. LFSRs are simple circuits constructed with registers and XOR gates that produce pseudo-random outputs when clocked. An example of an LFSR is depicted in Figure 15.14.

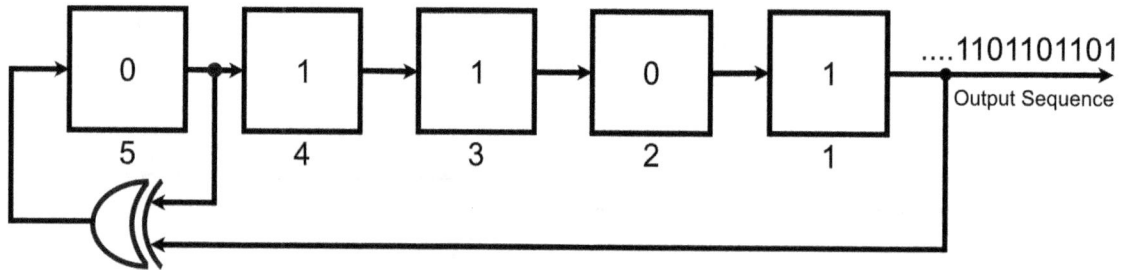

Figure 15.14: Linear feedback shift register circuit with polynomial: $z^5 + z^4 + 1$.

Some register outputs are tapped and XORed with other outputs. The position of the taps affect the output of the LFSR and are specified by a *generator polynomial*. The generator polynomial for the LFSR shown in Figure 15.5 is $z^5 + z^4 + 1$. Eventually the numbers output will begin to repeat, as the registers have a finite number of states. The time taken before repeating is called the cycle length and is determined by the generator polynomial and the initial state, or seed, of the registers. Note that the registers cannot all be initialised to zero, as this would result in a constant zero output.

For instance, the polynomial and seed employed in the LFSR in Figure 15.14 result in a cycle length of only 3. The maximum possible cycle length of an LFSR is $2^n - 1$ where n is the order of the LFSR (the number of registers). To achieve this maximum cycle length, one must use a primitive polynomial [357].

U_0 is a 12-bit number and is made from 12 independent LFSRs, and U_1 is an 8-bit number made from 8 independent LFSRs. The primitive generator polynomial used for all LFSRs is $z^{13} + z^{12} + z^{11} + z^8 + 1$ and each LFSR is randomly seeded, but ensuring the same seed is not used more than once.

Each random number is then input to a Look-up Table (LUT) that contain values of functions found in (15.1) and (15.2). Other methods exist for calculating these equations in hardware, which may be more resource efficient and accurate, but employing LUTs is the simplest to implement in practice.

The result is two normally distributed random numbers which form the real and imaginary parts of our complex noise signal. A final product stage that takes a value from the PS, *variance_sf*, allows for the noise to be scaled — thereby changing the resulting SNR. The calculation for this scaling factor will be covered in more detail when discussing the PS implementation in Section 15.3.

15.2.6. Log Likelihood Ratio (Soft Demodulation)

When performing LDPC decoding, the SD-FEC core only accepts soft bits in the form of Log Likelihood Ratios (LLRs) (see (14.10) on page 469) and so we must demodulate our noisy signal accordingly. Figure 15.15 shows the top level schematic of the LLR IP core.

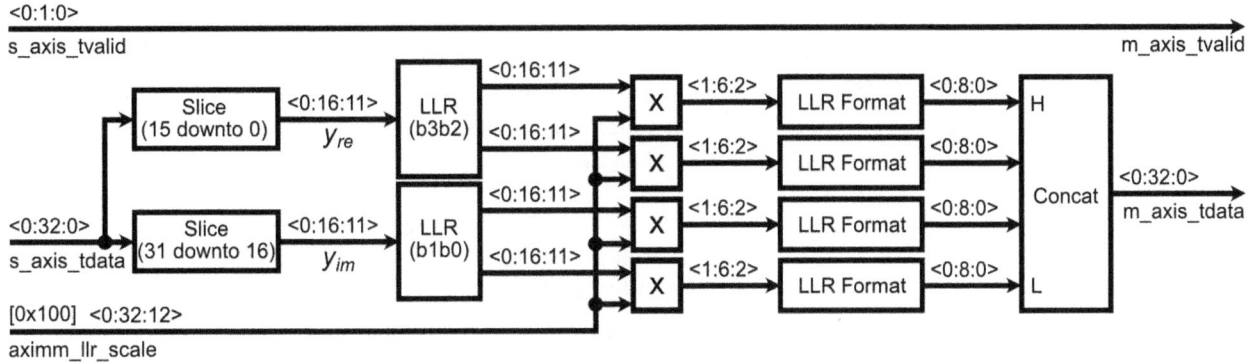

Figure 15.15: Top level block diagram of LLR IP core.

As this design only employs one modulation scheme, we can reduce the LLR calculations to a series of if-statements [311]. This greatly simplifies the hardware implementation required. Equations (15.3) to (15.6) provide the simplified calculations for obtaining LLR values for bit 3 to bit 0 of a symbol, respectively.

$$llr(b3) = \begin{cases} 2(y_{re} + 1) & : \text{if } y_{re} < -2 \\ y_{re} & : \text{if } -2 \leq y_{re} < 2 \\ 2(y_{re} - 1) & : \text{if } y_{re} > 2 \end{cases} \tag{15.3}$$

$$llr(b2) = -|y_{re}| + 2 \qquad \forall y_{re} \tag{15.4}$$

$$llr(b1) = \begin{cases} -2(y_{im} + 1) & : \text{if } y_{im} < -2 \\ -y_{im} & : \text{if } -2 \leq y_{im} < 2 \\ -2(y_{im} - 1) & : \text{if } y_{im} > 2 \end{cases} \tag{15.5}$$

$$llr(b0) = -|y_{im}| + 2 \qquad \forall y_{im} \tag{15.6}$$

The decoder SD-FEC block expects that the LLR values are signed 6-bit wordlengths with 2 fractional bits <1:6:2> that have been symmetrically saturated. Where regular saturation of this data type would be outside of the lower and upper bounds [-8, 7.75], symmetric saturation produces the lower and upper bounds [-7.75, 7.75]. If the data exceeds this range often, and is thereby incurring significant saturation, the performance of the decoder can be negatively impacted. Therefore, the LLR data can be scaled by a number less that 1 before performing the symmetric saturation, facilitated here with the AXI4-Lite register *aximm_llr_scale*.

Additionally, the LLR value should then be sign-extended to form a byte. Figure 15.16 illustrates how the LLR data is structured. The generated LLR data should first be converted to a signed fixed-point value (remembering to saturate to the range -7.75 and 7.75) with 1 sign bit, 3 integer bits and 2 fractional bits. The sign bit should then be extended to create an 8-bit fixed point value.

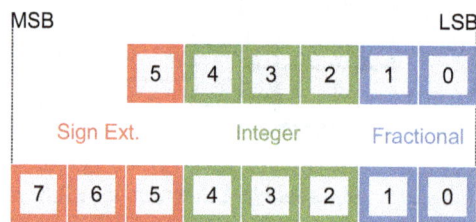

Figure 15.16: LLR data format for LDPC decoding.

The schematic shown in Figure 15.17 performs the fixed-point formatting required by the SD-FEC block when configured as a decoder. The data output from the scaling multipliers already performs regular saturation to [-8, 7.75]. A multiplexer is then used to check if the value drops below -7.75, in which case it outputs -7.75. The sign extension is achieved by slicing the most significant bit and performing a concatenation to form a byte.

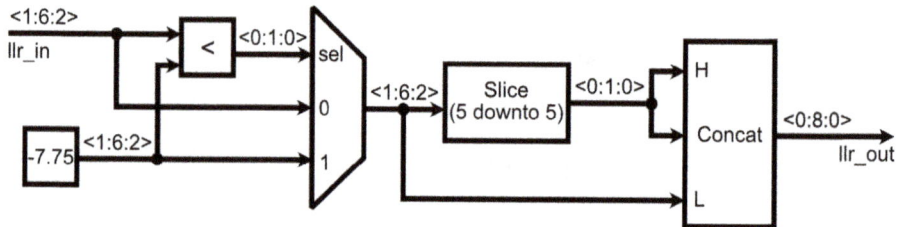

Figure 15.17: Block diagram of logic to format LLR values.

The four soft bits of a symbol, each represented by 8-bit words, are concatenated to form the data output to the decoder SD-FEC block, *m_axis_tdata*.

15.2.7. Decoder Configuration

Table 15.5 presents the SD-FEC core configuration to perform LDPC decoding using the DOCSIS 3.1 standard.

Table 15.5: SD-FEC IP core configuration for LDPC decoding.

Function		Interface		Runtime Loading	
Configuration		**Parameters (S AXI)**		Physical Utilization	N/A
Standard	DOCSIS 3.1	Interface	Runtime-Configured	Throughput Utilization	100
Turbo Decode Code Parameters		Code Parameters	N/A		
Turbo Decode	N/A	**DIN**			
Algorithm	N/A	Interface	Pre-Configured		
Scale	N/A	Lanes	1		
LDPC Decode Code Parameters		Word Configuration	Fixed		
LDPC Decode	Yes	Number of Words	4		
Support W>1	No	**DOUT**			
Code Definition	N/A	Interface	Pre-Configured		
Overrides	Disable	Lanes	1		
LDPC Encode Code Parameters		Word Configuration	Fixed		
LDPC Encode	No	Number of Words	1		
Code Definition	N/A				

This configuration is very similar to the encoder configuration, but is instead set up for decoding; it accepts four LLR values per cycle and so the *DIN* interface is configured for four words. The output of the decoder is set to one word, as this makes it easier to compare the Tx and Rx data buffers in the PS.

15.2.8. Rx Packet Creator

As noted before, the SD-FEC core outputs a *tlast* signal, indicating the last word of an encoded or decoded block. When reading data into the PS using a DMA, a high *tlast* signal coupled with a high *tvalid* signal indicates the end of a packet. Therefore, if multiple blocks are being grouped together into one larger packet to be sent, it is useful to discard the SD-FEC block's *tlast* signal and generate our own. Figure 15.18 presents some control logic to achieve this.

An AXI4-Lite register, *aximm_length*, provides a value to compare with a counter output. The counter counts the number of valid words output from the decoder. When the value of the comparator is reached, *tlast* is driven high and the counter is reset. In doing this, the number of words to be collected in the Rx buffer can be provided on *aximm_length*, and *tlast* will strobe accordingly.

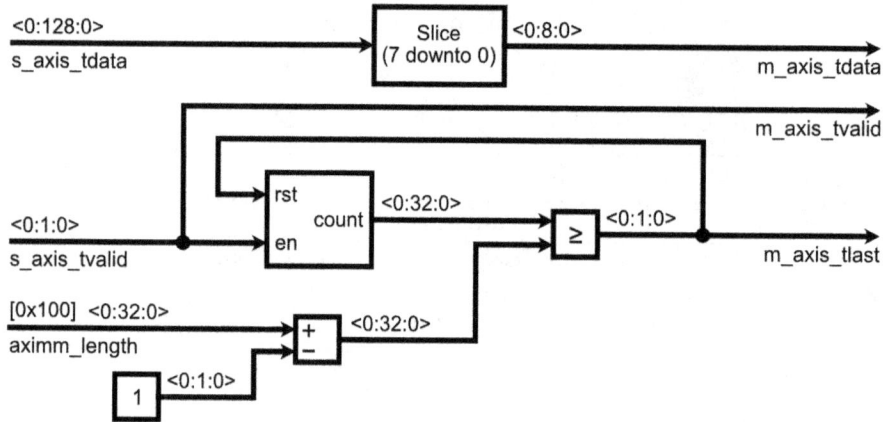

Figure 15.18: Block diagram illustrating the Rx Packet Creator IP core.

15.2.9. Vivado Block Design

Having discussed the various stages of the example design, which take the form of separate IP cores with AXI4-Lite and AXI-Stream interfaces, these should be connected up as outlined in Figure 15.3. When connecting the IP cores together in Vivado, the system has been tested where the SD-FEC IP cores are clocked at 666 MHz on *core_clk* and the rest of the design at 111 MHz. The DMAs should be configured to have 8-bit wordlengths and the FIFOs should be made large enough to accommodate the amount of data that is to be sent through the system.

Before generating the bitstream, it is important to constrain the core to achieve optimal timing results, as Vivado does not support timing-driven placement for SD-FEC instances. The code snippet in Figure 15.19 illustrates how this is achieved. The available locations are detailed in the SD-FEC Product Guide [45] in the *Placement Guidelines* table.

XDC

```
set_property LOC FE_X<x>Y<y> [get_cells */<ipinst_path>/inst/STD_OTHER_G.STD_OTHER_I/FE_I]
```

Figure 15.19: Constraining the SD-FEC instance.

15.3. Run-Time PYNQ Interaction

This section will present the Python code for configuring the hardware design, sending data to be encoded, receiving decoded data from the system, and performing BER analysis between the Tx and Rx data buffers.

A Python package for interacting with the SD-FEC core ships with PYNQ. Upon downloading an overlay to a RFSoC, if an SD-FEC core is detected in the design, a Python SD-FEC driver will bind to it. Upon binding, the Hardware Hand-off file (*.hwh) is scraped to obtain the LDPC parameters that are specified in the Vivado block design. A *.hwh file is required along with a bitstream (*.bit) when using PYNQ and contains information that PYNQ uses to identify various components of the Vivado block design.

The available LDPC parameters can then be inspected using the *available_ldpc_params()* function as shown in Figure 15.20.

Python

```python
from pynq import Overlay
from pynq import allocate
import numpy as np
import xsdfec
import math

ol = Overlay('ldpc_coding.bit')
ol.sd_fec_enc.available_ldpc_params()

>>  ['docsis_short_encode',
     'docsis_medium_encode',
     'docsis_long_encode',
     'docsis_init_ranging_encode',
     'docsis_fine_ranging_encode']
```

Figure 15.20: Downloading the overlay to the board and inspecting available LDPC codes.

The supported DOCSIS 3.1 standard that the SD-FEC blocks are configured with has five LDPC codes. We can load the code parameters into the SD-FEC block using the *add_ldpc_params()* function. This function takes five arguments: code ID, scale (SC) table offset, layer (LA) table offset, Quasi-Cyclic (QC) table offset, and the name of the LDPC code. The core has some configurable shared memory for holding the LDPC table parameters and can contain up to 128 different codes at once.

In the code snippet in Figure 15.21, a function is created that takes an SD-FEC core as an argument and ascertains the available LDPC codes. A loop then adds each of the codes to the SD-FEC block's internal memory. The table offsets are obtained using the *share_table_size()* function. The names provided must match one of the available LDPC codes as the function takes the code information provided in the *.hwh and packs it into a C struct to be used with the Bare-metal C code that the SD-FEC Python driver wraps around.

Python

```python
def add_all_ldpc_params(fec):
    ldpc_params = fec.available_ldpc_params()

    fec.CORE_ORDER = 0          # Maintain the order of blocks between input and output
    fec.CORE_AXIS_ENABLE = 0    # Disable all channels

    sc_offset = 0
    la_offset = 0
    qc_offset = 0
    for code_id in range(len(ldpc_params)):
        code_name = ldpc_params[code_id]
        table_sizes = fec.share_table_size(code_name)

        fec.add_ldpc_params(code_id, sc_offset, la_offset, qc_offset, code_name)

        sc_offset += table_sizes['sc_size']
        la_offset += table_sizes['la_size']
        qc_offset += table_sizes['qc_size']

fec.CORE_AXIS_ENABLE = 63        # Enable all channels
```

Figure 15.21: Python function to loop through available LDPC parameters and add them to the SD-FEC core.

Using this function, the LDPC parameters are loaded into the encoder and decoder SD-FEC blocks as demonstrated in Figure 15.22.

Python

```python
fec_enc = ol.sd_fec_enc
fec_dec = ol.sd_fec_dec
add_all_ldpc_params(fec_enc)
add_all_ldpc_params(fec_dec)
```

Figure 15.22: Add all available parameters to encoder and decoder SD-FEC cores.

Information about the codes is stored in a dictionary. Figure 15.23 illustrates how to use the dictionary to obtain information for a given code.

Python

```python
code_name = 'docsis_short_encode'
code_id = fec_enc.available_ldpc_params().index(code_name)

n = fec_enc._code_params.ldpc[code_name]['n']
k = fec_enc._code_params.ldpc[code_name]['k']
p = fec_enc._code_params.ldpc[code_name]['p']

print('Block Length (bits): %s\nInformation Bits: %s\nSub-Matrix Size: %s' % (n, k, p))

>>  Block Length (bits): 1120
    Information Bits: 840
    Sub-Matrix Size: 56
```

Figure 15.23: Accessing the Python dictionary containing LDPC code parameters.

The dictionary shows that for the DOCSIS Short LDPC code, the encoded block length in bits (n) is 1120, of which 840 are information bits (k). These values will be useful when configuring our hardware design for a given code. The variable *code_id* identifies each of the LDPC codes that have been loaded into the SD-FEC blocks. As these codes were loaded in the order given by *available_ldpc_params()*, we can obtain the ID by indexing the output of this function with the name of the desired code. The names given to the codes used by the encoder and decoder are not the same, however the order of the codes is the same, and therefore we only need to index the LDPC parameters of one of the SD-FEC cores and use the same code ID for both the encoder and decoder.

The code ID forms part of the control word that is sent to the SD-FEC block as detailed in Table 15.4. The control word comprises a number of fields and so it is worthwhile creating a function that takes in values for the fields and uses these to construct the code word. Figure 15.24 presents such a function. It accepts a dictionary containing the control word fields and their values as an argument. An additional dictionary is created inside the function with values initialised to zero for each of the fields, meaning that only fields with values that differ from this need included in the input argument. The control word is constructed by converting the field values into binary strings of the exact length of each field, as stipulated in Table 15.4. These values are then concatenated together in the correct order before being converted and returned as an integer.

Python

```python
def create_ctrl_word(ctrl_params_input):
    ctrl_params = {'id' : 0,
                   'max_iterations' : 0,
                   'term_on_no_change' : 0,
                   'term_on_pass' : 0,
                   'include_parity_op' : 0,
                   'hard_op' : 0,
                   'code' : 0}

    for key in ctrl_params_input:
        ctrl_params[key] = ctrl_params_input[key]

    id = '{0:08b}'.format(ctrl_params['id'])                              # (31:24) uint8
    max_iterations = '{0:06b}'.format(ctrl_params['max_iterations'])      # (23:18) uint6
    term_on_no_change = '{0:01b}'.format(ctrl_params['term_on_no_change']) # (17:17) bit1
    term_on_pass = '{0:01b}'.format(ctrl_params['term_on_pass'])          # (16:16) bit1
    include_parity_op = '{0:01b}'.format(ctrl_params['include_parity_op']) # (15:15) bit1
    hard_op = '{0:01b}'.format(ctrl_params['hard_op'])                    # (14:14) bit1
    reserved = '{0:07b}'.format(0)                                        # (13:7) uint7
    code = '{0:07b}'.format(ctrl_params['code'])                          # (6:0) uint7

    ctrl_word_bin = id + max_iterations + term_on_no_change + term_on_pass \
    + include_parity_op + hard_op + reserved + code

    return int(ctrl_word_bin,2)
```

Figure 15.24: Function to create control word given a dictionary containing parameters and values.

The two SD-FEC controller IP cores, designed in Section 15.2.3, can be configured by using this function to create the control word. Figure 15.25 details the Python code for setting up the controller IP cores.

Python

```python
reg_ctrl = 0x100
reg_len = 0x104

# Configure Encoder
enc_ctrl_params = {'code' : code_id}
enc_ctrl_word = create_ctrl_word(enc_ctrl_params)
data_len = int(k/8)

ol.fec_ctrl_enc.write(reg_ctrl, enc_ctrl_word)
ol.fec_ctrl_enc.write(reg_len, data_len)

# Configure Decoder
dec_ctrl_params = {'max_iterations' : 32,
                   'term_on_no_change' : 1,
                   'term_on_pass' : 1,
                   'include_parity_op' : 0,
                   'hard_op' : 1,
                   'code' : code_id}
dec_ctrl_word = create_ctrl_word(dec_ctrl_params)
data_len = int(n/4)

ol.fec_ctrl_dec.write(reg_ctrl, dec_ctrl_word)
ol.fec_ctrl_dec.write(reg_len, data_len)
```

Figure 15.25: Configuring the SD-FEC controllers.

For the encoder, we need only provide the code ID, as we will not be using the status word output from the encoder, and so can omit the *External Block ID* field. The length given to the controller dictates the rate at which the controller will send the control word to the SD-FEC encoder block. As the encoder takes information bits, k, with a wordlength of 8 bits, the length given to the controller is $k/8$. The control word and length are then written to the encoder controller at their address offsets.

The decoder uses more of the fields in the control word. Again, we can omit the *External Block ID* field. The maximum number of iterations is set to 32, and both terminate-on-no-change and terminate-on-pass are set to 1, meaning that if either of these conditions are met, the decoder will output early. We only want the decoded data in the form of hard bits to compare against the input data, and do not require the parity bits be returned. The decoder is configured to accept four soft bits per cycle on its *DIN* interface, and as such the length for the decoder controller is set to $n/4$, the encoded block length divided by the number of soft bits per cycle.

The symbol mapping IP core from Section 15.2.4 requires that a boolean *valid* signal is set in order to track the valid symbols. If the encoded block length divides evenly by 8 (the wordlength output by the encoder), then a value of zero is written to the register. If, however, the encoded block length produces a remainder when divided by 8, a value of one is written to the register. Figure 15.26 shows the symbol mapping IP core being configured in this manner.

```python
ol.qam_mapping.write(0x100, int(n % 8 > 0))
```

Figure 15.26: Configuring symbol mapping IP core.

To configure the AWGN channel IP core, we must provide a scaling factor that, when multiplied with the real and imaginary components of the Gaussian noise signal and added to the data signal, results in a specific Signal to Noise Ratio (SNR). One way of representing SNR is in terms of signal and noise variance, as expressed in (15.7).

$$SNR = \frac{\sigma^2_{signal}}{\sigma^2_{noise}} \tag{15.7}$$

SNR is the variable we would like to change, noise variance is our unknown, and we can calculate the signal variance using the formula for variance of a complex signal, i.e.

$$\sigma^2_{signal} = \frac{1}{N} \sum_{i=1}^{N} |(s_i - \bar{s})|^2 \tag{15.8}$$

where s_i is the value of the current symbol, $\bar{s}$ is the mean of all symbols (which for our case is zero) and N is the number of symbols, which is 16.

Using this equation, we calculate our signal variance to be 10. Rearranging (15.7), we can substitute in the value for signal variance along with a desired SNR, to obtain a value for noise variance. The variance of the Gaussian signal produced using the Box-Muller Transform is 1. Therefore our scaling factor is the square root of our desired variance divided by two, as our signal is complex. Figure 15.27 shows the Python code for configuring the AWGN channel IP core to supply the desired SNR. The variance scaling factor is multiplied by 2^{30} as only integer values can be written to AXI4-Lite registers from Python. We know, however, that the IP core will interpret this as an unsigned 32-bit number with 30 fractional bits. Therefore we bit shift our value 30 places to the left.

Python

```python
snr = 7                                    # desired SNR

signal_var = 10                            # variance of signal
noise_var = signal_var / 10**(snr/10)      # variance of noise
var_scale = math.sqrt(noise_var/2)         # calculate the variance scaling factor
var_scale = int(var_scale * pow(2,30))     # fixed-point <0:32:30>

ol.awgn_channel.write(0x100, var_scale)    # write the variance scaling factor
```

Figure 15.27: Configuring AWGN channel IP core.

The scaling of the LLR values prior to symmetric saturation can be configured by writing the bit-shifted scaling value to the appropriate register, as shown in Figure 15.28.

Python

```python
llr_scale = 1
aximm_llr_scale = int(np.uint32(llr_scale * pow(2,12))) # <0:32:12>

ol.soft_demodulation.write(0x100, aximm_llr_scale)
```

Figure 15.28: Configuring LLR IP core.

The scaling factor here should be modified depending on the SNR. Lower SNRs will mean that the conditions from (15.3) to (15.6) may exceed the bounds [-7.75, 7.75], and so scaling by a number less than one can reduce or prevent any saturation from occurring.

Having configured most of the IP cores in our system, all that is left to do is create the Tx and Rx data buffers for sending data to be encoded, and receiving the decoded data. The code snippet in Figure 15.29 shows how this is achieved. As the number of data blocks can be varied here, and therefore the size of the data buffers, this is also the where the packet creator IP core will be configured.

Python

```python
num_blocks = 6                          # Number of data blocks to be encoded
K = int(k/8)                            # Number of words in one data block
total_words = num_blocks * K            # Total number of words in a data buffer

# Generate random input data (Tx)
tx_buffer = allocate(shape=(total_words,), dtype=np.uint8)
for i in range(len(tx_buffer)):
    tx_buffer[i] = np.random.randint(0,256)     # Random numbers of the interval [0,256)

# Setup Rx
ol.packet_creator.write(0x100, total_words)
rx_buffer = allocate(shape=(total_words,), dtype=np.uint8)

# Perform transaction
ol.axi_dma_rx.recvchannel.transfer(rx_buffer)
ol.axi_dma_tx.sendchannel.transfer(tx_buffer)
ol.axi_dma_tx.sendchannel.wait()
ol.axi_dma_rx.recvchannel.wait()
```

Figure 15.29: Transmitting random data to be encoded and receiving decoded data.

The number of blocks to be transferred can be changed. From this, the size of the data buffers can be calculated by first calculating the number of words in one data block, and then multiplying by the desired number of data blocks. This value of *total_words* is used to create two data buffers using the *pynq.allocate()* function, which allocates contiguous memory and returns a *pynq.buffer* object where data can be written to or read from.

Random data is generated using the *numpy.random.randint()* function. The two arguments given indicate the minimum and maximum bounds of the random number returned. A maximum limit of 255 is set here as our wordlength is 8 bits. The data buffers are then sent and received. It is good practice to initiate the receive transfer first as, should the system have very low latency, there is the chance that initiating the send transfer first would mean some or all samples arrive at the receive side before the receive transfer is initiated.

We now have a Tx buffer containing our original data, and an Rx buffer which contains data that has been encoded, baseband modulated, introduced to some degree of channel noise, demodulated into soft bits and decoded. We can print these buffers out and visually compare by eye to see if they are the same, although this may prove more difficult for large data buffers.

A useful metric to assess our received buffer is BER, which is the rate of bit errors that have occurred as a result of channel or other noise. Figure 15.30 provides a code listing containing two functions. The first function, *serialise_data()*, serialises our data. That is, it converts our data buffer from having a wordlength of 8 into a larger array of individual bits. The second function, *calculate_ber()*, compares two buffers of serialised data and calculates the BER.

```python
def serialise_data(data):
    hard_data = data

    hard_binary = ''
    for hd in hard_data:
        h_bin = '{0:08b}'.format(hd)
        hard_binary += h_bin[::-1] # LSB first

    return hard_binary

def calculate_ber(tx, rx):
    tx_bits = np.asarray(list(serialise_data(tx)))
    rx_bits = np.asarray(list(serialise_data(rx)))

    compare = tx_bits == rx_bits
    error_bits = (compare == False).sum()

    ber = error_bits / len(tx_bits)

    return ber
```

Figure 15.30: Calculating BER.

We expect that, as we decrease the SNR, or in other words introduce more noise, the BER should increase. By using these functions and collecting the result over a range of SNRs, we are able to plot a BER curve. This is a very useful method for visualising the effects of noise in a system for a given LDPC code or modulation scheme.

Using the design outlined in this chapter, bit error analysis can be performed for the DOCSIS LDPC codes over a range of SNRs. A plot containing three BER curves for the Short, Medium and Long LDPC codes is shown in Figure 15.31.

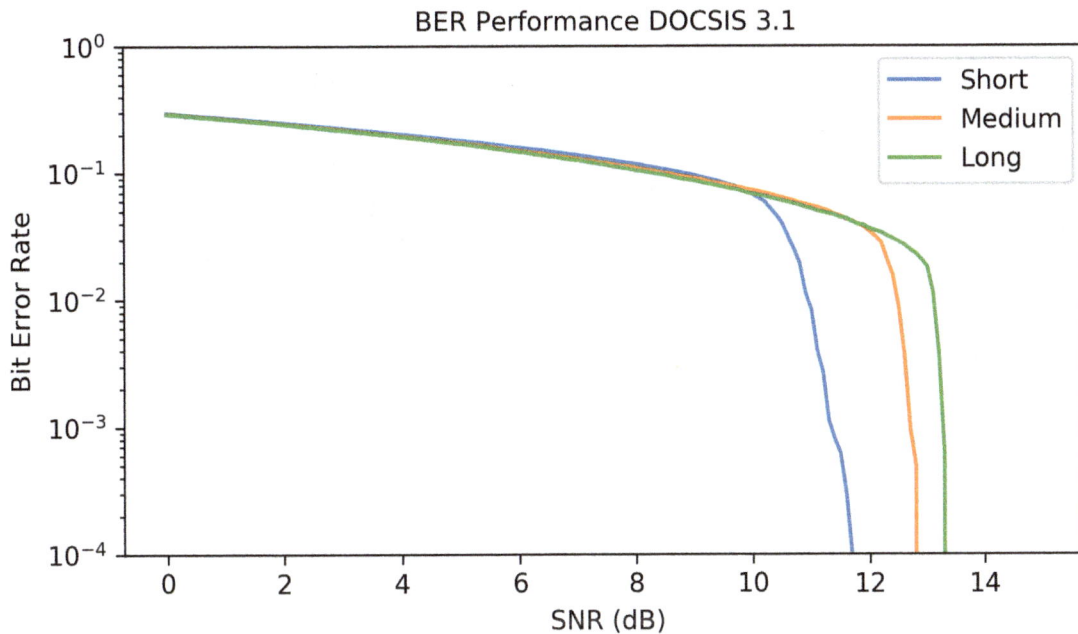

Figure 15.31: BER curves for DOCSIS Short, Medium and Long LDPC codes using 16-QAM.

15.4. Chapter Summary

This chapter has reviewed the practical use of the SD-FEC blocks on the RFSoC device, in the context of a larger system which simulates an AWGN channel with variable noise. A brief overview of the SD-FEC capabilities has been provided, alongside the IP core that is used to create an SD-FEC design. We have stepped through each stage of the hardware design, considering the impact of wordlengths and code rates, and configured the encoder and decoder accordingly. Additionally, we have seen how PYNQ can be employed to interact with the hardware design at run time, and produce BER plots for analysing the performance of various LDPC codes across a variety of SNRs.

Next, we introduce some accompanying practical notebooks that further demonstrate the use of the SD-FEC block.

Notebook Set H

Forward Error Correction

This series of notebooks investigate Forward Error Correction (FEC) in the context of RFSoC. We will use the hardened SD-FEC blocks that are integrated alongside the PL of some RFSoC devices. Although these notebooks provide some introductory material regarding FEC, much more detail of the underlying theory and practical design can be found in Chapters 14 and 15, respectively.

We begin by introducing FEC and parity check matrices, and then demonstrate how these matrices can be converted into the relevant parameters accepted by the SD-FEC block. Data is then generated in JupyterLabs and encoded using Low Density Parity Check (LDPC) codes on hardware using the SD-FEC IP core. Transmission of the encoded data is simulated in Jupyter, where the data is demodulated using soft-decision Log Likelihood Ratios (LLRs). The soft LLR data is then formatted and sent to an SD-FEC IP core, configured as a decoder, where it is decoded and hard values are returned. Finally, Bit Error Rate (BER) analysis is performed to illustrate the advantages of employing FEC.

There are five notebooks to investigate throughout this chapter using Jupyter Labs on your computer or RFSoC platform. The notebooks and their relative locations are listed as follows:

`ALL` 01_fec_first_principals.ipynb — *rfsoc_book/chapter_H/01_fec_first_principals.ipynb*

`RFSoC` 02_fec_encoding.ipynb — *rfsoc_book/chapter_H/02_fec_encoding.ipynb*

`ALL` 03_fec_channel_simulation.ipynb — *rfsoc_book/chapter_H/03_fec_channel_simulation.ipynb*

`RFSoC` 04_fec_decoding.ipynb — *rfsoc_book/chapter_H/04_fec_decoding.ipynb*

`RFSoC` 05_fec_bit_error_analysis — *rfsoc_book/chapter_H/05_fec_bit_error_analysis.ipynb*

H.1. First Principles FEC

In the first notebook, ***01_fec_first_principles.ipynb***, the reader is introduced to parity check matrices and how they are used for encoding and decoding data. The SD-FEC block supports LDPC coding, using both standardised and custom codes. This notebook will demonstrate how a parity check matrix, such as the one presented in Figure H.1, can be converted into the required YAML format that the SD-FEC IP core accepts, which enables the use of custom codes. YAML is a data-orientated language that is often used for application configuration files [364] .

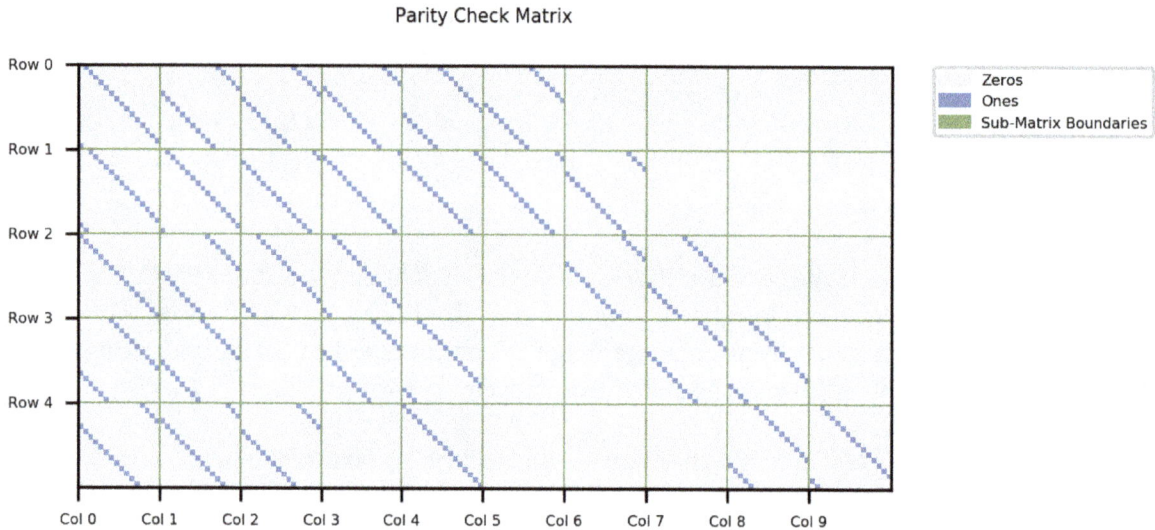

Figure H.1: *Parity check matrix and its sub-matrices.*

H.2. Using the Hardened SD-FEC Block for Encoding

The ***02_fec_encoding.ipynb*** notebook introduces the RFSoC's SD-FEC IP core and demonstrates how it can be configured to operate as an LDPC encoder. Figure H.2 shows the loop-back configuration of the SD-FEC core operating in the RFSoC. Notice that various DMAs are required to transfer data between the SD-FEC encoder and JupyterLabs.

The design is very simple and is intended to demonstrate how data input to the SD-FEC core should be formatted for Non-5G NR LDPC encoder configuration. In a similar vein, this notebook shows how data output from the SD-FEC core should be interpreted. Both of these aspects, formatting and interpreting, are achieved by setting up four buffers for transmission of data and control/status registers. The size of the data buffers (Tx, Rx) depend on the LDPC code employed. Multiple LDPC codes can be loaded into the SD-FEC block's internal memory and are easily switched between, using a control word.

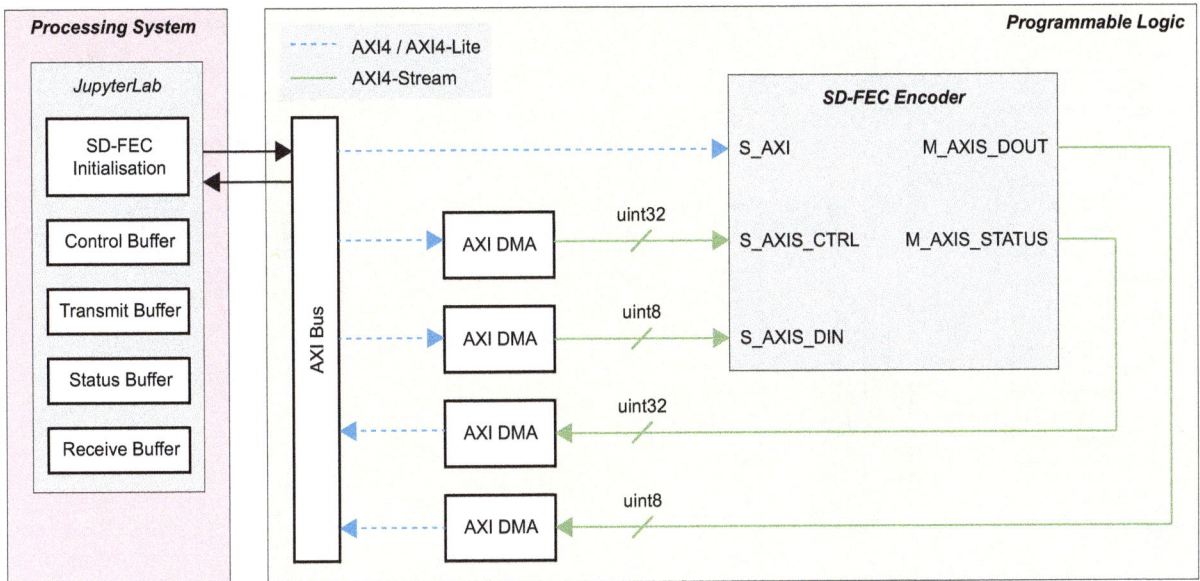

Figure H.2: Functional block diagram illustrating the loop-back implementation of the SD-FEC encoder.

H.3. Communications Channel Simulation

The data that was encoded by the SD-FEC core in the previous notebook now needs to be modulated using a digital communications scheme. The modulated data should then be transmitted through a communications channel. To begin modelling and simulating a communications channel, open the notebook named *03_fec_channel_simulation.ipynb*. This notebook applies Additive White Gaussian Noise (AWGN) to the signal to introduce noise and simulate a simple communications channel.

In this notebook, we will also highlight the type of data that the SD-FEC expects when configured as a decoder. We will specify the LLR values, which are soft decisions made during the demodulation process. A soft demodulator will be applied to the noisy encoded signal.

H.4. Using the Hardened FEC Block for Decoding

The decoding notebook in *04_fec_decoding.ipynb* follows a very similar process to the encoder, in terms of initialising the hardware and setting up the required data buffers. The block design is illustrated in Figure H.3. A significant difference between the decoder and encoder is the fixed-point representation of the data input to the decoder (fixed point formats are introduced in Section 4.4). The decoder expects the data input to be a signed 6 bit number with 2 fractional bits. This fixed-point number is symmetrically saturated to a range of -7.75 to 7.75 and then sign-extended to 8-bits. This notebook demonstrates how data communication is achieved using the LLR values acquired from the previous notebook.

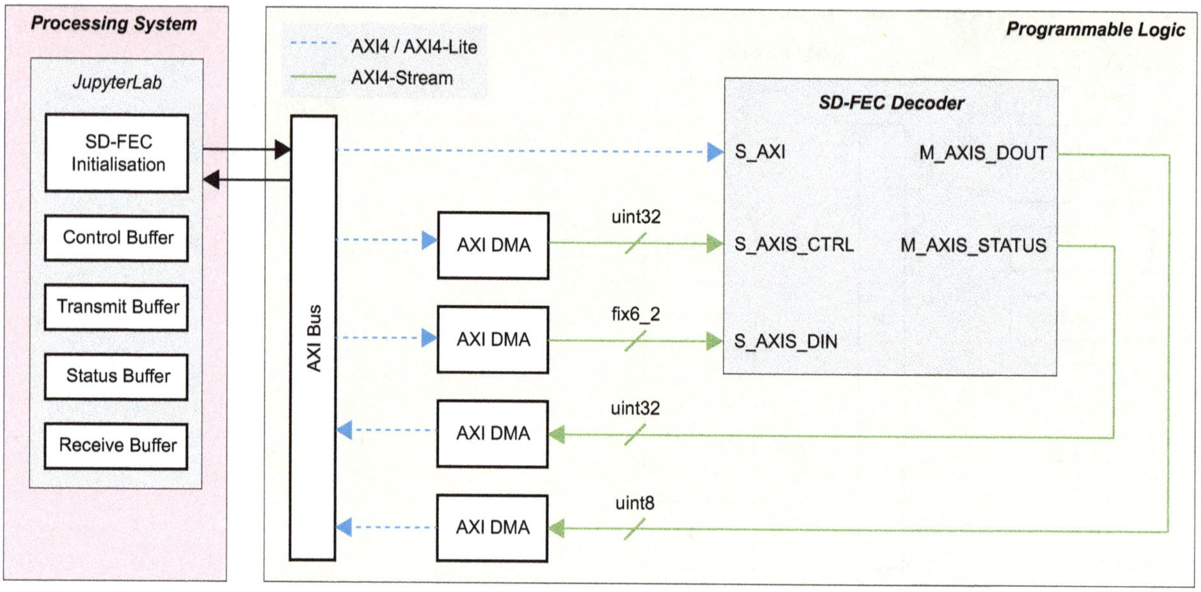

Figure H.3: Functional block diagram illustrating the loop-back implementation of the SD-FEC decoder.

H.5. Bit Error Rate and Analysis

The final notebook named ***05_fec_bit_error_analysis.ipynb*** will leverage the interactive JupyterLab environment to generate BER graphs, of similar form to Figure 15.31. These plots allow us to analyse the performance of various LDPC codes across a range of channel SNRs.

Chapter 16

OFDM: Orthogonal Frequency Division Multiplexing

Douglas Allan

This chapter introduces Orthogonal Frequency Division Multiplexing (OFDM), and provides the theoretical background to the *RFSoC OFDM* demonstration Jupyter notebooks that accompany this book. An overview of these examples is provided immediately following this chapter.

OFDM is a digital multi-carrier modulation method which allows for very efficient, reliable transmission and reception of data over wireless multipath channels. As such, it has become the modulation scheme of choice for a variety of wireless communications technologies and standards, including 4G LTE, 5G NR, Wi-Fi and both digital audio and video broadcasting, to name a few.

During transmission through the channel, a radio signal may experience non-linear gain across the frequency band that it occupies[1]. This is particularly true for wide-bandwidth signals, which can be subject to a complicated response across the occupied frequency band. In this case, the channel is said to be 'frequency selective' because the range of frequency components present within the signal experience different gains. Normally it is desirable to equalise the channel response (to achieve approximately linear gain across the entire signal bandwidth) by applying a compensating frequency response within the receiver. This equalising response is computed by measuring and adapting to the channel environment. Equalisation can be particularly difficult when the channel has a complicated frequency response, which changes over time (i.e. is 'time varying').

---

1. This effect can be attributed to *multipath propagation*, which is introduced in Figure 16.1.

OFDM addresses this problem by dividing the wideband, frequency selective channel into several parallel sub-channels. Each of these sub-channels is sufficiently narrow to ensure that they individually experience 'flat fading', meaning that the response across a sub-channel is a constant gain, or a simple linear response. As a result, sub-channels can be individually equalised using a very simple compensating response. The use of sub-channels significantly reduces the overall complexity of equalising a time varying multipath channel.

Figure 16.1 contrasts these two approaches — compare the frequency response of the wide-bandwidth signal, with the responses of individual sub-channels in an OFDM signal.

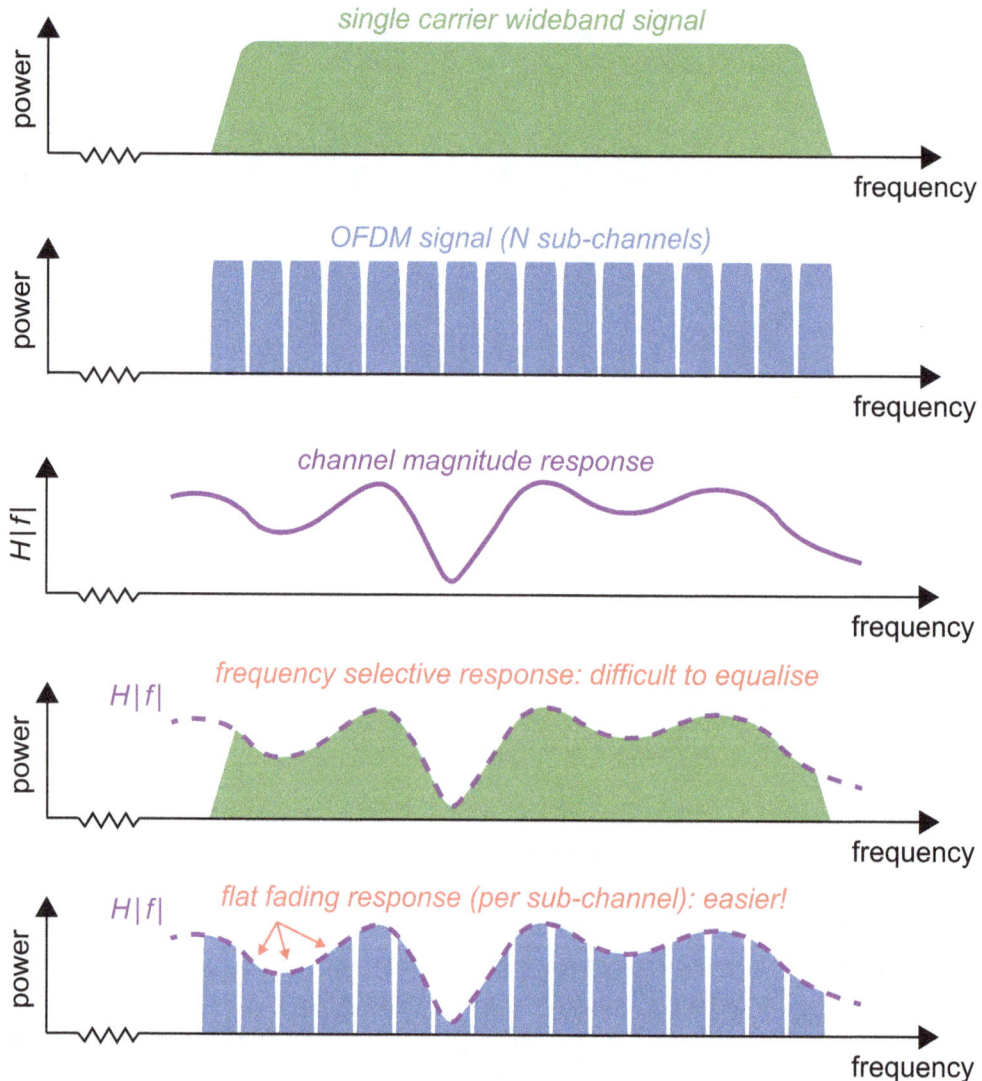

Figure 16.1: Frequency response comparison, highlighting use of sub-channels.

The data transmitted on each sub-channel modulates a different sub-carrier before being combined to form the final transmitted signal. The sub-carrier frequencies are orthogonal which allows the interfering sub-channels to be separated at the receiver, leading to improved spectral efficiency when compared to non-orthogonal *Multi-Carrier Modulation (MCM)*. The use of orthogonal sub-carriers is equivalent to the *Inverse Discrete Fourier Transform (IDFT)*, which means that the modulation and demodulation processes can be implemented efficiently using the *Fast Fourier Transform (FFT)* algorithm. In addition, the *Cyclic Prefix (CP)* maintains the orthogonality of sub-carriers in the multipath channel, provides a mechanism to prevent channel induced *Inter Symbol Interference (ISI)* and facilitates the *one-tap equaliser*.

16.1. Motivation for OFDM

In a wireless channel, the transmitted signal is reflected, refracted, diffracted and scattered by objects in the signal path. This leads to several delayed and scaled versions of the signal arriving at the receiver through multiple different paths. An example channel is shown in Figure 16.2, comprising a direct Line-of-Sight (LoS) path, and two Non-Line-of-Sight (NLoS) paths. The NLoS paths arise because some signal components are reflected by objects located between the transmitter and receiver.

Each resolvable multipath component can be modelled as having a time-varying complex amplitude (consisting of magnitude and phase components) and a time varying delay. This is expressed as

$$h_i(t) = a_i(t)e^{j\phi_i(t)}(t - \tau_i(t)),$$ (16.1)

where t denotes time, $a_i(t)$ is the time-varying amplitude of component i, $\phi_i(t)$ is the time-varying phase rotation associated with component i, $\tau_i(t)$ is the time varying delay of component i and $h_i(t)$ is the ith multipath component.

A resolvable multipath component is associated with one or more objects in the signal path. In general, each resolvable component comprises a large set of unresolvable components [176]. The time difference between the first and last significant resolvable components is the *delay spread*, denoted as d_s and measured in seconds.

The constructive and destructive interference of multipath components leads to a frequency selective channel, with different frequencies experiencing differing amounts of attenuation. The bandwidth over which the channel frequency response is approximately correlated is called the *coherence bandwidth*, denoted as B_c and measured in *Hz*. The delay spread and coherence bandwidth are approximately inversely proportional,

$$d_s \approx \frac{1}{B_c}.$$ (16.2)

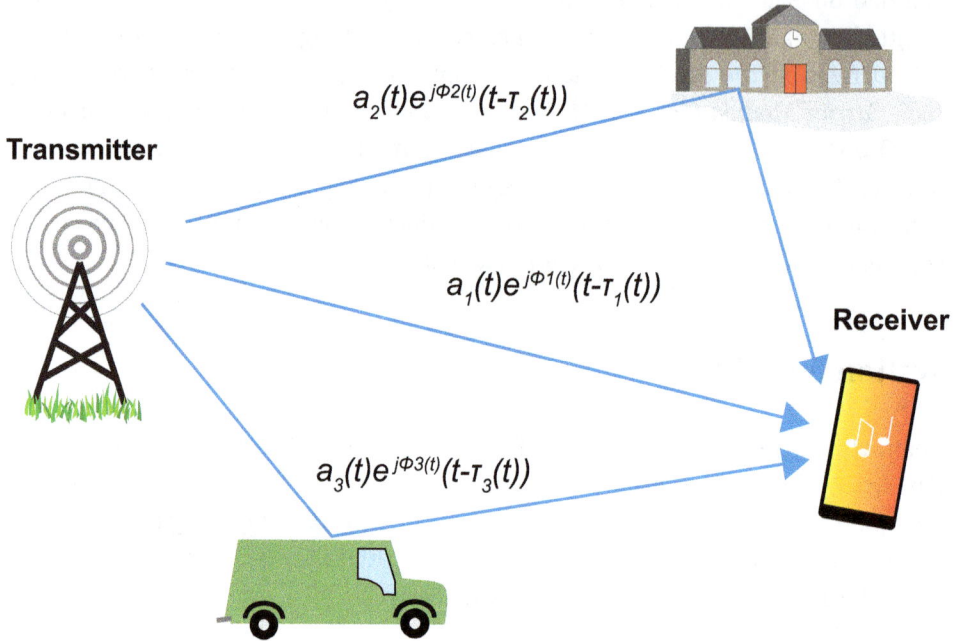

$a_2(t)e^{j\Phi2(t)}(t-\tau_2(t))$

Transmitter

$a_1(t)e^{j\Phi1(t)}(t-\tau_1(t))$

Receiver

$a_3(t)e^{j\Phi3(t)}(t-\tau_3(t))$

Figure 16.2: Illustration of a multipath channel.

The variation of signal power resulting from the multipath channel is known as *fading*. Depending on the characteristics of the transmitted signal, the channel can be classified as either *frequency selective fading* or *flat fading*. The channel is flat fading when the symbol period, T_{sym}, is long compared to d_s, i.e.

$$T_{sym} \gg d_s. \tag{16.3}$$

From a frequency domain point of view, the channel is flat fading when the signal bandwidth B is small compared to B_c, i.e.

$$B \ll B_c. \tag{16.4}$$

Due to the fact that the fading is correlated across all frequencies, and there is limited ISI between consecutive symbols, a flat fading channel is relatively easy to equalise in the receiver.

The channel is frequency selective when T_{sym} is small compared to d_s, i.e.

$$T_{sym} \ll d_s, \tag{16.5}$$

or equivalently in the frequency domain, where B is large compared to B_c,

$$B \gg B_c. \tag{16.6}$$

In contrast to the flat fading channel, in a frequency selective channel, different signal frequencies experience different amounts of fading and ISI is more significant, which makes equalising the channel more difficult. However, for high data rate communications in channels with significant delay spread, it becomes necessary to equalise a frequency selective channel.

If the transmitter and/or receiver and/or objects in the signal path are moving relative to one another, the channel varies with time. The channel *coherence time*, T_c, is defined as the time over which the channel response remains effectively constant. As the velocity of the transmitter and/or receiver increases, T_c decreases. The relative motion also generates a *Doppler shift* which differs for each multipath component due to the fact that each component arrives at the receiver from a different angle. The range of Doppler shifts associated with the multipath components is known as the *Doppler spread*, B_D, which is measured in *Hz*. T_c and B_D are approximately inversely proportional,

$$B_D \approx \frac{1}{T_c}. \qquad (16.7)$$

If $T_{sym} \ll T_c$, or equivalently if $B \gg B_D$, the channel is *slow fading*. Conversely, if $T_{sym} \gg T_c$ or equivalently $B \ll B_D$, the channel is characterised as *fast fading* [319].

In many cases, a channel will be *doubly dispersive* (i.e. will be characterised as either flat or frequency selective **and** either fast or slow fading). The term doubly dispersive refers to the fact that the channel is both *time dispersive* (due to the various delayed multipath components) and *frequency dispersive* (due to the Doppler shifts of the various multipath components).

At baseband, the multipath channel can be modelled as a discrete time FIR filter with complex coefficients of the form shown in (16.1). As such, the channel output is the *linear convolution* of the input signal and the channel *impulse response*,

$$x[n] = u[n]*h[n] \qquad (16.8)$$

where $x[n]$ is the discrete time output signal, $u[n]$ is the input signal, $h[n]$ is the complex impulse response, * denotes convolution and n is the sample index.

Figure 16.3 shows an illustration of the baseband channel filter. If there is no LoS component, the coefficients are drawn from a zero mean complex normal distribution, and the channel is said to be *Rayleigh fading*. Conversely, if there is a LoS component, the coefficients follow a complex normal distribution with non-zero mean and hence the channel is classed as *Rician fading*.

In the receiver, it is necessary to implement an equaliser to compensate for the effects of the channel. In single carrier systems such as QPSK and QAM, equalisers are commonly implemented in the time domain using *adaptive filters,* whose weights are updated using a training sequence known to both transmitter and receiver

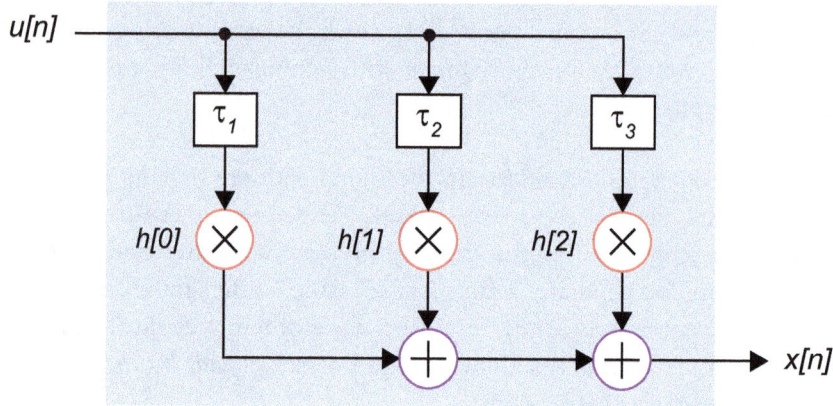

Figure 16.3: FIR filter representation of a multipath channel.

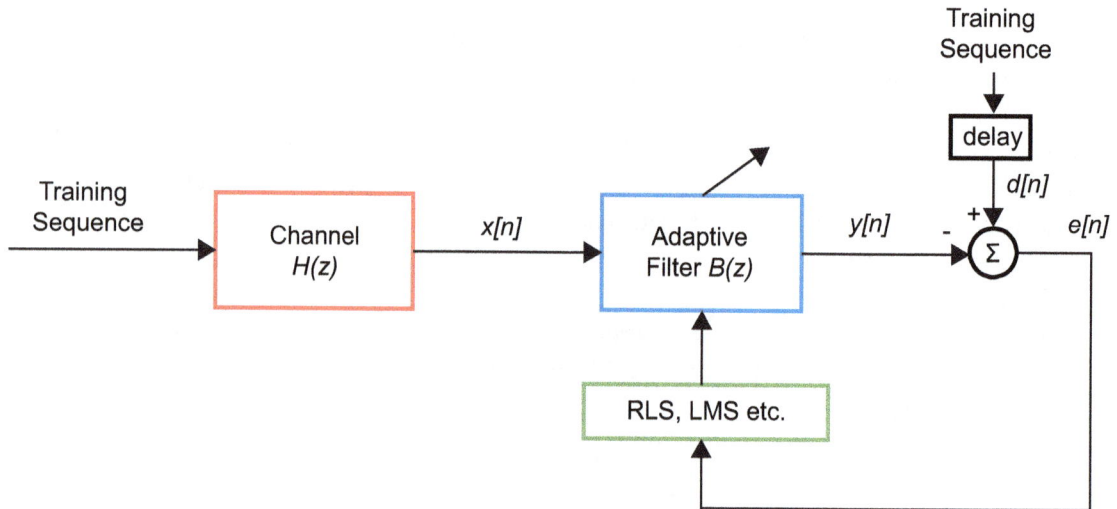

Figure 16.4: Linear Equaliser in training mode.

and an adaptive algorithm such as *Recursive Least Squares (RLS)* or *Least Mean Squares (LMS)*. Figure 16.4 shows the *Linear Equaliser* in training mode.

The adaptive algorithm (RLS, LMS etc.) converges towards a set of weights that minimises the error sequence, $e[n]$, given by

$$e[n] = d[n] - y[n], \qquad (16.9)$$

where $y[n]$ is the adaptive filter output, and $d[n]$ is the desired signal, i.e. the training sequence.

For the linear equaliser, the adaptive filter $B(z)$ would approximate the inverse of the multipath channel,

$$B(z) = H^{-1}(z). \tag{16.10}$$

In adaptive filtering terminology, the linear equaliser performs inverse system identification. The linear equaliser is also known as the *Zero Forcing* equaliser because it attempts to invert the channel and hence eliminate ISI. However, its main drawback is that it amplifies the noise at frequencies where there is a *deep fade*, i.e. great attenuation in the channel response, because it applies a large gain to compensate for the loss of signal power. As such, the linear equaliser is typically restricted to flat fading channels [248].

After the initial training mode, the equaliser is switched into *Decision Directed Mode (DDM)* to track variations in the channel between training periods. In DDM, a hard decision is made on the equaliser output, $y[n]$, to form the desired signal, $d[n]$. If the equalised symbols, $y[n]$, are close to their ideal values, $e[n]$ is small and hence the filter weights do not have to be updated significantly.

Another common form of equaliser is the *Decision Feedback Equaliser (DFE)*. The DFE uses a form of feedback filter whose input is the desired signal, $d[n]$. Figure 16.5 shows the DFE in training mode [324].

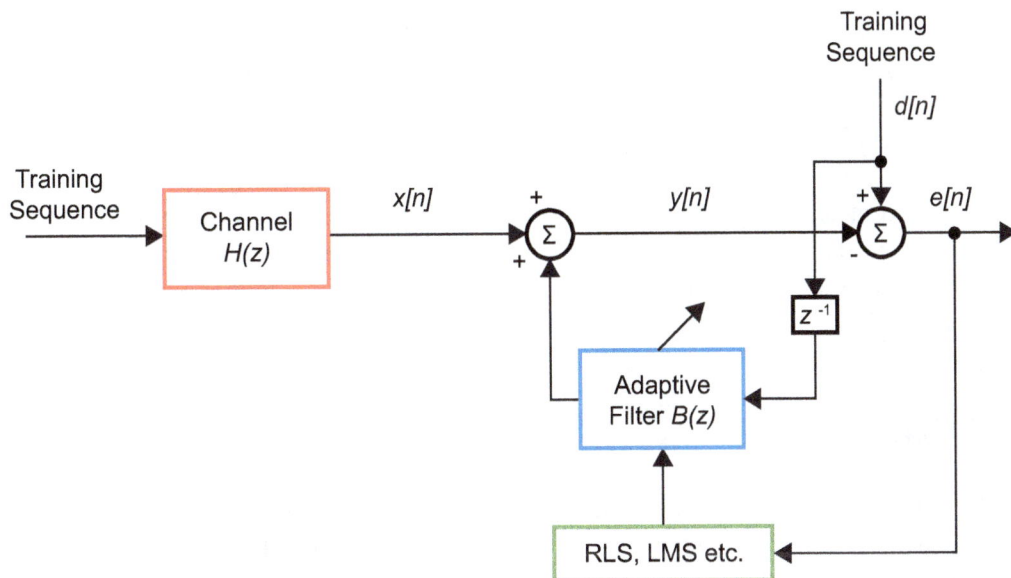

Figure 16.5: Decision Feedback Equaliser in training mode.

By feeding $d[n]$ into the adaptive filter, it converges towards the unknown system $H(z)$. Here, the adaptive filter performs system identification, rather than inverse system identification. The ideal filter response is

$$B(z) = 1 - H(z). \tag{16.11}$$

Since the DFE does not involve inverting the channel, it avoids the noise amplification problem associated with the linear equaliser and is therefore more suitable for equalising frequency selective channels. In addition, it requires fewer coefficients than its linear counterpart, which reduces the computational cost.

In general, the computational cost associated with time domain equalisation techniques such as the Linear or DFE equaliser increases with the Baud rate (or data rate) and the delay spread of the multipath channel. Moreover, as the velocity of the transmitter and/or receiver increases and coherence time decreases, the filter weights have to be updated more frequently.

In order to illustrate the computational cost with an example, consider that a QPSK signal operating at a Baud rate of R_{sym} = 10 Msym/s is passed through a multipath channel with d_s = 0.5μs. In this case, the symbol period is 5 times smaller then the delay spread, so the channel is frequency selective and ISI occurs. Assuming a DFE equaliser with 50 complex weights is used in the receiver, this amounts to $4 \times 50 \times 10$ million = 2 billion Multiply Accumulates per second (MAC/s) (assuming 4 real multipliers to perform one complex multiplication). If the Baud rate and hence bit rate were increased by a factor of 2, the cost would increase to $4 \times 50 \times 20$ million = 4 billion MACs/s. Moreover, if d_s increased, a larger number of weights would be required to properly equalise the channel, leading to an even greater computational cost.

From this simple example, it is clear that time domain equalisation becomes very expensive even for moderate data rates and delay spreads. In order to achieve R_b = 40 Mbps in a channel with d_s = 0.5 μs, the computational cost for the DFE equaliser was 4 billion MAC/s, assuming a 50 weight filter. In practice, an even longer filter would likely be required, which would further increase the computational overhead.

Since we are interested in achieving data rates on the order of 100's of Mbps (and approaching Gbps) in high delay spread environments, it is clear that more computationally efficient approaches to equalisation are necessary. This provides the motivation and rationale for the development of MCM techniques and more specifically OFDM.

16.2. Multi-Carrier Modulation

In MCM, the high-rate symbol stream is divided into several parallel low-rate streams, which each modulate a different sub-carrier. The number of sub-carriers, N, is chosen to ensure that the bandwidth of each sub-carrier is smaller than the channel coherence bandwidth, B_c. As a result of this, each sub-carrier experiences a flat fading channel and can therefore be equalised with low complexity in the receiver as illustrated in Figure 16.1. In this way, it is no longer necessary to equalise the frequency selective channel directly, reducing the overall complexity of the equalisation process. Figure 16.6 illustrates an MCM transmitter.

At the left hand side, the original symbol stream operating at R_{sym} Bd is passed into a 1-to-N serial-to-parallel converter which produces N parallel output streams, each operating at R_{sym}/N. These parallel streams are

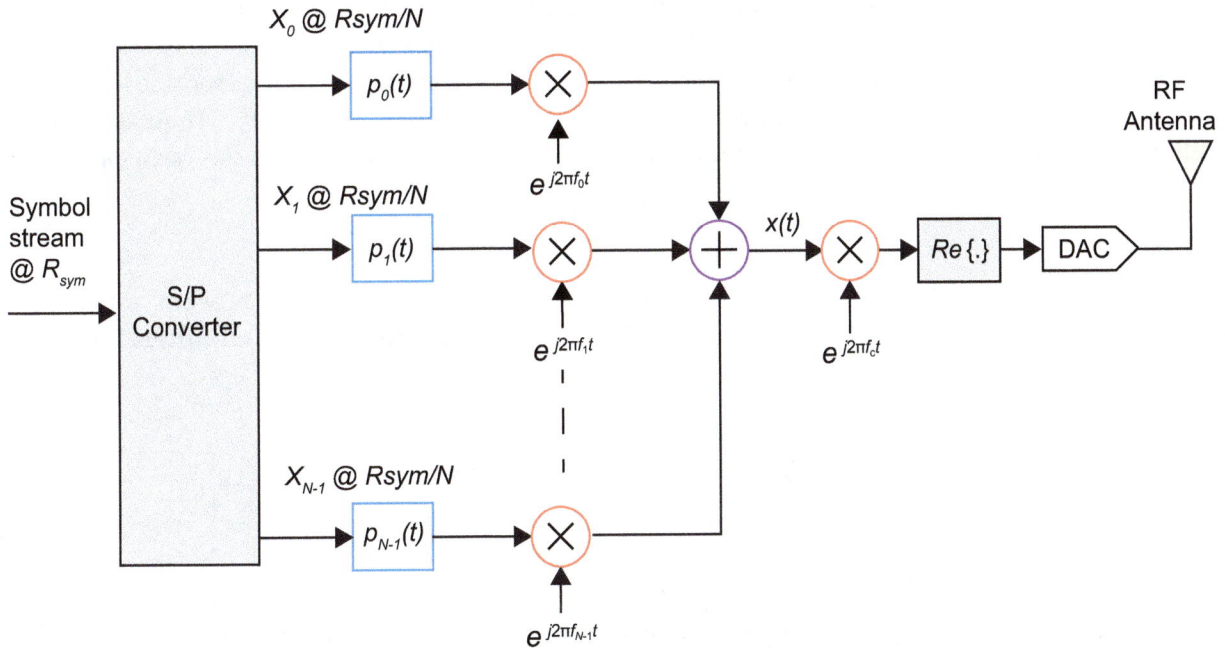

Figure 16.6: Multi-Carrier Modulation (MCM) transmitter.

then passed through pulse shaping filters $p_k(t)$ which limit their bandwidth to $B_n = R_{sym}/N$ Hz, where k is the sub-carrier index.

The streams then each modulate a different sub-carrier, $e^{j2\pi f_k t}$, where f_k is the frequency of sub-carrier k, before being summed together to produce the final output signal, $x(t)$. The $x(t)$ signal is then up-converted to Radio Frequency (RF) and transmitted across the channel.

The signal $x(t)$ can be expressed mathematically as,

$$x(t) = \sum_{k=0}^{N-1} X_k p_k(t) e^{j2\pi f_k t} \tag{16.12}$$

where x_k denotes the kth lower rate stream. In MCM, the sub-carrier spacing is chosen such that sub-carriers do not overlap or interfere in the frequency domain, allowing them to be separated successfully at the receiver. In an ideal system, the minimum sub-carrier spacing to avoid overlap is,

$$f_k = f_0 + kB_n \tag{16.13}$$

where f_k is the frequency of the kth sub-carrier. As such, the ideal system occupies a bandwidth of NB_n Hz, which is equal to the bandwidth occupied by an equivalent single-carrier operating at R_{sym} Bd.

The theoretical bandwidth of B_n Hz is impossible to achieve, however, as it would require "brick wall" pulse shaping filters with infinite length. Achieving an approximately band-limited signal requires filters with a large number of coefficients, which is prohibitively costly. Therefore, to keep the cost at a reasonable level and use filters with a practical number of weights, we cannot feasibly band-limit the signal to B_n Hz. As such, another method is used to ensure that the signals do not overlap: adding a guard band between sub-carriers,

$$f_k = f_0 + k(B_n + \alpha) \tag{16.14}$$

where α is the width of the guard band, in Hz. The necessity of adding guard bands between carriers means that the total occupied spectrum increases to $N(B_n + \alpha)$ Hz, which reduces the spectral efficiency compared to an equivalent single-carrier system.

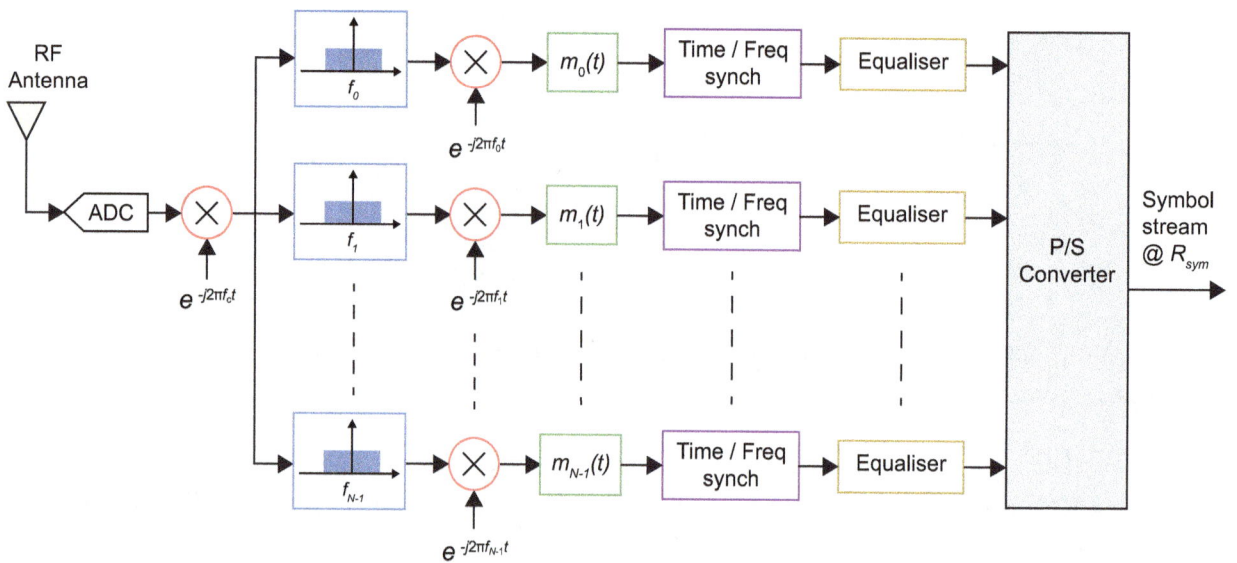

Figure 16.7: Multi-Carrier Modulation (MCM) receiver.

In the receiver, it is first necessary to recover the individual signals using a set of bandpass filters, each centred at frequencies corresponding to the set of sub-carrier frequencies. These individual signals are then demodulated using a complex oscillator and passed through a matched filter, $m_k(t)$, matched to the pulse shaping filters used in the transmitter. Finally, the signals are synchronised in time and frequency, and passed through a set of equalisation filters, before being recombined to produce the original symbol stream as shown in Figure 16.7. The requirement for N parallel bandpass filters, N complex demodulators and N separate baseband receive chains makes the generic MCM receiver very computationally complex.

It is clear that the potential benefits of the MCM scheme are outweighed by the additional overhead in terms of computation and spectrum usage.

16.3. OFDM Principles

In order to produce a more practical implementation of MCM, OFDM employs orthogonal sub-carriers. The property of orthogonality allows the sub-channels to overlap in frequency, thereby removing the need for guard-bands and expensive filters to keep the sub-channels separated. In addition, the use of orthogonal sub-carriers can be shown to be equivalent to the IDFT in the digital domain, allowing the modulation and demodulation processes to be implemented using the computationally efficient FFT algorithm.

16.3.1. OFDM Modulation and Demodulation

In OFDM, the sub-carriers are chosen to be orthogonal within a period of T_n seconds, where T_n denotes the symbol period for each sub-channel. Assuming a sampling rate of $f_s = R_{sym}$, this is equivalent to N samples. The orthogonality condition is expressed mathematically as,

$$\frac{1}{N} \sum_{n=0}^{N-1} e^{j2\pi(f_0/f_s)n} e^{-j2\pi(f_1/f_s)n} = 0 \tag{16.15}$$

where f_0 and f_1 denote two orthogonal frequencies and n is the sample index. In order to understand why this is useful, consider an OFDM system that employs two orthogonal sub-carriers. The transmitted signal $x[n]$ (before digital to analogue conversion) is given by,

$$x[n] = X_0 e^{j2\pi(f_0/f_s)n} + X_1 e^{-j2\pi(f_1/f_s)n} \tag{16.16}$$

where X_0 and X_1 denote the symbols transmitted on sub-carriers f_0 and f_1, respectively, within the period T_n. Using the relationship in (16.15), we can recover X_0 from the received mixture using the following receiver,

$$\frac{1}{N} \sum_{n=0}^{N-1} x[n] e^{-j2\pi(f_0/f_s)n}. \tag{16.17}$$

Expanding (16.17), we arrive at,

$$\frac{1}{N} \sum_{n=0}^{N-1} X_0 e^{j2\pi(f_0/f_s)n} e^{-j2\pi(f_0/f_s)n} + \frac{1}{N} \sum_{n=0}^{N-1} X_1 e^{j2\pi(f_1/f_s)n} e^{-j2\pi(f_0/f_s)n}, \tag{16.18}$$

and if we bring X_0 and X_1 to the outside of the summations, (16.18) becomes,

$$X_0 \left[\frac{1}{N} \sum_{n=0}^{N-1} e^{j2\pi(f_0/f_s)n} e^{-j2\pi(f_0/f_s)n} \right] + X_1 \left[\frac{1}{N} \sum_{n=0}^{N-1} e^{j2\pi(f_1/f_s)n} e^{-j2\pi(f_0/f_s)n} \right]. \qquad (16.19)$$

On the left-hand side of (16.19), the complex exponentials cancel to a value of 1, so the summation is simply the average of 1 over a period of N samples, which equals 1. On the right-hand side, the expression in square brackets is equivalent to (16.15). Therefore, (16.19) reduces to

$$X_0 \cdot 1 + X_1 \cdot 0 = X_0. \qquad (16.20)$$

Thus, by employing orthogonal sub-carriers, the signals can overlap or interfere in frequency and be separated at the receiver without expensive bandpass filters, unlike in generic MCM. Figure 16.8 illustrates the OFDM modulation process.

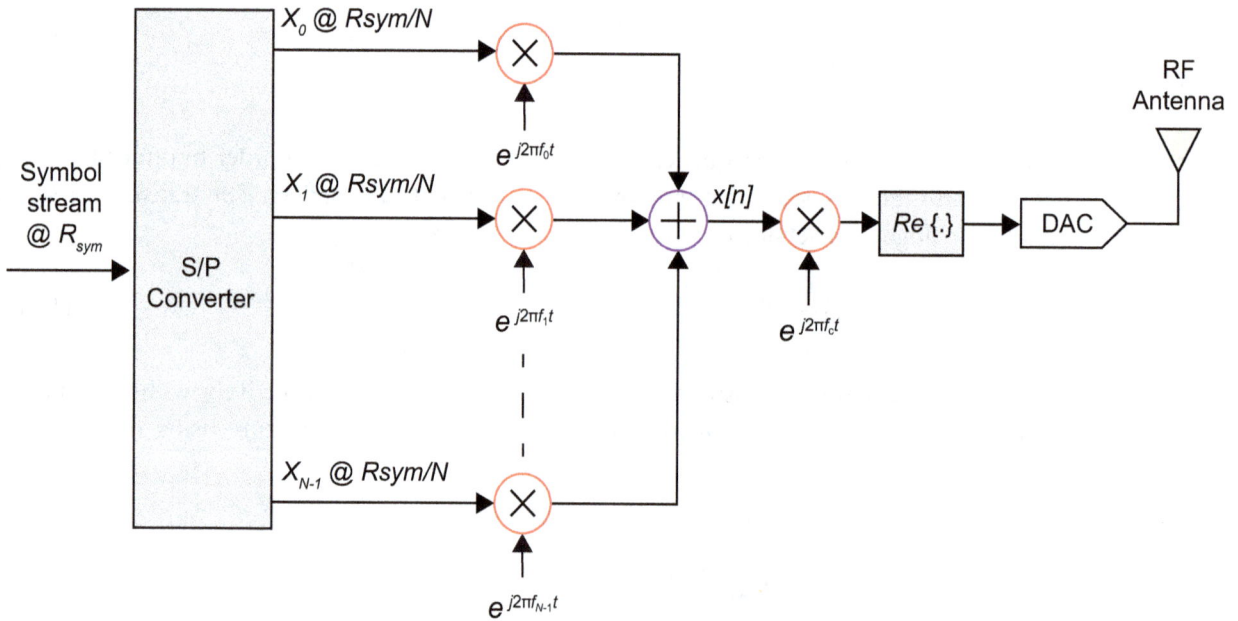

Figure 16.8: Model of an OFDM transmitter.

As the sub-carriers are orthogonal, it is not necessary to bandlimit the individual symbol streams using shaping filters, and they can simply be added together. This reduces the computational overhead significantly, and implies a rectangular pulse shape with period T_n. The use of a rectangular pulse shaping filter in the time domain results in each of the sub-carriers having a $(sin(x))/x$, or *sinc* frequency spectrum. Figure 16.9 shows the spectra of 5 OFDM sub-carriers.

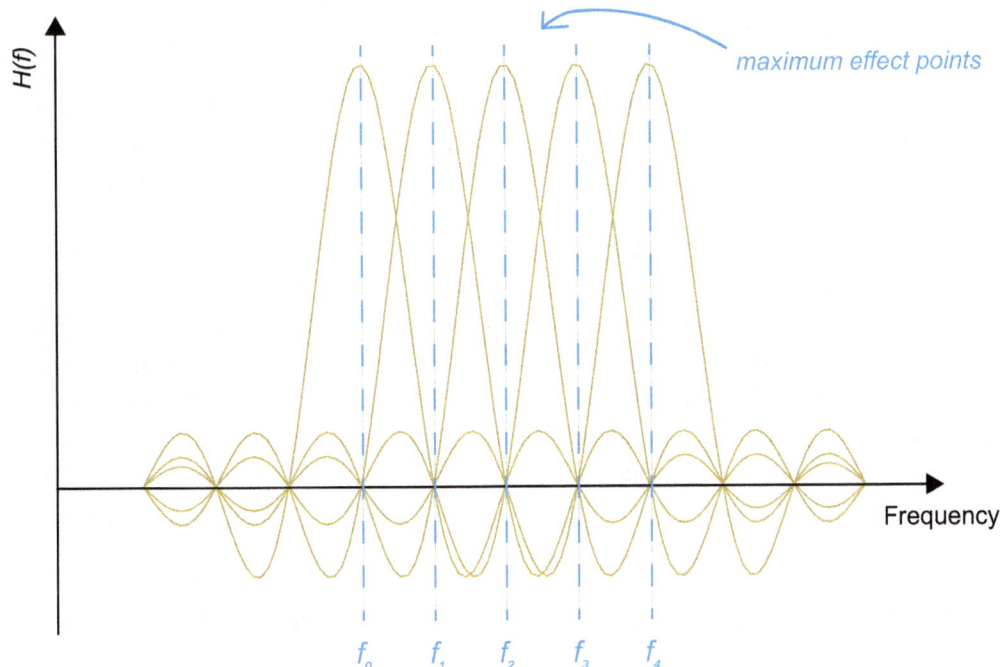

Figure 16.9: Frequency spectra of OFDM sub-carriers.

Due to the use of orthogonal frequencies, the sub-carriers do not interfere at their centre frequencies, thus allowing them to be separated at the receiver. The question remains, how to choose the sub-carrier frequencies? As it turns out, the relationship in (16.15) only holds for frequencies that have an integer number of periods within T_n [142]. Therefore, the sub-carrier frequencies are $1/T_n$ Hz and its integer multiples

$$f_k = \frac{k}{T_n} = \frac{k}{NT_s} = k\frac{f_s}{N} \, , \tag{16.21}$$

leading to a sub-carrier spacing of $\Delta f = 1/T_n$ Hz. The final OFDM signal can be expressed mathematically as

$$x[n] = \sum_{k=0}^{N-1} X_k e^{(2\pi kn)/N}. \tag{16.22}$$

Note, there is a DC (0 Hz) sub-carrier, but this is typically not used for data transmission. The expression in (16.22) is equivalent to the IDFT (minus the scaling factor), which means that OFDM modulation can be performed by taking the IDFT of a block of N complex symbols. The resulting signal, $x[n]$, is called an

OFDM symbol, and comprises the sum of N complex symbols, each modulating a different orthogonal sub-carrier. The overall data payload is carried across the wireless channel as a series of OFDM symbols.

As is well known, performing the IDFT directly requires $O(N^2)$ complex multiplications. The computational complexity can be reduced to $O((N/2)log2(N))$ by employing the Radix 2 FFT algorithm, and therefore the IFFT is used in practice [141]. As a result, the number of sub-carriers, N, is set to a power of 2.

The FFT is used to recover the transmitted symbols in the receiver. In an ideal scenario, this is equivalent to sampling the sub-carriers at the optimal frequencies, referred to as the *maximum effect points* (i.e. where there is no inter-carrier interference, and signal to noise ratio is maximised). As such, in OFDM, the transmitted pulses are sampled in the frequency domain rather than the time domain. In a realistic channel, however, frequency offsets (caused by oscillator mismatches and Doppler) cause the sub-carriers to drift from their ideal centre frequencies, meaning that the FFT does not sample them at their true maximum effect points. This leads to Inter Carrier Interference (ICI) which manifests as a form of additive noise, leading to an increase in symbol and bit errors in the receiver. However, the OFDM system can be designed to minimise the impact of ICI, as will be described later.

Now incorporating the IFFT, the OFDM transmitter can be redrawn as shown in Figure 16.10.

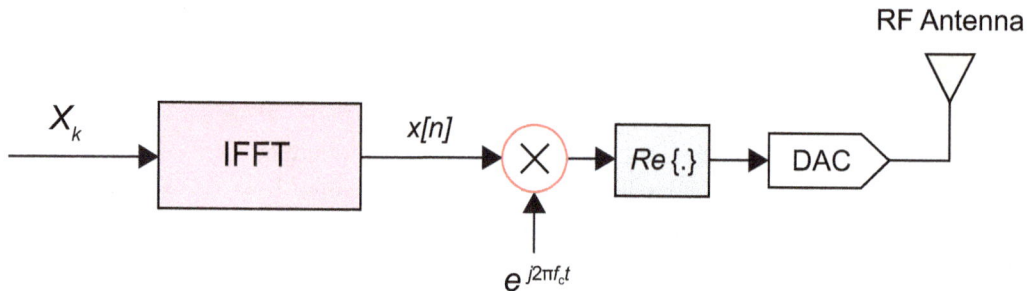

Figure 16.10: OFDM transmitter with IFFT.

With the replacement of the orthogonal modulation process by the IFFT, the OFDM transmitter is almost complete. However, there is one additional ingredient known as the Cyclic Prefix (CP). The CP is required to maintain the orthogonality of the sub-carriers through the multipath channel, eliminate ISI between OFDM symbols, and facilitate the one-tap equaliser.

16.3.2. The Cyclic Prefix

The purpose of the CP in OFDM can be understood by first analysing the effects of the multipath channel on the individual sub-carriers. The channel is equivalent to an L-tap FIR filter, and hence the output OFDM symbol, $y[n]$, is the linear convolution of the input OFDM symbol, $x[n]$, and the channel impulse response, $h[n]$.

The filter length, L, is dictated by the channel delay spread, d_s, with a longer delay spread requiring a longer filter. The linear convolution of an N sample signal with an L tap filter results in an output signal with length $N + L - 1$ samples, which can be divided into three stages: the *transient*, the *steady state* and the *decay* stages. These are illustrated in Figure 16.11, along with the duration of each stage.

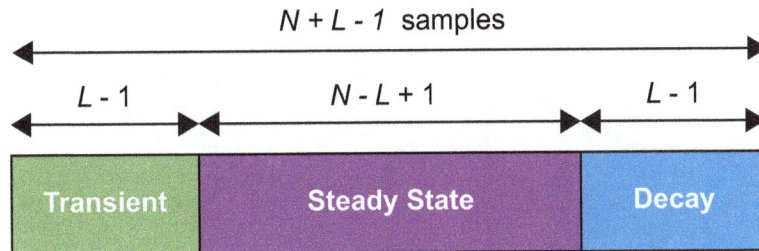

Figure 16.11: Channel filter output stages.

The transient stage occurs before the filter is fully occupied by input samples, and hence is $L - 1$ samples long. The steady-state stage starts as soon the filter memory is fully occupied by signal samples, and is $N - L + 1$ samples long. Finally, the decay stage starts when the filter is only occupied by past samples, and is thus also $L - 1$ samples long.

The transient stage distorts the initial portion of each sub-carrier, resulting in a loss of orthogonality and hence channel-induced ICI. Conversely, the decay stage causes sub-carriers from one symbol to bleed into sub-carriers in the next symbol, which causes ISI. Figure 16.12 shows the effects of the transient and decay stages on the real part (cosine) of a single unmodulated sub-carrier (note the effects are similar for all sub-carriers making up the OFDM symbol).

Both the transient and decay stages are shown in red, and the steady state stage is shown in blue. In order to maintain orthogonality and properly convey the symbol to the receiver, the amplitude and phase of the sub-carrier must be constant over the duration of N samples. However, due to the transient stage, the first $L - 1$ samples are distorted, which destroys the orthogonality. Similarly, the decay stage extends the sub-carrier beyond the N sample boundary, causing interference to the subsequent symbol.

Fortunately, both of these problems can be solved by the CP. The CP involves extracting a portion of the end of the OFDM symbol and appending it to the front. This has the effect of inserting a guard period between OFDM symbols. The length of the guard period is chosen to be greater than $L - 1$ samples, i.e. the duration of the transient and decay stages. The use of a CP is possible because the OFDM symbol period, N, is large compared to L, or equivalently the delay spread d_s of the channel, due to the fact that it comprises many low rate sub-carriers. Figure 16.13 illustrates the OFDM symbol with a CP.

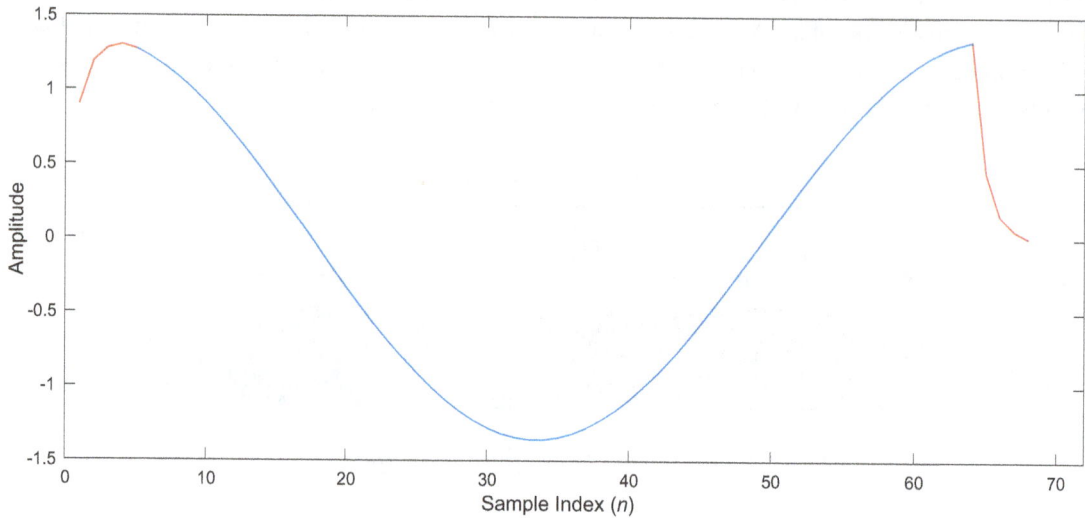

Figure 16.12: Plot of filtered sub-carrier.

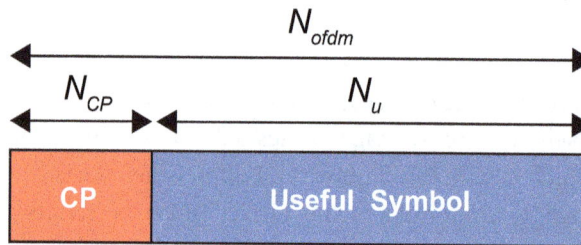

Figure 16.13: Addition of CP to OFDM symbol.

The overall OFDM symbol is now composed of the CP and the useful symbol (i.e. the portion that carries the data payload). Therefore, the overall length of the OFDM symbol, N_{ofdm}, is given by,

$$N_{ofdm} = N_u + N_{cp} \tag{16.23}$$

where N_u and N_{cp} are the lengths of the useful symbol and CP respectively. Note, $N_u = N$, is the number of samples in the useful symbol which is equivalent to the no. of sub-carriers since $N_u = 1/(f_s/N) = N \times T_s$. Figure 16.4 plots the real part of the sub-carrier with the CP attached.

It can be observed that, as the guard interval is formed using a portion of the OFDM symbol, there is no discontinuity between the CP and the beginning of the OFDM symbol. This is advantageous because any discontinuities between the CP and OFDM symbol would cause spurious out-of-band frequency components.

Note, there are still discontinuities between OFDM symbols, but these can be mitigated to some extent by applying windowing techniques, in a similar fashion to the windowing used in spectral analysis.

Furthermore, if we take the symbol to begin at the start of the CP, this is equivalent to the original symbol cyclically shifted to the right by N_{cp} samples. Therefore, we can take the beginning of the OFDM symbol to be anywhere within the CP, provided that we compensate for the resulting phase shift (this is compensated for by the equaliser). This is advantageous as it allows a greater margin of error for timing synchronisation in the receiver. However, it cannot be taken from any part of the CP affected by ISI. In most practical systems, this is achievable because the CP is made to be longer than the expected channel delay spread.

Figure 16.15 shows the sub-carrier with CP attached, after it has been passed through the multipath channel filter. On the left-hand side, the transient stage of the multipath channel now occurs during the CP, and therefore does not affect the sub-carriers. This means that the sub-carriers only enter the channel during the steady state stage, ensuring that orthogonality is maintained. On the right hand side, the decay stage is absorbed by the CP of the subsequent symbol (not shown), thus preventing ISI.

The last major benefit of the CP is that it facilitates the 'one-tap' equaliser, arising from the fact that periodic or circular convolution in the time domain is equivalent to multiplication in the frequency domain [141].

The addition of a CP which is at least L-1 samples long makes the OFDM symbol "appear" N periodic, i.e. being periodic with a period of N samples. As a result, after the FFT in the receiver, the received symbols Y_k are related to the original symbols X_k as,

$$Y_k = X_k H_k \tag{16.24}$$

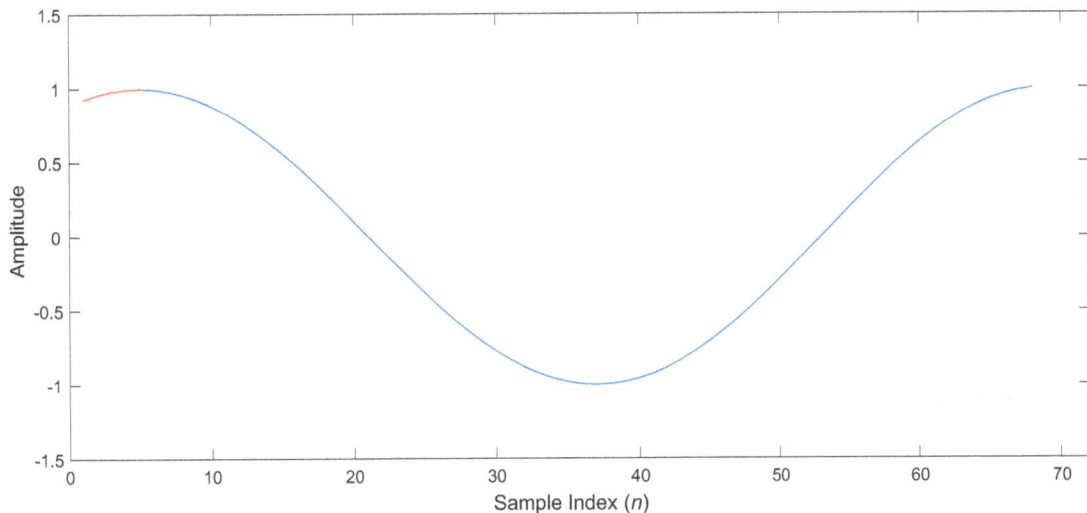

Figure 16.14: Plot of real part of sub-carrier with CP.

where H_k is the complex (magnitude and phase) frequency response at sub-carrier k. As a result of this, the channel can be equalised as,

$$\hat{X}_k = \frac{Y_k}{H_k} \qquad (16.25)$$

where $\hat{X}_k$ represents the estimated symbols after equalisation. More detail on how the addition of the CP leads to the one-tap equaliser can be found in the *OFDM Fundamentals* Jupyter notebook that accompanies this textbook (see the introduction to Notebook I on page 569).

It is clear that the one-tap equaliser used in OFDM represents a significantly simpler and more computationally efficient approach to equalisation than the time domain filters used in single carrier systems. Returning to the numerical example in Section 16.1, the d_s of 0.5 μs can be addressed with an equivalent OFDM system operating at $f_s = 20MHz$ with $N = 64$ sub-carriers and $N_{cp} = 16$. In this case, the total symbol duration is 4μs meaning that the computational complexity of equalisation is $4 \times 64 \times \frac{1}{4\mu s} = 64$ million MAC/s. This is a significant saving compared to the 4 billion MAC/s required for the 50 weight DFE in the QPSK system.

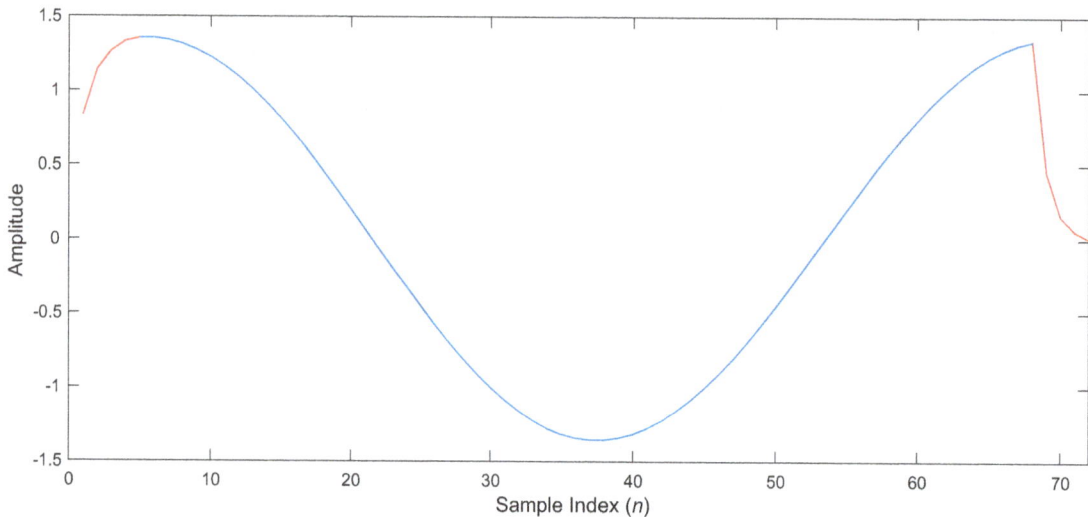

Figure 16.15: Filtered sub-carrier with CP.

16.4. OFDM Transmitter

Having introduced the IFFT and CP in the previous sections, we can now draw an updated diagram for the OFDM transmitter as shown in Figure 16.10, to include the CP insertion stage. The modified architecture is illustrated in Figure 16.16.

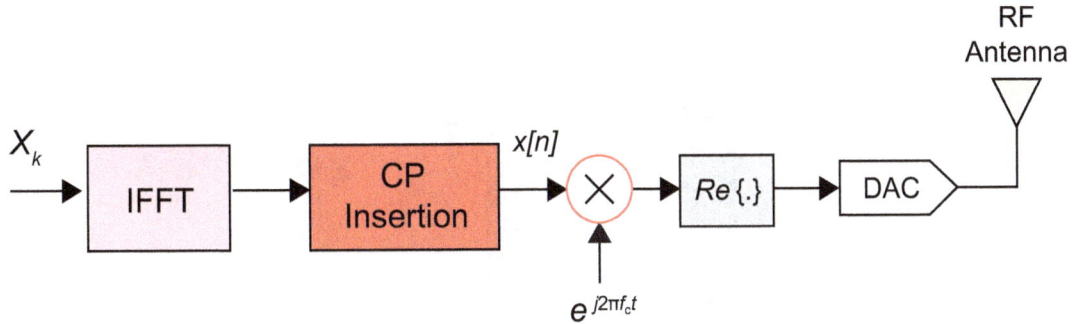

Figure 16.16: High level architecture for OFDM transmitter, including CP insertion.

It should be noted that not all sub-carriers are used for carrying data. In particular, both null sub-carriers and pilot sub-carriers are also transmitted. The null sub-carriers contain no data and are typically used at DC and the outer sub-carriers to relax the requirements for anti-aliasing and ant-imaging filters. Furthermore, the pilot sub-carriers carry special training symbols known as 'pilots' or 'reference signals' for example in 4G LTE and 5G NR, which can be used for synchronisation purposes, channel estimation, and signal power and channel quality measurements.

As an example, Figure 16.17 illustrates the composition of an OFDM symbol based on the IEEE 802.11-2012 standard [207]. The composition of the OFDM symbol is shown in a frequency versus time representation. This representation is used frequently in the 4G LTE and 5G NR standards, where it is known as the *resource grid*. These standards employ a multiple access method known as Orthogonal Frequency Division Multiple Access (OFDMA), wherein blocks of sub-carriers known as resource blocks are assigned to different users of the channel. The assignment of resource blocks to different users (in both the *downlink*, i.e. basestation to user; and the *uplink*, in the opposite direction) is controlled by a scheduling algorithm that runs in the Medium Access Control (MAC) Layer of the basestation (eNodeB in 4G, or gNodeB in 5G) [315].

In Figure 16.17, the null sub-carriers are highlighted in gold, the pilots in blue and the data sub-carriers in red. The OFDM symbol is drawn with the DC sub-carrier at the centre, with the spectrum extending from $-f_s/2$ to $f_s/2$ Hz. This structure is repeated for all other OFDM symbols in a frame. It can be observed that the null sub-carriers are on the outer edges of the band. As mentioned previously, this relaxes the requirements in terms of anti-imaging and anti-aliasing filters at the transmitter and receiver. In this case, the DC sub-carrier is a null sub-carrier, in order to avoid interference generated by radio artefacts including Local Oscillator Leakage (LOL) associated with Direct Conversion (RF to baseband) front end architectures [363]. Note, in 5G NR, the DC sub-carrier can be used for data transmissions [3].

There are four pilot sub-carriers per OFDM symbol, which are spread out across the signal bandwidth. In the IEEE 802.11-2012 standard, these are intended for phase tracking in the receiver. Channel estimation is handled using a special preamble appended to each data packet. This assumes that the channel does not change significantly over the duration of a single packet, which is a valid assumption for Wi-Fi, where mobility

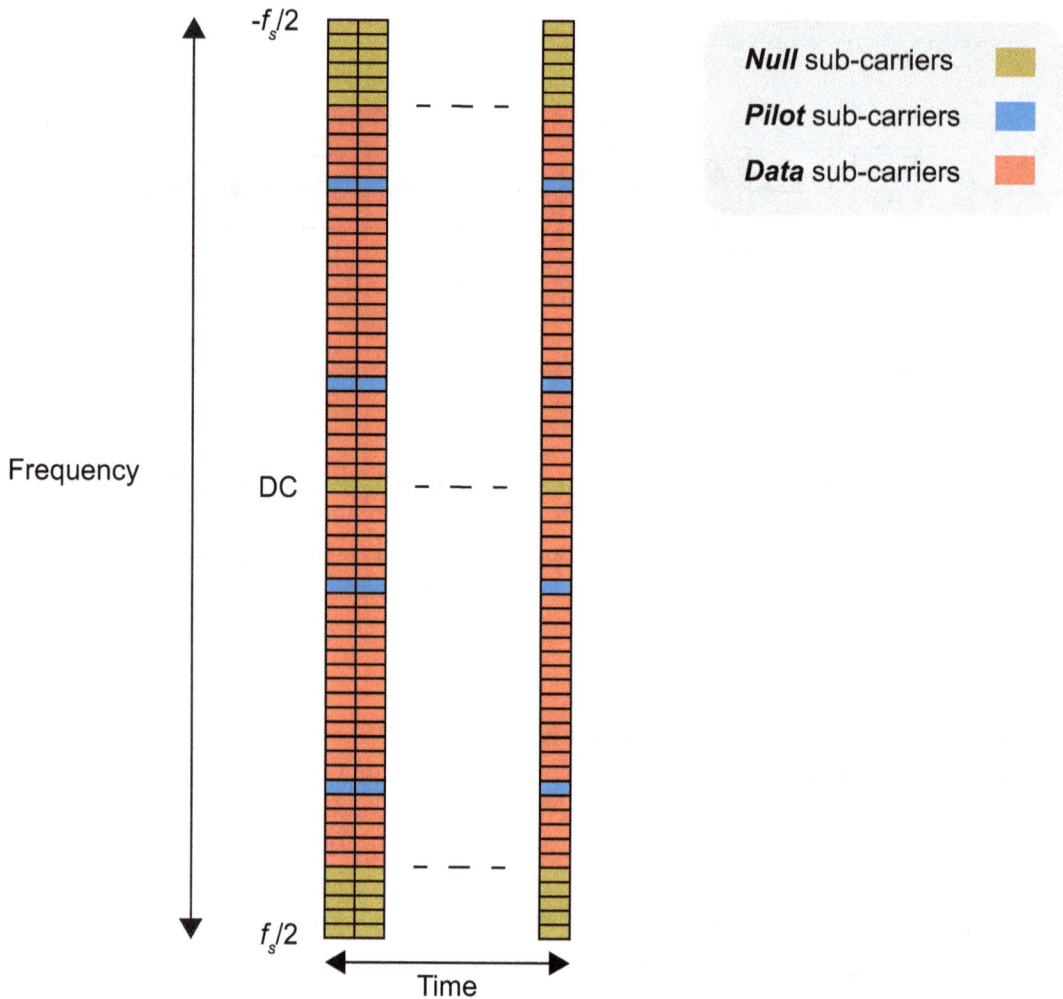

Figure 16.17: Composition of an example OFDM symbol.

is usually low. The phase error results from radio impairments such as residual frequency offset, oscillator phase noise and sampling frequency offset. Since this phase error varies from symbol to symbol, it has to be "tracked", and hence the pilots are repeated in each OFDM symbol. More details on the phase error will be provided in Section 16.5. The pilot arrangement shown in Figure 16.17 is known as *comb-type* [200].

In 4G and 5G, the transmission is continuous, so data is not transmitted in bursts, and consequently the channel cannot be assumed to remain constant over the duration of a transmission. Therefore, the channel is estimated using pilots or reference signals spread through frequency and time. For the sake of brevity, we will only consider the 4G case, since the reference signal configuration in 5G would require a more thorough treatment than is possible in this chapter. Figure 16.18 illustrates the pilot arrangement used in 4G.

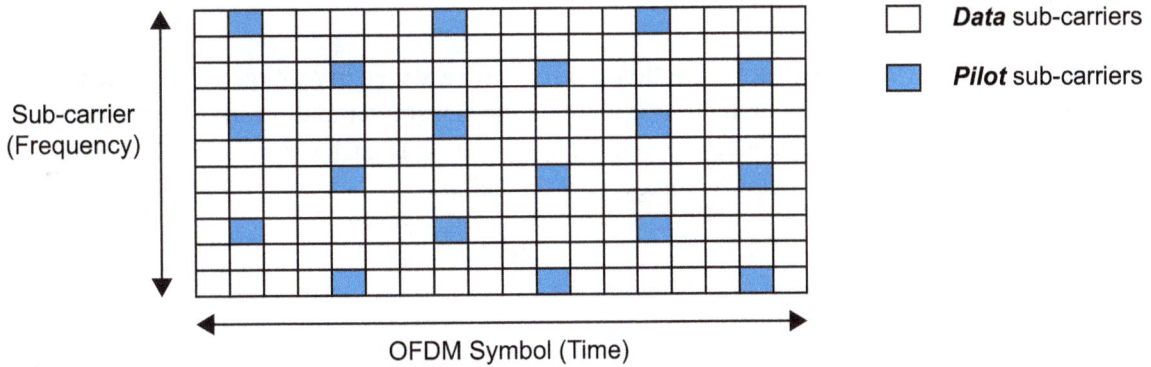

Figure 16.18: Pilot arrangement in 4G LTE.

The pilots are scattered throughout the resource grid, and thus it is referred to as a *scattered pilot* arrangement. The spacings of the pilots in frequency and time are determined by the expected coherence bandwidth and coherence time of the channel, respectively. In particular, the spacing in frequency, S_f, must be less than the coherence bandwidth B_c, i.e.

$$S_f < B_c. \tag{16.26}$$

Similarly, the spacing in time, S_t, must be less than the coherence time T_c,

$$S_t < T_c. \tag{16.27}$$

The requirement that $S_f < B_c$ implies that the sub-carrier spacing should be much smaller than B_c. Otherwise, a greater number of pilots would be required to estimate the channel, thereby reducing spectral efficiency. The channel estimates at the pilot sub-carriers are interpolated (usually linearly) in frequency and time to estimate the channel response at the data sub-carriers. These channel estimates can then be used to equalise the channel and recover the transmitted data symbols.

The null and pilot sub-carriers do not carry data, and therefore reduce the spectral efficiency that can be achieved in OFDM. However, with careful design, their use can be minimised as much as possible, while still guaranteeing the performance of the OFDM system.

16.5. Impairments in OFDM

In addition to the effects of the multipath channel, there are several radio imperfections which impair the OFDM signal and must be estimated and corrected for in the receiver. The main impairments are Timing Offset, Carrier Frequency Offset (CFO), Residual CFO and Phase Noise, and Sampling Frequency Offset. The details of these impairments are described in the following sections.

16.5.1. Timing Offset

The timing offset can be classified as either fractional or integer, measured in relation to f_s. A fractional timing offset is caused by a mismatch in sampling phase between the transmitter and receiver sampling clocks. The fractional timing offset causes a frequency dependent phase rotation to each symbol after the FFT [119]. This can be appreciated by recalling the following property of the DFT,

$$x[n - \tau] \leftrightarrow X[k]e^{-j\frac{2\pi kn}{N}} \tag{16.28}$$

where τ is the fractional offset normalised to f_s. The phase rotation is dependent on the sub-carrier index, k, and so each symbol experiences a different phase shift. This phase error does not vary with time, and is corrected by the channel equaliser.

The integer timing offset relates to an error in the estimated position of the beginning of an OFDM symbol, which causes the FFT to be performed over an incorrect block of N samples in the receiver. The impact of this error depends on whether the estimated starting position is earlier or later than the actual starting position. Figure 16.19 illustrates the two possible integer timing offsets.

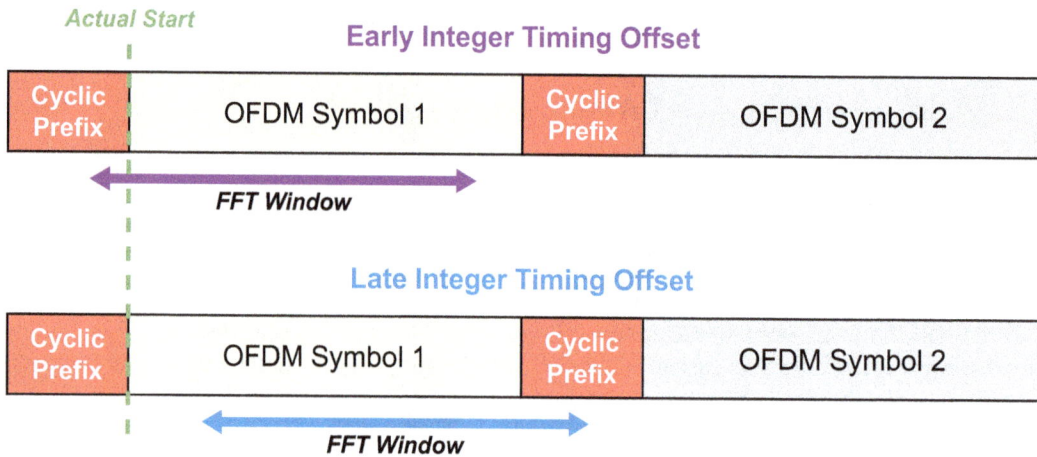

Figure 16.19: Early and late integer timing offsets.

In the early integer timing offset, the FFT window includes a portion of the CP of the current OFDM symbol. As described previously, this causes a simple phase shift to each symbol after the FFT, which can be corrected by the equaliser if the timing falls within a region of the CP not affected by ISI. Since the CP is typically longer than the delay spread, an early integer timing offset does not pose a problem, which relaxes the requirements in terms of timing synchronisation. In contrast, a late timing offset causes a portion of the CP belonging to the next OFDM symbol to be included in the FFT window for the current OFDM symbol. This leads to both ISI and ICI, and must be avoided [192].

16.5.2. Carrier Frequency Offset

Frequency offsets occur due to local oscillator mismatches between the transmitter and receiver, and Doppler shifts. The frequency offset causes the sub-carriers to drift from their ideal frequencies and, if not corrected prior to the FFT, results in a loss of orthogonality, and ICI. As mentioned earlier, ICI manifests as a form of additive noise which leads to an increase in symbol and bit errors in the receiver. In a similar fashion to timing offsets, the CFO comprises integer and fractional components, measured in relation to the sub-carrier spacing, Δf.

The effect of CFO on the received signal is expressed mathematically as,

$$x_r[n] = x_t[n]e^{j2\pi(f_o/f_s)n} \tag{16.29}$$

where $x_t[n]$ and $x_r[n]$ are the transmitted and received time domain OFDM signals, respectively, and f_o is the CFO. The CFO can be estimated using training signals, along with auto-correlation and cross-correlation techniques, as will be described in Section 16.6.

16.5.3. Residual Frequency Offset and Phase Noise

In practice, the frequency synchronisation process is not perfect, and therefore there is likely to be a residual frequency error after the FFT. This causes ICI, and a phase error which is common to all sub-carriers, but varies with time. If the residual offset is small compared to the sub-carrier spacing, the phase error will dominate compared to ICI. Since the phase error is common to all sub-carriers, it is known as Common Phase Error (CPE) [290]. In the case of residual frequency error, the CPE varies linearly with the OFDM symbol index, and therefore must be tracked using the pilot sub-carriers.

Phase noise is caused by oscillator imperfections and jitter in the clock that drives the oscillator. This is modelled as the addition of a time-varying phase term to the local oscillator output, i.e.

$$cos(2\pi f_c t + \phi(t)) \tag{16.30}$$

where $\phi(t)$ is the time-varying phase term. The phase noise causes the signal to spread in frequency, which introduces ICI. In addition, it causes a randomly varying CPE. If the sub-carrier spacing is large compared to the bandwidth of the phase noise, the CPE term will dominate compared to ICI [172]. At higher carrier frequencies, the phase noise bandwidth increases, which in turn necessitates a larger sub-carrier spacing. This is one of the reasons for the mixed numerology or sub-carrier spacing in 5G NR, which allows for a larger sub-carrier spacing to combat the effects of phase noise at higher carrier frequencies. In Frequency Range 2, or mmWave bands, a sub-carrier spacing of 120kHz is sued whereas 30kHz is typically used in mid-band (3.5GHz) deployments.

16.5.4. Sampling Frequency Offset

Sampling offset arises due to a mismatch in the sampling frequency between transmitter and receiver. This is equivalent to a sampling phase error that increases with time, and it means that the sub-carrier dependent phase error in (16.28) also varies with OFDM symbol index [320]. As with the CPE, this must be tracked using the pilot sub-carriers.

The sampling frequency offset also introduces an ICI term. In OFDM based standards such as IEEE 802.11 [207], there is often a worst case sampling frequency offset (and CFO) defined, which manufacturers must adhere to in order for their devices to be standard-compliant. This ensures that frequency offsets (sampling, carrier etc.) are manageable and can be estimated and corrected in the receiver.

16.6. OFDM Receiver

The OFDM receiver comprises all of the processing steps required to reverse the effects of the channel and radio impairments, and recover the transmitted bitstream. At a high level, the receiver stages are: timing synchronisation, frequency synchronisation, FFT demodulation, channel estimation, equalisation, and phase tracking. Figure 16.20 shows a high level illustration of an OFDM receiver.

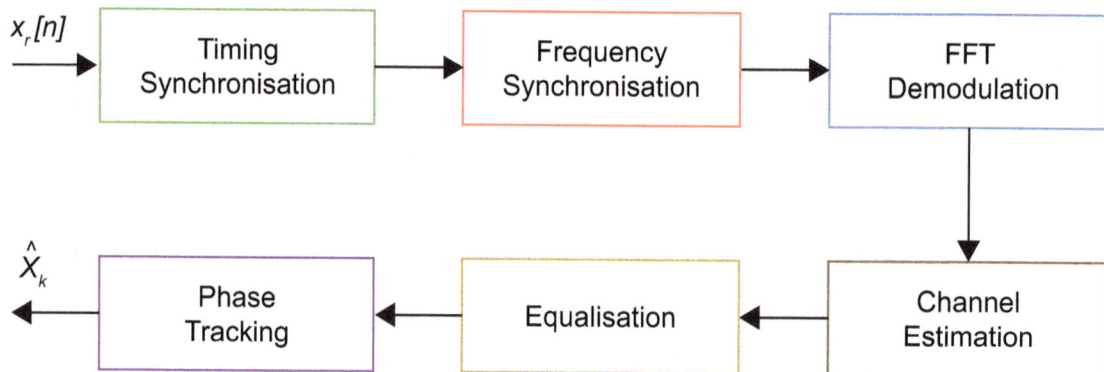

Figure 16.20: High level block diagram of an OFDM receiver.

At the top left of Figure 16.20, the received and digitised baseband signal, $x_r[n]$, is passed into the OFDM receiver. After the synchronisation, demodulation and equalisation stages, an estimate of the original symbol stream is generated, $\hat{X}_k$, as shown at the bottom left.

For the purposes of this chapter, which focuses on OFDM techniques, the channel decoding stages will not be considered as part of the OFDM receiver. Channel coding and decoding receives its own dedicated coverage, elsewhere in the book (please refer to Chapter 14, which covers the topic in detail).

It is useful to note, however, that channel coding and interleaving are important parts of many practical OFDM transceivers. An OFDM system that incorporates channel coding is known as Coded OFDM, or COFDM — this name is often used to refer to OFDM-based standards for broadcasting, such as Digital Video Broadcasting - Terrestrial (DVB-T).

The block diagram presented in Figure 16.20 is just intended as a high level overview of the main processing steps; the order of operations is not always as shown here, and some of the receiver tasks can be divided into several sub-tasks. Each of the main receiver stages will be reviewed over the following sections.

16.6.1. Timing Synchronisation

The first task that the OFDM receiver undertakes is to determine the starting point of the OFDM signal in the received sample stream. This is achieved using specially designed training sequences that are known to both transmitter and receiver, and which can be detected using auto-correlation and cross-correlation techniques in the receiver. These timing synchronisation algorithms are required to provide robust performance in multipath environments and low SNR conditions, and in the presence of radio impairments.

A prominent example of an auto-correlation-based timing synchronisation algorithm is the Schmidl & Cox (S&C) algorithm [140],[313]. This algorithm was designed for packet based OFDM protocols, specifically the IEEE 802.11 family of standards, and features in the *RFSoC OFDM* implementation accompanying this textbook (see page 569). It exploits the auto-correlation properties of a preamble which consists of repeated versions of a fixed length sequence. In the IEEE 802.11a/g standard, the preamble comprises 10 repetitions of a 16 sample sequence. The first seven repetitions are intended for signal detection, Automatic Gain Control (AGC) convergence, and diversity selection, while the last three repetitions are used for timing synchronisation and coarse frequency offset estimation [140]. The S&C algorithm itself only requires two repetitions.

The S&C algorithm computes the following timing metric [313],

$$M(d) = \frac{|P(d)|^2}{(R(d))^2} \tag{16.31}$$

where d denotes a sample index and $P(d)$ is an autocorrelation metric calculated as

$$P(d) = \sum_{m=0}^{L-1} (r_{d+m} r^*_{d+m+L}) . \tag{16.32}$$

where $(\ )^*$ denotes complex conjugation. The auto-correlation is performed at a shift of L, because there is a gap of L samples between identical samples in the training sequence. The result is then averaged across a window of L samples, which is equivalent to passing the auto-correlation through a length-L averaging filter. In the case of IEEE 802.11, $L = 16$. The auto-correlation metric is normalised by $R(d)$, which is given by,

$$R(d) = \sum_{m=0}^{L-1} \left| r_{d+m+L} \right|^2. \qquad (16.33)$$

This normalisation factor makes the timing metric, $M(d)$, independent of the signal and noise received power level, thus allowing for a fixed threshold to detect the timing metric. Figure 16.21 shows the timing metric for an IEEE 802.11a/g OFDM signal with no noise or multipath effects.

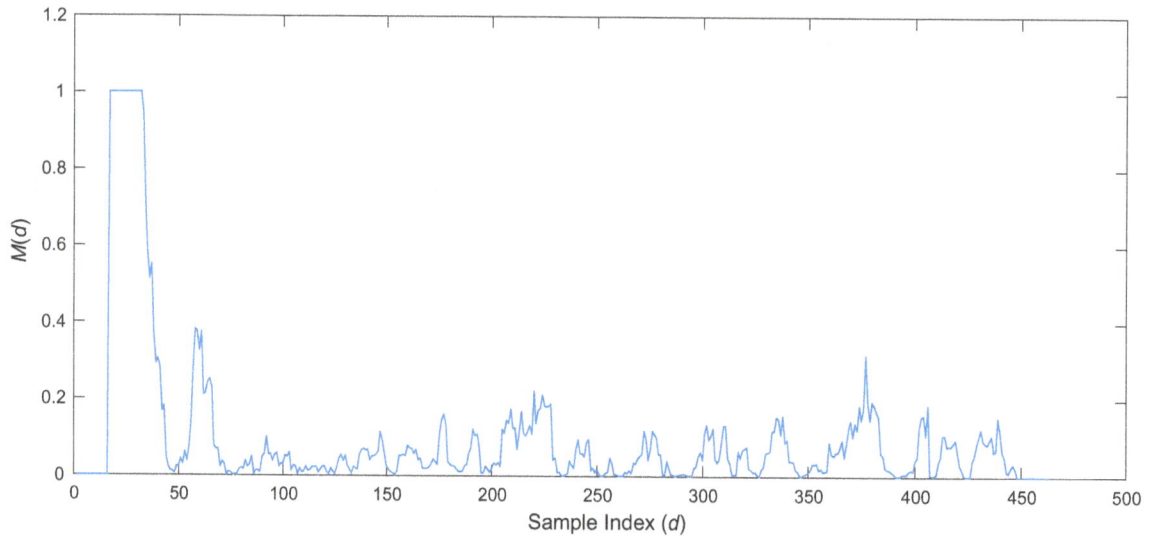

Figure 16.21: Schmidl & Cox timing metric.

In order to generate this timing metric, a data frame consisting of a preamble with two repetitions of the 16 sample training sequence, followed by a payload of 5 OFDM symbols, was passed to the Schmidl & Cox algorithm. The OFDM symbols were configured with parameters $N_u = 64$ and $N_{cp} = 16$.

The timing metric reaches a plateau when both halves of the preamble are aligned in the auto-correlation circuit. In an AWGN channel, the plateau has a length of L, corresponding to the length of each repetition. In multipath channels, the delay spread causes interference between the two halves of the preamble (since there is no CP between them) and therefore this reduces the length of the plateau by a factor equal to the delay spread [313]. In IEEE 802.11a/g, each repetition is 16 samples long (i.e. the length of the CP), hence the S&C algorithm is still effective in multipath environments.

Once the first OFDM symbol has been located, we automatically have the location of the remaining symbols. The location of the plateau can be determined by comparing the timing metric to a pre-defined threshold, such as 0.5. This is valid because the auto-correlation is low when the preamble signal is absent, as can be observed in Figure 16.21. The probability of a false alarm can be reduced by checking that the timing metric exceeds the threshold for several consecutive samples (say 5 or 10). The fact that a plateau is generated, rather than a clear peak, introduces a level of uncertainty, and therefore the S&C algorithm is usually used to provide a coarse estimate of the symbol timing. The coarse estimate is then refined with a further fine timing synchronisation stage, which makes use of cross-correlation techniques.

In cross-correlation based timing techniques, the input signal is passed through a filter whose coefficients are a locally stored copy of the synchronisation signal, known as a *matched filter*. The matched filter coefficients are a time-reversed and conjugated version of the synchronisation signal, since correlation is equivalent to convolution with the filter coefficients time-reversed. Once the synchronisation signal occupies the matched filter, a positive peak will be produced at the output, which can be compared to a threshold to detect the OFDM signal. Figure 16.22 illustrates the concept of a matched filter.

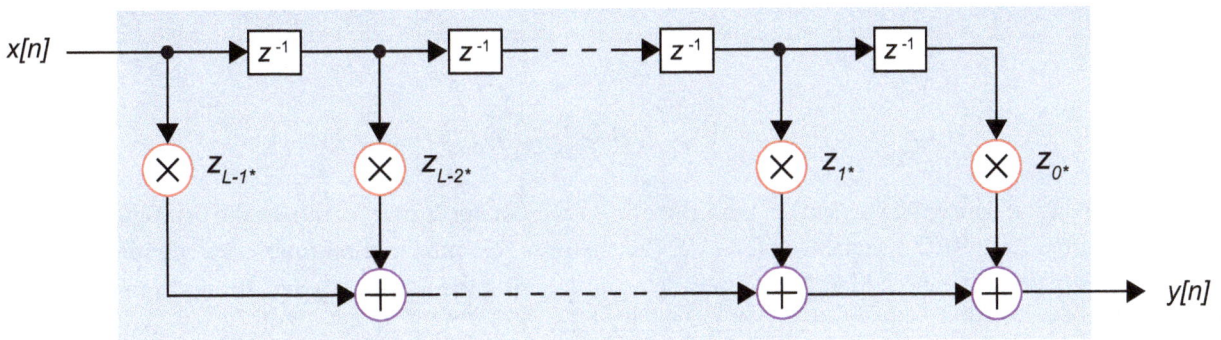

Figure 16.22: Matched filter for timing synchronisation.

Figure 16.23 shows the matched filter output for the Legacy Long Training Field (L-LTF) used in the IEEE 802.11 standards [207]. It can be observed that two main peaks are generated, which can be detected to refine the coarse timing estimate found using the S&C auto-correlation metric. It should be noted that matched filtering is sensitive to an uncorrected CFO (unless the matched filter contains a frequency shifted version of the synchronisation signal), and therefore it is necessary to perform frequency synchronisation prior to fine timing synchronisation.

In 4G LTE and 5G NR, a matched filter is used to detect the Primary Synchronisation Signal (PSS) to obtain downlink frame synchronisation during the initial cell search procedure. The PSS is based on a special sequence known as a Zadoff-Chu Sequence. A key property of these sequences is that they have zero correlation with cyclically shifted versions of themselves. As a result, multiple orthogonal sequences can be generated from a single base sequence, allowing them to be used for different purposes in the standard including PSS, the Physical Random Access Channel (PRACH) and certain reference signals [340].

There are three different PSS sequences, each corresponding to a unique cell identity within a cell identity group [340]. In the mobile User Equipment (UE), there will be three matched filters, each matched to one of the three PSS signals. When the PSS is received, only the filter matched to this particular PSS will produce a

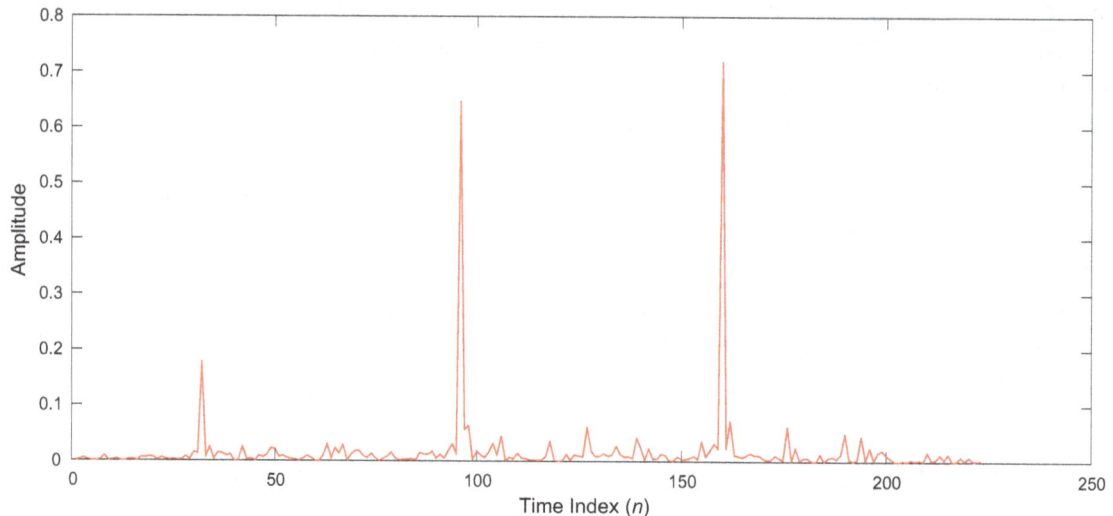

Figure 16.23: Matched filter output for L-LTF in IEEE 802.11 standards.

positive correlation, since the correlation with the other PSS sequences is zero, due to the orthogonality of the cyclically shifted Zadoff-Chu sequences. The UE can then use this information to select a cell identity within the cell identity group. The cell identity group is determined using the Secondary Synchronisation Signal (SSS) which allows identification of the final Physical Cell Identity (PCI). It is usually necessary to perform frequency synchronisation prior to PSS detection to reduce the vulnerability to CFO. However, the particular Zadoff-Chu sequences that are used for the PSS have a low sensitivity to an uncorrected CFO [340].

16.6.2. Frequency Synchronisation

The process of frequency synchronisation involves estimating the frequency offset between the transmitter and receiver local oscillators, and then frequency shifting the received signal in order to compensate for this offset.

The frequency offset can also be estimated with a preamble and an auto-correlation method such as the S&C algorithm. The auto-correlation, $r_{xx}[n]$ of the received signal, $x[n]$, at a lag of L samples is given by,

$$r_{xx}[n] = x[n]x^*[n-L] \tag{16.34}$$

In the presence of a frequency offset, f_o, the auto-correlation becomes,

$$r_{xx}[n] = x[n]e^{j2\pi(f_o/f_s)n}x^*[n-L]e^{-j2\pi(f_o/f_s)[n-L]}. \tag{16.35}$$

Expanding and re-arranging (16.35), we obtain,

$$r_{xx}[n] = x[n]x^*[n-L]e^{j2\pi(f_o/f_s)n}e^{-j2\pi(f_o/f_s)n}e^{j2\pi(f_o/f_s)L}, \tag{16.36}$$

and the first two complex exponential terms cancel, leaving

$$r_{xx}[n] = x[n]x^*[n-L]e^{j2\pi(f_o/f_s)L}. \tag{16.37}$$

In the noiseless case, when the two halves of the preamble are aligned in the auto-correlation circuit, the expression reduces to,

$$r_{xx}[n] = |x[n]|^2 e^{j2\pi(f_o/f_s)L}. \tag{16.38}$$

This is because multiplying a complex signal by its complex conjugate is equivalent to taking its magnitude squared. The frequency offset can be estimated as,

$$\hat{f}_o = \frac{arg(r_{xx}[n])}{2\pi LT_s} \tag{16.39}$$

where the *arg* operation denotes taking the argument or phase of $r_{xx}[n]$ and T_s is the sampling period. In the presence of noise, $r_{xx}[n]$ can be averaged over L samples to improve the estimate of f_o.

In the *RFSoC OFDM* implementation (introduced in Notebook I, see page 569), $arg(r_{xx}[n])$ is calculated using a Co-ordinate Rotation Digital Computer (CORDIC) processor operating in vectoring mode. CORDIC is a very computationally efficient method of computing trigonometric functions in hardware [354]; the details are beyond the scope of the current discussion.

Since the auto-correlation is performed between samples that are separated by LT_s seconds, it is only possible to unambiguously resolve frequencies offsets in the range,

$$\left(\frac{-1}{2LT_s}, \frac{1}{2LT_s}\right). \tag{16.40}$$

In the case of IEEE 802.11a/g, this is equivalent to offsets between $\pm 2\Delta f$ *Hz*. In systems without a preamble, such as 4G and 5G, autocorrelation can be performed at a shift of $L = N$ to exploit the fact that the CP and the end of the OFDM symbol are identical. This method is useful as it avoids the need to include a preamble in the OFDM signal, which increases spectral efficiency, although it limits the estimation range to $\pm\Delta f/2$ *Hz*.

In order to estimate integer frequency offsets beyond the limit of autocorrelation methods, the received signal can be cross-correlated with different frequency shifted versions of the synchronisation signal, i.e. the L-LTF or the PSS. The frequency shift that produces the largest positive correlation corresponds to the integer frequency offset. Due to the computational complexity of performing several different correlations for different frequency shifts, it is usually only feasible to test a limited range of integer frequency offsets. However, as alluded to previously, standards often specify a worst case frequency offset, which makes correcting for a large integer frequency offset unnecessary.

Once the CFO has been estimated, a correction is applied to shift the signal $x[n]$ to baseband, where the shifted version is given by

$$x_s[n] = x[n]e^{-j2\pi(\hat{f}_o/f_s)n}.$$

$$(16.41)$$

The complex sinusoid in (16.41) can be generated on an FPGA using a Numerically Controlled Oscillator (NCO) or a CORDIC processor operating in rotation mode. As mentioned earlier, in practice the frequency synchronisation process is not perfect, which leads to a residual frequency error that causes ICI and CPE. This is usually small enough that the CPE is the dominant effect, and therefore the performance degradation is not significant.

16.6.3. Channel Estimation and Equalisation

The channel is estimated using either a preamble for packet-based protocols, such as the IEEE 802.11 standards, or using pilots or reference signals in continuous transmission systems such as 4G and 5G. In packet-based protocols, the length of the packet is designed to be less than the expected coherence time (based on a maximum velocity assumption), which means that a channel estimate made at the beginning of the packet is usually valid for the entire packet. In continuous systems, it cannot be assumed that the channel will remain constant over the duration of a transmission, so the channel estimates need to be updated at regular intervals.

Regardless of the method employed for channel estimation, the channel estimate is calculated as,

$$\hat{H}_k = \frac{Y_k}{X_k}$$

$$(16.42)$$

where X_k is the training or pilot symbol transmitted on sub-carrier k, Y_k is the received training or pilot symbol on sub-carrier k, and $\hat{H}_k$ is the channel estimate at sub-carrier k. This channel estimation procedure is possible because the CP turns the linear convolution of the channel into an N periodic convolution, which is equivalent to multiplication after the FFT.

Once channel estimates are available for the data sub-carriers, the channel is equalised as,

$$\hat{X}_k = Y_k/\hat{H}_k$$

$$(16.43)$$

where Y_k is the received data symbol on sub-carrier k and $\hat{X}_k$ is the estimated data symbol at sub-carrier k. As mentioned, this is known as the one-tap equaliser. Since it is equivalent to inverting the channel estimate and multiplying the received symbol $\hat{Y}_k$ by the result, it is essentially performing the same task as the linear or zero forcing equaliser described in Section 16.1. As a result, the equaliser suffers from noise enhancement in channels with deep fades, which can lead to a significant degradation in performance. This problem is alleviated to some extent by the Minimum Mean Square Error (MMSE) equaliser. More information on this can be found in the *OFDM Transceiver* Jupyter notebook introduced in Notebook I on page 569.

16.6.4. Phase Tracking

The final synchronisation stage to consider is phase tracking. This corrects for the phase error resulting from residual frequency offset, phase noise and sampling frequency offset. The phase error for each OFDM symbol is a straight line, which is a function of the sub-carrier index k,

$$y = mk + c \qquad (16.44)$$

where y denotes the phase error, m is the gradient of the phase error, and c is the CPE. The gradient results from the sampling frequency offset and the CPE results from the residual frequency offset and phase noise. The m and c values vary with OFDM symbol index and therefore must be tracked continuously using the pilots. The phase error is illustrated in Figure 16.24.

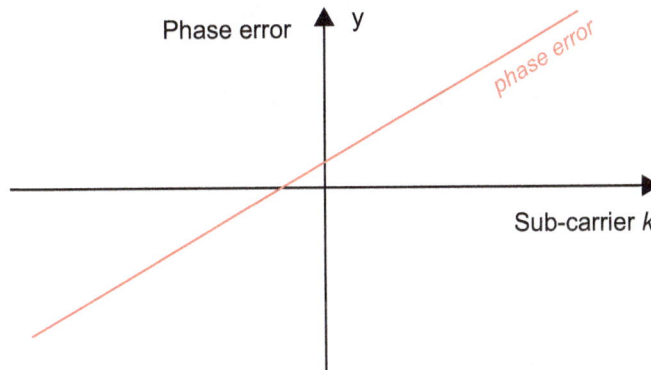

Figure 16.24: Phase error on a single OFDM symbol.

The linear phase error is estimated and corrected in two stages. In the first stage, we estimate the gradient of the straight line. By estimating the gradient of the line, the sub-carriers can then be rotated such that the phase error (straight line) is flat, i.e. has a gradient of zero as shown in Figure 16.25. After the first correction stage,

$$y = c, \qquad (16.45)$$

meaning that it just comprises the CPE term. This is estimated and corrected in the second stage, which reduces the phase error to zero (or as close as possible to zero), i.e. $y = 0$.

In packet-based systems, where the channel is assumed to remain constant over the duration of a transmission, the equaliser only compensates for the channel (and any constant phase error terms). This is because the channel estimate is performed once at the beginning of the packet. Therefore, an explicit phase tracking stage is required after the equaliser. In contrast, for continuous systems, the channel estimate is updated regularly and thus the phase error is also tracked and compensated by the equaliser. In 4G, the phase error has to be estimated through interpolation for symbols without reference signals. However, in 5G, a dedicated Phase

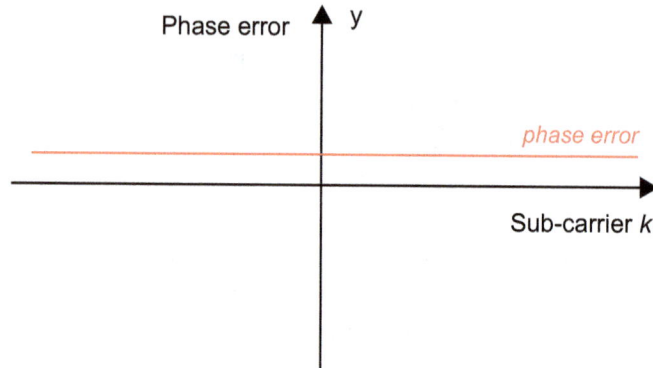

Figure 16.25: Phase error after the first correction stage.

Tracking Reference Signal (PTRS) has also been introduced to improve phase tracking performance. This is especially important for Frequency Range 2, where frequency offsets and phase noise effects are more pronounced.

The estimation and correction of phase errors during phase tracking in the FPGA requires CORDIC operating in vectoring and rotation modes [354].

16.7. Chapter Summary

This chapter has introduced the fundamental theory behind Orthogonal Frequency Division Multiplexing. OFDM is a digital multi-carrier modulation method that divides a high rate symbol stream into several parallel lower rate streams, which each modulate a different orthogonal sub-carrier. The use of orthogonal sub-carriers allows them to tightly overlap, which improves spectral efficiency compared to generic MCM, and makes the OFDM modulation and demodulation processes equivalent to the IDFT and DFT operations, respectively. These can both be implemented efficiently using the FFT algorithm.

The number of sub-carriers is chosen such that the bandwidth of each sub-carrier is small compared to the channel coherence bandwidth, and hence each experiences a flat fading channel, which can be equalised with a single complex tap after the FFT. This significantly reduces the computational complexity of equalisation when compared to the time domain equalisers used in traditional single-carrier systems. The use of many independent sub-carriers also gives rise to an efficient multiple access method, wherein several users are assigned different blocks of sub-carriers within a shared channel. The addition of a CP maintains the orthogonality of the sub-carriers, prevents ISI, and facilitates the one-tap equaliser. Due to its many advantages, OFDM has been deployed in the physical layer of many modern standards, including DAB, DVB-T, the IEEE 802.11 family of standards, 4G LTE, and most recently 5G NR.

Notebook Set I

Practical OFDM Design

The final notebook series of this book explores the design of an Orthogonal Frequency Division Multiplexing (OFDM) transceiver on RFSoC. These notebooks cover advanced OFDM topics and theory. Therefore, it is important that you have read Chapter 16, which motivates the need for OFDM in radio systems and describes the modulation scheme and theoretical background.

We will begin by exploring the design of an OFDM transmitter and receiver in Python and then investigate OFDM channel estimation and equalisation. Lastly, we will examine an OFDM transceiver implemented on RFSoC. We will be able to observe and plot various data paths in the OFDM radio design using PYNQ.

The are three notebooks to investigate throughout this chapter using Jupyter Labs on your RFSoC platform. The notebooks and their relative locations are listed as follows:

> `ALL`   01_ofdm_fundamentals.ipynb — ***rfsoc_book/notebook_I/01_ofdm_fundamentals.ipynb***

> `ALL`   02_ofdm_python_transceiver.ipynb — ***rfsoc_book/notebook_I/02_ofdm_python_transceiver.ipynb***

> `RFSoC`   03_rfsoc_ofdm_transceiver.ipynb — ***rfsoc_book/notebook_I/03_rfsoc_ofdm_transceiver.ipynb***

I.1. OFDM Fundamentals

This notebook explores the practical side of OFDM design using Python and can be accessed by opening the notebook named ***01_ofdm_fundamentals.ipynb***. Several topics are covered including OFDM baseband symbol generation, Inverse Fast Fourier Transform (IFFT) modulation, multipath channel effects, and the Cyclic Prefix (CP).

The notebook begins with an overview of the OFDM transmission process, which involves digital modulation and symbol mapping. The role of the IFFT / FFT for modulation and demodulation is established, and a brief summary of multipath channels is provided. Lastly, the purpose of the CP is discussed.

I.2. Channel Estimation and Equalisation

The next notebook in this series named ***02_ofdm_python_transceiver.ipynb*** explores an OFDM transceiver implemented in Python. In particular, an OFDM transmitter and receiver is designed from first principles and a wireless channel is also created to model multipath channel effects. The notebook includes a demonstration of channel estimation and investigates the one-tap, or Zero Forcing (ZF), equaliser.

Channel estimation is necessary to evaluate the properties of a wireless channel so that it is possible to adapt signal transmissions and improve communication data rates and reliability. It is important to perform channel estimation periodically as wireless channel conditions are rarely consistent over time. Channel estimation can be used to produce important information on channel effects such as scattering, fading, and the deterioration of signal power as it propagates through the channel.

When a signal is transmitted through a wireless channel and acquired at a receiver, it is subject to channel effects that distort the originally transmitted signal. The information produced from channel estimation can be used to suppress and remove the effects of fading and interference. This process is known as equalisation and can be used to restore the originally transmitted signal after it has undergone the effects of a wireless channel. Figure I.1 presents two constellation diagrams produced using a QPSK modulation scheme. The diagram on the left was created before equalisation, while the diagram on the right was created after equalisation (note the different scales). It is clear that channel estimation and equalisation are necessary to prevent signal distortion and bit errors.

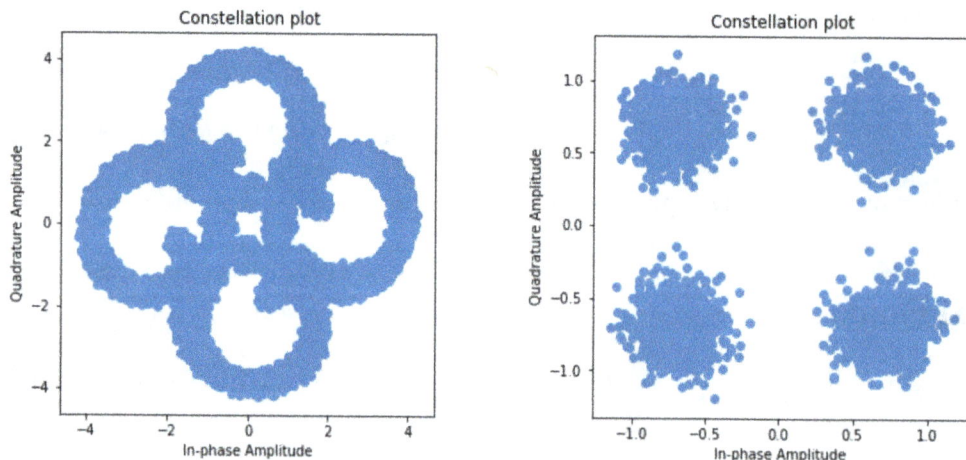

Figure I.1: The received constellation before applying equalisation (left) and after applying equalisation (right).

I.3. RFSoC OFDM Transceiver

The final notebook named ***03_rfsoc_ofdm_transceiver.ipynb*** presents an RFSoC OFDM transceiver design that is capable of transmitting and receiving OFDM waveforms. The radio system simply uses random data to demonstrate its OFDM transmission and reception capabilities. A simplified diagram of the OFDM radio system architecture is present in Figure I.2.

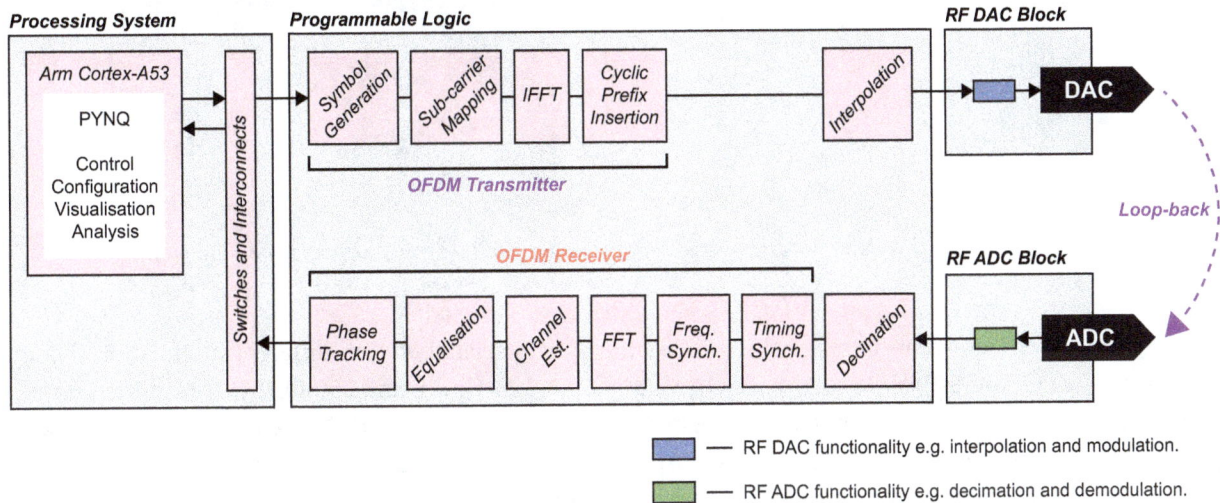

Figure I.2: Simplified architecture of the RFSoC OFDM demonstration system.

The OFDM transmitter consists of symbol generation logic that modulates random data using a digital modulation scheme. The symbols are then grouped into blocks and mapped to data sub-carriers. The final OFDM symbol is created by performing the IFFT and inserting the CP onto the resulting signal. The RF-DAC then transmits the signal, where it is received by the RF ADC using a loop-back connection.

Timing and frequency synchronisation are performed in the OFDM receiver. These processes acquire symbol timing and correct frequency offsets. The FFT is then performed to demodulate OFDM symbols and recover the underlying data symbols. Channel estimation is then performed to determine the frequency response of each sub-carrier, which allows the received data to be equalised. Phase tracking is then implemented to correct phase errors that remain in the signal. Lastly, the data is then passed to Jupyter for plotting and visualisation.

The RFSoC OFDM radio system provides users the opportunity to observe and plot various data paths and waveforms in real-time. This functionality is similar to the RFSoC radio demonstrator example presented in Notebook Set I. For example, Figure I.3 contains time and frequency domain plots of the received OFDM waveform after decimation. Introspection techniques such as observing the received waveform using plots are useful to validate the operation of the radio design.

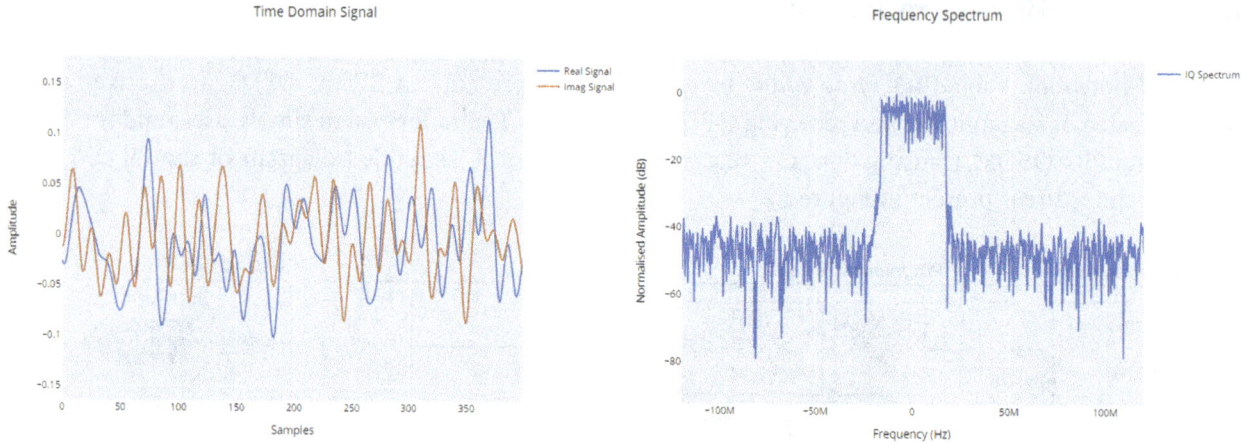

Figure I.3: RFSoC OFDM demonstration system time domain plot (left) and frequency domain plot (right).

The OFDM radio demonstration system also allows the user to select between different modulation schemes including BPSK, QPSK, 8-PSK, 16-QAM, 32-QAM, 64-QAM, 128-QAM, and 256-QAM. Transmitted symbols can be received using a loop-back connection between an RF DAC and RF ADC. The received symbols can be plotted on a constellation diagram for visual inspection. An example of plotting 16-QAM and 256-QAM symbols using the OFDM demonstration system is presented in Figure I.4.

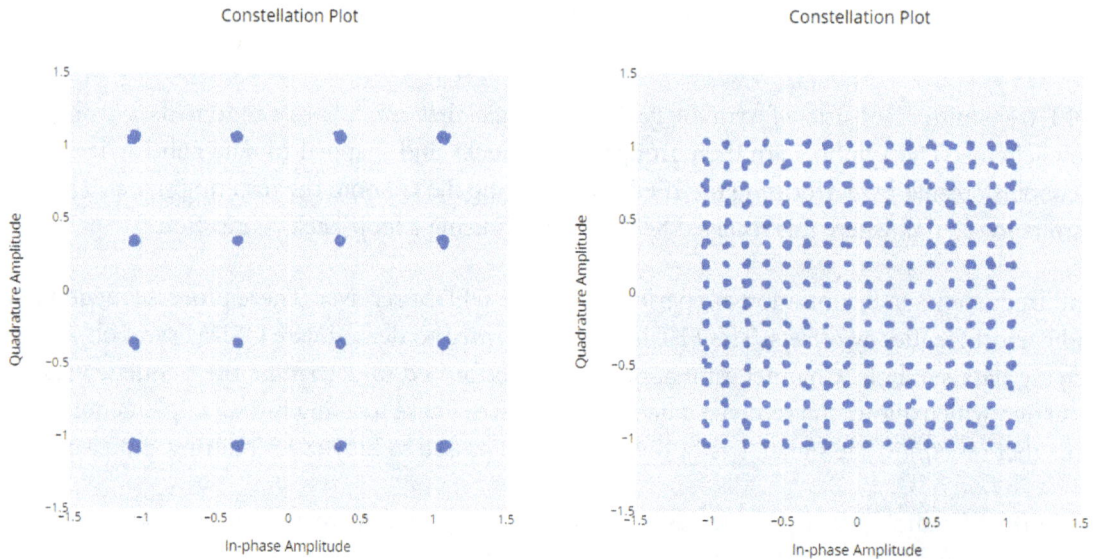

Figure I.4: 16-QAM constellation diagram (left) and 256-QAM constellation diagram (right).

Chapter 17

RFSoC Applications in Cellular Networks

Kenny Barlee

One of the primary use cases for the RFSoC family of products is in cellular networks. The high bandwidth, high performance and multi-channel RFDC SDR interfaces on products such as the RFSoC DFE [81] are well suited to the current 5G basestation implementations, and indeed as we evolve to 6G these RFSoC SDR interfaces will be even more important and prevalent. The RFSoC T1 and T2 Telco Accelerator cards [72],[73] can facilitate L1 (PHY layer) acceleration, and additionally, the T1 features high speed interfaces for use in OpenRAN networks which are becoming a key part of 5G deployments and setting new standards for interoperable components that will form part of the emerging and in-planning 6G standards and networks.

In this chapter, we review the evolving cellular network context, and discuss how both traditional and next-generation implementations can be enabled by RFSoC devices.

17.1. Introduction to 4G/5G Radio Access Networks (RANs)

A Radio Access Network (RAN) is the main part of the cellular network. It features the cellular basestations and a backhaul network that provides connectivity for remote User Equipment (UE, such as smartphones or mobile broadband modems) through to the core network and Wide Area Network (WAN), e.g. the Internet. 4G/5G UEs attach to basestations via 4G Long Term Evolution (LTE) and 5G New Radio (NR) air interface radio links. Figure 17.1 illustrates the primary components in a cellular network architecture.

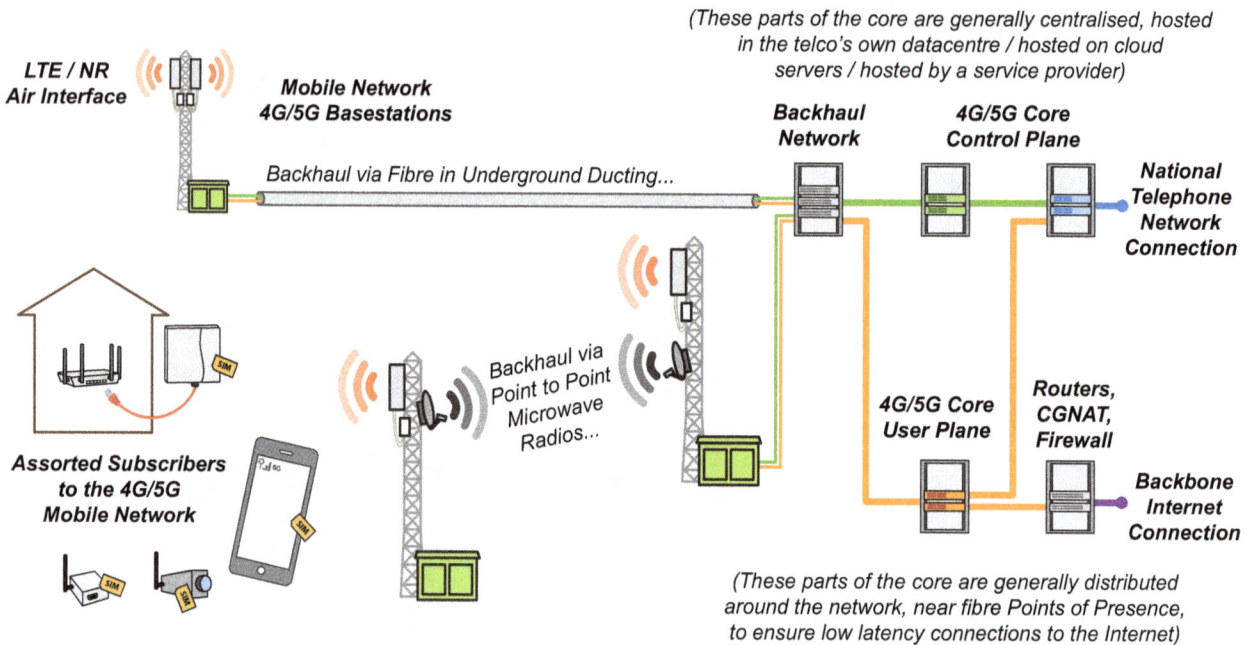

Figure 17.1: Basic components in a cellular network include the Basestation, the Backhaul, and the Core. The Radio Access Network (RAN) comprises the Basestation and Backhaul parts.

For 4G LTE, *long-term-evolution* to current 5G networks a key term is '*evolution*'. The new standards need to evolve from previous standards and equally importantly need to co-exist with other generations (2G and 3G are still out there) or indeed these radios might even be supported on the same RFSoC as 5G. Although 5+ years still in the future 6G as a standard that can enable a new set of applications, higher speeds, more efficiencies in spectrum usage, power and so on, will have many common features that will 'evolve' from 5G and current standards and designs. Therefore as we design and deploy for 5G, for OpenRAN, and for standalone and private SDR networks in the early to mid 2020s, this is in fact implementing with the baseline technologies on which we will now start to build 6G.

17.1.1. Traditional 4G/5G Basestation

A typical 'traditional' 4G/5G basestation site is depicted in Figure 17.2. It consists of a Baseband Unit (BBU), at least one Remote Radio Head (RRH), and an Antenna; alongside backhaul network components such as a cell site router. The primary components of the RAN are described in the following subsections.

Baseband Unit (BBU)

The BBU is fundamentally a high-performance computer or server, connected over the backhaul to the core network. The backhaul network carries control and user plane data packets (which are known as the S1-MME

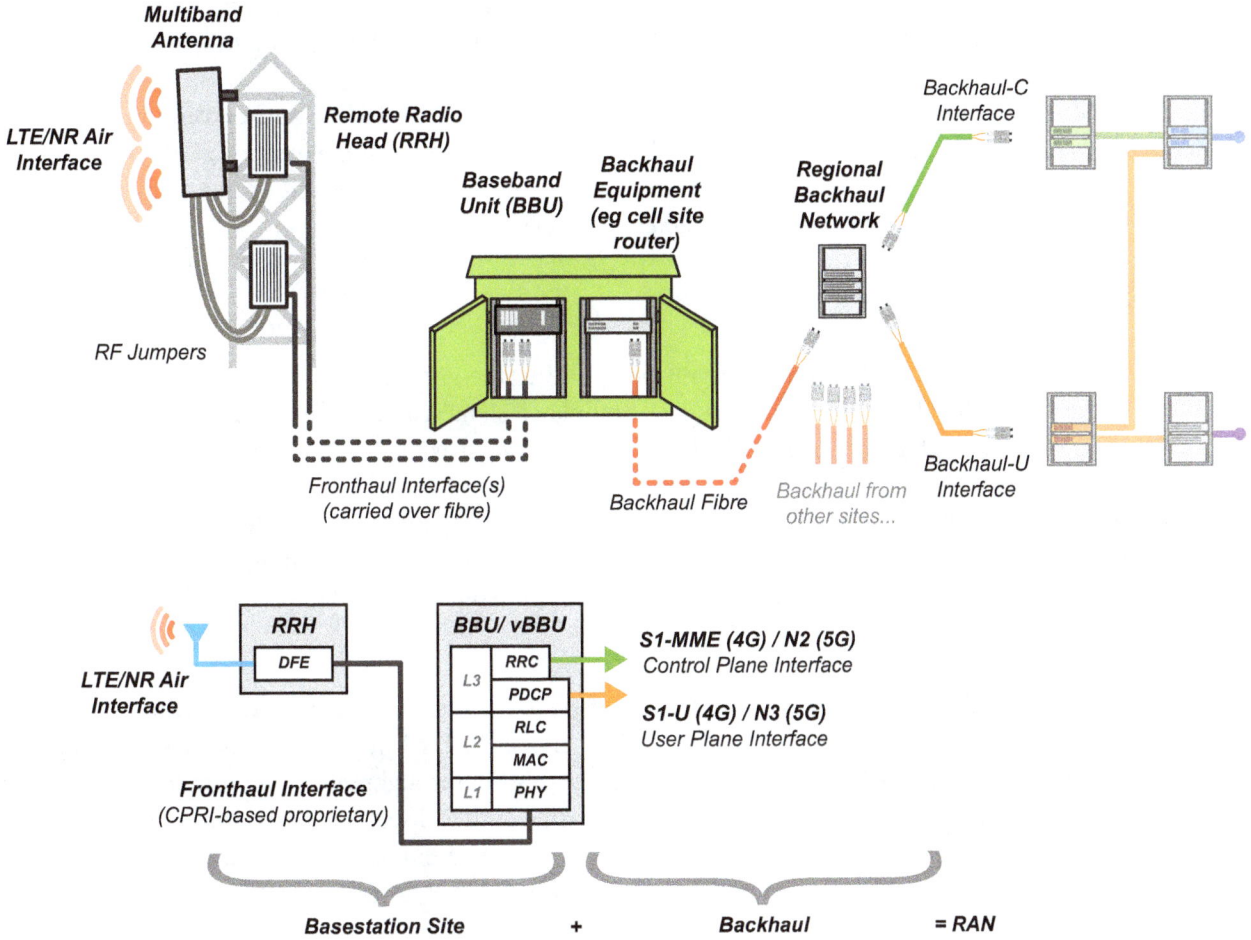

Figure 17.2: Basestation sites in 'traditional' 4G/5G RAN configurations feature a Baseband Unit (BBU), at least one Remote Radio Head (RRH), and an Antenna. The BBU will commonly reside in a cabinet at the base of the mast alongside backhaul network equipment such as a Cell Site Router.

and S1-U interfaces in 4G LTE; and the N2 and N3 interfaces in 5G NR). This connection is either made wirelessly (e.g. point-to-point microwave radio, Low Earth Orbit (LEO) satellite radio) or via cabling (e.g. copper line, fibre). The BBU is also connected to the RRH over a proprietary fronthaul fibre link, commonly using a CPRI interface. (The Common Public Radio Interface is a high speed serial communication interface used in fronthaul that carries radio control and management messages, radio synchronisation information, and raw IQ samples of radio signals [128]). The fronthaul fibre is typically short, running from the cabinet at the base of the mast up to the RRH, but can be up to 10km in length in certain circumstances.

The BBU runs the radio stack software (e.g. the LTE/NR software), and undertakes conversion of data between IP-encapsulated control and user plane data packets received from the core network, and digital baseband samples of modulated LTE/NR signals. The latest vRAN (virtualised RAN) radio stack software can be run on

COTS (commercial off the shelf) computer hardware and does not need special computer hardware provided by the stack vendor. LTE and NR stacks are very similar (considering NR is an extension of LTE).

Like most other communication protocols, there are multiple layers to the 4G/5G stack. From top to bottom, the stack layers in the BBU are: RRC, PDCP, RLC, MAC and PHY. The PHY implements an OFDMA modem, and contains a series of DSP components tasked with converting a stream of 1s and 0s to and from complex baseband modulated samples, as presented in Chapter 16 on OFDM.

Remote Radio Head (RRH)

The RRH is connected to the BBU over fronthaul fibre, and to the antenna over short RF coaxial cables. It converts the digital baseband samples to analogue RF modulated signals, and outputs them on a high powered RF carrier; and vice versa. This final stage of the basestation is known as the Digital Front End (DFE).

It is common for a basestation site to feature multiple RRHs, each operating in a different frequency band. Because RRHs feature high power amplifiers and appropriate analogue RF signal conditioning stages, they are fundamentally band locked. For instance, a RRH designed to operate in LTE Band 3 (1800MHz) cannot be used for LTE Band 28 (700MHz).

Depending on the standard specifications of the band, the RRH transmits and receives in either Frequency Division Duplex (FDD) or Time Division Duplex (TDD) mode. RRHs will generally use MIMO (Multiple Input Multiple Output) technology. MIMO essentially increases the throughput of the radio by adding more transmit and receive paths (MIMO is discussed further in Chapter 18). Most RRHs are configured as at least 2x2 MIMO (2 Tx, 2 Rx), but at the time of writing, 4x4 and 8x8 MIMO radios are also common. A 2x2 MIMO RRH uses two RF coaxial cables to connect to the antenna, while a 4x4 RRH uses four.

Standard bandwidths are defined for LTE and NR signals, and the radio will be configured to broadcast with one of these bandwidths. A base station site with one RRH has a single cell, while a site with 'n' RRHs has 'n' cells. Where multiple bands (and cells) are used simultaneously at a base station, subject to UE support, it is possible to use the technique of carrier aggregation to combine the cells together for greater throughput.

Antenna

The antenna is a passive object that is connected to the RRH, which radiates (transmits) and receives the electromagnetic waves that enable wireless communication. The radiation pattern and performance depends on the antenna size, shape, and band optimisations. The type of antenna chosen varies upon the RF band in use and the network use-cases.

It is common for a site to feature multiple RRHs, and these are either attached to individual antennae; or a multi-band (multi-port) antenna that serves multiple RRHs can be used. In line with the RRH MIMO point above, MIMO antennae are required when these features are used. The elements in a 2x2 MIMO antenna are offset by 90 degrees, giving one transmit path on a vertical plane, while the other is on a horizontal plane. Doing this increases the throughput, and helps to increase the signal strength and resilience. Increasingly, hardware vendors are supplying 'Active Antennae' for higher order MIMO systems (e.g. 64x64 MIMO). These are essentially all-in-one units comprising the RRH and (passive) antenna elements.

17.1.2. 4G and 5G Network Technology

4G networks are very capable; much more so than many people believe. A well kitted out 4G basestation site, with multiple high bandwidth RRH cells offering carrier aggregation, should be able to deliver real-world aggregate downlink data speeds in excess of 1Gbps. Of course, this total capacity must then be divided between all User Equipment (UE — i.e. phones, etc.) attached to the mast (as is also true for 5G networks). There are 4G networks in most countries around the world, and thousands of unique models of UE have been created to work on them. The heart of a 4G LTE network is the Evolved Packet Core (EPC). This features a control and a user plane, and is tasked with managing subscriber records, subscriber authentication, session management, mobility, charging etc.; and processing and outputting user data packets to the WAN.

The transition stage between 4G and 5G has resulted in a hybrid generation called 5G Non-StandAlone (NSA). 5G NSA networks combine 4G and 5G radios together, and can offer far higher aggregate data speeds than 4G alone could manage. In contrast, 'true' 5G is called 5G StandAlone (SA). This requires a 5G Core (5GC), and only uses the 5G radios for connectivity. 5G cells used for both NSA and SA networks run the upgraded version of LTE known as 5G NR. At the time of writing, the majority of NSA basestations connect to a 4G EPC, although with upgraded 4G components, it is possible to connect NSA basestations to a 5GC. The 5GC performs very similar tasks to the EPC, with the main architectural difference being that the network functions communicate with each other using different interfaces.

Another 4G/5G hybrid technology also exists, confusingly called Dynamic Spectrum Sharing (DSS). This is not spectrum sharing, but rather, radio resource sharing. A DSS radio can simultaneously output 4G LTE and 5G NR signals, in the same frequency channel, on a time division basis. The biasing between LTE and NR changes in real-time depending on UE demand. This makes it possible to run 4G and 5G SA networks simultaneously from single cell basestations. New 5G SA features and capabilities can be rolled out in areas where 4G is also required, and it costs less to implement compared to a two cell approach; which would also require two radio spectrum licences. DSS-enabled cells achieve lower overall throughputs than if they were configured for 4G alone, due to an increase in control signalling. DSS cells cannot be configured to operate in NSA mode with themselves; however, the 5G component could be connected with another local 4G cell to offer NSA (i.e. it would be possible to offer NSA at a site with x1 LTE cell and x1 DSS cell).

Finally, it is worth mentioning that *some* 5G radios can be connected both in 5G NSA and 5G SA at the same time. Therefore, on a base station site with x1 LTE cell and x1 NR cell, it would be possible to simultaneously support 4G, 5G NSA and 5G SA, and connect back to both an EPC and 5GC.

17.2. Evolution to 4G/5G OpenRAN Networks

OpenRAN (also referred to as Open RAN, O-RAN and ORAN) is a variant of the traditional RAN network architecture discussed in the previous section. It was brought about by supply chain diversification efforts and sees the layers of the LTE/NR stack 'split' into pieces. The new components are called the Centralised Unit (CU), the Distributed Unit (DU), and the Radio Unit (RU). New interfaces are introduced in a midhaul stage (a control plane interface *F1-C*, and a user plane interface *F1-U*), and what was a proprietary fronthaul in the 'traditional' RAN has been replaced with Open Fronthaul (control plane interface *F2-C*, and user plane interface *F2-U*).

The idea behind OpenRAN is that different vendors can provide hardware / software for each of the components, thanks to new standardised open interfaces. The open ideology also means that the software components in the CU and DU can be virtualised for maximum computing efficiency, and run on COTS computer hardware.

17.2.1. OpenRAN Architecture

The OpenRAN architecture is illustrated in Figure 17.3, and its major components are summarised as follows.

Centralised Unit (CU)

The CU software encompasses the RRC and PDCP stages of the 4G/5G stack. It is commonly split into two parts, with control plane CU-C and user plane CU-U components, linked by the E1 interface. These are not always co-located, and the CU-U may be deployed to the edge, alongside an edge 4G/5G core user plane server. CUs act as aggregators, and are connected to multiple DUs over the F1-C and F1-U midhaul interfaces. It is envisaged that the CU-C is hosted in the Telco's private cloud, next to the core network, potentially operated by a service provider.

Distributed Unit (DU)

The DU software normally exists on a server in the cabinet at the basestation site, but it can run on a server up to 10km away if desired. Each DU is connected to the (shared) CU over midhaul, and a single RU over the open eCPRI fronthaul F2-C and F2-U interface. (Enhanced-CPRI (eCPRI) fronthaul can be carried over routed packet switched networks[148], while the previous form of CPRI only supported direct dedicated point-to-point links). Multiple DUs can coexist on one server, using virtualisation software. The DU software features the RLC, MAC and PHY parts of the LTE/NR stack. Depending on the exact OpenRAN configuration, either the full PHY is implemented in the software and digital baseband samples are output over the fronthaul link to the RU; or only the High part of the PHY, denoted as PHY-H, is implemented in software. In this latter case, the Low part of the PHY, otherwise known as PHY-L, is implemented on the RU.

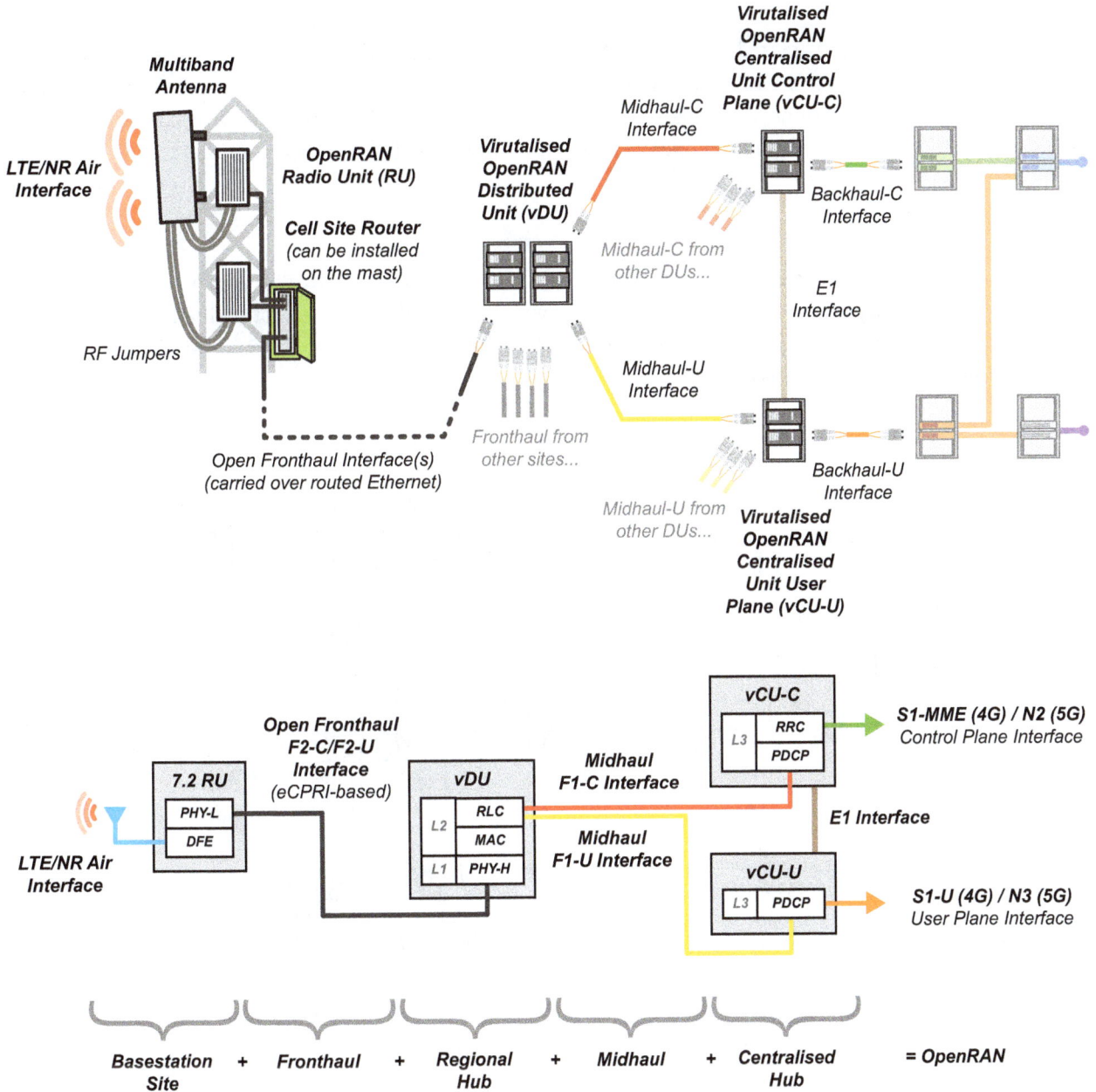

Figure 17.3: *Basestation sites in a 4G/5G OpenRAN network feature a cell site router, at least one Radio Unit (RU), and an Antenna. Distributed Units (DUs) that serve many RUs are hosted in regional hubs, and Centralised Units (CUs) in centralised hubs; i.e. co-located with the core control and user plane servers.*

Radio Unit (RU)

Akin to the RRH, the RU is tasked with converting the digital baseband samples to analogue RF modulated signals, and outputting these on a high powered RF carrier; and vice versa. In other words, it is the DFE. With some OpenRAN configurations, the RU also implements parts of the PHY (this will be discussed in the next section). The RU is often configured to operate in a MIMO mode, with 4x4 or 8x8 common for 5G NR cells.

17.2.2. OpenRAN Splits

There are various splits (implementations) that can exist in OpenRAN networks, which refer to the partitioning of functionality across the RU, DU and CU. There are a number of different defined ways to perform the split of the stack, and these have been numbered 1-8, as shown below. (Note the RLC, MAC and PHY have -H High and -L Low components).

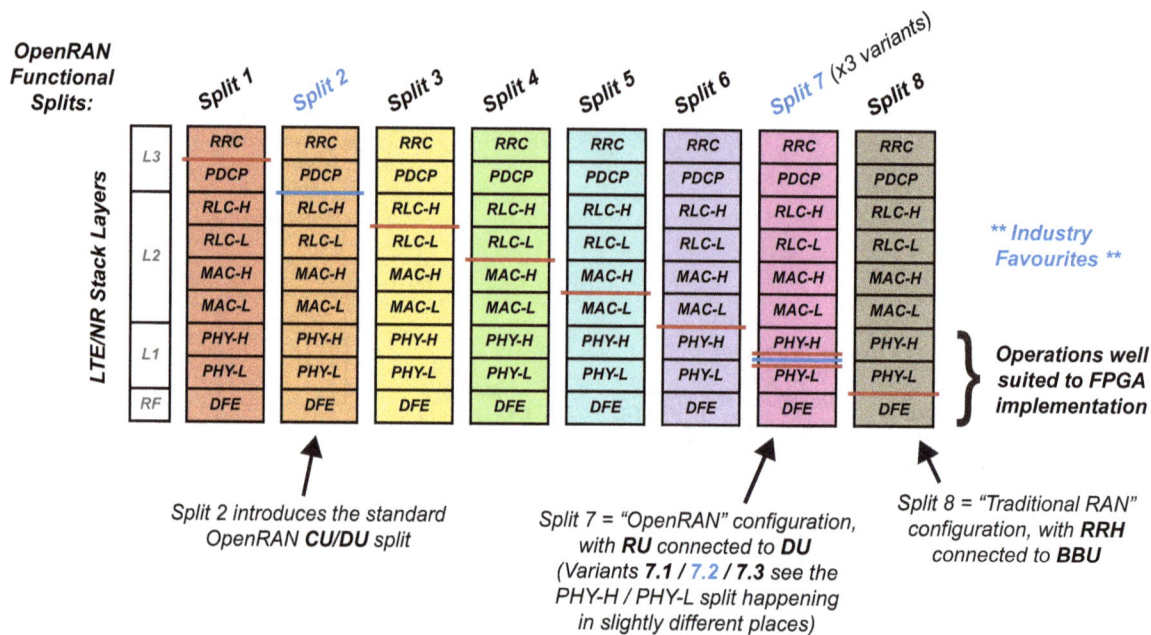

Figure 17.4: Diagram showing the numerous OpenRAN Splits.
Industry favourites are indicated by the blue text.

Splits are commonly combined. For example, if the network was being designed with distinct CUs and DUs rather than an all-in-one CUDU processing unit, Split 2 will likely be used. Another split is used between the software stack and RF stage. Split 8 can support traditional RRHs, and therefore can be used as a convenient 'stepping stone' during OpenRAN network upgrades. Compared to Split 8, Split 7 offers a distinct advantage, as part of the PHY-L is pushed onto the RU. As will be talked about in the next section, this will reduce the throughput requirements on the fronthaul link. There are three variants of Split 7, each with a minor difference in the location of the PHY-H / PHY-L split. Split 7.2 (the industry favourite variant) sees the symbol mapping, iFFT/FFT, cyclic prefix, synchronisation and Digital Pre-Distortion (DPD) stages implemented on the RU; and these DSP operations are perfectly suited to FPGA acceleration.

The most commonly discussed OpenRAN implementations are Split 7.2 (all-in-one CUDU + RU), and the disaggregated RAN model with Split 2 + Split 7.2 (CU + DU + RU). The latter is the architecture presented in Figure 17.3.

17.2.3. Disaggregated RAN: Motive for OpenRAN Split 7.2

As described above, OpenRAN Split 7.2 is where the Low PHY is implemented in the RU. The primary reason for doing so is to achieve reduced data rates on the fronthaul connection between the DU and RU. The best way to explain this is with an example:

Imagine a traditional RAN configuration with a RRH broadcasting a 20MHz LTE cell. The RRH receives a continuous feed of complex digital baseband samples from the BBU. Each sample has a digital wordlength, and both I and Q components. These complex samples are normally stored in a 32-bit binary word. As the sample rate for a 20MHz cell is 30.72Msps, this means the throughput on both the uplink and downlink paths from BBU to RRH is equal to 983.04Mbps per SISO channel. Therefore, if the cell operates in 4x4 MIMO mode, the throughput on the fronthaul paths would be 3,932.16Mbps in each direction.

An OpenRAN 20MHz LTE cell operating with Split 7.2 is fed a stream of QAM symbols from the DU, rather than complex digital basesband samples. The maximum Modulation Coding Scheme (MCS) supports 256 QAM, so each symbol can be represented in an 8-bit binary word. The iFFT in the RU has 2048 bins. Only 1334 data symbols are required for each iFFT operation (as the remaining 714 are for a guard band), and the iFFT sample rate remains at 30.72MSps. Therefore, the overall data throughput per SISO path from DU to RU is equal to 160.08Mbps per SISO channel; a fronthaul capacity reduction of around 80%! (Additional OpenRAN control signalling is required, however this is insignificant in the overall context). For complete comparison, if the cell was to operate in 4x4 MIMO, the data rate on the fronthaul paths would be 640.32Mbps in each direction.

This all plays into the disaggregated RAN model, where there are DU 'hubs' distributed geographically, connected into centralised CUs, as presented in Figure 17.3. Each of the virtualised DU software stacks (vDU) serve an RU in the area around them. A small number of high performance servers can be used to run the numerous vDU software stacks, and eCPRI fronthaul connections carried over high speed packet switched networks will connect the vDUs to the basestation sites. Due to the significant reduction in fronthaul capacity requirements per vDU to RU link, it is possible to aggregate multiple 4G/5G RU fronthaul eCPRI connections onto a single Ethernet connection (e.g. a Tx/Rx fibre). This can greatly reduce the cost of deploying basestation sites, as fewer 'pipes' are required for fronthaul connections. Cell site routers are installed at each basestation site, and these then connect to each of the RUs to complete the routed eCPRI fronthaul interfaces.

With this eCPRI aggregation, it would be possible to connect a basestation site with a fibre connection at a line speed of 10Gbps, and simultaneously support five 4x4 MIMO 20MHz LTE cells, a 4x4 MIMO 50MHz NR cell, and an 8x8 MIMO 50MHz NR cell (and have capacity to spare!). To achieve a similar 'basestation hotel' style configuration in a traditional Split 8 Centralised RAN (C-RAN, where BBUs are grouped together, away from the basestation sites), six 10Gbps fibres and one 25Gbps fibre would likely be required. As the aggregated eCPRI link is carried over a packet switched network, it would be possible to use point-to-point microwave radios, or even LEO satellite radios instead of fibre in the fronthaul, for example in rural scenarios where laying kilometres of fibre in trenches to connect remote sites is not economically viable.

Another advantage of removing compute-intensive DU software from the basestation site and virtualising it in a distributed 'hub' configuration is a large power requirement reduction at the site. In place of a multi-kW electricity grid connection (which can cost $15,000+ to install and commission in rural areas, with ongoing energy costs), it may be possible to run the RUs, cell site routers and any backhaul radios from renewable energy sources, without a mains connection; e.g. with some solar panels, a small wind turbine, an inverter and a bank of batteries. This can greatly reduce the cost to deploy and run greenfield rural basestation sites.

17.3. RFSoC Products for Applications in Cellular Networks

As mentioned in the introduction to this chapter, one of the primary use cases for the RFSoC family of products is in cellular networks. Example product placements for the AMD Zynq RFSoC DFE and T1/T2 Telco Accelerator cards are shown in Figure 17.5.

17.3.1. RFSoC DFE for RRH and RU Applications

The RFSoC DFE has high performance, high bandwidth, multi-channel (8T8R) RFDC SDR interfaces that support Direct RF conversion for all FR1 bands up to 7.125 GHz [81]. It features a number of Hard IP cores that can efficiently accelerate PHY-L and DFE operations, such as: FEC cores, iFFT/FFT, programmable filters, Crest Factor Reduction (CFR), complex equalisation, DPD, DUC/DDC and mixers. These are highlighted in Figure 17.6. The remaining stages of the 4G/5G PHY-L stack (such as the cyclic prefix, windowing, synchronisation, AGC) can be implemented using Soft IP cores on the adaptable logic fabric. This RFSoC DFE product, therefore, is designed to accelerate the development and deployment of next generation RRHs and OpenRAN RUs for 4G/5G networks [114].

RRH/ RU Implementation

An example RFSoC DFE based RU hardware implementation is presented in Figure 17.6. As illustrated, the transmit and receive chains of an OpenRAN RU OFDMA modem implemented on the DFE comprise a mixture of Hard and Soft IP core components, alongside software management and control systems running on the processor. Some of the Soft IP components required in the OFDMA modem are provided pre-assembled with the DFE, while others require custom implementation. Many individual basestation radio

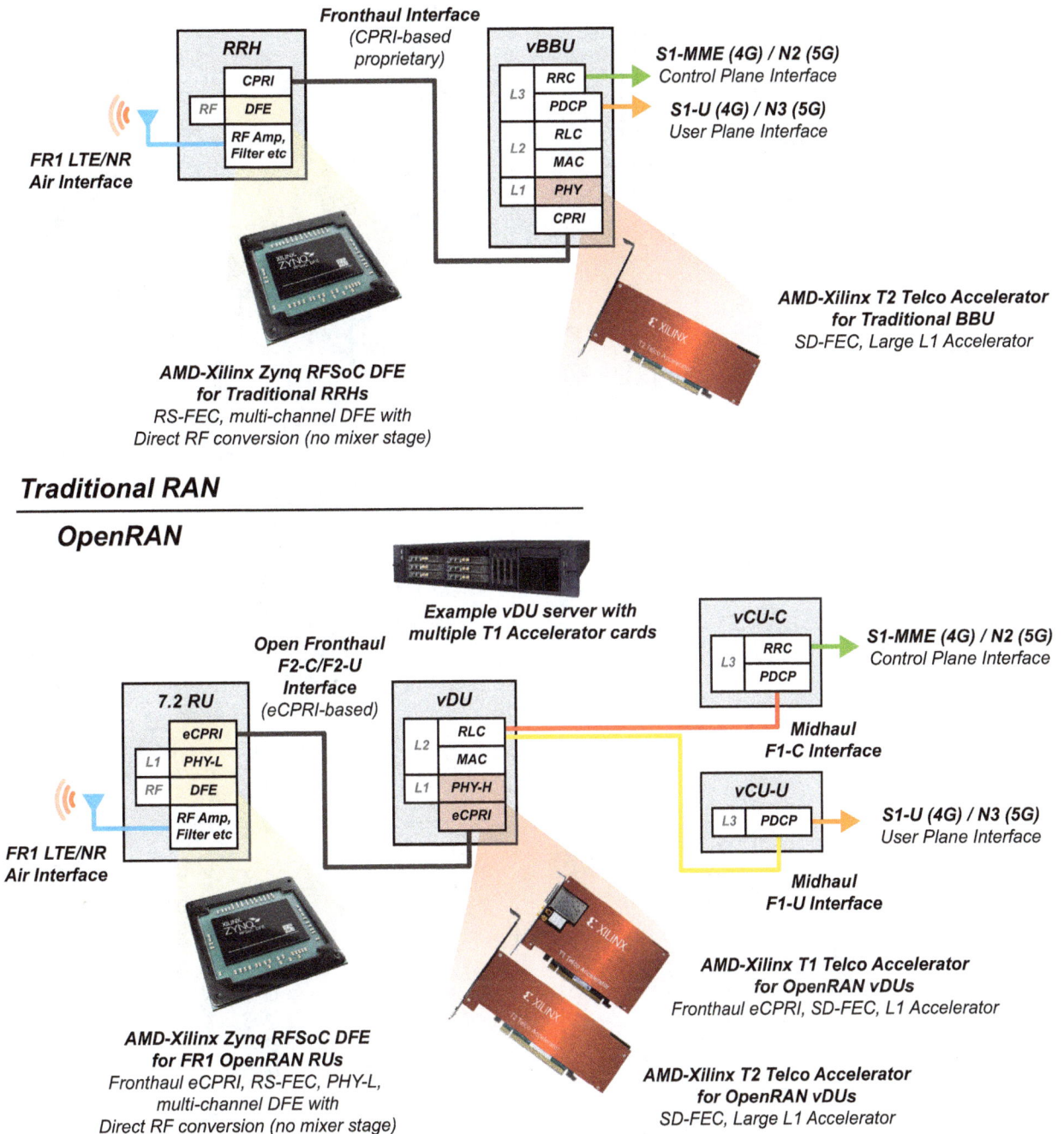

Figure 17.5: RFSoC DFE and T2 Telco Accelerator use in Traditional RAN Networks, RFSoC DFE and T1/ T2 Telco Accelerator use in OpenRAN Networks

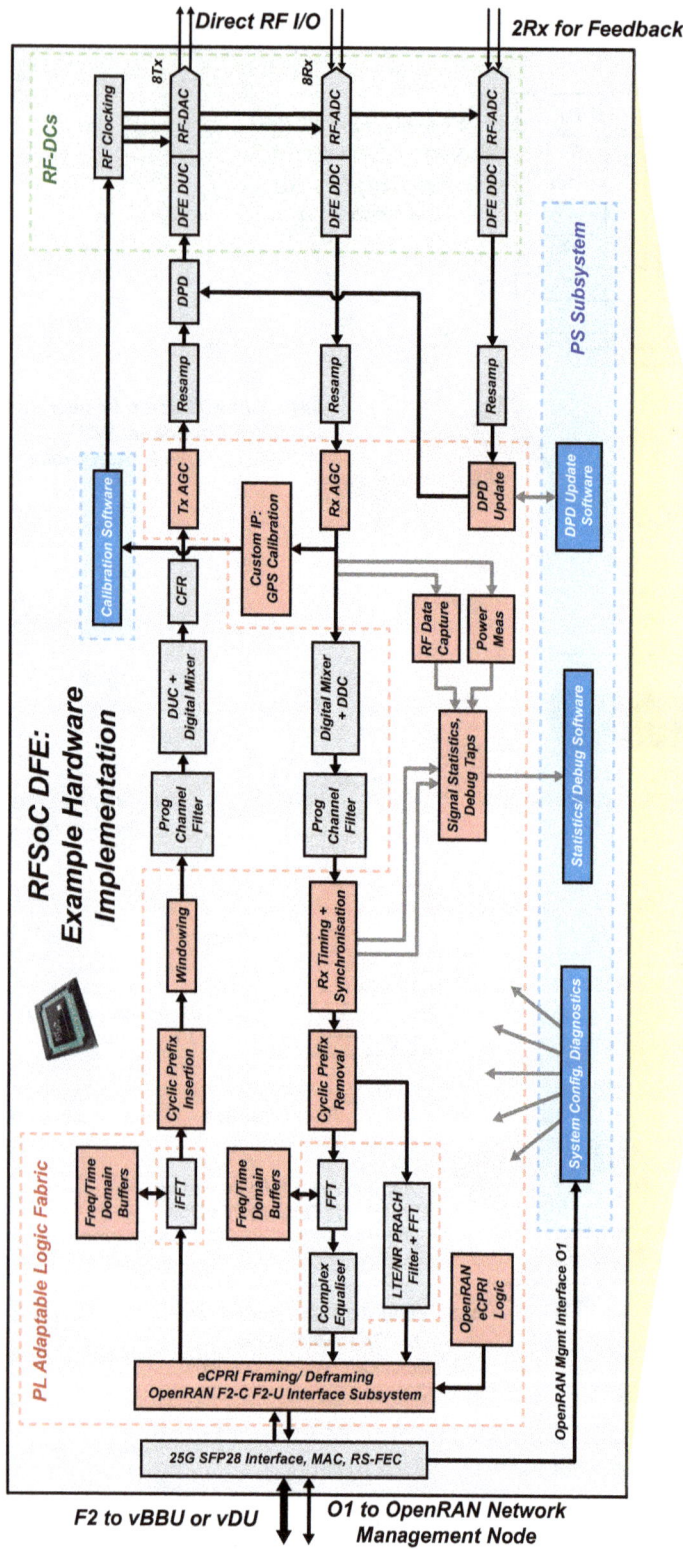

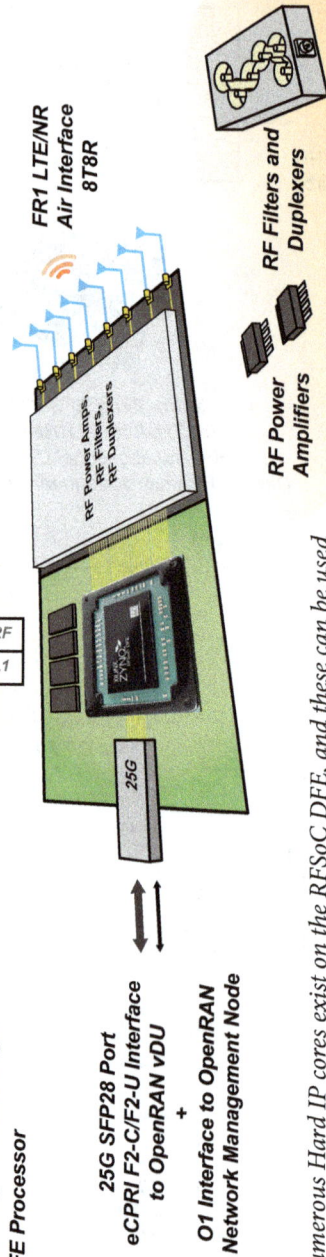

Figure 17.6: *Numerous Hard IP cores exist on the RFSoC DFE, and these can be used alongside adaptable logic fabric Soft IP cores and software on the PS to implement key parts of next gen RRHs and OpenRAN RUs. This diagram shows an example configuration for an OpenRAN 7.2 RU, and highlights where the DFE chip sits inside the physical radio.*

manufacturers have 'value add' features (such as novel synchronisation systems, neighbour discovery, or unique implementations of components they own patents for), so the flexibility to implement custom IP is very attractive in this regard, and likely offers significant advantages over fixed ASIC-based solutions.

As has been highlighted, an RF signal conditioning stage is required after the RFDCs. When the RFSoC DFE performs Direct RF conversion and outputs analogue RF signals on their RF carriers, the signal has an extremely low power level. If this was connected directly to an antenna, the 'radio' would not provide any significant coverage; and would probably be limited to a few metres at most. Therefore, RF Power Amplifiers (PAs) are needed to increase the signal power. In turn, analogue RF Bandpass Filters are required in order to ensure the PAs do not transmit unwanted energy at other frequencies. And finally, in order to combine pairs of transmit and receive ports together, RF Duplexers are used.

These components are band specific. A wideband RF PA, for example, might only offer a 'linear' power gain of 20dBm over a 200MHz window of the spectrum, between 3800-4000MHz. (Note that in practice, the PA will not be perfectly linear, and there will be slight amplitude and phase distortions across the 200MHz window). Outside this frequency range, the performance of the amplifier is likely to rapidly decline. The gain here will be 'non-linear', which would cause problems when used for wideband multicarrier waveforms such as LTE/NR; and high power harmonics may be introduced, affecting the signal quality. Therefore, this PA could not be used for a radio broadcasting at 3700MHz, as the signal being amplified by the PA would be 'damaged' in the process.

The frequency response of a PA is similar to that of a bandpass filter. There is a linear pass band, with transition bands above and below it. *(Normally a digital FIR bandpass filter is designed to have tight transition bands and high attenuation in the stop bands; and this is easy to achieve by using a large number of filter coefficients).* The difference is that the non-linear 'transition bands' in the analogue PA can be hundreds of MHz wide, and a significant amount of power will be applied throughout them. Therefore, an accurately designed RF Bandpass Cavity Filter is required to isolate only the band of interest.

Cavity Filters are large, slightly strange-looking units, generally manufactured by accurately milling holes into a block of aluminium. These holes are 'cavities', and they contain rods referred to as 'resonators'. Well designed Cavity Filters have a low insertion loss in the passband, and excellent frequency selectivity (extremely tight transition bands, and sometimes >100dB rejection in the stopband). Furthermore, they can operate with very high signal power levels, as would be found in a macro basestation radio. In the world of cellular standards, spectral emissions limits have been specified that state acceptable levels of Adjacent Channel Leakage Ratio (ACLR, a measure of relative power) — for example, in the TS 38.104 specification for 5G NR basestations [155] — and the Cavity Filters must be designed so that the radio as a whole meets these requirements.

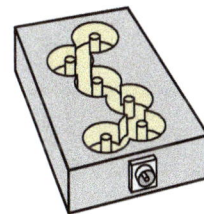

Figure 17.7: Illustration of a cavity filter, milled from a block of Aluminium

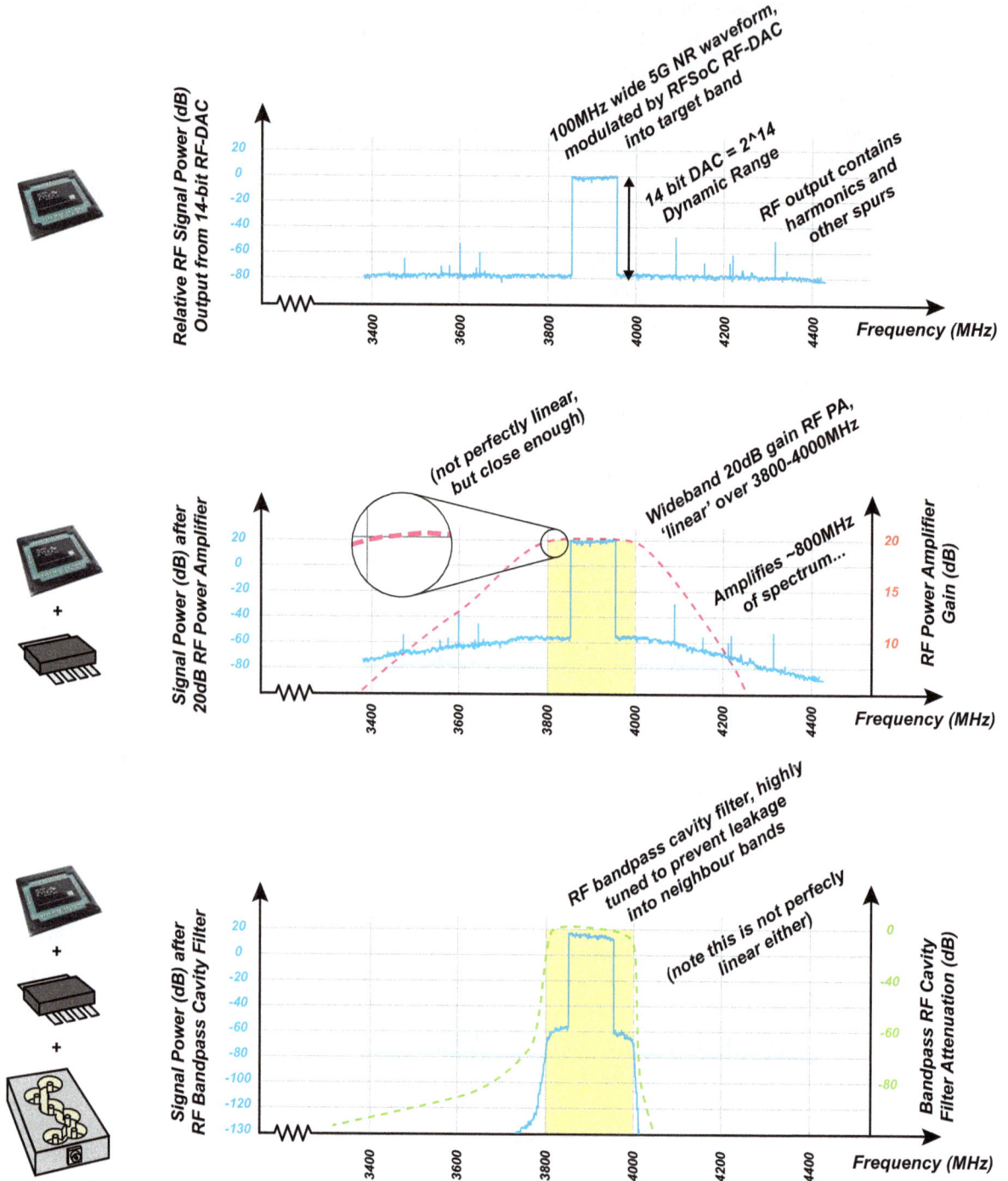

Figure 17.8: *An RF signal conditioning stage is required to amplify and 'clean up' the output of the SDR, and ensure that it only transmits energy in the band of interest. This is a common requirement for radio spectrum licence applications.*

Commonly, RF Duplexer functionality is integrated into the cavity filter. Duplexers are components which enable bi-directional communication (Tx, Rx) over a single RF path and antenna element. *(As highlighted in Figure 17.6, an 8T8R RU RF interface will likely only feature 8 physical RF ports — the duplexers allow each of the ports to transmit and receive simultaneously).* The Cavity Filter/ Duplexer designer will ensure there is a significant attenuation between transmit and receive ports, so that the transmit path does not saturate the receive path of the SDR. This is especially important in FDD radios, which operate Tx and Rx paths simultaneously.

Each of these stages is presented in Figure 17.8. All of these components: the RFSoC DFE with the RF-DCs, the RF PA, and the RF Cavity Filter/ Duplexer, generate a significant amount of thermal energy. As a result, a cooling system is required in order to keep components operating within temperature tolerances. Radios designed for outdoor installation (e.g. at the top of a radio mast) must be watertight, and as a therefore need to be passively cooled. As illustrated in Figure 17.3, this is achieved using radiator 'fins' on the radio chassis; heat is lost by cold air falling vertically through the fins. Radios designed for indoor use, or installation in watertight cabinets, may use active cooling systems with fans and vents.

Multicarrier Transceiver

A single 8T8R RFSoC DFE chip can support multiple cells across multiple 4G and 5G cellular bands at the same time, as shown in Figure 17.9. All of these cells could be used simultaneously by an operator to allow greater throughputs by leveraging 4G/5G carrier aggregation, or 5G NSA dual connectivity.

Alternatively, an interesting model here is a multi-operator basestation, where each of the cells is connected back to a different cellular network. This "neutral host" radio configuration can enable very cost and power efficient multi-operator deployments, and could have applications in both indoor small cell networks (e.g. airports, train stations, sports stadiums, high rise office blocks) and macro outdoor networks.

mmWave Modulation

While the RFSoC DFE supports Direct RF conversion for all FR1 bands, it is also possible to use it in an Intermediate Frequency SDR configuration, as part of an FR2 mmWave radio. (The FR2 bands span from 24.25 GHz to 52.6 GHz). This architecture is presented in Figure 17.10.

Additional RF hardware is required to achieve this. Instead of outputting an analogue RF signal from the RF-DAC to the RF signal conditioning stage, as with the FR1 mode of operation, the analogue IF signal (i.e. the mmWave signal on a sub 7 GHz IF carrier) is mixed up to the correct carrier frequency, then amplified, filtered and output from the transmitter. The wide carrier bandwidths that can be achieved with RFSoC mean that the radio can support the headline 400MHz Ultra Wideband (UWB) FR2 waveforms.

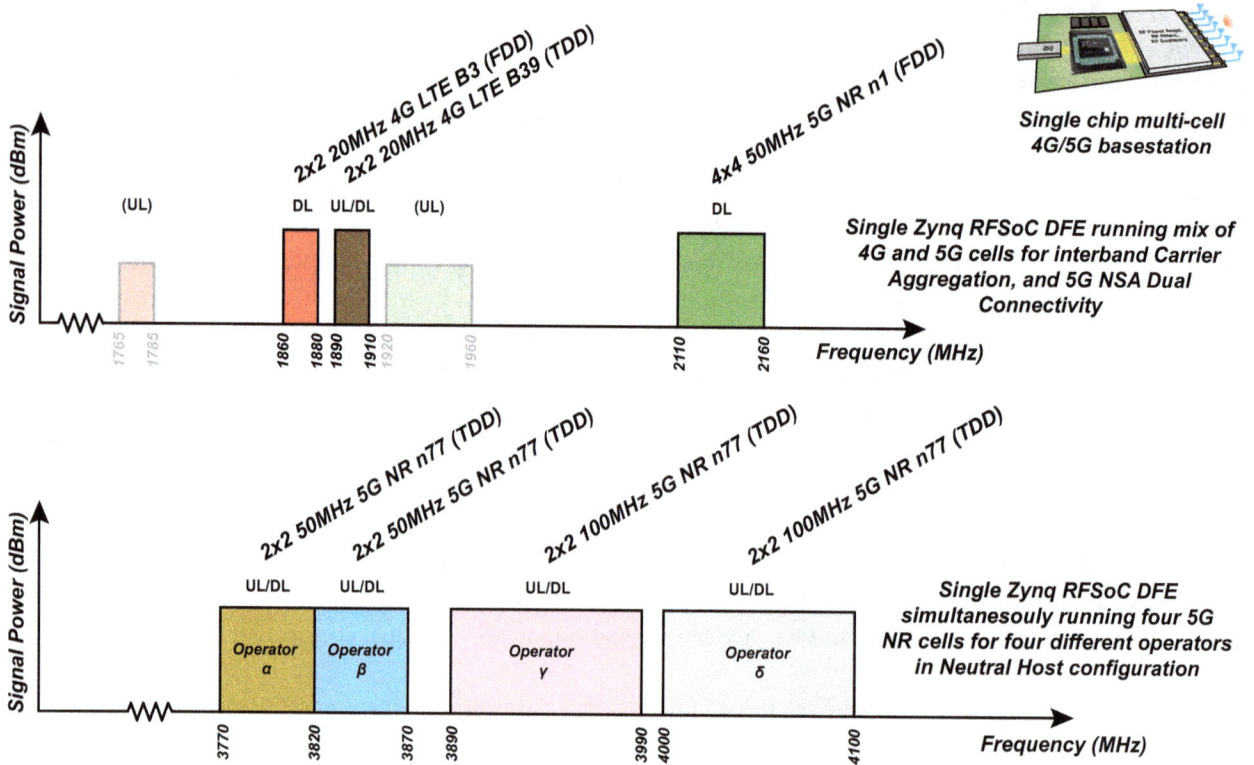

Figure 17.9: A single 8T8R RFSoC DFE based radio can be used as a multi-cell 4G/5G basestation; or even be combined with other DFE chips to create massive MIMO systems

17.3.2. RFSoC Telco Accelerator Cards for vBBU and vDU

The most compute-intensive processes in the PHY-H stack are the real-time Channel Encode and Decode stages. With the transition to using General Purpose Processors (GPPs) and standard server hardware to implement vBBU and vDU software, these stages are a significant bottleneck when it comes to achievable throughputs, and the number of visualised stacks that can simultaneously run on each server.

The T1 and T2 Telco Accelerator cards have been designed to facilitate high speed L1 PHY-H CPU offload and hardware acceleration in the vBBU/vDU. Additionally, the T1 features SFP28 eCPRI interfaces for use in OpenRAN fronthaul. An example vDU implementation is presented in Figure 17.11. The components highlighted in red are the parts of the stack that can be accelerated by the reference design provided with the board.

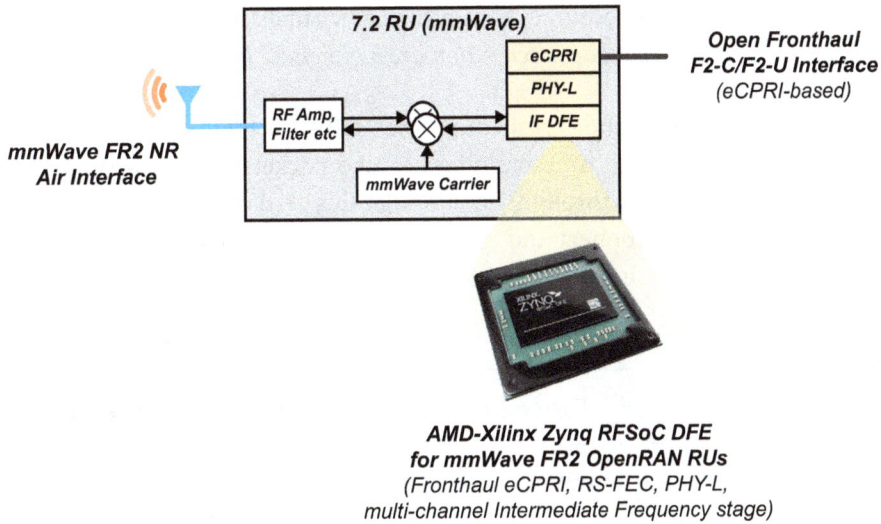

Figure 17.10: RFSoC DFE can be used as an Intermediate Frequency SDR stage in FR2 mmWave radios

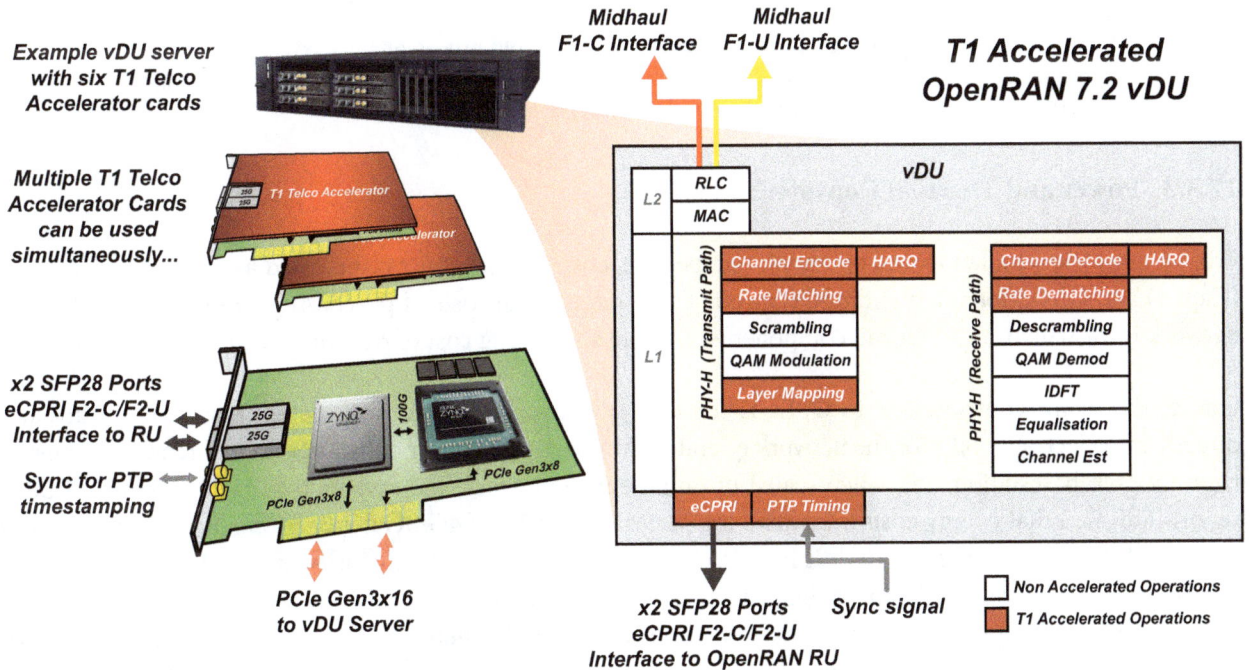

Figure 17.11: Compute intensive L1 PHY-H functions in the vDU can be accelerated by the T1/T2 Telco Accelerator cards. The T1 features an eCPRI fronthaul interface, while the T2 has a larger L1 accelerator.

Tests have shown that shifting the Encode and Decode stages onto a T1/T2 Telco Accelerator card drastically increases performance, freeing up GPP resources to run further virtualised vBBU and vDU stacks in parallel. *(Test results demonstrating T1 Encode/Decode acceleration are highlighted in Table 17.1).*

The T1 card is capable of accelerating and terminating (via eCPRI) a total of 16 TRx layers with 100MHz bandwidth, and supports IEEE 1588 PTP timestamping of radio packets with sub-ns accuracy. The T2 card offers roughly double the L1 acceleration throughput, although this card does not feature the fronthaul eCPRI interfaces. With both, it is possible to connect and use multiple cards simultaneously on a single high performance host server, increasing visualisation efficiencies and scalability.

Table 17.1: Live measurement results of T1 hardware acceleration

L1 Stage	GPP only [1]		GPP + T1 Telco Accelerator [2]	
	Throughput	Latency	Throughput	Latency
Channel Encode	0.718Gbps	45µs	17.7Gbps (x24 higher)	14.15µs (x3.2 lower)
Channel Decode	0.183Gbps	62.7µs	7.8Gbps (x42 higher)	16.21µs (x3.8 lower)

1. Single thread performance measured on Intel® Xeon® Gold processor;
2. Single thread performance measured on Intel Xeon Gold processor with T1 Telco Accelerator card; *both running FlexRAN software on Dell™ R740 server [71].*

17.3.3. Power and Thermal Considerations

From a network installation and operation perspective, costs are generally categorised as Capital Expenditure (CapEx) and Operational Expenditure (OpEx). These refer to the costs of purchasing and installing all of the network infrastructure and related components, and the continuing cost of running the network, respectively.

One of the major contributions to OpEx is the energy consumed by network equipment, which translates directly into electricity bills for the network operator. Any saving in power consumption will reduce the cost of running each basestation, and when scaled up to an entire network, the savings may be significant. A 2019 report indicated that energy costs accounted for 21% of the OpEx of 4G/5G network operators in the UK and developed Asia-Pacific regions [91]. It has also been established that the majority of mobile network energy consumption is in the RAN (approximately 73%, with the remainder attributed to the network core, data centres, and other operations [184]). As well as costs, naturally there are also environmental reasons for seeking to reduce energy consumption [276].

With these factors in mind, there is a clear motivation to optimise the power efficiency of each component within the network infrastructure; even though the electronics might not be the biggest power-consumer (power amplification and radiated power being more significant), savings are still worthwhile. Therefore, the power-saving attributes of the RFSoC previously discussed in Section 3.5.5 (i.e. savings due to integration, hardening of computationally intensive blocks, support for DPD to linearise power amplifier gain, and the power management features of the PMU) are extremely relevant to cellular network operations. The RFSoC DFE is particularly optimal in terms of hardening, with its functionality tailored to the requirements of 5G NR mobile networks [82].

Improving the thermal performance of the network infrastructure, particularly at the RRH, is significant both in terms of CapEx and OpEx. The RRH is situated close to the antenna and is exposed to the weather, and therefore requires to be in a sealed unit; pushing hot air out via fans is therefore not practicable, and instead, heatsinks must be used to dissipate heat. The use of lidless packaging for RFSoC devices (mentioned in Section 3.5.5) helps to achieve a lower operating temperature than with a conventional package [300], and this implies that smaller heatsinks can be used, which is beneficial in terms of materials and manufacturing costs, as well as ease of installation. In terms of OpEx, lower operating temperature is advantageous because components can typically enjoy longer lifetimes, easing maintenance requirements and reducing downtime.

17.4. Chapter Summary

This chapter has introduced the general architectures of current and next-generation OpenRAN cellular networks, and introduced the various components that comprise a RAN. We have also highlighted candidate cellular use case cases for some of the products in the RFSoC family, and commented on the energy and cost considerations.

Chapter 18

MIMO and Beamforming

James Craig and Blair McTaggart

One of the most significant features of RFSoC is its multiple RF-ADCs and RF-DACs, which enables systems with several transmit and receive channels to be implemented. As noted in Chapter 3, most RFSoC devices have either 8 or 16 RF-DACs and RF-ADCs, which can either be used individually (for real signals) or in pairs (for complex signals). This multi-channel capability enables applications such as multi-standard radio systems, wherein several different bands and protocols are serviced on the same chip — for instance, an access point serving cellular, Wi-Fi, GPS, and Bluetooth connectivity. Two other notable applications, which form the basis of this chapter, are Multiple-Input-Multiple-Output (MIMO) systems, and beamforming.

In MIMO systems, the transmitter and receiver both have multiple signal paths and antennas, which contrasts with the conventional single transmitter, single receiver configuration (also known as Single-Input-Single-Output, or SISO[1]). As will be reviewed over the coming pages, MIMO creates a more diverse set of signal paths through the radio channel, which can be exploited to increase overall data rate.

Beamforming involves the use of multiple antennas at either the transmitter or receiver. By applying DSP techniques to the set of input/output signals, the beam can be electronically steered in a desired direction. This has the advantage that a transmitter can concentrate its transmitted power towards a desired target, or that a receiver can be most sensitive in a particular direction, all without any mechanical steering of antennas. Beamforming and MIMO are key technologies for 5G networks, and will be even more prevalent for 6G.

Both of these topics would be worthy of books on their own — the intent of this chapter is to provide an introductory overview of each, noting RFSoC implementation aspects, with suggested sources for further reading.

---

1. Multiple-Input-Single-Output (MISO) and Single-Input-Multiple-Output (SIMO) are two (less common) variations.

18.1. Introduction to MIMO Systems

MIMO is a technique that improves the data throughput and reliability of communication systems by using multiple (typically two to eight) transmit and receive antennas each to simultaneously send several data streams. Since there are many different paths between each set of transmitter and receiver antennas, the signals experience different channel effects. By using the various transmitter and receiver processing techniques discussed later in this chapter, the impact of these channel effects can be mitigated and even exploited, providing a significant improvement over SISO systems in many cases.

The concept of MIMO can be traced as far back as the 1970's, with research papers discussing digital transmission systems using multiple channels [223], and now it has become an essential technique in 4G Long Term Evolution (LTE) and 5G New Radio (NR) communications. Mobile data traffic is increasing at an exponential rate [266], and to keep up with this growth, there must be a corresponding increase in the achievable throughput of wireless networks. Network throughput is a measure of the data per second in a given area, and can be split into three main factors, i.e.

Area throughput [bits/s/km^2] = Bandwidth [Hz] · Density [cells/km^2] · Spectral Efficiency [bits/s/Hz/cell]

In the past, approaches to increasing capacity (or throughput) have focused on allocating more bandwidth and the densification of base stations; however, further growth in either sense is unlikely to be practical, or indeed sufficient. The amount of RF spectrum that can be used for mobile communications is limited, for both physical and regulatory reasons, and therefore allocating more bandwidth is not generally possible. When considering the base station costs for site acquisition, backhaul links, and construction, as well as maintenance and environmental factors, and the elevated interference levels implied, then it is clear that increasing the density of basestations is not a viable solution in the long term either.

Attention then turned to improving the spectral efficiency of cells, which can be achieved in a variety of ways, such as increasing the transmit power or modulation order [109]. One of the key technologies in enhancing spectral efficiency is the use of multiple antennas to simultaneously transmit over the same band of frequencies, but with spatial or directional separation. For example, via the use of sectored antennas pointing away from the mast at regular angles. Another approach is to use arrays of transmitters and receivers to generate diversity in the transmission paths taken through the radio channel — in other words, MIMO. This usually causes interference or degradation of SNR, but by exploiting the concepts of spatial multiplexing and spatial diversity then the individual transmission paths can be resolved at the receiver to improve the data throughput and reliability of the system, and thus improve spectral efficiency.

18.2. Spatial Multiplexing and Spatial Diversity

MIMO can provide many benefits, depending on the configuration of transmit and receive antennas, and the processing undertaken in the transmitter and receiver. Broadly speaking, there are two approaches: spatial multiplexing and spatial diversity.

In the most basic spatial multiplexing case, the data to be transmitted is demultiplexed into several lower-rate, independent data streams, which are transmitted simultaneously from the set of transmit antennas, and subsequently arrive at the set of receive antennas, having taken different paths. An example of such a configuration (in this case, 4×4 MIMO) is shown in Figure 18.1.

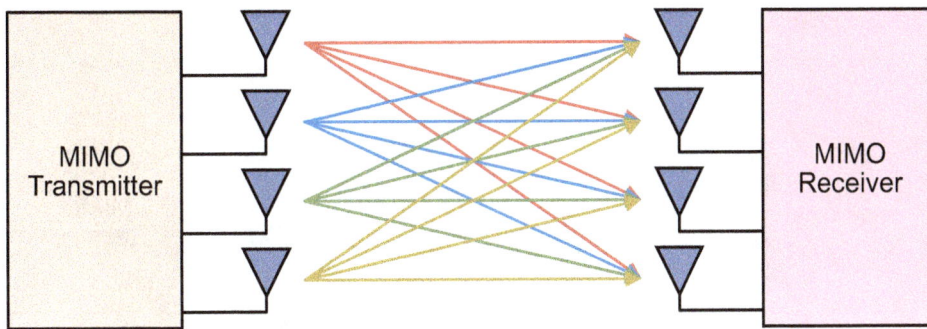

Figure 18.1: A simple example of spatial multiplexing for a 4 x 4 MIMO system.

Alternatively, we can consider that the baud rate at each transmitter antenna path remains the same. Therefore with multiple transmit antenna paths, the overall data rate is increased. In both cases, however, there is no improvement in the reliability of the transmission.

Turning to spatial diversity, it is possible to employ transmitter diversity and receiver diversity. These techniques provide enhanced reliability by combining copies of the same data stream that travel different paths between the transmitter and receiver antennas. Transmitter diversity is where the same data stream is sent from multiple transmit antennas, whereas receiver diversity is when the same data stream is combined across multiple receive antennas. Due to the diverse paths taken by the various copies of the data stream, they experience different channel effects, which can be cancelled out to recover the original data stream. This can provide an increase in reliability, but offers no throughput benefit compared to a SISO system.

Examples of transmitter and receiver diversity systems are shown in Figure 18.2.

Depending on the channel conditions, a focus on either spatial multiplexing or spatial diversity may be preferable, or a hybrid solution combining both approaches can be adopted.

(a)

(b)

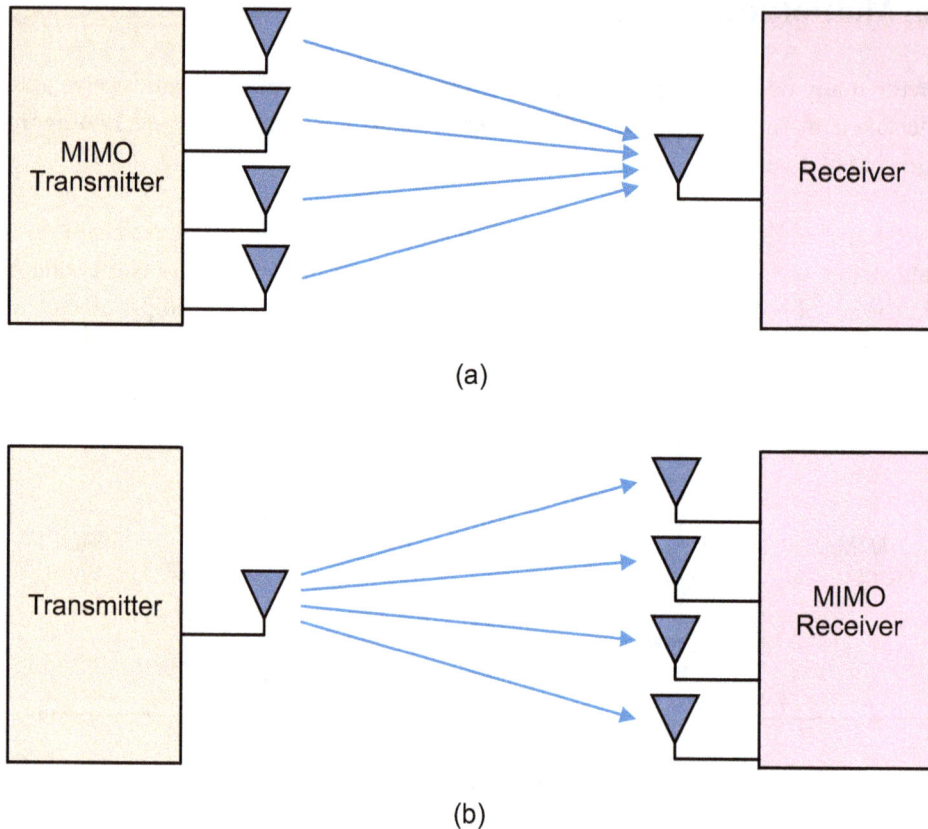

Figure 18.2: (a) transmitter diversity; and (b) receiver diversity.

18.3. MIMO Channel Representation

As mentioned previously, in MIMO systems the transmitter and receiver comprise an array of antennas, and so the transmitted and received signals can be represented by vectors. These vectors are composed of symbols, expressed as complex values. Since there can be a different path between every transmit and receive antenna, the channel can be represented as a matrix, where each matrix element corresponds to the channel gain for a distinct path. Figure 18.3 clarifies how these quantities relate to the MIMO system.

By including the effect of noise on the channel as an additional vector, the equation for a basic MIMO channel can be expressed as

$$y = Hx + n \tag{18.1}$$

where y is a vector of the received symbols, x is a vector of the transmitted symbols, H is the channel matrix, and n is the noise vector.

$$\begin{bmatrix} x_1 \\ x_2 \end{bmatrix} \qquad \begin{bmatrix} h_{11} & h_{12} \\ h_{21} & h_{22} \end{bmatrix} \qquad \begin{bmatrix} y_1 \\ y_2 \end{bmatrix}$$

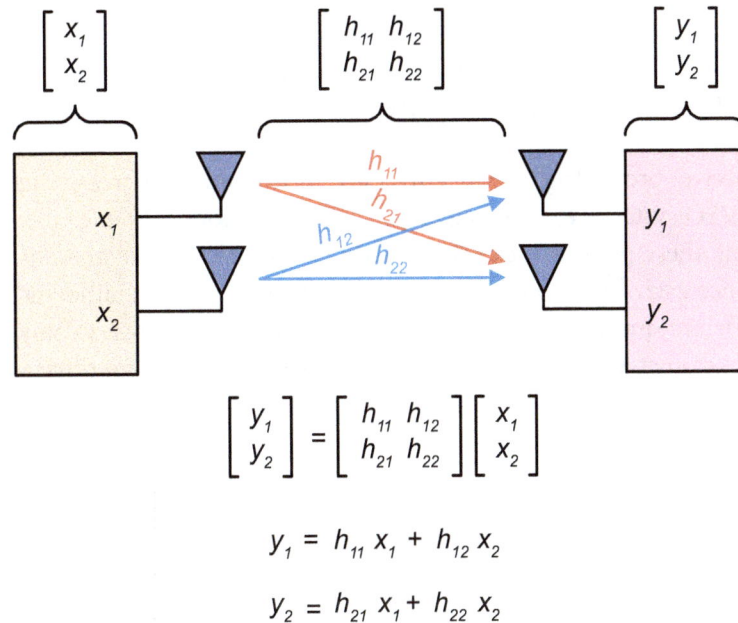

$$\begin{bmatrix} y_1 \\ y_2 \end{bmatrix} = \begin{bmatrix} h_{11} & h_{12} \\ h_{21} & h_{22} \end{bmatrix} \begin{bmatrix} x_1 \\ x_2 \end{bmatrix}$$

$$y_1 = h_{11} x_1 + h_{12} x_2$$

$$y_2 = h_{21} x_1 + h_{22} x_2$$

Figure 18.3: Matrix-vector representation of a 2 x 2 MIMO system.

To gain an estimate of the transmitted symbols at the receiver, processing must be undertaken to remove the effects of the channel and noise from the received symbols. This can be done at either the transmitter, receiver, or both. Over the next two sections, we will consider receiver and transmitter processing techniques, respectively, in a little more detail.

18.4. Receiver Processing Techniques

To obtain the greatest benefit from having multiple antennas, MIMO receiver processing techniques must be employed to recover the transmitted information. Various forms of receiver processing are discussed in this section, focusing on simple techniques such as Switch Diversity, and linear receivers such as Zero Forcing (ZF) and Linear Minimum Mean Square Error (L-MMSE) detectors. A brief overview is provided of techniques beyond these three examples, covering non-linear processing techniques such as Maximum Likelihood (ML) and Sphere decoders, as well as extensions of the basic techniques mentioned previously, with references to further reading on these topics.

18.4.1. Switch Diversity

Switch diversity is a simple form of receiver processing which uses the multiple receive antennas as a form of redundancy to increase the chance that the transmitted signal will be correctly received.

The simplest form of receiver processing is switch diversity, which aims to increase channel reliability through the use of MIMO. In a SISO system experiencing fading, there are short periods of intense attenuation known as *deep fades*, and during these periods it can be difficult to acquire the transmitted symbol. The benefit of using MIMO is that, since each path from transmitter to receiver is slightly different, they experience these deep fades at slightly different times. The core concept behind switch diversity is to only accept the signal at the receiver with the highest gain at a given moment, so it will switch to the best receive antenna [124]. A simple switch diversity system is sketched in Figure 18.4.

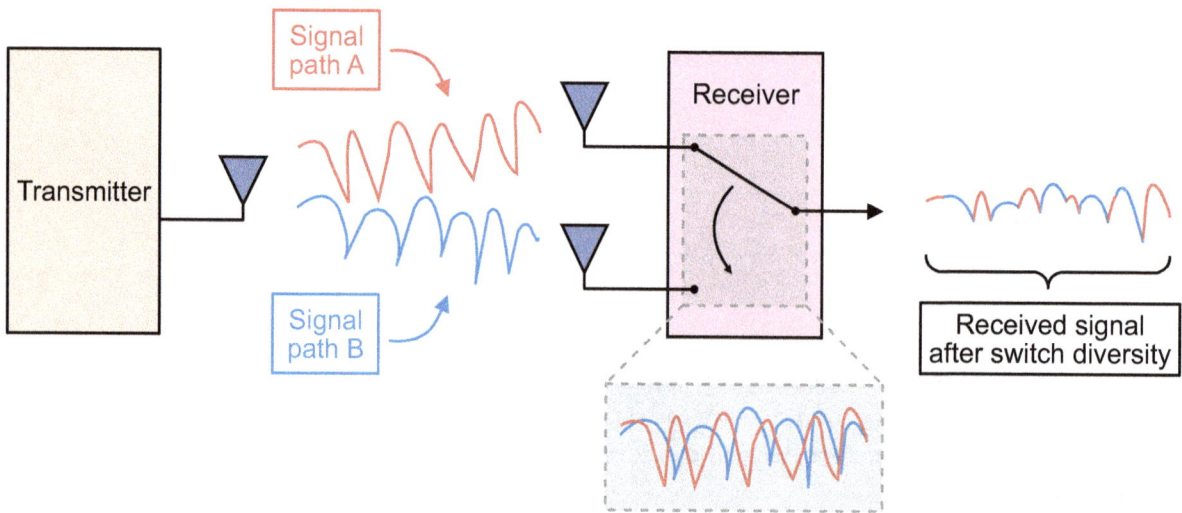

Figure 18.4: MIMO switch diversity.

MIMO switch diversity is a very limited technique because it discards all other received signals, which may still contain useful information.

18.4.2. Zero Forcing

The technique of Zero Forcing (ZF) seeks to position nulls at the incident direction(s) of other transmissions that are received by the system, to minimise interference. It achieves this by simply multiplying the received symbols by the inverse of the channel to cancel its effects. This requires an estimate of the channel at the receiver, which can be obtained through the transmission of pilot symbols. The ZF technique can be described mathematically as

$$\hat{x} = H^{-1}(Hx + n) = x + H^{-1}n, \tag{18.2}$$

where $\hat{x}$ is the estimated transmitted symbol vector.

The ZF technique gives a good estimate of the transmitted symbols if there is minimal noise present, however, there are a few potential drawbacks. The channel inverse is computationally difficult to calculate, and it may not even exist! Additionally, the channel inverse can be ill-conditioned, meaning that it is highly sensitive to small errors such as those caused by noise. This can cause the effects of noise to be greatly amplified when multiplied by the channel inverse [124].

18.4.3. Linear Minimum Mean Square Error (L-MMSE)

An alternative approach is to use Minimum Mean Square Error (MMSE) processing at the receiver. This method builds upon the Zero Forcing receiver by adding a regularisation term, which reduces the sensitivity of the receiver to the conditioning of the channel matrix. When this regularisation term is equal to N_t/SNR, where N_t is the number of transmit antennas, this is known as the Linear MMSE (L-MMSE) detector, and generates an estimate of the transmitted signals given by

$$\hat{x} \;=\; H^H\!\left(HH^H + \left(\frac{N_t}{SNR}I \right) \right)^{-1}(Hx+n), \tag{18.3}$$

where H^H represents the conjugate transpose of the channel matrix, H, also known as the Hermitian transpose, I is the identity matrix, and all other symbols are as previously defined. When the SNR increases then the regularisation term is reduced, and the solution approaches the Zero Forcing receiver [238].

18.4.4. Additional Receiver Processing Techniques

In addition to the previously mentioned methods, non-linear processing techniques can also be used. Maximum Likelihood (ML) techniques enable optimal detection to be achieved by finding the minimum Euclidean distance between the transmitted and received signal vectors. This method is more demanding than linear approaches, but can result in higher data rates for a given channel and is more resilient to antenna correlation (where correlation refers to the degree of independence between signals received at adjacent antennas). The complexity of this ML approach increases exponentially with the number of transmit antennas, which makes it impractical for most applications.

To reduce the complexity, the Sphere Decoding algorithm confines the search range of the ML algorithm to a sphere of set radius, which can still achieve comparable performance [124]. This radius can be adjusted to achieve a point in the trade-off between complexity and performance, with increasing radius approaching the ML solution.

18.5. Transmitter Processing

As well as receiver-based processing, signals can also be processed in the transmitter to help mitigate the effects of the channel. Systems which feed back channel information to the transmitter are known as *closed-loop MIMO*, and this approach can greatly simplify the processing required in the receiver, which is very useful for wireless mobile applications. Just as in the receiver, Zero Forcing can be performed in the transmitter, which removes the risk of noise amplification. Alternatively, Singular Value Decomposition (SVD) can be used to split the channel into subchannels, which can then have individual gains applied to them. These two techniques are discussed further in the following sections, with references provided for further reading.

18.5.1. Zero Forcing

If an estimate of the channel is available at the transmitter, either through feedback or channel estimation, then precoding can be performed in the transmitter to negate the effects of the channel at the receiver. Similar to ZF in the receiver, ZF can be performed in the transmitter by precoding the transmitted signal with the inverse of the channel. The outcome is to cancel out the effects of the channel, leaving only the transmitted signal at the receiver.

Unlike ZF at the receiver, this method does not amplify additive noise from the channel, since only the transmitted symbols are multiplied by the inverse channel matrix, however, it still experiences a problem if the inverse does not exist. Further, the technique requires the channel estimate to be available at the transmitter, necessitating some form of feedback.

Transmitter-based ZF can be expressed as

$$\hat{x} = HH^{-1}x + n = x + n. \tag{18.4}$$

18.5.2. Singular Value Decomposition

Another method is to use Singular Value Decomposition (SVD) to split the channel into three matrices, which are labelled U, S, and V^H, i.e.

$$H = USV^H \tag{18.5}$$

The matrix S contains the singular channel gain values, ordered highest to lowest, as a diagonal matrix,

$$S = \begin{bmatrix} g_1 & 0 & \dots & 0 \\ 0 & g_1 & \dots & 0 \\ \dots & \dots & \dots & \dots \\ 0 & 0 & \dots & g_n \end{bmatrix} \tag{18.6}$$

The matrices U and V^H are unitary matrices which represent rotations in the matrix space, however, these can be negated using careful processing choices in the transmitter and receiver. By interpreting the channel using SVD, the symbols can be precoded with V in the transmitter and with U^H in the receiver. Since $U^H U$ and $V^H V$ both equal the identity matrix, I, the effect is to negate the rotation in the transmitted symbols, leaving only the scaling factor of S.

This can be expressed as

$$x = U^H U S V^H V x + U^H n = I S I x + U^H n = S x + U^H n \qquad (18.7)$$

The SVD technique essentially splits the communication channel into subchannels with gains listed as the diagonal values in the matrix S, which can then be adjusted by adding a power matrix. A different gain can be applied to each subchannel in order to improve performance. For cases when there is minimal noise, it is preferred to apply equal power across each subchannel, which is known as *waterfilling* [317]. Alternatively, where there is a low SNR, it is better to focus power into the subchannel with the highest singular value. This is also known as beamforming, which will be discussed further in the second part of this chapter.

Codebook Selection

The problem with using SVD is that the precoding matrix, V, must be known at the transmitter, but the channel estimation is only available at the receiver, having been developed from the received pilot signals. The overhead to transmit the entire matrix would be excessive, and therefore to combat this problem, the concept of codebooks was introduced as part of the 4G LTE 3GPP standard [2]. Codebooks are a set of predefined matrices that are known at both the transmitter and receiver; the receiver must simply select the closest matrix and send only the index of that matrix back to the transmitter.

The concept of codebooks was developed further in the 5G NR 3GPP standard with the introduction of Type-I and Type-II codebooks. Type-I uses the same logic as the 4G LTE codebooks but with a more diverse range of matrices to choose from. Type-II codebooks can select a linear combination of matrices and then use amplitude scaling for improved precision.

18.6. MIMO Performance Metrics

Despite all of the benefits it can provide, not all scenarios are ideal for MIMO deployment. For example, using spatial multiplexing in situations with a strong Line of Sight (LoS) component will not provide any capacity benefit, as there is no rich scattering environment. Since the MIMO channel is represented by a complex matrix, then continuous analysis of this entire matrix would be extremely computationally challenging, especially for larger MIMO deployments. However, performance metrics such as Rank Indicator and Condition Number can be used to gain a simple mathematical understanding of the channel matrix, and so evaluate MIMO performance [312].

18.6.1. Rank Indicator

The Rank Indicator (RI) is a representation of how many independent communication channels are available in a MIMO system. It is based on the mathematical term *rank index*, referring to how many linearly independent vectors make up a matrix. For effective MIMO communications, a high RI is desirable, since it implies that the MIMO system is successfully spatially multiplexing the data into independent channels. On the other hand, if a User Equipment (UE) reports a RI of 1, this means that the system operation is equivalent to a SISO system. When using SVD for a MIMO channel, the RI is simply the number of non-zero elements in the singular matrix, S [317].

18.6.2. Condition Number

The Condition Number (CN) is measure of how easily the received symbols can be reconstructed, and is based on the mathematical concept of *condition*, which indicates how easily the linear vectors that make up a matrix can be solved. Ideally, the CN should be as close to 1 as possible, however, if the CN is high, this may imply that the channel is ill-conditioned, which can make it very difficult to recover the transmitted symbols.

CN can be expressed as

$$\kappa(H) = \frac{g_{max}}{g_{min}}, \tag{18.8}$$

where g_{max} and g_{min} are the maximum and minimum singular channel gain values, respectively. Together with the RI, these can be used to analyse all configurations of MIMO channels, regardless of the number of transmit and receive antennas [317].

18.7. MIMO Communications

MIMO has been vital in improving the throughput and reliability of wireless systems for years and is included in numerous wireless standards such as Wi-Fi, 4G, and 5G. It can be used in various configurations, and in conjunction with other communications waveforms and protocols. In this section, we briefly review some of the notable use-cases for MIMO.

18.7.1. MIMO OFDM

In cases where the delay spread is high relative to the symbol length, such as in outdoor transmissions, frequency selective fading can occur. This is a type of multipath interference where the attenuation experienced by a signal varies with frequency, and therefore the frequency response of the channel is not flat.

The presence of a frequency selective channel would normally require channel equalisation, which may be relatively complex in a computational sense. To achieve the high bandwidth communications that MIMO enables, OFDM can instead be used to split the channel into subchannels that each have an approximately flat frequency response. When MIMO is combined with OFDM in this way, the MIMO aspect partitions the signal in the spatial domain, while OFDM distributes the signal components across the frequency domain, thus providing the benefits of both techniques [124]. An overview of this scheme is provided in Figure 18.5.

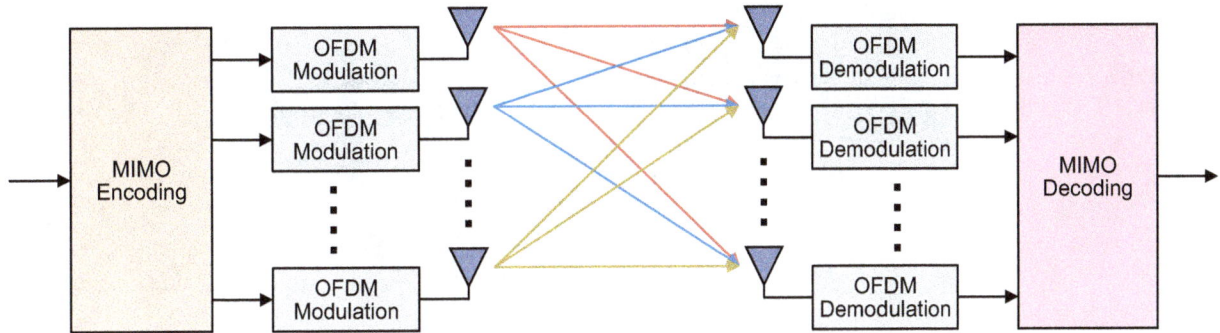

Figure 18.5: Simplified MIMO OFDM block diagram.

18.7.2. Massive MIMO

Massive MIMO, as the name implies, is a type of MIMO that uses a large number of antennas. The use of large MIMO arrays is attractive because they provides greater control and flexibility than a smaller array. There is no strict definition of how many antennas are needed for a MIMO system to be designated "massive"; however, Massive MIMO systems typically have 64, 256, or more transmit and receive antennas, as compared to more traditional MIMO systems which may have 2, 4, or 8 transmit and receive antennas. An example 64-element beamformer is depicted in Figure 18.6, demonstrating that it can steer the azimuth and elevation of the beam.

The main motivation behind Massive MIMO lies in the frequencies used. Low frequencies (which here we will consider as < 1 GHz) are favoured for their wide coverage. However, limited bandwidth is available at these frequencies, and therefore with increasing demand for high data rates, wireless data communications systems generally adopt higher frequencies (for instance, Wi-Fi at 2.4 GHz and 5.8 GHz, with systems using even higher frequencies, such as 'mmWave' bands, under development).

As the transmission frequency increases, the optimal size of a radiating element decreases. As such, the power that each antenna element can emit becomes limited, and consequently more antennas must be added, thus leading to Massive MIMO [109], [308]. The major benefit of Massive MIMO is that the large number of antennas available allows much more precise control over signal directionality, as well as access to advanced techniques such as MU-MIMO, which is covered next.

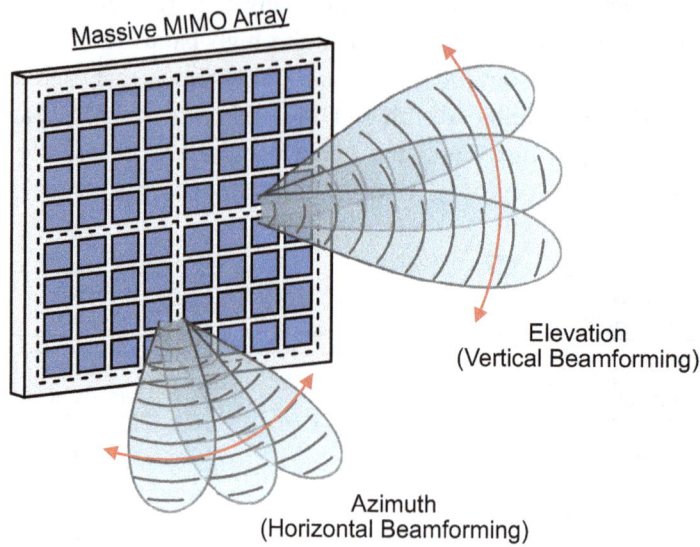

Figure 18.6: Massive MIMO array, showing directions of beam steering.

18.7.3. Single-User MIMO and Multiple-User MIMO

Single-User MIMO (SU-MIMO), introduced in the 802.11n wireless standard, allows multiple data streams to form a link between a single transmitter and receiver pair, providing the benefits discussed earlier (such as increased data rates and reliability). Multiple-User MIMO (MU-MIMO) was then incorporated in the Wi-Fi 5 (802.11ac) standard, enabling multiple data streams to target multiple different users simultaneously in the downlink. Uplink support was later added in the Wi-Fi 6 (802.11ax) standard. Both SU- and MU-MIMO are also supported as part of the 4G LTE and 5G NR standards.

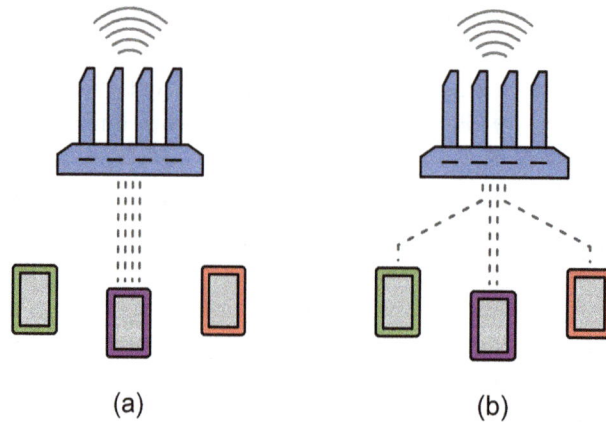

Figure 18.7: (a) Single user MIMO; and (b) Multiple-user MIMO.

The number of data streams, or layers, that can be created is limited by the number of antennas (typically one stream can be generated per antenna). As the number of antennas increases, such as in Massive MIMO, the number of supported layers also increases, allowing more data streams and more users to be supported, and resulting in a great improvement in wireless network throughput. MIMO can also be combined with OFDM to separate transmissions to different users across the frequency and spatial domains [109].

18.8. Phased Array Beamformers

Next, and for the remainder of this chapter, we switch the focus from MIMO to beamforming. This section will give a brief overview of phased array antennas and beamforming, and establish the context for why they are necessary in many modern digital communications systems.

18.8.1. Antenna Directionality

With traditional forms of signal transmission and reception, a singular omnidirectional antenna would be used to radiate or detect radio signals uniformly across all directions [103]. For many applications such as FM radio, this is a perfectly viable method as each radio station is modulated to its own frequency band. This allows for one antenna to transmit the FM radio signal containing all radio stations, modulated to different frequency bands, and one antenna to receive the FM radio signals and perform digital filtering to demodulate the signal and listen to the chosen radio station [326].

As technology advances, more and more applications (such as Wi-Fi, RADAR and 5G) require the transmitter and receiver antennas to transmit or receive signals from one or more specific directions, and filter out signals from all other directions [115]. These modern antennas must also enable the chosen 'look' direction to be easily and quickly adapted in response to a dynamic wireless environment.

Figure 18.8 presents an intuitive comparison between omnidirectional and directional transmit antennas. In this diagram, the viewpoint is from above: the ideal omnidirectional antenna emits energy equally in all directions, whereas a directional antenna concentrates its energy within one particular range of angles. For a given power output, directional antennas can transmit further than omnidirectional ones, because they concentrate the emitted power within a limited angular range.

One method of an antenna producing a narrow beam that can be directed in a specific direction is with a dish antenna, an example of which is illustrated in Figure 18.9. These antennas use a parabolic reflector, whose size must be larger than the signal wavelength in order to reflect the signal into a singular focal point. In this way, these dish antennas can produce a narrow beam or region of sensitivity that can be oriented in a desired direction for transmission or reception. The downside to these types of antennas is that they need to be physically moved in order to change the sensitive direction. Depending of the wavelength of the signals, dish antennas can also become very large and heavy, making them difficult and slow to reorient [339].

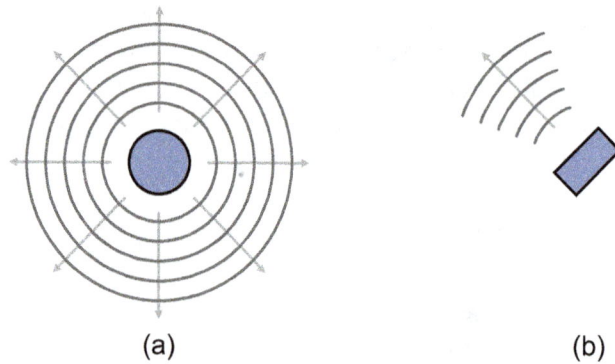

Figure 18.8: *Transmitter radiating signal: (a) uniformly across all directions, (b) in one specific direction.*

Figure 18.9: *Illustration of a parabolic dish antenna.*

18.8.2. Array Antennas

Another method of producing a directional signal is through the use of an array antenna, and the technique of *beamforming*. An array transmit antenna has two or more antenna elements, and the signals emitted by these elements intersect and combine to form a relatively narrow beam in a particular direction. Each of the antenna elements in the array transmits the same signal, causing constructive or destructive interference to occur, depending on the angle with respect to the array.

In the example shown in Figure 18.10, identical copies of the signal are transmitted simultaneously, and a narrow beam is generated. This occurs because constructive interference combines the signal energy perpendicular to the array, while destructive interference cancels out the signals emitted in other directions [361]. More formally, we note that the sensitivity (gain) of the array antenna varies with angle, and this is known as a *beam pattern* or *radiation pattern*.

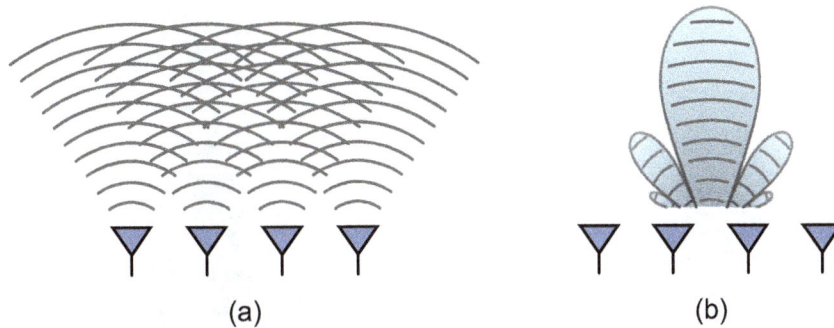

Figure 18.10: Signals from a linear array: (a) overlapping and causing interference, (b) forming a narrow beam.

The technique of beamforming exploits the interference generated by antenna arrays to control the radiation pattern, and in particular the narrow beam where constructive interference occurs (the *main lobe*). Our initial example considered the simple case where all transmit array elements emit identical and synchronised signals, however the radiation pattern can be customised by controlling the phases of signals emitted from the array elements, thus 'steering' the beam in a desired direction.

Note that beamforming is equally applicable to array receivers, in which case it defines the direction(s) that the receiver is most sensitive to.

18.8.3. Phased Array Antennas

As depicted in Figure 18.11, a phased array includes an electronically controlled phase shifter for each antenna element. By applying individual phase offsets to the signals emitted by the array elements, the pattern of constructive and destructive interference can be controlled, thereby steering the main lobe to a desired angle. As the phase offsets are applied digitally and can be reprogrammed, the phased array antenna becomes dynamically steerable, and the look direction can be altered as quickly as new phase offsets can be calculated.

The main benefit of using a phased array is that the direction of the narrow beam can be digitally steered without the need for any physical movement of the array. Another advantage is response time: the phase delays required for the set of antenna elements can be readily re-calculated and applied digitally, and therefore the beam can be steered incredibly quickly [99].

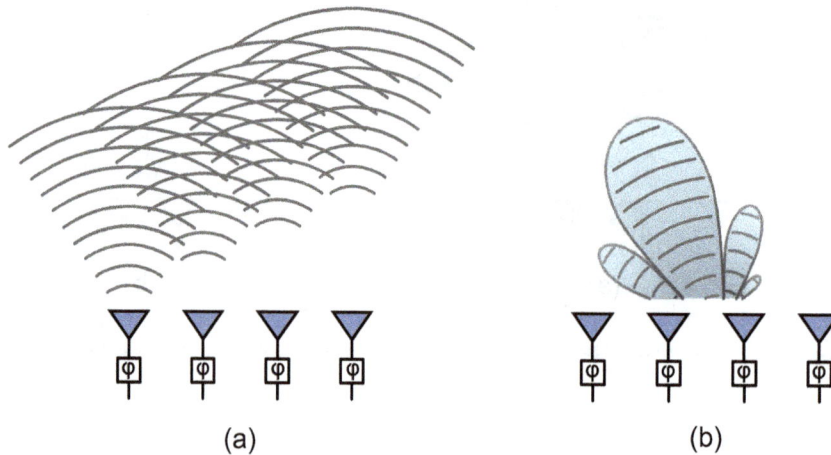

Figure 18.11: Signals from a linear array with phase delay:
(a) overlapping and causing directed interference, (b) forming a directed narrow beam.

18.8.4. Phased Array Antenna Orientations

Phased array antennas can be categorised into three main types: linear arrays, planar arrays and frequency scanning arrays. These categories are determined by the positioning of each antenna element and the number of phase shifters used. Each type is now briefly reviewed.

Linear Array

The antenna elements of a linear array are arranged in a straight line with a progressive phase shift applied to them. The antenna elements are supplied with increasing multiples of a base phase shift to accurately perform the required beam steering. A linear array is the simplest and cheapest form of a phased array that allows for the steering of the antenna array to be dynamically controlled. The disadvantage is that beamforming can only take place in a two-dimensional plane that is perpendicular to the line of antenna elements.

Planar Array

A planar array is composed of antenna elements arranged in a two dimensional pattern, which is typically square or rectangular, but can also be another shape, e.g. circular or triangular. As illustrated in Figure 18.13, each antenna element has its own individual phase shifter, which allows beam steering to operate in three-dimensional space, and thus the planar array exhibits more beamforming flexibility than the linear array. However, with a greater number of antenna elements, each requiring a unique phase shifter and independent phase offset calculation (rather than multiples of the same phase offset), the planar array is considerably more complex and expensive to implement than a linear array.

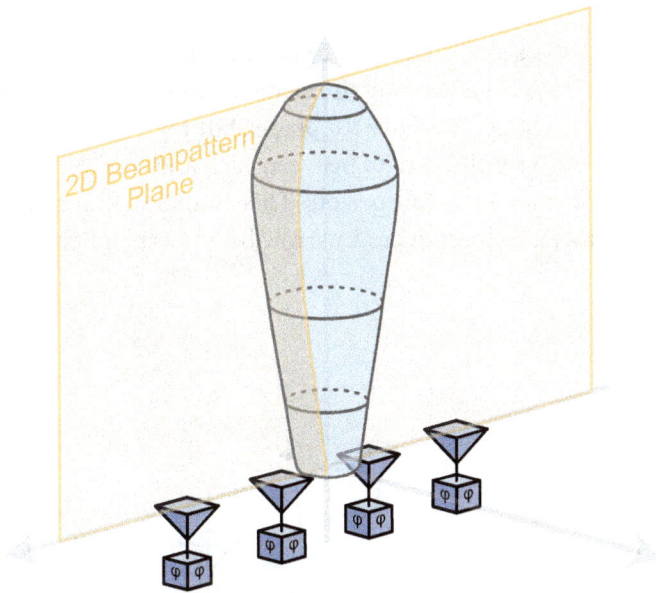

Figure 18.12: Linear array.

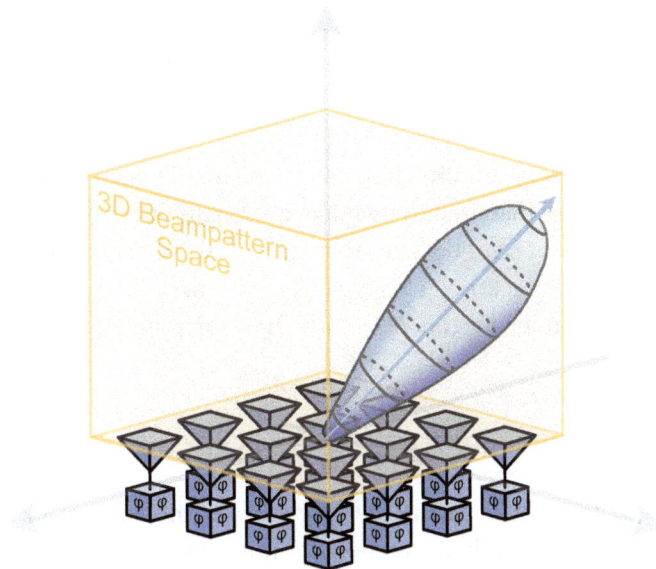

Figure 18.13: Planar array.

Frequency Scanning Array

A frequency scanning array uses the signal frequency to perform the beamsteering, instead of phase shifters. The signal is fed along a *serpentine feed* connecting the elements in the antenna array, as illustrated in Figure 18.14. These feed lines are designed to apply a full 360° phase shift for a specific frequency. Thus, the transmit signal must travel along a wire with a specific length between each antenna element, with the length calculated to delay that specific signal frequency by a full period. This results in the signal arriving at each antenna element being completely in phase, which produces a narrow beam perpendicular to the antenna array.

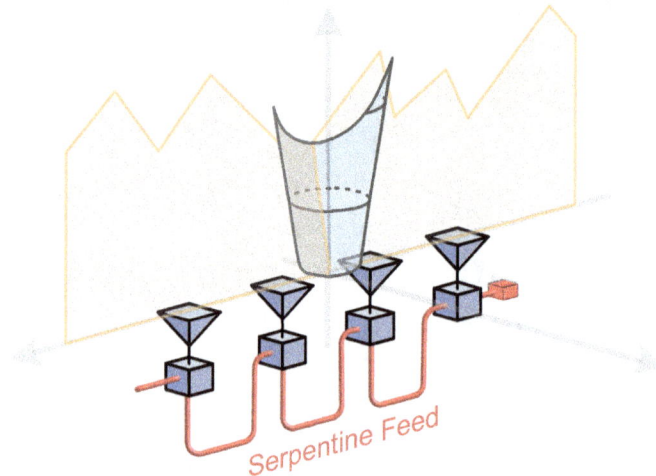

Figure 18.14: Frequency scanning array.

In the frequency scanning array, the beam direction can be steered by altering the frequency of the signal. If the signal frequency is different than that used to calculate the length of the feed lines, the signal arrives at each antenna element with a phase offset (i.e. it deviates from the ideal 360°), giving the narrow beam a direction that is not perpendicular to the antenna array. This type of phased array is very simple, but it lacks flexibility as it is limited to only a few predetermined frequencies, which in turn provide only a few predetermined beam steering directions.

18.8.5. Passive and Active Antenna Arrays

Phased array antennas have played a huge part in meeting the ever-increasing capacity and coverage demands of modern wireless communications. Early adaptations of this technology, in the form of "passive" antenna arrays, combined multiple antenna elements to create directional beams through phase shift manipulation. Passive antenna arrays allow for bespoke radiation patterns to be designed based on specific applications, however, these patterns are static, and unable to adapt or manoeuvre in response to a dynamic environment. This is where "active" antenna arrays offer a distinct advantage, as they have the ability to electronically alter their radiation patterns. The active antenna array gives a higher degree of flexibility and control than the passive array, and in particular, offers the ability to dynamically steer the direction of the beam in response to changing parameters [9].

Passive Electronically Steered Arrays

Passive antenna arrays are seen as the first generation of phased arrays, and are typically used in radio communications. All antenna array elements are connected to a single Transmit / Receive Module (TRM), as illustrated in Figure 18.15, with the amplitude and phase delay needed to achieve beamforming being applied in a passive manner.

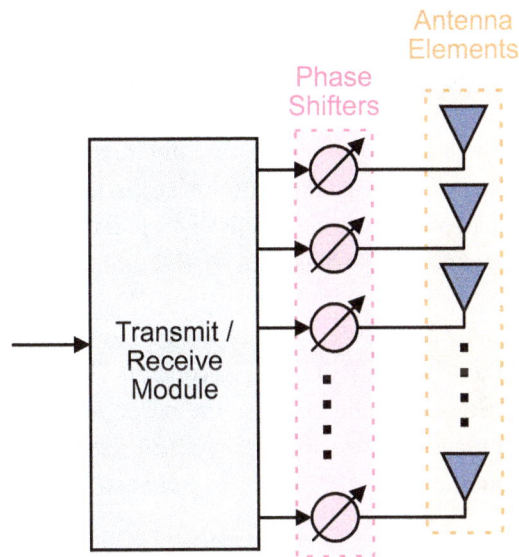

Figure 18.15: Passive electronically steered array.

Early iterations of the passive antenna array were constructed using a common feed, with transmission line technology applying the amplitude and phase needed at each antenna element to form a narrow beam. However, the resulting radiation pattern has fixed characteristics (such as gain and beam width) as well as a fixed direction.

More dynamic iterations of the passive antenna array were then introduced, integrating variable phase shifters into the design to allow the phased array direction to be electronically steered. This can be done directly at the antenna, or wirelessly through the use of a Remote Electrical Tilt (RET) actuator.

The number of variable phase shifters implemented varies according to the design requirements. A phase shifter can be applied to each antenna element individually, or to a group of elements in a sub-array. There is a trade-off in terms of performance versus cost: the fewer phase shifters used, the lower the cost, but at the expense of performance (the number of sub-arrays determines the overall beamforming degrees of freedom and flexibility).

For passive antenna arrays to be implemented efficiently, they need to be designed with the specific application requirements and characteristics in mind. On the other hand, an active antenna array is more flexible and can adapt to dynamic environments.

Active Electronically Steered Arrays

Active antenna arrays are considered to be the second generation of phased array antennas. They are designed with active components that deliver a flexible antenna, capable of producing a wide variety of radiation patterns. These characteristics make active antenna arrays a much more suitable solution for a dynamic wireless environment than the static characteristics of a passive antenna array.

The phase and amplitude manipulation for these active antenna arrays are controlled from a singular beamforming unit, capable of electronically calculating the phase/amplitude values needed at each transceiver to produce the desired radiation pattern. These electronically calculated values can be reapplied continuously, such that the radiation pattern is dynamically adjusted.

The optimal implementation of an active antenna array is to have a TRM for every antenna element, as depicted in Figure 18.16, giving the maximum amount of control. This enables the antenna array to have the highest degrees of freedom and flexibility, allowing for the narrowest beams to be formed with the widest range of steerability. The disadvantage is that this approach can be very costly (and for some applications, perhaps even infeasible) due to manufacturing constraints such as space, weight, heat dissipation, and so on. If

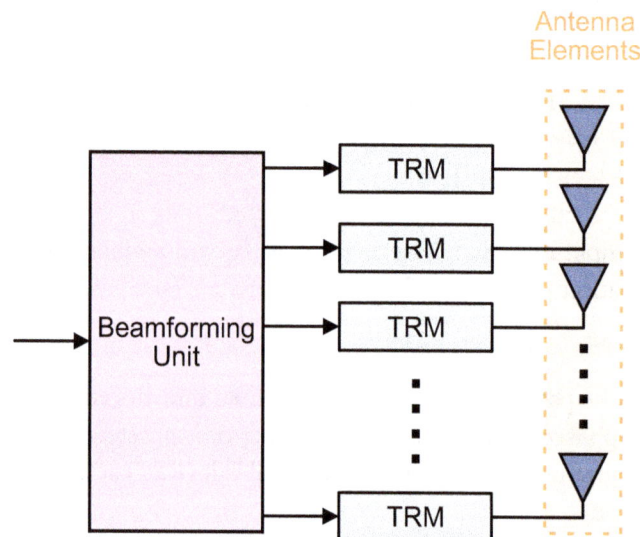

Figure 18.16: Active electronically steered array.

these are limiting factors, then a more viable solution can be implemented by reducing the number of transceivers and connecting them to a group of antenna elements. This method retains the "active" characteristics of the antenna array, but with a reduced degree of flexibility.

18.9. Beamforming Techniques

There are three broad categories of beamforming techniques used in phased array antennas: analogue, digital and hybrid beamforming [187], all of which are illustrated in Figure 18.17. These three techniques provide methods for steering an antenna array with differing levels of control, complexity and cost [222].

18.9.1. Analogue Beamforming

In an analogue beamformer, all processing relating to the beamforming operation is undertaken in the analogue domain. All antenna elements are driven by a singular RF source that is split and passed through analogue phase shifters that apply the appropriate phase offset to steer the antenna array in a desired direction.

The original analogue beamformer used phase shifters with fixed delays to produce a static radiation pattern. Improvements on this design saw multiple phase shifters with different fixed delays, selectable via a switch, which added the option to select different radiation patterns. Eventually, adjustable phase shifters were introduced at each antenna element to create a flexible antenna array.

The limitation that comes with analogue beamforming is the inability to produce multiple beams. An analogue beamformer is only capable of producing a single directional beam. For multi-beam applications, such as MIMO, a digital or hybrid beamforming implementation is required.

18.9.2. Digital Beamforming

In digital beamforming, all signal processing operations required to steer an antenna array are performed in the digital domain. Each antenna element has its own RF signal chain that is electronically controlled via a digital beamforming unit, and therefore analogue phase shifters are not required.

The digital beamforming unit implements a beamforming algorithm that calculates the appropriate adjustments to phase and amplitude, in the form of complex weight values, at each antenna element. Doing so allows the radiation pattern of the antenna array to be electronically shaped and steered.

A fully digital beamformer is able to produce multiple directional beams at multiple frequencies simultaneously, and these can each be independently controlled. The number of beams that can be produced is directly related to the number of RF chains applying complex weight values (calculated by the beamforming unit). For a truly digital beamformer, each antenna element has its own digitally controlled amplitude and phase shift, thus maximising the degree of flexibility and control available.

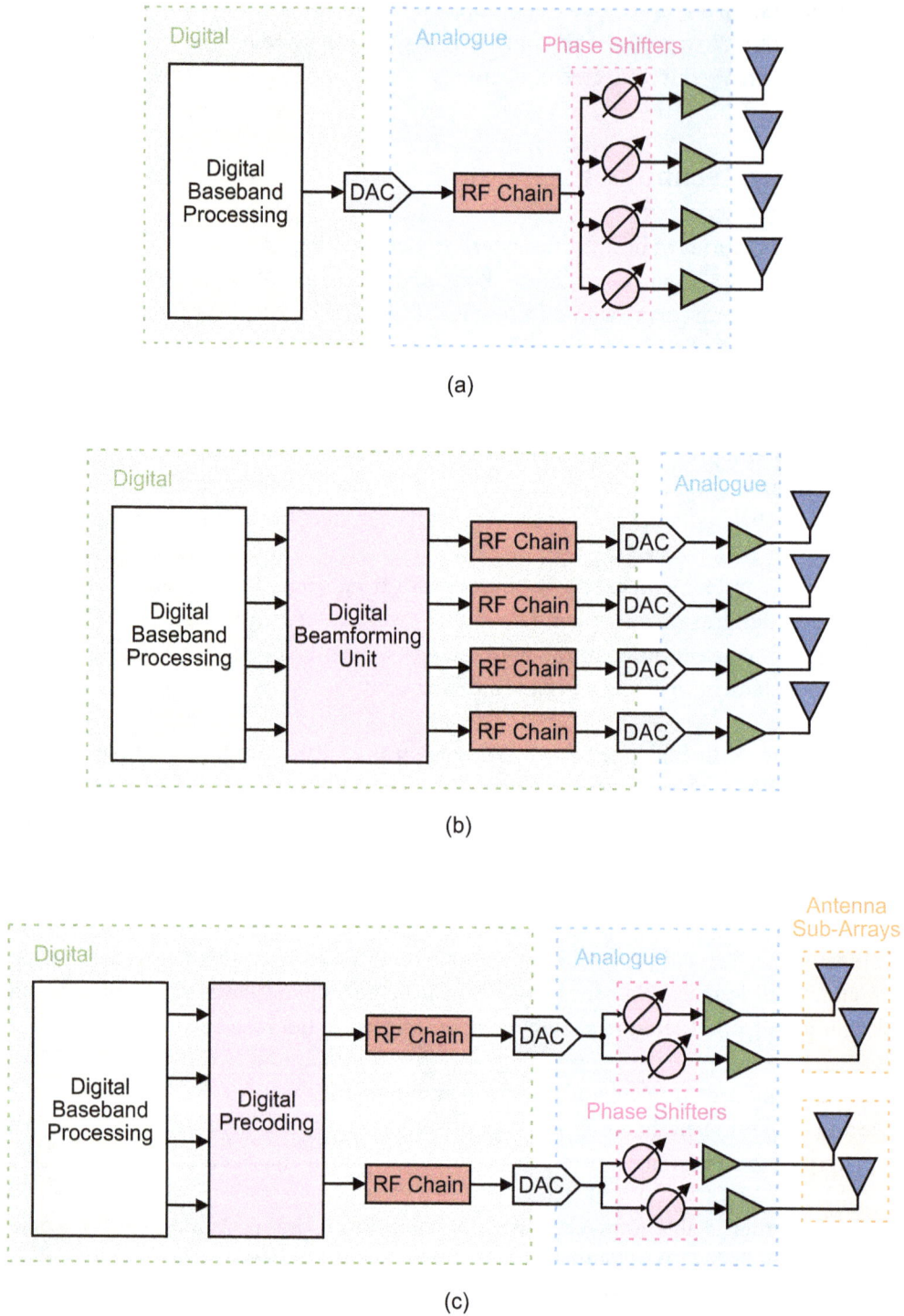

Figure 18.17: Three categories of beamformer implementation: (a) analogue, (b) digital, and (c) hybrid.

Digital beamforming can be very complex and costly to implement, however, requiring significant hardware resources and computation, and resulting in a higher power consumption than analogue beamformers. With the AMD RFSoC platform, fully digital beamforming implementations can be achieved due to the integration of ADCs and DACs inside the chip, and the vast amount of PL resources available. If however, there are limited PL resources available for beamforming in a specific application (perhaps due to the demands of other parts of the system), an alternative solution is hybrid beamforming, where a combination of analogue and digital beamforming is used to reduce hardware complexity.

18.9.3. Hybrid Beamforming

Hybrid beamforming provides a flexible antenna array, capable of generating multiple directional beams, at a reduced hardware resource and processing power cost compared to digital beamforming. It achieves this by grouping antenna elements into sub-arrays and using a combination of analogue and digital beamforming techniques.

As in analogue beamforming, each antenna element is connected to an analogue phase shifter. However, all elements are not driven by the same RF signal chain. Instead, multiple pairs of antennas and phase shifters are grouped together in subsets of arrays (sub-arrays), with each sub-array driven by its own RF chain and data converter. Every individual RF chain requires a level of digital precoding to achieve multi-beam functionality, in a similar manner to a digital beamformer but to a lesser degree.

The advantages and limitations of hybrid beamforming both arise from the number of RF chains and data converters used. The advantage being the ability to produce a radiation pattern of more than one beam at a reduced cost to digital beamforming. The limitation is the number of beams that can be produced, and therefore the overall degree of flexibility of the antenna array being less than that of a digital beamformer.

18.10. Beamforming Implementations

Beamforming is a versatile technique that can be applied to many different applications to electronically steer a phased antenna array. If the direction in which the phased array is to be steered is known, then a conventional beamformer can be used, as it calculates the phase shifts based on a known direction. If the direction is not known, then an adaptive beamformer is required. With an adaptive beamformer, a replica of the received signal should be available, or at least a portion of that signal should be known in advance. The beamformer then uses an adaptive algorithm to calculate complex weight values that, when applied to the phased array antenna data, will best recover the desired signal. We will now go on to discuss how conventional and adaptive beamformers are implemented.

18.10.1. Conventional Beamforming

In conventional beamforming, shown in Figure 18.18, the phase shifts are electronically calculated based on a known signal direction or location. Meaning the conventional beamformer needs to be manually steered towards a desired signal. This is a simpler beamforming implementation and is ideal for applications where the positioning of both transmitter and receiver are known.

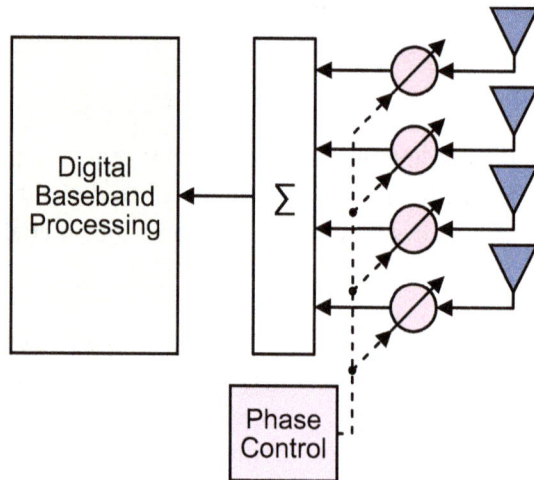

Figure 18.18: Conventional beamformer implementation.

Conventional Beamformer Example

In conventional beamformers, the phase shift value applied to each antenna element of a phased array antenna can be calculated using simple mathematics, provided that specific information relating to the phased array and RF signals are known.

Figure 18.19 gives a simplified diagram for a phased array receiver antenna with the antenna elements arranged in a linear array. It is worth noting here that even though this example is of a receiver antenna, the same properties apply to a transmitting antenna, due to the principle of reciprocity. The principle of reciprocity states that the direction of sensitivity of a receiver antenna is the same as the radiation pattern direction of a transmitting antenna.

The signals incident at the phased array antenna originate from a far field signal, so it can be assumed that a plane wave arrives at each antenna. Each antenna element has been spaced at the optimal separation distance,

$$d = \frac{\lambda}{2},$$

(18.9)

for this example scenario, where λ is the wavelength of the incoming RF signal.

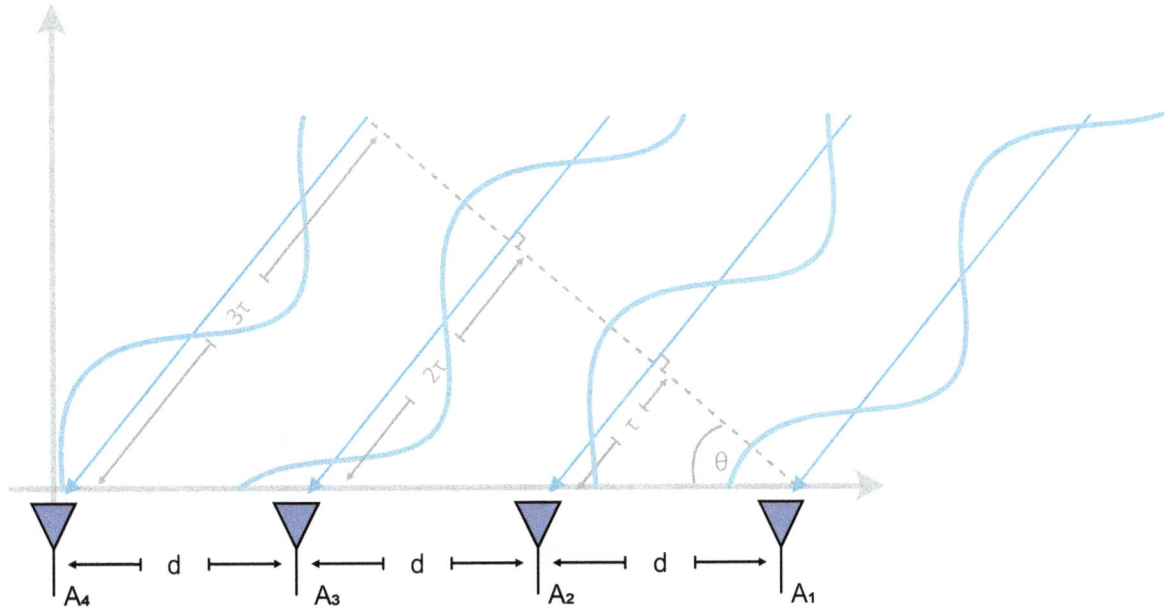

Figure 18.19: *Example of a signal arriving at a phased array.*

Each signal arrives at an antenna element at a slightly different time (τ) due to the angle at which the far field signal is positioned relative to the antenna array (θ). By applying the appropriate phase shifts to each antenna element, the antenna arrays sensitivity can be steered towards the far field signal, aligning the received signals.

Figure 18.20, overleaf, provides a visualisation of how the phase shifts are used to steer the antenna array in the direction of the desired signal.

The signal received at the second antenna element, A_2, can be represented by

$$y(t) = sin(2\pi f(t + \tau)) \qquad (18.10)$$

where f is the centre frequency of the desired signal. Eq. (18.10) can be simplified to

$$y(t) = sin(2\pi ft + \varphi) \qquad (18.11)$$

where $\varphi = 2\pi f\tau$.

The value of τ can be calculated if the angle of the received signal relative to the phased array (θ) and the distance between antenna elements (d) are both known, i.e.

$$\tau = \frac{d \cdot sin(\theta)}{c} \qquad (18.12)$$

where c is the RF signal propagation speed, which is equal to the speed of light.

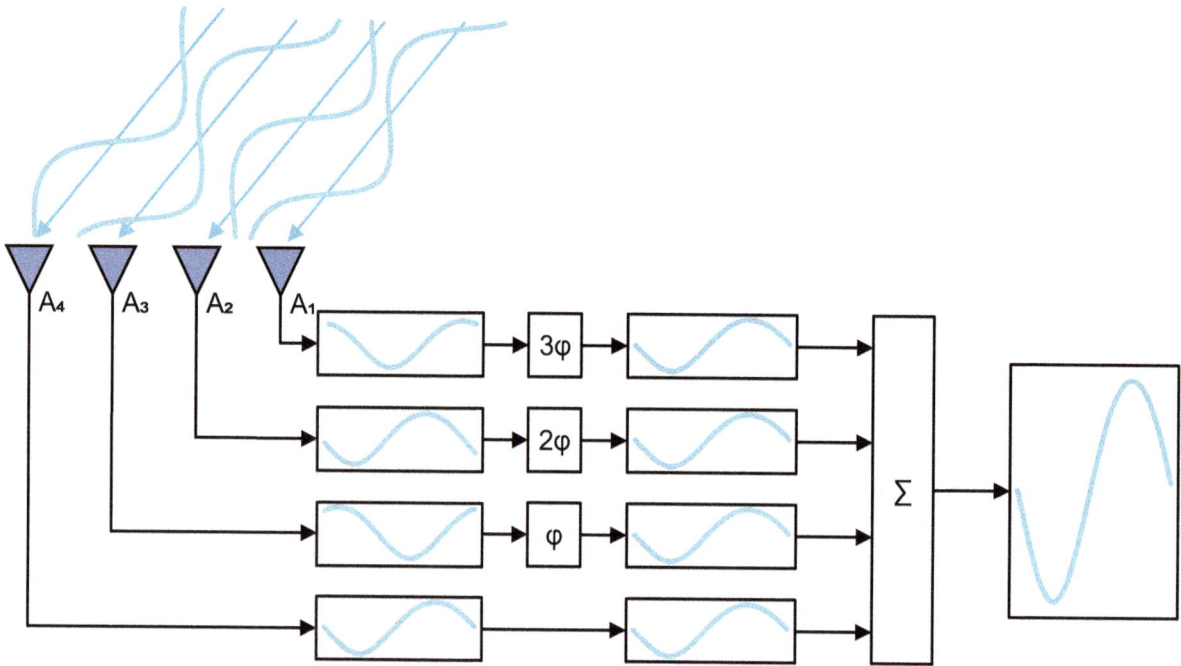

Figure 18.20: Example of steering a phased array with phase shifts.

The phase delay, φ, can now be calculated using the value of τ,

$$\varphi = 2\pi f \cdot \frac{d \sin(\theta)}{c}. \tag{18.13}$$

Given the relationship between the antenna spacing and signal wavelength expressed in (18.9), we can substitute for d, yielding

$$\varphi = 2\pi f \cdot \frac{(\lambda/2)\sin(\theta)}{c} = \pi f \cdot \frac{\lambda \sin(\theta)}{c}. \tag{18.14}$$

A further simplification then arises from the relationship $\lambda = c/f$, which gives

$$\varphi = \pi f \cdot \frac{(c/f)\sin(\theta)}{c} = \pi \sin(\theta). \tag{18.15}$$

Having determined the phase delay needed to steer the phased array in the direction of the desired signal (φ), it can be applied to the receiver antennas as seen in Figure 18.20. Note that since each antenna element is equally spaced, antenna elements are applied with integer multiples of the same phase shift.

18.10.2. Adaptive Beamforming

Adaptive beamforming calculates the phase shifts without explicit knowledge of the desired signal direction or location, instead using information derived from the received signal itself. Adaptive algorithms are used to find the optimal phase shifts needed to steer a phased array antenna towards a desired signal and attenuate any interfering signals.

In order to achieve this, the adaptive algorithm requires prior knowledge of some aspect of the desired signal. A common approach is to embed a signal component that is known to the both transmitter and receiver, and a local copy of the desired signal is stored at the receive side to act as an input to the adaptive beamforming algorithm. Such a configuration is shown in Figure 18.21.

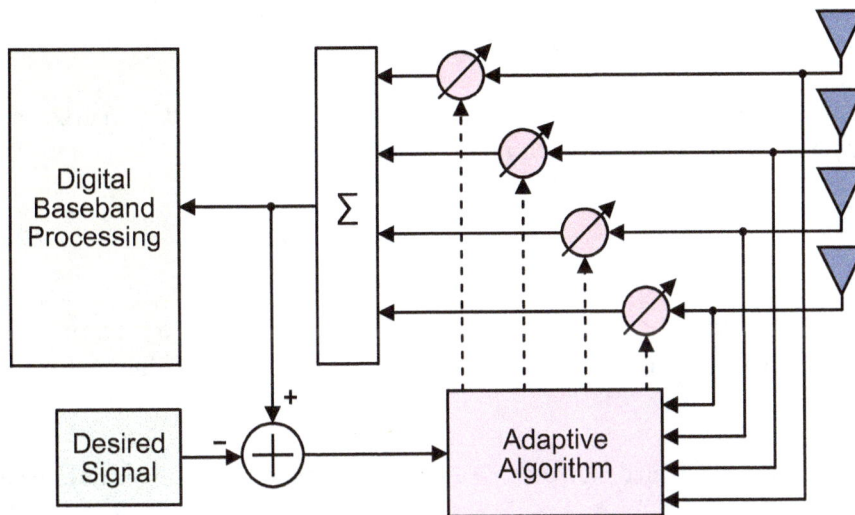

Figure 18.21: Adaptive beamformer implementation.

18.10.3. Adaptive Beamforming Example

For an adaptive beamformer, the phase shift values needed to steer the antenna array in a specific direction are calculated without knowing the direction required. They are instead calculated using known information about the desired signal and applied with complex weight values. There are several different methods of performing adaptive beamforming, but for the purposes of this example, a QR Decomposition (QRD) based adaptive beamforming algorithm will be discussed [181].

QRD is a method of performing matrix inversion to solve a set of simultaneous equations that is optimised for FPGA implementation. An advantage of the QRD method is that it maintains a limited dynamic range between values, allowing the arithmetic wordlengths to be kept within a reasonable range, and resulting lower

FPGA resource consumption than other methods [93]. The QRD splits a data matrix (X) into two separate matrices, an upper triangular matrix (R), and an orthogonal matrix (Q).

To provide an illustrative example for a data matrix of dimensions 4×4, the QRD can be expressed in matrix notation as

$$X = QR \tag{18.16}$$

$$\begin{bmatrix} x_{00} & x_{01} & x_{02} & x_{03} \\ x_{10} & x_{11} & x_{12} & x_{13} \\ x_{20} & x_{21} & x_{22} & x_{23} \\ x_{30} & x_{31} & x_{32} & x_{33} \end{bmatrix} = \begin{bmatrix} q_{00} & q_{01} & q_{02} & q_{03} \\ q_{10} & q_{11} & q_{12} & q_{13} \\ q_{20} & q_{21} & q_{22} & q_{23} \\ q_{30} & q_{31} & q_{32} & q_{33} \end{bmatrix} \cdot \begin{bmatrix} r_{00} & r_{01} & r_{02} & r_{03} \\ 0 & r_{11} & r_{12} & r_{13} \\ 0 & 0 & r_{22} & r_{23} \\ 0 & 0 & 0 & r_{33} \end{bmatrix} \tag{18.17}$$

Since Q is an orthogonal matrix (meaning that $Q^T Q = I$, where I is an identity matrix) the equation can be rearranged in terms of R without the use of matrix division.

$$R = Q^T X \tag{18.18}$$

This is significant because division is a costly operation to implement, particularly for complex numbers, and can add considerably to the critical path. The complex weight values needed to steer the antenna array in the direction of a desired signal can thus be computed using the transpose operation and matrix multiplication implied by (18.18).

As shown in Figure 18.22, the far field signal arrives at the antenna array from an unknown angle. The data collected at each antenna element forms the data matrix X, and is passed into the QRD-based adaptive algorithm along with a copy of the originally transmitted desired signal (D). The aim of the adaptive algorithm is to calculate a weight vector (W) which, when applied to the antenna data, will reproduce the desired signal with the least degree of error.

$$XW = D \tag{18.19}$$

Therefore, to calculate the weight values, the desired signal, D, is simply divided by the data matrix, X.

$$W = \frac{D}{X} \tag{18.20}$$

Unfortunately, it is not as simple as it sounds, as matrix division is a complex and costly operation to perform. Instead, the QRD can be applied to the data matrix X, turning the weight value calculation into a more convenient operation.

$$QRW = D \tag{18.21}$$

$$RW = Q^T D \tag{18.22}$$

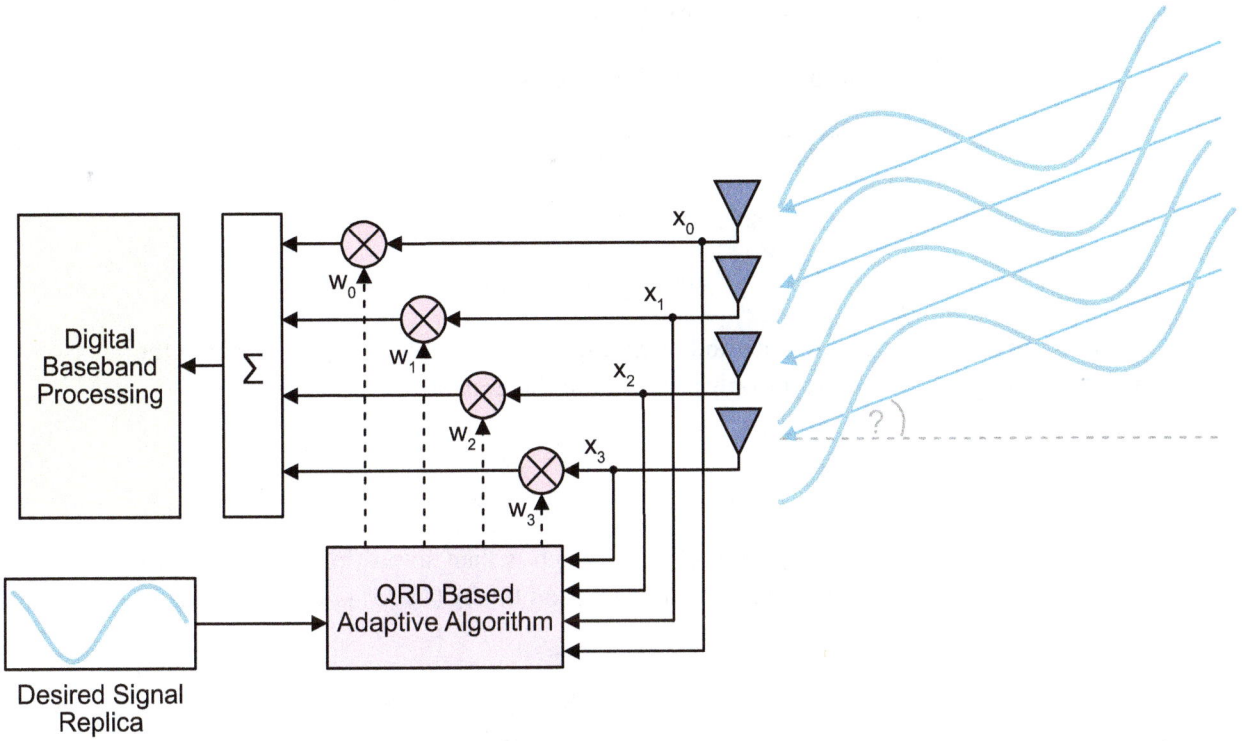

Figure 18.22: QRD-based adaptive beamforming example.

Eq. (18.22) can be expressed in matrix form as

$$
\begin{bmatrix} r_{00} & r_{01} & r_{02} & r_{03} \\ 0 & r_{11} & r_{12} & r_{13} \\ 0 & 0 & r_{22} & r_{23} \\ 0 & 0 & 0 & r_{33} \end{bmatrix} \cdot \begin{bmatrix} w_0 \\ w_1 \\ w_2 \\ w_3 \end{bmatrix} = \begin{bmatrix} q_{00} & q_{10} & q_{20} & q_{30} \\ q_{01} & q_{11} & q_{21} & q_{31} \\ q_{02} & q_{12} & q_{22} & q_{32} \\ q_{03} & q_{13} & q_{23} & q_{33} \end{bmatrix} \cdot \begin{bmatrix} d_0 \\ d_1 \\ d_2 \\ d_3 \end{bmatrix}
\tag{18.23}
$$

which can be simplified by substituting $D' = Q^T D$, giving

$$
\begin{bmatrix} r_{00} & r_{01} & r_{02} & r_{03} \\ 0 & r_{11} & r_{12} & r_{13} \\ 0 & 0 & r_{22} & r_{23} \\ 0 & 0 & 0 & r_{33} \end{bmatrix} \cdot \begin{bmatrix} w_0 \\ w_1 \\ w_2 \\ w_3 \end{bmatrix} = \begin{bmatrix} d'_0 \\ d'_1 \\ d'_2 \\ d'_3 \end{bmatrix}.
\tag{18.24}
$$

The matrix equation from (18.24) can be expressed as a set of four simultaneous equations, i.e.

$$r_{00}w_0 + r_{01}w_1 + r_{02}w_2 + r_{03}w_3 = d'_0 \tag{18.25}$$

$$r_{11}w_1 + r_{12}w_2 + r_{13}w_3 = d'_1 \tag{18.26}$$

$$r_{22}w_2 + r_{23}w_3 = d'_2 \tag{18.27}$$

$$r_{33}w_3 = d'_3 \tag{18.28}$$

These equations can be solved using the method of *backward substitution*. Using this method, the value of w_3 is calculated first, using (18.28), as it is the only unknown in this equation:

$$w_3 = \frac{d'_3}{r_{33}} \tag{18.29}$$

The calculated value of w_3 is substituted into (18.27), which is then solved for w_2. The process repeats by substituting both w_3 and w_2 into (18.26) to calculate w_1, and finally (18.25) is solved to find w_0 in a similar manner.

The calculated weight values (w_0, w_1. w_2, and w_3) are then applied to the receiver antenna data to recover the desired signal and steer the antenna array in the required direction.

18.10.4. QR Decomposition Implementation

The QRD of an adaptive beamformer can be implemented as a signal systolic array architecture, which maps to the PL of the RFSoC device. As mentioned earlier, the QRD method keeps the numerical values bounded within a limited numerical range, and therefore the implementation can use fixed-point arithmetic in preference to floating-point, which is required for numerically less-well-conditioned approaches (floating-point provides larger dynamic range, but at the cost of more resource-intensive operations, and so the fixed-point arithmetic of the QRD approach is advantageous).

The systolic array architecture used to implement the QRD is built using two types of repeating cells (denoted as *Boundary* and *Internal* cells, that in combination perform complex-valued Givens rotations [134], [358]. Figure 18.23 provides a diagram of this systolic array structure, highlighting the Boundary and Internal cells, for a QRD implementation with four antenna inputs. All input data is assumed to be complex valued.

The systolic array takes in the data received from each antenna element (X_n) along with a replica of the desired signal (d). The triangle formed from Boundary and Internal cells (highlighted in red in Figure 18.23) is used to construct the upper triangular matrix R, while the column of Internal cells (highlighted in blue in Figure 18.23) is used to construct the D' vector, as in (18.24). The set of operations performed by the Boundary and Internal cells will now be described.

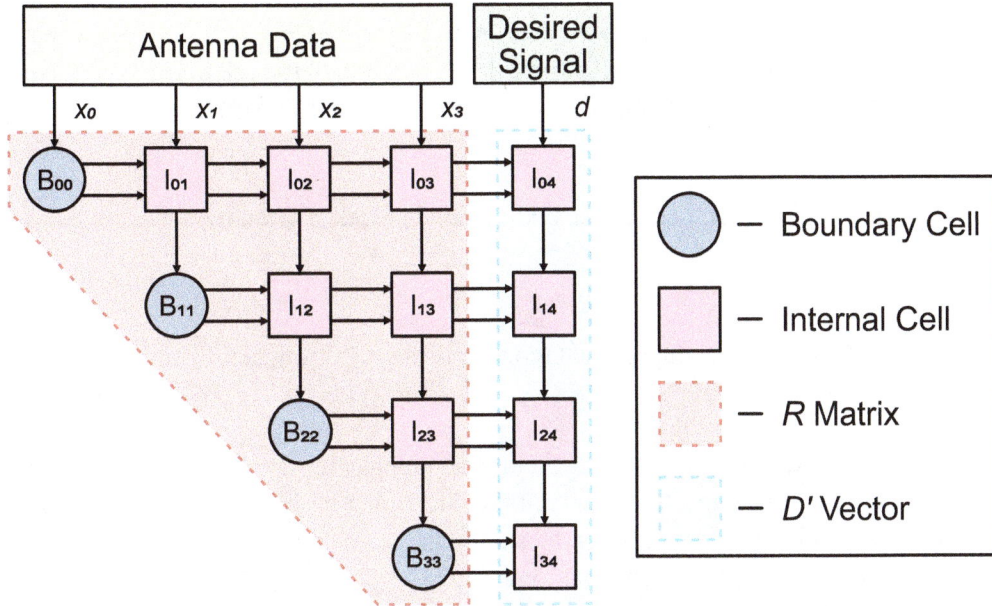

Figure 18.23: QR Decomposition systolic array architecture.

Boundary Cells

A boundary cell begins each row and performs three main operations: a pre-rotation, a cell local memory update, and a Givens rotation calculation. These operations are performed on the data sample input to each individual cell, which we denote here as x_{in}. The first step is the pre-rotation which calculates the magnitude and angle of the complex input data sample, and a new data sample, denoted by x_{new}, is then assigned with the computed magnitude. These operations can be expressed as

$$x_{new} = \sqrt{\Re(x_{in})^2 + \Im(x_{in})^2} \tag{18.30}$$

$$\theta = tan^{-1}\left(\frac{\Im(x_{in})}{\Re(x_{in})}\right), \tag{18.31}$$

where $\Re(\ )$ and $\Im(\ )$ denote the real and imaginary parts, respectively.

The purpose of the pre-rotation is to remove the imaginary component of the input data by rotating it in the complex plane, as the values on the diagonal of the upper triangular R matrix should be real-valued. Therefore, x_{new} is the input data sample, x_{in}, after rotation to remove its imaginary component, and θ is the angle of this rotation.

The next step is a memory update. Each Boundary and Internal cell stores a local r value which is updated on each new input data sample. With reference to the architecture shown in Figure 18.23, these values are indexed in the form r_{ij}, where i and j represent the row and column position of the cell in the systolic array structure, respectively. The r_{ij} values correspond to elements of the R matrix and D' vector of the QRD, as given in Eq. (18.24).

The local memory update calculation performed by the Boundary cell is given by

$$r_{new} = \sqrt{x_{new}^2 + (\mu r_{prev})^2} \tag{18.32}$$

where x_{new} is the input data sample after pre-rotation, μ is the forgetting factor, and r_{prev} is the value of r from the previous iteration.

The forgetting factor is an exponential weighting factor that emphasises the most recent data, i.e. it gradually 'forgets' previous data. This is an important aspect of the QRD adaptive beamforming algorithm as it enables the antenna array to respond to changes in observed signal characteristics; without it, the antenna array would not be able to automatically adapt to a changing environment. For example, the beamformer can track a moving signal effectively because information arising from its previous positions is gradually forgotten, and the calculated beam pattern is weighted towards the most recent data. More generally, the beamformer can constantly update to adapt its radiation pattern in response to dynamic environments.

The final Boundary cell operation is to calculate the Givens rotations angle (φ). This angular value is used to set an individual elements of a data matrix to zero, thus forming the R and Q matrices.

$$\varphi = tan^{-1}\left(\frac{\mu r_{prev}}{x_{new}}\right) \tag{18.33}$$

The Givens rotations angle, φ, and the pre-rotation angle, θ, are passed on to all of the Internal cells in the same row. The r_{new} value is stored locally in the cell, as well as being passed into the Backward Substitution Unit (BSU) to update the complex weight values.

Internal Cells

After the Boundary cell, the remainder of each row in the systolic array is composed of Internal cells, as shown in Figure 18.23. Each Internal cell also performs three operations, and these are: a pre-rotation, a local memory update, and an application of the Givens rotation calculated by the Boundary cell.

The first operation, as in the Boundary cell, is the pre-rotation. However, unlike the Boundary cell, this pre-rotation does not remove the imaginary component, but instead applies the same rotational value calculated by the Boundary cell to the input of the Internal cell.

The real and imaginary component are calculated as follows:

$$\Re(x_{new}) = \Re(x_{in})\cos\theta + \Im(x_{in})\sin\theta \qquad (18.34)$$

$$\Im(x_{new}) = \Im(x_{in})\cos\theta - \Re(x_{in})\sin\theta. \qquad (18.35)$$

Like the Boundary cell, the Internal cell also performs a local memory update. However, unlike the Boundary cell, the Internal cells holds a complex value in local memory, meaning that the update is performed on both the real and imaginary components, i.e.

$$\Re(r_{new}) = \Re(x_{new})\cos\varphi + \Re(r_{prev})\mu\sin\varphi \qquad (18.36)$$

$$\Im(r_{new}) = \Im(x_{new})\cos\varphi + \Im(r_{prev})\mu\sin\varphi. \qquad (18.37)$$

Similar to the Boundary Cell, x_{new} is the input to the Internal cell input after pre-rotation, μ is the forgetting factor, and r_{prev} is the value of r calculated at the previous iteration. Also, φ represents the Givens rotations value calculated by the Boundary cell at the beginning of the row. The r_{new} value is stored locally within the cell for use at the next iteration, and is also passed into the BSU.

The final operation of the Internal cell is to apply the Givens rotation to the pre-rotated input data, x_{new}.

$$\Re(x_{out}) = \Re(r_{prev})\mu\cos\varphi - \Re(x_{new})\mu\sin\varphi \qquad (18.38)$$

$$\Im(x_{out}) = \Im(r_{prev})\mu\cos\varphi - \Im(x_{new})\mu\sin\varphi \qquad (18.39)$$

The complex sample, x_{out}, calculated in (18.38) and (18.39) is passed as input data to the Boundary and Internal cells in the row below.

QR Systolic Array Implementation

Systolic arrays are named as such because of the regular pulsing of data through the architecture, in a manner similar to blood pumping through a heart. The regular cell structure of the QR systolic array, and its inherent parallelism, means that it is ideally mapped to the PL portion of the RFSoC device, with its arithmetic support, vast routing resources, and concurrent processing capabilities.

The significance of using this method of implementing the QRD-based adaptive beamforming algorithm is that all of the Boundary and Internal cell calculations (Eq. (18.30) to (18.39)) can be performed entirely using the CORDIC (CoOrdinate Rotation DIgital Computer) technique [354]. CORDIC is able to perform a variety of trigonometric functions almost entirely using shift and add operations, which are inexpensive to implement. This is ideal for PL hardware implementation on the RFSoC, as the QRD can be fully implemented using efficient methods, using the logic fabric. A further benefit is that the systolic array is constructed of repeating cells that perform the same operation. Therefore, the potential exists to reduce the footprint of the design using a scheduling algorithm to time-share a smaller number of Boundary and Internal cells.

With this systolic array implementation of the QRD, backward substitution needs to be performed separately. The dynamic range of values begins to increase at the backward substitution stage, resulting in larger wordlengths and thus greater FPGA resource utilisation for arithmetic implementation and storage. However, the BSU can only calculate new weight values as fast as it is presented new data, meaning that its rate is limited to the throughput of the QR systolic array. The BSU can also be configured to run at a much slower rate than the QRD array, if application requirements allow, which can be exploited to reduce the FPGA resources needed (recall from Section 18.10.3 that backward substitution involves the hardware intensive operations of division and complex multiplication).

18.11. RFSoC Support for MIMO and Beamforming

RFSoC platforms are well suited for tackling the design problems introduced by MIMO and beamforming, both in 5G NR and other applications, and can eliminate the need for many components. Traditional architectures require several different ICs for processing tasks such as transmission and calibration, however, by handling all tasks on a single chip the RFSoC can provide a cost effective and low footprint solution for massive MIMO designs. This is due in large part to the integration of ADCs and DACs inside the RFSoC chip, negating the need for communication with external data converters (e.g. via a JESD204 interface). The integration of data converters not only reduces the overall component footprint, but also the total power consumption required for a phased array implementation. The reduction in form factor and power consumption also extends the frequency operating range, as antenna elements can be placed physically closer together — this accommodates smaller antenna element separations, corresponding to the shorter wavelengths of high frequency carrier signals [325].

Another advantage of using RFSoC for applications such as 5G NR massive MIMO systems is its Multi Tile Synchronisation (MTS) functionality [90]. Using this feature, each RFSoC device can synchronise all of its channels from a single clock, *and* multiple RFSoC devices can be synchronised via a single external clock. A single RFSoC can implement up to 16 channels of RF-ADCs and RF-DACs, meaning that a single RFSoC chip can be used for a 16-element MIMO antenna. Designs combining multiple RFSoCs are required to achieve larger MIMO configurations, and this is directly supported using MTS functionality.

For massive MIMO antennas, RF path alignment is very important, especially if the antenna needs to deploy beamforming algorithms, and therefore it is important that all elements in the array are accurately time synchronised. To achieve a tightly synchronised MIMO system involving multiple RFSoCs, a *SYSREF* signal is interfaced to all devices to provide a common timing reference. This method can be used to support MIMO arrays of various sizes [46]. The example shown in Figure 18.24 depicts four RFSoC chips, all synchronised from one reference clock, with each supporting 16 antenna elements to create a 64-element MIMO array.

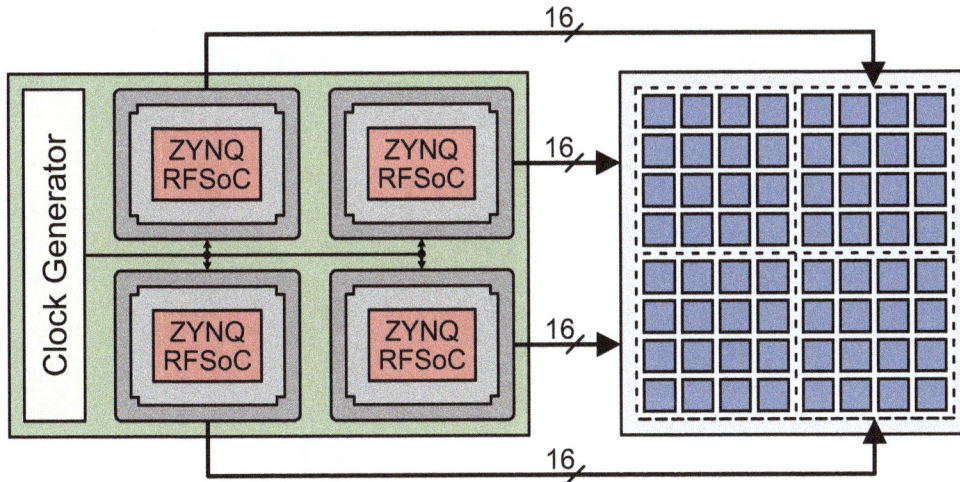

Figure 18.24: Four RFSoC devices synchronised to a common clock and supporting a 64-element antenna array.

18.12. Chapter Summary

This chapter has reviewed two important technologies for modern communications systems such as 5G and *beyond-5G*. MIMO exploits spatial diversity to improve the spectral efficiency of wireless transmissions, and this can be used to improve reliability and/or increase capacity. Beamforming — the ability to electronically steer the transmit beam, or the sensitive region of a receiver, in a desired direction — can be used to optimise transmitter and receiver performance in the spatial domain, and is also a key capability for military and radar systems. The fundamental principles of both techniques were reviewed, and the multi-channel support provided by the RFSoC was highlighted as being particularly relevant for implementation of MIMO and beamforming systems.

Chapter 19

Dynamic Spectrum Access and Cognitive Radio

Tawachi Nyasulu, Graeme Fitzpatrick, Andrew Maclellan, Ehinomen Atimati, and David Crawford

As wireless applications evolve and the demand for RF spectrum grows, it is becoming increasingly important to develop new spectrum access methods that make better use of spectrum resources. For example, the opportunities and benefits of private, pop-up networks are expected to lead to widespread deployment by industry and community organisations large and small, and this will require flexible and adaptive access to spectrum for a wide variety of use cases in different locations and operating environments. With the competition for spectrum from 5G, satellite, private networks and the expectation of frequencies being assigned for the future 6G, its very clear that efficient management, sensing, and sharing of spectrum is very important.

In this chapter, we review relevant background in spectrum regulation and the emergence of Dynamic Spectrum Access (DSA) as an alternative, agile method of accessing spectrum. The motivations and drivers for DSA are outlined, and the frameworks that have been developed in the UK and USA are presented as examples.

We also investigate ways in which the effectiveness of DSA may be enhanced by Cognitive Radio, where radio terminals have built-in intelligence which allows them to alter their behaviour in response to observed conditions. Following a brief review of the underpinning fundamentals, we explore why the RFSoC is a suitable platform for implementing DSA and cognitive radio, and we take a look forward to new challenges and opportunities in this area.

19.1. Spectrum Regulation

Radio spectrum is a natural resource that requires regulation to control access and enforce rules for its use. This is mainly because, at any given time and place, when a portion of spectrum is in use by one transmitting station, another radio station cannot be allowed to transmit at the same frequency, otherwise the transmissions would interfere with one another. Spectrum regulation refers to the process of overseeing the usage of RF spectrum in a given location, with the goal of preventing radios from interfering with one another while allowing optimum usage of the spectrum.

19.1.1. Hierarchy of Spectrum Regulation

The International Telecommunication Union Radio Communication Sector (ITU-R) of the United Nations (UN) is the global regulator of RF spectrum. It ensures coordination of spectrum management among member states to avoid interference between countries. The ITU-R has divided the world into three ITU regions for the purposes of managing global radio spectrum: Region 1 covers Africa, the Middle East and Europe; Region 2 covers the Americas and Greenland; Region 3 comprises Asia (see Figure 19.1).

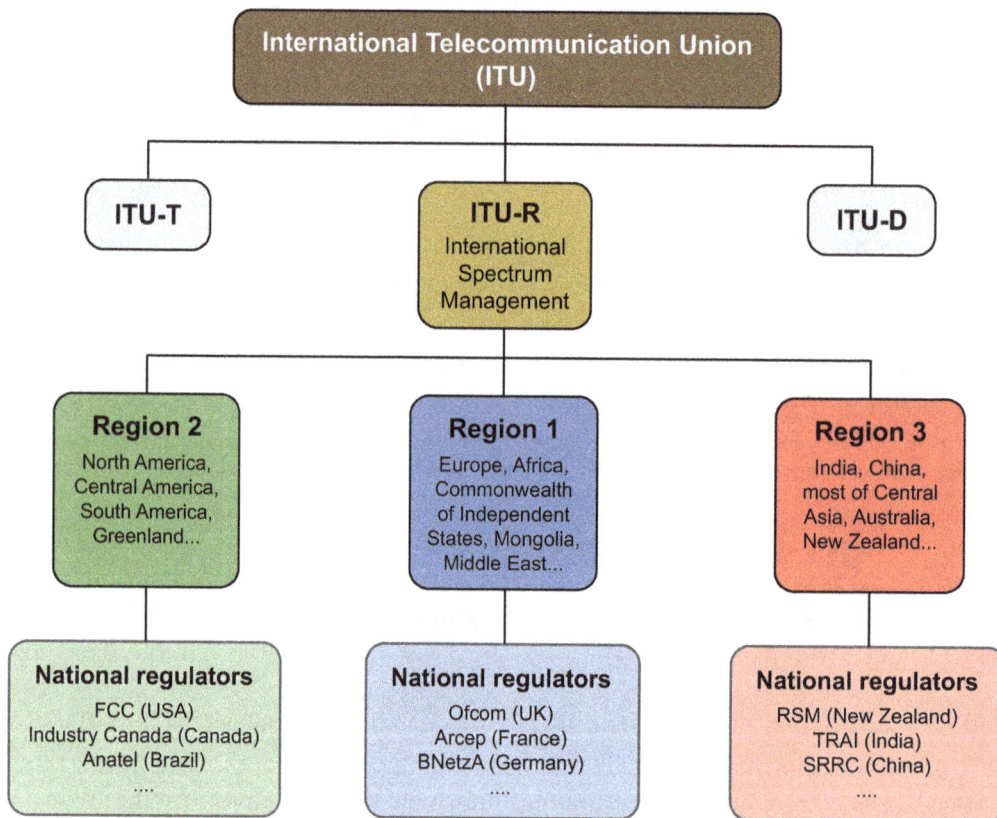

Figure 19.1: Hierarchy of spectrum management and regulation.

At country level, every nation has its own national regulator to oversee spectrum management within its borders and to coordinate interference management with its neighbouring nations. For example, the Office of Communications (Ofcom) and the Federal Communications Commission (FCC) are the national regulators for the United Kingdom and the United States of America, respectively. National regulators coordinate their spectrum regulation activities at ITU regional level and at continental level to ensure technical interoperability of radio systems. For example, in Europe, national regulators work together through the European Telecommunications Standards Institute (ETSI).

19.1.2. Spectrum Allocation

Spectrum allocation involves the allocation of portions of RF spectrum to specific radio usage in line with international radio regulations, such as ITU-R rules, and in accordance with technical propagation characteristics and the potential for interference.

National development priorities and policies are also taken into consideration during spectrum allocation.

19.1.3. Spectrum Licensing

Spectrum licensing involves authorising access to spectral resources for exclusive use by radio communication stations under conditions that are specified in the licence. The goal of spectrum licensing is to ensure proper use of the radio spectrum and to achieve efficient use of spectrum through spectrum reuse, etc. Compliance is enforced by the regulator through monitoring of the radio spectrum and implementation of measures to deter unauthorised use.

The frequencies in a spectrum band can be assigned to different licensees; for example, spectrum reserved for mobile phone use can be allocated to different Mobile Network Operators (MNOs) to serve subscribers (i.e. mobile phone users). The assignment of frequencies is done in such a way as to avoid harmful interference between the radios of different users. Traditionally, spectrum assignment is static: regulators assign a licence which guarantees "exclusive rights" to use the assigned frequencies for a large geographical area and for a specific term, e.g. 15 years. This means that no other operator is allowed to use the assigned frequencies in the specified geographical location, even where and when the licence holder is not using them. This results in wastage of spectrum resources.

19.1.4. Spectrum Engineering

Spectrum engineering is a regulatory function that is responsible for the development of technical standards for radios to access a particular spectrum band for a particular communication service. Technical standards can be broadly divided into two categories: 1) ***Spectrum operating standards***, which describe procedural rules for how spectrum is to be used; 2) ***Radio equipment standards***, which specify the minimum requirements of radio hardware, such as transmit power, operating centre frequency, bandwidth, adjacent channel leakage ratio, and so on.

Specifications contained in the radio equipment standard form the basis for certification of radio equipment to determine compliance. There are various standards development organisations. In Europe, for example, ETSI produces harmonised technical standards for the EU single market. The Institute of Electrical and Electronics Engineers Standards Association (IEEE SA) develops global standards for various industries, including wireless communications under the 802 LAN/MAN standards committee. The 3rd Generation Partnership Project (3GPP) is a global consortium of standards organisations which develop protocols for mobile telecommunications.

19.1.5. Improving Spectrum Utilisation through Spectrum Sharing

Use of wireless-based technologies and applications is continuously increasing, owing to growth in the sales of smart phones, tablets, and other connected devices, and to the socio-economic development opportunities that these devices provide. To support this increasing demand for wireless connections, there is need for efficiency in the use of spectrum. While more efficient radio transceiver technologies have been developed to improve spectrum utilisation by sending more information per unit of spectrum, further improvement can be realised through efficient spectrum management approaches. Exclusively licensed spectrum guarantees certainty of spectrum availability for sizeable long-term investment in communications infrastructure and for the provision of services at a guaranteed Quality of Service (QoS) level. However, spectrum occupancy measurements have shown that most licensed spectrum is not in use at all times and in all places [349]. Therefore, some regulators have moved towards implementing regulatory frameworks for the sharing of spectrum.

Spectrum under-utilisation can result if, for example, a national licence holder chooses not to roll out a network to the entire geographical area covered by the licence at once, or chooses not to cover areas that are deemed not to be economically viable. This type of wastage could be addressed through shared spectrum, by offering affordable short-term leases of unused spectrum to small-scale network operators, e.g. for up to 3 years. This is illustrated in Figure 19.2 where, for example, a local licence-holder (User 1), has been granted rights to Band 1 in Area A, whereas another local licence-holder (User 3) has been granted a licence for the same band (Band 1), but in a different area (Area C).

The spectrum sharing approach will support deployment of private 5G/6G mobile networks, such as private voice and data networks and machine-to-machine (M2M) communications [283]. For example, in the UK, some spectrum bands have been allocated for use under a Shared Access Licence (SAL) framework introduced by Ofcom in 2019, allowing SAL licence holders to share spectrum geographically.

Additionally, the Local Access Licence (LAL) framework allows for shared use of mobile spectrum (i.e. spectrum that is listed in the Mobile Trading Regulations) in geographical areas where the national licensee (a mobile operator) is not using the spectrum and has no immediate plans to use it.

Another 'sharing' spectrum utilisation approach can be seen in New Zealand, where the Interim Maori Spectrum Commission manage allocated bands as part a government allocation (2022) to Maori for 5G radio.

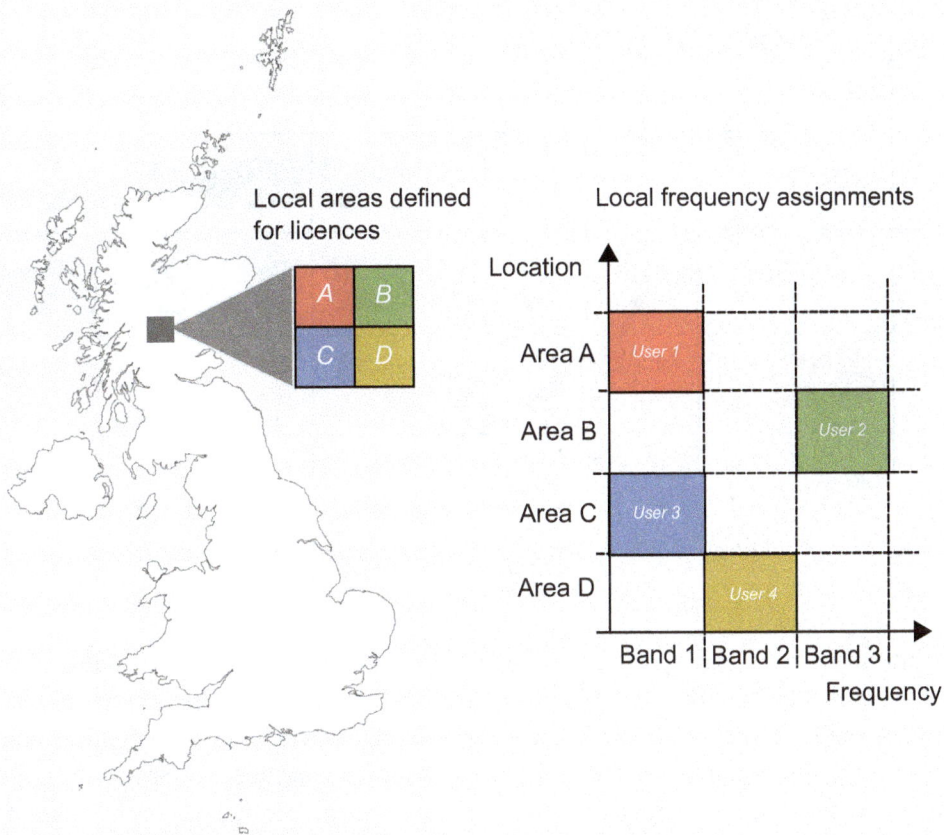

Figure 19.2: Illustration of geographic spectrum sharing through short-term leasing in specific areas.

19.2. Dynamic Spectrum Access (DSA)

Spectrum holes, or spectrum opportunities, are frequency bands that are allocated to a particular type of spectrum use but are under-utilised by licence holders in some locations and/or at certain times. Spectrum utilisation could be improved by identifying the vacant spectrum and allowing other users to make use of it, perhaps even for a service that is different from that which is allocated to the band.

In this section, an overview of the developments towards regulation of dynamic spectrum management is given. Figure 19.3 illustrates dynamic spectrum access. The user who holds the long-term licence to a particular frequency is referred to as the Primary User (PU), while the opportunistic user is called a Secondary User (SU). In period 2T, SU1 can choose to occupy any of the vacant spectrum, in this case either Band 2 or Band 3. In period 3T, PU2, which is assigned Band 2, begins transmission and SU1 moves to the vacant spectrum in Band 3.

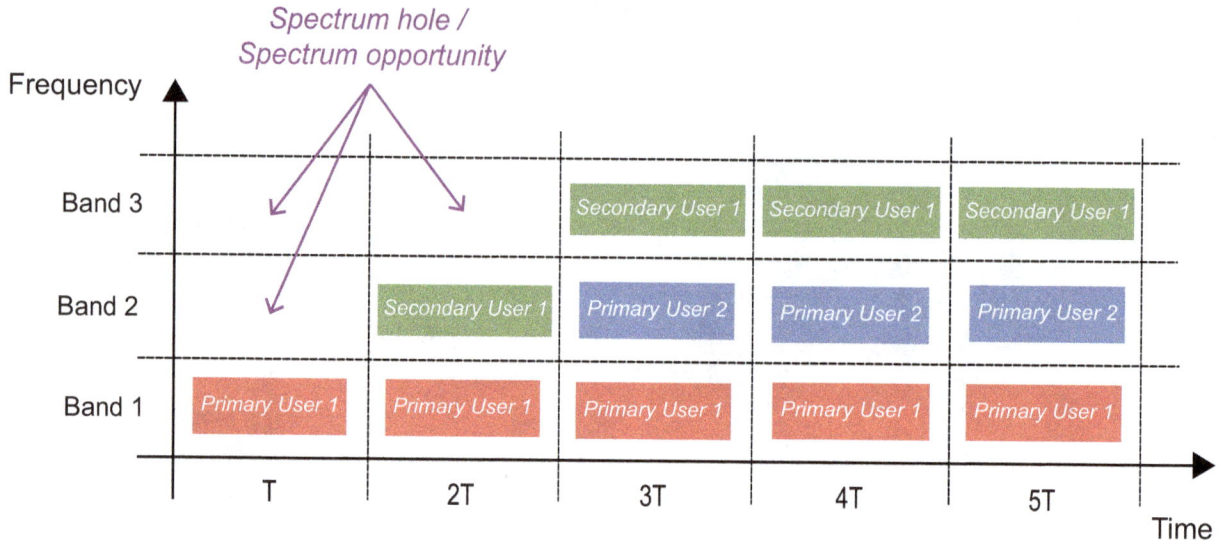

Figure 19.3: *Illustration of Dynamic Spectrum Access between primary and secondary users.*

The terms 'white space' and 'grey space' have been used to classify the occupancy status of spectrum. *White space* refers to spectrum that is not in use at a particular location, while *grey space* refers to spectrum that is in use only some of the time, e.g. by a radio that transmits only when there is payload or control data to be conveyed, resulting in the assigned spectrum not being occupied continuously. Spectrum can also be referred to as grey space in areas which are 'blind spots' for the radio transmitter that has been assigned that frequency. While spectrum utilisation can be improved through spectrum sharing as discussed in Section 19.1.5, it can be improved even further if radios can access spectrum dynamically to exploit spectrum usage needs in both space and time. This can be achieved via a combination of techniques which may involve additional radio capabilities such as spectrum awareness and frequency-agile modulation techniques.

19.2.1. 'Dynamic vs Static' and 'Automated vs Manual'

Experience has shown that the nuances of 'dynamic' vs 'static' spectrum access and 'automatic' vs 'manual' mechanisms for providing access to spectrum are, in general, not understood in a consistent manner, even amongst those directly involved in spectrum policy discussions. With this in mind, it is worth clarifying the terminology and concepts in this regard.

Dynamic vs Static

The IEEE Standards Association defines Dynamic Spectrum Access (DSA) as the process of making continuously-changing spectrum assignments to radio access networks within a composite wireless network that is operating in a given location and time [205]. The frequency assignments are adapted in near-real-time in response to: changes in the environment, such as changes in spectrum availability and types of application;

changes in the radio state, such as its geolocation; and changes in objectives and external constraints, such as operational policies for spectrum usage, QoS, and energy conservation. The physical aspects of spectrum utilisation that can be adjusted include the frequency range that can be accessed and transmission characteristics such as transmission power. Thus, the DSA model enables flexibility in spectrum allocation and licensing, since the user and the type of use can be changed dynamically. This contrasts with 'static' spectrum allocation in which assignments are made on a relatively long-term basis with fixed conditions in force during the term of the licence.

Automated vs Manual

Traditionally, access to licensed spectrum is managed on a manual basis: a licence application gets submitted to the regulator, who manually reviews it and makes a decision about whether or not to issue a licence. (This does not happen for access to licence-exempt spectrum such as Wi-Fi or car key fobs, of course, where the regulator stipulates device compliance requirements but does not get directly involved in specific spectrum usage decisions.)

If the radio network operating environment, radio states, or policy objectives change slowly enough, then a manual approach to licensing may suffice. This has typically been the case for traditional applications such as TV broadcasting and public mobile services. However, as the use of pop-up private networks becomes more widespread, it is increasingly likely that spectrum access decisions will need to be made more quickly for each radio involved in the network, and the use of adaptive, automated spectrum access mechanisms will be required. Figure 19.4 illustrates the potential relationship between manual and automated spectrum management, and the temporal characteristics of the radio network.

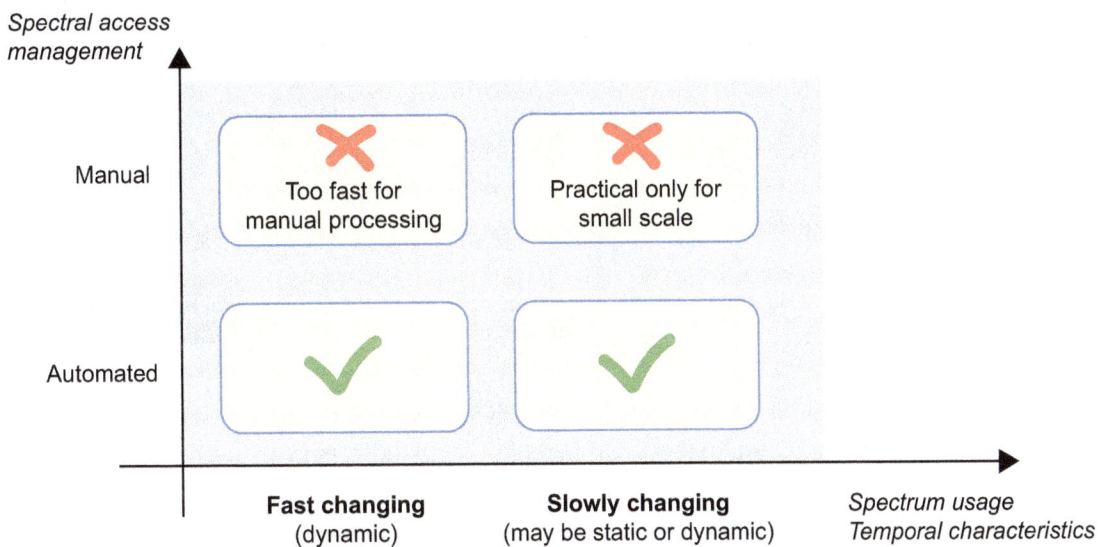

Figure 19.4: Spectrum access management approach vs temporal characteristics.

19.2.2. DSA Regulation

Implementation of spectrum sharing using dynamic spectrum access techniques requires certain conditions to be met through spectrum regulation. These can be summarised as:

- The spectrum allocation function is responsible for drawing up regulations to authorise spectrum sharing in specified bands and the types of use, with the objective of improving spectrum availability for current and future demands. The spectrum allocation function may also outline commercial incentives for the spectrum sharing initiative.

- The spectrum licensing function, in consultation with all stakeholders, has the role of determining a suitable framework for implementing spectrum sharing. The key component of the framework is the set of measures that must be implemented to ensure users share the spectrum effectively, according to the spectrum access priorities and rules, and without causing harmful interference to each other. This includes specification of the method used to determine the availability of spectrum. The spectrum licensing function also provides the rules for implementation of spectrum analysis to determine the spectrum holes, such as the sensing threshold requirements in spectrum sensing.

- The spectrum engineering function is responsible for developing the technical specifications required for radio hardware to operate effectively within the spectrum sharing framework. Besides the traditional specifications for control and data communication functions, DSA requires specification for the cognitive functions in all layers of the protocol stack.

19.2.3. DSA Frameworks

Initiatives towards regulation of spectrum sharing using dynamic spectrum access started in 2008 when the FCC voted to approve licence-exempt use of unused spectrum in the Television (TV) band [162]. DSA frameworks have so far been approved by regulatory bodies in some frequency bands, such as the TV band. New DSA models for spectrum sharing in 5G networks have also been proposed in the literature. The following are some of the approaches for implementation of dynamic sharing.

Concurrent Shared Access

In concurrent shared access, also known as commons-use, the spectrum band is open to any user as long as they use certified equipment in accordance with prevailing regulations. All users have equal rights to access the shared spectrum and are covered by general licence-exemption. This model is already in use in the Industrial, Scientific, and Medical (ISM) band which is used for Wi-Fi and Bluetooth applications and car keyfobs, etc. Medium Access Control (MAC) protocols that are inherent in the radio system are used to coordinate access to the shared medium.

Tiered Shared Access

Tiered Shared Access is used to implement spectrum sharing among users that are allocated to the same spectrum band but have different spectrum access rights. Implementation of tiered access requires a coordination system to enforce spectrum access priorities, to establish where and when spectrum is available, and to determine the technical usage conditions such as transmission power levels, so that the likelihood of interference is minimised.

When licence-exempt users are allowed to make opportunistic use of licensed spectrum when it is not in use by licence holders, a two-tiered spectrum access system can be used in which the licence holder operates in the top tier and must be protected from any harmful interference that may be caused by licence-exempt users, whereas licence-exempt users operate in the bottom tier and are not guaranteed any interference protection from the licence holder or from other licence-exempt users.

A three-tier access system can also be implemented as follows: licence-holders operate in the top tier and must be protected at a given location and time; users that acquire a short-term licence for priority and exclusive access in a given location and time operate in the middle tier and are protected from other priority users and licence-exempt users. Licence-exempt users operate on the bottom tier and are not guaranteed any interference protection. (For note, CBRS operates on such a three-tier model.)

Dynamic Licensed Shared Access

In traditional Licensed Shared Access (LSA), a licence-holder is given exclusive rights for a geographical area, which is considered as an exclusion zone in which the licensee is protected, usually in the form of a circular area defined by its radius as measured from the geolocation of the base station [154]. However, for light users, the spectrum could be wasted during periods of time when the spectrum is unused. Spectrum utilisation can be improved if spectrum is shared temporally through DSA methods. This is illustrated in Figure 19.5, which shows temporal sharing of Band 1 by four users in Area A.

In this regard, proposals have been made for architecture for LSA-type spectrum sharing that is based on DSA, and targeted for 5G mobile networks [269],[287]. Spectrum auctioning and licensing can be implemented using an automated Real-Time Secondary Spectrum Market (RTSSM) [112]. The users' radio equipment would communicate directly with the RTSSM system to request spectrum access, and the RTSSM system would grant spectrum access rights to the radio equipment.

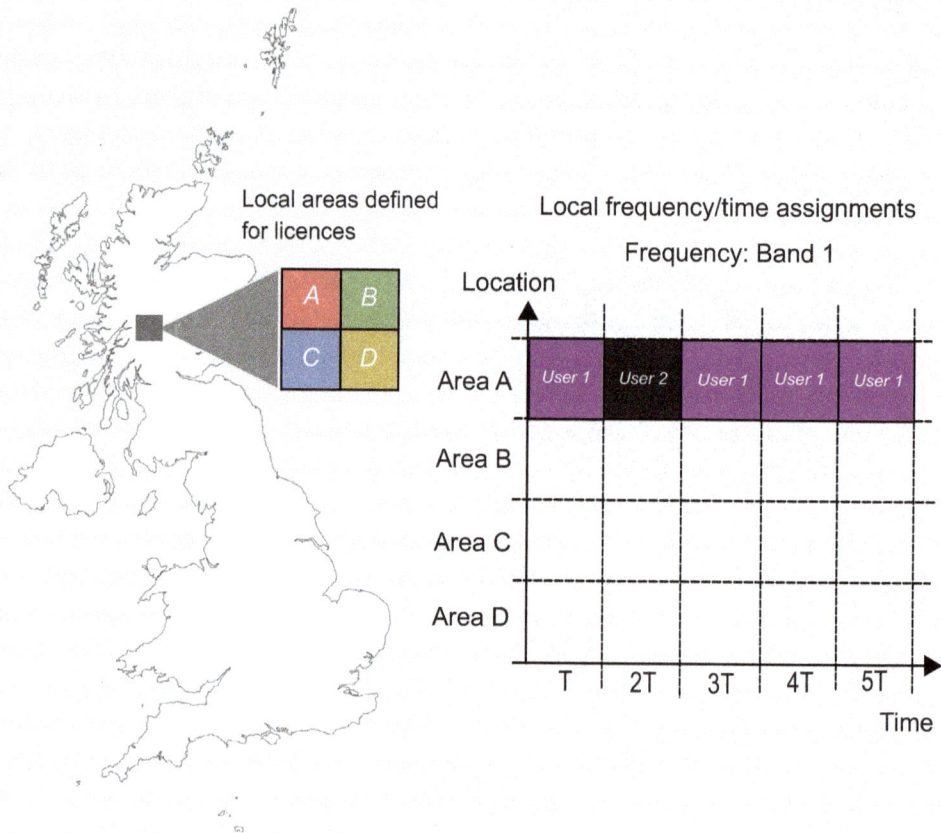

Figure 19.5: Illustration of Dynamic Licensed Shared Access.

19.3. DSA Frameworks in Practice

'Fully dynamic' spectrum access is where spectrum usage is shaped to permit fully flexible access: shared both intelligently and automatically. A number of DSA frameworks have already been defined and implemented in various parts of the world. In this section, we review the two foremost examples: TV White Space, and the Citizens Broadband Radio Service (CBRS).

19.3.1. TV White Space

Traditional access to TV services is via terrestrial broadcast, in which TV signals are transmitted from a national network of high-power TV transmitters distributed across the entire country. This results in large bands of fixed (static) licensed spectrum being reserved for television broadcasting. The transition to Digital TV resulted in more spectrally efficient TV broadcasts, revealing considerable portions of unused TV spectrum in different locations. These large chunks of unused TV spectrum (so-called 'White Space' spectrum) can be used by other applications in those geographical areas. In 2010, the FCC proposed the use of dynamic access to TV White Spaces (TVWS) through automated frequency coordination using a centralised database. This allows for the operation of localised, licence-exempt networks — on the condition that they do not interfere with licensed TV transmissions in the same spectrum band [145].

One of the main attractions of TV White Space is that it has a longer wavelength than other commonly used spectrum (such as Wi-Fi at 5 GHz); this means that TVWS signals can travel further, and can better penetrate trees, buildings, and other obstacles, and can diffract better around/over objects such as hills. For example, a 200MHz TVWS signal used for Wi-Fi can cover an area over five times larger than that of a 5 GHz Wi-Fi signal, while having improved resilience to obstacles. In many circumstances, using TVWS can replace the need for additional cabling or infrastructure. As well as reducing the cost of deploying connectivity to rural areas and remote IoT networks, another potential benefit is faster deployment [145].

Implementation

Figure 19.6 shows a high-level overview of a typical TVWS system. In most, if not all, TVWS implementations, fixed 'basestation' devices register with a geolocation spectrum database, which provides the GPS location along with other relevant information. Once registered and authenticated, they submit regular spectrum availability requests to the database, which uses various pieces of information to perform calculations and provide a list of currently available channels to the TVWS device. Using these available channels, the TVWS 'basestation' device can set up its own wireless network with a number of 'user' devices connected to it.

The frequency of spectrum availability requests depends on the implementation and regulations in the area. In the UK, TVWS devices must poll the database every 15 minutes, while in the USA the interval can be as long as every 24 hours [163].

19.3.2. Citizens Broadcast Radio Service (CBRS)

In 2015, the FCC took some key concepts associated with TV White Space and applied them to establish the Citizens Broadband Radio Service (CBRS) band. This is an example of a tiered shared access DSA framework. It makes use of the 3.5 to 3.7 GHz band, which was previously completely reserved for military use across the entire country, despite being used predominantly in coastal areas. CBRS allows tiered access to this spectrum as long as there is no interference to higher priority users, the highest priority still being reserved for the US Military [145].

Neutral Hosts Capability

Another advantage of CBRS is that it allows for 'Neutral Hosts'. These are single LTE deployments that can serve subscribers from different operators, letting mobile network operators share certain infrastructure and

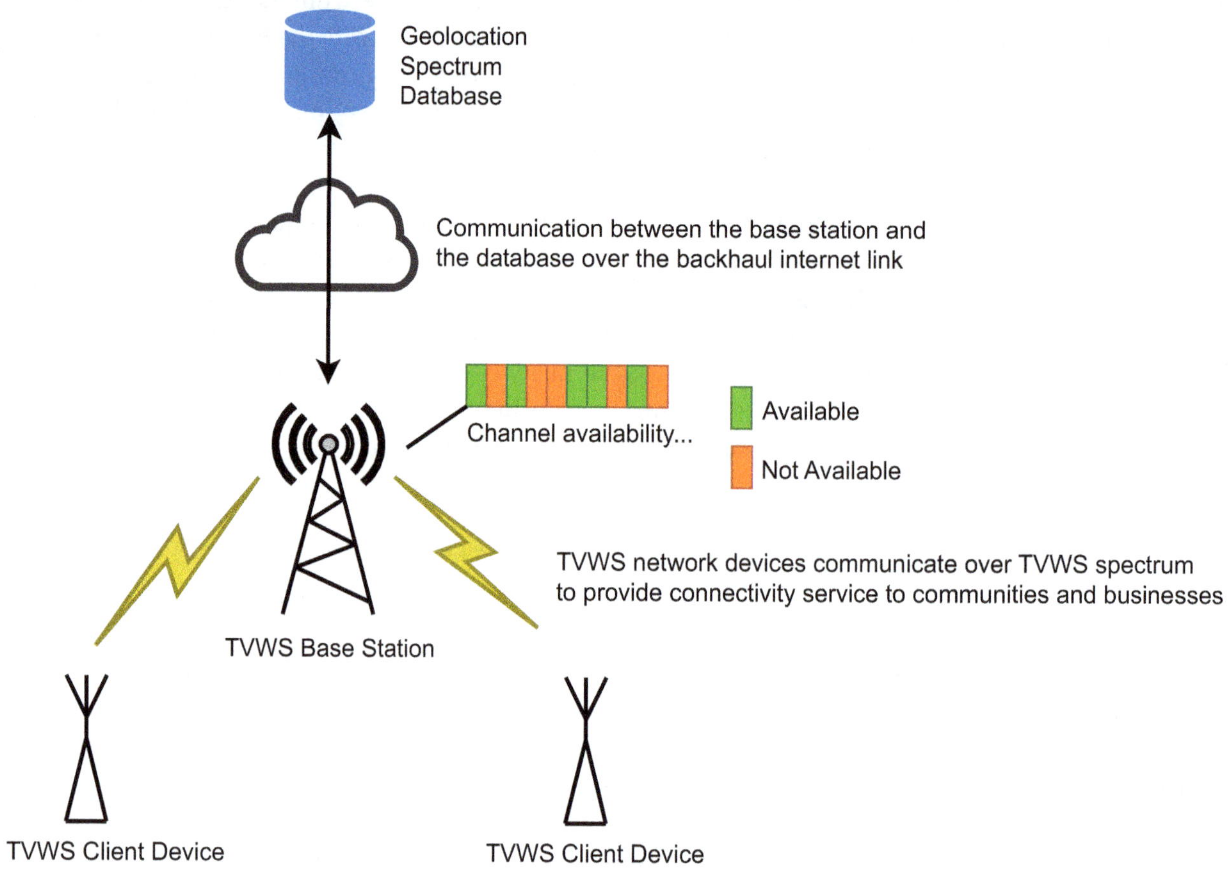

Figure 19.6: High level overview of TVWS networks.

spectrum to improve user experience and coverage. This is particularly useful for places where many people who subscribe to different network operators are gathered, such as airports and public stadiums.

The ability to provide the same QoS, regardless of which MNO provides a particular customer's broadband, could be an attractive selling point for public events [174]. From the perspective of an MNO, instead of each MNO having to manage its own individual network infrastructure to provide service, several MNOs can leverage a single antenna and single access point (basestation) to provide mobile service to their subscribers. This has multiple benefits, such as reducing infrastructure costs, and offloading some of the indoor traffic from their own traditional infrastructure onto some of the newly introduced CBRS infrastructure [272].

Implementation

CBRS has a three-tier, priority-based system which allows lower tier users to access spectrum in a particular location where higher priority users are not using it. Highest priority (Tier 1) users are the incumbent users —

those such as the military who have existing licences to use these frequencies. Tier 2 is for Priority Access License (PAL), which gives users a higher priority service for a relatively small price. Tier 3 is for General Authorized Access (GAA), which is the default and represents free-to-use public access to the CBRS band [349].

The tiered nature of the CBRS framework means that its implementation becomes relatively complex, as there is a need to ensure that higher-priority users experience little to no interference or disruption in their services due to lower priority users. This is made difficult as each potential user has information only about the interference they can potentially cause to other users, and not necessarily about the interference other individual users can cause. If a higher priority user is experiencing relatively small amounts of interference from multiple parties, this can add up to produce unacceptable levels. Therefore, as in TV White Space — a centralised system is required, with the additional computational requirement to calculate and manage the level of interference that is experienced by each user. A centralised system can manage *how* each user is allowed to transmit (e.g. at which frequencies in the spectrum, and how powerful their signal is allowed to be) in order to maximise efficient use of the band and the QoS provided for each priority of user.

Previously, this kind of system would have been complex and costly to implement, but with the advent of services such as cloud computing — which is scalable and provides immense information management capability — we can rely on such services to provide sufficient computational power to manage spectrum in this way. The cloud computing service that manages the CBRS band is known as the 'Spectrum Access System' (SAS). All users give information to the SAS, which then ensures that the interference experienced by higher priority users is minimal, and access to the spectrum is shared efficiently and effectively [117].

Spectrum Sensing in CBRS

Another key component of CBRS is Environmental Sensing Capability (ESC), which is depicted in Figure 19.7. This feature uses spectrum sensing to ensure that if any incumbent (Tier 1) users or PAL (Tier 2) users are detected, then the SAS can be alerted to clear those channels and protect higher priority users. In practice, ESC tends to be deployed around the coast to account for the possible usage of navy radar systems under the incumbent license priority [145].

Effective spectrum sensing entails capturing and monitoring the entire CBRS band, and measuring the power of different signals in the band accurately. Currently, due to the bandwidth occupied by the CBRS band, this is done by inspecting slices of spectrum, one at a time, and combining the data. However, recent technological advancements have enabled the direct digital sampling of much larger signal bandwidths. A prime example of this is the RFSoC spectrum analyser application presented in Notebook Set C. This application shows that RFSoC Devices can be used to directly inspect the radio spectrum up to 6 GHz (i.e. with no external analogue demodulation stage), which is well suited to CBRS applications [329].

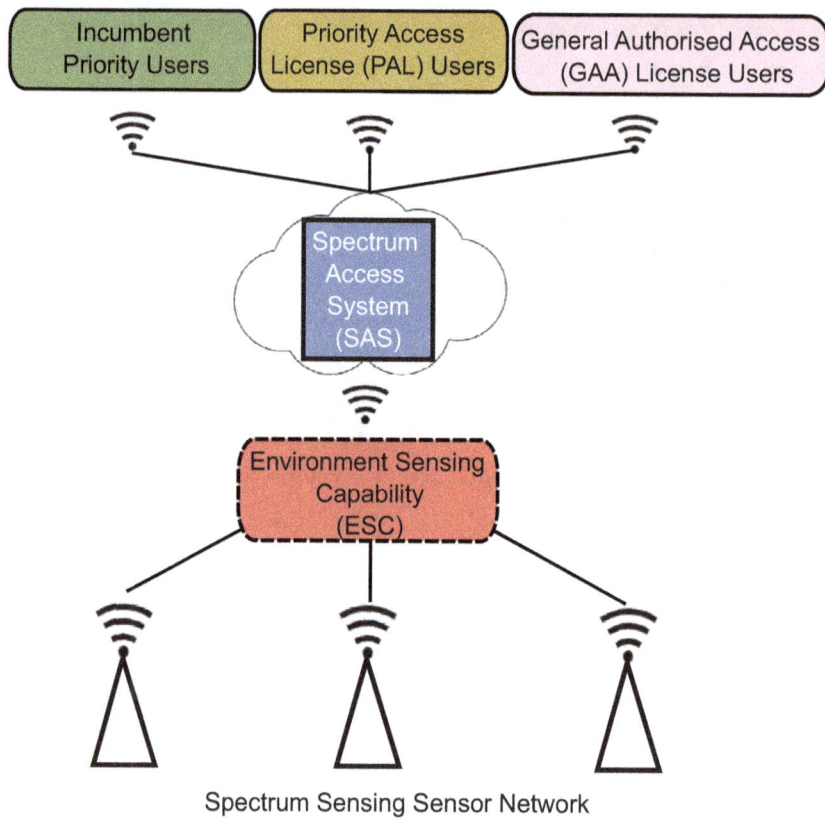

Figure 19.7: High level overview of CBRS.

19.4. Deployment of Fully Dynamic Spectrum Access

Fully dynamic DSA requires flexibility of radio resource usage and coexistence management, which is supported by intelligent and adaptive wireless transceivers. A Cognitive Radio (CR) is a wireless communication system that is aware of its ambient environment, is able to learn from the environment and to adapt its operation parameters according to changes in the environment, operating conditions, or user requirements, in order to improve utilisation of spectrum and to achieve reliable and efficient wireless communication [199]. Spectrum agility, that is, the ability to transition quickly from one operating frequency to another, is a key component in implementing cognitive radio architectures. This is achieved through SDR — wireless communication systems in which the transmission parameters can be controlled and reconfigured dynamically through software-controlled signal processing algorithms.

It is clear from the definitions of DSA and CR that the effective design of dynamic access methods for cognitive radio networks requires technical knowledge and skills spanning science, engineering and economics. It also requires mathematical optimisation to implement efficient radio resource allocation, machine learning and Artificial Intelligence (AI) to implement cognitive capabilities, and economies of scale to control spectrum costs and/or revenue.

In this section, we will review methods of operating DSA techniques with reference to SDR technology, and specifically the capabilities of the RFSoC. Spectrum sensing for shared spectrum is introduced, and CR techniques for more efficient spectrum management are discussed. The concepts of Machine Learning (ML) and intelligent radio are outlined as enablers for the next stage of CR evolution.

19.4.1. Software Defined Radio for Dynamic Spectrum Access

DSA, discussed in Section 19.2, can improve the overall spectrum utilisation of a wireless communication network. DSA techniques include smart wireless spectrum access techniques where multiple users can share the same portion of the spectrum with minimal interference. To facilitate fully dynamic spectrum access, radio transceivers must evolve to scan a spectrum and make decisions based on these observations.

Traditional radio communication systems typically comprise hardware components, such as mixers, filters, amplifiers, etc., which are configured to perform a specific task without the flexibility to change their operation or to make decisions. SDR devices, however, implement and/or control core aspects of a radio communication system using software running on an embedded processor, or an externally-connected processing device (recall the discussion of SDR architectures from Chapter 8). This coupling of a processor to a radio communication system allows for an SDR device to make dynamic decisions, given the observation of signals received via the antenna(s). Therefore SDRs are a great fit for DSA applications.

RFSoC Capabilities

The RFSoC device is well-suited for SDR implementations, including those for DSA-based applications. The RF front end supports the entire 5G sub-6GHz band, and its multi-GHz ADCs facilitate wideband spectrum sensing (for instance, the entire TVWS and CBRS bands can be simultaneously digitised).

As reviewed in Chapter 3, the RFSoC's onboard features include a processing system equipped with a quad-core Arm Cortex processor and DDR memory, FPGA programmable logic, and multiple channels of RF data converters (ADCs and DACs, with associated DDCs and DUCs, respectively). Of particular note to spectrum sensing, up to 16 channels of ADCs are available, and likewise up to 16 DACs. The PS provides plenty of resources for implementing software control of the radio, including decision-making algorithms for DSA. There is also scope to hardware-accelerate algorithms that work with radio data close to the DACs and ADCs, by leveraging the PL. The tight coupling between the PL and PS enables flexible radio operation, with low latency spectrum sensing and decision making, while maintaining high data throughput characteristics.

The RFSoC platform provides the necessary resources for implementing DSA applications, particularly as flexible software can be developed that benefits from hardware acceleration assistance from the PL. The ADCs that interface directly into the PL enable broad sensing of the frequency spectrum, while mixing frequency and decimation factor can be programmatically changed — this enables spectrum sensing even in very dynamic scenarios where frequency bands over a broad range of frequencies must be monitored. The extensive resources on the chip can also allow for the development of CR systems where the radio device takes decisions to maximise the throughput on the channel. In a DSA scenario, CR techniques could be used to allocate channels based on a selection of performance metrics rather than following a rigid set of rules.

19.4.2. Cognitive Radio

Traditional radio communications systems transmit and receive data on allocated channels of the spectrum in order to minimise collisions with one another and to ensure consistent transmissions. With the number of connected devices growing exponentially, and the RF spectrum being increasingly allocated, the spectrum resources available to support more devices are quickly diminishing [271],[284],[362]. Some spectrum users are faced with expensive licence fees even though they want to transmit only intermittently. However, studies have shown that allocated spectrum is not consistently occupied by primary user transmissions, and that actually some portions of the spectrum are extremely under-used [349]. For radio devices to exploit vacant spectrum opportunities, they must evolve to dynamically change their transmission parameters based on knowledge of other devices that are transmitting around them.

CR techniques can enable radios to manage access to spectrum in a dynamic manner, with the aim of improving the overall quality of transmissions within a communications network. A CR has the ability to sense the spectrum and identify possible ways to optimise its own transmissions via various metrics, discussed in the following sections. CR and DSA techniques can be integrated into wireless communications networks to improve the overall spectrum performance of the network, as well as maximise bandwidth utilisation. These can be ideal tools for SU radios — which do not have priority access to the spectrum but aim to transmit around idle PU radios.

The Cognitive Cycle

The cognitive cycle, illustrated in Figure 19.8, refers to the tasks performed by CRs [271],[367]. It consists of several stages, which include:

- **Spectrum Sensing** — The CR scans the available spectrum and captures usage and vacancies.

- **Spectrum Decision** — The CR chooses the best vacant channel for the SU's communication given prior analysis.

- **Spectrum Mobility** — The CR detects a change in the channel conditions. Either the quality of the channel has degraded, or the PU has appeared, and the CR must change its channel of operation to a different vacant portion of the spectrum.

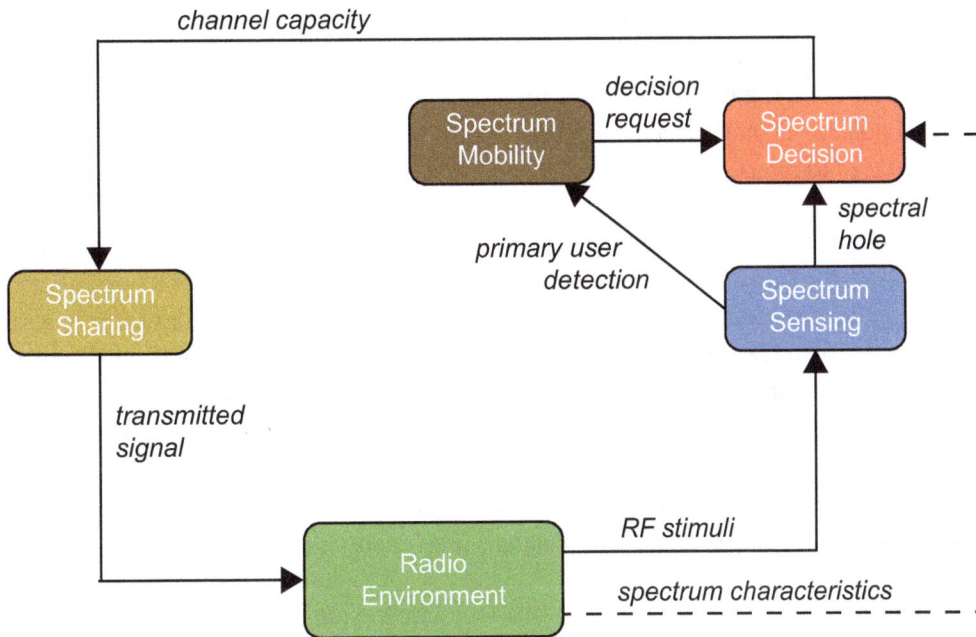

Figure 19.8: The cognitive cycle.

- **Spectrum Sharing** — The CR coordinates with other CR user's transmissions to avoid collisions when accessing the same portion of vacant spectrum.

The radio makes decisions, typically through finite state machine or ML-based techniques, based on information obtained during both the sensing and mobility stages of the cycle.

Spectrum Sensing Techniques

The spectrum sensing task aims to identify the PUs and other SUs of the spectrum while also seeking out unused portions of spectrum. There are several ways to sense signals in the spectrum. These include:

- Energy detection

- Matched filter detection

- Cyclostationary detection

- Covariance based detection

Energy detection is the most basic form of signal detection, and it simply compares the magnitude of received samples, when scanning, to a prior known noise floor. If the sample magnitudes are above the threshold, then a signal is assumed to be present; otherwise, it is assumed that no signal is present [368]. This method of detection assumes that the CR has knowledge of the noise level prior to sensing; however, even with a prior estimate, the actual noise level may vary due to other channel factors, and this could cause misdiagnosis of the spectrum and could lead to interference occurring.

Matched filter detection is an alternative technique that can be more robust. This method of detection requires prior information about the transmitted signal, such as pilot sequences, modulation schemes, spreading codes, preambles and packet formats. Such knowledge is required because the matched filter technique correlates one or more of these known patterns with the observed signal to detect the presence of the PU transmissions. For example, an SU radio could aim to transmit on a channel that is known to be allocated to a 5G NR signal, and uses the matched filter detection technique to correlate known synchronisation signals with the observed signal while searching [5]. If the matched filter detects a correlation, the 5G NR signal is assumed to be present; otherwise, it is assumed to be absent. This technique does not require knowledge of the noise floor and can be a faster and more robust way to detect signals than energy detection [271].

Cyclostationary detection is another option, and this technique exploits the periodic nature of many transmitted signals. Wireless communications signals are loaded with repeating codes, sinusoidal signals, cyclic prefixes, and hopping sequences, giving them cyclostationary characteristics, i.e. some of their statistical properties exhibit periodicity. Cyclostationary detectors are able to identify communications signals by measuring these periodic statistics and comparing them to thresholds.

Finally, covariance-based detection was introduced to overcome the dependency on prior knowledge of the noise level or signal parameters [368]. This method takes advantage of the statistical covariances between signals and noise, which are usually different. Through manipulation of covariance matrices, this method can compare the statistical differences between noise (assuming no signal presence) and signals to determine whether a PU is occupying the channel space.

The detection techniques described here represent a few of the more popular ways to detect PU activity on a channel. They range from simple implementations (energy detection) to more complex deployments (matched filter and covariance-based detection) where an SDR capable of performing complex algorithms can be used. Furthermore, these detection techniques will require their parameters to be altered depending on what type of signal is being detected (synchronisation patterns, pilot symbols, etc.), and therefore a reprogrammable architecture is required for processing these algorithms.

Exploiting Vacancies in All Dimensions

PU vacancies in the spectrum can occur within frequency channels, but can also occur in the form of temporal and spatial gaps. Towards achieving optimal spectral usage in the future, SU radios can find ways to use an already occupied channel but exploit the empty space in time and spatial separation. The detection algorithms mentioned in the previous section can be taken a step further and used in conjunction with estimation models to predict gaps in temporal signals — the CR can then transmit with a good level of confidence that the PU will not also transmit at the same time [350]. SU radios could also learn to exploit the spatial separation between the PU and SU, and transmit at power levels that are low enough to avoid causing interference to PU activity [138],[355].

Common Cognitive Radio Problems

In a wireless communications network, the PU radio is usually part of a separate system, and therefore will probably not be cooperating with SU radios. A possible problem is that, when a PU is transmitting, the SU CR may not correctly sense the presence of a PU transmitter, for instance if it is outside the SU detection range. This could result in an SU transmission that interferes with PU network activity. The problem is known as the 'hidden node problem' and it can become an issue for SU radios if they repeatedly interfere with PU radios.

Many factors can contribute to the hidden node problem, including extreme multipath fading, and shadowing from nearby structures [149],[271],[299]. Possible solutions include cooperative spectrum sensing, where two SU radios share the same channel and cooperatively sense the PU activity and relay the information to each other [171]. This cooperation can scale to more devices while increasing the robustness of PU transmission sensing. Similarly, the hidden node problem extends to networks with extremely unstable channel conditions where transmitted signals may fade faster than expected at unpredictable times, resulting in the SU mis-sensing the PU. Again, cooperative sensing can assist in mitigating this issue.

The flexibility of CR is promising for achieving improved spectral utilisation via DSA; however, this flexibility is not possible with traditional hardware-based radios. For a radio to sense, analyse, and change its own parameters dynamically, a more adjustable radio architecture must be used. SDRs are an excellent resource for implementing CR because of their reconfigurable architectures: parameters such as operating frequency, modulation scheme, number of antennas used, bandwidth, and transmit power are among the many parameters that can be dynamically adjusted on SDRs.

19.4.3. Intelligent Radio

CR is a promising paradigm for SUs spectrum access. Using CR techniques, SU terminals can access spectrum without the need for a licence, and share vacant spectrum with the PUs through sensing transmitter activity and transmitting opportunistically. DSA using CR does have pitfalls, however; the possibility of mis-detection in the spectrum can result in interference and disruption of service for both PU and SU radios. As the number of devices using the radio spectrum grows, the likelihood of collisions and interference also increases, and it becomes more difficult to successfully detect vacant spectrum. Smarter solutions for CR must be adopted and basic sensing techniques, such as energy detection, may not suffice. CR can integrate Artificial Intelligence (AI) into its sensing and decision making to address these more complex scenarios [239].

Introducing Artificial Intelligence to Cognitive Radio

AI is the study of building machines capable of performing tasks that typically require human intelligence. This can entail building machines to perform tasks that were typically designed by humans; it can also be taken even further, to state that AI machines can perform tasks that humans are unable to design for.

The terms *Artificial Intelligence*, *Machine Learning*, and *Deep Learning (DL)* can at times seem confusing, especially as they may even be used interchangeably. To clarify what we mean when using each of these terms, it can be useful to think of them as encompassing one another, as illustrated in Figure 19.9. ***Artificial Intelligence (AI)*** is the term given to the science of developing and deploying intelligent systems. ***Machine Learning (ML)*** is a subset of AI, and involves the design and training of mathematical algorithms to learn how to make decisions based on data features, observations, and experiences provided to a model during training. ***Deep Learning (DL)*** is a subset of ML wherein the mathematical models used to train the ML model receive raw, unprocessed data which gets passed into a DL model to learn decision-making tasks alongside feature extraction techniques.

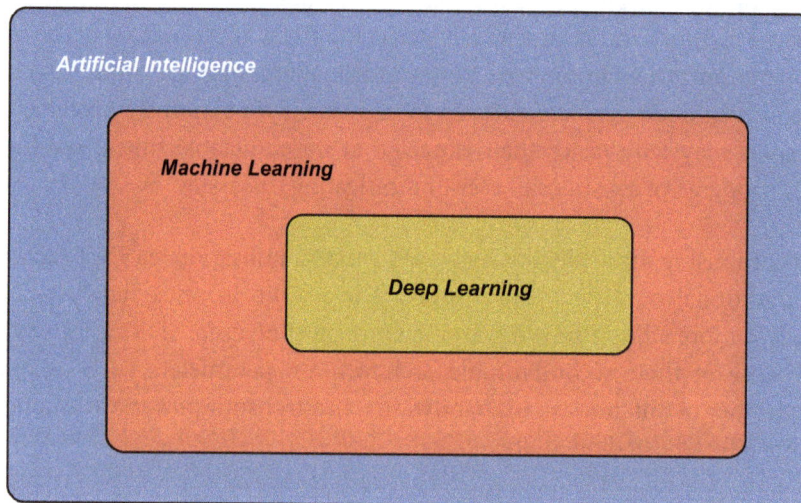

Figure 19.9: AI/ML/DL distinction.

DL techniques typically seen in research areas such as image processing and natural language processing have made their way into the wireless communications field, and have been demonstrated as capable of solving problems that were otherwise difficult to design for. DL differs compared to ML, which uses more classical statistical techniques such as Bayesian ML and statistical algorithms, and has removed the need to pre-process the input data and pass features into the AI model. Instead, DL allows for the AI model to learn how to extract features from raw data, leading to smarter and more efficient feature extraction methods. This approach to learning has greatly assisted in the research areas of image processing and natural language processing, where features within an image or a word may not be immediately obvious, and it now provides benefits in wireless communications as DL algorithms can act directly upon the received data.

The benefits of DL for CR can be seen in many areas of PHY layer wireless communications, ranging from channel estimation and signal detection, to a whole PHY layer radio stack built from a Neural Network (NN) [139]. Additionally, the field of wireless communications has an abundance of accurate simulation and generation tools that make it possible to produce an infinite amount of labelled data for use in DL. With this benefit, DL models for wireless communications can be trained more easily, without the laborious effort normally involved in building datasets (such as labelling the content of photographs in image processing applications).

Basic Deep Learning Architectures

The forthcoming review of DL in CR requires that the reader has a basic understanding of common DL architectures, and therefore as a precursor, this section introduces the building blocks of NNs and their layers, and outlines how models can be trained to operate on cognitive tasks.

The **Neuron** is a fundamental building block of all Deep Neural Network (DNN) architectures. The artificial 'neuron' originated from the biological concept of human neurons within the brain. An artificial neuron is a mathematical construct that sums all connected weighted inputs, before passing them to the output via a non-linear activation function. Connecting a network of neurons together is known as an Artificial Neural Network (ANN), where every neuron has its own connected weighted inputs and activation function.

Figure 19.10 illustrates an artificial neuron. Its output can be expressed as

$$y = f\left(\sum_{n=0}^{N-1} x_n w_n + b \right) \tag{19.1}$$

where N is the total number of weights, x_n represents the nth input, and w_n represents the nth weight. The terms used in this equation are further defined and described next.

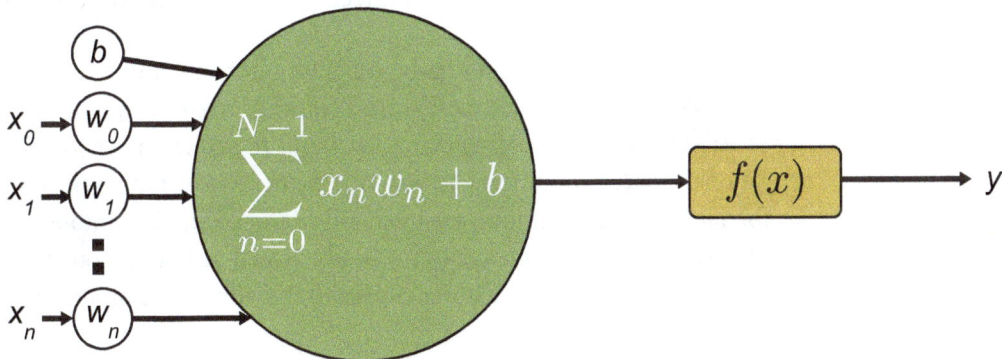

Figure 19.10: An artificial neuron.

The **Bias** (*b*) is an addition made alongside the summed weighted inputs to provide a means of shifting the decision boundary of a neuron.

An **Activation Function** (*f*) is the non-linear function applied at the output of each neuron in an NN. They are essential in assisting DNNs in learning non-linear data structures during training. Figure 19.11 shows the formulae and resulting plots for the most common activation functions.

Sigmoid	ReLU	Tanh
$\dfrac{1}{1+e^{-x}}$	$max(0, x)$	$\dfrac{2}{1+e^{-2x}} - 1$

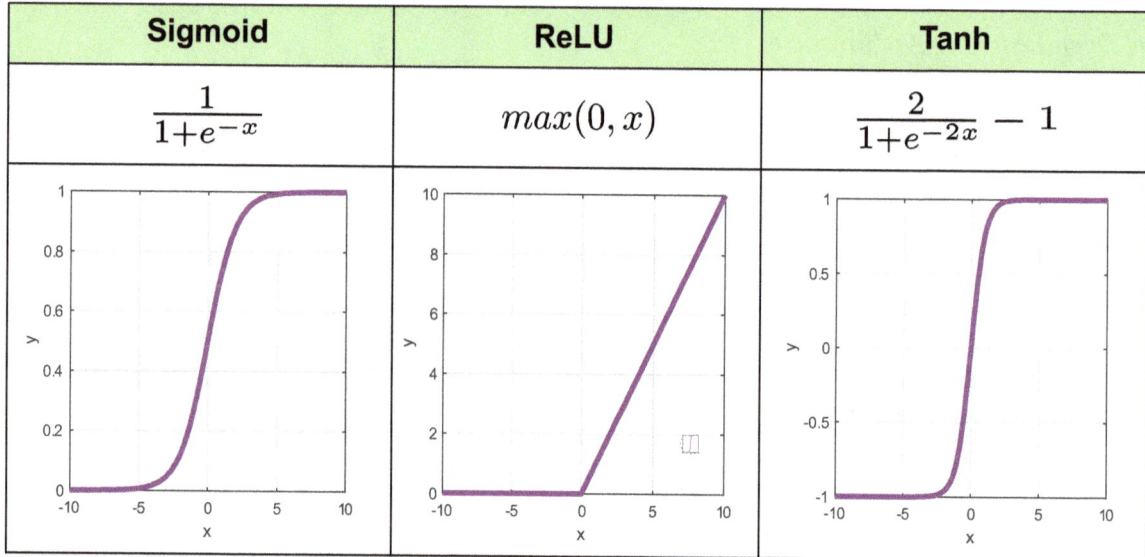

Figure 19.11: Plots of common activation functions.

Multi-layer Perceptron (MLP) is a term that can loosely mean any feed-forward ANN that consists of multiple layers of connected neurons. MLPs, historically, were some of the first NN structures, and gave way to other architectures such as Convolutional Neural Networks (CNN) and Recurrent Neural Networks (RNN).

Convolutional Neural Networks (CNN) are structured similarly to MLPs, but some layers in the network are replaced with convolutional layers. These types of networks are primarily used within image and video processing, due to their ability to pass a kernel of weights across an input image/frame. Kernels in convolutional layers can find local correlations between input samples without the need to weight every sample to each neuron. This not only allows for larger inputs without exponentially increasing the number of weighted connections, but also focuses learning on identifying features in locally placed samples. Convolutional layers only store the weights required for the kernels, which are independent of the number of input samples.

For a given input sample in the convolutional layer at location i, j, the kernel K of dimensions $M \times N$ correlates each kernel weight with the samples that the kernel overlaps, and forms a sum. Finally, the bias, b, is added to the result. The following formula is an example of a two-dimensional cross-correlation and is repeated over the entire input, X.

$$Y_{i,j} = \sum_{m=0}^{M-1} \sum_{n=0}^{N-1} (X_{i+m,\,j+n} \, K_{m,n}) + b \qquad (19.2)$$

This is how most DL libraries implement 2D convolutional layers [182].

Figure 19.12 illustrates how a CNN operates for a given input image. The first convolutional layer passes its kernels over the input image and calculates the output for each kernel in the convolutional layer. This output is then passed to the second layer, where the same operation occurs with its respective kernels. The output from the second convolutional layer is flattened and passed through the first and second fully-connected layers to obtain an output classification.

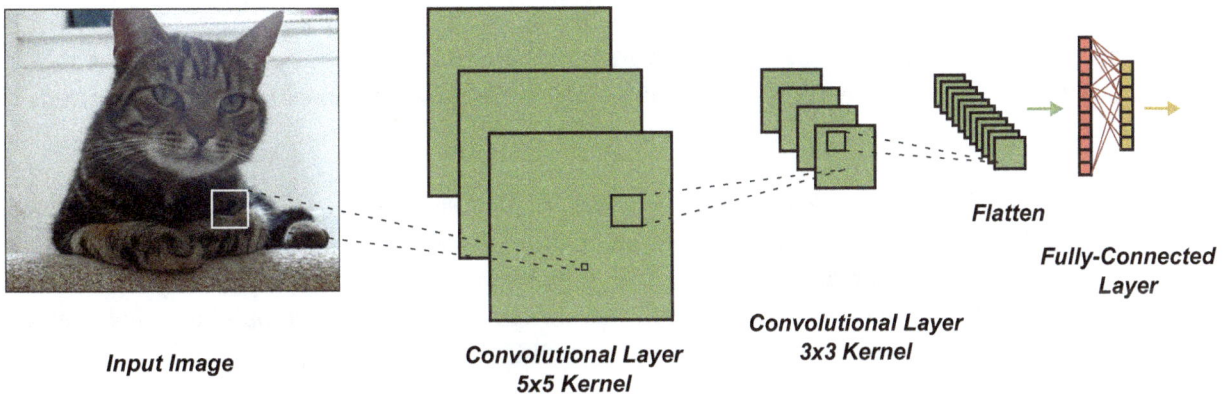

Figure 19.12: Depiction of a CNN operation on an input image.

For extended coverage of DL concepts, the reader may be interested to follow up with a popular textbook such as [182]. There are also DL-focused chapters in our previous book on Zynq UltraScale+ MPSoC, which are directly relevant to RFSoC implementations [131].

Deep Learning in Cognitive Radio: Intelligent Radio

DL is a tool that has been shown to perform well in the areas of image and video processing due to its ability to self-learn decisions and feature extraction methods purely on the labelled data that it receives. For DSA, DL could further enhance CR operations to improve tasks such as identifying spectrum vacancies, learning the transmit patterns of PU radios, correcting channel degradation affecting received signals, and implementing smarter decoding algorithms. DL provides a powerful means for radios to make smart decisions and adapt to complex transmission scenarios.

Signal Detection with Deep Learning

The purpose of signal detection and classification is to accurately maintain an understanding of the current spectrum environment. This involves: detecting other users of the spectrum, isolating sources of interference, learning PU radios' transmission patterns, and identifying spectral vacancies around PU activity. Signal detectors can loosely fall into two categories of detector:

- **General Method of Detection** — This does not require prior information about the signal being detected; an example is energy detection (previously mentioned in Section 19.4.2). However, these detectors have poor Constant False Alarm Rate (CFAR) performances and do not provide information about the signal type that is present [149],[362].

- **Specialised Method of Detection** — These are detectors for specific signal types. Examples of specialised methods of detection include matched filters, cyclostationary detectors, and machine learning models. These detectors provide insight into the signal type being detected; however, they are not very scalable as the detector would require updating for any new signal types that are introduced [149].

A task of DSA is to reliably detect signals while also identifying the signal type, which would enable practical and reliable transmission decisions to be made (bearing in mind that an SU cognitive radio has to avoid interfering with any priority user of the spectrum). If this problem were to be approached from a non-ML perspective, the statistical properties of each signal would need to be analysed to build a reliable detector, which could make deploying such a detector laborious and difficult.

A new class of ML-based radio waveform detectors that leverages the power of NNs has the potential to improve signal detection and classification performance. The ability of CR to identify and differentiate between radio broadcasts, mobile phone carriers, and other sources of potential radio interference is a powerful tool for approaching the spectrum access problem, where each transmission has different behaviours and requirements. Modulation recognition is a step towards understanding what transmission sources exist in the vicinity, by recognising and classifying the modulation type of each radio signal. This can be approached as an N-class decision problem where an input of complex-valued time samples is input to the ML model, and a classification vector of size $1 \times N$ is generated, indicating which modulation scheme was detected. Figure 19.13 illustrates the modulation classification setup for CRs.

Digital modulation is a method of applying discrete modulation levels onto an carrier signal, such that digital data is conveyed. A variety of different digital modulation schemes are available, including phase modulation (BPSK, QPSK, 8-PSK etc.), and quadrature amplitude modulation (e.g. 16-QAM, 64-QAM, etc.), as was previously reviewed in Chapter 6. If a cognitive radio can classify and understand the modulation scheme that a transmitter in the vicinity is using, it can better understand what kind of transmission the source is emitting [149],[362].

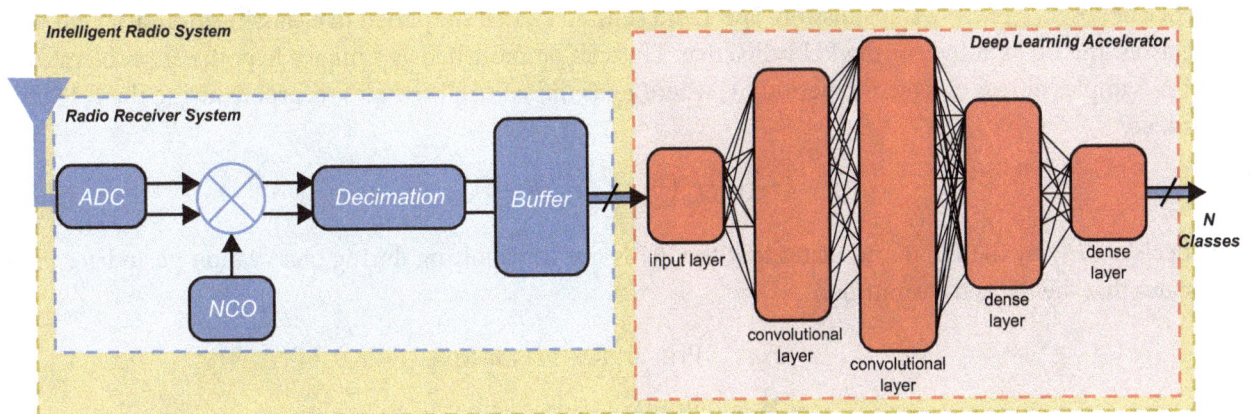

Figure 19.13: Modulation recognition DL block diagram.

In Figure 19.13, a DL model trained to classify N modulation schemes has been implemented as part of a CR. The radio receives signals from the air through the antenna and converts them into the digital domain through an ADC. The digitised signal is then mixed with an oscillator to bring it down to baseband, before the signal of interest is decimated (low-pass filtered and down-sampled). The resulting samples are then passed into the trained CNN, where the model deduces the modulation scheme that has been used. This information can then be used further to improve understanding of spectrum usage in the vicinity of the radio. Combining modulation classification with other spectrum sensing tasks provides a detailed outlook of the spectrum usage in the area, and this can further help in making optimal transmission decisions.

We will return to the topic of modulation classification again a little later, in Section 19.5.1.

Primary User Activity Monitoring with Deep Learning

A further example of intelligent radio for DSA is in the case of PU radio monitoring and prediction. Traditionally, feature-based detection techniques such as cyclostationary detection and matched filtering, although accurate and reliable methods for detecting PU signals, require prior knowledge of the PU transmission characteristics for the method to work properly. In a real spectrum sensing scenario, it is not practical to know the characteristics of all possible PUs, and therefore such traditional methods would require careful implementation for very specific monitoring scenarios. For a vision of fully autonomous DSA, CRs would have to evolve to intuitively detect PU activity without prior knowledge of their transmission characteristics. Intelligent radio advancements using DL can provide a tool towards implementing fully autonomous DSA [362].

The task of sensing a signal without prior knowledge of its characteristics is known as 'blind sensing', and a DL-based system can be trained to perform well in these types of scenario. The self-learning of feature extraction methods that DL is capable of would otherwise be difficult to implement if designed by a human.

In the following example, a DL model is used for spectrum sensing, and specifically the task of detecting the presence or absence of a PU transmission. Spectrum sensing can be simplified to a classification problem with two classes: the PU is active, or the PU is inactive. This can be modelled as a binary hypothesis problem for a set of N samples during a detection period, u, where r_u is the set of all samples recorded during the detection period, i.e.

$$r_u = [r_u(0), r_u(1), ..., r_u(N-1)] . \tag{19.3}$$

The decision H_0 by the DL model states that the PU is not transmitting during observation period u, while H_1 states that the PU is transmitting:

$$f(r_u) = \begin{cases} H_0 & : \ \text{Primary User is inactive} \\ H_1 & : \ \text{Primary User is active} \end{cases} \tag{19.4}$$

Figures 19.14 and 19.15 show the DL model in two modes of operation: Figure 19.14 represents the training mode, where the NN processes off-line training samples and compares its result y with label Z; Figure 19.15 shows the trained NN deployed at the output of an ADC and decimation chain, predicting the PU activity through analysis of real-world samples received by the antenna.

For the DL model to successfully detect the PU with no prior knowledge of the signal characteristics, it must first be trained to do so. The DL model in this example is therefore trained with synthetic data off-line before it is deployed in a real-life scenario. As mentioned previously, functions that exist in wireless communications software development tools can be used to generate data for training the DL models and to provide the associated label for each sample. Additionally, these tools can also mimic interference and degradation effects introduced by the radio channel, which can include: multipath fading, Doppler effects, frequency offset, phase noise, AWGN, and channel attenuation. Training the DL model with these impairments can teach it to detect not only a PU transmission in an ideal environment, but also where there are channel impairments, while avoiding mis-classification.

In the example shown in Figure 19.14, the DL model contains three distinct parts: *convolutions, BiLSTM and Self Attention*, and *Dense layers*.

The *convolutions* portion of the model allows it to learn feature extraction techniques through localised kernels that pass over the input data to extract local features. These features are then concatenated with the input to compensate for some of the features lost during transmission. Convolutional layers are good at finding features in a local 'instantaneous' set of data, but lack the ability to detect temporal features that extend over long time periods. Long Short Term Memory (LSTM) refers to layers that can effectively identify hidden connections among a sequence of features captured by the convolutional layers. Typically, LSTM layers can only analyse features in one direction, from current to past samples, so instead the example DL model uses a *BiLSTM layer* that analyses connections between features that extend both forward and backward in time.

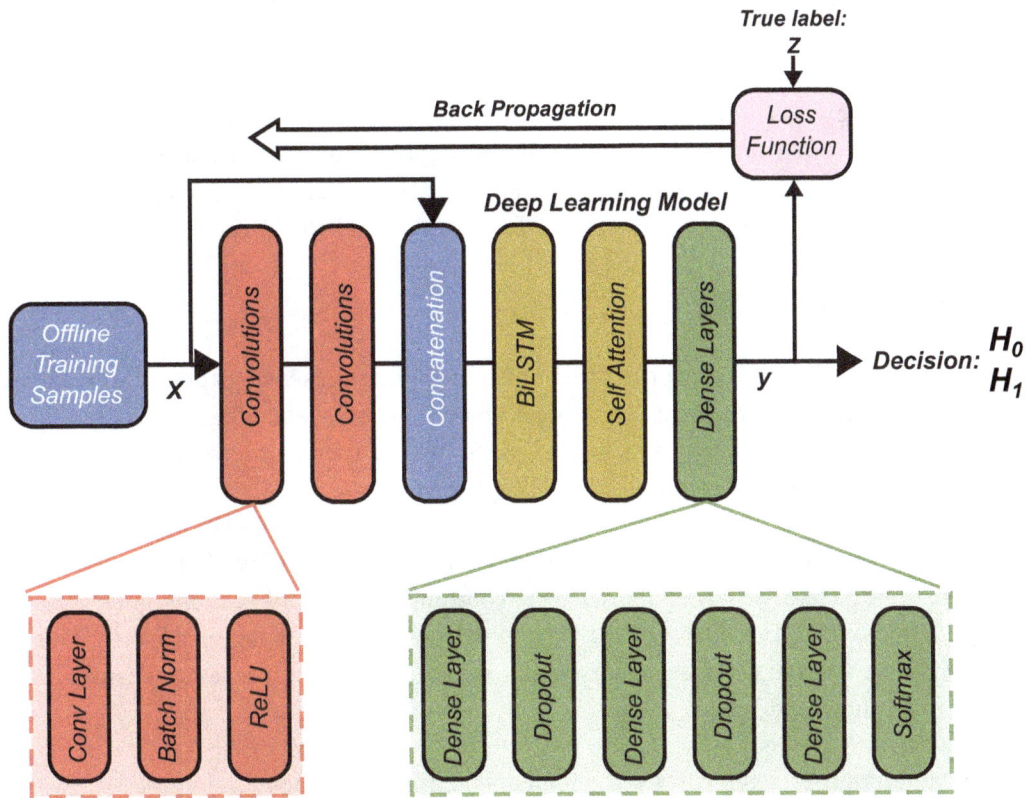

Figure 19.14: Block diagram representation of PU monitoring DL in training mode.

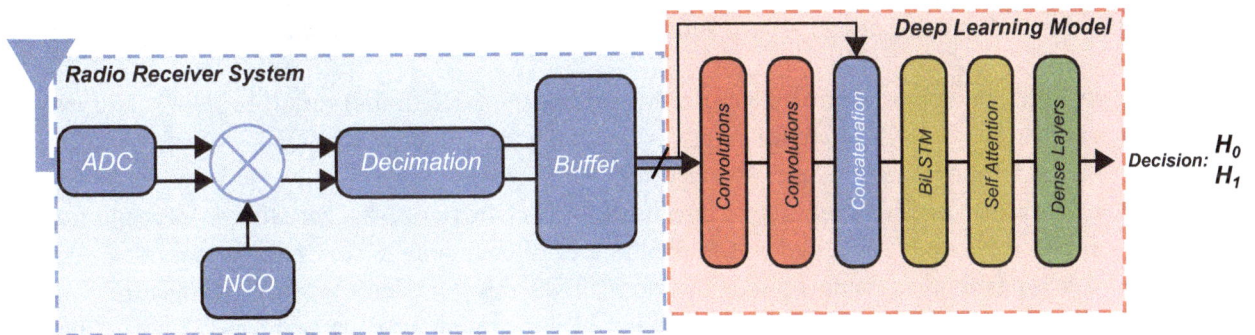

Figure 19.15: Block diagram representation of PU monitoring DL in deployed mode.

The *Self Attention (SA)* layer allows the inputs to interact with each other ('self') and work out what other samples they should be more associated with ('attention'). The output of the layer is the aggregation of these attention scores.

The *dense layers* (also known as *fully-connected layers*) form the classification stages of the model. The dropout layers help the network avoid overfitting during training and the softmax layers convert the output classes into a propability distribution. The dense, dropout, and softmax layers use the local, temporal, and dependant deatures from the previous layer to calculate the likelihood of the PU being present or absent.

Deploying Deep Learning on SDRs

While computer-created DL models are proving successful in spectrum sensing applications, actually running them on an SDR and deploying them in a real-life scenario can be challenging. DNNs require parallelised computational resources, typically trained using GPUs due to their vector processing capabilities, and devices that are able to pass a great number of samples quickly between layers.

An RFSoC-based SDR can implement receiver PHY layer processing such as filtering on the PL, in parallel form. Where DL models are deployed for wireless communications applications, and they complement existing SDR functions, the DL models should be implemented on the same logic fabric. This ensures that bottlenecks due to data transfers (for instance, passing samples to / from a separate device) can be avoided, thus optimising throughput and latency.

The RFSoC contains ample PL resources for the implementation of DL models alongside other CR functions. To support such designs, AMD provides the Vitis AI development environment, a development platform for AI inference on AMD hardware devices [66]. The platform consists of optimised IP, libraries, models, and example designs. Vitis can convert existing DL models made with various deep learning frameworks such as PyTorch, Tensorflow, and Caffe into Quantised Neural Networks (QNNs) that are deployable on hardware through architectures like the Deep Learning Processor Unit (DPU) [26]. Vitis also assists with identifying bottlenecks in the implemented design to help with optimising the DL model inference speed [27].

Another notable project supported and maintained by AMD is the FINN framework [110],[347]. This experimental framework explores DNN inference on FPGAs by designing and deploying QNNs. Furthermore, it has an emphasis on generating dataflow-like architectures. The framework is not intended to generate generic DNN models, but rather highly optimised architectures for specific networks. For wireless communication, streams of samples are already processed and filtered in a dataflow manner (i.e. where samples flow sequentially through a series of processing stages); therefore FINN can complement such architectures by also deploying the NN in a dataflow style. FINN translates NNs built using a PyTorch library, 'Brevitas', that has additional functions for quantisation-aware training.

19.5. Advancing DSA Networks

The architecture of DSA networks can be made more intelligent by enabling the network to learn from itself. This has been termed by some as *Intelligent Networks*. Intelligent networks can be referred to as a group of intelligent radios or cognitive radios capable of learning from their environment, and an intelligent network can be established using cooperative or distributive learning. In the former, there is a central control system that coordinates learning, while in the latter, individual smart radios learn and decide on network properties independently. A network can combine both methods to improve its overall performance.

Intelligent networks have become important as the demand for wireless communication such as the internet of things, wireless devices, and services are on the increase. Therefore, the learning ability of a network can be used to optimise spectral resources, predict network behaviour, and optimise the use of resources [365]. Key DSA network coordination functions that have been improved by learning algorithms are channel allocation and prediction, spectrum sensing and predictions, device clustering, adaptive routing, and resource management. Improvements range from more efficient spectral utilisation, effective detection of spectrum availability, forecasting of network behaviour, and the selection of optimal network parameters. A major benefit of learning algorithms in DSA networks is the level of flexibility and adaptability it gives to the network such as mobile user equipment, heterogeneous networks, and dense networks. Therefore, simulated and implemented intelligent DSA networks are explored in this section.

In most of these cases, the centralised architecture of the DSA network is replaced with the individual learning of nodes or access points in making resource-sharing decisions. A decentralised system therefore needs a very good view of the network environment through sensing and a means of sharing decisions made by nodes in the network. A close example is the environment sensing module in a CBRS network. Both centralised and decentralised architectures are shown in Figure 19.16.

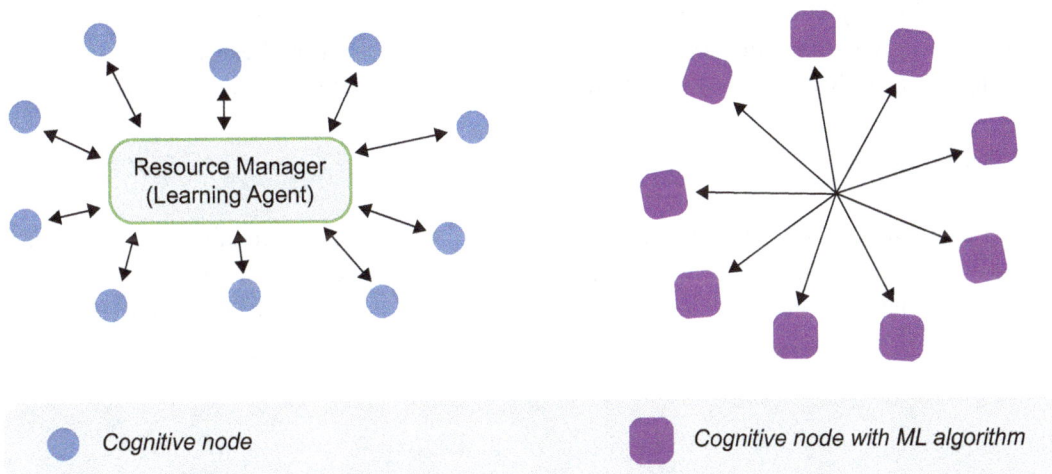

Cognitive node

Cognitive node with ML algorithm

Figure 19.16: Centralised network management (left) and decentralised network management with ML (right).

Nodes or cognitive radios, therefore, need excellent sensing abilities in detecting the presence of incumbent or SU signals and provide the Spectrum Access System (SAS) with real-time information on spectral usage. Suggested improvement and learning have been to use a semi-centralised system like the CBRS or a fully decentralised system where each cognitive radio autonomously makes spectral occupancy decisions without consulting a centralised network or controller.

Each type of network architecture has its own challenge: decentralised systems suffer from hidden node issues, while centralised systems have considerable packet overheads for network control. DSA's future of better network resource management is such that the information generated from different base stations and nodes in real-time is used for resource allocation and network assessment and to inform the adaptable reconfiguration of the network.

19.5.1. Supervised, Unsupervised and Reinforcement Learning

Machine Learning refers to the use of algorithms to make sense of, or extract knowledge from, data. It provides outstanding results in the prediction of future events based on learned patterns. Adopting Samuel's idea of machine learning, 'it is the field of study that gives computers the ability to learn without explicit programming' [309]. Thus, a machine or device or an algorithm can learn from a set of data fed to it. It extracts sequences, trends, or patterns from this information and can make better decisions by correctly predicting future occurrences of such patterns.

ML has been successfully explored extensively in facial recognition, text completion, image classification, gaming, and robotics. Each of these broad applications makes use of different kinds of learning algorithms to achieve their aims. Some make use of multiple learning structures; it may therefore prove difficult to categorise all machine learning algorithms into classes. However, generally, they have been categorised into three very broad groups: Supervised, Unsupervised, and Reinforcement Learning, with each group comprising multiple algorithms that are the subject of ongoing research and development. Supervised and Reinforcement learning are very popular in predicting and controlling wireless networks, hence these are discussed extensively. Unsupervised learning has been scarcely used to date, and as such a brief introduction is presented.

Supervised Learning

The term *supervised learning* refers to learning from large data sets that have already expected outcomes. That is to say, the algorithm is trained with a large set of observed features with known results, and the trained model is used to predict the outcomes for newly observed features. It is termed 'supervised', as the patterns in the input data (with its labelled outcomes) guide the model in making future predictions.

A general workflow for supervised learning (irrespective of the algorithm) is shown in Figure 19.17. In this example, labelled images[1] of four modulations schemes are used to solve a modulation classification problem similar to that presented in Section 19.4.3.

As shown in Figure 19.17, the data preparation phase entails, proper labelling (data cleaning) and augmenting of the dataset. It also includes the partitioning of the entire dataset to training and testing sub sets. The model extracts signal features automatically (with DL) or manually (using mathematical tools) from the training data subsets, and uses this to generate an error function that tunes its weights until the output is similar to the results from the training data subset, as detailed in Section 19.4.3.

The trained model's prediction performance is evaluated by feeding the unseen test data subset through the model. The model's prediction of the test dataset (in the prediction phase) is compared with the ground truth of the test data subset in the prediction assessment phase. When deployed in a DSA network, the image of a PU's waveforms can be correctly classified, and used to train a model that can later be used to determine the presence of PUs when any sensed waveform is presented to it [116].

The overall performance of a model is judged by its ability to generalise effectively (in other words, the extent to which the model can adapt to predict new and unseen data). A model that achieves a prediction accuracy of 80% on a training data subset may exhibit better generalisation than an algorithm that *appears* to have very high accuracy, but which has memorised the training data points (overfitting) and cannot correctly predict new examples.

Supervised learning algorithms include:

- **ML algorithms:** k-Nearest Neighbour, Linear Regression, Logistic Regression, Support Vector Machines (SVM), Decision Trees and Random Forests;

- **DL algorithms:** NNs: Artificial, Convolutional, and Recurrent, etc.

Unsupervised Learning

In unsupervised learning there is no prior knowledge of outcomes. There are no results or ground truths with which to label a dataset, as was the case in supervised learning. Unsupervised learning creates its own outcome based on the perceived structure in the dataset. An unsupervised learning algorithm extracts data patterns with no prior knowledge of what they ought to be. Therefore, it searches for similarities in the features of the input data and categorises them into clusters.

Clustering involves the grouping of unsorted data into sets based on similarity indices, such that elements in a cluster have similar features, and dissimilar features from other clusters. Clustering has been used in DSA

---

1. Modulation classification can also be undertaken based directly on received samples, rather than by working with images.

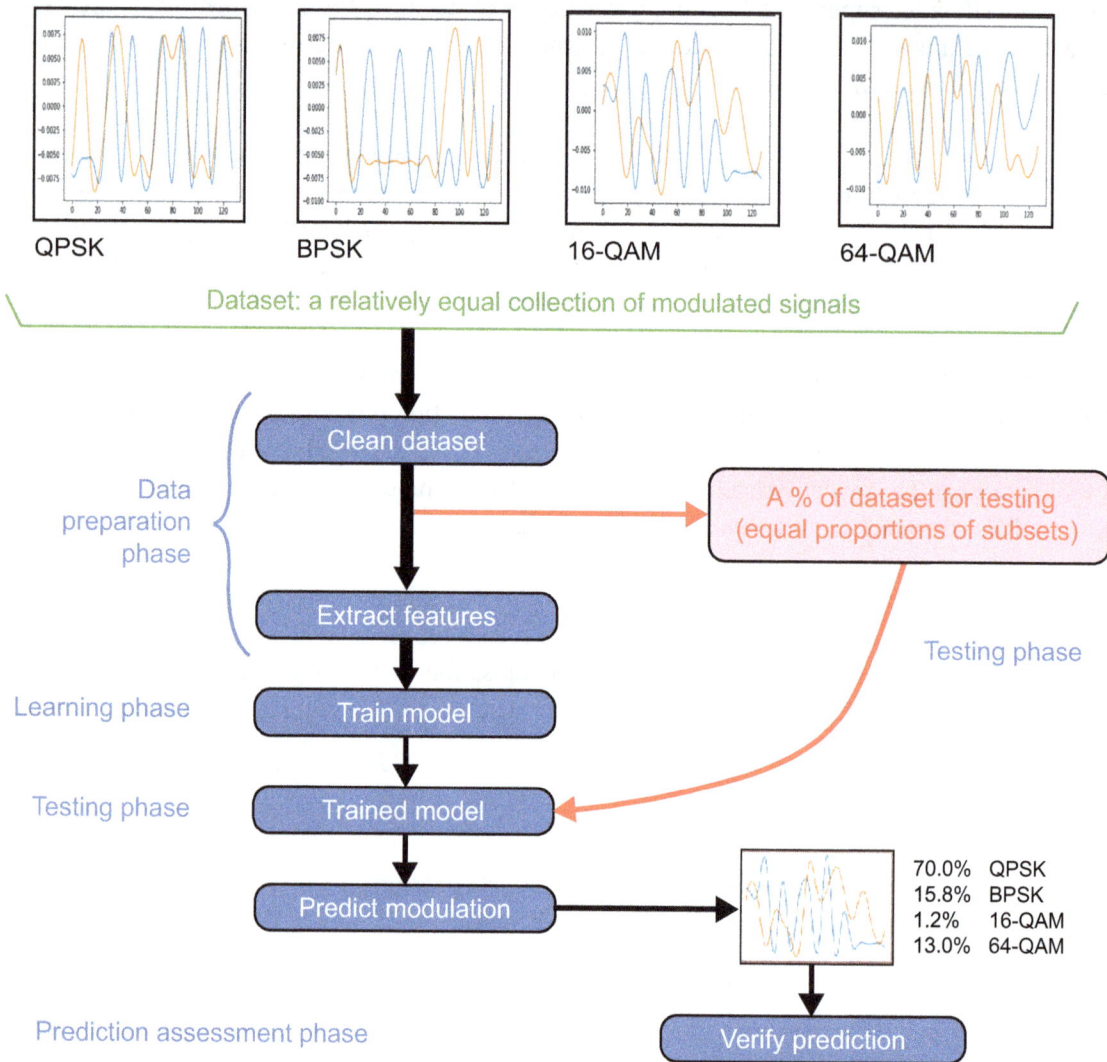

Figure 19.17: Multiple level classification of modulation signals showing supervised learning workflow.

systems to group network nodes based on certain criteria such as distance from a transmitter [199]. It is also used for dimensionality reduction of state space in reinforcement learning algorithms.

Examples of unsupervised learning algorithms are:

- **Clustering:** *k* Means, Hierarchical Cluster Analysis, and Expectation Maximisation;

- **Dimension reduction algorithms:** Principal Component Analysis (PCA), Kernel PCA, Locally Linear Embedding (LLE), and *t*-distributed Stochastic Neighbour Embedding (*t* SNE).

This type of learning algorithm is necessary for reducing large datasets, or grouping datasets that were not labelled. Thus, unlabelled large datasets can be classified and labelled using an unsupervised learning algorithm. However, evaluating the correctness of these classifications becomes an issue due of the absence of ground truth in the unstructured dataset. Techniques to measure the correctness or closeness of cluster elements are used to assess the performance of such algorithms.

Reinforcement Learning

Reinforcement Learning (RL) is a type of learning that seeks to find a set of optimal decisions or policies. The first two categories of machine learning algorithms (supervised and unsupervised) are trained by labelled or unlabelled data, respectively. They are also used predominantly for classification, pattern recognition, and data reduction. These use cases are quite different from an RL algorithm, which is implemented by an agent, to assist in creating a policy that guides the agent's actions to obtain the best cumulative reward. The RL structure, therefore, needs an agent and an environment which the agent acts upon. The effect of these actions on the environment is measured by a reward function, and may trigger a change in the environment's state, as illustrated in Figure 19.18.

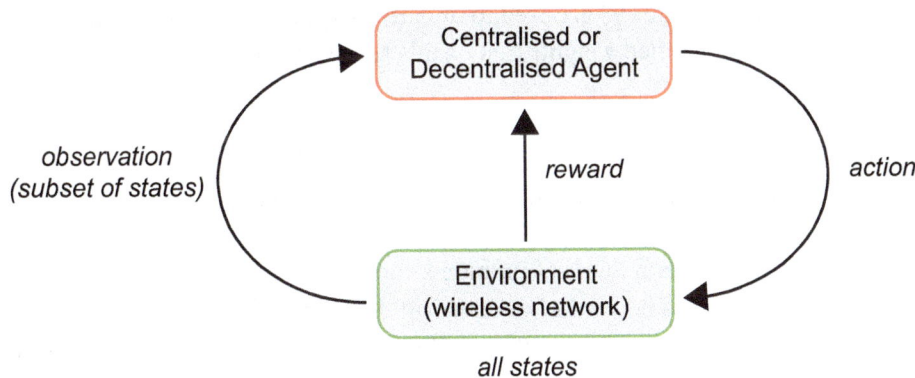

Figure 19.18: Reinforcement learning workflow.

RL consists of two elements: trial and error, where different actions are tried/explored in the environment; and the optimisation stage, where the best trial and error actions are chosen [330]. The kinds of problems that are based on trial and error can be optimised by formulating them as Markov Decision Problems (MDPs). Each trial is assessed by a reward system that measures how good or bad the trial was. The manner of conducting these trials and errors over the entire environment can be categorised as *exploration* or *exploitation*.

In *exploration*, trials are conducted on the entire environment. Different actions are tested on the entire environment to keep track of all possible reactions, or rewards the agent will receive from the environment for its actions. However, if the agent is constantly exploring, it will never decide on a sequence of actions to optimise. On the other hand, if it performs more *exploitation*, and is fixed on optimising these set of actions only, it may overlook actions that it may have taken on other parts of the environment (and which could have led to a better-accumulated reward).

It is desirable to find a trade-off between exploitation and exploration that achieves an optimal result, and this can be defined by choosing an appropriate policy. Training an RL algorithm requires a timed sequence of data, which makes it suitable for a DSA network. Hence in training an RL algorithm, framed as an MDP, mathematical MDP solutions are used to achieve optimisation and fast convergence.

One such popular solution is Bellman's equation, which assigns a value to an action (a), state (s) pair [330]. An extension of this solution is the Temporal Difference (off policy) control or Q-Learning algorithm, which is known for its fast convergence when solving optimization problems. The Q-learning update equation,

$$TewQ(s, a) \; = \; Q(s, a) + \alpha[R((s, a) + (\gamma \max Q'(s', a') - Q(s, a))). \tag{19.5}$$

updates a state action matrix which stores these values as q-values ($Q(s, a)$), such that for each action a taken by an agent from a state s, a reward ($R(s, a)$) is obtained. This reward, the previously stored q-value from the q-table for a state-action pair $Q(s, a)$, the next state's maximum rewards for all future actions ($\max Q'(s', a')$), the learning rate (α) and the discount rate (γ) are used to update the q-table with a set of new q-values. The q-table, therefore serves as a matrix of learned weights that enable the agent to choose the right action from any state.

The Q-learning algorithm is independent of the policy adopted [5]. Thus (19.5) provides a set of $Q(s, a)$, which maps the value of taking a certain action from a specific state in a state-action matrix.

We have seen that in a DSA system, several levels of problems can be framed and solved using RL algorithms. Decision-making applications in wireless networks have employed RL algorithms such as Q-learning, Deep Q-learning algorithms, Double Deep Q-learning algorithms, Temporary Difference, and State-Action-Reward-State-Action (SARSA) methods. The use of these algorithms is well explained in [330].

A pre-requisite for applying RL to solve problems is formulating such a problem into an MDP. The following questions can serve as a guide to aid achieving this:

1. What decisions/actions does the agent or decision-maker need to learn?

2. What indices or indicators can the agent use to make these decisions?

3. Criteria for knowing that the right decision was made.

The first question above can be used to determine the actions that the agent would need to take, or learn to take. The second question reveals the observable states that can either be influenced directly by an agent's action or follow a predefined sequence. Finally, question three provides a reward function that helps the agent know that the sequence of actions taken resulted in achieving the right outcome. In each of the RL applications discussed next, understanding how these questions were used to fix specific DSA network issues is addressed.

19.5.2. Applications of Machine Learning in Advancing DSA

The database's role of spectrum detection, access, and allocation in a DSA network can been improved upon using ML or DL algorithms. Some examples of published research in this area are:

- **Intelligent Channel Access** — Access to available channel slots can be managed by ML-based DSA control systems. These algorithms learn from the historical usage of channel slots, the demand from SUs, and the current state of the available channels, to predict future channel slot availability [307]. Decentralised secondary users can learn to select unoccupied channels using reinforcement learning algorithms [270].

- **Channel sensing, allocation, and security** — Channel availability and sensing for PU signals are fundamental aspects of DSA systems. The assessment of a channel can be aided by training ML techniques to identify PU signals (even in the presence of impairments). By extending this idea to train on a larger dataset of modulated signals, it can be determined whether a channel is occupied by a PU, an SU, or an unidentified / potentially malicious user, or is available for use [318].

- **Interference Mitigation** — It is vital to protect the communications of incumbent users, or PUs, in a DSA system, and this is achieved by ensuring the level of interference is managed appropriately. ML methods can help to achieve this goal, by learning the transmission patterns of incumbent transmitters, and sharing these learned patterns with other SUs, enabling them to avoid PUs active periods [116].

- **Power Control in Interference Mitigation** — Interference between SUs and PUs can further be managed through power control, i.e. appropriate selection of a transmit power level, in conjunction with channel selection. RL based methods, together with other ML techniques, can be applied to this problem [168],[175],[360].

- **Spectrum Mobility** — In DSA systems, spectrum handoff refers to the transfer of a transmission from one channel to another, based on changing conditions. If interference worsens, a decision must be made to transfer channel to another available channel, or to stay in the existing channel and wait for conditions to improve, or adjust transmission parameters. The decision making process can be implemented using RL techniques [224].

19.5.3. Challenges of Machine Learning in Network Management

The application of ML to network management provides new and exciting opportunities, and is an area of active research. There are a number of challenges, as outlined as follows.

Data Collection and Cleaning

Prediction of PU patterns requires a large amount of data to train a supervised learning algorithm. Storage space for the volume of data remains a huge challenge in developing an intelligent network. Labelling and cleaning data automatically without much human intervention becomes necessary in a constantly varying environment, and sometimes relies on error-prone unsupervised clustering for labelling. Further, missing or incomplete data can impede the efficacy of predictions [120]. Another challenge is the collation of data from different network components (sensors, PUs, SUs), in real time, for automatic network management and coordination.

Computational Power and Time

ML algorithms can be computationally expensive, especially NNs with several layers. The backward propagation for updating the weights of each of the many neurons requires high computational power and long training time. A proffered solution is the use of pre-trained models whose weights can be seldom updated. These models can then be stored on the cloud for DSA networks to use; this however, may not be convenient for low latency communication networks. In such low latency communication networks, intelligent CRs may be preferred, however, they too will struggle with the power and time consumption required for training models.

Convergence Issues

When RL is applied to a problem with a large set of actions or observations (states), it may exhibit poor or slow convergence behaviour. The environment for optimisation should be defined carefully in order to manage the optimisation state space; there is a tendency for the algorithm to struggle to achieve convergence when there are a large number of nodes / access points. Deep RL offers a possible solution, but requires very high computational power (a typical issue with NNs).

Real-time Updates

Signals are time and space dependent. While RL and RNN can handle time-dependent data, they each have different issues that have to be addressed if adopted. RNNs require a level of memory to retain previous key features, while RL may suffer convergence issues when they have a large state space. Implementing supervised learning in managing DSA networks requires that trained models are retrained/updated with real-time data from the deployed network after a period of time. Determining the frequency of these updates can be challenging, especially in dynamic networks with mobile SUs [14],[116].

19.6. Chapter Summary

This chapter has reviewed the topics of DSA and CR, and provided the necessary regulatory and theoretical background.

Firstly, the motivation to develop new spectrum access paradigms was established; namely, to make better use of the radio spectrum, which is a finite resource under pressure from growing demand for wireless connectivity. DSA principles were described and existing frameworks were reviewed, in particular TVWS and CBRS.

We noted that the ability to sense the ambient radio environment and make spectrum access decisions are both important enablers for DSA — a combination of capabilities that RFSoC can support due to the integrated nature of the device.

Taking DSA a step further requires advanced radio terminals with cognitive capabilities. A CR has the ability to sense and understand its local radio environment, including the types of transmissions emitted by other terminals in the vicinity. We discussed the importance of AI, and in particular DL, as a means for radios to generate such understanding, and considered how these algorithms would be implemented, and integrated into an RFSoC-based SDR.

Finally, the potential applications of DL at the wireless network level were considered, for instance in managing the interference that is generated by transmitting nodes. The potential of RL for network management tasks was highlighted.

"The wireless telegraphy is one of the most wonderful inventions the world has ever seen. I think it will be of great commercial use some day and as I have seen it demonstrated on the ship in which I have just arrived I can say that it is very marvellous indeed."

— WILLIAM THOMSON, LORD KELVIN

(1824 - 1907)

List of References

All references last accessed December 2022.

[1] *3rd Generation Partnership Project (3GPP)* website.
Available: https://www.3gpp.org/

[2] 3GPP, "LTE and MIMO test challenges" webpage.
Available: https://www.3gpp.org/news-events/press-clippings/1323-LTE-and-MIMO-test-challenges

[3] 3GPP, TR 38.912, "5G; Study on new radio access technology," *3rd Generation Partnership Project; Technical Report Group Radio Access Network.*

[4] 3GPP, TS 36.212, "LTE; Evolved Universal Terrestrial Radio Access (E-UTRA); Multiplexing and Channel Coding," *3rd Generation Partnership Project; Technical Specification Group Radio Access Network.*

[5] 3GPP, TS 38.101-1, "NR; User Equipment (UE) radio transmission and reception; Part 1: Range 1 Standalone," *3rd Generation Partnership Project; Technical Specification Group Radio Access Network.*

[6] 3GPP, TS 38.212, "NR; Multiplexing and channel coding," *3rd Generation Partnership Project; Technical Specification Group Radio Access Network.*

[7] 3GPP, TS 38.321, "5G NR; Medium Access Control (MAC) protocol specification," *3rd Generation Partnership Project; Technical Specification Group Radio Access Network.*

[8] 4G.co.uk, "4G and 5G frequency bands" webpage.
Available: https://www.4g.co.uk/4g-frequencies-uk-need-know/.

[9] 5G Americas, "Advanced Antenna Systems for 5G," white paper, August 2019.
Available: https://www.5gamericas.org/wp-content/uploads/2019/08/5G-Americas_Advanced-Antenna-Systems-for-5G-White-Paper.pdf

[10] S. Abrantes, "From BCJR to turbo decoding: MAP algorithms made easier," Faculty of Engineering, University of Porto, Porto, Portugal, 2004.
Available: https://paginas.fe.up.pt/~sam/textos/From%20BCJR%20to%20turbo.pdf

[11]   S. Ahmadi, *5G NR: Architecture, Technology, Implementation, and Operation of 3GPP New Radio Standards*, Academic Press, 2019.

[12]   A. M. A. Ali, *High Speed Data Converters*, IET Digital Library, 2016.
       DOI: 10.1049/PBCS026E

[13]   P. E. Allen and D. R. Holberg, *CMOS Analog Circuit Design*, Oxford University Press, 3rd Edition, 2011.

[14]   M. A. Alsheikh, S. Lin, D. Niyato, and H. P. Tan, "Machine Learning in Wireless Sensor Networks: Algorithms, Strategies, and Applications," *IEEE Communications Surveys and Tutorials*, Volume 16, Issue 4, Fourth Quarter 2014, pp. 1996–2018.
       DOI: 10.1109/COMST.2014.2320099

[15]   Amaranth, "Amaranth HDL GitHub".
       Available: https://github.com/amaranth-lang/amaranth

[16]   AMD, "4G and 5G Wireless Radio Examples Using the Zynq UltraScale+ RFSoC," video, 2019.
       Available: https://www.xilinx.com/video/soc/4g-5g-wireless-radio-examples-using-zynq-ultrascale-plus-rfsoc.html

[17]   AMD, "AIRRAYS Massive MIMO Antenna Reference Design based on Zynq® UltraScale+™ RFSoC," video, 2019.
       Available: https://www.xilinx.com/video/events/airrays-massive-mimo-antenna-reference-design-zynq-ultrascale-plus-rfsoc.html

[18]   AMD, "AXI DMA LogiCORE IP Product Guide", PG021 (v7.1), April 2022.
       Available: https://docs.xilinx.com/r/en-US/pg021_axi_dma/AXI-DMA-v7.1-LogiCORE-IP-Product-Guide

[19]   AMD, "AXI Interconnect", PG059 (v2.1), May 2022.
       Available: https://docs.xilinx.com/r/en-US/pg059-axi-interconnect

[20]   AMD, "AXI Reference Guide," UG1037 (v4.0), July 2017.
       Available: https://docs.xilinx.com/v/u/en-US/ug1037-vivado-axi-reference-guide

[21]   AMD, "CIC Compiler v4.0: LogiCORE IP Product Guide," PG140, February 2021.
       Available: https://docs.xilinx.com/v/u/en-US/pg140-cic-compiler

[22]   AMD, "CLK104 RF Clock Add-on Card," UG1437 (v1.0), March 2020.
       Available: https://docs.xilinx.com/v/u/en-US/ug1437-clk104

[23]   AMD, "Clocking Wizard LogiCORE IP Product Guide," PG065 (v6.0), April 2022.
       Available: https://docs.xilinx.com/r/en-US/pg065-clk-wiz/Clocking-Wizard-v6.0-LogiCORE-IP-Product-Guide

[24]   AMD, "Configuration Security Unit", UG1137 (v2021.1), 2021.
       Available: https://docs.xilinx.com/r/2021.1-English/ug1137-zynq-ultrascale-mpsoc-swdev/Configuration-Security-Unit

[25]   AMD, "Designing IP Subsystems Using IP Integrator", UG994 (v2022.2), October 2022.
       Available: https://www.xilinx.com/content/dam/xilinx/support/documents/sw_manuals/xilinx2022_2/ug994-vivado-ip-subsystems.pdf

[26]   AMD, "DPU for Convolutional Neural Network," webpage.
       Available: https://www.xilinx.com/products/intellectual-property/dpu.html

[27]   AMD, "DPUCZDX8G for Zynq UltraScale+ MPSoCs," Product Guide PG338 (v4.0), June 2022.
       Available: https://www.xilinx.com/content/dam/xilinx/support/documents/ip_documentation/dpu/v4_0/pg338-dpu.pdf

[28]   AMD, "Fast Fourier Transform v9.1 LogiCORE IP Product Guide," PG109, May 2022.
       Available: https://docs.xilinx.com/r/en-US/pg109-xfft/Fast-Fourier-Transform-v9.1-LogiCORE-IP-Product-Guide

[29]   AMD, "freertos10_xilinx" GitHub repository.
       Available: https://github.com/Xilinx/embeddedsw/tree/master/ThirdParty/bsp/freertos10_xilinx

[30]   AMD, "Integrated SD-FEC in Zynq UltraScale+ RFSoCs for Higher Throughput and Power Efficiency," WP498 (v1.1), May 2018.
       Available: https://www.xilinx.com/content/dam/xilinx/support/documents/white_papers/wp498-sdfec.pdf

[31] AMD, "Intellectual Property" webpage.
Available: https://www.xilinx.com/products/intellectual-property.html

[32] AMD, "JESD204C LogiCORE IP Product Guide," PG242 (v4.2), June 2022.
Available: https://docs.xilinx.com/r/en-US/pg242-jesd204c

[33] AMD, "Mechanical and Thermal Design Guidelines for Lidless Flip-Chip Packages," XAPP1301 (v1.6), February 2021.
Available: https://docs.xilinx.com/v/u/en-US/xapp1301-mechanical-thermal-design-guidelines

[34] AMD, "MicroBlaze Soft Processor Core," webpage.
Available: https://www.xilinx.com/products/design-tools/microblaze.html

[35] AMD, "Petalinux Tools," webpage.
Available: https://www.xilinx.com/products/design-tools/embedded-software/petalinux-sdk.html

[36] AMD, "PetaLinux Tools Documentation," Reference Guide, UG1144 (v2021.1), June 2021.
Available: https://docs.xilinx.com/r/2021.1-English/ug1144-petalinux-tools-reference-guide/Overview

[37] AMD, "Platform Management Unit Firmware," UG1137 (v2021.1), July 2021.
Available: https://docs.xilinx.com/r/2021.1-English/ug1137-zynq-ultrascale-mpsoc-swdev/Platform-Management-Unit-Firmware

[38] AMD, "PYNQ Overlays," documentation page.
Available: https://pynq.readthedocs.io/en/latest/pynq_overlays.html

[39] AMD, "PYNQ: Python Productivity," website.
Available: http://www.pynq.io/

[40] AMD, "PYNQ: Python Productivity for Xilinx Platforms, Read the Docs", web documentation.
Available: https://pynq.readthedocs.io/en/latest/

[41] AMD, "RFSoC 2x2 Kit," webpage.
Available: https://www.xilinx.com/support/university/xup-boards/RFSoC2x2.html#overview

[42] AMD, "RFSoC 4x2 Overview," webpage.
Available: http://www.rfsoc-pynq.io/rfsoc_4x2_overview.html

[43] AMD, "RFSoC Frequency Planner Quick Start Guide," Version 2.0, April 2022.
Available: https://www.xilinx.com/content/dam/xilinx/publications/quick-start/rfsoc-frequency-planner-quick-start-guide.pdf

[44] AMD, "RFSoC-PYNQ," website.
Available: https://www.rfsoc-pynq.io/

[45] AMD, "Soft-Decision FEC Integrated Block v1.1 LogiCORE IP Product Guide," PG256 (v1.1), October 2022.
Available: https://docs.xilinx.com/r/en-US/pg256-sdfec-integrated-block

[46] AMD, "Synchronization of Signal Processing in Multiple RF Data Converters Subsystems," XAPP1349 (v1.0), February 2022.
Available: https://xilinx.eetrend.com/files/2022-03/wen_zhang_/100558157-244514-xapp1349-rfdc-subsystems.pdf

[47] AMD, "UltraFast Design Methodology Timing Closure Quick Reference Guide," UG1292 (v2022.1), June 2022.
Available: https://www.xilinx.com/content/dam/xilinx/support/documents/sw_manuals/xilinx2022_2/ug1292-ultrafast-timing-closure-quick-reference.pdf

[48] AMD, "UltraRAM: Breakthrough Embedded Memory Integration on UltraScale+ Devices," WP477 (v1.0), June 2016.
Available: https://docs.xilinx.com/v/u/en-US/wp477-ultraram

[49] AMD, "UltraScale Architecture Configurable Logic Block User Guide," UG574 (v1.5), February 2017.
Available: https://docs.xilinx.com/v/u/en-US/ug574-ultrascale-clb

[50] AMD, "UltraScale Architecture GTY Transceivers," UG578 (v1.3.1), September 2021.
Available: https://docs.xilinx.com/v/u/en-US/ug578-ultrascale-gty-transceivers

[51] AMD, "UltraScale Architecture Memory Resources", UG573 (v1.13), September 2021.
Available: https://docs.xilinx.com/v/u/en-US/ug573-ultrascale-memory-resources

[52] AMD, "UltraScale+ Devices Integrated 100G Ethernet Subsystem LogiCORE IP Product Guide," PG203 (v3.1), November 2022.
Available: https://docs.xilinx.com/r/en-US/pg203-cmac-usplus

[53] AMD, "Understanding Key Parameters for RF-Sampling Data Converters," WP509, February 2019.
Available: https://docs.xilinx.com/v/u/en-US/wp509-rfsampling-data-converters

[54] AMD, "Using the MicroBlaze Processor to Accelerate Cost-Sensitive Embedded System Development," WP469 (v1.0.1), June 2016.
Available: https://docs.xilinx.com/v/u/en-US/wp469-microblaze-for-cost-sensitive-apps.

[55] AMD, "Using Xilinx SDK: Board Support Packages (SDK)," webpage.
Available: https://www.xilinx.com/htmldocs/xilinx2018_1/SDK_Doc/SDK_concepts/sdk_c_bsp_internal.html

[56] AMD, "Vitis_Libraries," Github page.
Available: https://github.com/Xilinx/Vitis_Libraries

[57] AMD, "Vitis Unified Software Platform," webpage.
Available: https://www.xilinx.com/products/design-tools/vitis/vitis-platform.html

[58] AMD, "Vitis Unified Software Platform Documentation," UG1416 (v2022.2), May 2022.
Available: https://docs.xilinx.com/v/u/en-US/ug1416-vitis-documentation

[59] AMD, "Vitis Unified Software Platform Documentation: Application Acceleration Development," UG1393 (v2022.2), October 2022.
Available: https://docs.xilinx.com/r/en-US/ug1393-vitis-application-acceleration/Getting-Started-with-Vitis

[60] AMD, "Vitis Unified Software Platform Documentation: Embedded Software Development," UG1400 (v2022.2), October 2022.
Available: https://docs.xilinx.com/r/en-US/ug1400-vitis-embedded

[61] AMD, "Vivado 2022.2 - Using IP Integrator," design hub webpage.
Available: https://www.xilinx.com/support/documentation-navigation/design-hubs/dh0009-vivado-using-ip-integrator-hub.html

[62] AMD, "Vivado Design Suite User Guide: Designing with IP," UG896 (v2022.2), October 2022.
Available: https://docs.xilinx.com/r/en-US/ug896-vivado-ip/Using-the-IP-Catalog.

[63] AMD, "Vivado Design Suite User Guide: Dynamic Function eXchange," UG909 (v2022.1), June 2022.
Available: https://docs.xilinx.com/r/en-US/ug909-vivado-partial-reconfiguration/Introduction

[64] AMD, "Vivado Implementation," webpage.
Available: https://www.xilinx.com/products/design-tools/vivado/implementation.html

[65] AMD, "Vivado Overview," webpage.
Available: https://www.xilinx.com/products/design-tools/vivado.html

[66] AMD, "Vitis AI," webpage.
Available: https://www.xilinx.com/products/design-tools/vitis/vitis-ai.html

[67] AMD, "Vitis High-Level Synthesis User Guide," UG1399 (v2022.2), October 2022.
Available: https://docs.xilinx.com/r/en-US/ug1399-vitis-hls/Introduction-to-Vitis-HLS

[68] AMD, "Vitis Model Composer," webpage.
Available: https://www.xilinx.com/products/design-tools/vitis/vitis-model-composer.html

[69] AMD, "Vitis Model Composer User Guide," UG1483 (v2022.2), October 2022.
Available: https://docs.xilinx.com/r/en-US/ug1483-model-composer-sys-gen-user-guide/Overview

[70] AMD, "Xilinx Standalone Library Documentation: BSP and Libraries Document Collection," UG643 (v2022.2), October 2022.
Available: https://docs.xilinx.com/r/en-US/oslib_rm

[71]   AMD "Xilinx 5G Telco Accelerator Cards," presentation, Xilinx Wired and Wireless Group.
       Available: https://www.xilinx.com/publications/presentations/xilinx-5g-telco-accelerator-cards.pdf

[72]   AMD, "Xilinx T1 Telco Accelerator Card," solution brief, 2021.
       Available: https://www.xilinx.com/content/dam/xilinx/publications/product-briefs/xilinx-t1-product-brief.pdf

[73]   AMD, "Xilinx T2 Telco Accelerator Card," solution brief, 2021.
       Available: https://www.xilinx.com/content/dam/xilinx/publications/product-briefs/xilinx-t2-product-brief.pdf

[74]   AMD, "Xrfclk Package," 2021.
       Available: https://github.com/Xilinx/PYNQ/tree/master/sdbuild/packages/xrfclk/package

[75]   AMD, "ZCU111 Evaluation Board User Guide," UG1271 (v1.2), October 2018.
       Available: https://docs.xilinx.com/v/u/en-US/ug1271-zcu111-eval-bd

[76]   AMD, "ZCU1285 Characterization Board User Guide," UG1348 (v1.0), July 2019.
       Available: https://docs.xilinx.com/v/u/en-US/ug1348-zcu1285-char-bd

[77]   AMD, "ZCU208 Evaluation Board User Guide," UG1410 (v1.0), July 2020.
       Available: https://docs.xilinx.com/v/u/en-US/ug1410-zcu208-eval-bd

[78]   AMD, "ZCU216 Evaluation Board User Guide," UG1390 (v1.1), July 2020.
       Available: https://docs.xilinx.com/v/u/en-US/ug1390-zcu216-eval-bd

[79]   AMD, "ZCU670 Evaluation Board User Guide," UG1532 (v1.0), March 2022.
       Available: https://docs.xilinx.com/r/en-US/ug1532-zcu670-eval-bd/Introduction

[80]   AMD, "Zynq-7000 SoC Data Sheet: Overview," DS190 (v1.11.1), July 20.
       Available: https://docs.xilinx.com/v/u/en-US/ds190-Zynq-7000-Overview

[81]   AMD, "Zynq RFSoC DFE," product brief, 2021.
       Available: https://www.xilinx.com/publications/product-briefs/xilinx-zynq-dfe-product-brief.pdf

[82]   AMD, "Zynq RFSoC DFE Backgrounder," 2021.
       Available: https://www.xilinx.com/content/dam/xilinx/publications/backgrounders/zynq-rfsoc-dfe-backgrounder.pdf

[83]   AMD, "Zynq UltraScale+ Device Technical Reference Manual," UG1085 (v2.3), September 2002.
       Available: https://docs.xilinx.com/r/en-US/ug1085-zynq-ultrascale-trm/Zynq-UltraScale-Device-Technical-Reference-Manual

[84]   AMD, "Zynq UltraScale+ MPSoC Data Sheet: Overview," DS891 (v1.9), May 2020.
       Available: https://docs.xilinx.com/v/u/en-US/ds891-zynq-ultrascale-plus-overview

[85]   AMD, "Zynq UltraScale+ MPSoC Product Tables and Product Selection Guide," XMP104 (v2.6).
       Available: https://docs.xilinx.com/v/u/en-US/zynq-ultrascale-plus-product-selection-guide

[86]   AMD, "Zynq UltraScale+ MPSoC Software Developer Guide," UG1137 (v. 2021.2), October 2021.
       Available: https://docs.xilinx.com/r/en-US/ug1137-zynq-ultrascale-mpsoc-swdev

[87]   AMD, "Zynq UltraScale+ RFSoC Data Sheet: DC and AC Switching Characteristics," DS926 (v1.10), April 2022.
       Available: https://docs.xilinx.com/r/en-US/ds926-zynq-ultrascale-plus-rfsoc/Summary

[88]   AMD, "Zynq UltraScale+ RFSoC Data Sheet: Overview", DS889 (v1.13), January 2022.
       Available: https://docs.xilinx.com/v/u/en-US/ds889-zynq-usp-rfsoc-overview

[89]   AMD, "Zynq UltraScale+ RFSoC Product Tables and Product Selection Guide".
       Available: https://docs.xilinx.com/v/u/en-US/zynq-usp-rfsoc-product-selection-guide

[90]   AMD, "Zynq UltraScale+ RFSoC RF Data Converter v2.6 Gen 1/2/3 Product Guide", April 2022.
       Available: https://www.xilinx.com/content/dam/xilinx/support/documents/ip_documentation/usp_rf_data_converter/v2_6/pg269-rf-data-converter.pdf

[91]   Analysys Mason, "What Are Key Considerations for 5G Sites?," white paper, September 2019.
Available: https://www.analysysmason.com/globalassets/x_migrated-media/media/analysys_mason_5g_key_considera-tions_white_paper_oct20192.pdf

[92]   A. Antoniou, *Digital Filters: Analysis, Design, and Signal Processing Applications*, McGraw Hill, 2018.

[93]   J. A. Apolinário Jr, *QRD-RLS Adaptive Filtering*, Springer, 2009.

[94]   Arm, "AMBA 4 AXI4-Stream Protocol Specification", Issue A, 2010.
Available: https://developer.arm.com/documentation/ihi0051/a.

[95]   Arm, "AMBA AXI and ACE Protocol Specification AXI3, AXI4, and AXI4-Lite ACE and ACE-Lite," Issue E, February 2013.
Available: https://developer.arm.com/documentation/ihi0022/e

[96]   Arm, "Arm Cortex-A53 MPCore Processor Technical Reference Manual," Issue J, Revision r0p4, June 2018.
Available: https://developer.arm.com/documentation/ddi0500/latest/

[97]   Arm, "Arm NEON," webpage.
Available: https://www.arm.com/technologies/neon

[98]   Arm, "Cortex-R5 Technical Reference Manual," Issue D, Revision r1p2, 2011.
Available: https://developer.arm.com/documentation/ddi0460/d/?lang=en

[99]   Avnet Abacus, "5G beamforming: an engineer's overview," webpage.
Available: https://www.avnet.com/wps/portal/abacus/solutions/markets/communications/5g-solutions/5g-beamforming/

[100]   Avnet, "Qorvo 2-Channel RF Front-end 1.8 GHz Card Hardware User's Guide," v2.0, 2020.
Available: https://www.avnet.com/opasdata/d120001/medias/docus/203/$v2/Qorvo1800MHz-HW-User-Guide-ver2.0.pdf

[101]   W. Ayoub, A. E. Samhat, F. Nouvel, M. Mroue, and J-C Prevotet, "Internet of Mobile Things: Overview of LoRaWAN, DASH7, and NB-IoT in LPWANs Standards and Supported Mobility," *IEEE Communications Surveys and Tutorials*, Vol. 21, Issue 2, Second quarter 2019, pp. 1561 - 1581.
DOI: 10.1109/COMST.2018.2877382

[102]   L. Bahl, J. Cocke, F. Jelinek and J. Raviv, "Optimal decoding of linear codes for minimizing symbol error rate (Corresp.)," in *IEEE Transactions on Information Theory*, vol. 20, no. 2, pp. 284-287, March 1974.
DOI: 10.1109/TIT.1974.1055186

[103]   C. A. Balanis, *Antenna Theory Analysis and Design*, Hoboken, New Jersey: John Wiley & Sons, Inc., 2005.

[104]   K. W. Barlee, R. W. Stewart, L. H. Crockett and N. C. MacEwen, "Rapid Prototyping and Validation of FS-FBMC Dynamic Spectrum Radio With Simulink and ZynqSDR," in *IEEE Open Journal of the Communications Society*, vol. 2, pp. 113-131, January 2021.
DOI: 10.1109/OJCOMS.2020.3039928

[105]   G. Baudoin, O. Venard, and D-K. Germain Pham, "Digital Pre-Distortion," in *Digitally Enhanced Mixed Signal Systems*, edited by C. Jabbour, P. Desgreys, and D. Dallet, IET Digital Library, 2019.
DOI: 10.1049/PBCS040E_ch3

[106]   A. Benksy, *Short-Range Wireless Communication*, 3rd Edition, Newnes, 2019.

[107]   C. Berrou, A. Glavieux and P. Thitimajshima, "Near Shannon limit error-correcting coding and decoding: Turbo-codes. 1," *Proceedings of ICC '93: IEEE International Conference on Communications*, Geneva, Switzerland, May1993, vol.2, pp. 1064-1070.
DOI: 10.1109/ICC.1993.397441

[108]   M. Bertocco, C. Narduzzi, P. Paglierani and D. Petri, "Accuracy of Effective Bits Estimation Methods," in *IEEE Instrumentation and Measurement Technology Conference*, Brussels, Belgium, 1996.
DOI: 10.1109/ICC.1993.397441

[109]   E. Björnson, J. Hoydis and L. Sanguinetti, "Massive MIMO Networks: Spectral, Energy, and Hardware Efficiency," *Foundations and Trends in Signal Processing*, Vol. 11: No. 3-4, November 2017, pp 154-655.
DOI: 10.1561/2000000093

[110] M. Blott, T. Preusser, N. Fraser, G. Gambardella, K. O'Brien, and Y. Umuroglu, "FINN-R: An End-to-End Deep-Learning Framework for Fast Exploration of Quantized Neural Networks," *ACM Transactions on Reconfigurable Technology and Systems*, Volume 11, Issue 3, Article No. 16, September 2018.
DOI: 10.1145/3242897

[111] N. Bonello, S. Chen and L. Hanzo, "Low-Density Parity-Check Codes and Their Rateless Relatives," in *IEEE Communications Surveys & Tutorials*, vol. 13, no. 1, First Quarter 2011, pp. 3-26.
DOI: 10.1109/SURV.2011.040410.00042

[112] A. Bourdena, E. Pallis, E. Karditsis, G. Mastorakis, and G. Kormentzas, "A radio resource management framework for TVWS exploitation under the RTSSM policy," in *International Conference on Telecommunications and Multimedia*, Ottawa, Canada, July 2012, pp. 1-6.
DOI: 10.1109/ICC.2012.6364713

[113] G. E. P. Box and M. E. Muller, "A Note on the Generation of Random Normal Deviates," *The Annals of Mathematical Statistics*, Vol. 29, Issue 2, June 1958, pp. 610 - 611.
DOI: 10.1214/aoms/1177706645

[114] D. Brubaker and A. Collins, "Zynq RFSoC DFE: The Next Frontier for Mass 5G NR Deployment," AMD webinar, 21st April 2022.

[115] P. v. Butovitsch, D. Astely, A. Furuskär, B. Göransson, B. Hogan, J. Karlsson and E. Larsson, "Advanced antenna systems for 5G networks" Ericsson White Paper, November 2018.
Available: https://www.ericsson.com/4a8a87/assets/local/reports-papers/white-papers/10201407_wp_advanced_antenna_system_nov18_181115.pdf

[116] M. Camelo et al., "An AI-Based Incumbent Protection System for Collaborative Intelligent Radio Networks," *IEEE Wireless Communications*, vol. 27, no. 5, pp. 16–23, October 2020.
DOI: 10.1109/MWC.001.2000032

[117] Celona, "CBRS SAS: Simple Explanation of the Spectrum Access System," webpage.
Available: https://www.celona.io/cbrs/cbrs-sas

[118] V. Chandrasetty and S.M. Aziz, "LDPC Decoding Algorithms," in *Resource Efficient LDPC Decoders: From Algorithms to Hardware Architectures*, Academic Press, 2017.

[119] Q. Chaudhari, "Effect of Timing Mismatch in OFDM Systems," webpage.
Available: https://wirelesspi.com/effect-of-timing-mismatch-in-ofdm-systems/

[120] M. Chen, U. Challita, W. Saad, C. Yin, and M. Debbah, "Artificial Neural Networks-Based Machine Learning for Wireless Networks: A Tutorial," *IEEE Surveys and Tutorials*, Volume 21, Issue 4, Fourth Quarter 2019.
DOI: 10.1109/COMST.2019.2926625

[121] Z. Chen et al, "A Survey on Terahertz Communications", *IEEE China Communications*, Vol. 6, Issue 2, February 2019.
DOI: 10.12676/j.cc.2019.02.001

[122] Chisel, "Chisel/FIRRTL Hardware Compiler Framework," webpage.
Available: https://www.chisel-lang.org/

[123] S-Y. Chung, G. D. Forney, T. J. Richardson and R. Urbanke, "On the design of low-density parity-check codes within 0.0045 dB of the Shannon limit," in *IEEE Communications Letters,* vol. 5, no. 2, pp. 58-60, Feb 2001.
DOI: 10.1109/4234.905935

[124] I. Collings, "Prof Iain Collings (IEEE Fellow) - Digital Communications," Iaincollings.com webpages.
Available: https://www.iaincollings.com/digital-communications

[125] A. Colvin, "CSMA with Collision Avoidance," Computer Communications, Vol. 6, Issue 5, October 1983, pp. 227-235.
DOI: 10.1016/0140-3664(83)90084-1

[126] P. G. Cook and W. Bonser, "Architectural Overview of the SPEAKeasy System," *IEEE Journal on Selected Areas in Communications*, Vol. 17, Issue 4, April 1999, pp. 650 - 661.
DOI: 10.1109/49.761042

[127] J. W. Cooley and J. W. Tukey, "An Algorithm for the Machine Calculation of Complex Fourier Series," *Mathematics of Computation*, Vol. 19, No. 90, 1965, pp. 297-301.
DOI: 10.2307/2003354

[128] CPRI, "CPRI Specification Overview," webpage.
Available: http://www.cpri.info/spec.html

[129] R. E. Crochiere and L. R. Rabiner, *Multirate Digital Signal Processing*, Prentice Hall, 1983.

[130] L. H. Crockett, R. A. Elliot, M. A. Enderwitz, and R. W. Stewart, *The Zynq Book: Embedded Processing with the ARM Cortex-A9 on the Zynq-7000 All Programmable SoC*, Strathclyde Academic Media, 2014.

[131] L. H. Crockett, D. Northcote, C. Ramsay, F. Robinson, and R. W. Stewart, *Exploring Zynq MPSoC: with PYNQ and Machine Learning Applications*, Strathclyde Academic Media, 2019.

[132] G. C. Danielson and C. Lanczos, "Some improvements in practical Fourier analysis and their application to X-ray scattering from liquids," Journal of the Franklin Institute, Vol. 233, Issue 4, 1942, pp. 365-380.
DOI: 10.1016/S0016-0032(42)90767-1

[133] T. Darwish and M. Bayoumi, "Trends in Low-Power VLSI Design", in *The Electrical Engineering Handbook*, Academic Press, 2005, pp. 263-280.

[134] C. Dick, F. Harris, M. Pajic and D. Vuletic, "Real-Time QRD-Based Beamforming on an FPGA Platform," *Proceedings of the 40th Asilomar Conference on Signals, Systems and Computers*, Asilomar, CA, USA, 2006, pp. 1200 - 1204.
DOI: 10.1109/ACSSC.2006.354945

[135] Digilent, Inc., "What is a Constraints File?" webpage.
Available: https://digilent.com/reference/programmable-logic/guides/vivado-xdc-file

[136] P. Diniz, *Adaptive Filtering: Algorithms and Practical Implementation*, 5th Edition, Springer, 2019.

[137] D. Divsalar and E. Pollara, "Turbo Codes for Deep-Space Communications," *JPL TDA Progress Report 42-120*, Feb. 15, 1995.

[138] T. Do and B. L. Mark, "Joint Spatial-Temporal Spectrum Sensing for Cognitive Radio Networks," *IEEE Transactions on Vehicular Technology*, vol. 59, Issue 7, May 2010, pp. 3480–3490.
DOI: 10.1109/TVT.2010.2050610

[139] S. Dorner, S. Cammerer, J. Hoydis, and S. T. Brink, "Deep Learning Based Communication over the Air," *IEEE Journal on Selected Topics in Signal Processing*, Vol. 12, Issue 1, February 2018, pp. 132–143.
DOI: 10.1109/JSTSP.2017.2784180

[140] DSP Illustrations, "The Schmidl and Cox Synchronization Technique for OFDM," webpage.
Available: https://dspillustrations.com/pages/posts/misc/schmidlcox-synchronization-for-ofdm.html

[141] DSP Related, "Fast Fourier Transform (FFT) Algorithms," webpage.
Available: https://www.dsprelated.com/freebooks/mdft/Fast_Fourier_Transform_FFT.html

[142] DSP Related, "Orthogonality of Sinusoids," webpage.
Available: https://www.dsprelated.com/freebooks/mdft/Orthogonality_Sinusoids.html

[143] *DVB* website.
Available: https://dvb.org/

[144] *Dynamic Spectrum Alliance* website.
Available: http://dynamicspectrumalliance.org/

[145] Dynamic Spectrum Alliance, "Automated Frequency Coordination: An Established Tool for Modern Spectrum Management," research report, March 2019.
Available: https://dynamicspectrumalliance.org/wp-content/uploads/2019/03/DSA_DB-Report_Final_03122019.pdf

[146] Electronics notes, "What is Frequency Modulation?" webpage.
Available: https://www.electronics-notes.com/articles/radio/modulation/frequency-modulation-fm.php

[147] P. Elias, "Coding for noisy channels," *IRE Convention Records,* 3(4):37-46, 1955.

[148] Ericsson AB, Huawei Technologies Co. Ltd, NEC Coroporation and Nokia, "eCPRI Transport Network Requirements Specification".
Available: http://www.cpri.info/downloads/Requirements_for_the_eCPRI_Transport_Network_V1_2_2018_06_25.pdf

[149] T. Erpek, T. J. O'Shea, Y. E. Sagduyu, Y. Shi, and T. C. Clancy, "Deep Learning for Wireless Communications," *Studies in Computational Intelligence*, vol. 867, pp. 223–266, May 2020.
DOI: 10.1007/978-3-030-31764-5_9

[150] Ettus Research, "RFNoC (RF Network on Chip)," webpage.
Available: https://www.ettus.com/sdr-software/rfnoc/

[151] Ettus Research, "USRP Hardware Driver and USRP Manual," Version 4.3.0.0-1-g8cb669977.
Available: https://files.ettus.com/manual/page_properties.html

[152] Ettus Research, "USRP X410" webpage.
Available: https://www.ettus.com/all-products/usrp-x410/

[153] European Telecommunications Standards Institute (ETSI), "EN 300 175-1 V2.9.1 (2022-03) — Digital Enhanced Cordless Telecommunications (DECT); Common Interface (CI); Part 1: Overview", ETSI Standard, March 2022.

[154] European Telecommunications Standards Institute (ETSI), "TR 103 113 - V1.1.1 - Electromagnetic Compatibility and Radio Spectrum Matters (ERM); System Reference Document (SRdoc); Mobile Broadband Services in the 2300 MHz - 2400 MHz Frequency Band Under Licensed Shared Access Regime", Technical Report, July 2013.

[155] European Telecommunications Standards Institute (ETSI), "TS 38.104".
Available: https://www.etsi.org/deliver/etsi_ts/138100_138199/138104/15.14.00_60/ts_138104v151400p.pdf

[156] everythingRF, "Xilinx Demonstrates Digital RF Solutions to Address 5G Radio Challenges", webpage, 7th August 2020.
Available: https://www.everythingrf.com/News/details/10667-xilinx-demonstrates-digital-rf-solutions-to-address-5g-radio-challenges

[157] K. Fall and W. Stevens, *TCP/IP Illustrated: The Protocols, Volume 1*, 2nd Edition, Addison-Wesley, 2011.

[158] C. W. Farrow, "A Continuously Variable Digital Delay Element", *IEEE International Symposium on Circuits and Systems*, Espoo, Finland, June 1998, pp. 2641 - 2645.
DOI: 10.1109/ISCAS.1988.15483

[159] *Federal Communications Commission (FCC)* website.
Available: https://www.fcc.gov/

[160] Federal Communications Commission (FCC), "3.5GHz Overview," webpage.
Available: https://www.fcc.gov/35-ghz-band-overview

[161] Federal Communications Commission (FCC), "Amateur Radio Service", webpage.
Available: https://www.fcc.gov/wireless/bureau-divisions/mobility-division/amateur-radio-service

[162] Federal Communications Commission (FCC), 'FCC-08-260A1', November 2008.
Available: https://docs.fcc.gov/public/attachments/FCC-08-260A1.pdf.

[163] A. B. Flores, R. E. Guerra, E. W. Knightly, P. Ecclesine and S. Pandey, "IEEE 802.11af: A Standard for TV White Space Spectrum Sharing", *IEEE Communications Magazine*, vol. 51, no. 10, October 2013, pp. 92-100.
DOI: 10.1109/MCOM.2013.6619571

[164] G. Forney, R. Gallager, G. Lang, F. Longstaff and S. Qureshi, "Efficient Modulation for Band-Limited Channels," in *IEEE Journal on Selected Areas in Communications*, vol. 2, no. 5, September 1984, pp. 632-647.
DOI: 10.1109/JSAC.1984.1146101

[165] J. Fourier, "Théorie analytique de la chaleur," (in French) Paris: Firmin Didot Père et Fils, 1822.
Available: https://archive.org/details/bub_gb_TDQJAAAAIAAJ

[166] Free Software Foundation, "GNU Operating System" website.
Available: https://www.gnu.org/home.en.html

[167] H. T. Friis, "A Note on a Simple Transmission Formula," *Proceedings of the I.R.E. and Waves and Electrons,* Vol. 34, Issue 5, May 1946, pp 254–256.
DOI: 10.1109/JRPROC.1946.234568

[168] A. Galindo-Serrano and L. Giupponi, "Distributed Q-learning for Aggregated Interference Control in Cognitive Radio Networks," *IEEE Transactions on Vehicular Technology*, Volume 59, Issue 4, May 2010, pp. 1823–1834.
DOI: 10.1109/TVT.2010.2043124

[169] R. G. Gallager, "Low Density Parity Check Codes", D. Sc. Thesis, Dept. Elect. Eng. and Comp. Sci., MIT, Cambridge, MA, 1960.

[170] R. G. Gallager, *Low Density Parity Check Codes*, 1963.
Available: https://web.stanford.edu/class/ee388/papers/ldpc.pdf

[171] G. Ganesan and Y. Li, "Cooperative Spectrum Sensing in Cognitive Radio, Part I: Two User Networks," *IEEE Transactions on Wireless Communications,* vol. 6, issue 6, June 2007, pp. 2204–2212.
DOI: 10.1109/TWC.2007.05775

[172] A. Garcia Armada, "Understanding the effects of phase noise in orthogonal frequency division multiplexing (OFDM)," in *IEEE Transactions on Broadcasting*, vol. 47, no. 2, June 2001, pp. 153-159.
DOI: 10.1109/11.948268

[173] Z. Geng and S. N. Simrock, "Evaluation of Fast ADCs for Direct Sampling RF Field Detection for the European XFEL and ILC", in *XXIV Linear Accelerator Conference*, Victoria, British Columbia, Canada, 2008.

[174] R. Ghai, "CBRS Use-Cases With Focus on Localized Indoor Mobile Access (LIMA), Mobility and Service Continuity", white paper, Technicolor, 2018.
Available: https://www.technicolor.com/sites/default/files/whitepapers/2018-cbrs-use-cases.pdf

[175] L. Giupponi, A. Galindo-Serrano, P. Blasco, and M. Dohler, "Docitive Networks: An Emerging Paradigm for Dynamic Spectrum Management," *IEEE Wireless Communications*, Vol. 17, Issue 4, August 2010, pp. 47–54.
DOI: 10.1109/MWC.2010.5547921

[176] A. Goldsmith, *Wireless Communications,* 1st ed. Cambridge; Cambridge University Press, 2005.

[177] GNU, "The GNU C Library (glibc)," webpage.
Available: https://www.gnu.org/software/libc/

[178] *GNU Radio* website.
Available: https://www.gnuradio.org/

[179] GNU Radio, "Embedded Python Block," wiki page.
Available: https://wiki.gnuradio.org/index.php/Embedded_Python_Block

[180] GNU Radio, "Guided Tutorial GRC," wiki page.
Available: https://wiki.gnuradio.org/index.php/Guided_Tutorial_GRC

[181] G. H. Golub and C. F. Van Loan, *Matrix Computations*, 4th Edition, John Hopkins University Press, 2013.

[182] I. Goodfellow, Y. Bengio, and A. Courville, *Deep Learning*. MIT Press, 2016.

[183] A. Graps, "An introduction to wavelets," *IEEE Computational Science and Engineering*, Vol. 2, No. 2, Summer 1995, pp. 50-61.
DOI: 10.1109/99.388960

[184] GSMA Intelligence, "Going green: benchmarking the energy efficiency of mobile", technical report, June 2021.
Available: https://data.gsmaintelligence.com/api-web/v2/research-file-download?id=60621137&file=300621-Going-Green-efficiency-mobile.pdf

[185] S. Guitchev, K. Moessner, D. Thilakawardana, T. Dodgson, and R. Tafazolli, "Evaluation of Software Defined Radio Technology," technical report, Centre for Communication Systems Research, University of Surrey, February 2006.
Available: https://www.ofcom.org.uk/__data/assets/pdf_file/0028/27388/eval.pdf

[186] J. Hagenauer and P. Hoeher, "A Viterbi Algorithm With Soft-Decision Outputs and Its Applications," *IEEE Global Telecommunications Conference and Exhibition*, Dallas, Texas, November 1989, pp. 1680-1686 vol.3.
DOI: 10.1109/GLOCOM.1989.64230

[187] M. N. Hamdy, "Beamformers Explained," Commscope white paper, 2020.
Available: https://www.commscope.com/globalassets/digizuite/542044-beamformer-explained-wp-114491-en.pdf

[188] f. j. harris, *Multirate Signal Processing for Communications Systems*, 2nd Edition, River Publishers, 2022.

[189] f. j. harris, "On the use of windows for harmonic analysis with the discrete Fourier transform," *Proceedings of the IEEE*, Vol. 66, No. 1, Jan. 1978, pp. 51-83.
DOI: 10.1109/PROC.1978.10837

[190] M. Hata, "Empirical Formula for Propagation Loss in Land Mobile Radio Services," *IEEE Transactions on Vehicular Technology*, Vol. VT-29, No. 3, August 1980.
DOI: 10.1109/T-VT.1980.23859

[191] S. Haykin, *Adaptive Filter Theory*, 5th Edition, Pearson, 2013.

[192] J. Heiskala and J. Terry, *OFDM Wireless LANs: A Theoretical and Practical Guide*, 1st Ed. Sam Publishing, 2002.

[193] J. Henkel, "A low power hardware/software partitioning approach for core-based embedded systems," in *Proceedings of the 1999 Design Automation Conference*, New Orleans, LA, USA, 1999.
DOI: 10.1109/DAC.1999.781296

[194] J. Heo, "Analysis of scaling soft information on low density parity check code", *IEEE Electronics Letters*, 2003.

[195] E. Hewitt and R. E. Hewitt, "The Gibbs-Wilbraham Phenomenon: An Episode in Fourier Analysis," *Archive for History of Exact Sciences*, Vol. 21, 1979, pp. 129-160.
DOI: 10.1007/BF00330404

[196] T. Hinamoto and W-S. Lu, *Digital Filter Design and Realization*, River Publishers, 2017.

[197] E. Hogenauer, "An Economical Class of Digital Filters for Decimation and Interpolation", *IEEE Transactions on Acoustics, Speech, and Signal Processing*, Vol. 29, Issue 2, April 1981, pp. 155 - 162.
DOI: 10.1109/TASSP.1981.1163535

[198] G. Hosoya, M. Hasegawa and H. Yashima, "LLR calculation for iterative decoding on fading channels using Padé approximation," *International Conference on Wireless Communications and Signal Processing (WCSP)*, Huangshan, China, October 2012, pp. 1-6.
DOI: 10.1109/WCSP.2012.6542931

[199] E. Hossain et al., "Introduction to Cognitive Radio," in *Dynamic Spectrum Access and Management in Cognitive Radio Networks*, Cambridge University Press, 2009, pp 41-74.

[200] M-H. Hsieh and C-H. Wei, "Channel estimation for OFDM systems based on comb-type pilot arrangement in frequency selective fading channels," in *IEEE Transactions on Consumer Electronics*, vol. 44, no. 1, Feb. 1998, pp. 217-225.
DOI: 10.1109/30.663750

[201] F. Huang, "Evaluation of Soft Output Decoding for Turbo Codes," Masters dissertation, Virginia Tech., Blacksburg, VA, 1997.

[202] G. A. Hufford, A. G. Longley, and W. A. Kissick, "A Guide to the Use of the ITS Irregular Terrain Model in the Area Prediction Mode", *NTIA Report 82-100, National Telecommunications and Information Administration,* April 1, 1982.

[203] E. Ifeachor and B. Jervis, *Digital Signal Processing: A Practical Approach*, 2nd Edition, Pearson, 2001.

[204] *Institute of Electrical and Electronics Engineers (IEEE) website.*
Available: https://www.ieee.org/

[205] IEEE, "IEEE standard for definitions and concepts for dynamic spectrum access: Terminology relating to emerging wireless networks, system functionality, and spectrum management - red- line," IEEE Std 1900.1-2019 (Revision of IEEE Std 1900.1-2008) - Redline, pp. 1-144, April 2019.

[206] IEEE, "Standard for Ethernet", IEEE Std 802.3-2022, July 2022.
DOI: 10.1109/IEEESTD.2022.9844436

[207] IEEE, "Standard for Information technology — Telecommunications and information exchange between systems Local and metropolitan area networks — Specific requirements Part 11: Wireless LAN Medium Access Control (MAC) and Physical Layer (PHY) Specifications," in IEEE Std 802.11-2020 (Revision of IEEE Std 802.11-2016), February 2021.
DOI: 10.1109/IEEESTD.2012.6178212

[208] IEEE, "IEEE Standard for SystemVerilog — Unified Hardware Design, Specification, and Verification Language," IEEE Standard 1800-2009, December 2009.
DOI: 10.1109/IEEESTD.2009.5354441

[209] IEEE, "IEEE Standard for Verilog Hardware Description Language," IEEE standard 1364-2001, September 2001.
DOI: 10.1109/IEEESTD.2001.93352

[210] IEEE, "IEEE Standard for VHDL Language Reference Manual," IEEE standard 1076-1993, June 1994.
DOI: 10.1109/IEEESTD.1994.121433

[211] IEEE, "IEEE Standard for VHDL Language Reference Manual," IEEE standard 1076-2008, January 2009.
DOI: 10.1109/IEEESTD.2009.4772740

[212] International Standards Organisation (ISO) and International Electrotechnical Commission (IEC), ISO/IEC 7498-1:1994 standard, "Information Technology — Open Systems Interconnection — Basic Reference Model: The Basic Model," 2nd edition, 1994.
Available: https://standards.iso.org/ittf/PubliclyAvailableStandards/s020269_ISO_IEC_7498-1_1994(E).zip

[213] International Standards Organisation (ISO) and International Electrotechnical Commission (IEC), "Information technology — Telecommunications and information exchange between systems — Local and metropolitan area networks — Specific requirements — Part 2: Logical Link Control," International Standard ISO/IEC 8802-2:1998, May 1998.
DOI: 10.1109/IEEESTD.1998.8684692

[214] International Telecommunication Union (ITU), "Radiowave Propagation Recommendations (P series)," webpage.
Available: https://www.itu.int/rec/R-REC-P/en

[215] Internet Engineering Task Force (IETF), "Internet Protocol, Version 6 (IPv6) Specification," *Request for Comments: 8200*, July 2017.
Available: https://www.rfc-editor.org/info/rfc8200

[216] Internet Engineering Task Force (IETF), "Transmission Control Protocol," *Request for Comments: 793*, prepared for DARPA by Information Sciences Institute, University of Southern California, September 1981.
Available: https://www.ietf.org/rfc/rfc793.txt

[217] Internet Engineering Task Force (IETF), "Transmission Control Protocol (TCP)," *Request for Comments 9293*, August 2022.
Available: https://www.rfc-editor.org/info/rfc9293

[218] *Javascript* website.
Available: https://www.javascript.com/

[219] JEDEC, "JESD204C standard".
Available: https://www.jedec.org/

[220] M. Jones, "Anatomy of the Linux kernel," *IBM Developer* website, June 2007.
Available: https://developer.ibm.com/articles/l-linux-kernel/

[221] jupyter-widgets, "GitHub Source Code Repository for the IPywidgets Python Library," webpage.
Available: https://github.com/jupyter-widgets/ipywidgets

[222] M. Kappes, "All-Digital Antenna for mmWave Systems," *Microwave Journal*, pp. 8-11, 2019.
Available: https://www.microwavejournal.com/articles/32449-all-digital-antennas-for-mmwave-systems

[223] A. Kaye and D. George, "Transmission of Multiplexed PAM Signals Over Multiple Channel and Diversity Systems," *IEEE Transactions on Communication Technology*, vol. 18, no. 5, pp. 520-526, October 1970.
DOI: 10.1109/TCOM.1970.1090417

[224] A. M. Koushik, F. Hu, and S. Kumar, "Intelligent Spectrum Management Based on Transfer Actor-Critic Learning for Rateless Transmissions in Cognitive Radio Networks," *IEEE Transactions on Mobile Computing*, Volume 17, Issue 5, May 2018, pp. 1204–1215.
DOI: 10.1109/TMC.2017.2744620

[225] C. M. Kozeirok, *TCP/IP Guide: A Comprehensive, Illustrated Internet Protocols Reference*, No Starch Press, 2005.

[226] S. Y. Kung, "On Supercomputing with Systolic / Wavefront Array Processors", *Proceedings of the IEEE*, Vol. 72, Issue 7, July 1984, pp. 867 - 884.
DOI: 10.1109/PROC.1984.12944

[227] J. F. Kurose and K. W. Ross, *Computer Networking: A Top Down Approach*, 8th Edition, Pearson, 2021.

[228] B. Leiner, *LDPC Codes - a brief Tutorial,* 2005.
Available: http://www.bernh.net/media/download/papers/ldpc.pdf

[229] Z. Li, L. Chen, L. Zeng, S. Lin and W. H. Fong, "Efficient encoding of quasi-cyclic low-density parity-check codes," in *IEEE Transactions on Communications*, vol. 54, no. 1, Jan. 2006, pp. 71-81.
DOI: 10.1109/TCOMM.2005.861667

[230] F. Ling, *Synchronization in Digital Communication Systems*, Cambridge University Press, 1st Edition, 2017.

[231] Linux manual, "Linux Kernel System Call Interface API," documentation.
Available: https://man7.org/linux/man-pages/dir_section_2.html

[232] Linx Technologies, "LCW Series Low Cost 2.4 GHz Dipole Antenna", webpage.
Available: https://linxtechnologies.com/wp/product/lcw-series-low-cost-2-4ghz-dipole-antenna/

[233] S. Louwsma, E. van Tuijl and B. Nauta, *Time-interleaved Analog-to-Digital Converters*, Springer, 2010.

[234] R. Lyons, *Understanding Digital Signal Processing*, 3rd Edition, Pearson, 2010.

[235] R. J. McEliece and L. Swanson, "Reed-Solomon Codes and the Exploration of the Solar System," NASA technical report, 1994.
Available: https://trs.jpl.nasa.gov/handle/2014/34531

[236] D. J. C. MacKay and R. M. Neal, "Near Shannon limit performance of Low Density Parity Check Codes," in *Electronics Letters,* vol.32, no. 18, pp. 1645-1646, Aug 1996.
DOI: 10.1049/el:19961141

[237] D. J. C. MacKay, S. T. Wilson and M. C. Davey, "Comparison of constructions of irregular Gallager codes," in *IEEE Transactions on Communications*, vol. 47, no. 10, Oct. 1999, pp. 1449-1454.
DOI: 10.1109/26.795809

[238] M. R. McKay, I. B. Collings and A. M. Tulino, "Achievable Sum Rate of MIMO MMSE Receivers: A General Analytic Framework," *IEEE Transactions on Information Theory*, vol. 56, no. 1, January 2010, pp. 396-410.
DOI: 10.1109/TIT.2009.2034893

[239] M. M. Mabrook, H. A. Khalil, and A. I. Hussein, "Artificial Intelligence Based Cooperative Spectrum Sensing Algorithm for Cognitive Radio Networks," *Procedia Computer Science*, vol. 163, January 2019, pp. 19–29.
DOI: 10.1016/j.procs.2019.12.081

[240] C. Mack, "The Multiple Lives of Moore's Law," *IEEE Spectrum*, Vol. 52, Issue 4, April 2015, pp. 31 - 37.
DOI: 10.1109/MSPEC.2015.7065415

[241] G. Manganaro, *Advanced Data Converters*, Cambridge University Press, 2011.

[242] A. S. Margulies and J. Mitola III, "Software Defined Radios: A Technical Challenge and a Migration Strategy," *Proceedings of the IEEE 5th International Symposium on Spread Spectrum Technologies and Applications*, September 1998, Sun City, South Africa, pp. 551 - 556.
DOI: 10.1109/ISSSTA.1998.723845

[243] J. Markel, "FFT pruning," *IEEE Transactions on Audio and Electroacoustics*, Vol. 19, No. 4, December 1971, pp. 305-311.
DOI: 10.1109/TAU.1971.1162205

[244] T. L. Marzetta, "Massive MIMO: An Introduction," *Bell Labs Technical Journal,* vol. 20, pp. 11-22, March 2015.

[245] Massachusetts Institute of Technology (MIT), *Chapter 8: Viterbi Decoding of Convolutional Codes*, lecture notes, February 2012.
Available: http://web.mit.edu/6.02/www/s2012/handouts/8.pdf

[246] MathWorks, "ARM Cortex A Support from Embedded Coder," webpage.
Available: https://uk.mathworks.com/hardware-support/arm-cortex-a.html

[247] MathWorks, "Embedder Coder," webpage.
Available: https://uk.mathworks.com/products/embedded-coder.html

[248] MathWorks, "Equalization," webpage.
Available: https://uk.mathworks.com/help/comm/ug/equalization.html

[249] MathWorks, "HDL Coder," webpage.
Available: https://uk.mathworks.com/products/hdl-coder.html

[250] MathWorks, *"LLR vs. Hard Decision Demodulation in Simulink,"* webpage.
Available: https://uk.mathworks.com/help/comm/ug/llr-vs-hard-decision-demodulation-in-simulink.html

[251] MathWorks,"MATLAB," webpage.
Available: https://uk.mathworks.com/products/matlab.html

[252] MathWorks, "MATLAB Coder," webpage.
Available: https://uk.mathworks.com/products/matlab-coder.html

[253] MathWorks, "Simulink," webpage.
Available: https://uk.mathworks.com/products/simulink.html

[254] MathWorks, "Simulink Coder," webpage.
Available: https://uk.mathworks.com/products/simulink-coder.html

[255] MathWorks, "SoC Blockset," webpage.
Available: https://uk.mathworks.com/products/soc.html

[256] MathWorks, "SoC Builder," webpage.
Available: https://uk.mathworks.com/help/soc/ref/socbuilder.html

[257] *Matplotlib* website.
Available: https://matplotlib.org/

[258] Measurement Computing, "Analog to Digital Conversion," webage.
Available: https://www.mccdaq.com/PDFs/specs/Analog-to-Digital.pdf.

[259] Microsoft, "Azure RTOS," webpage.
Available: https://azure.microsoft.com/en-us/products/rtos/

[260] Microwave Journal, "AMD's Latest RFSoC Added to Evenstar O-RAN Reference Designs," 11th May 2022.
Available: https://www.microwavejournal.com/articles/38122-amds-latest-rfsoc-added-to-evenstar-o-ran-reference-designs

[261] L. Milic, *Multirate Filtering for Digital Signal Processing: MATLAB Applications*, IGI Global, 1993.

[262] Mini-Circuits, "VLF-1800+ LTCC Low Pass Filter, DC - 1800 MHz, 50Ω, Connector Type: SMA," webpage.
Available: https://www.minicircuits.com/WebStore/dashboard.html?model=VLF-1800%2B

[263] Mini-Circuits, "VBF-2435+ LTCC Band Pass Filter, 2340 - 2530 MHz, 50Ω, Connector Type: SMA," webpage.
Available: https://www.minicircuits.com/WebStore/dashboard.html?model=VBF-2435%2B

[264] J. Mitola III, "The Software Radio Architecture", *IEEE Communications Magazine*, May 1995, pp. 26 - 38.
DOI: 10.1109/35.393001

[265] J. Mitola III, "Software Radios: Survey, critical evaluation and future directions", *IEEE Aerospace and Electronic Systems Magazine*, Vol. 8, Issue 4, April 1993, pp. 25 - 36.
DOI: 10.1109/62.210638

[266] Moblitciti, "The Exponential Rise of Mobile Data Usage," webpage.
Available: https://www.mobliciti.com/the-exponential-rise-of-mobile-data-usage/

[267] T. Moon, *Error Correction Coding: Mathematical Methods and Algorithms*, 2nd ed.Wiley, 2021.

[268] G. E. Moore, "Cramming More Components onto Integrated Circuits," *Electronics*, Vol. 38, No. 8, April 1965.

[269] A. Morgado et al., "Dynamic LSA for 5G networks the ADEL perspective," *European Conference on Networks and Communications (EuCNC)*, Paris, France, 2015, pp. 190-194.
DOI: 10.1109/EuCNC.2015.7194066

[270] S. Mosleh, Y. Ma, J. D. Rezac, and J. B. Coder, "Dynamic Spectrum Access with Reinforcement Learning for Unlicensed Access in 5G and beyond," *91st IEEE Vehicular Technology Conference (VTC2020-Spring)*, Antwerp, Belgium, May 2020.
DOI: 10.1109/VTC2020-Spring48590.2020.9129381

[271] N. Muchandi, and R. Khanai, "Cognitive radio spectrum sensing: A survey," *International Conference on Electrical, Electronics, and Optimization Techniques (ICEEOT)*, Chennai, India, March 2016, pp. 3233-3237.
DOI: 10.1109/ICEEOT.2016.7755301

[272] K. Mun, "CBRS: New Shared Spectrum Enables Flexible Indoor and Outdoor Mobile Solutions and New Business Models," Mobile Experts white paper, 2017.
Available: https://federatedwireless.com/wp-content/uploads/2017/09/Mobile-Experts-CBRS-Overview.pdf

[273] musl, "git index: musl," webpage.
Available: https://git.musl-libc.org/cgit/musl

[274] MyHDL, "MyHDL" webpage.
Available: https://www.myhdl.org/

[275] National Instruments, "Wireless Industry's New Gambit: Terahertz Communication Bands", *IEEE Spectrum Magazine*, 17th April 2019.
Available: https://spectrum.ieee.org/wireless-industrys-newest-gambit-terahertz-communication-bands

[276] Next Generation Mobile Networks (NGMN) Alliance, "NGMN Defines Industry Standard for a "Global Green Networks Benchmark"", press release, 18th January 2022.
Available: https://www.ngmn.org/highlight/the-next-generation-mobile-networks-alliance-ngmn-alliance-today-announces-the-establishment-of-methodologies-and-kpis-for-a-global-green-networks-benchmark.html

[277] NooElec, "Nooelec - LaNA Wideband Ultra Low-Noise Amplifier (LNA) Module," webpage.
Available: https://www.nooelec.com/store/lana.html

[278] NooElec, "Nooelec RaTLSnake M6 v2 - Premium 3-Antenna Bundle for SMA-Input SDRs," webpage.
Available: https://www.nooelec.com/store/ratlsnake-m6.html

[279] NooElec, "Nooelec - SMA Attenuator Kit - 1dB to 42 dB of Attenuation in 1 dB Increments," webpage.
Available: https://www.nooelec.com/store/attenuator-bundle.html

[280] NooElec, "Nooelec - VeGA Barebones - Ultra Low-Noise Variable Gain Amplifier (VGA) Module for RF & Software Defined Radio (SDR). Highly Linear & Wideband 30MHz-4000MHz Frequency Capability," webpage.
Available: https://www.nooelec.com/store/vega-barebones.html

[281] *NumPy* website.
       Available: https://numpy.org/

[282] *Office of Communications (Ofcom)* website.
       Available: https://www.ofcom.org.uk/

[283] Office of Communications (Ofcom), U.K., "Enabling wireless innovation through local licensing", *statement*, 25th July 2019.
       Available: https://www.ofcom.org.uk/__data/assets/pdf_file/0033/157884/enabling-wireless-innovation-through-local-licensing.pdf

[284] Office of Communications (Ofcom), U.K., "Mobile networks and spectrum: Meeting future demand for mobile data", Discussion Paper, February 2022.
       Available: https://www.ofcom.org.uk/__data/assets/pdf_file/0017/232082/mobile-spectrum-demand-discussion-paper.pdf.

[285] Office of Communications (Ofcom), U.K., "UK Frequency Allocation Table," 8 January 2020.
       Available: http://static.ofcom.org.uk/static/spectrum/fat.html

[286] *pandas* website.
       Available: https://pandas.pydata.org/

[287] G. K. Papageorgiou et al., "Advanced Dynamic Spectrum 5G Mobile Networks Employing Licensed Shared Access", in *IEEE Communications Magazine*, vol. 58, no. 7, July 2020, pp. 21-27.
       DOI: 10.1109/MCOM.001.1900742

[288] S. S. Pereira, L. Almeida, A. S. R. Oliveira, N. B. Carvalho, and P. P. Monteiro, "Multi-band, Multi-technology Remote Unit (RU) Based on RFSoC", *50th European Microwave Conference*, Utrecht, Netherlands, January 2021.
       DOI: 10.23919/EuMC48046.2021.9338153

[289] F. Pérez and B. E. Granger, "IPython: A System for Interactive Scientific Computing," *Computing in Science and Engineering*, vol. 9, no. 3, May/June 2007, pp. 21-29.
       DOI: 10.1109/MCSE.2007.53

[290] D. Petrovic, W. Rave and G. Fettweis, "Common phase error due to phase noise in OFDM-estimation and suppression," *IEEE 15th International Symposium on Personal, Indoor and Mobile Radio Communications*, Barcelona, Spain, September 2004, pp. 1901-1905 Vol.3.
       DOI: 10.1109/PIMRC.2004.1368329

[291] *Plotly* website.
       Available: https://plotly.com/

[292] J. Postel, "User Datagram Protocol", *Request for Comments: 768*, August 1980.
       Available: https://www.ietf.org/rfc/rfc768.txt

[293] J. Proakis and M. Salehi, *Digital Communications*, 5th Edition, McGraw-Hill, 2007.

[294] *Project Jupyter* website.
       Available: https://jupyter.org/

[295] *Python* website.
       Available: https://www.python.org/

[296] QBayLogic, "Clash," webpage.
       Available: https://clash-lang.org/

[297] *QEMU* website.
       Available: https://www.qemu.org/

[298] H. Qi and N. Goertz, "Low-Complexity Encoding of LDPC Codes: A new algorithm and its performance," Institute for Digital Communications, Joint Research Institute for Signal & Image Processing, School of Engineering and Electronics, The University of Edinburgh.

[299] QinetiQ Ltd., "Cognitive Radio Technology: A Study for Ofcom — Summary Report", technical report, 2007.
Available: https://www.ofcom.org.uk/__data/assets/pdf_file/0017/40364/cograd_summary.pdf.

[300] G. Refai-Ahmed, H. Do, B, Philofsky, and A. Torza, "Extending the Cooling Limit of Remote Radio Head (RRH) Systems Based on Level 1 Thermal Management," 20th IEEE Electronics Packaging Conference (EPTC), December 2018, Singapore.
DOI: 10.1109/EPTC.2018.8654306

[301] M. Rice, *Digital Communications: A Discrete-Time Approach*, independently published, 2018.

[302] T. J. Richardson and R. L. Urbanke, "Efficient encoding of low-density parity-check codes," in *IEEE Transactions on Information Theory*, vol. 47, no. 2, Feb 2001, pp. 638-656.
Available: 10.1109/18.910579

[303] P. Robertson, P. Hoeher and E. Villebrun, "Optimal and Sub-Optimal Maximum A Posteriori Algorithms Suitable for Turbo Decoding," *European Transactions on Telecommunications*, Vol. 8, 1997, pp. 119-125.
DOI: 10.1002/ett.4460080202

[304] P. Robertson, E. Villebrun and P. Hoeher, "A comparison of optimal and sub-optimal MAP decoding algorithms operating in the log domain," *Proceedings of the IEEE International Conference on Communications (ICC)*, Seattle, WA, USA, June 1995, pp. 1009-1013, vol.2.
DOI: 10.1109/ICC.1995.524253

[305] P. A. Roncagliolo, J. G. Garcia, and C. H. Muravchik, "Optimized Carrier Tracking Loop Design for Real-Time High Dynamics GNSS Receivers", *International Journal of Navigation and Observation*, Volume 2012, Article ID 651039, 18 pages, 2012.
DOI: 10.1155/2012/651039

[306] H. R. Sadjadpour, N. J. A. Sloane, M. Salehi and G. Nebe, "Interleaver design for turbo codes," in *IEEE Journal on Selected Areas in Communications*, vol. 19, no. 5, May 2001, pp. 831-837.
DOI: 10.1109/49.924867

[307] A. Sahoo, "A Machine Learning Based Scheme for Dynamic Spectrum Access," *IEEE Wireless Communications and Networking Conference (WCNC)*, Nanjing, China, March/April 2021.
DOI: 10.1109/WCNC49053.2021.9417401

[308] Samsung Electronics, "Massive MIMO for New Radio", white paper, December 2020.
Available: https://images.samsung.com/is/content/samsung/assets/global/business/networks/insights/white-papers/1208_massive-mimo-for-new-radio/MassiveMIMOforNRTechnicalWhitePaper-v1.2.0.pdf

[309] A. L. Samuel, "Some Studies in Machine Learning Using the Game of Checkers," IBM J. Res. Dev., vol. 3, no. 3, 1959, pp. 210–229.

[310] K. Sankar, "Convolutional Code," webpage, 2009.
Available: http://www.dsplog.com/2009/01/04/convolutional-code/

[311] K. Sankar, "Softbit for 16-QAM," webpage, 2009.
Available: http://www.dsplog.com/2009/07/05/softbit-16qam/

[312] S. Schindler and H. Mellein, "Assessing a MIMO Channel," white paper, Rohde and Schwarz, 2012.
Available: https://www.rohde-schwarz.com/uk/applications/assessing-a-mimo-channel-white-paper_230854-15677.html

[313] T. M. Schmidl and D. C. Cox, "Robust frequency and timing synchronization for OFDM," in *IEEE Transactions on Communications*, vol. 45, no. 12, Dec. 1997, pp. 1613-1621.
DOI: 10.1109/26.650240

[314] *SciPy* webpage.
Available: https://scipy.org/

[315] S. Sesia et al., *LTE - The UMTS Long Term Evolution: From Theory to Practice,* 2nd Ed. John Wiley & Sons, 2011.

[316] J. S. Seybold, *Introduction to RF Propagation*, Wiley-Interscience, 2007.

[317] ShareTechNote, "5G/NR - MIMO DL," webpage.
Available: https://www.sharetechnote.com/html/5G/5G_MIMO.html

[318] Y. Shi, K. Davaslioglu, Y. E. Sagduyu, W. C. Headley, M. Fowler, and G. Green, "Deep Learning for RF Signal Classification in Unknown and Dynamic Spectrum Environments," *2019 IEEE International Symposium on Dynamic Spectrum Access Networks (DySPAN)*, Newark, NJ, USA, November 2019.
DOI: 10.1109/DySPAN.2019.8935684

[319] B. Sklar, *Digital Communications: Fundamentals and Applications*, 2nd Ed. Prentice Hall, 2001.

[320] M. Sliskovic, "Sampling frequency offset estimation and correction in OFDM systems," 8th IEEE International Conference on Electronics, Circuits and Systems (ICECS), Malta, 2001, vol.1, pp. 437-440.
DOI: 10.1109/ICECS.2001.957773

[321] S. W. Smith, *The Scientist and Engineer's Guide to Digital Signal Processing*, California Technical Publishing, 1997 - 2001.
Available: http://www.dspguide.com/

[322] T. Socolofsky and C. Kale, "A TCP/IP Tutorial", *Request for Comments: 1180*, January 1991.
Available: https://www.rfc-editor.org/rfc/rfc1180

[323] Spectrum Instrumentation, "ADC and Resolution," webpage.
Available: https://spectrum-instrumentation.com/support/knowledgebase/hardware_features/ADC_and_Resolution.php

[324] Steepest Ascent Ltd (acquired by MathWorks, 2013), *DSPedia Notes 8: Equalisation,* 2004.

[325] K. Steiner and M. Yeary, "Least-Squares Equalizer Demonstrations Using a Full-Digital Bandwidth Sub-Nyquist-Sampled Wideband Beamformer on an RFSoC," *IEEE Transactions on Aerospace and Electronic Systems*, May 2022.
DOI: 10.1109/TAES.2022.3176844

[326] R. W. Stewart, K. W. Barlee, D. S. W. Atkinson, and L. H. Crockett, *Software Defined Radio using MATLAB & Simulink and the RTL-SDR*, 1st Ed., Strathclyde Academic Media, 2015.

[327] StrathSDR, "GitHub Source Code Repository for the RFSoC Book Practical Exercises (rfsoc_book)", webpage.
Available: https://github.com/strath-sdr/rfsoc_book/

[328] StrathSDR, "GitHub Source Code Repository for the RFSoC Spectrum Analyser Module (rfsoc_sam)," webpage.
Available: https://github.com/strath-sdr/rfsoc_sam/

[329] StrathSDR, "Spectrum Analyser on PYNQ," Github repository.
Available: https://github.com/strath-sdr/rfsoc_sam

[330] R. S. Sutton and A. G. Barto, *Reinforcement Learning: An Introduction*, 2nd Ed., The MIT Press, Cambridge, Massachusetts, 2018.

[331] A. H. Syed, *Adaptive Filters*, 1st Ed., Wiley-IEEE Press, 2008.

[332] A. Tanenbaum, N. Feamster, and D. Wetherall, *Computer Networks*, 6th Ed., Pearson, 2021.

[333] R. Tanner, "A recursive approach to low complexity codes," in *IEEE Transactions on Information Theory*, Vol. 27, No. 5, September 1981, pp. 533-547.
DOI: 10.1109/TIT.1981.1056404

[334] S. Taylor and J. Metzler, "Why it's time to let the OSI model die", Network World website, September 2008.
Available: https://www.networkworld.com/article/2276158/why-it-s-time-to-let-the-osi-model-die.html

[335] Tcl Developer Xchange, "About Tcl/Tk Language," webpage.
Available: https://www.tcl-lang.org/about/language.html

[336] Texas Instruments, "Spurs Analysis in the RF Sampling ADC," Application Report SLAA824, February 2018.
Available: https://www.ti.com/lit/an/slaa824/slaa824.pdf

[337] Texas Instruments, "TICSPRO-SW," webpage.
Available: https://www.ti.com/tool/TICSPRO-SW

[338] Texas Instruments, "TIDA-01161 1-GHz Signal Bandwidth RF Sampling Receiver Reference Design," webpage.
Available: https://www.ti.com/tool/TIDA-01161

[339] L. Q. Thao, D. Q. Loc and N. X. Tuyen, "Study Comparative of Parabolic and Phased Array Antenna", *Journal of Science: Mathematics - Physics*, vol. 30, no. 3, 2014, pp. 31-36.

[340] F. Tomatis and S. Sesia, "Synchronization and Cell Search," in *LTE — The UMTS Long Term Evolution: From Theory to Practice*, Wiley, 2011, pp.151-164.
DOI: 10.1002/9780470978504.ch7

[341] L. B. Torvalds, "Free minix-like kernel sources for 386-AT," Google Groups forum post, 5th October 1991.
Available: https://groups.google.com/g/comp.os.minix/c/4995SivOl9o/m/GwqLJlPSlCEJ?pli=1

[342] L. Torvalds, *Linux kernel source tree*.
Available: https://git.kernel.org/pub/scm/linux/kernel/git/torvalds/linux.git/about/

[343] F. Tosato and P. Bisaglia, "Simplified soft-output demapper for binary interleaved COFDM with application to HIPERLAN/2," *IEEE International Conference on Communications*, New York, NY, USA, April/May 2002, pp. 664-668 vol.2.
DOI: 10.1109/ICC.2002.996940

[344] S. Trimberger, "Three Ages of FPGAs: A Retrospective on the First Thirty Years of FPGA Technology," *Proceedings of the IEEE*, Vol. 103, No. 3, March 2015, pp. 318 - 331.
DOI: 10.1109/JPROC.2015.2392104

[345] M. Tropea and F. A. De Rango, "Comprehensive Review of Channel Modeling for Land Mobile Satellite Systems", *Electronics*, 11, 820, March 2022.
DOI: 10.3390/electronics11050820

[346] D. Tse and P. Viswanath, *Fundamentals of Wireless Communication*, Cambridge University Press, 2005.

[347] Y. Umuroglu, N. J. Fraser, G. Gambardella, M. Blott, P. Leong, M. Jahre, and K. Vissers, "FINN: A Framework for Fast, Scalable Binarized Neural Network Inference," *Proceedings of the 2017 ACM/SIGDA International Symposium on Field-Programmable Gate Arrays (FPGA)*, Monterey, CA, USA, February 2017, pp. 65–74.
DOI: 10.1145/3020078.3021744

[348] P. Vaidyanathanm, *Multirate Systems and Filter Banks*, Prentice Hall, 1993.

[349] V. Valenta, R. Marsalek, G. Baudoin, M. Villegas, M. Suarez and F. Robert, "Survey on spectrum utilization in Europe: Measurements, analyses and observations", *Proceedings of the Fifth International Conference on Cognitive Radio Oriented Wireless Networks and Communications (CROWNCOM)*, Cannes, France, June 2010, pp. 1-5.
DOI: 10.4108/ICST.CROWNCOM2010.9220

[350] M. Venkatesan, A. V. Kulkarni, and R. Menon, "Stochastic Time Series Learning Scheme for Throughput Prediction in Cognitive Radio System," *Proceedings of the 4th International Conference on Computing, Communication Control and Automation (ICCUBEA)*, Pune, India, August 2018.
DOI: 10.1109/ICCUBEA.2018.8697672

[351] Vic Myers Associates, "Deploying Xilinx Zynq UltraScale+ RFSoC," white paper.
Available: https://www.vicmyers.com/deploying-xilinx-zynq-ultrascale-rfsoc/

[352] A. Viterbi, "Error bounds for convolutional codes and an asymptotically optimum decoding algorithm," in *IEEE Transactions on Information Theory*, Vol. 13, No. 2, April 1967, pp. 260-269.
DOI: 10.1109/TIT.1967.1054010

[353] C. Vogel and H. Johansson, "Time-Interleaved Analog-to-Digital Converters: Status and Future Directions," *IEEE International Symposium on Circuits and Systems*, Kos, Greece, May 2006.
DOI: 10.1109/ISCAS.2006.1693352

[354] J. E. Volder, "The CORDIC Trigonometric Computing Technique", *IRE Transactions on Electronic Computers*, Vol. EC-8, no. 3, 1959, pp. 330-334.
DOI: 10.1109/TEC.1959.5222693

[355] Z.Wei, Z. Feng, Q. Zhang, and W. Li, "Three Regions for Space-Time Spectrum Sensing and Access in Cognitive Radio Networks," *IEEE Global Communications Conference (GLOBECOM)*, Anaheim, California, USA, 2012, pp. 1283-1288. DOI: 10.1109/GLOCOM.2012.6503290

[356] Wi-Fi Alliance, "Global Economic Value of Wi-Fi: 2021-2025," report, 2021. Available: https://www.wi-fi.org/file/global-economic-value-of-wi-fi-2021-2025

[357] Wolfram MathWorld, "Primitive Polynomial," webpage. Available: https://mathworld.wolfram.com/PrimitivePolynomial.html

[358] R. Woods, J. McAllister, G. Lightbody, and Y. Ying, *FPGA-based Implementation of Signal Processing Systems*, 2nd Edition, Wiley, 2017.

[359] *Worlddab* website. Available: https://www.worlddab.org/

[360] C. Wu, K. Chowdhury, M. Di Felice, and W. Meleis, "Spectrum Management of Cognitive Radio Using Multi-agent Reinforcement Learning Categories and Subject Descriptors," *Proceedings of the 9th International Conference on Autonomous Agents and Multiagent Systems (AAMAS 2010)*, Toronto, Canada, May 2010, pp. 1705–1712.

[361] A. Wulff, *Beginning Radio Communications: Radio Projects and Theory*, Cambridge, MA, USA, Apress Media, 2019.

[362] H. Xing, H. Qin, S. Luo, P. Dai, L. Xu, and X. Cheng, "Spectrum sensing in cognitive radio: A deep learning based model," *Transactions on Emerging Telecommunications Technologies*, 33(1), January 2022. DOI: 10.1002/ETT.4388

[363] S. Yamada, O. Boric-Lubecke, and V. M. Lubecke, "Cancellation Techniques for LO Leakage and DC offset in Direct Conversion Systems," *IEEE MTT-S International Microwave Symposium Digest*, Atlanta, GA, USA, June 2008. DOI: 10.1109/MWSYM.2008.4633271

[364] *YAML* website. Available: https://yaml.org/

[365] H. Yang, X. Xie, and M. Kadoch, "Machine Learning Techniques and A Case Study for Intelligent Wireless Networks," *IEEE Network*, vol. 34, no. 3, May/June 2020, pp. 208–215. DOI: 10.1109/MNET.001.1900351

[366] *Yocto Project* website. Available: https://www.yoctoproject.org/

[367] M. Zareei, A. K. Muzahidul Islam, S. Baharun, C. Vargas-Rosales, L. Azpilicueta, and N. Mansoor, "Medium access control protocols for cognitive radio ad hoc networks: A survey," *Sensors,* MDPI journal, September 2017. DOI: 10.3390/s17092136

[368] Y. Zeng and Y. C. Liang, "Covariance Based Signal Detections for Cognitive Radio", *2nd IEEE International Symposium on New Frontiers in Dynamic Spectrum Access Networks*, Dublin, Ireland, April 2007, pp. 202 - 207. DOI: 10.1109/DYSPAN.2007.33

[369] *Zephyr Project* website. Available: https://www.zephyrproject.org/

[370] Y. Zhou, L. Liu, H. Du, L. Tian, X. Wang and J. Shi, "An overview on intercell interference management in mobile cellular networks: From 2G to 5G," *IEEE International Conference on Communication Systems,* Macau, China, November 2014, pp. 217-221. DOI: 10.1109/ICCS.2014.7024797

List of Acronyms

123

4G	4th Generation
5G	5th Generation
5G NR	5th Generation New Radio
6G	6th Generation

A

AC	Alternating Current
ACAP	Adaptive Compute Acceleration Platform
ACLR	Adjacent Channel Leakage Ratio
ACS	Add Compare Select
ADC	Analogue to Digital Converter
AES	Advanced Encryption Standard
AGC	Active Gain Control
AI	Artificial Intelligence
AKE	Asymmetric Encryption Key
ALU	Arithmetic Logic Unit
AM	Amplitude Modulation
AMBA	Advanced Microcontroller Bus Architecture
AMD	Advanced Micro Devices, Inc.
AMP	Asymmetric Multi Processing
ANN	Artificial Neural Network
AP	Access Point

API	Application Programming Interface
APU	Application Processing Unit
ARM	Advanced RISC Machines
ARQ	Automatic Repeat ReQuest
ASIC	Application-Specific Integrated Circuit
ASK	Amplitude Shift Keying
AWGN	Additive White Gaussian Noise
AXI	Advanced eXtensible Interface
AXIS	AXI Stream

B

BB	Baseband
BBU	Baseband Unit
BCH	Broadcast Control Channel
BCJR	Bahl, Cocke, Jelinek and Raviv algorithm
BER	Bit Error Rate
BGM	Base Graph Matrix
BiLSTM	Bidirectional Long Short Term Memory
BM	Branch Metric
BOM	Bill Of Materials
BPF	Bandpass Filter
BPSK	Binary Phase Shift Keying
BRAM	Block RAM
BSP	Board Support Package
BSU	Backward Substitution Unit
BUF	Buffer

C

CA	Collision Avoidance
CAN	Controller Area Network
CBRS	Citizen Band Radio Service
CCI	Cache Coherent Interconnect
CD	Compact Disc
CDC	Clock Domain Crossing

CFAR	Constant False Alarm Rate
CFO	Carrier Frequency Offset
CFR	Crest Factor Reduction
CFS	Completely Fair Scheduler
CGNAT	Carrier-Grade Network Address Translation
CIB	Cryptography Interface Block
CIC	Cascade Integrate Comb
CLB	Configurable Logic Blocks
CM	Common Mode
CMAC	Centralised Media Access Controller (MAC)
CN	Condition Number
CNN	Convolutional Neural Network
COFDM	Coded OFDM
CORDIC	Co-ordinate Rotation Digital Computer
COTS	Commercial Off the Shelf
CP	Cyclic Prefix
CPD	Critical Path Delay
CPE	Common Phase Error
CPM	Circulant Permutation Matrix
CPRI	Common Public Radio Interface
CPU	Central Processing Unit
CR	Cognitive Radio
CRC	Cyclic Redundancy Check
CSMA	Carrier Sense Multiple Access
CSU	Configuration Security Unit
CTRL	Control Signal
CU	Centralised Unit

D

DAB	Digital Audio Broadcast
DAC	Digital to Analogue Converter
DARPA	Defence Advanced Research Projects Agency (DARPA)
DC	Direct Current
DCI	Downlink Control Information

DCLK	Data Clock
DDC	Digital Down-Converter
DDM	Decision Directed Mode
DDR	Double Data Rate
DECT	Digitally Enhanced Cordless Telecommunications
DEMUX	De-multiplexer
DFE	Digital Front End
DFT	Discrete Fourier Transform
DFX	Dynamic Function eXchange
DIN	Data In
DIP	Dual In-line Package
DL	Deep Learning
DMA	Direct Memory Access
DNN	Deep Neural Network
DoD	Department of Defence
DOUT	Data Out
DPD	Digital Pre-Distortion
DPU	Deep Learning Processing Unit
DRC	Design Rule Check
DSA	Dynamic Spectrum Access
DSB	Double Sideband
DSP	Digital Signal Processing
DSS	Dynamic Spectrum Sharing
DU	Distributed Unit
DUC	Digital Up-Converter
DVB	Digital Video Broadcast
DVD	Digital Video Disc

E

EIRP	Effective Isotropic Radiated Power
ENOB	Effective Number Of Bits
EPC	Evolved Packet Core
ERO	Elementary Row Operation
ESC	Environmental Sensing Capability

ETSI	European Communications Standards Institute
EU	European Union
EVM	Error Vector Magnitude

F

FCC	Federal Communications Commission
FD	Frequency Domain
FDD	Frequency Division Duplex
FEC	Forward Error Correction
FF	Flip Flop
FFT	Fast Fourier Transform
FIFO	First In First Out
FINN	Fast, Scalable Binarised Neural Network Inference
FIR	Finite Impulse Response
FLOSS	Free/Libre and Open Source Software
FM	Frequency Modulated
FMC	FPGA Mezzanine Card
FPGA	Field Programmable Gate Array
FPU	Floating Point Unit
FreeRTOS	Free Real Time Operating System
FS	Full Scale
FSK	Frequency Shift Keying
FWA	Fixed Wireless Access

G

GAA	General Authorised Access
GB	GigaByte
GCM	Galios Counter Mode
GIC	General Interrupt Controller
GigE	Gigabit Ethernet
GMSK	Gaussian Minimum Shift Keying
GNSS	Global Navigation Satellite System
GNU	GNU is Not Unix (recursive acronym!)
GPIO	General Purpose Input Output

GPP	General Purpose Processor
GPS	Global Positioning System
GPU	Graphical Processing Unit
GSM	Global System for Mobile
GSps	Giga samples per second
GTIS	Gain/Time Interleaving Spurs
GTY	Gigabit Transceivers
GUI	Graphical User Interface

H

HARQ	Hybrid Automatic Repeat Request
HDD	Hard Decision Decoding
HDL	Hardware Description Language
HDMI	High-Definition Multimedia Interface
HLS	High Level Synthesis
HTTP	Hyper Text Transfer Protocol

I

IC	Integrated Circuit
ICI	Inter Carrier Interference
ID	IDentity
IDE	Integrated Development Environment
IDFT	Inverse Discrete Fourier Transform
IEC	International Electrotechnical Commission
IEEE	Institute of Electrical and Electronics Engineers
IF	Intermediate Frequency
IFFT	Inverse Fast Fourier Transform
IIC or I^{2}C	Inter Integrated Circuit
IIR	Infinite Impulse Response
ILA	Integrated Logic Analyser
IMD	Inter Modulation Distortion
IMR	IMage Rejection
IO	Input Output
IoT	Internet of Things

IP	Intellectual Property
IPI	IP Integrator
IQ	In-phase Quadrature
ISI	Inter-Symbol Interference
ISM	Industrial Scientific and Medical
ISO	International Standards Organisation
ITU	International Telecommunications Union

J

JEDEC	Joint Electronic Device Engineering Council
JTAG	Joint Test Action Group
JTRS	Joint Tactical Radio Systems

L

LA	Layer
LAL	Local Access License
LAN	Local Area Network
LDPC	Low Density Parity Check
LED	Light Emitting Diode
LEO	Low Earth Orbit
LFSR	Linear Feedback Shift Register
LLC	Logical Link Control
LLE	Locally Linear Embedding
LLR	Log Likelihood Ratio
LMS	Least Mean Squares
LNA	Low Noise Amplifier
LO	Local Oscillator
LOL	Local Oscillator Leakage
LoS	Line of Sight
LPF	Low Pass Filter
LSA	License Shared Access
LSB	Least Significant Bit
LSTM	Long Short Term Memory
LTE	Long Term Evolution

| L-LTF | Legacy Long Training Field |
| LUT | Look Up Table |

M

MAC	Media Access Control / Controller
MAN	Metropolitan Area Network
MCM	Multi-Carrier Modulation
MCS	Modulation Coding Scheme
MDP	Markov Decision Process
MIMO	Multiple Input Multiple Output
MIO	Multiplexed Input/Output
MISO	Multiple Input Signal Output
ML	Machine Learning
MLP	Multi-Layer Perceptron
MMCM	Mixed-Mode Clock Manager
MME	Mobility Management Entity
MMIO	Memory Mapped Input/Outputs
MMITS	Modular Multifunction Information Transfer Systems
MMMC	Massive Machine-to-Machine type Communications
MMSE	Minimum Mean Square Error
MMU	Memory Management Unit
MNO	Mobile Network Operator
MPE	Media Processing Engine
MPSoC	Multi Processor System on Chip
MPU	Memory Protection Unit
MSB	Most Significant Bit
MSK	Minimum Shift Keying
MTS	Multi-Tile Synchronisation
MU	Multi User
MUX	Multiplexer

N

| NB | Narrow Band |
| NBIoT | Narrow Band Internet of Things |

NCO	Numerically Controlled Oscillator
NFC	Near Field Communication
NLoS	Non Line of Sight
NN	Neural Network
NR	New Radio
NSA	Non-Stand-Alone
NSD	Noise Spectral Density

O

OFDM	Orthogonal Frequency Division Multiplexing
OFDMA	Orthogonal Frequency Division Multiple Access
OLED	Organic Light Emitting Diode
OOK	On/Off Keying
ORAN	Open Radio Access Network
OS	Operating System
OSI	Open Systems Interconnection
OTA	Over The Air

P

PA	Power Amplifier
PAL	Priority Access License
PAN	Personal Area Network
PAPR	Peak to Average Power Ratio
PC	Personal Computer
PCA	Principal Component Analysis
PCB	Printed Circuit Board
PCH	Paging Control Channel
PCI	Physical Cell Identity
PCS	Physical Coding Sublayer
PDCP	Packet Data Convergence Protocol
PDF	Portable Document Format
PDF	Probability Density Function
PDU	Protocol Data Unit
PHY	Physical Layer

PL	Programmable Logic
PLL	Phase Locked Loop
PM	Path Metric
PMA	Physical Medium Attachment
PMBUS	Power Management Bus
PMU	Platform Management Unit
PRACH	Physical Random Access Channel
PS	Processing System
PSK	Phase Shift Keying
PSS	Primary Synchronisation Signal
PTP	Precision Time Protocol
PTRS	Phase Tracking Reference Signal
PU	Primary User

Q

QAM	Quadrature Amplitude Modulation
QC	Quasi Cyclic
QEMU	Quick EMUlator
QMC	Quadrature Modulation Correction
QNN	Quantized Neural Network
QoS	Quality of Service
QSPI	Queued Serial Peripheral Interface
QPSK	Quadrature Phase Shift Keying

R

RADAR	RAdio Detection And Ranging
RAM	Random Access Memory
RAN	Radio Access Network
RC	Raised Cosine
ReLU	Rectifier Linear Unit
RET	Remote Electrical Tilt
RF	Radio Frequency
RF-ADC	Radio Frequency Analogue-to-Digital Converter
RF-DAC	Radio Frequency Digital-to-Analogue Converter

RFDC	Radio Frequency Data Converter
RFMC	Radio FPGA Mezzanine Card
RFNoC	Radio Frequency Network on Chip
RFSoC	Radio Frequency System on Chip
RI	Rank Indicator
RL	Reinforcement Learning
RLC	Radio Link Control
RLS	Recursive Least Squares
RMS	Root Mean Square
RNN	Recurrent Neural Network
ROM	Read Only Memory
RPU	Real time Processing Unit
RRC	Root Raised Cosine
RRH	Remote Radio Head
RS	Reed Solomon
RSA	Rivest-Shamir-Adleman encryption
RSC	Recursive Systematic Convolutional
RSFEC	Reed Solomon Forward Error Correction
RSM	Radio Spectrum Management
RTOS	Real Time Operating System
RTSSM	Real Time Secondary Spectrum Market
RU	Radio Unit
Rx	Receiver

S

SA	Stand-Alone
SA	Standards Association
SAL	Shared Access License
SARSA	State Action Reward State Action
SAS	Spectrum Access System
SATA	Serial Advanced Technology Attachment
SC	Suppressed Carrier
SCH	Shared Channel
SCU	Snoop Control Unit

SDAP	Service Data Adaptation Protocol
SDD	Soft Decision Decoding
SDK	Software Development Kit
SDR	Software Defined Radio
SDSoC	Software Defined System on Chip
SER	Symbol Error Rate
SFDR	Spurious Free Dynamic Range
SHA	Secure Hash Algorithms
SIHO	Soft Input Hard Output
SIMD	Single Instruction Multiple Data
SIMO	Single Input Multiple Output
SINAD	Signal to Noise and Distortion Ratio
SIOU	Serial Input Output Unit
SISO	Single Input Single Output
SL	Scallop Loss
SMA	SubMinature type-A
SMP	Synchronous Multi Processing
SNDR	Signal to Noise and Distortion Ratio
SNE	Stochastic Neighbour Embedding
SNR	Signal to Noise Ratio
SoC	System on Chip
SOVA	Soft Output Viterbi Algorithm
SPB	Secure Processing Block
SPI	Serial Peripheral Interface
SQNR	Signal to Quantisation Noise Ratio
SRL	Shift Register
SRRC	State Radio Regulation of China
SSR	Super Sample Rate
SSS	Secondary Synchronisation Signal
STFT	Short Time Fourier Transform
SU	Secondary User
SVD	Singular Value Decomposition
SVM	Support Vector Machines
SYNC	Synchronisation signal

T

TCM	Tightly-Coupled Memory
TCP	Transmission Control Protocol
TD	Time Domain
TDD	Time Division Duplex
TDMA	Time Division Multiple Access
THD	Total Harmonic Distortion
TRAI	Telecom Regulatory Authority of India
TRM	Transmit/Receive Module
TV	Television
TVWS	TV White Space
Tx	Transmitter

U

UART	Universal Asynchronous Receiver/Transmitter
UCI	Uplink Control Information
UDP	User Datagram Protocol
UE	User Equipment
UHD	Ultra High Definition
UHF	Ultra High Frequency
UK	United Kingdom
UL	Up-Link
UN	United Nations
URLLC	Ultra Reliable Low Latency Communications
US	United States
USA	United States of America
USB	Universal Serial Bus
USD	United States Dollar
USRP	Universal Software Radio Peripheral
UWB	Ultra Wide Band

V

VGA	Variable Gain Amplifier
VHDL	VHSIC Hardware Description Language (see VHSIC)

VHSIC	Very High Speed Integrated Circuit
VHF	Very High Frequency
VOP	Variable Output Power

W

WAN	Wide Area Network
WLAN	Wireless Local Area Network

X

XDC	Xilinx Design Constraints
XML	Extensible Markup Language
XRT	Xilinx RunTime

Y

YAML	YAML Ain't Markup Language (recursive acronym!)

Z

ZF	Zero Forcing
ZOH	Zero Order Hold

List of Equations and Trigonometric Identities

Sum to Difference Trigonometric Rules:

$$cos(u \pm v) = cos(u)cos(v) \mp sin(u)sin(v) \tag{a.1}$$

$$sin(u \pm v) = sin(u)cos(v) \pm cos(u)sin(v) \tag{a.2}$$

$$tan(u \pm v) = \frac{tan(u) \pm tan(v)}{1 \mp tan(u)tan(v)} \tag{a.3}$$

Product to Sum Trigonometric Rules:

$$cos(u)cos(v) = \frac{1}{2}\left[cos(u-v) + cos(u+v)\right] \tag{a.4}$$

$$sin(u)sin(v) = \frac{1}{2}\left[cos(u-v) - cos(u+v)\right] \tag{a.5}$$

$$sin(u)cos(v) = \frac{1}{2}\left[sin(u+v) + sin(u-v)\right] \tag{a.6}$$

$$cos(u)sin(v) = \frac{1}{2}\left[sin(u+v) - sin(u-v)\right] \tag{a.7}$$

Additional Trigonometric Rules

$$cos^2(u) = \frac{1}{2}\Big(1 + cos(2u)\Big) \tag{a.8}$$

$$sin^2(u) = \frac{1}{2}\Big(1 - cos(2u)\Big) \tag{a.9}$$

$$sin(2u) = 2\,sin(u)\,cos(u) \tag{a.10}$$

Euler's Formula

$$e^{j\omega t} = cos(\omega t) + j\,sin(\omega t), \text{ where } j = \sqrt{-1} \tag{a.11}$$

Complex Representations of Trigonometric Functions

$$cos(\omega t) = \Re\Big[e^{j\omega t}\Big] = \frac{e^{j\omega t} + e^{-j\omega t}}{2} \tag{a.12}$$

$$sin(\omega t) = \Im\Big[e^{j\omega t}\Big] = \frac{e^{j\omega t} - e^{-j\omega t}}{2j} \tag{a.13}$$

Discrete Fourier Transform Pair:

$$X[k] = \sum_{n=0}^{N-1} x[n]\,W_N^{nk} \qquad k = 0, 1, 2, \dots, N-1 \tag{a.14}$$

$$x[n] = \frac{1}{N}\sum_{k=0}^{N-1} X[k]\,W_N^{-nk} \qquad n = 0, 1, 2, \dots, N-1 \tag{a.15}$$

Index

S

www.ingramcontent.com/pod-product-compliance
Lightning Source LLC
Chambersburg PA
CBHW051747200326
41597CB00025B/4468